Principles of Combustion

Principles of Combustion

Third Edition

Allan T. Kirkpatrick
Department of Mechanical Engineering
Colorado State University
USA

Kenneth K. Kuo
Department of Mechanical Engineering
The Pennsylvania State University
USA

WILEY

Published by John Wiley & Sons, Inc., Hoboken, New Jersey.
Published simultaneously in Canada.

Library of Congress Cataloging-in-Publication Data

Names: Kirkpatrick, Allan T., author. | Kuo, Kenneth K., author.
Title: Principles of combustion / Allan T. Kirkpatrick, Kenneth K. Kuo.
Description: Third edition. | Hoboken, New Jersey : Wiley, [2025] |
Includes index.
Identifiers: LCCN 2024014114 (print) | LCCN 2024014115 (ebook) | ISBN 9781394187065 (hardback) | ISBN 9781394187089 (adobe pdf) | ISBN 9781394187072 (epub)
Subjects: LCSH: Combustion engineering.
Classification: LCC TJ254.5 .K57 2025 (print) | LCC TJ254.5 (ebook) | DDC 621.402/3–dc23/eng/20240403
LC record available at https://lccn.loc.gov/2024014114
LC ebook record available at https://lccn.loc.gov/2024014115

Cover Design: Wiley
Cover Images: © Bikash Mallick/Getty Images, © Jose A. Bernat Bacete/ Getty Images, © AVN Photo Lab/Shutterstock, © Oleg Pankeev/Shutterstock, © Lianov/Shutterstock

Set in 9.5/12.5pt STIXTwoText by Straive, Chennai, India

Printed in Singapore

M WEP335348 010125

Contents

About the Authors

Dr. Allan T. Kirkpatrick is Emeritus Professor at Colorado State University. He earned his bachelor's and PhD degrees in mechanical engineering from the Massachusetts Institute of Technology. He taught for 42 years at Colorado State (1980–2022), where he was Professor of Mechanical Engineering and a former Department Head. Dr. Kirkpatrick is an internationally recognized authority in the applied thermal-fluid sciences. He has authored four books and numerous journal and conference publications on the internal combustion engine, combustion instability, fluid jets, and engineering education areas. He is a Fellow of the American Society of Mechanical Engineers (ASME) and is the recipient of numerous awards, including the Ben Sparks Medal by ASME.

Dr. Kenneth K. Kuo was awarded his bachelor's degree by the National Taiwan University. He earned a master's degree in mechanical engineering from the University of California, Berkeley, and earned a PhD in mechanical and aerospace engineering from Princeton University. Dr. Kuo taught for 39 years at Penn State (1972–2011), where he was Professor of Mechanical Engineering and Director of the High Pressure Combustion Laboratory. He was an internationally recognized authority in chemical propulsion and propellant combustion and authored four books and numerous publications on energetic material combustion and chemical propulsion. An elected fellow of the American Institute of Aeronautics and Astronautics (AIAA), the ASME and the International Ballistics Society, Kuo received numerous awards from various professional societies and from U.S. government agencies.

Preface

The third edition of this combustion textbook builds on the foundation established by the two previous editions (1986, 2004). For over 30 years, these editions have documented the continuing development of the field of combustion. The editions have demonstrated the application of the principles of thermodynamics, heat and mass transfer, and chemical kinetics to combustion processes and reflected the balance between engineering analysis and numerical computation in improving our understanding of combustion. The third edition continues to place emphasis on the development of idealized models to represent actual features of a combustion process.

Better understanding of the fundamental principles of combustion processes will provide engineers and scientists with the expertise needed to address a wide range of combustion issues. To reduce greenhouse gas emissions, there is a need to both increase the use of biofuels and carbon-free fuels such as hydrogen and ammonia and develop more efficient combustion technologies.

Since the publication of the second edition, major advancements have been made in theoretical, computational, and experimental aspects of combustion. Increased computational power has allowed engineers and scientists to simulate detailed reaction mechanisms and transport processes. Highly comprehensive model simulations can now be conducted with in-depth consideration of many complex thermochemical processes. At the same time, newly developed nonintrusive diagnostic techniques have allowed researchers to explore detailed phenomena associated with ignition and combustion processes. However, a note of caution should be sounded. As the capacity of computers increase, there can be a temptation to rely exclusively on numerical computation. Engineering insight is also required. It is a sense of a "feel for the answer," which is developed through engineering analysis and modeling.

There are now 11 chapters in the third edition, each of which focuses on a particular area of combustion. Content from the second edition has been extensively reorganized and updated to reflect suggestions from instructors and students. The topics of chemical kinetics and reaction mechanisms are now in separate chapters. The chapters on transport processes, premixed combustion, and nonpremixed combustion now each include turbulence and turbulent combustion modeling. Three new chapters on fuels (including carbon-free fuels), emissions, and solid fuel combustion have been added. The number of examples and homework problems has been increased to illustrate solution techniques for various combustion problems.

This edition is aimed toward advanced undergraduate and graduate engineering students who have an interest in combustion. It is also designed for combustion engineers and scientists to be used as a reference book in their research, development, and design work.

August 24, 2024

Allan T. Kirkpatrick
Fort Collins, Colorado

Acknowledgments

It should be noted that Kenneth Kuo, the author of the first two editions of the combustion book, passed away in 2016. Dr. Kuo should be recognized for his contributions to combustion education and research. As noted below, Dr. Kuo was an internationally recognized authority in chemical propulsion and propellant combustion. This edition is dedicated to Professor Kuo, who sparked an interest in combustion for hundreds of students over the course of his academic career.

Discussions with Professors Anthony Marchese at the University of Rhode Island on combustion modeling have been very helpful. Thanks are due to Professor Peyman Givi at the University of Pittsburgh, and Dr. Ondrej Masek at the University of Edinburgh for supplying selected combustion figures. Former CSU graduate students Aron Dobos and Kevin Knappmiller deserve a heartfelt thanks for their contributions to the computational sections of the nonpremixed combustion chapters.

Many thanks are due to the editorial staff at John Wiley & Sons, Inc., for their work on the third edition. Ms. Lauren Poplawski, Mr. Ashik Melvin, and Mr. Sindhu Raj Kuttappan deserve special acknowledgment for their editorial assistance with this project. Allan T. Kirkpatrick thanks his wife Susan and his extended family, Anne, Matt, Maeve, Michael, Rob, Kristin, Thomson, Charlotte, and Theo, for their unflagging support while this third edition was being written.

About the Companion Website

This book is accompanied by a companion website:

www.wiley.com/go/kirkpatrick/principlesofcombustion3e

This website includes:

- Combustion Matlab Files
- Combustion Solutions
- Lecture Notes

1
Introduction to Combustion

1.1 Introduction

The goal of this book is to provide insight into the modeling and analysis of combustion processes. The term combustion encompasses many high-temperature chemical reactions that involve oxygen. Combustion is a chemical reaction between a fuel and an oxidizer producing heat and light. It is a rapid energy conversion process that converts molecular energy stored in the fuel to thermal energy. The book covers the basic principles of combustion with a focus on the application of the chemical and thermal sciences to better understand combustion processes.

Our society relies heavily on the thermal energy released by combustion to meet a wide variety of needs, as most (~85%) of the world's energy use comes from combustion processes. This energy from combustion is primarily used for three purposes: to supply heat, to provide power, and to produce manufactured materials. We use petroleum in the form of gasoline, diesel, or kerosene for transportation, natural gas for heating our buildings and industrial processes, and coal and natural gas for electricity generation.

The mix of U.S. energy production and consumption has changed over time. Fossil fuels have dominated the U.S. energy mix for more than 100 years, but the mix has changed. Currently, in 2023, about a third of the U.S. energy production is petroleum based and another third is natural gas, with the remaining third a combination of renewables, coal, and nuclear energy (Smil 2017).

There are a number of private and governmental initiatives underway to decarbonize, i.e., reduce the amount of greenhouse gas emissions, primarily CO_2, from combustion devices. About three-quarters of the worldwide greenhouse gas emissions are due to the combustion of fossil fuels. The goal of decarbonization is to limit global warming to no more than 2.0 °C by 2050. The initiatives include increased use of biofuels and carbon-free fuels such as hydrogen and ammonia and improved combustion and process efficiency. Two technologies have dominated policy discussions about mitigating climate change: renewable energy generation and carbon capture and storage. Renewable technologies for electricity generation now being deployed widely are wind turbines and solar cells. Carbon capture and storage technologies include a process of converting carbon dioxide into fuel. In a two-stage process, the CO_2 gas is chemically captured and turned into a solid form as calcium carbonate, then heated to drive off the CO_2 and convert it to carbon monoxide.

Principles of Combustion, Third Edition. Allan T. Kirkpatrick and Kenneth K. Kuo.

Companion website: www.wiley.com/go/kirkpatrick/principlesofcombustion3e

1.2 Combustion Applications

Electrical power generation: Natural gas and coal are burned in the furnaces of electrical power stations to produce steam in order to generate electricity. For many years, most power plants used coal as a fuel, but due to health, economic, and climate change considerations, coal is being replaced by natural gas and also renewable energy sources.

Transportation: Liquid and gaseous fuels are used as an energy source for transportation by automobiles, trucks, aircraft, and ships due to their high energy density. Gasoline and diesel fuels are the major fuels used in internal combustion engines, and kerosene and natural gas are the major fuels used in gas turbines. There have been significant developments since the late 1960s in internal combustion engine and gas turbine technology in order to reduce their emissions and increase thermal efficiency.

Process industry for materials manufacturing: Manufacturers of primary metals (iron and steel), nonferrous materials (glass, ceramics, cement, and nanoparticles), chemicals, petroleum products, food, and paper produce these materials through heating and combustion processes. The combustion devices used in materials manufacturing include boilers, furnaces, ovens, and blast furnaces. For example, cement is produced by sintering ground limestone to 1450 °C in large kilns. In the Haber process, hydrogen and nitrogen are reacted together at high temperatures and pressures to produce ammonia, NH_3.

Domestic and industrial heating: Heating of residential homes, commercial buildings, industrial factories, offices, and various types of buildings such as schools and hospitals is accomplished through combustion of natural gas. There is an extensive natural gas network in the United States and many countries which delivers natural gas directly to a domestic and industrial end user.

Cooking More than 2.5 billion people, about 1/3 of the world's population, rely on solid biomass, kerosene, or coal as their primary cooking fuel. The burning of solid fuels in households for cooking and heating can lead to very low indoor air quality and health issues.

1.3 Organization of Text

In Chapter 1, we introduce various applications of combustion processes, fuels, and emissions from various combustion applications. As discussed below, these applications include electrical power generation, transportation, industrial processes, and heating of domestic and commercial buildings. The major fuels for these combustion processes include natural gas, liquid hydrocarbons (gasoline and diesel fuels), and solids (wood and coal).

Chapter 2 covers chemical thermodynamics, including thermodynamic properties of fuel–air mixtures and equilibrium thermodynamics of multicomponent reacting systems. Chapter 3 reviews chemical kinetics, reaction rates, and the rate of production of combustion products. Chapter 4 introduces detailed reaction mechanisms of various fuels. In Chapter 5, the general conservation equations of mass, momentum, energy, and species for reacting systems are presented. Chapter 5 also introduces turbulence models and turbulent combustion.

In Chapter 6, premixed laminar flames are modeled to determine burning velocities and flammability limits. Chapter 7 examines subsonic deflagration waves and supersonic detonation waves of premixed gases. Chapter 8 covers non-premixed or diffusion flames, including jet mixing and flame length. In Chapter 9, the evaporation and burning of fuel droplets are presented. Chapter 10 discusses the emissions, including nitrogen oxides, carbon monoxide, carbon dioxide, and particulates from combustion processes. The final chapter, Chapter 11, introduces the processes of solid fuel (biomass, i.e., wood and coal) combustion. A Homework Problem section is included at the end of each chapter.

1.4 Solution Methods for Combustion Problems

A combustion problem can be solved using various combinations of analytical, computational, and experimental approaches. The text introduces a number of solution techniques and gives background references for more information. Although the design of many modern combustion systems leans toward calculation and simulation rather than extensive hardware fabrication and testing, diagnostic techniques are still very important.

Indeed, the rise in computational fluid dynamics (CFD) is beginning to allow the evaluation of complex flow and combustion systems over a wide range of design parameters more quickly, inexpensively, and thoroughly than can be accomplished with actual hardware tests. Reactive flows including multistep kinetic considerations have become computationally tractable. For example, there are three major categories of numerical analysis, Reynolds average Navier–Stokes (RANS), large-eddy simulation (LES), and direct numerical simulation (DNS). The predictions of these numerical solution techniques for a diffusion flame are shown in Figure 1.1. A discussion of these methods is provided in Chapter 5.

However, these computational models often require adequate input of kinetic information in the form of sub-models, which are based on specific measurements or basic understanding of the chemical and physical mechanisms derived from experimental observations. In particular, experimental validation of computed results will be necessary to prove the predictive capability of theoretical models and computer codes before they are used for guiding actual designs.

There are a number of software packages available for numerical solution of combustion problems, from commercial, national laboratory, and academic sources. A popular open-source program is Cantera, which is a suite of software tools for problems involving chemical kinetics, thermodynamics, and transport processes. Since Cantera is open source, users can incorporate detailed chemical kinetics and transport models into their calculations. Cantera can be used from Python and Matlab, or in applications written in C/C++ and Fortran 90.

A popular commercial combustion software tool is Chemkin-Pro. Many researchers work with combustion mechanisms in Chemkin format, which has become a de facto standard in

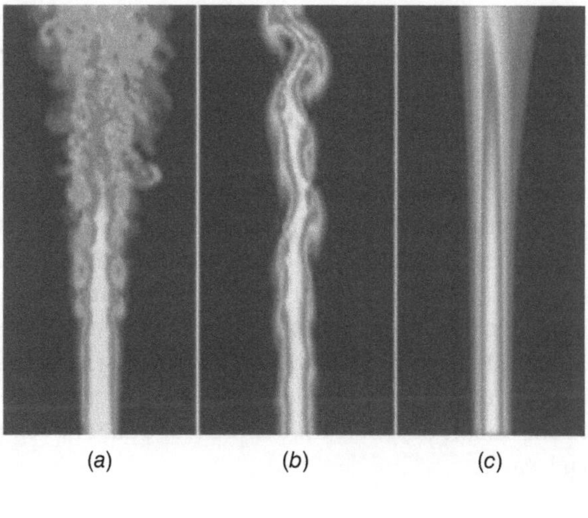

Figure 1.1 Predicted results for a diffusion flame using (*a*) DNS, (*b*) LES, and (*c*) RANS. (Courtesy of Prof. P. Givi.)

the combustion community. The Chemkin software was originally developed in the U.S. Sandia National Laboratories and has since become a standard computational program for combustion engineering as well as in other fields such as fluid dynamics and chemistry. The Chemkin-Pro package is a set of programs which incorporate complex chemical kinetics into simulations of gas-phase and gas-surface reacting flow. The software includes application programs for stirred reactors, plug-flow reactors, premixed and diffusion flames, and opposed-flow diffusion flames. For specific applications such as combustion in internal combustion engines and gas turbines, commercial software packages such as Converge, WAVE, OpenFOAM, Fluent, STAR-CD, and KIVA are also used by the combustion research community.

1.5 Fuels Background

In this and the next few sections, we introduce a range of conventional and alternative fuels used in combustion devices. We discuss the chemical structure of fuels and oxidizers, their thermodynamic properties, and their use in various combustion devices, such as boilers, internal combustion engines, gas turbines, and rockets.

A typical hydrocarbon-based liquid fuel may consist of 100 different hydrocarbons and another 100–200 trace species. Gasoline and diesel fuels are primarily obtained by distillation from petroleum oil. Petroleum oil has a relatively low cost and a high energy density. Since petroleum contains carbon, its combustion produces carbon dioxide, a greenhouse gas contributing to global warming and climate change.

Petroleum is a fossil fuel composed from ancient organic materials. Formation of petroleum and natural gas reservoirs occurs underground during the pyrolysis of hydrocarbons in a variety of endothermic reactions at high temperature and/or pressure. Wells are drilled into oil reservoirs to extract the crude oil. In 1858, Edwin Drake drilled the first U.S. oil well, a 21-m deep well in Titusville, Pennsylvania. He is credited with inventing the technique of drilling inside a pipe casing to prevent water seepage. Innovations in the technology for oil recovery have allowed deeper and deeper wells to be drilled, both on land and in the oceans. In Europe, petroleum oil is currently extracted from reservoirs located about 3000 m below the North Sea seabed.

The identified worldwide crude oil reserves are estimated by the American Petroleum Institute to be about 1 trillion barrels, with 0.6 trillion barrels remaining to be identified. At present consumption rates, at about 100 million barrels per day, it is estimated that petroleum reserves will last for 60–95 years. Technological advances in extraction have created continual increases in the size of the worldwide petroleum reserves. In 1950, the identified worldwide petroleum reserves were estimated to be about 0.09 trillion barrels, so in the last 60 years, the identified petroleum reserves have increased 10-fold. To put the consumption of petroleum into perspective, about 0.7 trillion barrels of petroleum have been consumed since the advent of the Industrial Revolution. The recent invention and commercialization of hydraulic fracturing, commonly known as "fracking," have enabled greater production of petroleum and natural gas from shale and related geological formations.

Crude oil is classified by its geographical origin, its American Petroleum Institute (API) gravity (light, heavy), and its sulfur content (low sulfur is labeled as sweet, and high sulfur is labeled as sour). Light crude oil produces a higher gasoline fraction. Sweet crude oil is more valuable than sour crude oil because it requires less refining to meet sulfur standards. Crude oil contains a very large number of different hydrocarbons. For example, 25,000 different compounds have been found in one sample of petroleum-derived crude oil. The compounds range from gases to viscous liquids and waxes.

Solid fuels like wood and coal are heterogenous hydrocarbon compounds that do not burn completely. Their combustion residue is in the form of inorganic ash particles. Gasification is a combustion technique that has been developed to convert a solid fuel into a clean gaseous fuel that can be used in combustion devices. Coal and woody biomass are broken down at high temperature (between 700 and 800 °C) and high pressure in a low-oxygen environment, a process called pyrolysis. Coal and wood gasification has many applications, as it has been used for the generation of synthetic natural gas ("syngas"), hydrogen, ammonia, and specialty chemicals.

A gasifier differs from a typical combustor as it uses a controlled air or oxygen supply and steam. With coal or wood gasification, only drying, distillation, and pyrolysis processes occur. The pyrolysis reactions are controlled by the level of oxygen or air made available for the reactions. A volatile flame is not established. The products are low molecular mass hydrocarbons, CO, H_2, CO_2, H_2O, N_2, and tar. The tar, char, and ash particles and any vaporized inorganic compounds are filtered from the pyrolysis products before use in an engine.

Coal gas was the primary source of gaseous fuel in the United States until replaced by natural gas in the 1940s. Coal gas is obtained by coking, i.e., partial pyrolysis of coal, similar to the process of producing charcoal from wood. The pyrolysis process drives off the volatile constituents in the coal. Coal gas is typically 50% hydrogen, 35% methane, 10% carbon monoxide, and other trace gases such as ethylene. The earliest internal combustion engines in the late 1800s were fueled with coal gas.

Biogas, a mixture of methane and carbon dioxide, is produced from the decay of organic matter. When organic matter, such as kitchen scraps and farm or livestock waste, decays anaerobically at low temperature with no oxygen present, either in nature or under controlled conditions in a sealed tank, it ferments, producing biogas. Feedstocks can include residues from the harvest of wheat, maize, rice, sugar beet, sugar cane, soybean, and other crops. The methane content of biogas typically ranges from 45% to 75% by volume, with most of the remainder being CO_2. The Assyrians may have used biogas to heat bath water, and Marco Polo reported that in the 13th century, the Chinese extracted energy from covered sewage pots. Biogas from sewage treatment powered English streetlights in the 1890s. In the United States, the primary pathway for biogas has been through landfill gas collection.

1.6 Hydrocarbon Fuel Chemistry

Hydrocarbon fuels are composed of blends of hydrocarbons, grouped into families of hydrocarbon molecules termed paraffins, olefins, naphthenes, and aromatics. The hydrocarbon families each have characteristic carbon–hydrogen bond structures and chemical formulae.

Paraffins (alkanes) are molecules in which carbon atoms are chained together by single bonds. The remaining bonds are with hydrogen. They are called saturated hydrocarbons because there are no double or triple bonds. The general formula for the paraffin family is $C_nH_{(2n+2)}$. The number of carbon atoms is specified by a prefix:

1-meth	2-eth	3-prop	4-but
5-pent	6-hex	7-hept	8-oct
9-non	10-dec	11-undec	12-dodec

Paraffin is designated as an alkane by the suffix *-ane*. Examples of paraffins are methane, CH_4, and octane, C_8H_{18}, as shown schematically in Figure 1.2. Compounds with straight chains are also labeled as normal or *n*-. Octane is sometimes called normal octane or *n*-octane, and straight-chain heptane, C_7H_{16}, is labeled *n*-heptane.

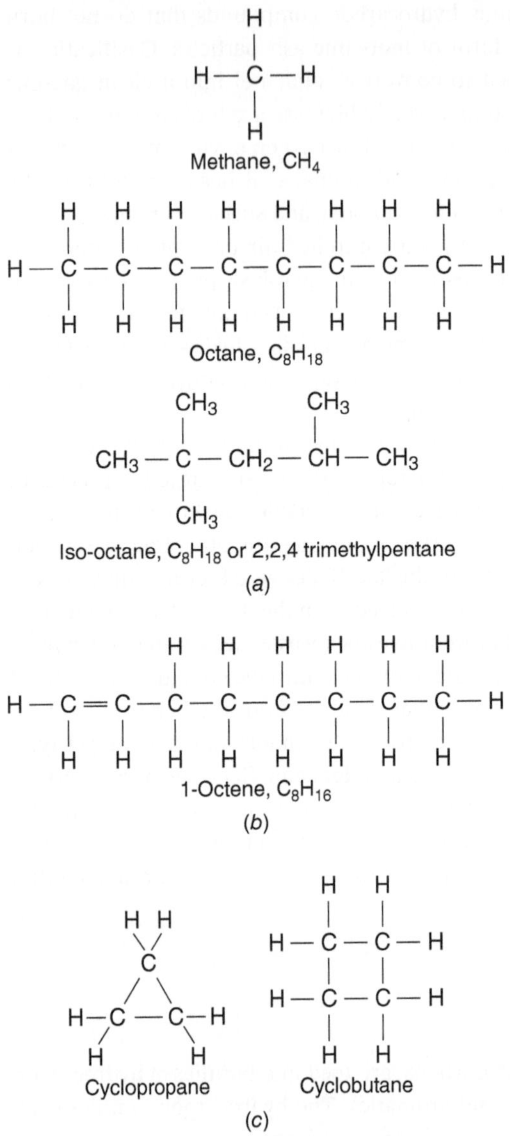

Figure 1.2 (*a*) Paraffins, (*b*) olefins, and (*c*) naphthenes.

Iso-octane, shown in Figure 1.2, is an example of a highly branched chain isomer of octane. That is, it has the same number of carbon atoms as octane but not in a straight chain. The group CH_3 attached to the second and fourth carbons from the right is called a methyl radical, *meth* because it has one carbon atom, *yl* because it is of the alkyl radical family $C_nH_{(2n+1)}$, and *radical* because it is a molecule that contains at least one unpaired valence electron, which makes it very reactive. Iso-octane is more properly called 2,2,4-trimethyl pentane, 2, 2, 4 because methyl groups are attached to the second and fourth carbon atoms, *trimethyl* because three methyl radicals are attached, and *pentane* because the straight chain has five carbon atoms.

Olefins (alkenes) are molecules with one or more carbon=carbon double bonds. Mono-olefins have one double bond, the general formula C_nH_{2n}, and their names end with *-ene*. The molecule 1-octene, C_8H_{16} is shown in Figure 1.2. Isomers are possible not only by branching the chain with the addition of a methyl radical but also by shifting the position of the double bond without changing the carbon skeleton. Olefins with more than one carbon=carbon double bond are undesirable components of fuel that lead to storage problems. Consequently, they are refined out and the only olefins of significance in diesel fuel or gasoline fuel are mono-olefins.

Naphthenes (cycloalkanes) have the same general formula as olefins, C_nH_{2n}, but there are no double bonds. They are called *cyclo* because the carbon atoms are in a ring structure. Two examples are cyclopropane and cyclobutane, shown in Figure 1.2. Cycloalkane rings having more than six carbon atoms are not as common.

Aromatics are hydrocarbons with carbon=carbon double bonds internal to a ring structure. The most common aromatic is benzene, shown schematically in Figure 1.3. Benzene is a regulated toxic compound, as it is a known carcinogen. Notice that the double bonds alternate in position between the carbon atoms. This makes the molecule's bonds difficult to break, so that a greater temperature is required to initiate combustion. As a result, aromatics are desirable in gasoline since they increase the octane number. Aromatics are undesirable components of diesel fuels. Some common aromatics (toluene, ethylbenzene, and styrene) have groups such as methyl radicals substituted for hydrogen atoms, and others (biphenyl) have more than one ring. Finally, there are polycyclic aromatic hydrocarbons (PAHs) which are aromatics with two carbon atoms shared between more than one ring (naphthalene and anthracene).

An alcohol is a partially oxidized hydrocarbon, formed by replacing a hydrogen atom with the hydroxyl radical OH. If the hydrogen atom attached to an aromatic ring is replaced by the hydroxyl radical, the molecule is called a phenol. Ethers are isomers of alcohol with the same number of carbon atoms. Some examples, shown in Figure 1.4, are methanol, ethanol, phenol, and methyl ether.

Nitromethane CH_3NO_2 is formed from a paraffinic hydrocarbon by replacing a hydrogen atom with a NO_2 group, as shown in Figure 1.4. It has twice the bound oxygen as monohydric alcohols

Figure 1.3 Aromatics.

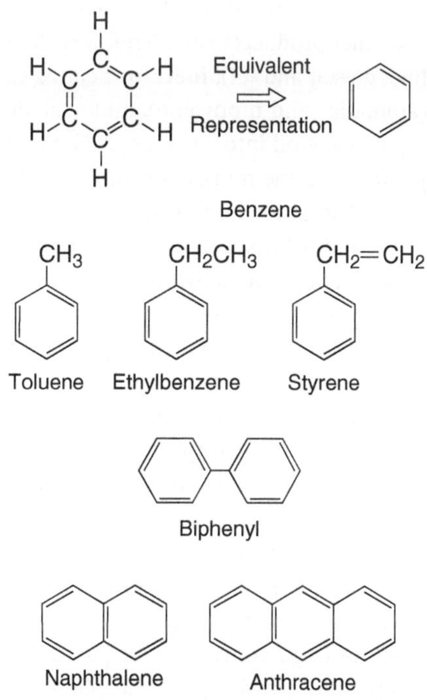

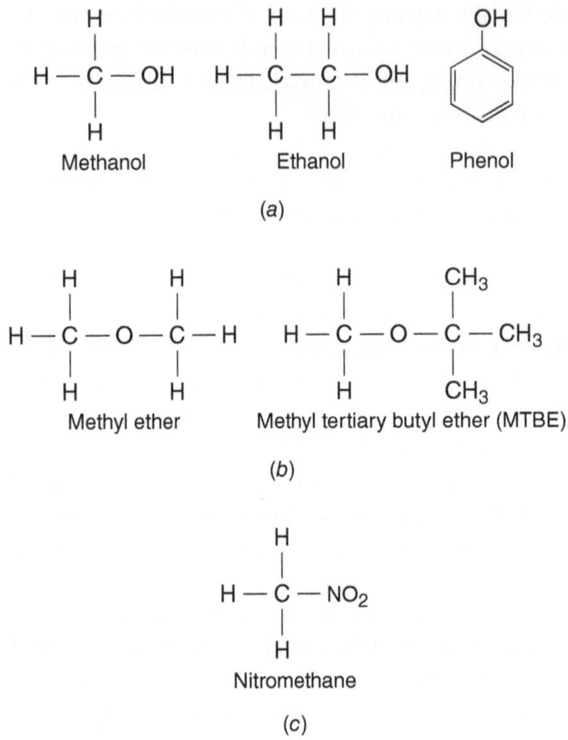

Figure 1.4 (*a*) Alcohols, (*b*) ethers, and (*c*) nitroparaffins.

and can combust without air. At ambient temperature, it is a liquid, and it is widely used as a drag racing fuel.

Crude oil is physically separated into various fractions, and then chemically processed into fuels and other products in an oil refinery. A full-scale crude oil refinery produces fuels for engines (gasoline, diesel, and jet), fuels for heating (burner, coke, kerosene, and residual), chemical feedstock (aromatics and propylene), and asphalt. On the average, a refinery will refine about 40% of the input crude oil into gasoline, 20% into diesel and heating fuel, 15% into residual fuel oil, 5% into jet fuel, and the remainder into the other listed hydrocarbons. The fraction separation process is called distillation and the device employed is a fractionating column.

Alkylation is used to increase the molecular mass and octane number of gasoline by adding alkyl radicals to a gaseous hydrocarbon molecule. Light olefin gases are reacted with iso-butane using a catalyst. Iso-octane results from reacting butene with iso-butane. This process requires relatively low temperature (275 K) and pressure (300 kPa) and, therefore, consumes relatively less energy than other refining processes.

Catalytic cracking uses activated catalysts to break the molecular chains of a distillate to produce naphthas. Naphtha is a liquid mixture consisting of straight-chained and cyclic aliphatic hydrocarbons having from five to nine carbon atoms per molecule. The naphtha products of catalytic cracking are blended with other hydrocarbons to produce high octane number gasolines. The reactions are at high temperature (700–800 K) and at low to moderate pressure (200–800 kPa). Considerable energy is consumed in the process.

Reforming refers to reactions designed to alter the molecular structure to yield higher octane gasoline (e.g., conversion of paraffins into aromatic hydrocarbons). This is often done using catalysts in a hydrogen atmosphere at high temperature (800 K) and high pressure (3000 kPa).

Considerable hydrogen is produced as a result of the reaction:

$$C_nH_{2n+2} \longrightarrow C_nH_{2n-6} + 4H_2$$

Coking is the process used to convert heavy reduced crude fractions to the more usable naphtha and distillate fractions. Upon heating, the molecules undergo pyrolytic decomposition and recombination. The average molecular mass of the fraction remains the same, but a greater spectrum of components is produced. The heaviest component, called coke, is a solid carbon material similar to charcoal.

Example 1.1 ***Fuel Chemical Structure***

What is the chemical structure of (a) 3-methyl-3-ethylpentane?

Solution

This pentane has methyl CH_3 and an ethyl C_2H_5 on the third carbon. A pentane paraffin has five single-bonded carbon atoms. Its chemical formula is C_8H_{18}, so it is an isomer of octane. Its structure is:

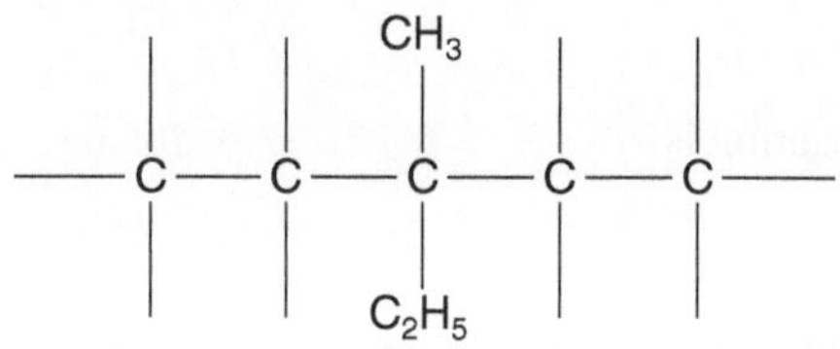

1.7 Gaseous and Liquid Fuels for Combustion Engines

In this section, we introduce the gaseous and liquid fuels used in combustion devices such as internal combustion engines, gas turbines, and thermal power generation. Spark ignition engine fuels include natural gas, propane, hydrogen, gasoline, ammonia, methanol, and ethanol.

Table 1.1 compares the thermal properties of a variety of gaseous fuels. The lower heat of combustion q_{lhc} or heating value is defined as the state where all of the water in the products is vapor (the quality $\chi = 1$), and the higher heat of combustion, q_{hhc}, or heating value is defined

Table 1.1 Thermodynamic Properties of Common Gaseous Fuels.

Fuel	Hydrogen H_2	Methane CH_4	Ammonia NH_3	Propane C_3H_8
Boiling Temperature (°C) at 1 atm	−253	−161	−33.4	−42.1
Condensation Pressure (atm) at 25 °C	n/a	n/a	9.90	9.40
Lower Heat of Combustion q_{lhc} (MJ/kg)	120	50	18.6	46.4
Flammability Limits ϕ				
(Lean–Rich Equivalence Ratios)	0.10–7.1	0.50–1.7	0.63–1.40	0.51–2.5
Adiabatic Flame Temperature $T_{f,ad}$ (K)	2383	2240	1800	2268
Maximum Laminar Flame Speed S_L (m/s)	2.91	0.37	0.07	0.43
Minimum Autoignition Temperature (°C)	520	630	650	450

as the state where all of the water in the products is liquid (the quality $\chi = 0$). At $T = 298$ K, the enthalpy of vaporization of water $h_{fg} = 2442$ kJ/kg on a mass basis, and $\overline{h}_{fg} = 43.99$ kJ/mol on a molar basis. Thermal properties of fuels and associated measurements are discussed further in Chapter 2, and tables of transport properties of fuels are provided in the Appendix E.

In the United States, petroleum gasoline is currently the dominant fuel for internal combustion engines, as it comprises about 70% of the vehicular fuel consumption, followed by diesel fuel at 23%, ethanol at 5%, and biodiesel at 0.5%. The performance of spark ignition engines is limited by the occurrence of an autoignition process called knock, which constrains the maximum compression ratio and thus the overall engine efficiency. Compression ignition engines have separate fuel and air streams that combust as they are mixed together at a temperature greater than the autoignition temperature, a process called diffusion combustion. The performance of compression ignition engines is limited by emissions of unburned hydrocarbons such as soot.

Example 1.2 ***Hydrogen Heat of Combustion***

If the mass lower heat of combustion q_{lhc} of hydrogen H_2 is 120 MJ/kg, what is its molar high heat of combustion $\overline{q}_{hhc}$?

Solution

The hydrogen combustion reaction is

$$H_2 + \frac{1}{2}O_2 \longrightarrow H_2O$$

where 1 mol of hydrogen yields 1 mol of water. Converting q_{lhc} to a molar basis, where M_{H_2} is the molar mass (kg/kmol) of hydrogen,

$$\begin{aligned} \overline{q}_{hhc} &= q_{lhc}\, M_{H_2} + n_{H_2O}\, \overline{h}_{fg} \\ &= (120)(2.016) + (1)(43.99) \\ &= 286\ \text{MJ/kmol} \end{aligned}$$

1.7.1 Methane and Natural Gas

Natural gas is a naturally occurring gaseous fuel found in oil fields. It is primarily composed of about 90–95% methane (CH_4), with small amounts of additional compounds such as 0–4% nitrogen, 4% ethane, and 1–2% propane. Due to its structure, methane is a greenhouse gas and, over a 20-year period, has a global warming potential about 80 times that of carbon dioxide.

Since methane has a lower carbon to hydrogen ratio relative to gasoline, the CO_2 emissions from the combustion of methane are about 22–25% lower when compared to gasoline combustion. Natural gas has been used for many years for electric power generation, industrial processes, and commercial and residential heating. An extensive distribution network of natural gas pipelines exists to meet the need for natural gas for applications.

Natural gas has slow ignition kinetics and a low cetane number. The combustion of methane is different from that of higher order hydrocarbon combustion since only carbon–hydrogen bonds are involved, and no carbon=carbon bonds, so the combustion process is more likely to be more complete, producing less non-methane hydrocarbons. Optimal thermal efficiency occurs at rich conditions with equivalence ratios from 1.3 to 1.5. The combustion reactions of methane produce intermediate molecules, such as formaldehyde, a criteria pollutant. Since it is spark ignited, the particulate emissions of methane are very low relative to diesel fuel. Methane has a lower adiabatic flame temperature (2240 K) than that of gasoline (2310 K), due to its greater product water content. Operation under lean conditions will also lower the peak combustion temperature.

Natural gas is stored in a compressed (CNG) state at room temperature and also in a liquid (LNG) form at −160 °C. Natural gas is pressurized to 20 MPa in vehicular storage tanks and has about one-third of the volumetric energy density of gasoline. The storage pressure is about 20 times that of propane.

Natural gas-fueled vehicles have been in use since the 1950s, and conversion kits are available for both spark and compression ignition natural gas and gasoline or diesel fuel. Like propane, natural gas is delivered to an internal combustion engine through a pressure regulator, either through a mixing valve located in the intake manifold, port fuel injection at about 750 kPa, or through direct injection into the cylinder. With intake manifold mixing or port fuel injection, the engine's volumetric efficiency and power are reduced due to the displacement of about 10% of the intake air by the natural gas, and the loss of evaporative charge cooling. Natural gas does not require mixture enrichment for cold starting, reducing potential cold start HC and CO emissions. Natural gas has an octane number (RON) of about 120, so that natural gas-fueled engines can operate at a compression ratio greater than that of gasoline fueled engines.

1.7.2 Propane

Propane (C_3H_8) is a saturated alkane hydrocarbon. When blended with butane C_4H_{10} or ethane C_2H_6, it is also designated as liquefied petroleum gas (LPG). A common LPG blend is P92, which is 92% propane and 8% butane. Propane is a by-product of crude oil refining and natural gas processing. In the United States, about one-half of the LPG supply is obtained from the lighter hydrocarbon fractions produced during crude oil refining, and the other half from heavier components of wellhead natural gas.

Propane has been used as a vehicular fuel since the 1930s. As of 2020, there are about 10 million LPG vehicles operating worldwide. There is a relatively extensive refueling network for propane. For example, there are about 15,000 refueling stations available in North America. There are a number of original equipment manufacturers that currently sell propane-fueled vehicles, primarily light- and medium-duty fleet vehicles, such as pickup trucks and vans.

In vehicles, propane is stored as a compressed liquid, at about 1 MPa (10 atm) pressure. Its evaporative emissions are essentially zero, since it is used in a sealed system. A pressure regulator controls the supply of propane to the engine and converts the liquid propane to a gas through a throttling process. As the pressure is reduced in the pressure regulator, the propane vaporizes into a gas. Propane gas can be injected into the intake manifold, into the ports, or directly into the cylinder. Propane has an octane number of 112 RON, so vehicular applications of propane can operate at a raised compression ratio.

1.7.3 Hydrogen

Hydrogen (H_2) can be produced from many different feedstocks, including natural gas, coal, biomass, and water. The production processes include steam reforming of natural gas, presently the most economical method, gasification of coal, and electrolysis of water. The various methods used to produce hydrogen as a fuel are classified by color. "Green" hydrogen is produced using renewable energy (wind and solar) to split H_2O into H_2 and O_2, which generates no greenhouse gases. Currently, most hydrogen is produced from natural gas and is denoted "gray" hydrogen. "Blue" hydrogen also uses natural gas with the added step of capturing and storing the carbon-dioxide emissions. The steam reforming process using natural gas to produce "gray" or "blue" hydrogen is given in Eq. (1.1):

$$CH_4 + 2H_2O \longrightarrow CO_2 + 4H_2 \tag{1.1}$$

Hydrogen is a colorless, odorless, and nontoxic gas, and hydrogen flames are invisible and smokeless. The global warming potential of hydrogen is insignificant in comparison to hydrocarbon-based fuels since the combustion of hydrogen produces no carbon-based compounds such as HC, CO, and CO_2. At present, the largest user of hydrogen fuel is the aerospace community for rocket fuel.

The combustion characteristics of hydrogen are very different from gasoline, as the laminar flame speed of a hydrogen–air mixture is about 3 m/s, about 10 times that of methane and gasoline, and the adiabatic flame temperature is about 100 °C higher than gasoline and methane. Since it has wide flammability limits (5–75% by volume), preignition and backfiring in an internal combustion engine can be a problem. The flammability limits correspond to equivalence ratios of 0.10–7.1. Water injection into the intake manifold of an engine is used to mitigate preignition and provide cooling. The relatively high octane rating of hydrogen of 106 RON allows the use of an increased compression ratio.

There have been a number of hydrogen vehicular demonstration projects, but the relatively high cost of hydrogen fuel has hindered adoption as an alternative fuel. Hydrogen produced by onboard electrolysis of water can also be injected into the intake manifold of diesel engines to reduce soot emissions.

One of the major obstacles related to the use of hydrogen fuel is the lack of a manufacturing, distribution, and storage infrastructure. The most economical method would be to distribute hydrogen through pipelines, similar to natural gas distribution. It has been proposed to blend relatively low (5–15%) concentrations of hydrogen with natural gas and use the natural gas distribution infrastructure for distribution of the hydrogen-natural gas blend. This concept is not new, since coal gas, a blend of hydrogen and methane, was used in gas distribution systems in the 1800s until replaced by natural gas.

The three methods used to store hydrogen are: (1) in a liquid form at −253 °C in cryogenic containers; (2) as a metal hydride, such as iron–titanium hydride FeTi H_2, or (3) in a pressurized gaseous form at 20–70 MPa. The metal hydride releases hydrogen when heated by a heat source, such as a vehicle exhaust system. The most common storage methods are liquid and hydride storage, which have comparable volumetric storage capabilities, both requiring about 10 times the space required by an equivalent 5-gallon gasoline tank.

Compressed hydrogen at 70 MPa has one-third the energy density by volume of compressed natural gas, and liquid hydrogen has one-fourth the energy density by volume of gasoline. The use of liquid hydrogen has an additional energy cost, as liquefaction of hydrogen to a temperature of 20 K requires an expenditure of energy approximately equal to the energy content of the liquid hydrogen. If mixed with air in the intake manifold of an engine, the volume of hydrogen is about 30% of the intake mixture volume at stoichiometric, decreasing the volumetric efficiency. The techniques used to reduce NO_x levels in hydrogen combustion are retarding the spark ignition timing, exhaust gas recirculation, and leaner operation to offset the effect of higher flame speeds and combustion temperatures.

1.7.4 Ammonia

Ammonia (NH_3) is a carbon-free fuel formed using the Haber–Bosch process, a catalytic synthesis reaction of hydrogen and nitrogen, invented in 1913. Since the combustion of ammonia produces no carbon-based compounds, ammonia has the potential to contribute to the decarbonization of engine fuels. The production and transport technologies of ammonia are mature. Ammonia has been mainly used for the past hundred years as a chemical raw material, for the production of fertilizer, and as a working fluid in refrigeration cycles. It is also used as a reducing agent in selective catalytic reduction (SCR) of NO_x in thermal power generation and in internal combustion engines.

Ammonia is a fuel that can be obtained from fossil fuels, biomass, or other renewable sources. Some advantages of ammonia with respect to hydrogen are its lower cost per unit of stored energy, higher volumetric energy density, easier production, handling and distribution, and better commercial viability. Ammonia has a narrow flammability range and is therefore generally considered nonflammable when transported.

Ammonia concentrations in air are regulated due to its high level of toxicity. The OSHA 8 hour NH_3 exposure limit is 25 ppm, and the immediately dangerous to life and health (IDLH) limit is 300 ppm.

The thermal properties of ammonia, such as condensation pressure and boiling temperature, are very similar to that of propane. At 25 °C, it can be stored as a liquid at pressures greater than 10 bar, with a density of 603 kg/m^3. The lower heat of combustion q_{lhc} = 18.6 MJ/kg of ammonia is 40% of the heat of combustion of propane. The stoichiometric laminar flame speed at atmospheric pressure is 7.0 cm/s, a factor of about five slower than that of hydrocarbon fuels. The slow flame speed results in poor combustion quality in engines, so combustion promoters such as hydrogen, kerosene, gasoline, or diesel fuels are required. The adiabatic flame temperature of ammonia is 1800 K, lower than that of hydrocarbon fuels, and the autoignition temperature is relatively high at about 650 K.

1.7.5 Gasoline

Gasoline is a hydrocarbon fuel composed of a blend of light distillate hydrocarbons, including paraffins, olefins, naphthenes, and aromatics. It has a hydrogen to carbon ratio varying from 1.6 to 2.4. A typical formula used to characterize gasoline is C_8H_{15}, with a molecular mass of $M = 111$. A high hydrogen content gasoline is C_7H_{17}. Gasoline properties of interest for internal combustion engines include the octane number, volatility, gum content, viscosity, specific gravity, and sulfur content. The American Society for Testing and Materials (ASTM) has established a set of gasoline specifications for each property.

Gasoline has been the major vehicular liquid fuel for spark ignition engines since the early 1900s. It has a very high volumetric energy density, a relatively low cost, and reliable cold-weather startup properties. It has the highest fueling station accessibility of engine fuels. Gasoline-fueled engines are also adaptable to alternative fuels such as propane and natural gas.

1.7.6 Methanol

Methanol (CH_3OH) is an alcohol fuel formed from natural gas, coal, or biomass feedstock. Methanol has been used as a vehicular fuel since the early 1900s and is also used as a fuel for diesel engines and fuel cells. It is also called wood alcohol. It is a liquid at ambient conditions. Its chemical structure is a hydrocarbon molecule with a single hydroxyl (OH) radical. The hydroxyl radical increases the polarity of the hydrocarbon, so that methanol is miscible in water and has a relatively low vapor pressure. Since oxygen is part of the chemical structure, less air is required for complete combustion. Methanol is very toxic, and ingestion can cause blindness and death.

To form methanol, the feedstock, such as natural gas, is steam reformed to make CO and H_2,

$$CH_4 + H_2O \longrightarrow CO + H_2 \tag{1.2}$$

Methanol is then synthesized by the subsequent reaction of CO and H_2,

$$CO + 2H_2 \longrightarrow CH_3OH \tag{1.3}$$

Pure methanol is labeled M100, and a mix of 85% methanol and 15% gasoline is labeled M85, with an octane rating of 102. The octane rating of methanol M100 of 111 RON allows the use of an increased compression ratio. Methanol is partially miscible with gasoline. The low volatility of methanol requires priming agents such as gasoline for improving cold starting performance. Adding gasoline to methanol provides more volatile components that can vaporize more easily at low temperatures. Satisfactory cold starting with M85 requires a rich mixture so that enough volatiles are present to form a combustible mixture.

Methanol has been adopted as a racing fuel, both for performance and for safety reasons. Since methanol is highly miscible with water, a methanol fire can be extinguished with water, which is not the case with gasoline. The relatively high enthalpy of evaporation (1215 kJ/kg) of methanol relative to gasoline (310 kJ/kg) produces greater intake aircooling and a corresponding increase in volumetric efficiency relative to gasoline. The energy density by volume of methanol is about half that of gasoline. However, because of its oxygen content, it has a higher stoichiometric energy density (3.09 MJ/kg_{air}) relative to gasoline (2.96 MJ/kg_{air}). For maximum power, a rich equivalence ratio of $\phi = 1.6$ is used.

1.7.7 Ethanol

Ethanol (C_2H_5OH) is an alcohol fuel formed from the fermentation of sugar and grain stocks, primarily corn and sugar cane, which are renewable energy sources. Its properties and combustion characteristics are very similar to those of methanol. Ethanol is also called "grain" alcohol. It is a liquid at ambient conditions and nontoxic at low concentrations.

Gasohol (E10) is a gasoline–ethanol blend with about 10% ethanol by volume. E85 is a blend of 85% of ethanol and 15% of gasoline. In the United States, the primary source of ethanol is currently from starch feedstocks, such as corn. As a result of the 2005 Renewable Fuel Standard which mandates that gasoline contain 10% biofuels, in 2016, about 40% of the U.S. corn production was used to produce ethanol. In Brazil, about half of the vehicles use an ethanol-based fuel "alcool," primarily E93, produced from sugar cane. There are efforts underway to produce ethanol from cellulosic feedstocks such as corn fiber, forestry waste, and switch grass.

The energy density by volume of ethanol is relatively high for an alternative fuel, about two-thirds that of gasoline. The octane rating of ethanol of 111 RON allows the use of an increased compression ratio. The cetane number of ethanol is low, at about 8, and it can be used in compression ignition engines with diesel fuel pilot ignition.

1.7.8 Kerosene

Kerosene is a liquid fuel composed of a mixture of hydrocarbon molecules with a range of 9–16 carbon atoms per molecule. The major components are branched and straight-line alkanes and naphthenes. Kerosene is often approximated by dodecane ($C_{12}H_{26}$). It is produced by the fractional distillation of petroleum and condenses at a temperature between diesel fuel and gasoline. Accordingly, at standard conditions, its volatility, i.e., vapor pressure of 0.7 kPa, is between a diesel fuel vapor pressure of 0.007 kPa and a gasoline vapor pressure of 60 kPa. Kerosene is mainly used as an aviation gas turbine fuel (Jet A) and as a rocket fuel (RP-1). Other uses include home heating and cooking.

1.7.9 Octane Number

To provide a standard measure of a spark ignition fuel's knock characteristics, a scale has been devised in which fuels are assigned an octane number, *ON*. A knock scale was developed in the 1920s by the American Society for Testing Materials (ASTM) Cooperative Fuel Research Committee (CFR). There are two fuels used, one with a low resistance to knock and the other with a high resistance to knock. In the ASTM octane scale, the low primary reference fuel is straight-chain *n*-heptane, assigned an octane number of 0, with a very short ignition delay. The high primary reference fuel is branched chain iso-octane, assigned an octane number of 100, with a much longer ignition delay. These two fuels were chosen as they also have very similar thermodynamic properties, with densities of 0.68 kg/m^3, boiling points of 99 °C, and vapor pressures of 5.3 kPa at 25 °C.

The ASTM CFR Committee worked with the Waukesha Motor Company to develop a standardized CFR fuel research engine and specific operating conditions to measure the octane number of a fuel. The first CFR engine was designed and built in 1929 and is still the standard by which liquid fuel octane measurements are made today. An antiknock index (AKI) is the average of the research (ASTM D2699) and motored (ASTM D2700) octane numbers and is displayed on gasoline pumps at service stations (e.g., 85, 87, and 91). In general, it has been found that the octane number is increased by reducing the straight-chain length of the hydrocarbon fuel. This can be accomplished by reducing the total number of carbon atoms or by rearranging them into a branch chain structure.

Many compounds have been tested for use as octane improvers in gasoline. Tetraethyl lead was the primary octane improver in general use from 1923 to 1975. Its use in motor vehicles was prohibited in 1995 due to its toxicity and its adverse effect on catalytic converters and oxygen sensors. Currently, lead is only used in aviation and off-road racing gasolines. Thomas Midgley (1889–1944), a mechanical engineer from the General Motors Research Laboratory, discovered lead additives in 1921. Lead deactivates the free radicals produced during the combustion process, thus decreasing the rate of heat release. As an aside, Midgley also was the inventor of Freon (F-12), a refrigerant initially developed for automotive air conditioning systems. Freon was the most widely used refrigerant in the world until the mid-1990s when it was determined that the ultraviolet decomposition of Freon in the stratosphere releases chlorine, causing depletion of the stratospheric ozone layer. The manufacturing of Freon in the United States was prohibited in 1998.

1.7.10 Methane Number

The methane number *MN* is a measure of the tendency for a gaseous fuel to knock. A fuel's methane number limits the maximum compression ratio and thus the theoretical engine efficiency. A blend of 20% hydrogen and 80% methane has $MN = 80$. As indicated in Table 1.2, Malenshek and Olsen (2009) found a linear relationship between the maximum compression ratio and the methane number, for a variety of gaseous fuels, including coal gas, wood gas, digester gas, and landfill gas. An engine optimized to operate on natural gas with a methane number of about 90 is susceptible to knock when operated on gases that have a lower methane number, such as coal gas which has a methane number of 24. The octane number of methane is 120 RON, one of the highest values for hydrocarbon fuels.

Table 1.2 Critical Compression Ratio versus Methane Number.

Gas	Compression Ratio	Methane Number
Coal Gas	8.0	23.9
Steam Reformed Natural Gas	10.5	62.4
Wood Gas	10.3	70.2
Natural Gas		78–98
Methane	14.4	100
Digester Gas	17.6	139.1
Landfill Gas	17.6	139.6

Source: Adapted from Malenshek and Olsen (2009).

Example 1.3 *Fuel Mixtures*

A flexible fuel engine operates with a mixture of 35% gasoline and 65% methanol, by volume. If the combustion is to be stoichiometric, what should the mass air–fuel ratio be?

Solution

The volume or mole fraction $X_i = V_i/V$, and the mass fraction $Y_i = m_i/m$ are reviewed in Section 2.2, and stoichiometry is reviewed in Section 2.10. The volume fractions X_i are to be converted to mass fractions. Writing the mass fraction Y_i in terms of density ρ_i and X_i.

$$Y_i = \frac{m_i}{m} = \frac{\rho_i \cdot V_i}{\sum \rho_i \cdot V_i} = \frac{\rho_i \cdot V_i/V}{\sum \rho_i \cdot V_i/V} = \frac{\rho_i \cdot X_i}{\sum \rho_i \cdot X_i}$$

Using data from Table 1.3,

Fuel	Formula	X_i	ρ_i	$\rho_i X_i$	Y_i	$AF_{st,i}$
Gasoline	C_8H_{18}	0.35	710	248.5	0.324	15.04
Methanol	$CH_3\ OH$	0.65	796	517.4	0.675	6.43
Sum		1.0		765.9	1	

The mass air–fuel ratio is the weighted sum of the stoichiometric air–fuel ratios of the two fuels.

$$\begin{aligned} \mathrm{AF_{st}} &= \sum \mathrm{AF}_{\mathrm{st},i} \cdot x_i \\ &= 15.04 \cdot 0.324 + 6.43 \cdot 0.675 \\ &= 9.22 \end{aligned}$$

1.7.11 Diesel Fuels

Diesel fuel for compression ignition engines consists of a mixture of light distillate hydrocarbons that have molecular masses from about 170 to 200 kg/kmol with a corresponding range of 12–20 carbon atoms per molecule. Diesel fuels vaporize in the range between about 180 and 360 °C, higher than gasoline. It is estimated that there are more than 10,000 isomers in diesel fuel. Like gasoline, diesel fuels are mixtures of paraffinic, olefinic, naphthenic, and aromatic hydrocarbons, but

Table 1.3 Thermodynamic Properties of Spark Ignition Liquid Fuels.

	Gasoline	Ethanol	Methanol	Kerosene
Molecular Mass, M (kg/kmol)	~110	46.07	32.04	~170
Stoichiometric Air–Fuel Ratio, AF_s	15.04	8.94	6.43	14.92
Lower Heat of Combustion, q_{lhc} (MJ/kg)	44.5	26.8	19.9	43.1
Liquid Density, ρ_l (kg/m^3) at 15 °C	710	794	796	800
Enthalpy of Vaporization, h_{fg} (kJ/kg) at 298 K	310	850	1215	251
Adiabatic Flame Temperature (K)	2310	2197	2151	2094
Autoignition Temperature (°C)	280–370	365	470	220
Flammability Limits (% Volume)	1.4–7.6	3.5–26	5.5–26	0.6–4.9
Research Octane Number (RON)	90–98	111	112	20-50
Motor Octane Number (MON)	80–90	92	91	20-50
Vapor Pressure (kPa) at 32 °C	62–90	17	32	0.7
Boiling Point T_{sat} (°C) at 1 bar	30–225	−42	−160	200–260

Source: Adapted from Owen and Coley (1995).

their relative proportions are different. In CFD simulations, diesel fuel is often approximated by tetradecane (n-$C_{14}H_{30}$) for computation of spray breakup, evaporation, and mixing. The ignition and combustion characteristics of diesel fuel are approximated by n-heptane (n-C_7H_{16}), as its cetane number of 56 is approximately equal to that of Diesel 2-D.

Diesel fuels have about an 8% greater energy density by volume than gasoline and are the primary fuel used by heavy duty vehicles. Diesel fuel is also much less flammable than gasoline, as the flashpoint temperature of diesel 2-D is 52 °C, much higher than that of gasoline, which is about −40 °C. The flash point is defined as the lowest temperature at which a vapor–air mixture will ignite when an ignition source is applied, in contrast to the autoignition temperature, which has no ignition source.

Diesel fuels are named after Rudolph Diesel (1858–1913) who in 1897 was the first to develop an engine designed for compression ignition. In a compression ignition engine, a low volatility fuel must be converted from a liquid state into a finely atomized state, vaporized, mixed with air, and its temperature raised to a point to support autoignition. The autoignition temperature of vehicular diesel fuel (2-D) is about 530 K, about 110 K less than that of gasoline.

Diesel fuels are classified by cetane number (*CN*) and compared using an ignition delay metric. The cetane number characterizes the ability of the fuel to autoignite and is the opposite of the octane number. The range of cetane numbers *CN* for vehicular diesel fuels is from about 40–60. As the cetane number is increased, the ignitability of the fuel increases, with a decrease in the ignition delay time. If the cetane number is low, the ignition delay is long, and fuel will not ignite until late in the injection process. In this situation, the fuel is well mixed, so that once combustion is initiated, the burning rate is controlled by the chemical kinetics.

Aromatic hydrocarbons and alcohols have strong chemical bonds, resulting in a long ignition delay. If these fuels mix completely with air before autoignition occurs, when ignition does occur, they will all burn rapidly in a premixed phase, producing a large rate of change of pressure and a high peak pressure. If the cetane number is too low, the engine will misfire and exhibit poor cold start behavior.

At higher cetane numbers, with shorter ignition delays, combustion is initiated while the fuel is being injected. The chemical bonds of some fuels, such as alkanes (straight-chain paraffins), are easily broken. Relatively little fuel burns in the premixed phase and most of the fuel burns in a diffusion phase at a rate limited by the rate of mixing with the cylinder air. If the cetane number is too high, combustion can occur before the fuel and air are completely mixed, resulting in increased soot formation.

One component of diesel fuel that has attracted particular regulatory attention is sulfur, due to the adverse impact that sulfur-containing particulate emissions have on air quality. During combustion, the sulfur in the diesel fuel forms sulfates (SO_4^-), and sulfur dioxide (SO_2), which reacts with water to form sulfuric acid, a component of acid rain. Ultralow sulfur diesel (ULSD) has a sulfur content of 15 ppm or less. As part of the Euro V emission standards, ULSD is required in Europe. After December 1, 2014, all highway, non-road, locomotive and marine diesel fuel produced and imported in the United States was mandated to be ultralow sulfur.

1.7.12 Alternative Compression Ignition Fuels

A variety of fuels have been developed as an alternative for petroleum-based diesel fuels. Alternative diesel fuels have a higher cost, and lower volumetric energy density than fossil-based diesel fuel, but do produce lower CO and particulate emissions. Biodiesel fuels are manufactured from their corresponding vegetable oils, through the trans-esterification process, where one ester is converted into another. Biodiesel is produced from vegetables such as soybean, castor, canola, sunflower, cotton, palm, coconut, jatropha, and algae. Interest in algae-based biodiesel has been increasing, due to the far greater sunlight–oil conversion efficiency of algae relative to land-based crops; however, the production costs remain relatively noncompetitive.

Biodiesel fuels are produced through the trans-esterification of the triglycerides in vegetable oil using a low molecular mass alcohol, such as methanol. The esterification process decreases the kinematic viscosity of the neat vegetable oil from about 30×10^{-6} to about 4×10^{-6} m^2/s, for satisfactory operation of the fuel injectors. Typical chemical syntheses of biodiesel fuels depend on the distribution of the triglycerides in their corresponding vegetable oils.

Trans-esterification is basically a substitution of methanol (a monohydric alcohol) for the triglycerides. The methyl ester is obtained through a process in which methyl alcohol and a catalyst (such as sodium hydroxide or potassium hydroxide) chemically breaks down the triglyceride molecule into a fatty acid methyl ester (FAME) of the oil and a glycerin byproduct. Due to chemical equilibrium considerations, not all of the tri-, di-, and mono-glycerides can be removed in the esterification process.

Biodiesel is not a single chemical compound, as the triglycerides in vegetable oils are a variable mixture of unsaturated and saturated fatty acids. As shown in Table 1.4, a representative biodiesel formula is $C_{19}H_{35}O_2$. The iodine number is used as a measure of the degree of unsaturated bonds in biodiesel fuel, with higher iodine numbers indicating more unsaturated carbon=carbon double bonds in the fuel molecule. The test method used to determine the iodine number is ASTM D1959.

Biodiesels are fully miscible with conventional diesel fuel and can be blended at various levels denoted by the volume percent of biodiesel. Biodiesel fuels are designated with the prefix B, so a mixture of 20% biodiesel is labeled B20. Diesel engines are rated for a maximum percentage of biodiesel, typically from B5 to B20. The most common blend in the United States is B20. In Europe, diesel fuel is blended with 7% biodiesel to produce B7. If 5% ethanol is added to B20, the resulting blend is labeled B20E5.

Table 1.4 Comparison of Thermodynamic Properties of Various Compression Ignition Fuels.

	Diesel (2D–4D)	Dimethyl Ether (DME)	Bio-diesel (B100)
Chemical Formula	$C_{15}H_{25}$	CH_3 OC H_3	$C_{19}H_{35}O_2$
Molecular Mass (kg/kmol)	170–200	46.07	295
Cetane Number	40–60	55–60	48–58
Liquid Density (kg/m^3)	775–860	668	882
Autoignition Temperature (°C)	256	350	375–450
Stoichiometric A/F Ratio	14.5–14.4	9.0	11.2–12.5
Kinematic Viscosity (m^2/s) at 40 °C	2.8×10^{-6}	2.2×10^{-7}	4.1×10^{-6}
Lower Heating Value[a)], Mass (MJ/kg_{fuel})	43.2–42.8	28.4	37.7
Lower Heating Value[a)], Volume (MJ/l_{fuel})	33.7–36.8	19.0	33.25
Thermal Expansion Coefficient (%/K)	8.4×10^{-4}		
Boiling Point at 1 bar (°C)	180–360	−25	328–366
Vapor Pressure at 38 °C (bar)	0.0069	8	1×10^{-8}

a) Equivalent to the lower heat of combustion.

In cold climates, it can be a challenge to fuel vehicles with high blends of biodiesel because the biodiesel cloud points and pour points occur at higher temperatures relative to conventional diesel. The cloud point of soybean biodiesel is about 1 °C, whereas the cloud point for No. 1 diesel is about −40 °C. The difference between the cloud point and the pour point for soybean biodiesel is about 1 °C, whereas the difference for No. 1 diesel is about 20 °C.

Two additional compression ignition fuels with high cetane numbers are dimethyl ether (DME) and Fischer–Tropsch (F–T) fuel. DME, CH_3OCH_3 (see Table 1.4), is an oxygenated fuel produced by dehydration of methanol, from lignocellulosic biomass, or from synthesis gas. It has a relatively low autoignition temperature and a short ignition delay, as indicated by its cetane number $CN = 55$–60. DME is composed of two methyl groups linked by an oxygen atom and has the same atomic composition as ethanol, with a greater adiabatic flame temperature. It burns with a visible blue flame, similar to that of natural gas. It has low lubricity, so its use in engines requires lubricity additives, such as a fatty acid lubricity improver. Since its vapor pressure is between propane and butane, it is a gas at standard conditions. In vehicles, a fuel tank pressurized to about 5 bar, similar to pressures used for LPG storage, is required to store DME in liquid form. It is noncorrosive to metals but does deteriorate some elastomers. The volumetric energy density (MJ/l) of liquid DME is relatively low, about half that of diesel fuel, so adjustment of the fuel injection duration is needed for conversion to DME.

F–T synthetic diesel fuel is produced from a mixture of carbon monoxide and hydrogen gas using a high-temperature gas to liquid catalytic reforming process. It is a polymerization process resulting in long straight-chain hydrocarbon alkanes. A variety of catalysts can be used, but the most commonly used are cobalt, iron, and ruthenium.

The F–T reaction is:

$$n\text{CO} + (2n+1)\text{H}_2 \longrightarrow C_nH_{2n+2} + n\text{H}_2\text{O} \tag{1.4}$$

The F–T process was first developed by two German scientists, Franz Fischer and Hans Tropsch, in 1925. The source CO is generated by pyrolysis of coal or woody materials, such as switchgrass. The F–T process has been successfully scaled up and used in large-scale natural gas-to-liquid and coal-to-liquid facilities worldwide. It is also used in CO_2 capture schemes, in conjunction with a reverse water–gas reaction.

1.7.13 Cetane Number

The standard measure of the autoignition characteristics of compression ignition fuels is the cetane number *CN*. It is a dimensionless number with a scale that varies from 15 to 100. The upper and lower limits of the cetane number are defined by two primary alkane reference fuels. The upper reference fuel is *n*-cetane (hexadecane), an unbranched $C_{16}H_{34}$, with a reference cetane value of 100. The lower reference fuel is an isomer *iso*-cetane (heptamethylnonane), a highly branched $C_{16}H_{34}$, with a reference cetane value of 15. Additives such as nitrate esters, for example, 2-ethyl hexyl nitrate, can be used to increase the cetane number. The cetane number is measured using a standard Waukesha CFR research engine with a pre-chamber and a variable compression ratio. The cetane number *CN* and the octane number *ON* are inversely correlated, as gasoline is a poor diesel fuel and vice versa.

1.8 Fuels for Rocket Engines

Rocket engines are propulsion devices that carry their own fuel and oxidizer and do not need air for combustion. A comparative measure of different rocket fuels is the specific impulse I_{sp}, defined as the thrust T per unit weight flow, $\dot{m}g$, with units of seconds. The specific impulse indicates for how long a time that engine can exert a continuous force, i.e., thrust, until fully burning the propellant mass. Historically, the first rockets used solid propellants, such as gunpowder. R. Goddard (1882–1945), an American engineer, built the first liquid-fueled rocket in 1926, using gasoline and liquid oxygen (LOX). Kerosene has also been used as a rocket fuel with LOX as the oxidizer. Similar rocket fuel research was conducted at the same time in Russia by K. Tsiolkovsky (1857–1935) and in Germany by H. Oberth (1894–1989).

1.8.1 Cryogenic Fuels

A cryogenic fuel is one that is at a temperature less than $T = 183\,\text{K}$ ($-90\,°\text{C}$). Cryogenic fuels include liquid hydrogen (LH_2) at $T = 20\,\text{K}$ and liquid methane at $T = 100\,\text{K}$. LOX at 50–90 K is used as the oxidizer. Liquid hydrogen fuel has a very high specific impulse of 450 seconds. These cryogenic mixtures require a spark for initial ignition.

1.8.2 Hypergolic Propellants

Hypergolic propellants are fuels that ignite upon contact with one another. An example of a hypergolic rocket fuel is monomethylhydrazine (MMH) CH_3NHNH_2 with the oxidizer nitrogen tetroxide N_2O_4. The specific impulse of hypergolic fuels ranges from 260 to 310 seconds. Rocket engines with hypergolic fuels are usually simple and reliable because they need no ignition system. The most common hypergolic fuels are in a liquid state at ambient temperatures and pressures. The corrosivity, toxicity, and carcinogenicity of traditional hypergolic fuel require extensive safety precautions.

1.8.3 Solid Propellants

With solid propellants, the fuel and oxidizer are premixed and cast in solid form. An example of a solid propellant is a mixture of aluminum powder as a fuel and ammonium perchlorate NH_4ClO_4 as an oxidizer. The specific impulse of solid propellants ranges from 200 to 300 seconds. The combustion process is dependent on the surface area of the fuel. Once ignited, a solid propellant will burn to completion, so it is difficult to throttle a solid propellant for rocket thrust control.

1.9 Solid Fuels

The two main categories of solid fuels are biomass and coal. Biomass, a renewable organic material, includes wood, agricultural residues, and organic waste, with the largest biomass energy source being wood. Coal is the solid fossilized residue of plant biomass, with common varieties such as lignite, bituminous, and anthracite coal.

Wood has a complex heterogeneous cellular structure and is composed of three main components: cellulose, hemicellulose, and lignin. The main component of wood is cellulose, which is often used as a surrogate for biomass. The inner cavity of a wood cell is called the lumen, and the surrounding structural layer is the cell wall, which consists mainly of cellulose. Cellulose is a carbohydrate polymer composed of long linear chains of glucose ($C_6H_{12}O_6$) units linked by β-1,4-glycosidic bonds. Hemicellulose is also a carbohydrate polymer, shorter in length than cellulose, with an amorphous branched structure, and is located between the cellulose and lignin. Lignin is a polymer composed mainly of linked phenylpropane ($C_6H_5CH_2$ $CH{=}CH_2$) units which hold the cellulose structure together.

Coal is primarily composed of carbon with variable amounts of other elements such as hydrogen, sulfur, oxygen, and nitrogen. During the fossilization or coalification process, the hydrocarbon aliphatic chains are replaced by aromatic rings, which then fuse into polyaromatic linked rings, with a corresponding increase in the percentage of carbon and decrease in the percentage of hydrogen and oxygen, as indicated in Table 1.5. Coal is classified in terms of rank and grade. The rank is a measure of the carbon content and heating value. In increasing rank are lignite,

Table 1.5 Comparison of Representative Wood and Coal Components and Heating Value (Dry, Ash-free).

	Fir	Pine	Anthracite Coal	Bituminous Coal
Proximate analysis (wt%)				
Volatile matter	86.2	87.0	6.4	17.7
Fixed carbon	13.7	12.8	81.4	71.9
Ultimate analysis (wt%)				
Carbon	52.3	59.0	82.1	80.1
Hydrogen	6.3	7.2	2.3	4.3
Oxygen	40.5	32.7	2.0	2.2
Nitrogen	0.1	—	0.8	1.3
Sulfur	—	—	0.6	1.7
Ash	0.8	1.1	12.2	10.4
Higher heating value q_c (MJ/kg)	21.5	24.2	30.8	32.5

Source: Adapted from Tillman et al. (1981).

bituminous, and anthracite coal. The grade indicates the quality of the coal, which depends on the ash and sulfur content.

In recent years, coal has supplied about a quarter of the world's primary energy. However, the CO_2 emissions from the combustion of coal since the advent of the Industrial Revolution have been the major contributor to global climate change, leading many countries to reduce or eliminate their use of coal. The CO_2 generated by biomass combustion is eventually recycled back to biomass through photosynthesis. Therefore, renewable fuels such as biomass can be CO_2 neutral if sourced using sustainable cultivation practices.

The composition of solid fuels is reported with a proximate analysis or an ultimate analysis. Proximate analysis is a standardized method to determine the moisture, volatile matter, char, and ash content of a given solid fuel. Volatile matter is defined as the volatile gases that are produced and driven off by heating the solid fuel. The volatile gases burn more rapidly than the remaining char particles and contribute to flame ignition and stability. Char is a measure of the amount of nonvolatile carbon remaining in a sample. Ash is the assortment of oxidized inorganic mineral compounds remaining after the combustion process is complete. An ultimate analysis is a standardized method which provides the chemical element composition of a solid fuel, namely, the molecular carbon, hydrogen, nitrogen, oxygen, and sulfur content, and is reported on a dry, ash-free weight percent basis.

The analysis procedures for the proximate analysis, ultimate analysis, and heating value of wood are given in ASTM E870, and the procedures for the proximate and the ultimate analysis of coal are given in ASTM D3172 and ASTM D3176, respectively. With the ASTM D3172 proximate analysis procedure, a sample of coal is crushed and dried in an oven at 105–110 °C to determine residual moisture. The sample is then heated in a covered crucible at 900 °C to a constant weight. This weight loss is the volatile matter. The remaining sample is then placed in an oven at 750 °C with the crucible cover removed to allow the sample to combust. The weight loss on completion of combustion is termed fixed carbon or char. The remaining residue is defined as ash. For ultimate analysis procedure, a coal sample is combusted in an elemental analyzer, which measures the weight percent of the elements carbon, hydrogen, nitrogen, sulfur, and ash. The oxygen percentage is determined by the remaining percentage.

Representative compositions of the softwoods fir and pine, anthracite coal, and bituminous coal are compared in Table 1.5. Wood has about 60–70% of the heating value of coal. Wood has substantially more volatile matter while having less fixed carbon than coal, so that about 60% of the heating value of wood is produced by the volatile matter. The large amount of volatile matter associated with wood is a result of the high number of functional groups (alkanes, etc.) and the low number of aromatic structures in the wood. The mass hydrogen/carbon ratio H/C of the softwood fir is about 0.12, and bituminous coal has a mass H/C ratio = 0.054. Coal also has substantially larger amounts of sulfur and nitrogen, which create environmental issues.

Wood is classified as either a hardwood or softwood, each having different physical structures. Hardwoods are deciduous trees such as maple, oak, and cherry. Softwoods are generally conifers such as pine, cedar, and fir trees. The primary distinguishing feature between softwoods and hardwoods is that hardwoods, such as oak and teak, have pores that transport water throughout the wood. The dry, ash-free high heating value (q_c) of hardwoods ranges from 19 to 21 MJ/kg and that of softwoods ranges slightly greater at 20–22 MJ/kg. Hardwoods generally have a higher density, higher hemicellulose, and a lower lignin content, and sometimes a slightly higher cellulose content.

Wood and coal are hygroscopic and absorb water, which is contained in both free and bound form. Free water is water that is absorbed by capillary uptake and resides in the wood or coal pores. Bound water is water that is adsorbed into the wood or coal cell wall, and is held by hydrogen bonds. Adsorption is characterized by the release of thermal energy due to the formation of hydrogen bonds and the interaction of Van der Waal's forces. Therefore, the heating value of a given wood species is strongly dependent on its free and bound water content. As such, the energy required to dry a given species of wood depends on the energy needed to heat the wood and the subcooled water to 373 K, the heat of vaporization (~ 2400 kJ/kg) of the free water, and the heat of sorption (~200–300 kJ/kg)of the bound water, all of which also affect the rate of combustion. The moisture content of wood can range from 5% for dried wood to 50% for green wood.

The composition of solid fuels is reported in three ways, as-received (ar), dry (*d*), or dry, ash-free (daf). The as-received composition is therefore

$$m_{\text{ar}} = m_{\text{daf}} + m_w + m_{\text{ash}} \tag{1.5}$$

where m_w is the mass of water and m_{ash} is the mass of ash in the fuel. The heating value of the fuel q_c will be reduced when the non-combusting moisture and ash in the fuel is accounted for:

$$q_{\text{c,ar}} = q_{\text{c,daf}} \frac{m_{\text{daf}}}{m_{\text{ar}}} \tag{1.6}$$

Example 1.4 ***Solid Fuel Heating Value***

From proximate analysis, a sample of pinewood has 25% moisture, 6% ash, and a high heating value of 20 MJ/kg on a dry, ash-free basis. What is the high heating value of an as-received (ar) basis?

Solution

The as-received mass m_{ar} is

$$\begin{aligned} m_{\text{ar}} &= m_{\text{daf}} + m_w + m_{\text{ash}} \\ &= m_{\text{daf}} + 0.25\, m_{\text{daf}} + 0.06\, m_{\text{daf}} \\ &= 1.31\, m_{\text{daf}} \end{aligned}$$

The heating value on an as-received basis is

$$\begin{aligned} q_{\text{c,ar}} &= q_{\text{c,daf}} \frac{m_{\text{daf}}}{m_{\text{ar}}} \\ &= 20 \frac{m_{\text{daf}}}{1.31\, m_{\text{daf}}} \\ &= 15.3\,\text{MJ/kg} \end{aligned}$$

1.10 Additional Reading

The references at the end of this introductory chapter contain a listing of both past and current books that will provide further information about both combustion science and engineering. These sources will give the reader a deeper appreciation of how our understanding and modeling of combustion processes has advanced in the last 50 years. In chronological order, these books are Kanury (1975), Gaydon et al. (1979), Williams (1985), Lewis and von Elbe (1987), Linan and Williams (1994), Fristrom (1995), Drysdale (1999), Law (2006), Warnatz et al. (2006), McAllister et al. (2011), Kuo and Acharya (2012), Glassman et al. (2014), Turns and Haworth (2020), Kirkpatrick (2021), Poinsot and Veynante (2022), and Bryden et al. (2022).

In addition, scholarly journals including

Combustion and Flame,
Combustion Science and Technology,
Journal of Chemical Physics,

Journal of Physical Chemistry,
Journal of Propulsion and Power,
Progress in Energy and Combustion Science, and
Proceedings of the Combustion Institute

provide up to date research findings in the various combustion areas.

References

Bryden, K., K. Ragland, and S. Kong (2022), *Combustion Engineering*, CRC Press, Boca Raton, FL.

Drysdale, D. (1999), *An Introduction to Fire Dynamics*, 2nd ed., John Wiley and Sons, New York.

Fristrom, R. (1995), *Flame Structure and Processes*, Oxford University Press, New York.

Gaydon, A., H. Wolfhard, and S. Penner (1979), *Flames, Their Structure, Radiation and Temperature*, 4th ed., Chapman and Hall, London, England.

Glassman I., R. Yetter, and N. Glumac (2014), *Combustion*, 5th ed., Academic Press. Cambridge, MA.

Kanury A. (1975), *Introduction to Combustion Phenomena*, Gordon and Breach, New York.

Kirkpatrick, A. (2021), *Internal Combustion Engines: Applied Thermosciences*, 4th ed., John Wiley and Sons, New York.

Kuo, K. and R. Acharya (2012), *Fundamentals of Turbulent and Multiphase Combustion*, John Wiley and Sons, New York.

Law, C. K. (2006), *Combustion Physics*, Cambridge University Press, New York.

Lewis, G. and B. von Elbe (1987), *Combustion, Flames, and Explosions of Gases*, 3rd ed., Elsevier, New York.

Linan, A. and F. Williams (1994), *Fundamental Aspects of Combustion*, Oxford University Press, Oxford, UK.

Malenshek, M. and D. Olsen (2009), Methane number testing of alternative gaseous fuels, *Fuel*, Vol. 88, pp. 650–656.

McAllister, S., J. Chen, and A. Fernandez-Pello (2011), *Fundamentals of Combustion Processes*, Springer, New York.

Owen, K. and T. Coley (1995), *Automotive Fuels Reference Book*, Society of Automotive Engineers, Warrendale, PA.

Poinsot, T. and D. Veynante (2022), *Theoretical and Numerical Combustion*, 3rd ed., CNRS.

Smil, V. (2017), *Energy Transitions: Global and National Perspectives*, Praeger, Santa Barbara, CA.

Tillman, D. A., A. J. Rossi, and W. D. Kitto (1981), *Wood Combustion-Principles, Processes, and Economics*, Academic Press, New York.

Turns, S. and D. Haworth (2020), *An Introduction to Combustion*, McGraw-Hill, New York.

Warnatz, J., U. Maas, and R. Dibble (2006), *Combustion*, Springer-Verlag, Berlin, Germany.

Williams, F. (1985), *Combustion Theory*, 2nd ed., CRC Press, Boca Raton, FL.

Homework Problems

1.1 What is the chemical structure of 2,4-diethylpentane?

1.2 What is the chemical structure of (a) *n*-hexane, (b) methylnaphthalene, and (c) heptamethylnonane?

1.3 An internal combustion engine fuel is a mixture of 70% octane and 30% nitromethane, by volume. What should the mass air–fuel ratio be to run rich at an equivalence ratio $\phi = 1.24$?

1.4 If a hydrocarbon fuel is represented by the general formula C_xH_{2x}, what is its stoichiometric mass air–fuel ratio?

1.5 A fuel has the following composition by mass: 10% pentane, 35% heptane, 30% octane, and 25% dodecane. If its general formula is of the form C_xH_y, (a) find x and y; (b) What is its stoichiometric combustion equation?

1.6 What is the percentage difference in the heat of combustion for methanol when used as a liquid at 25 °C and when used as a vapor at 25 °C? (Hint: Review Tables 2.1 and 2.9.)

1.7 What are the normalized CO_2 emissions (g_{CO_2}/MJ_{fuel}) resulting from the stoichiometric combustion of propane, methane, methanol, ethanol, and gasoline?

1.8 Pinewood has a high heating value (ash-free dry basis) of 24.2 MJ/kg. What is the ash-free high heating value and lower heating value as a function of moisture content at 0, 20, 40, and 60%?

2

Chemical Thermodynamics

2.1 Introduction

In this chapter, we introduce and discuss some of the fundamental concepts of chemical thermodynamics and their application to combustion problems. The chapter includes multicomponent ideal-gas properties, followed by the first and second laws, stoichiometry, combustion modeling, and the properties of combustion products. We will develop relationships for the thermodynamic properties of both fuel–air mixtures and combustion products as a function of pressure, temperature, and species mole fractions.

Chemical thermodynamics treats matter as a macroscopic system and does not address detailed molecular structures and microscopic events. Unlike kinetic theory, which can provide information about the rates of chemical reactions, chemical thermodynamics, if given the initial state of a thermodynamic system, allows the determination of the final equilibrium state of the system.

When discussing the laws of chemical thermodynamics, it is useful to classify systems according to the exchanges of energy (heat and work) and mass that can take place across the boundary of the thermodynamic system:

- *Isolated* systems exchange neither energy nor mass with their surroundings, i.e., $\delta Q = 0$ and $\Delta m = 0$.
- *Closed* systems exchange energy but no mass with their surroundings.
- *Open* systems exchange both energy and mass with their surroundings.

It is also useful to define *thermodynamic equilibrium* by considering the following three distinct kinds of equilibrium:

- *Mechanical equilibrium* exists when there are no unbalanced forces in the interior of a system or between a system and its surroundings.
- *Thermal equilibrium* exists when all parts of a system are at the same temperature, which is the same as that of the surroundings.
- *Chemical equilibrium* exists when a system has no tendency to undergo a spontaneous change in chemical composition, no matter how slow.

When all three kinds of equilibrium are satisfied, the system is said to be in a state of *thermodynamic equilibrium*. Under such conditions, the analysis becomes simpler, since the state variables do not change with respect to time. The state of complete equilibrium can then be described in terms of macroscopic coordinates. It is found that the thermodynamic coordinates that are independent and appropriate for combustion studies are the pressure in the system p, the volume V, and n_i, which represents the moles of all chemical species in different phases present in the system.

Principles of Combustion, Third Edition. Allan T. Kirkpatrick and Kenneth K. Kuo.

Companion website: www.wiley.com/go/kirkpatrick/principlesofcombustion3e

The approach to be taken here will be that of irreversible (nonequilibrium) thermodynamics rather than classical. It is useful to understand the major difference between the two and the reason for using the nonequilibrium approach in combustion studies. When a process is followed by means of classical thermodynamics, it must be regarded as consisting of a succession of states of thermodynamic equilibrium, that is, as taking place infinitely slowly. Such processes are necessarily reversible. A *reversible* process is one that is performed in such a way that the system and its surroundings can both return to their initial states. Such a process must be carried out very slowly so that the system remains in equilibrium throughout. In the real world, we must deal with irreversible processes that are a succession of nonequilibrium states.

When we come to the description of a system in chemical nonequilibrium, we shall assume that the system is in mechanical and thermal equilibrium and that it is homogeneous in space. We assume the system to have a definite volume V; since it is in mechanical and thermal equilibrium, we can assign to it a definite pressure p. Furthermore, since the system is homogeneous, its composition can be specified by giving the number of moles, n_i, of each of the constituent chemical species, $\mathcal{A}_i$. The unit g-mole is commonly used for n_i; one g-mole designates the quantity of substance $\mathcal{A}_i$ whose mass in grams is equal to the molecular mass of the substance. One mole of any substance contains 6.02252×10^{23} molecules; these molecules could be in the form of atoms, ions, radicals, or undissociated molecules. This number is called Avogadro's number in honor of the Italian physicist Amedeo Avogadro (1776–1856).

2.2 Properties of Gas Mixtures

In this section, we review the properties of fuel–air mixtures and combustion products. An *intensive* property is one that is unchanged when the size of the system is increased by adding to it any number of systems that are identical to the original system. Some intensive properties are density ρ, pressure p, temperature T, chemical potential of jth species μ_j, and mass fraction of kth species Y_k. An *extensive* property is one that increases in proportion to the size of the system in such a process. Some extensive properties are volume V, mass m, total stored energy E, enthalpy H, and Gibbs free energy G.

In the modeling that follows, we treat the fuel–air (or fuel–oxidant) mixture and the products of combustion as ideal-gas mixtures. A few words about the atmosphere are in order. The properties of air vary geographically, with altitude and with time. In this text, we will assume that air is 21% oxygen and 79% nitrogen by volume, i.e., for each mole of O_2, there are 3.76 moles of N_2. Selected physical properties of air, oxygen, and nitrogen are given in the Appendix C. The extension of our analyses to different air mixtures encountered in practice is straightforward. The most frequent differences accounted for are the presence of water and argon in air.

For an ideal gas, the familiar relationships between pressure p, temperature T, and volume V are given by the ideal-gas equation:

$$\begin{aligned} pV &= nR_u T \\ pV &= mRT \\ pv &= RT \end{aligned} \tag{2.1}$$

The mass, m (kg), of a gas mixture is the sum of the mass of all k components

$$m = \sum_{i=1}^{N} m_i \tag{2.2}$$

The mass fraction, Y_i, of any given species is defined as

$$Y_i = \frac{m_i}{m} \tag{2.3}$$

and it should be clear that

$$\sum_{i=1}^{N} Y_i = 1 \tag{2.4}$$

The total number of moles, n, of a mixture is the sum of moles of all N components

$$n = \sum_{i=1}^{N} n_i \tag{2.5}$$

and the mole fraction X_i of any species is the fraction of the total number of moles

$$X_i = \frac{n_i}{n} \tag{2.6}$$

Using the rule of additive pressures for an ideal-gas mixture, formulated by John Dalton (1766–1844), an English scientist, in which the partial pressure p_i is the pressure resulting from a gas species occupying the entire volume at the given temperature, the mole fraction for low pressures can also be defined as

$$X_i = \frac{p_i}{p} \tag{2.7}$$

The molecular mass M (kg/kmol) of a mixture is the conversion factor required between molar intensive and mass intensive units:

$$M = \sum_{i=1}^{N} X_i M_i \tag{2.8}$$

The relationship between Y_i, the mass fraction, and X_i, the mole fraction, is

$$Y_i = \frac{m_i}{m} = \frac{n_i M_i}{nM} = X_i \frac{M_i}{M} \tag{2.9}$$

If there are just two components in a gas mixture, say fuel (F) and air (A), then the fuel mass fraction Y_F is related to the fuel mole fraction X_F by

$$Y_F = \frac{1}{1 + (1/X_F - 1)M_A/M_F} \tag{2.10}$$

The mass-specific gas constant R is related to the molar universal gas constant R_u by

$$R = \frac{R_u}{M} \tag{2.11}$$

where $R_u = 8.314$ kJ/kmol-K. The molar concentration $[M]$ (kmol/m^3) of a gas mixture is

$$[M] = \frac{p}{R_u T} \tag{2.12}$$

We adopt capital letters for extensive variables and reserve lowercase letters for intensive, i.e., specific (per unit mass or mole) variables. Accordingly, the internal energy U (kJ) of a mixture is the sum of the internal energy u_i of all k components

$$U = \sum_{i=1}^{N} m_i u_i \tag{2.13}$$

The specific internal energy u (kJ/kg) is

$$u = \sum_{i=1}^{N} Y_i \, u_i \tag{2.14}$$

The internal energy U of a mixture can also be written on a molar basis as

$$U = \sum_{i=1}^{N} n_i \, \bar{u}_i \tag{2.15}$$

where the molar-intensive properties are denoted with an overbar. The specific molar internal energy (kJ/kmol) is the mole fraction weighted sum of the component internal energies:

$$\bar{u} = \sum_{i=1}^{N} X_i \, \bar{u}_i \tag{2.16}$$

$$\bar{u}_i = M_i \, u_i \tag{2.17}$$

Analogous relations for the enthalpy H (kJ) and specific enthalpy h (kJ/kg) are

$$H = \sum_{i=1}^{N} m_i \, h_i \tag{2.18}$$

$$h = \sum_{i=1}^{N} Y_i \, h_i \tag{2.19}$$

Note that the enthalpy h_i of a component is evaluated at the total pressure p

$$h_i = u_i + p v_i$$

On a molar basis, the enthalpy is

$$H = \sum_{i=1}^{N} n_i \, \overline{h}_i \tag{2.20}$$

$$\overline{h} = \sum_{i=1}^{N} X_i \, \overline{h}_i \tag{2.21}$$

$$\overline{h}_i = M_i \, h_i \tag{2.22}$$

The constant pressure and constant volume-specific heats are defined as follows:

$$c_p = \left(\frac{\partial h}{\partial T} \right)_p \tag{2.23}$$

$$c_v = \left(\frac{\partial u}{\partial T} \right)_v \tag{2.24}$$

Useful relationships between ideal-gas-specific heats c_v, c_p, and the gas constant R can be developed from the definition of enthalpy h:

$$\begin{aligned} h &= u + pv \\ &= u + RT \end{aligned} \tag{2.25}$$

in differential form,

$$\begin{aligned} dh &= du + R\,dT \\ c_p\,dT &= c_v\,dT + R\,dT \end{aligned} \tag{2.26}$$

so

$$R = c_p - c_v$$
$$\frac{c_v}{R} = \frac{1}{\gamma - 1} \tag{2.27}$$
$$\frac{c_p}{R} = \frac{\gamma}{\gamma - 1}$$

The entropy S (kJ/K) of a mixture is the sum of the entropy of each component

$$S = \sum_{i=1}^{N} m_i s_i = \sum_{i=1}^{N} n_i \bar{s}_i \tag{2.28}$$

The Gibbs equation relates entropy to temperature and pressure, so for an ideal gas,

$$ds = c_p \frac{dT}{T} - R\frac{dp}{p} \tag{2.29}$$

Upon integration, the mass and molar-specific entropy, s_i (kJ/kg-K) and $\bar{s}_i$ (kJ/kmol-K), of component i are

$$s_i(T, p_i) = s_i^o(T) - R_i \ln(p_i/p_o) \tag{2.30}$$

$$\bar{s}_i(T, p_i) = \bar{s}_i^o(T) - R_u \ln(p_i/p_o) \tag{2.31}$$

where standard molar entropy $\bar{s}_i^o(T)$ is defined as

$$\bar{s}_i^o(T) = \int_{T_o}^{T} \bar{c}_p \frac{dT}{T} \tag{2.32}$$

The standard molar entropy $\bar{s}_i^o$ of a component depends only on temperature T and is evaluated at the reference pressure p_o, i.e., $p_i = p_o = 1$ bar. Tabulated molar standard entropy data for ideal gases at $p_o = 1$ bar are given in the Appendix C. At 298 K, the molar standard entropy of carbon dioxide, CO_2, is 213.794 kJ/kmol-K.

The third law of thermodynamics, postulated by W. Nerst (1864–1941), sets the entropy of pure crystalline elements and compounds to zero at a temperature of 0 K, which has been chosen as the reference state for entropy. Values of the specific entropy at other states are determined relative to 0 K. It follows that a substance that does not have a perfect crystalline structure at absolute zero but instead has a degree of randomness (such as a solid solution or a glassy solid) has a finite value of entropy at absolute zero. The experimental evidence on which the third law rests is primarily data on chemical reactions at low temperatures and measurements of the heat capacity at temperatures approaching absolute zero.

Substitution of Eq. (2.30) into Eq. (2.28) yields convenient relations for the mixture mass (s) and molar ($\bar{s}$)-specific entropies with separate pressure- and temperature-dependent terms:

$$s = \frac{S}{m} = \sum Y_i s_i = -R \ln(p/p_o) + \sum_{i=1}^{N} Y_i (s_i^o - R_i \ln X_i) \tag{2.33}$$

$$\bar{s} = \frac{S}{N} = \sum X_i \bar{s}_i = -R_u \ln(p/p_o) + \sum_{i=1}^{N} X_i (\bar{s}_i^o - R_u \ln X_i) \tag{2.34}$$

The molar-specific Gibbs free energy for a component is defined as

$$\bar{g}_i = \bar{h}_i - T\bar{s}_i \tag{2.35}$$

We separate the temperature and pressure terms in $\bar{g}_i$ by defining the standard state Gibbs free energy $\bar{g}_i^o$ evaluated at p_o:

$$\bar{g}_i^o = \bar{h}_i^o - T\bar{s}_i^o \tag{2.36}$$

where $\bar{h}_i^o = \bar{h}_i$, since for an ideal gas, the enthalpy h is a function of temperature only. Upon substitution of Eqs. (2.31) and (2.36) into Eq. (2.35),

$$\begin{aligned}\bar{g}_i(T, p_i) &= \bar{h}_i(T) - T[\bar{s}_i^o(T) - R_u \ln(p_i/p_o)] \\ &= \bar{g}_i^o + TR_u \ln(p_i/p_o)\end{aligned} \tag{2.37}$$

The Gibbs free energy G of an ideal-gas mixture is defined as

$$\begin{aligned}G = H - TS &= \sum_{i=1}^{N} n_i \bar{g}_i \\ &= \sum_{i=1}^{N} n_i (\bar{h}_i - T\bar{s}_i)\end{aligned} \tag{2.38}$$

and the molar-specific Gibbs free energy for an ideal-gas mixture is

$$\bar{g} = \frac{G}{n} = \sum X_i \bar{g}_i = \bar{h} - T\bar{s} \tag{2.39}$$

Example 2.1 ***Properties of Ideal-Gas Mixtures***

Compute the molecular mass M, mass-specific gas constant R, molar concentration $[M]$, mass-specific enthalpy h (kJ/kg), mass-specific entropy s (kJ/kg-K), and mass-specific Gibbs free energy g (kJ/kg) of a mixture of combustion products at $p = 2000$ kPa and $T = 1000$ K. The constituents and their mole fractions are:

Species	X_i
CO_2	0.109
H_2O	0.121
N_2	0.694
CO	0.0283
H_2	0.0455

Solution

Using the tabular ideal-gas data in the Appendix, the following table of component properties can be generated and the mixture properties computed:

Species	X_i	M_i (kg/kmol)	$\bar{h}_{f,i}^o$ (kJ/kmol)	$\Delta\bar{h}(T)$ (kJ/kmol)	$\bar{s}_i^o$ (kJ/kmol-K)	$X_i(\bar{s}_i^o - R_u \ln X_i)$ (kJ/kmol-K)
CO_2	0.109	44.01	−393, 522	33,397	269.30	31.36
H_2O	0.121	18.015	−241, 826	26,000	232.74	30.28
N_2	0.694	28.013	0	21,463	228.17	160.05
CO	0.0283	28.01	−110, 527	21,686	234.54	7.48
H_2	0.0455	2.016	0	20,663	166.22	8.73

1. Molecular mass: $M = \sum X_i M_i = 27.3$ kg/kmol.
2. Mass-specific gas constant $R = R_u/M = 8.314/27.3 = 0.305$ kJ/kg-K.
3. Molar concentration $[M] = p/R_u T = 2000/(8.314)(1000) = 0.240$ kmol/m^3.
4. Molar-specific enthalpy: $\overline{h} = \sum X_i\, \overline{h}_i = -52{,}047$ kJ/kmol.
5. Molar-specific entropy: $\overline{s} = -R_u \ln(p/p_o) + \sum X_i\,(\overline{s}_i^o - R_u \ln X_i) = 213.5$ kJ/kmol-K.
6. Molar-specific Gibbs free energy: $\overline{g} = \overline{h} - T\overline{s} = -265{,}550$ kJ/kmol.

Therefore,

$$h = \overline{h}/M = -1906 \text{ kJ/kg}$$

$$s = \overline{s}/M = 7.82 \text{ kJ/kg-K}$$

$$g = \overline{g}/M = -9727.0 \text{ kJ/kg}$$

Comment: The entropy is calculated using the partial pressure, and the values of enthalpy and Gibbs free energy are relatively low, since this is a mixture of the products of combustion. The gas constant $R \approx 0.30$ for a mixture of combustion products is greater than that for air $R = 0.287$.

2.2.1 Specific Heat of Gas Mixtures

If the composition, i.e., mole fractions, of a gas mixture are known, the thermodynamic properties h, u, s, and v of the mixture are found by the application of the above property relations. The constant pressure-specific heat c_p of a gas mixture requires a more detailed analysis. The constant pressure-specific heat depends not only on the change in enthalpy but also on the change in the mixture composition as a function of temperature. It is defined as

$$c_p = \left(\frac{\partial h}{\partial T}\right)_p \tag{2.40}$$

Since $h = \overline{h}/M$,

$$\frac{\partial h}{\partial T} = \frac{1}{M}\frac{\partial \overline{h}}{\partial T} + \overline{h}\frac{\partial}{\partial T}\left(\frac{1}{M}\right) \tag{2.41}$$

Differentiating the molar-specific heat with respect to temperature,

$$\overline{h} = \sum_{i=1}^{N} X_i\, \overline{h}_i \tag{2.42}$$

$$\frac{\partial \overline{h}}{\partial T} = \sum_{i=1}^{N} X_i \frac{\partial \overline{h}}{\partial T} + \sum_{i=1}^{N} \overline{h}_i \frac{\partial X_i}{\partial T} \tag{2.43}$$

results in

$$c_p = \frac{1}{M}\left[\sum_{i=1}^{N} X_i \overline{c}_{pi} + \sum_{i=1}^{N} \overline{h}_i \frac{\partial X_i}{\partial T}\right] - \frac{\overline{h}}{M^2}\frac{\partial M}{\partial T} \tag{2.44}$$

The molecular mass is $M = \sum X_i M_i$, so upon substitution,

$$c_p = \frac{1}{M}\left[\sum_{i=1}^{N} X_i \overline{c}_{pi} + \sum_{i=1}^{N} \overline{h}_i \frac{\partial X_i}{\partial T} - \frac{\overline{h}}{M}\sum_{i=1}^{N} M_i \frac{\partial X_i}{\partial T}\right] \tag{2.45}$$

The frozen-specific heat, $c_{p,f}$, is computed holding the composition constant. It is defined as

$$c_{p,f} = \frac{1}{M}\sum_{i=1}^{N} X_i \bar{c}_{p,i} \tag{2.46}$$

Note the important role that the changes in mole fraction with respect to pressure and temperature, $\partial X_i/\partial p$ and $\partial X_i/\partial T$, have in the determination of the mixture-specific heat.

Since $c_p \sim \Delta h/\Delta T$, interpolation of the tabulated enthalpies $h(T)$ in the Appendix for a given temperature T can provide an estimate of $c_p(T)$ for ideal gases. For computer calculations, the specific heats of various species have been curve-fitted to polynomials by minimizing the least squares error (Gordon and McBride 1994). The function we will employ for any given species is

$$\frac{c_p}{R} = \frac{\bar{c}_p}{R_u} = a_1 + a_2 T + a_3 T^2 + a_4 T^3 + a_5 T^4 \tag{2.47}$$

For an ideal gas, $dh = c_p(T)dT$ and $ds = (c_p(T)/T)dT$, so the enthalpy and standard entropy at atmospheric pressure can be expressed as

$$\frac{h}{RT} = \frac{\bar{h}}{R_u T} = \frac{1}{RT}\int c_p(T)dT = a_1 + \frac{a_2}{2}T + \frac{a_3}{3}T^2 + \frac{a_4}{4}T^3 + \frac{a_5}{5}T^4 + \frac{a_6}{T} \tag{2.48}$$

$$\frac{s^o}{R} = \frac{\bar{s}^o}{R_u} = \int (c_p(T)/RT)dT = a_1 \ln T + a_2 T + \frac{a_3}{2}T^2 + \frac{a_4}{3}T^3 + \frac{a_5}{4}T^4 + a_7 \tag{2.49}$$

where a_6 and a_7 are constants of integration determined by matching the enthalpy and entropy to a zero datum at some reference temperature. As discussed above, the reference temperature for enthalpy is chosen to be 298.15 K with the enthalpy of H_2, O_2, N_2, and C(s) set to zero.

Values of the curve-fit constants for several species of interest in combustion, CO_2, H_2O, N_2, O_2, CO, H_2, H, O, OH, and NO, are given in the Appendix D for the temperature ranges 300–1000 K and 1000–3000 K. Similar curve-fit coefficients for several fuels are also given in the Appendix.

The mass-specific enthalpies of CO_2 and H_2O given by Eq. (2.48) are plotted versus temperature in Figure 2.1. At 298 K, the enthalpy of CO_2 is −8942 kJ/kg, and the enthalpy of H_2O vapor is −13,424 kJ/kg, consistent with the definition of enthalpy of formation. Note that the slope of

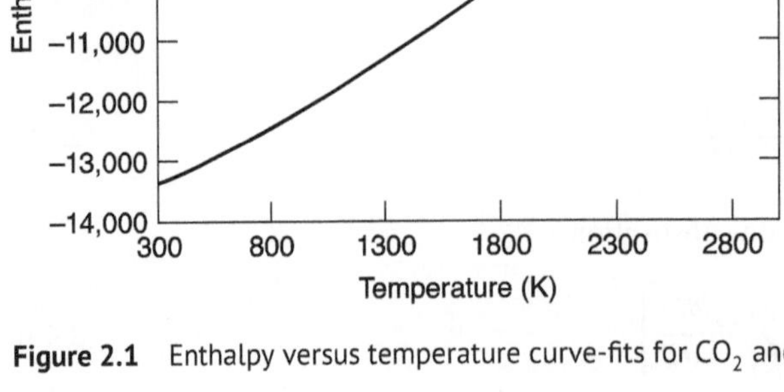

Figure 2.1 Enthalpy versus temperature curve-fits for CO_2 and H_2O.

the H_2O curve is steeper than that of the CO_2 curve, as a result of the greater specific heat of the water vapor.

Thermodynamic data for elements, combustion products, and many pollutants are also available in a compilation published by the National Institute of Standards and Technology (NIST) called the JANAF Tables. For single-component fuels, the data presented by Stull et al. (1969) are in the same format as that of the JANAF Tables.

For several decades, S. Gordon and B. McBride at the NASA Glenn Research Center provided least-square coefficients of thermodynamic property data for use in computer programs. A representative listing of their publications, McBride et al. (1993) and McBride et al. (2002), is given in the chapter references. In addition to these references, a compilation by Rossini (1953) is useful for hydrocarbon fuels at temperatures as high as 1500 K.

2.3 Liquid–Vapor–Gas Mixtures

Since liquid fuel–air mixtures are composed of more than one thermodynamic phase, the thermodynamics of fuel injection, vaporization, and condensation are complicated, as liquid fuel–air mixtures are composed of more than one thermodynamic phase. Fortunately, we can make some simplifications that are quite accurate for our intended use.

First, let us consider a pure substance in terms of its compressed liquid, saturated liquid, saturated vapor, and superheated vapor states. The simplifications that we will introduce are the following:

1. Compressed and saturated liquids are incompressible.
2. Saturated and superheated vapors are ideal gases.

For an incompressible substance, it can be shown that the internal energy and entropy depend only on temperature. Hence, the approximation for compressed liquids can be

$$u \simeq u_f(T) \tag{2.50}$$

$$s \simeq s_f(T) \tag{2.51}$$

where the notation $u_f(T)$ and $s_f(T)$ denote the internal energy and entropy of saturated liquid at the temperature T. The enthalpy of a compressed liquid depends on pressure, and it is consistent with Eqs. (2.50) and (2.51) to assume that

$$h = h_f^o + (h - h_f^o) + (p - p_{atm})\,v \tag{2.52}$$

where h_f^o is the enthalpy of formation of the compressed liquid at standard atmospheric pressure ($p_o = 101.25$ kPa) and temperature ($T_o = 298$ K). The only property remaining to be prescribed is the specific volume v. Let us choose it to be the specific volume of compressed liquid at atmospheric pressure p_o as these data are readily available:

$$v = v_o(T) \tag{2.53}$$

To introduce the enthalpy of vaporization into our equations of state for the liquids is convenient, since these data are usually easier to find than the saturated liquid data. We then have

$$h = h_g - h_{fg} + (p - p_{atm})\,v \tag{2.54}$$

Unlike specific volume, data for the enthalpy of vaporization at saturation pressure are readily available. Hence, we choose

$$h_{fg} = h_{fg}(T) \tag{2.55}$$

Typically

$$|p - p_{\text{atm}}|\,v \ll |h_{fg}| \quad \text{and} \quad |pv| \ll |h| \tag{2.56}$$

so that in many cases, we can write for liquids

$$h \approx h_g - h_{fg} = h_f^o + (h - h_f^o) - h_{fg} \tag{2.57}$$

$$u \approx h \tag{2.58}$$

Table 2.1 gives the molar enthalpy of formation for the vapor and liquid state, molar enthalpy of vaporization, the saturation pressure, and the specific volume of compressed liquid, for several liquid fuels and water at $T = 298$ K. The enthalpy of vaporization is the difference between the enthalpies of formation of the liquid and the gaseous states. Table 2.2 gives most of the same information for octane but as a function of temperature. At $T = 298$ K, the molar enthalpy $\overline{h}$ of liquid octane is $\overline{h} = \overline{h}^o_{f,\text{gas}} - \overline{h}_{fg} = -208.45 - 41.51 = -249.96$ J/mol.

The saturated vapor pressure is needed for the calculations of liquid fuel droplet evaporation. The Clausius–Clapeyron equation for the saturated vapor pressure as a function of temperature is

$$\frac{dp_{\text{sat}}}{dT} = \frac{h_{fg}}{Tv_{fg}} \tag{2.59}$$

Table 2.1 Ideal-Gas Enthalpy of Formation, Enthalpy of Vaporization, Saturation Vapor Pressure, and Specific Volume of Some Hydrocarbon Fuels at $T = 298$ K.

Formula	Name	$\overline{h}^o_{f,\text{gas}}$ (kJ/mol)	$\overline{h}^o_{f,\text{liquid}}$ (kJ/mol)	$\overline{h}_{fg}$ (kJ/mol)	p_{sat} (bar)	v_o (m^3/kg)
CH_3NO_2	Nitromethane	−74.73	−113.1	38.37	0.050	0.879×10^{-3}
CH_4O	Methanol	−201.17	−239.09	37.92	0.186	1.264×10^{-3}
C_2H_6O	Ethanol	−234.81	−277.15	42.34	0.084	1.267×10^{-3}
$C_4H_{10}O$	Ethyl ether	−252.21	−278.74	26.53	0.733	1.413×10^{-3}
C_5H_{12}	Pentane	−146.00	−172.44	26.44	0.710	1.597×10^{-3}
C_6H_{14}	Hexane	−167.19	−198.68	31.49	0.242	1.514×10^{-3}
C_6H_6	Benzene	+82.93	+48.91	34.02	0.129	1.138×10^{-3}
C_7H_{17}	Gasoline	−267.12	−305.63	38.51		1.449×10^{-3}
C_8H_{18}	Octane	−208.45	−249.96	41.51	0.022	1.423×10^{-3}
C_8H_{18}	Isooctane	−224.14	−259.25	35.11	0.071	1.445×10^{-3}
C_8H_{10}	Ethylbenzene	+29.29	−12.72	42.01	0.013	1.153×10^{-3}
$C_{12}H_{26}$	Dodecane	−290.87	−229.55	61.32	< 0.001	1.336×10^{-3}
$C_{14}H_{30}$	Tetradecane	−332.13	−403.37	71.24	< 0.001	1.311×10^{-3}
$C_{14.4}H_{24.9}$	Diesel fuel	−100.00	−174.08	74.08		1.176×10^{-3}
$C_{16}H_{34}$	Hexadecane (cetane)	−373.34	−454.48	81.14	< 0.001	1.293×10^{-3}
$C_{19}H_{40}$	Nonadecane	−435.14	−530.17	95.03	< 0.001	1.286×10^{-3}
H_2O	Water	−241.83	−285.85	43.99	0.0317	1.000×10^{-3}

Source: p_{sat} and v_o are from the Lide (2022); $\overline{h}_{fg}$ is from Vargaftik (1975); and $\overline{h}^o_f$ is from Stull et al. (1969).

Table 2.2 Enthalpy of Vaporization, Saturation Vapor Pressure, and Specific Volume of Octane, C_8H_{18}.

T (K)	$\bar{h}_{fg}$ (kJ/mol)	p_{sat} (bar)	v_o (m³/kg)
298	41.51	0.018	1.432×10^{-3}
325	39.99	0.075	1.479×10^{-3}
350	38.22	0.211	1.527×10^{-3}
375	36.45	0.503	1.579×10^{-3}
400	34.46	1.058	1.639×10^{-3}
425	32.35	2.003	1.708×10^{-3}
450	30.07	3.500	1.789×10^{-3}
475	27.49	5.707	1.890×10^{-3}
500	24.32	8.837	2.022×10^{-3}
550	20.35	13.210	2.214×10^{-3}

Source: Vargaftik (1975)/with permission of Hemisphere Pub. Corp.

Table 2.3 Curve-Fit Coefficients for Antoine's Equation for Saturation Vapor Pressure p_{sat} (bar).

Formula	Name	*a*	*b*	*c*
CH_3NO_2	Nitromethane	4.1135	1229.6	−76.221
CH_4O	Methanol	5.1585	1569.6	−34.846
C_2H_6O	Ethanol	4.9253	1432.5	−61.819
C_8H_{18}	Octane	4.0487	1355.1	−63.633
$C_{12}H_{26}$	Dodecane	4.1055	1625.9	−92.839
$C_{14}H_{30}$	Tetradecane	4.1373	1739.6	−105.62
$C_{16}H_{34}$	Hexadecane (cetane)	4.1731	1845.7	−117.05
H_2O	Water	5.4022	1838.67	−31.737

Source: https://webbook.nist.gov/.

For conditions away from the critical point, we can assume ideal-gas behavior $v_{fg} \simeq v_g \simeq RT/p_{sat}$, so Eq. (2.59) becomes

$$\frac{dp_{\text{sat}}}{p_{\text{sat}}} = \frac{1}{R}\frac{h_{fg}(T)}{T^2}dT \tag{2.60}$$

The saturated vapor pressure is typically found from an integrated form of the Clausius–Clapeyron equation, called Antoine's equation. Antoine's equation (2.61) has empirical coefficients, *a*, *b*, and *c*, that are curve-fits for a given fluid. Table 2.3 lists the curve-fit coefficients for Antoine's equation for several liquid fuels and water, and temperature *T* in Kelvin:

$$\log_{10}(p_{\text{sat}}) = \left[a - \frac{b}{T+c}\right] \tag{2.61}$$

For example, Eq. (2.61) predicts that the saturated vapor pressure p_{sat} of a cetane droplet at $T = 500$ K is 0.226 bar.

Table 2.4 Curve-Fit Coefficients for Enthalpy of Vaporization $\bar{h}_{fg}$ (kJ/mol).

Formula	Name	A (kJ/mol)	α	β	T_c (K)	p_c (bar)
CH_3NO_2	Nitromethane	53.33	0.2732	0.2732	588.0	58.7
CH_4O	Methanol	45.30	−0.3100	0.4241	512.6	81.0
C_2H_6O	Ethanol	50.43	−0.4475	0.4989	513.9	63.0
C_8H_{18}	Octane	58.46	0.1834	0.3324	568.8	24.9
$C_{14}H_{30}$	Tetradecane	95.66	0.2965	0.2965	694.0	14.4

Source: https://webbook.nist.gov/.

To integrate the Clausius–Clapeyron equation for the saturation pressure, one needs to know the temperature dependence of the enthalpy of vaporization, $\bar{h}_{fg}$. Equation (2.62) is a curve-fit for the molar enthalpy of vaporization $\bar{h}_{fg}$ as a function of the temperature ratio T/T_c, where T_c is defined as the temperature at the critical point where $\bar{h}_{fg} = 0$. Table 2.4 lists the curve-fit coefficients for Eq. (2.62) for several liquid fuels, and critical temperatures for various gases are listed in Table 2.7:

$$\bar{h}_{fg} = A\, e^{(-\alpha T/T_c)}(1 - T/T_c)^{\beta} \tag{2.62}$$

2.3.1 Quality of Exhaust Products

The products from a combustion process can include water in the vapor phase and also in the liquid phase, depending on the partial pressure of the water. If the partial pressure is greater than the saturation pressure, condensation will occur, reducing the partial pressure. The parameter that characterizes the amount of water vapor in the combustion products is the quality χ, which has its origins in the steam power industry. A value of $\chi = 0$ indicates no water vapor, and a value of $\chi = 1$ indicates a maximum value. To obtain the equilibrium water quality χ_{eq} of combustion products, we assume the partial pressure of the water vapor is equal to the saturation pressure corresponding to the mixture temperature. The saturation pressure is determined by the water vapor mole fraction $X_{H_2O,g}$, i.e.,

$$\frac{p_{sat}}{p} = X_{H_2O,g} = \frac{n_{H_2O,g}}{n_g} \tag{2.63}$$

where n_g is the number of moles of exhaust products in the gas phase. The total number of moles of exhaust products, n, is the sum of the gas-phase and liquid-phase moles,

$$n = n_g + n_{H_2O,f} \tag{2.64}$$

and since

$$n_{H_2O} = n_{H_2O,f} + n_{H_2O,g} \tag{2.65}$$

we have

$$n_{H_2O,g} = (n - n_{H_2O})\left(\frac{p}{p_{sat}} - 1\right)^{-1} \tag{2.66}$$

Therefore, for an equilibrium mixture of products,

$$\chi_{eq} = \frac{n_{H_2O,g}}{n_{H_2O}} = \left(\frac{1}{X_{H_2O}} - 1\right)\left(\frac{p}{p_{sat}} - 1\right)^{-1} \tag{2.67}$$

If the combustion is stoichiometric, using the chemical formula of the fuel represented by $C_aH_bO_cN_d$, the equilibrium quality χ_{eq} can be expressed as

$$\chi_{eq} = \left(\frac{9.52a + 1.88b - 3.76c + d}{b}\right)\left(\frac{p}{p_{sat}} - 1\right)^{-1} \tag{2.68}$$

Example 2.2 ***Quality of Exhaust Products***

Compute the quality χ_{eq} of the water in exhaust products for the stoichiometric combustion of cetane. Assume the exhaust products are at standard reference conditions $p = p_o = 1$ bar and $T_o = 298.15$ K.

Solution

The stoichiometric combustion equation per mole of cetane ($a = 16$, $b = 34$, $c = 0$, and $d = 0$) is

$$C_{16}H_{34} + 24.5(O_2 + 3.76N_2) \longrightarrow 16CO_2 + 17H_2O + 92N_2$$

At 298.15 K, the saturation pressure P_{sat} of water vapor is 3.169 kPa, so upon substitution in Eq. (2.67), we have

$$\chi_{eq} = \left(\frac{9.52(16) + 1.88(34) - 0 + 0}{34}\right)\left(\frac{101.325}{3.169} - 1\right)^{-1} = 0.205$$

The enthalpy of the system can be written as

$$H = (1 - \chi)\, m_1\, (h_{g,1} - h_{fg}) + \chi\, m_1 h_{g,1} + \sum_{i=2}^{N} m_i h_i \tag{2.69}$$

or

$$h = \sum_{i=1}^{N} Y_i h_i - (1 - \chi) Y_1 h_{fg} \tag{2.70}$$

The indexing is chosen so that $i = 1$ corresponds to the condensable substance and $i = 2, \dots, N$ corresponds to all other gases. Likewise, the system volume is

$$V = (1 - \chi)\, m_1\, (v_{g,1} - v_{fg}) + \chi\, m_1\, v_{g,1} + \sum_{i=2}^{N} m_i\, v_i \tag{2.71}$$

and the specific volume is

$$v = \sum_{i=1}^{N} Y_i\, v_i - (1 - \chi) Y_1\, v_{fg} \tag{2.72}$$

where the v_i are computed at the total pressure P in accordance with the Amagat Law of Additive Volumes, postulated by E. Amagat (1841–1915), a French scientist, which states that the total volume of a gas mixture is equal to the sum of the volumes of the component gases, if their temperature and pressure are the same.

Using the ideal-gas equation, the specific volume v_i of a component in a gas mixture is therefore

$$v_i = \frac{TR_u}{pM_i} \tag{2.73}$$

2.4 Standard Enthalpies of Formation

The enthalpy h is defined using a standard reference state, $T = 298.15$ K and $p_o = 1$ bar, and has two parts, the enthalpy related to the chemical bond energy needed to form the substance from its elements, defined as the sum of the enthalpy of formation $\overline{h}_f^o$ and the enthalpy $\Delta\overline{h}(T)$ related to the temperature T. The superscript o is used to designate the standard state enthalpy. Values of specific enthalpy at other states are determined relative to the enthalpy of formation at $T = 298.15$ K and $p_o = 1$ bar, as shown by Eq. (2.74):

$$\overline{h}(T) = \overline{h}_f^o + \Delta\overline{h}(T) \tag{2.74}$$

The standard state of an element is the stable form of that element at room temperature and 1 bar. By convention, each *element* in its standard state is assigned an enthalpy of zero or has a standard heat of formation of zero. For *compounds*, the enthalpy of formation is the enthalpy required to form the compound from its elements in their standard states. When energy is released (*exothermic reaction*) in the formation of the compound, then the $\overline{h}_f^o$ value of that compound is a negative quantity, since energy must be removed from the system in order to maintain an isothermal reaction process.

For example, the enthalpy of formation of CO_2 is $-393,522$ kJ/kmol $= -8942$ kJ/kg. As we shall see, due to the difference in the bond energies, the enthalpy of CO_2 (-8942 kJ/kg) at the standard reference state is consequently less than the enthalpy of its elements C and O_2 (0 kJ/kg) at the same reference state. Similarly, the enthalpy of formation of H_2O vapor is $-241,826$ kJ/kmol $= -13,424$ kJ/kg.

The enthalpy of formation of a hydrogen atom H from a hydrogen molecule H_2 is a positive value equal to 218.08 kJ/mol. This is due to the fact that external energy must be supplied for the reaction to proceed (*endothermic reaction*). When energy is absorbed in the formation of a compound, the heat of formation of that compound is a positive quantity, since heat must be added to the system. It is useful to note that substances with large positive standard heats of formation are usually more chemically active species. Tabular molar-specific enthalpy data for elemental gases and combustion products are given in the Appendix.

In 1780, A. L. Lavoisier and P. S. Laplace developed the following law, stated as follows:

The quantity of heat, which must be supplied to decompose a compound into its elements, is equal to the heat evolved when the compound is formed from its elements. A more general form of this law states that *the heat change accompanying a chemical reaction in one direction is exactly equal in magnitude, but opposite in sign, to that associated with the same reaction in the reverse direction.*

In 1840, G. H. Hess developed empirically what is known as the law of constant heat summation. This law states that *the resultant enthalpy change, at constant pressure or constant volume, in a given chemical reaction is the same whether it takes place in one or in several stages.* This means that the net heat of reaction depends only on the initial and final states, as expected, since enthalpy is a thermodynamic property.

Hess's law can be used to determine the overall energy required for a chemical reaction by dividing it into steps that are individually easier to characterize. As a result of Hess's law, thermochemical equations can be added and subtracted just like algebraic equations. The following example demonstrates the application of this law.

Example 2.3 ***Enthalpy of Formation***

Show that the standard enthalpy of formation $\overline{h}_f^o$ of CO_2 of 393.68 kJ/mol can be obtained from the known enthalpy of formation of carbon monoxide CO and the energy released from the following reaction:

$$CO(g) + \frac{1}{2}O_{2(g)} \xrightarrow{298.15K} CO_{2(g)} - 283.09\ kJ$$

Solution

Note that the energy released in the above exothermic reaction is not the standard enthalpy of formation for 1 mole of $CO_{2(g)}$, since the reactant is not an element in its standard state. In the formation reaction of $CO_{(g)}$, we have

$$C_{(s)} + \frac{1}{2}O_{2(g)} \xrightarrow{298.15K} CO_{(g)} - 110.59\ kJ$$

Adding the above two thermochemical equations and subtracting the common CO term according to Hess's law, we have

$$C_{(s)} + O_{2(g)} \xrightarrow{298.15K} CO_{2(g)} - 393.68\ kJ$$

Note that the two species (carbon solid and oxygen gas) on the reactant side in the above equation are elements in their respective standard states. Therefore, this thermochemical equation is the formation reaction of carbon dioxide, and the standard enthalpy of formation of $CO_{2(g)}$ is

$$(\overline{h}_f^o)_{CO_{2(g)}} = -393.68\ kJ/mol$$

2.5 Relationship Between Bond Energy and Enthalpy of Formation

The bond energy for dissociation, $D^o(R-X)$, is also known as the strength of a chemical bond between the components R and X of the molecule RX. When a molecule is dissociated into two parts by the reaction $RX \rightarrow R + X$, the energy required to break the bond at $T = 298.15$ K is related to the enthalpies of formation of the original molecule and its dissociated components by

$$D^o_{298}(R-X) = \overline{h}_{f,R}^o + \overline{h}_{f,X}^o - \overline{h}_{f,RX}^o \tag{2.75}$$

Physically, the bond energy for dissociation represents the average amount of energy, per mole, required to break that bond in a molecule and separate the two components far apart. In general, the energy required to pull two molecular components (R and X) apart is a function of the distance between R and X, as shown in Figure 2.2. The bond energy corresponds to the difference between the energy level at infinity and the minimum of the potential energy curve. Bond energy values are an average of the bond-dissociation energies within a given molecule. Some bond energies selected are tabulated in Table 2.5. These values are given at 298.15 K.

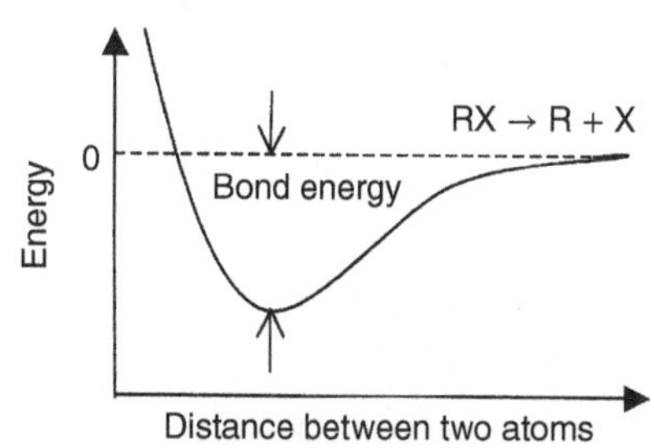

Figure 2.2 Potential energy between two molecular fragments as a function of interatomic separation distance.

Table 2.5 Mean Bond Energies.

Bond	Energy (kJ/mol)	Bond	Energy (kJ/mol)
Br—Br	193	H—F	565
C—C	348	H—H	432
C=C	614	H—I	295
C≡C	839	H—P	318
C—Br	280	H—S	339
C—Cl	326	I—I	151
C—F	485	N—H	368
C—H	411	N—N	167
C—I	268	N≡N	946
C—N	305	O—H	459
C≡N	879	O—N	201
C—O	358	O—O	142
C=O	799	O=O	498
C—S	272	P—Br	268
Cl—Cl	240	P—Cl	326
F—F	155	P—P	201
H—Br	362	S—Cl	250
H—Cl	432	S—S	226

Source: Adapted from Lide (2022).

It is useful to note that the bond energies are based on data for substances in the gaseous state and, therefore, should be used for reactions involving gases only. Also, the bond energy has the negative value of the energy required to form a particular bond. As we shall see, bond energies can be used to estimate the heats of formation of certain chemical compounds. The mean bond energies between some selected atoms are listed in Table 2.5.

In addition to the bond energy, one should take into account the possibility of resonance in some molecules in order to estimate their enthalpy of formation. Table 2.6 gives the resonance energies

Table 2.6 Resonance Energies.

Compound	Resonance Energy (kJ/mol)
Benzene, C_6H_6	205
—COOH Group (Carboxyl Group)	117
CO_2	138
Naphthalene, $C_{10}H_8$	368
Aniline, $C_6H_5NH_2$	291

Source: Adapted from Lide (2022).

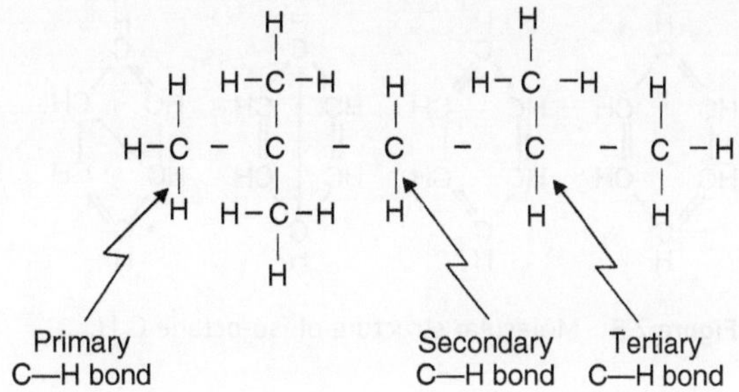

Figure 2.3 Different molecular structures of the benzene (C_6H_6) molecule.

of some molecules for which different molecular structures exist; their resonance energies must be considered in the estimation of their enthalpies of formation. Bond and resonance energies can also be used to estimate the enthalpy of combustion of a given reaction.

For example, the benzene molecule C_6H_6 can resonate between the five structures shown in Figure 2.3. As a result of resonance, the enthalpy of formation $\overline{h}^o_{f,C_6H_6}$ is much larger than the sum of three C=C, three C—C, and six C—H bond energies. The additional energy is termed the resonance energy; it must be considered in addition to the bond energies to calculate the actual heat of formation of a given chemical compound.

Example 2.4 ***Determination of Enthalpy of Formation from Bond Energy***

Use the tabulated bond energies to estimate the enthalpy of formation $\overline{h}^o_f$ of ethanol, $C_2H_6O_{(g)}$. The molecular structure of $C_2H_6O_{(g)}$ is shown in Figure 2.4

Solution

The sum of the bond energies of one mole of $C_2H_6O_{(g)}$ is

$$\begin{aligned}&5(C-H)+(C-C)+(C-O)+(O-H)\\&\quad=5\times(411)+(348)+(358)+(459)=3220\,\text{kJ/mol}\end{aligned}$$

The bond energy is the negative value of the energy required to form a particular bond, so the energy required to form one mole of $C_2H_6O_{(g)}$ is

$$6H_{(g)}+O_{(g)}+2C_{(g)}\rightarrow C_2H_6O_{(g)}-3220\,\text{kJ/mol}$$

Since the standard enthalpy of formation is a measure of the energy released or consumed when one mole of a substance is created from its standard state elements ($H_{2(g)}$, $C_{(s)}$, etc.), we next utilize Hess's law and the standard state thermochemical equations,

$$3H_{2(g)}\rightarrow 6H_{(g)}+(6\cdot 218.08=1308.50)\,\text{kJ}$$

$$\tfrac{1}{2}O_{2(g)}\rightarrow 1O_{(g)}+(1\cdot 249.31=249.31)\,\text{kJ}$$

$$2C_{(s)}\rightarrow 2C_{(g)}+(2\cdot 717.0=1434.0)\,\text{kJ}$$

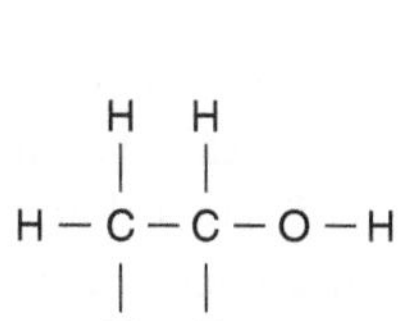

Figure 2.4 Molecular structure of $C_2H_6O_{(g)}$.

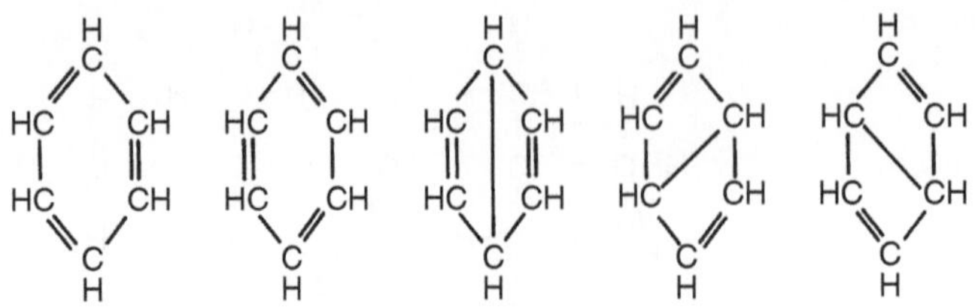

Figure 2.5 Molecular structure of iso-octane C_8H_{18}.

Summing up the above reaction equations, we have

$$3H_{2(g)} + \tfrac{1}{2}O_{2(g)} + 2C_{(s)} \rightarrow C_2H_6O_{(g)} + (-3220 + 2991.8)\text{kJ/mol of } C_2H_6O$$

Therefore, the estimated value for the enthalpy of formation $\overline{h}_f^0$ of ethanol, $C_2H_6O_{(g)}$, is $(-3220 + 2991.8) = -228.2$ kJ/mol. The actual value of the enthalpy of formation is -234.9 kJ/mol; the difference is attributed to the fact that the estimated value is based on average bond energies.

The previous example showed that the estimated heat of formation is very close to the actual value. However, this is not always true. It is important to note that the bond energy between the two atoms can vary significantly, depending on their respective locations in the molecule. For example, the molecular structure of iso-octane is shown in Figure 2.5,

The bond strength ratios between primary, secondary, and tertiary bonds of iso-octane are 13, 5, and 1, respectively. Therefore, the enthalpy of formation estimated by the mean bond energy could differ significantly from the actual value. The reason for the difference in bond strengths is due to the fact that the tertiary C–H bond has a higher probability of forming antibonding orbitals than the primary and secondary C–H bonds. In antibonding orbitals, the electrons help pull the nuclei apart rather than helping to pull them together.

2.6 Fugacity and Real-Gas Equations of State

When the combustion system pressure is at an elevated level, for example, in a rocket engine, the ideal-gas law can no longer be applied. Various forms of real-gas equations of state for high-pressure considerations are given in the Appendix. In addition to these real-gas equations of state, it is useful to consider a thermodynamic parameter called fugacity.

Fugacity is defined as the pressure of an ideal gas which has the same temperature and molar Gibbs free energy as the real gas. For example, nitrogen gas at $T = 0\,°C$ and $p = 100$ atm has a fugacity $\mathbf{f} = 97.03$ atm.

G. N. Lewis (1875–1946) in 1901 introduced the concept of *fugacity*, which has proved of great value in representing the actual behavior of real gases, especially at high pressures. Lewis was an American scientist known for his many contributions to chemical thermodynamics, including the discovery of the covalent bond and the concept of electron pairs.

Accurate calculations of chemical equilibrium for real gases at high pressure should use the fugacity $\mathbf{f}$ rather than the pressure. For ideal gases at a constant temperature, the change in the Gibbs free energy with pressure is

$$(dG)_T = nR_uT\,d(\ln p)_T \tag{2.76}$$

For a gas that departs from ideal behavior, the above equation will not hold, but the fugacity may be defined in such a manner that the relationship

$$(dG)_T \equiv nR_uT\, d(\ln \mathbf{f})_T \tag{2.77}$$

is always satisfied, irrespective of whether the gas is ideal or not. Integrating Eq. (2.77) from thermodynamic state 1 to state 2, we have

$$G_2 - G_1 = nR_uT \ln \frac{\mathbf{f}_2}{\mathbf{f}_1} \tag{2.78}$$

The fugacity of the ith species may be considered to be the corrected partial pressure for that species, where $\mathbf{f}_i$ has units of pressure,

$$\frac{\mathbf{f}_i}{p_i} = \Gamma \tag{2.79}$$

and Γ is the fugacity coefficient.

For an ideal gas,

$$\frac{\mathbf{f}}{p} = 1 \tag{2.80}$$

As the pressure of a real gas is decreased, its behavior approaches that of an ideal gas. Therefore, low pressure is chosen as the reference state, and it is assumed that the ratio $\mathbf{f}/p$ then approaches unity,

$$\lim_{p\to 0} \frac{\mathbf{f}}{p} = 1 \tag{2.81}$$

In terms of specific quantities, Eq. (2.77) can be written as

$$(dg)_T = RT\,(d \ln \mathbf{f})_T \tag{2.82}$$

For a real gas, an equation of state is $pv = ZRT$, so

$$(dg)_T = ZRT(d \ln p)_T \tag{2.83}$$

where Z is the compressibility factor of the gas. Therefore, for an isothermal process at temperature T, the fugacity can be determined from the compressibility factor Z and the pressure p by the following relationship:

$$Z\, d(\ln p)_T = d(\ln \mathbf{f})_T \tag{2.84}$$

Integrating Eq. (2.84) along an isothermal line from $p = 0$ to a nondimensional reduced pressure p_r, we have

$$\begin{aligned} \ln \Gamma = \ln \frac{\mathbf{f}}{p} &= \int_0^{p_r} (Z-1)\, d(\ln p_r)_{T_r} \\ &= \int_0^{p_r} \frac{(Z-1)}{p_r}\, dp_r \end{aligned} \tag{2.85}$$

The reduced pressure (p_r) and reduced temperature (T_r) of a gas are defined as the ratios of temperature and pressure to their critical values,

$$p_r \equiv \frac{p}{p_c} \tag{2.86}$$

$$T_r \equiv \frac{T}{T_c} \tag{2.87}$$

Table 2.7 Critical Temperatures and Pressures of Selected Gases.

Gas	Formula	Critical Temperature T_c (K)	Critical Pressure p_c (MPa)
Air	—	132.5	3.766
Methane	CH_4	190.56	4.592
Acetylene	C_2H_2	308.3	6.138
Ethylene	C_2H_4	282.34	5.041
Ethane	C_2H_6	305.32	4.872
Propane	C_3H_8	369.83	4.248
n-Butane	C_4H_{10}	425.12	3.796
Isopentane	C_5H_{12}	460.45	3.38
n-Heptane	C_7F_{16}	474.8	1.62
Chlorine	Cl_2	416.9	7.991
Carbon Dioxide	CO_2	304.13	7.375
Carbon Monoxide	CO	132.91	3.499
Fluorine	F_2	144.13	5.172
Helium	He	5.19	0.227
Hydrogen	H_2	32.97	1.293
Neon	Ne	44.40	2.760
Nitrogen	N_2	126.20	3.390
Hydrazine	N_2H_4	653.0	14.7
Nitric Oxide	NO	180.0	6.48
Nitrogen Dioxide	NO_2	430.95	10.132
Nitrous Oxide	N_2O	309.57	7.255
Oxygen	O_2	154.58	5.043
Ozone	O_3	261.1	5.57
Water	H_2O	646.99	22.064
Xenon	Xe	289.73	5.840

Source: Adapted from Lide (2022).

The critical temperatures (T_c) and pressures (p_c) of selected gases are shown in Table 2.7. At the critical point, the saturated liquid and saturated vapor states are identical, and the fluid exists in a single state. Note that hydrocarbons have a critical temperature at or above 273 K, and diatomic gases such as H_2 generally have critical temperatures below 150 K.

A generalized compressibility chart for some selected gases is shown in Figure 2.6. The compressibility factor Z is plotted as a function of reduced pressure p_r and reduced temperature T_r. The figure indicates that as the reduced temperature increases, the compressibility factor increases. The generalized chart does not apply specifically to any one substance but is instead an approximation for many gases that are fairly simple in molecular structure.

An equation of state which takes the finite volume of the molecules in a system into account is

$$p(v - b) = RT \tag{2.88}$$

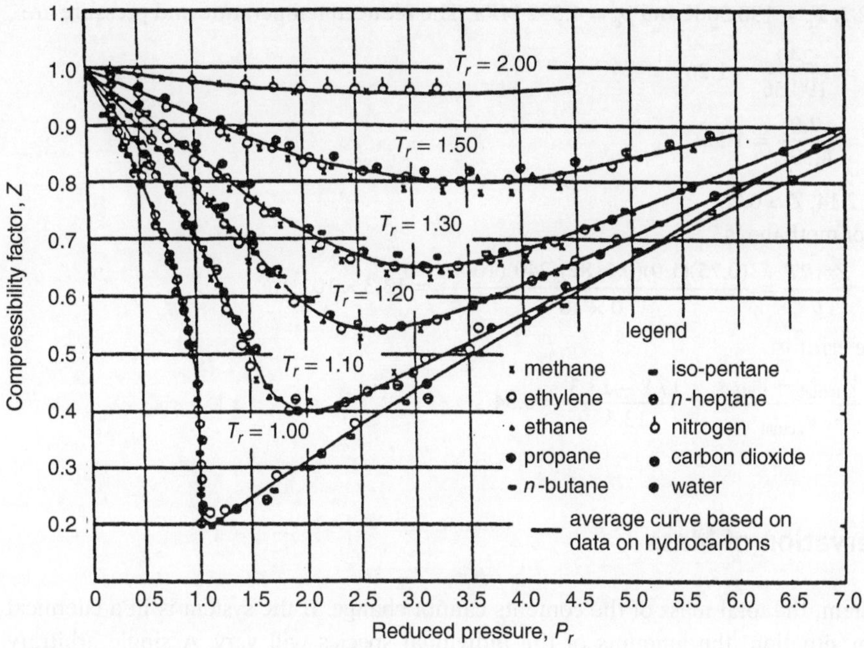

Figure 2.6 Generalized compressibility chart for selected gases. (Su 1946)/with permission of American Chemical Society.

and the corresponding compressibility factor Z is

$$Z = \frac{pv}{RT} = 1 + \frac{bp}{RT} \tag{2.89}$$

For this equation of state,

$$\begin{aligned} \ln \Gamma = \ln \frac{f}{p} &= \int_0^{P_r} \frac{(Z-1)}{P_r} \, dp_r \\ &= \int_0^{p} \frac{b}{RT} \, dp \\ &= \frac{bp}{RT} \end{aligned} \tag{2.90}$$

and the fugacity coefficient Γ is

$$\Gamma = e^{(bp/RT)} \tag{2.91}$$

Example 2.5 ***Compressibility Factor***

Using the equation of state $pV = ZmRT$, calculate the volume V of 1.0 kg of methane CH_4 at a temperature $T = 240$ K and pressure $p = 7.0$ MPa. What is the relative error if the volume V were calculated from the ideal-gas equation? The methane gas constant $R = 0.5182$ kJ/kg-K.

Solution

The methane volume from the ideal-gas law is

$$V_{\text{ideal}} = \frac{mRT}{p} = \frac{(1.0)(0.5182)(240)(1000)}{7.0 \times 10^6} = 17.8 \times 10^{-3}\ \text{m}^3$$

From Table 2.7, $T_c = 190.56$ K and $p_c = 4.592$ MPa. The reduced temperature and pressure are

$$T_r = \frac{T}{T_c} = \frac{240}{190.56} = 1.26$$

$$p_r = \frac{p}{p_c} = \frac{7.0}{4.592} = 1.52$$

From Figure 2.14, $Z \simeq 0.75$

The volume of methane is

$$V_{\text{actual}} = \frac{ZmRT}{p} = \frac{(0.75)(1.0)(0.5182)(240)(1000)}{7.0 \times 10^6} = 13.3 \times 10^{-3}\,\text{m}^3$$

and the relative error is

$$\text{error} = \frac{V_{\text{actual}} - V_{\text{ideal}}}{V_{\text{actual}}} = \frac{17.8 - 13.3}{13.3} = 0.34$$

2.7 Conservation of Mass

In a closed system, the total mass of the contents cannot change. If the system is in a chemical nonequilibrium situation, the amounts of the individual species will vary. A single arbitrary chemical reaction can be written as

$$\sum_{i=1}^{N} \nu_i \mathcal{A}_i \to \sum_{i=1}^{N} \nu_i' \mathcal{A}_i \tag{2.92}$$

where ν_i is the coefficient for species i appearing as a reactant, ν_i' is the coefficient for species i appearing as a product, and $\mathcal{A}_i$ represents the chemical symbol for species i. Species that are not reactants have $\nu_i = 0$, while those that do not appear as products have $\nu_i' = 0$.

The reaction equation (2.92) indicates a relationship between the changes in the number of moles of each species. Equation (2.92) implies that when $(\nu_i' - \nu_i)$ moles of $\mathcal{A}_i$ are formed, $(\nu_j - \nu_j')$ moles of $\mathcal{A}_j$ disappear due to the chemical reaction. The number of moles of products is not necessary equal to the number of moles of reactants, as due to the creation and consumption of species, so that $j \neq i$.

Example 2.6 ***Mole Fractions***

What is the change in the number of moles in the oxidation of CO to form CO_2, according to

$$CO + \frac{1}{2}O_2 \to CO_2 \tag{2.93}$$

Solution

Let

$$\mathcal{A}_1 = CO \tag{2.94}$$

$$\mathcal{A}_2 = O_2 \tag{2.95}$$

$$\mathcal{A}_3 = CO_2 \tag{2.96}$$

Then,

$$\nu_1 = 1 \qquad \nu_1' = 0 \tag{2.97}$$

$$\nu_2 = \frac{1}{2} \qquad \nu_2' = 0 \tag{2.98}$$

$$\nu_3 = 0 \qquad \nu_3' = 1 \tag{2.99}$$

When

$$\nu_3' - \nu_3 = 1 \text{ mol of } CO_2 \text{ is formed, } \Delta n_3 = 1 \tag{2.100}$$

$$\nu_1 - \nu_1' = 1 \text{ mol of CO disappears, } \Delta n_1 = -1 \tag{2.101}$$

$$\nu_2 - \nu_2' = \frac{1}{2}\text{mol of } O_2 \text{ disappears, } \Delta n_2 = -\frac{1}{2} \tag{2.102}$$

Example 2.6 shows that we obtained the following relationship between the changes of moles of reactants and the product species:

$$\frac{\Delta n_1}{\nu_1' - \nu_1} = \frac{\Delta n_2}{\nu_2' - \nu_2} = \frac{\Delta n_3}{\nu_3' - \nu_3} \tag{2.103}$$

Now let us generalize it to any elementary reaction involving an infinitesimal change. Introducing the dimensionless single-reaction progress variable ε, where

$$dn_i = (\nu_i' - \nu_i)\, d\varepsilon \quad i = 1, 2, \dots, N \tag{2.104}$$

If $n_{i,r}$ denotes the number of moles of the various species at the same initial or reference condition at which ε is zero, the above equation can be integrated to obtain

$$n_i - n_{i,r} = (\nu_i' - \nu_i)\,\varepsilon \quad i = 1, 2, \dots, N \tag{2.105}$$

If m_i is the mass of the ith species and M_i the molecular mass of that species, then from Eq. (2.104)

$$dm_i = (\nu_i' - \nu_i) M_i\, d\varepsilon \quad i = 1, 2, \dots, N \tag{2.106}$$

Since the total mass of a closed system is constant,

$$m = \sum_{i=1}^{N} m_i = \text{constant} \tag{2.107}$$

We therefore have

$$\sum_{i=1}^{N} dm_i = 0 \tag{2.108}$$

Substituting Eq. (2.106) into Eq. (2.108), we have

$$\sum_{i=1}^{N} [(\nu_i' - \nu_i) M_i]\, d\varepsilon = 0 \tag{2.109}$$

If the degree of the reaction is not zero (i.e., $d\varepsilon \neq 0$), then

$$\sum_{i=1}^{N} [(\nu_i' - \nu_i)] M_i = 0 \tag{2.110}$$

which is known as the *stoichiometric equation*. If Eq. (2.104) is differentiated with respect to time, we obtain the rate equation,

$$\frac{dn_i}{dt} = (\nu_i' - \nu_i)\frac{d\varepsilon}{dt} \tag{2.111}$$

2.8 The First Law of Thermodynamics

This law states that there exists an extensive function called the stored energy E, composed of (1) internal energy U, (2) kinetic energy KE, and (3) potential energy PE:

$$E = U + \mathrm{KE} + \mathrm{PE} \tag{2.112}$$

where

$$E = E(p, V, n_i) \tag{2.113}$$

The function E has the property that, for a closed system (no mass exchange with its surroundings), the heat added to the system in an infinitesimal process is related to the change of E and the work done by the system (δW), according to

$$\delta Q = dE + \delta W \tag{2.114}$$

where δ indicates an inexact differential.

The first law of thermodynamics states that during any cycle a system undergoes, the cyclic integral of the heat is equal to the cyclic integral of work, i.e.,

$$\oint \delta Q = \oint \delta W \tag{2.115}$$

where $\oint \delta Q$ is the cyclic integral of the energy transfer or the net heat transferred during the cycle, and $\oint \delta W$ is the cyclic integral of the work or the net work done during the cycle.

The physical significance of the property E is that it represents all the energy of the system in the given state. This energy can be present in many forms, including thermal energy, kinetic energy, potential energy (with respect to the chosen coordinate frame), energy associated with the motion and position of molecules, energy associated with atomic structure, chemical energy (e.g., in a storage battery), electrostatic energy (e.g., in a charged capacitor), and so on.

In the study of thermodynamics, it is convenient to consider the bulk kinetic energy and potential energy separately and then to lump all other forms of energy of the system in a single property, which we shall call *internal energy* and give the symbol U. Therefore,

$$E = \text{internal energy} + \text{kinetic energy} + \text{potential energy} = U + \mathrm{KE} + \mathrm{PE} \tag{2.116}$$

where

$$\mathrm{KE} = \frac{1}{2}m|\mathbf{v}|^2 \quad \text{and} \quad \mathrm{PE} = mgz \tag{2.117}$$

In differential form, this is $dE = dU + d(\mathrm{KE}) + d(\mathrm{PE})$, which from Eqs. (2.114) and (2.117) may be written as

$$\delta Q = dU + d\left(\frac{1}{2}m|\mathbf{v}|^2\right) + d(mgz) + \delta W \tag{2.118}$$

Assuming g is a constant, the above equation may be integrated between states 1 and 2 to give

$${}_1Q_2 = (U_2 - U_1) + m\left[\frac{|\mathbf{v}_2|^2 - |\mathbf{v}_1|^2}{2}\right] + mg(z_2 - z_1) + {}_1W_2 \tag{2.119}$$

For a closed system undergoing an infinitesimal reversible process, the first law is usually expressed in classical (i.e., equilibrium) thermodynamics by the equation

$$dU = \delta Q - p\,dV \tag{2.120}$$

A positive value of δQ represents the amount of heat the system receives from its surroundings, and a positive value of $p\,dV$ represents the work done by the system. Here, p and U are state functions related to the other variables V and T by state equations of the form

$$p = p(V, T), \quad U = U(V, T) \tag{2.121}$$

For studying chemical nonequilibrium problems, the law of conservation of energy can be applied. It is necessary, however, to redefine the state functions p and U as

$$p = p(V, T, n_1, n_2, \dots, n_N), \quad U = U(V, T, n_1, n_2, \dots, n_N) \tag{2.122}$$

In nonequilibrium thermodynamics, as in equilibrium thermodynamics, it is useful to define the enthalpy by the equation

$$H \equiv U + pV \tag{2.123}$$

For chemical nonequilibrium cases, H is also given by a state relation of the form

$$H = H(V, T, n_1, n_2, \dots, n_N) \tag{2.124}$$

It is useful to note that the first law places no restrictions on the direction of flow of heat and work. We know from our experience that a proposed cycle that does not violate the first law does not ensure that the cycle will actually occur. It is this kind of experimental evidence that has led to the formulation of the second law of thermodynamics.

2.9 The Second Law of Thermodynamics

Let us recall that a *reversible process* for a system is defined as a process that, once having taken place, can be reversed and, in so doing, leaves no change in either the system or surroundings. In general, factors that render processes irreversible include friction, heat transfer, mixing of different gases, combustion, unrestrained expansion, i^2R loss, etc.

The second law of thermodynamics postulates the existence of a state function called the entropy S and defines the basic properties of this function. Entropy is the measure of a system's energy that is *unavailable* for work. It is used in the analysis of both equilibrium and nonequilibrium systems.

For a closed system that undergoes a change from one state of thermodynamic equilibrium 1 to another state 2, the change in entropy is given by

$$S_2 - S_1 = \int_1^2 \left(\frac{\delta Q}{T}\right)_{\text{rev}} \tag{2.125}$$

where *rev* implies any reversible path between 1 and 2, δQ is the heat transfer to or from the system, and T is the corresponding absolute temperature of the boundary between the reservoir and the system. The important point to note here is that since the change in the entropy of a substance is path independent, it is also the same for all processes. Entropy is a derived property, so its value must be determined by calculation, not experiment.

The equation given here enables us to find the change in entropy only along a reversible path, but once evaluated, the magnitude of change is the same for all processes between these two states. We may also note here that T takes the role of an integrating factor in that it has converted the inexact differential δQ to the exact differential $(\delta Q/T)_{\text{rev}}$.

If the same system undergoes an irreversible or real process between the same two equilibrium end states 1 and 2, we have

$$S_{gen} = (S_2 - S_1)_{irrev} - \int_1^2 \left(\frac{\delta Q}{T}\right)_{irrev} > 0 \tag{2.126}$$

where a positive δQ is the heat added to the system in the particular process, and S_{gen} is the entropy generation, which is path dependent.

Given the existence of the state variable entropy S, the change in entropy, dS, in any closed system undergoing any process can be split into two parts:

$$dS = dS_e + dS_i \tag{2.127}$$

where dS_e is the change in entropy resulting from interaction between the system and its surroundings (i.e., heat transfer to or from the system), and dS_i is the change in entropy resulting from a process taking place within the system (e.g., chemical reaction, constant-pressure mixing of different gases). These differential quantities may also be thought of, respectively, as the flow of entropy into the system from the surroundings and the production of entropy by irreversible processes within the system.

The entropy change dS_i is never negative:

$$dS_i = 0\,\text{(reversible process)} \tag{2.128}$$

$$dS_i > 0\,\text{(irreversible process)} \tag{2.129}$$

For a *closed* system undergoing any process, reversible or irreversible, dS_e, is given by

$$dS_e = \frac{\delta Q}{T} \tag{2.130}$$

If mass is added to an *inert open reversible* system, the entropy associated with the additional mass can be considered to be a part of the entropy flow into the system from the surroundings. We then have

$$dS = dS_e = \frac{\delta Q}{T} + \sum_{j=1}^{N} s_j dn_{e,j} \tag{2.131}$$

where s_j is the entropy per mole of the jth species added, and $dn_{e,j}$ is the change in the number of moles of the jth species flowing into the system. With mass addition, the first law becomes

$$dU = \delta Q - p\,dV + \sum_{j=1}^{N} h_j dn_{e,j} \tag{2.132}$$

where h_j is the enthalpy per mole associated with the jth species flowing into the system.

Let us define a new parameter μ_j in which

$$\mu_j \equiv h_j - Ts_j \tag{2.133}$$

where μ_j represents the chemical potential, which is equal to the change of the Gibbs free energy per mole of the jth species. Dividing Eq. (2.132) by T, we obtain

$$\frac{\delta Q}{T} = \frac{dU}{T} + \frac{p\,dV}{T} - \frac{1}{T}\sum_{j=1}^{N} h_j dn_{e,j} \tag{2.134}$$

Substituting Eqs. (2.134) and (2.133) into Eq. (2.131) gives

$$dS = d_e S = \frac{1}{T}\,dU + \frac{p}{T}\,dV - \frac{1}{T}\sum_{j=1}^{N} \mu_j dn_{e,j} \tag{2.135}$$

for an *inert open reversible* system.

Equation (2.135) is based on the assumption that all the species involved were chemically inert; hence, there exist no irreversible processes resulting from chemical reactions within the system.

However, when chemical reactions occur, the mole number of jth species may change both by mass addition from outside the system and by chemical reaction inside. Then, an infinitesimal process involving change in p, T, and n_j will, in general, be *irreversible*, since the system will not be in chemical equilibrium. Therefore,

$$dn_j = dn_{i,j} + dn_{e,j} \tag{2.136}$$

where $dn_{i,j}$ represents internal differential change of the moles of the jth species due to chemical reaction occurring inside the system, while $dn_{e,j}$ represents differential change of the moles of the jth species due to the exchange of jth material with the surroundings. If we further consider each species j to occupy a separate subsystem of the total system, then each subsystem can be regarded as an inert open system. This means that for each species

$$dS_j = \frac{1}{T}\,dU_j + \frac{p_j}{T}\,dV - \frac{1}{T}\mu_j\,dn_j \tag{2.137}$$

where p_j is the partial pressure due to species j, and U_j is the internal energy associated with species j. Summing the preceding equation overall species (i.e., overall subsystems) and noting that

$$p = \sum_{j=1}^{N} p_j \tag{2.138}$$

$$U = \sum_{j=1}^{N} U_j \tag{2.139}$$

$$S = \sum_{j=1}^{N} S_j \tag{2.140}$$

we obtain for the total system

$$dS = \frac{1}{T}\,dU + \frac{p}{T}\,dV - \frac{1}{T}\sum_{j=1}^{N}\mu_j\,dn_j \tag{2.141}$$

which is the fundamental equation of chemical thermodynamics for *open irreversible chemical processes*.

From the equation of state, we have

$$S = S(U, V, n_1, n_2, \ldots, n_N) \tag{2.142}$$

or in differential form

$$dS = \left(\frac{\partial S}{\partial U}\right)_{V,n_j} dU + \left(\frac{\partial S}{\partial V}\right)_{U,n_j} dV + \sum_{j=1}^{N}\left(\frac{\partial S}{\partial n_j}\right)_{U,V,n_j'} dn_j \tag{2.143}$$

where the symbol n_j' in the last term indicates that all mole numbers except n_j itself are held constant when the derivative is taken. By comparing Eqs. (2.141) and (2.143), we have

$$\left(\frac{\partial S}{\partial U}\right)_{V,n_j} = \frac{1}{T} \tag{2.144}$$

$$\left(\frac{\partial S}{\partial V}\right)_{U,n_j} = \frac{p}{T} \tag{2.145}$$

and

$$-T\left(\frac{\partial S}{\partial n_j}\right)_{U,V,n_j'} = \mu_j \tag{2.146}$$

Since enthalpy is defined as

$$H \equiv U + pV \tag{2.147}$$

the differential form is,

$$dU = dH - p\,dV - V\,dp \tag{2.148}$$

Substituting Eq. (2.148) into Eq. (2.141), we obtain

$$dS = \frac{1}{T}\,dH - \frac{V}{T}\,dp - \frac{1}{T}\sum_{j=1}^{N}\mu_j\,dn_j \tag{2.149}$$

Similarly, we can obtain

$$\mu_j = -T\left(\frac{\partial S}{\partial n_j}\right)_{H,p,n_j'} \tag{2.150}$$

Combining Eq. (2.146) with Eq. (2.150) gives

$$\left(\frac{\partial S}{\partial n_j}\right)_{H,p,n_j'} = \left(\frac{\partial S}{\partial n_j}\right)_{U,V,n_j'} \tag{2.151}$$

As introduced earlier in this chapter, the Gibbs free energy G, in differential form, is

$$dG = dH - T\,dS - S\,dT \tag{2.152}$$

which upon substitution into Eq. (2.149) gives

$$dG = V\,dp - S\,dT + \sum_{j=1}^{N}\mu_j\,dn_j \tag{2.153}$$

Since

$$dG = \left(\frac{\partial G}{\partial p}\right)_{T,n_j} dp + \left(\frac{\partial G}{\partial T}\right)_{p,n_j} dT + \sum_{j=1}^{N}\left(\frac{\partial G}{\partial n_j}\right)_{P,T,n_j'} dn_j$$

by comparing the coefficients of the above two equations, we can therefore write

$$V = \left(\frac{\partial G}{\partial p}\right)_{T,n_j} \tag{2.154}$$

$$S = -\left(\frac{\partial G}{\partial T}\right)_{p,n_j} \tag{2.155}$$

$$\mu_j \equiv \left(\frac{\partial G}{\partial n_j}\right)_{p,T,n_j'} = -T\left(\frac{\partial S}{\partial n_j}\right)_{H,p,n_j'} \tag{2.156}$$

where Eq. (2.156) can be taken as the definition of μ_j, which is known as the chemical potential and plays an important role in chemical thermodynamics. The chemical potential, an intensive variable, is in general a function of the state of the system, as given by p, T, and all n_j.

2.10 Stoichiometric Chemical Reactions

A stoichiometric chemical reaction is a reaction in which the fuel burns to completion and the only products for hydrocarbon fuels are carbon dioxide and water. The word *stoichiometric* is from the Greek, meaning "element measure." This reaction is used as a reference, and we will treat a wider variety of combustion products in subsequent sections. The product composition resulting from fuel–air combustion depends on the stoichiometry. If the reactants are fuel rich, there is not enough oxygen to react completely with the fuel, resulting in the formation of additional products such as CO and H_2, and if the mixture is fuel lean, there is not enough fuel to consume the oxygen, so there will be unburned oxygen in the product mixture.

Let us represent the chemical formula of a fuel as $C_aH_bO_cN_d$ and assume the following stoichiometric reaction:

$$C_aH_bO_cN_d + a_{st}(1O_2 + 3.762\ N_2) \longrightarrow n_1CO_2 + n_2H_2O + n_3N_2 \tag{2.157}$$

and solve for the stoichiometric molar air–fuel ratio = $4.762 \cdot a_{st}$, and the moles n_i $(i = 1, 2, 3)$ that describe the product composition. Note that the stoichiometric reaction is expressed per mole of fuel. The constant 3.762 comes from the ratio of approximately 79% nitrogen to 21% oxygen by volume in air.

We know that atoms are conserved, so using an atomic species balance, we can write

$$\begin{aligned}
&\mathrm{C}: a = n_1\\
&\mathrm{H}: b = 2n_2\\
&\mathrm{O}: c + 2a_{st} = 2n_1 + n_2\\
&\mathrm{N}: d + 2 \times 3.76 \times a_{st} = 2n_3
\end{aligned} \tag{2.158}$$

Solution of these four equations gives

$$\begin{aligned}
a_{st} &= a + \frac{b}{4} - \frac{c}{2}\\
n_1 &= a\\
n_2 &= \frac{b}{2}\\
n_3 &= \frac{d}{2} + 3.76a_{st}
\end{aligned} \tag{2.159}$$

The mass air–fuel ratio (A/F) for a gas mixture is defined as

$$(A/F) \equiv \frac{\text{mass of air}}{\text{mass of fuel}} = \frac{m_A}{m_F} \tag{2.160}$$

The mass oxidant–fuel ratio (O/F) is

$$(O/F) \equiv \frac{\text{mass of oxidant}}{\text{mass of fuel}} = \frac{m_O}{m_F} \tag{2.161}$$

The stoichiometric mass air–fuel ratio $s_A \equiv (A/F)_{st}$ can be determined from a_{st} and the fuel–air mixture molecular mass. Since $M_i = m_i/n_i$, we have

$$\begin{aligned}
s_A \equiv (A/F)_{st} &= \left(\frac{m_A}{m_F}\right)_{st} = \left(\frac{M_A\, n_A}{M_F\, n_F}\right)_{st}\\
&= \frac{28.97\,[(4.762)a_{st}]}{(12.01\,a + 1.008\,b + 16.00\,c + 14.01\,d)}
\end{aligned} \tag{2.162}$$

For example, the mass air–fuel ratio s_A for the stoichiometric combustion of hydrogen in air is

$$s_A = \left(\frac{m_A}{m_F}\right)_{st} = \frac{n_A M_A}{n_{H_2} M_{H_2}} = \frac{0.5(4.76)(28.97)}{1(2.016)} = 34.20 \tag{2.163}$$

Similarly, the stoichiometric mass oxidant–fuel ratio $s_O \equiv (O/F)_{st}$ can be determined from a_{st} and the fuel–oxidant mixture molecular mass. For oxygen (O_2) as the oxidant,

$$\begin{aligned} s_O \equiv (O/F)_{st} &= \left(\frac{m_O}{m_F}\right)_{st} = \left(\frac{M_{O_2} n_{O_2}}{M_F n_F}\right)_{st} \\ &= \frac{32.00\, a_{st}}{(12.01\, a + 1.008\, b + 16.00\, c + 14.01\, d)} \end{aligned} \tag{2.164}$$

For example, the mass oxidant–fuel ratio s_O for the stoichiometric combustion of hydrogen in oxygen is

$$s_O = \left(\frac{m_O}{m_F}\right)_{st} = \frac{n_{O_2} M_{O_2}}{n_{H_2} M_{H_2}} = \frac{0.5(32)}{1(2.016)} = 7.94 \tag{2.165}$$

The relationship between s_O and s_A is

$$s_O = \left(\frac{m_O}{m_F}\right)_{st} = \frac{m_O}{m_A}\frac{m_A}{m_F} = [Y_O]_A\, s_A \tag{2.166}$$

where $[Y_O]_A$ is the mass fraction of an oxidant in air. If the air is composed entirely of oxidant, $[Y_O]_A = 1$ and $s_O = s_A$.

A dimensionless measure of the air–fuel ratio is the equivalence ratio, ϕ, and defined in Eq. (2.167) as the stoichiometric air–fuel ratio $(A/F)_{st}$ divided by the actual air–fuel ratio (A/F):

$$\phi = \frac{(A/F)_{st}}{(A/F)} = \frac{(F/A)}{(F/A)_{st}} = (F/A)\, s_A \tag{2.167}$$

The equivalence ratio has the same value on a mole or mass basis. If $\phi < 1$, the mixture is lean, if $\phi > 1$, the mixture is rich, and if $\phi = 1$, the mixture is stoichiometric. Similar considerations apply to an oxidant–fuel (O/F) reaction.

Typical stoichiometric mass air–fuel ratio, s_A, values of various fuels are given in Table 2.8. For hydrocarbons, i.e., ($c = d = 0$), the mass stoichiometric air–fuel ratio is approximately 15 and not a strong function of the number of carbon atoms. Table 2.8 also lists the molar stoichiometric air–fuel ratio, a_{st}, the CO_2, and H_2O product mole fractions, at a reference temperature of 298.15 K.

The mole and mass fractions of fuel in a fuel–air mixture are as follows:

$$X_F = \frac{n_F}{n_F + n_A} = \frac{1}{1 + 4.76\, a_{st}/\phi} \tag{2.168}$$

$$Y_F = \frac{m_F}{m_F + m_A} = \frac{1}{1 + (A/F)} \tag{2.169}$$

Example 2.7 ***Stoichiometry of Air–Fuel Mixtures***

(a) What is the molecular mass M_{mix} (kg/kmol) and the gas constant R (kJ/kg-K) of a stoichiometric mixture of octane (C_8H_{18}) and air?
(b) If the mixture fills an engine cylinder with a volume V of 9.0×10^{-4} m^3 at a pressure $P =$ 100 kPa and temperature $T = 300$ K, what is the mass of the fuel–air mixture in the cylinder?
(c) If the mass fraction Y_F of octane at the lean limit for combustion is 0.051, what is the fuel mole fraction X_F, air–fuel ratio (A/F), and the equivalence ratio ϕ at the lean limit?

Table 2.8 Molecular Mass, Stoichiometric Air–Fuel Ratios, and Product Mole Fractions.

Fuel	Chemical Formula	M (kg/kmol)	s_A	a_{st}	X_{CO_2}	X_{H_2O}
Hydrogen	H_2	2.016	34.20	0.50	0.000	0.347
Methane	CH_4	16.04	17.20	2.00	0.095	0.190
Ammonia	NH_3	17.03	6.05	0.75	0.000	0.311
Methanol	CH_4O	32.04	6.43	1.50	0.116	0.231
Propane	C_3H_8	44.09	15.57	5.00	0.116	0.155
Ethanol	C_2H_6O	46.07	8.94	3.00	0.123	0.184
Nitromethane	CH_3NO_2	61.04	1.69	0.75	0.158	0.237
Hexane	C_6H_{14}	86.17	15.14	9.5	0.123	0.144
Gasoline	C_7H_{17}	101.21	15.27	11.25	0.121	0.147
Octane	C_8H_{18}	114.22	15.03	12.50	0.125	0.141
Diesel	$C_{14.4}H_{24.9}$	198.04	14.30	20.63	0.138	0.119
Tetradecane	$C_{14}H_{30}$	198.39	14.54	21.50	0.127	0.110
Hexadecane (Cetane)	$C_{16}H_{34}$	226.44	14.56	24.5	0.128	0.136

Solution

(a) From the Appendix, the molecular mass of air is $M_A = 28.97$ kg/kmol. From Table 2.8, the molecular mass M_F of octane is 114.22 kg/kmol, the stoichiometric molar air–fuel ratio $a_{st} = 12.50$, and the stoichiometric mass air–fuel ratio $(A/F)_{st} = 15.03$. Therefore,

$$X_F = \frac{1}{1 + 4.76\,a_{st}} = 0.0166$$

$$X_A = 1 - X_F = 0.9834$$

$$M_{mix} = X_A M_A + X_F M_F = (0.9834)(28.97) + (0.0166)(114.22) = 30.39\,\text{kg/kmol}$$

$$R = R_u/M_{mix} = 8.314/30.39 = 0.274\,\text{kJ/kg-K}$$

(b) The mass of the mixture in the cylinder is

$$m_{mix} = \frac{PV}{RT} = \frac{(100)(9.0 \times 10^{-4})}{(0.274)(300)} = 1.097\,\text{g}$$

(c) Rearranging Eq. (2.10),

$$X_F = \left[\left(\frac{1}{Y_F} - 1\right)\frac{M_F}{M_A} + 1\right]^{-1} = 1.34 \times 10^{-2}$$

$$(A/F) = \frac{m_A}{m_F} = \frac{1 - Y_F}{Y_F} = \frac{1 - 0.051}{0.051} = 18.60$$

$$\phi = \frac{(A/F)_{st}}{(A/F)} = \frac{15.03}{18.60} = 0.81$$

Comment: For an automotive class engine, the molecular mass of the fuel–air mixture is of the order of 30 kg/kmol, there is typically a gram of combustible mixture in a cylinder, and the equivalence ratio at the lean limit is about 0.8. Note that the addition of fuel to air in this example will reduce the gas constant R from 0.287 to 0.274.

2.11 Mixture Fraction *f*

The mixture fraction f (also defined with the symbol Z) is an extremely useful variable in the study of combustion. It is defined as the mass fraction of fuel or fuel species both burned and unburned in a gaseous mixture, i.e.,

$$f \equiv \frac{\text{mass of material having its origin in the fuel stream}}{\text{mass of mixture}} \tag{2.170}$$

The parameter f is also a conserved scalar parameter, since it is neither created nor destroyed by chemical reactions. A conserved scalar is defined as a scalar with no sources or sinks. As will be discussed in later chapters, the species differential equation in terms of the mixture fraction f will not have any source or sink terms, such as a reaction rate. The mixture fraction concept is very useful in the solution of reacting flow problems, since the mixture fraction is conserved throughout the flow field. One application is in the flamelet model, as an important parameter in the flamelet model is the mixture fraction f.

In the case of a hydrocarbon fuel ($C_aH_bO_cN_d$) burning with pure oxygen, the products will consist of many species, such as H_2O, CO_2, and CO. If the mass fractions of all combustion product species in a control volume are known, the amount of the burned fuel can be determined from the product species. By adding all these amounts of mass originally coming from the fuel, we have the burned amount of fuel in the control volume. There also could be unburned portions of the fuel species $C_aH_bO_cN_d$. Adding this amount to the burned hydrocarbons in the control volume, we can determine the numerator in Eq. (2.170). By dividing this number by the total mass of the mixture in the control volume, the mixture fraction f can be determined.

More generally, let us consider a stoichiometric combustion process of two streams of reactants, a fuel (F) stream and an air (A) stream, resulting in Products P as shown in the sketch in Figure 2.7.

In terms of mass balance, the global reaction can be written as

$$[m_F]\ \text{Fuel} + [m_A]_{\text{st}}\ \text{Air} \rightarrow [m_F + m_A]\ \text{Products} \tag{2.171}$$

and per unit mass of fuel,

$$1 \cdot F + \left(\frac{m_A}{m_F}\right)_{\text{st}} \text{Air} \rightarrow \left(1 + \frac{m_A}{m_F}\right) \text{Products} \tag{2.172}$$

$$1 \cdot F + s_A\ \text{Air} \rightarrow (1 + s_A)\ \text{Products} \tag{2.173}$$

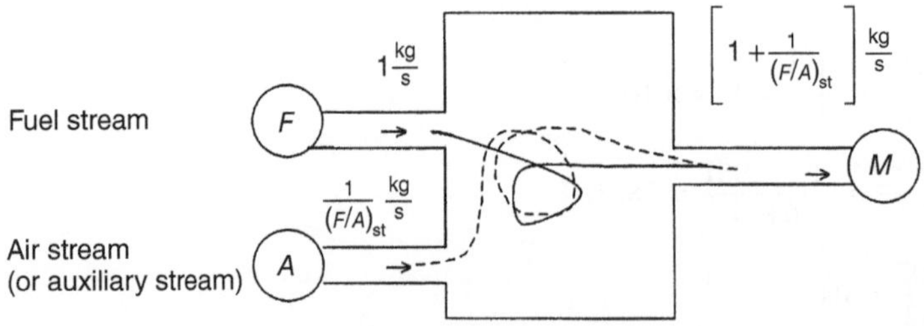

Figure 2.7 Mixing and combustion of fuel and air streams in a combustion chamber.

The reaction may not be complete due to poor mixing or low-temperature conditions. Some portion of the unburned fuel and oxidizer are contained in the "mixture" (M) with all three types of gases (fuel, air, and product). Assume we know the local mass fractions Y of these three gases at a downstream location within the mixture. We now would like to find out the fraction f of the original material from the fuel stream in a small local control volume in the mixture stream and then call this fraction the mixture fraction.

The next step is to determine the relationship between f, Y_F, Y_P, and s_A. Since no fuel is present in the air stream, the mass fraction f of any partially burned material having its origin in the fuel stream can be written as the sum of two parts, namely, the fraction of unburned fuel in the mixture and the fraction of burned fuel in the mixture.

Accordingly, the mixture fracture f is expressed as

$$f = 1 \cdot Y_F + \left[\frac{1}{1+s_A}\right] Y_P \tag{2.174}$$

With stoichiometric combustion in air, the mass fraction of fuel $Y_F = 0$ and the mass fraction of product $Y_P = 1$, so the stoichiometric mixture fraction f_{st} is

$$f_{st} = \frac{1}{1+s_A} = \frac{1}{1+(A/F)_{st}} \tag{2.175}$$

Now let us consider the air stream that consists of oxidizer only; then we have

$$f = 1 \cdot Y_F + \left[\frac{1}{1+s_O}\right] Y_P \tag{2.176}$$

It should be noted that any linear combination of conserved properties, ζ_i, is also a conserved property, where $a_0, a_1, \ldots, a_n$ are constants.

$$a_0 + a_1\zeta_1 + a_2\zeta_2 + \cdots + a_n\zeta_n \tag{2.177}$$

In general, in a two-stream combustion process, the mixture fraction can be expressed as

$$f = \frac{\zeta_M - \zeta_A}{\zeta_F - \zeta_A} \tag{2.178}$$

where ζ represents any conserved property free from sources and sinks. It is useful to prove that the following properties are conserved scalars in this sense:

- Y_{inert} – The mass fraction of a chemically inert mixture component.
- $[Y_F - Y_O/s_O]$ – A "composite" mass fraction made of Y_F and Y_O.
- $\left[Y_F + \frac{1}{1+s_O} Y_P\right]$ – A "composite" mass fraction, where Y_P is the mass fraction of combustion products P.
- $\left[\frac{1}{s_O} Y_O + \frac{1}{1+s_O} Y_P\right]$ – Another "composite" mass fraction.

Using the composite mass fraction $[Y_F - Y_O/s_O]$ to represent the ζ for a two-stream mixing/combustion process, the fuel and oxidant mass fractions are then linked with f as follows:

$$f = \frac{[Y_F - Y_O/s_O]_M - [Y_F - Y_O/s_O]_A}{[Y_F - Y_O/s_O]_F - [Y_F - Y_O/s_O]_A} \tag{2.179}$$

If the F stream contains only fuel and the A stream contains oxidizer but no fuel, then we have

$$[Y_F]_A = 0, \quad [Y_F]_F = 1, \quad [Y_O]_F = 0 \tag{2.180}$$

$$f = \frac{[Y_F - Y_O/s_O]_M + [Y_O/s_O]_A}{1 + [Y_O/s_O]_A}$$

If the chemical reaction is complete within the mixing chamber, either the fuel or the oxidant will have zero concentration in the mixture M state. Therefore, the stoichiometric value of f is

$$f_{\text{st}} = \frac{[Y_O/s_O]_A}{1 + [Y_O/s_O]_A} \tag{2.181}$$

and

$$\text{if} \quad f < f_{\text{st}} \quad \text{(fuel-lean case)} \tag{2.182}$$

$$f = \frac{-[Y_O/s_O]_M + [Y_O/s_O]_A}{1 + [Y_O/s_O]_A}$$

$$\text{if} \quad f > f_{\text{st}} \quad \text{(fuel-rich case)}$$

$$f = \frac{[Y_F]_M + [Y_O/s_O]_A}{1 + [Y_O/s_O]_A}$$

Upon rearrangement,
for $f < f_{\text{st}}$:

$$[Y_F]_M = 0$$
$$[Y_O]_M = [Y_O]_A \frac{f_{\text{st}} - f}{f_{\text{st}}} \tag{2.183}$$

and for $f > f_{\text{st}}$:

$$[Y_O]_M = 0$$
$$[Y_F]_M = \frac{f - f_{\text{st}}}{1 - f_{\text{st}}} \tag{2.184}$$

If the air stream contains some diluent gas, then we have the following relationships for any f:

$$[Y_D]_M = (1 - f)\,[Y_D]_A \tag{2.185}$$

$$[Y_P]_M = 1 - [Y_D]_M - [Y_O]_M - [Y_F]_M$$

where $[Y_D]_M$ represents the mass fraction of diluent in the mixture M state. The linear relationships in Eqs. (2.183) and (2.184) between Y_O, Y_F, Y_P, Y_D and f are shown on the graph in Figure 2.8 as f varies from 0 to 1 on the burned plane. The minimum values of Y_O and Y_F in the mixture occur at stoichiometric mixture fraction f_{st}. As the mixture is made more lean ($f < f_{\text{st}}$), Y_O increases, and Y_P decreases. As the mixture is made more rich ($f > f_{\text{st}}$), Y_F increases, and again Y_P decreases.

It should be noted that this graph is not the most general form; the F stream could have some diluent, and both streams might also be contaminated with product. It would be a useful exercise

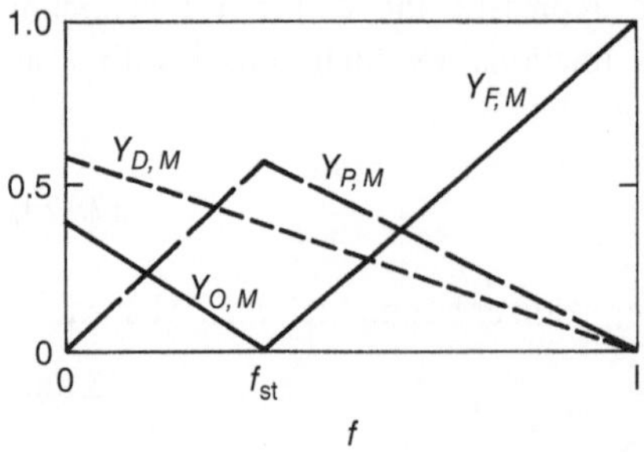

Figure 2.8 Mass fractions as functions of mixture fraction.

for readers to construct an unburned plane of Y_O, Y_F, Y_P, and Y_D versus f and to demonstrate that the dependence of Y_i on f is totally linear. It is important to note that the mixture fraction f should be distinguished quite clearly from the mass fraction of fuel Y_F.

Example 2.8 ***Mixture Fraction of Elements***

Separate streams of methane (F) and oxygen (O) are injected into a combustor as indicated in Figure 2.3. They combust with the stoichiometric reaction

$$1CH_4 + 2O_2 \longrightarrow 1CO_2 + 2H_2O$$

What is the mixture fraction of H atoms in the mixture (M) just before and after combustion?

Solution

Let $\zeta = Y$, the mass fraction of H atoms. With Mixture (M), fuel (F), and oxidizer (O).

$$f = \frac{Y_M - Y_O}{Y_F - Y_O} = \frac{Y_M}{Y_F}$$

Since there are no H atoms in the oxidizer stream, $Y_O = 0$.

Before combustion,

$$Y_M = \frac{m_H}{m_M} = \frac{4(1)}{1(16) + 2(32)} = 0.05$$

$$Y_F = \frac{m_H}{m_F} = \frac{4(1)}{1(16)} = 0.25$$

$$f = \frac{Y_M}{Y_F} = \frac{0.05}{0.25} = 0.20$$

After combustion,

$$Y_M = \frac{m_H}{m_M} = \frac{4(1)}{1(44) + 2(18)} = 0.05$$

$$Y_F = \frac{m_H}{m_F} = \frac{4(1)}{1(16)} = 0.25$$

$$f = \frac{Y_M}{Y_F} = \frac{0.05}{0.25} = 0.20$$

The mixture fraction of H atoms is a conserved scalar, since elements are neither created nor destroyed by chemical reactions.

2.12 First Law Analysis of Combustion

With chemical thermodynamics modeling, we are able to predict the equilibrium state that results from burning a fuel–air mixture as a function of initial conditions, such as pressure, temperature, and equivalence ratio. In this section and the following Section 2.13, we apply the first law of thermodynamics to fuel–air combustion processes and determine the heat and work interactions between the fuel–air mixture and the environment. We apply the open system energy equation for the determination of the heat of reaction, heat of combustion, and the adiabatic flame temperature.

In reciprocating internal combustion engines, there is residual gas mixed with the fuel and air, since not all of the combustion gases leave the cylinder. The residual gas fraction f_r is the ratio of the gas mass left from the previous cycle to the total gas mass in the cylinder. Typical values of the residual gas fraction f_r are in the 0.03–0.12 range, with smaller values for Diesel cycle engines due to their greater compression ratio in comparison with Otto cycle engines.

2.12.1 Open System Energy Equation

For a reactor or engine modeled as an open system, we adopt a control volume approach. The steady flow control volume energy equation indicates that the change in enthalpy, the difference between the product (p) enthalpy and the reactant (r) enthalpy, is equal to the rate of heat transfer $\dot{Q}$ *into* the system, minus the rate of shaft work $\dot{W}$ *out* of the system. By convention, the heat energy transferred out of the system is defined as negative, and the work performed by the system is defined as positive.

$$\dot{Q} - \dot{W} = +\dot{m}_F(h_p - h_r) \tag{2.186}$$

where $\dot{m}_F$ is the fuel flow rate, and the enthalpy h is expressed on a per kg of fuel basis.

Example 2.9 ***Open System Energy Equation***

An internal combustion engine uses propane as a fuel. The propane and air enter the engine at 298 K, and the propane mass flowrate is 0.01 kg/s. The combustion is rich with $\phi = 1.11$, so there is CO and H_2 in the exhaust, which exits at 1000 K. The heat loss from the engine to the cooling water is 120 kW. What is the power output (kW) and specific fuel consumption (sfc) of the engine?

Solution

The combustion equation per mole of propane with $\phi = 1.11$ is

$$1C_3H_8 + 4.50(O_2 + 3.76N_2) \longrightarrow 2.7CO_2 + 0.3CO + 3.3H_2O + 0.7H_2 + 16.92N_2$$

The steady open system energy equation is

$$\dot{Q} - \dot{W} = +\dot{m}_F(h_p - h_r)$$

From Table 2.9, the enthalpy of formation of propane $\overline{h}_f^o = -103.9$ kJ/mol. The enthalpies of the product species at 1000 K are given in the Appendix. The enthalpy change during combustion of the propane per kmol of fuel, assuming no water condensation in the exhaust products, is

$$\begin{aligned}\overline{h}_r - \overline{h}_p &= \sum_r n_r\overline{h}_r - \sum_p n_p\overline{h}_p = \sum_r n_r\overline{h}_{f,r}^o - \sum_p n_p(\overline{h}_{f,p}^o + \Delta\overline{h})\\ &= +1(-103.9) - 2.7(-393.5 + 33.4) - 0.3(-110.5 + 21.7)\\ &\quad - 3.3(-241.8 + 26.0) - 0.7(0 + 20.7) - 16.92(0 + 21.5)\\ &= -103.9 + 1333.6 = +1229.7\ \text{MJ/kmol}\end{aligned}$$

The molecular mass of propane is $M = 44.09$ kg/kmol. The mass-specific enthalpy change $h_r - h_p$ is therefore

$$h_r - h_p = \frac{\overline{h}_r - \overline{h}_p}{M} = \frac{1229.7}{44.09} = 27.89\ \text{MJ/kg}_{\text{fuel}} = 27,890\ \text{kJ/kg}_{\text{fuel}}$$

The power output of the engine is

$$\begin{aligned}\dot{W} &= \dot{Q} + \dot{m}_F(h_r - h_p)\\ &= -120 + 0.01(27,890) = 158.9\ \text{kW}\end{aligned}$$

Table 2.9 Enthalpy of Formation, Entropy, Lower/Higher Heat of Combustion, and Maximum Available Energy of Combustion[a].

FUEL		$\bar{h}_f^o$ (MJ/kmol)	$\bar{s}_{298}^o$ (kJ/kmol K)	q_{lhc} (MJ/kg)	q_{hhc} (MJ/kg)	a_c (MJ/kg)
CH_4 (g)	Methane	−74.9	186.2	50.01	55.5	52.42
CH_4O (l)	Methanol	−239.09	126.8	19.91	22.7	22.68
C_2H_6O (l)	Ethanol	−277.15	160.7	26.82	29.7	29.71
CH_3NO_2 (l)	Nitromethane	−113.1	171.8	10.54	11.6	12.43
C_3H_8 (g)	Propane	−103.9	269.9	46.36	50.3	49.16
C_7H_{17} (l)	Gasoline[b]	−305.63	345.8	44.51	46.3	47.87
C_8H_{18} (l)	Octane	−249.96	360.8	44.30	47.9	47.67
C14.4 H24.9 (l)	Diesel[b]	−174.08	525.9	42.94	45.5	45.73
$C_{16}H_{34}$ (l)	Cetane	−454.48	586.0	43.99	47.26	47.22
H_2 (g)	Hydrogen	0.0	130.6	119.95	141.6	119.52
C_2H_2 (g)	Acetylene	226.7	200.8	48.22	49.9	48.58
C_2N_2 (g)	Cyanogen	309.1	241.5	21.06	21.0	21.29
NH_3 (g)	Ammonia	−45.7	192.6	18.61	22.5	20.29
C_6H_6 (l)	Benzene	48.91	173.0	40.14	41.9	42.14
$C_{10}H_8$ (s)	Naphthalene	78.1	166.9	38.86	40.3	40.84
C (s)	Graphite	0.0	5.7	32.76		33.70

a) Based on equilibrium water quality, lean combustion at $\phi = 0.01$, $T_o = 298$ K, $P_o = 1.013$ bar, and unmixed reactants.
b) Estimated for typical fuel.

The sfc is

$$\text{sfc} = \frac{\dot{m}_F}{\dot{W}} = \frac{(0.01)(3600)}{158.9} = 0.226\,\text{kg/kWh}$$

2.12.2 Heats of Reaction and Combustion

The *heat of reaction* q_r is defined as the thermal energy transferred out of a system during a chemical reaction from a given initial temperature T_1 to a given final temperature T_2, with no work done other than that required for flow. The *heat of combustion* q_c is a special case of the heat of reaction. The heat of combustion, also termed heating value, is defined as the thermal energy transferred out of a system when the initial and final states are at the reference temperature and pressure, $T_o = 298.15$ K and $p_o = 101$ kPa.

We write combustion equations on a per mole or per mass of fuel basis and define $\overline{q}_c$ as the heating value with units of kJ/kmol$_{\text{fuel}}$ and q_c as the heating value with units of kJ/kg$_{\text{fuel}}$. Again, by convention, the heat transferred out of a system is defined as negative, and since the heat of combustion is defined as a positive number, that is, $q_c = -q$, so

$$q_c = h^o_{f,r} - h^o_{f,p} \tag{2.187}$$

when the heat of combustion referred to without qualification, constant pressure combustion is implied.

Furthermore, with hydrocarbon heats of combustion, the combustion process is assumed to be complete, with the fuel reacting with oxygen to produce carbon dioxide and water. Since the products are at low temperature ($T_o < 1000$ K), the analyses that led to Table 2.9 for the lean or stoichiometric case can be used to compute the heat of combustion q_c on a mass basis and $\overline{q}_c$ on a molar basis. In this case, it can be shown that Eq. (2.187) per kmol of fuel becomes

$$\begin{aligned}\overline{q}_c &= \sum_r n_r \overline{h}^o_{f,r} - \sum_p n_p \overline{h}^o_{f,p} \\ &= 1 \cdot \overline{h}_{\text{fuel}} - n_{CO_2}\overline{h}^o_{f,CO_2} - n_{H_2O}[\overline{h}^o_{f,H_2O} - (1-\chi)\overline{h}_{fg,H_2O}]\ \text{kJ/kmol}_{\text{fuel}}\end{aligned} \tag{2.188}$$

since the enthalpies of oxygen O_2 and nitrogen N_2 are assigned to be zero at the reference temperature $T_o = 298.15$ K.

Two values of q_c are recognized: (1) the lower heat of combustion q_{lhc} is defined as the state where all of the water in the products is vapor (the quality $\chi = 1$), and (2) the higher heat of combustion, q_{hhc} is defined as the state where all of the water in the products is liquid (the quality $\chi = 0$). If the water quality is not specified, one usually assumes that $q_c = q_{\text{lhc}}$.

The lower q_{lhc} and higher heat of combustion q_{hhc} of several gaseous and liquid fuels are given in Table 2.9. The heat of combustion is used primarily in two ways: (1) in some cases, such as in gas cycles or when solving reacting Navier–Stokes equations, it is desirable to relax the rigor of the thermodynamics by using the heat of combustion to define an equivalent energy release; and (2) for practical fuels, the enthalpy at $T_o = 298.15$ K can be determined inexpensively by measurement of the heat of combustion. Table 2.9 also lists the enthalpy of formation $\overline{h}^o_f$; the absolute entropy at 298 K, $\overline{s}^o_{298}$; and a_o, the maximum available energy of combustion.

Example 2.10 ***Heat of Combustion***

Compare the lower, q_{lhc}, and higher, q_{hhc}, heat of combustion of cetane $C_{16}H_{34}$ (l) to the equilibrium heat of combustion q_{eq} with equilibrium water quality in the product species.

Solution

Assume standard reference conditions $p = p_o = 1$ atm and $T_o = 298.15$ K for reactants and products, and the air and fuel enter unmixed. The molecular mass M of cetane is 226.44 kg/kmol. The equilibrium water quality χ_{eq} for this reaction is 0.208. The fuel enthalpy is obtained from Table 2.9. The nitrogen N_2 enthalpy is zero for both the reactants and products and is not included in the energy equation computation.

The stoichiometric combustion equation per mole of cetane is

$$1C_{16}H_{34} + 24.5O_2 \longrightarrow 16CO_2 + 17H_2O$$

The first law, Eq. (2.187), per kmol of fuel, is

$$\begin{aligned}\overline{q}_c &= \sum_r n_r \overline{h}^o_{f,r} - \sum_p n_p \overline{h}^o_{f,p} \\ &= 1 \cdot \overline{h}_{\text{fuel}} - n_{CO_2}\overline{h}^o_{f,CO_2} - n_{H_2O}[\overline{h}^o_{f,H_2O} - (1-\chi)\overline{h}_{fg,H_2O}] \\ &= +1(-454.48) - 16(-393.52) - 17[(-241.83 - (1-\chi)(43.99))] \\ &= +9952.9 + (747.8)(1-\chi)\ \text{MJ/kmol}\end{aligned}$$

Therefore,

$$q_{\text{lhc}} = \frac{\overline{q}_c(\chi = 1)}{M} = 9952.9/226.4 = 43.96\,\text{MJ/kg}$$

$$q_{\text{eq}} = \frac{\overline{q}_c(\chi = 0.208)}{M} = 10,545.2/226.4 = 46.58\,\text{MJ/kg}$$

$$q_{\text{hhc}} = \frac{\overline{q}_c(\chi = 0)}{M} = 10,700.8/226.4 = 47.26\,\text{MJ/kg}$$

2.12.3 Adiabatic Flame Temperature

Another useful combustion parameter is the adiabatic flame temperature T_f. It is defined as the temperature of the combustion products when completely burned with no shaft work ($W = 0$) and heat transfer ($Q = 0$) to the surroundings. The adiabatic flame temperature represents the maximum temperature of a combustion process, since any heat transfer from the reaction and incomplete combustion will lower the temperature of the products. The adiabatic flame temperature can be controlled with the use of excess air or diluent. The adiabatic flame temperature for non-dissociating products is normally called the adiabatic frozen flame temperature.

For constant pressure combustion, the first law is

$$H_p = H_r \tag{2.189}$$

The initial state of the reactants is assumed to be at the reference temperature and pressure, $T_o = 298.15$ K and $p_o = 101.3$ kPa. Since the product temperature is generally unknown in the first law combustion calculations, iteration with an initial temperature estimate is required to determine the product mixture enthalpy. An assumption also needs to be made about the amount of dissociation in the combustion products. The adiabatic flame temperature with dissociation of the combustion products will be less than the computed value for combustion without dissociation.

For constant volume combustion, the first law of thermodynamics with no work ($W = 0$) and heat transfer ($Q = 0$) to the surroundings reduces to a constant internal energy U,

$$U_p = U_r \tag{2.190}$$

The constant volume adiabatic flame temperature is greater than the constant pressure adiabatic flame temperature, since the $p\,dV$ work is zero in a constant volume process.

The stoichiometric adiabatic flame temperature in air of several fuels is listed in Table 2.10. The results tabulated assume initial conditions of $T_o = 298.15$ K, $p_o = 101.3$ kPa, the residual gas fraction $f_r = 0.0$, and equivalence ratio $\phi = 1.0$. There is little dependence on fuel type among the hydrocarbons, which have similar adiabatic flame temperatures at about 2250 K.

Employing a simple thermodynamic model where the specific heats c_p of all the reactant and product species are constant and equal allows us to express the heat of combustion q_c in terms of the unburned mixture temperature T_u and the adiabatic flame temperature T_f, as follows. A global reaction between a fuel (F) and air (A) with products (P) can be expressed as

$$1 \cdot \text{F} + s_A \cdot \text{A} \rightarrow (1 + s_A) \cdot \text{P} \tag{2.191}$$

where s_A is the mass air/fuel ratio. If the reaction is adiabatic with no shaft work, the enthalpy is constant,

$$H_r = H_p \tag{2.192}$$

and per unit mass of fuel,

$$1 \cdot h_F + s_A h_A = (s_A + 1) h_P \tag{2.193}$$

Table 2.10 Stoichiometric Adiabatic Flame Temperature in Air of Various Fuels.

Formula	Fuel	T_f(K)
CH_4 (g)	Methane	2227
C_3H_8 (g)	Propane	2268
C_8H_{18} (l)	Octane	2266
$C_{15}H_{32}$ (l)	Pentadecane	2269
$C_{20}H_{40}$ (g)	Eicosane	2291
CH_4O (l)	Methanol	2151
C_2H_6O (l)	Ethanol	2197
CH_3NO_2 (l)	Nitromethane	2545
H_2 (g)	Hydrogen	2383
C_2N_2 (g)	Cyanogen	2596
NH_3 (g)	Ammonia	2076
C_2H_2 (g)	Acetylene	2540
$C_{10}H_8$ (s)	Naphthalene	2328

$p_o = 101.3$ kPa, $T_o = 298$ K.

Expressing the enthalpy in terms of the enthalpy of formation and the specific heat,

$$h = h_f^o + c_p(T - T_{\text{ref}}) \tag{2.194}$$

Upon substitution, Eq. (2.193) is

$$\begin{aligned}&[h_f^o + c_p(T_u - T_{\text{ref}})]_F + s_A[h_f^o + c_p(T_u - T_{\text{ref}})]_A\\&= (s_A + 1)[h_f^o + c_p(T_f - T_{\text{ref}})]_P\end{aligned} \tag{2.195}$$

Rearranging,

$$\begin{aligned}&h_{f,F}^o + s_A h_{f,A}^o - (s_A + 1)h_{f,P}^o\\&= -c_p(T_u - T_{\text{ref}}) - s_A c_p(T_u - T_{\text{ref}}) + (s_A + 1)c_p(T_f - T_{\text{ref}})\\&= (s_A + 1)c_p(T_f - T_u)\end{aligned} \tag{2.196}$$

The heat of combustion q_c (per unit mass of fuel) for reaction (2.193) is

$$q_c = h_{f,F}^o + s_A h_{f,A}^o - (s_A + 1)h_{f,P}^o \tag{2.197}$$

Therefore

$$q_c = (s_A + 1)c_p(T_f - T_u) \tag{2.198}$$

Example 2.11 ***Adiabatic Flame Temperature***

A stoichiometric mixture of gasoline C_7H_{17}, air, and residual gas is burned at constant pressure. Given that $T_1 = 298$ K, $p_1 = 101.3$ kPa, and residual gas fraction $f_r = 0.10$, what is the constant pressure adiabatic flame temperature?

Solution

The text website contains a listing of the program `AdiabaticFlameTemp.m` that computes the constant pressure adiabatic flame temperature. The program uses the fuel–air–residual gas routine

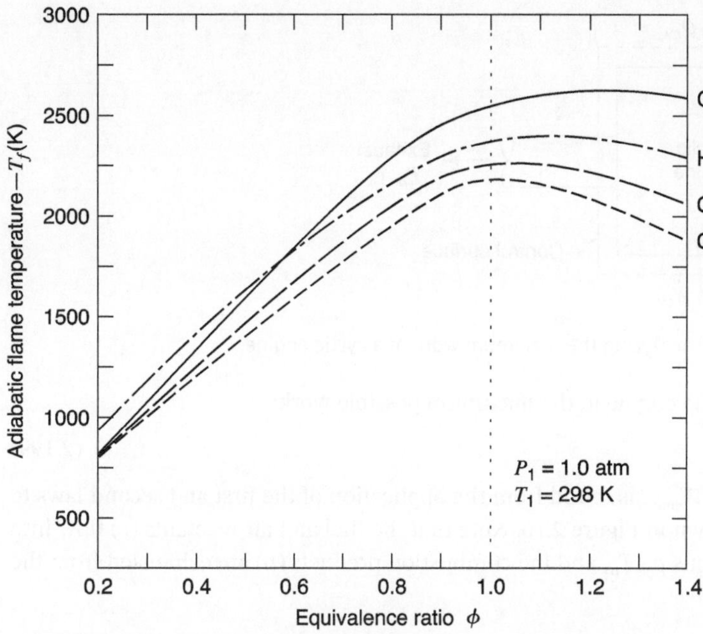

Figure 2.9 Adiabatic flame temperature of some fuels initially at atmospheric pressure and temperature.

`farg.m`, the equilibrium combustion routine `ecp.m`, and Newton–Raphson iteration for constant pressure, with inputs the pressure p, initial temperature T, fuel–air equivalence ratio ϕ, and residual gas fraction f_r. The available fuel types are methane (CH_4), gasoline (C_7H_{17}), diesel ($C_{14.4}H_{24.9}$), methanol (CH_3OH), ethanol (C_2H_6O), and nitromethane (CH_3NO_2).

The above information is entered into the program, and the resulting adiabatic flame temperature is computed to be $T = 2094$ K. The adiabatic flame temperature has decreased from the 2250 K range as a result of the presence of the residual gas.

The effect of equivalence ratio is plotted in Figure 2.9, which indicates that the adiabatic flame temperature is maximum near stoichiometric. This is consistent with the effect of equivalence ratio on enthalpy, as shown earlier in this chapter, since the combustion is incomplete with lean or rich mixtures on either side of stoichiometric. For hydrocarbons, the maximum flame temperature occurs slightly on the fuel-rich side of stoichiometric owing to the presence of diatomic molecules, particularly H_2. Additional calculations show the adiabatic flame temperature decreases with increasing residual gas fraction and slightly increases with pressure. Note that of the fuels plotted, ethanol (C_2H_6O) consistently has the lowest and nitromethane (CH_3NO_2) has the highest adiabatic flame temperature over most of the equivalence ratio range.

2.13 Second Law Analysis of Cyclic Engines

The efficiency of cyclic engines, such as internal combustion engines, can be examined from the perspective of the second law of thermodynamics. While the first law takes the energy transfers to the surroundings by way of the coolant and the exhaust into account, it does not consider the maximum possible work of the engine. The second law definition of engine efficiency, η_{II}, is the

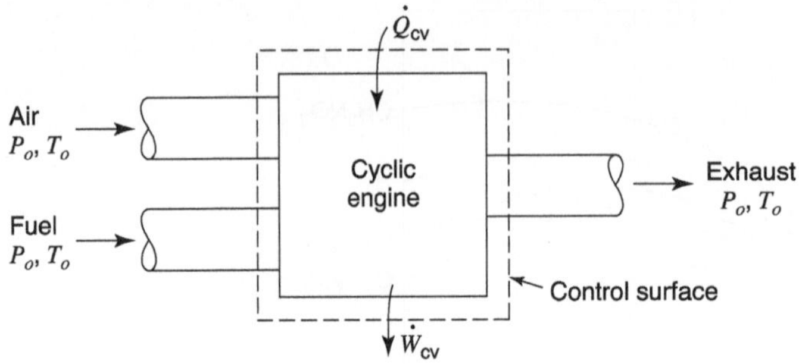

Figure 2.10 A control volume for analyzing the maximum work of a cyclic engine.

ratio of the net work done by the engine to the maximum possible work:

$$\eta_{II} = W_{cv}/W_{max} \tag{2.199}$$

The maximum possible work, W_{max}, is found from the application of the first and second laws to the control volume (*cv*) as shown in Figure 2.10. Note that the fuel and air reactants (*r*) flow into the engine at standard conditions p_o, T_o, and the combustion products (*p*) are exhausted from the engine also at p_o, T_o.

$$\dot{Q}_{cv} - \dot{W}_{cv} = \left(\frac{dE}{dt}\right)_{cv} + \sum_p \dot{m}h - \sum_r \dot{m}h \tag{2.200}$$

Let us integrate over one period of the engine's cycle

$$Q_{cv} - W_{cv} = \sum_p mh - \sum_r mh \tag{2.201}$$

The maximum work is obtained only if the process is reversible, in which case, the second law applied to the control volume is an equality:

$$Q_{cv} = T_o\left(\sum_p ms - \sum_r ms\right) \tag{2.202}$$

The only way in which the reversible heat transfer of Eq. (2.202) can occur between an engine and its surroundings is via an intervening Carnot engine. Upon substitution of Eq. (2.202) into Eq. (2.200), the maximum work is

$$\begin{aligned} W_{max} &= \sum_r mh - \sum_p mh + T_o\left(\sum_p ms - \sum_r ms\right) \\ &= (H_r - H_p) - T_o(S_r - S_p) \\ &= G_r - G_p \end{aligned} \tag{2.203}$$

The maximum work can also be expressed as the change in the Gibbs free energy, *G*. The available energy of combustion a_c is defined as the maximum work per unit mass (or mole) of fuel,

$$\begin{aligned} a_c &= W_{max}/m_f \\ &= \frac{1}{m_f}[(H_r - H_p) + T_o(S_p - S_r)] \end{aligned} \tag{2.204}$$

and the second law of efficiency can be expressed as

$$\eta_{II} = \frac{W_{cv}}{m_f a_c} = \frac{w_{cv}}{a_c} \tag{2.205}$$

The difference between the available energy of combustion, $\Delta(H - T_o S)$, and the heat of combustion, ΔH, is that the available energy of combustion takes into account the change in entropy due to the changes in composition of reactants. Including the residual gas fraction f_r and the equivalence ratio ϕ in the analysis,

$$\eta_{II} = \frac{W_{cv}}{m_F\, a_c} = \frac{w_{cv}(m_A + m_F + m_r)}{m_F\, a_c} \tag{2.206}$$

$$f_r = \frac{m_r}{m_A + m_F + m_r} \tag{2.207}$$

so

$$1 - f_r = \frac{m_A + m_F}{m_A + m_F + m_r} \tag{2.208}$$

Since $m_F = m_A\,\phi\,(F/A)_s = m_A\phi/s_A$,

$$1 - f_r = \frac{m_A(1 + \phi/s_A)}{m_A + m_F + m_r} \tag{2.209}$$

$$\begin{aligned} m_A + m_F + m_r &= \frac{m_A(1 + \phi/s_A)}{1 - f_r} \\ &= \frac{m_F(1 + \phi/s_A)}{\phi/s_A(1 - f_r)} \end{aligned} \tag{2.210}$$

and the second law efficiency becomes

$$\eta_{II} = \frac{w_{cv}(1 + \phi/s_A)}{\phi/s_A(1 - f_r)\, a_c} \tag{2.211}$$

Figure 2.11 is a plot of the available energy of combustion and the heat of combustion as a function of equivalence ratio at $T_o = 298$ K, $p_o = 1$ atm. The curves are relatively insensitive to equivalence ratio for $\phi < 1$. This is because the oxygen O_2 in the products does not contribute to the product enthalpy as its enthalpy of formation is zero. The main effect on the available energy of combustion and the heat of combustion is through the change in the product mole fractions and thus the quality. The curves decrease with increasing equivalence ratio for $\phi > 1$, due to the reduced concentration of CO_2 and increased concentrations of carbon monoxide CO and hydrogen H_2. It should be noted that CO is at a negligible concentration and H_2 has a zero enthalpy of formation at the product temperature of 298 K, so the main influence on the available energy of combustion and the heat of combustion for $\phi > 1$ is the reduced concentration of CO_2 in the product gases.

The maximum work is attained only if the exhaust is in equilibrium with environment at the state T_o, p_o. This also implies that each exhaust species such as CO_2 and H_2O is at the chemical potential, partial pressure, and concentration that it exists in the environment. The mole fraction of water vapor is evaluated by setting the partial pressure of the water vapor equal to the saturation vapor pressure at $T = T_o$. At $T_o = 298$ K, the saturation water vapor pressure is 3.17 kPa, so the environmental mole fraction of water vapor $X_{H_2O} = 3.17/101.3 = 3.12 \times 10^{-2}$. Note that the atmospheric concentration of CO_2 is increasing at a rate of a few ppm per year and is currently (2020) about 405 ppm, so the environmental mole fraction of carbon dioxide $X_{CO_2} = 4.05 \times 10^{-4}$.

It is evident from Figure 2.11 that more energy is available per unit mass of fuel if an engine is fueled lean than if it is fueled rich. By inspection of Figure 2.11, one would use the stoichiometric heat of combustion, where the heat of combustion is maximum. Arguments can also be made for the use of either the lower or the higher heat of combustion. Additionally, from Table 2.9, notice that for the most part, there is little difference between the higher heat of combustion q_{hhc} and the maximum available energy a_c.

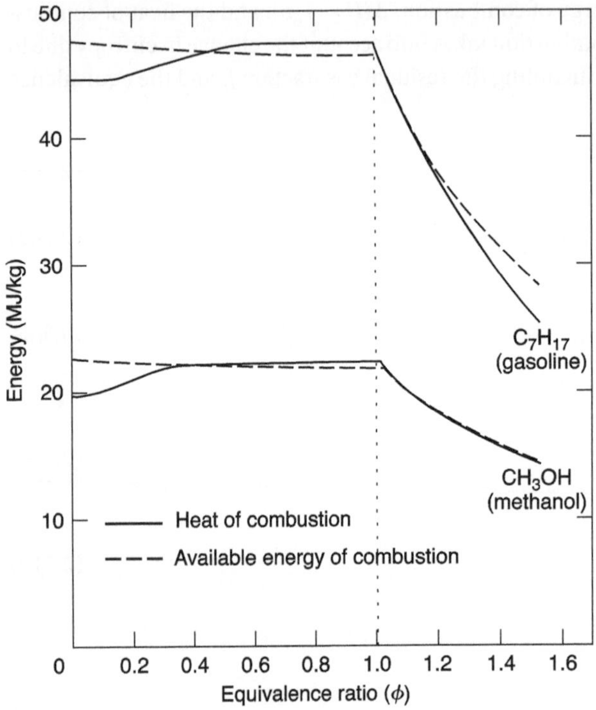

Figure 2.11 Comparison of the available energy a_c and the equilibrium heat of combustion $q_{c,\text{eq}}$ for liquid gasoline and methanol. Fuel and air are unmixed. Products are mixed $p_o = 1.013$ bar, $T_o = 298$ K.

Example 2.12 ***Heat of Combustion and Available Energy of Combustion***

Compare the available energy of combustion, a_c with the lower heat of combustion, q_{lhc}, for the stoichiometric combustion of methane at standard reference conditions.

Solution

Assume standard reference conditions $p = p_0 = 1$ bar and $T_o = 298.15$ K, the air and fuel enter unmixed, with mole fractions $X_{CH_4} = 1$ and $X_{O_2} = 0.21$, and that the products mix with the atmosphere, with mole fractions $X_{H_2O} = 3.12 \times 10^{-2}$ and $X_{CO_2} = 4.05 \times 10^{-4}$. The molecular mass M of methane is 16.04 kg/kmol. Since the nitrogen enthalpy is zero for both the reactants and products, and the environmental nitrogen mole fraction is a constant, it is not included in the computation. The combustion equation per mole of methane is

$$1CH_4 + 2O_2 \longrightarrow CO_2 + 2H_2O$$

From the tabular data for $\overline{h}_f^o$ and $\overline{s}_i^o$ given in Table 2.9,

$$\begin{aligned}\overline{h}_r &= \sum_{r_i} n_i \overline{h}_i = n_{CH_4}\overline{h}_{CH_4} + n_{O_2}\overline{h}_{O_2} \\ &= (1)(-74.87) + 2(0) = -74.87\ \text{MJ/kmol} \\ \overline{h}_p &= \sum_{p_i} n_i \overline{h}_i = n_{CO_2}\overline{h}_{CO_2} + n_{H_2O}\overline{h}_{H_2O} \\ &= (1)(-393.5) + (2)(-241.8) = -877.17\ \text{MJ/kmol}\end{aligned}$$

$$\bar{s}_r = \sum_{r_i} n_i[\bar{s}_i^o - R_u \ln X_i)] = n_{CH_4}[\bar{s}^o_{CH_4} - R_u \ln X_{CH_4}] + n_{O_2}[\bar{s}^o_{O_2} - R_u \ln X_{O_2}]$$

$$= (1)[186.2] + (2)[205.15 - 8.314 \ln(0.21)] = 622\,\text{kJ/K-kmol} = 0.622\,\text{MJ/K-kmol}$$

$$\bar{s}_p = \sum_{p_i} n_i[\bar{s}_i^o - R_u \ln X_i] = n_{CO_2}[\bar{s}^o_{CO_2} - R_u \ln X_{CO_2}] + n_{H_2O}[\bar{s}^o_{H_2O} - R_u \ln X_{H_2O}]$$

$$= (1)[213.79 - 8.314 \ln(4.05 \times 10^{-4})] + (2)[188.83 - 8.314 \ln(3.12 \times 10^{-2})]$$

$$= 714\,\text{kJ/K-kmol} = 0.714\,\text{MJ/K-kmol}$$

Therefore,

$$\bar{q}_c = \bar{h}_r - \bar{h}_p$$
$$= -74.87 - (-877.17) = 802.3\,\text{MJ/kmol}$$
$$q_c = \bar{q}_c/M = 802.3/16.04 = 50.01\,\text{MJ/kg}$$
$$\bar{a}_c = (\bar{h}_r - \bar{h}_p) - T_o(\bar{s}_r - \bar{s}_p)$$
$$= 802.3 - (298)(0.622 - 0.714) = 830.3\,\text{MJ/kmol}$$
$$a_c = \bar{a}_c/M = 830.3/16.04 = 51.76\,\text{MJ/kg}$$

Comment: The available energy of combustion a_c is about 3% greater than the lower heat of combustion q_{lhc} for stoichiometric combustion. For very lean conditions, $\phi = 0.01$, a_c will increase slightly to 52.42 MJ/kg, as listed in Table 2.9.

2.14 Equilibrium Constant K_p

The first and second laws of thermodynamics determine the conditions for the progress of a chemical reaction and equilibrium of the product species. The first and second laws are:

$$\delta Q - p\,dV = dU$$
$$dS \geq \frac{\delta Q}{T} \tag{2.212}$$

we have

$$dU + p\,dV - T\,dS \leq 0 \tag{2.213}$$

The Gibbs free energy G is defined as $G = H - TS$, so for a constant T and p,

$$dG = dH - T\,dS - S\,dT$$
$$= dU + p\,dV + V\,dp - T\,dS - S\,dT \tag{2.214}$$
$$= dU + p\,dV - T\,dS$$

Upon substitution of Eq. (2.213), the condition for progress of a chemical reaction is

$$dG \leq 0 \tag{2.215}$$

Therefore, a chemical reaction will proceed in the direction of a decreasing overall Gibbs free energy and establish chemical equilibrium of the product species when the Gibbs free energy is at a minimum. At equilibrium, individual product species will react at equal rates of production and consumption, so the change in the overall Gibbs free energy is zero, with no net change in the mole fractions of the product species.

The condition for equilibrium is therefore

$$dG = \sum_i \overline{g}_i dn_i = 0 \tag{2.216}$$

For a three-component ideal-gas mixture, with species A, B, and C,

$$dG = \overline{g}_A\, dn_A + \overline{g}_B\, dn_B + \overline{g}_C\, dn_C = 0 \tag{2.217}$$

Consider this reactive mixture to have an elementary reaction represented by

$$\upsilon_A A + \upsilon_B B \rightleftharpoons \upsilon_C C \tag{2.218}$$

where the υs are the stoichiometric coefficients for the balanced chemical equation. Note that since Eq. (2.218) represents the elementary reaction at the molecular level, the values of υ_i are constants and not related to the actual number of moles of each species in the reactive mixture.

The changes in the number of moles of each species dn_i during the reaction are not independent of each other, since the change in the moles of reactants and products is proportional to the stoichiometric coefficients. Therefore, if we define the degree of forward reaction by ϵ, then for a differential change in the elementary reaction,

$$\begin{aligned} dn_A &= -d\epsilon\, \upsilon_A \\ dn_B &= -d\epsilon\, \upsilon_B \\ dn_C &= +d\epsilon\, \upsilon_C \end{aligned} \tag{2.219}$$

The A and B species decrease, and the C species increases as the forward reaction progresses. Upon substitution of Eq. (2.219) into Eq. (2.217),

$$dG = -\upsilon_A \overline{g}_A - \upsilon_B \overline{g}_B + \upsilon_C \overline{g}_C = 0 \tag{2.220}$$

This is called the equation of reaction equilibrium, as it indicates the interrelationship between the properties of the reactants and products for a given reaction. Incorporating the Gibbs free energy equation, Eq. (2.37), for each component, and choosing $p_o = 1$ atm, we obtain

$$-\upsilon_A(\overline{g}_A^o + TR_u \ln p_A) - \upsilon_B(\overline{g}_B^o + TR_u \ln p_B) + \upsilon_C(\overline{g}_C^o + TR_u \ln p_C) = 0 \tag{2.221}$$

The standard state terms are grouped together into the overall standard state Gibbs free energy change, ΔG^o (kJ/kmol):

$$\Delta G^o = -\upsilon_A \overline{g}_A^o - \upsilon_B \overline{g}_B^o + \upsilon_C \overline{g}_C^o \tag{2.222}$$

Therefore

$$\begin{aligned} \Delta G^o &= -R_u T(-\upsilon_A \ln p_A - \upsilon_B \ln p_B + \upsilon_C \ln p_C) \\ &= -R_u T \ln\left[\frac{p_C^{\upsilon_C}}{p_A^{\upsilon_A} p_B^{\upsilon_B}}\right] \end{aligned} \tag{2.223}$$

We define an equilibrium constant K_p for chemical equilibrium in terms of partial pressures as

$$K_p = \frac{p_C^{\upsilon_C}}{p_A^{\upsilon_A} p_B^{\upsilon_B}} \tag{2.224}$$

Upon substitution and rearrangement, Eq. (2.223) becomes

$$\ln K_p = \frac{-\Delta G^o}{R_u T} \tag{2.225}$$

The equilibrium constant K_p can be determined from the standard state Gibbs free energy change. Once K_p is known for a given reaction, it can be used to find the equilibrium composition of a reactive product mixture by using Eq. (2.7) to express the partial pressures of the reacting species in terms of their mole fractions.

$$p_i = X_i p \tag{2.226}$$

Upon substitution, the equilibrium constant K_p is given by Eq. (2.227), where the total mixture pressure p is in units of atm,

$$K_p = \frac{X_C^{\upsilon_C}}{X_A^{\upsilon_A} X_B^{\upsilon_B}} p^{(\upsilon_C - \upsilon_A - \upsilon_B)} = \frac{n_C^{\upsilon_C}}{n_A^{\upsilon_A} n_B^{\upsilon_B}} \left(\frac{p}{n}\right)^{(\upsilon_C - \upsilon_A - \upsilon_B)} \tag{2.227}$$

Equation (2.227) has been derived for a three-component reaction with two reactants and one product. It can be generalized to reactions with any number of given reactants and products.

The equilibrium constant K_p is a measure of the completeness of a reaction, and the larger the value of K_p, the more complete the reaction. Values of K_p above 10^3 indicate a reaction that is mostly complete and composed primarily of products, and conversely, values of K_p below 10^{-3} indicate a reaction that is composed mostly of reactants. We will see in Chapter 3 that the equilibrium constant K can also be expressed in terms of concentration and identified as K_c.

It should be noted that K_p is independent of system pressure, since K_p is related to ΔG^o and T by Eq. (2.225). Accordingly, K_p is a function of temperature only and usually increases with increasing temperature for endothermic reactions and decreases for exothermic reactions. Various curve-fit equations are available, for example, a simple two parameters curve-fit for K_p is given by

$$K_p(T) = \exp(a_1 + a_2/T) \tag{2.228}$$

Example 2.13 ***Equilibrium Mole Fractions***

Determine the equilibrium mole fractions of CO_2, CO, and O_2 in an exhaust gas at $p = 10$ atm and $T = 2500$ K for the CO_2 dissociation reaction

$$CO_2 \longrightarrow aCO_2 + bCO + cO_2$$

Assume the equilibrium constant K_p is given by the following curve-fit equation:

$$K_p(T) = \exp(a_1 + a_2/T)$$

where $a_1 = 9.9283$ and $a_2 = -33,808$.

Solution

There are three unknown product moles, a, b, and c, and the total number of product moles is $n = a + b + c$. The carbon atom balance gives:

$$1 = a + b$$

and the oxygen balance is:

$$2 = 2a + b + 2c$$

The third equation is from the equilibrium constant equation, Eq. (2.227), for the elementary stoichiometric reaction

$$1CO_2 \rightleftharpoons 1CO + \frac{1}{2}O_2$$

where

$$v_{CO} = 1, \quad v_{O_2} = 1/2, \quad v_{CO_2} = 1$$

The equilibrium constant for this equation is

$$K_p = \frac{n_{CO}^{v_{CO}} n_{O_2}^{v_{O_2}}}{n_{CO_2}^{v_{CO_2}}} \left(\frac{p}{n}\right)^{(v_{CO}+v_{O_2}-v_{CO_2})}$$

Upon substitution, the fourth equation becomes

$$\exp(a_1 + a_2/T) = \frac{b^1 c^{1/2}}{a^1} \left(\frac{p}{a+b+c}\right)^{1+1/2-1}$$

The four equations for a, b, c, and n constitute a nonlinear equation set which needs to be solved with the aid of a computer. The resulting mole fractions are listed below.

```
Mole fractions
CO2 = 0.924
O2 = 0.0252
CO = 0.0505
```

2.15 Chemical Equilibrium Using Lagrange Multipliers

In general, we often consider a combustion problem that has many product species. At the start of combustion, the fuel is initially mixed with air with an equivalence ratio ϕ. After combustion, the products of the reaction are assumed to be in equilibrium at temperature T and pressure p. The composition and thermodynamic properties of the product mixture are to be determined. The overall combustion reaction per mole of fuel is:

$$\begin{aligned} &C_aH_bO_cN_d + \frac{a_{st}}{\phi}(O_2 + 3.76N_2) \longrightarrow \\ &n_1CO_2 + n_2H_2O + n_3N_2 + n_4O_2 + n_5\ CO + n_6H_2 \\ &+ n_7H + n_8O + n_9\ OH + n_{10}\ NO + n_{11}N + n_{12}C(s) \\ &+ n_{13}NO_2 + n_{14}CH_4 + \cdots \end{aligned} \tag{2.229}$$

This condition for equilibrium is usually stated in terms of thermodynamic functions such as the minimization of the Gibbs or Helmholtz free energy. If temperature and pressure are used to specify a thermodynamic state, the Gibbs free energy G is most easily minimized since temperature and pressure are its fundamental variables.

For a product mixture of N species, from Eq. (2.38), the Gibbs free energy is

$$G = \sum_{j=1}^{N} n_j \bar{g}_j \tag{2.230}$$

The equilibrium state can be determined by a Lagrangian multiplier approach, i.e., minimizing the Gibbs free energy subject to constraints. In this case, the constraint is the conservation of the number of atoms of each reacting species, b_i':

$$b_i' = \sum_{j=1}^{N} a_{ij} n_j \tag{2.231}$$

or

$$b_i - b'_i = 0 \tag{2.232}$$

where the index $i = 1, \ldots, l$, the integer l is the number of atom types, a_{ij} is the number of atoms of element i in species j, b'_i is the number of atoms of element i in the reactants, and b_i is the number of atoms of element i in the products.

$$b_i = \sum_{j=1}^{N} a_{ij} n_j \tag{2.233}$$

Using the Lagrangian optimization procedure, we first define the parameter B:

$$B = G + \sum_{i=1}^{l} \lambda_i (b_i - b'_i) \tag{2.234}$$

where the λ_i are the Lagrangian multipliers, one for each element. The variational condition, $\delta B = 0$, for equilibrium is

$$\delta B = \sum_{j=1}^{N} \left(\bar{g}_j + \sum_{i=1}^{l} \lambda_i a_{ij} \right) \delta n_j + \sum_{i=1}^{l} (b_i - b'_i) \delta \lambda_i = 0 \tag{2.235}$$

Treating the variations δn_j and $\delta \lambda_i$ as independent

$$\bar{g}_j + \sum_{i=1}^{l} \lambda_i a_{ij} = 0 \quad j = 1, \ldots, N \tag{2.236}$$

For ideal gases, the molar Gibbs free energy $\bar{g}_j$ is

$$\bar{g}_j = \bar{g}_j^o + R_u T \ln (n_j/n) + R_u T \ln (p/p_o) \tag{2.237}$$

where the standard state Gibbs free energy at T_o, p_o is

$$\bar{g}_j^o = \bar{h}_j^o - T_o \bar{s}_j^o \tag{2.238}$$

Therefore,

$$\frac{\bar{g}_j^o}{R_u T} + \ln (n_j/n) + \ln (p/p_o) + \sum_{i=1}^{l} \pi_i a_{ij} = 0 \quad j = 1, \ldots, N \tag{2.239}$$

where the dimensionless Lagrange multiplier, π_i, is

$$\pi_i = \lambda_i / R_u T \tag{2.240}$$

To determine the equilibrium composition using the Lagrange multiplier approach, we have to solve a set of $N + l + 1$ equations. For a given temperature and pressure (T, p), Eq. (2.239) is a set of N equations for the N unknowns n_j, l unknowns π_i, and n. Equation (2.232) provides the additional l equations and we close the set with an equation for the total number of moles n.

$$n = \sum_{j=1}^{N} n_j \tag{2.241}$$

Once the composition of the product species has been determined, we can now compute the thermodynamic properties of the equilibrium mixture. Recall that any two of the independent properties T, p, H, S, U, and V specify the thermodynamic state. For constant pressure combustion, the enthalpy is known instead of the temperature. For this case, we include an equation for the known enthalpy to our set of equations,

$$H = \sum_{j=1}^{N} n_j \bar{h}_j \tag{2.242}$$

For an isentropic compression or expansion, or expansion to a specified pressure, the entropy is given instead of enthalpy or temperature. In this case, we have

$$S = \sum_{j=1}^{N} n_j (\bar{s}_j^o - R_u \ln(n_j/n) - R_u \ln(p/p_o)) \tag{2.243}$$

Finally, if in any case specific volume rather than pressure is known, then we have to minimize the Helmholtz free energy. In this case, a similar analysis (Gordon and McBride 1994) shows that Eq. (2.239) is replaced by

$$\frac{\bar{g}_j^o}{R_u T} + \ln(n_j/n) + \ln(RT/p_o v) + \sum_{i=1}^{l} \pi_i a_{ij} = 0 \quad j = 1, \ldots, k \tag{2.244}$$

For constant volume combustion, the internal energy is known, so we include

$$U = \sum_{j=1}^{N} n_j (\bar{h}_j - R_u T) \tag{2.245}$$

For an isentropic expansion or compression to a specified volume v, we include

$$S = \sum_{j=1}^{N} n_j (\bar{s}_j^o - R_u \ln(n_j/n) - R_u \ln(RT/p_o v)) \tag{2.246}$$

Solution of these problems for practical application requires numerical iteration on a computer. Fortunately, there are now several computer programs available. Thermodynamic properties and equilibrium compositions can be computed using a classic NASA program called CEA (Chemical Equilibrium with Applications) Gordon and McBride (1994). This program uses the Lagrangian multiplier approach, i.e., minimizing the Gibbs free energy subject to constraints, for computation of the composition of the product species.

Results illustrating composition shifts with temperature and equivalence ratio are given in Figures 2.12 and 2.13 for the combustion of C_8H_{18} at $p = 50$ bar. Composition as a function of temperature is shown in Figure 2.12. The largest mole fractions are N_2, H_2O, and CO_2. As the reaction temperature is increased above 1500 K, there is an exponential rise in product species such as CO, NO, OH, O_2, O, H_2, and H. For lean ($\phi < 1$) conditions, the O_2 fraction is relatively insensitive to temperature. For rich conditions, the H_2 mole fraction first decreases and then increases with increasing temperature.

Notice that at high temperatures, there is a significant amount of nitric oxide (NO). If any gas in an engine cylinder is raised to these high temperatures, that gas will tend toward equilibrium at a rate determined by chemical kinetics. Since the chemistry for most species that contribute to the thermodynamic properties is fast enough relative to engine time scales, in many cases, local equilibrium may be assumed. Nitric oxide, however, is significant even though its concentrations are relatively low because it is an air pollutant. Unlike the species of thermodynamic importance, its chemistry is not fast enough to assume that it is in equilibrium concentrations. Likewise, once formed, its concentration "freezes" during the expansion stroke of an internal combustion engine so that even in the low-temperature exhaust gases, nitric oxides are found. This will be discussed more fully when we deal with emissions in the next chapters.

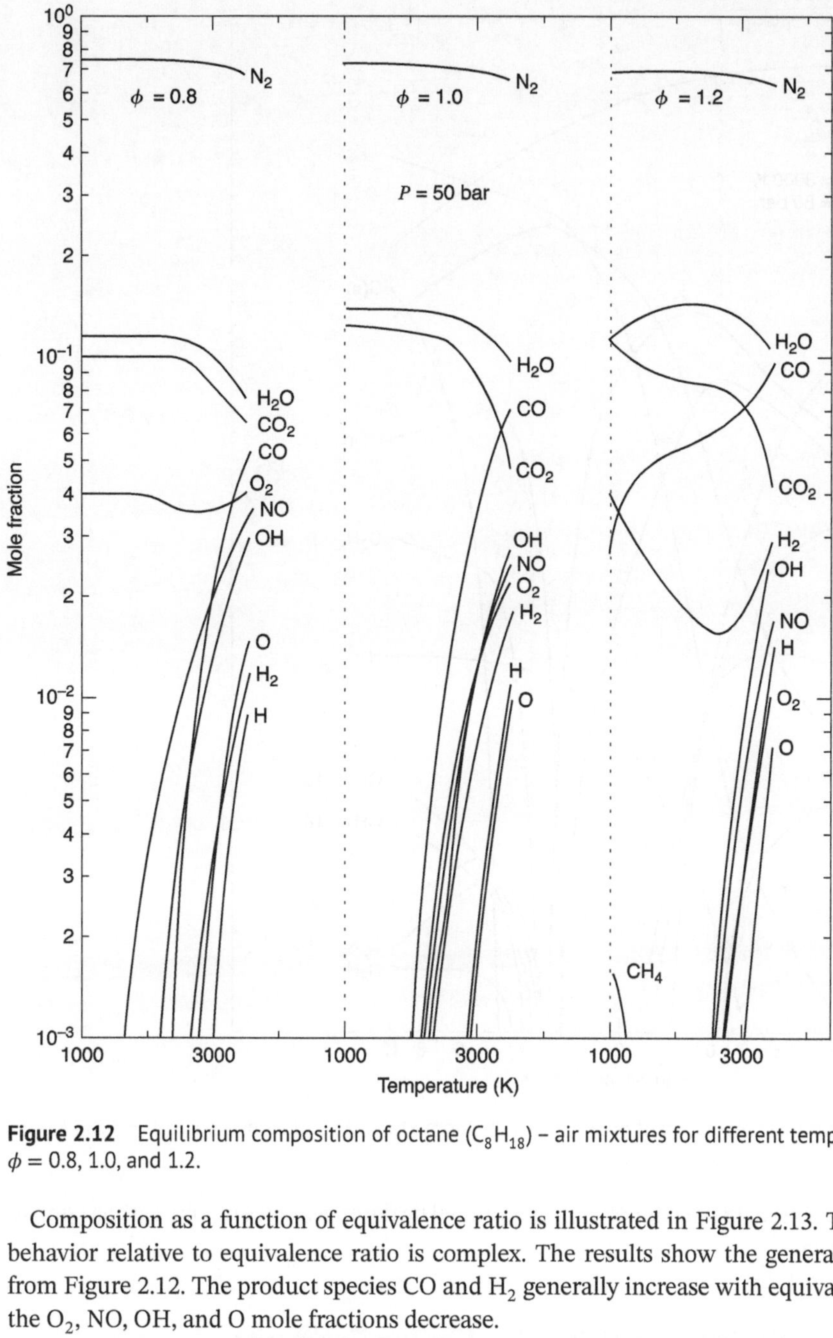

Figure 2.12 Equilibrium composition of octane (C_8H_{18}) – air mixtures for different temperatures at $\phi = 0.8$, 1.0, and 1.2.

Composition as a function of equivalence ratio is illustrated in Figure 2.13. The mole fraction behavior relative to equivalence ratio is complex. The results show the general trends expected from Figure 2.12. The product species CO and H_2 generally increase with equivalence ratio, while the O_2, NO, OH, and O mole fractions decrease.

If the equivalence ratio ϕ is greater than about 4, the product species list becomes quite large and includes solid carbon, C(s); hydrogen cyanide, HCN; acetylene, C_2H_2; and methane, CH_4. Thus, if anywhere in the combustion chamber, there are fuel–air pockets where $\phi > 3$, such as in diesel or stratified charge internal combustion engines, there will be a tendency for these species to form. Similar to nitric oxides, their concentration may freeze when mixed with leaner pockets or when the temperature drops, so these species can appear in the exhaust. With diesel engines, even though

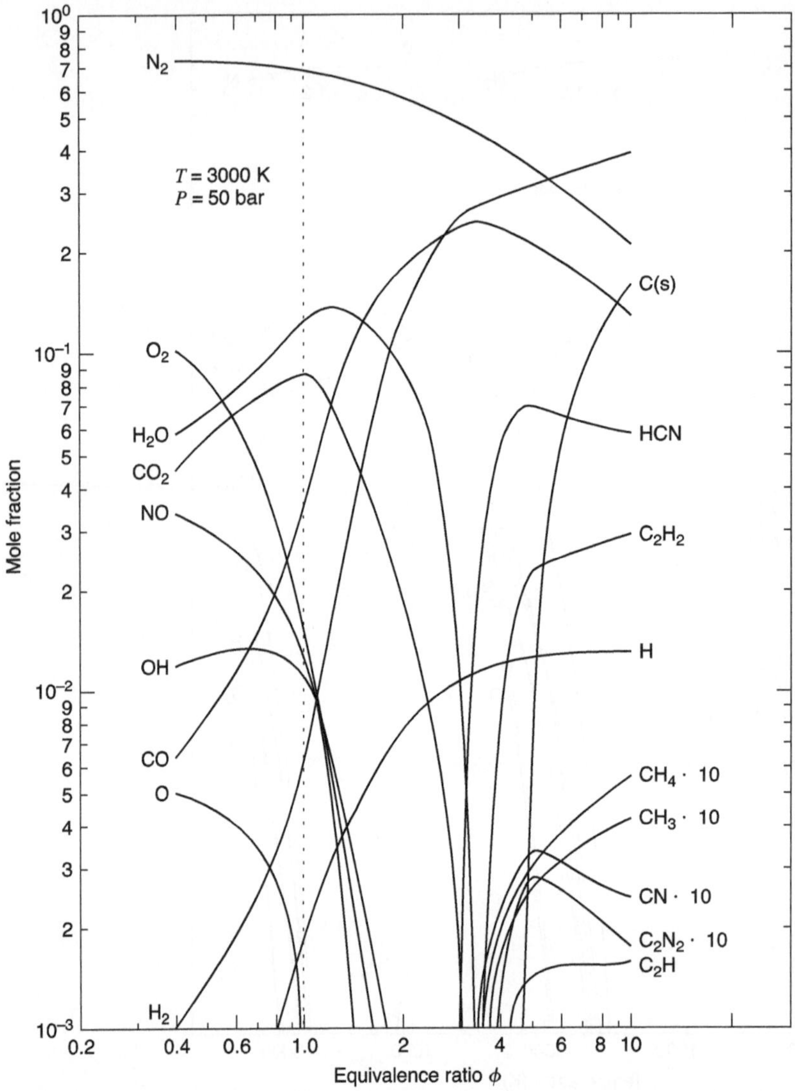

Figure 2.13 Equilibrium composition of octane (C_8H_{18}) – air mixtures as a function of ϕ at $T = 3000$ K, and $p = 50$ bar.

the engine is running lean, the maximum power is limited by the appearance of solid carbon (smoke and soot) in the exhaust.

2.16 Chemical Equilibrium Using Equilibrium Constants

This section presents a numerical solution for the properties of equilibrium combustion products based on an equilibrium constant method applied by Olikara and Borman (1975) to the gas phase products of combustion of hydrocarbon fuels. The use of equilibrium constants is also based on the minimization of the Gibbs free energy of the product species mixture; however, it is algebraically less complex than the Lagrange multiplier approach of Section 2.15 when considering restricted species lists. The equilibrium constant method does require, however, that the equilibrium reactions between the product species be specified.

A more complete reaction calculation, such as done in Section 2.15, needs to be performed first to determine the significant product species to include in the equilibrium constant analysis. Inspection of Figures 2.12 and 2.13 shows that if $\phi < 3$, the only product species of importance resulting from dissociation are O, H, OH, and NO. Therefore, the species list in Eq. (2.229) can be terminated at $i = 10$, that is, we need to consider only 10 species. Therefore, let us consider the following reaction:

$$\begin{aligned} &C_aH_bO_cN_d + \frac{a_{st}}{\phi}(O_2 + 3.76N_2) \longrightarrow \\ &n_1CO_2 + n_2H_2O + n_3N_2 + n_4O_2 + n_5CO + n_6H_2 \\ &+ n_7H + n_8O + n_9OH + n_{10}NO \end{aligned} \tag{2.247}$$

Atom balancing yields the following four equations:

$$\begin{aligned} &C : a = (X_1 + X_5)\,n \\ &H : b = (2X_2 + 2X_6 + X_7 + X_9)\,n \\ &O : c + 2a_{st}/\phi = (2X_1 + X_2 + 2X_4 + X_5 + X_8 + X_9 + X_{10})\,n \\ &N : d + 7.52a_{st}/\phi = (2X_3 + X_{10})\,n \end{aligned} \tag{2.248}$$

where n is the total number of moles. By definition, the mole fractions sum to 1:

$$\sum_{i=1}^{10} X_i = 1 \tag{2.249}$$

From these equations, three constants are defined:

$$d_1 = \frac{b}{a}$$

$$d_2 = \frac{c}{a} + 2\frac{a_{st}}{\phi a}$$

$$d_3 = \frac{d}{a} + \frac{7.52a_{st}}{\phi a}$$

Upon substitution into the atom balance equations, and with some rearrangement,

$$\begin{aligned} 2X_2 + 2X_6 + X_7 + X_9 - d_1X_1 - d_1X_5 &= 0 \\ 2X_1 + X_2 + 2X_4 + X_5 + X_8 + X_9 + X_{10} - d_2X_1 - d_2X_5 &= 0 \\ 2X_3 + X_{10} - d_3X_1 - d_3X_5 &= 0 \\ \sum X_i &= 1 \end{aligned} \tag{2.250}$$

We now introduce six product species equilibrium reactions each with an associated equilibrium constant, $K_{p,i}(i = 1, 6)$, based on partial pressure. These reactions include the the dissociation of hydrogen, oxygen, water, and carbon dioxide, and the formation of OH and NO:

$$\begin{aligned}
&\frac{1}{2}H_2 \rightleftharpoons H && K_{p,1} = \frac{X_7 p^{1/2}}{X_6^{1/2}}\\
&\frac{1}{2}O_2 \rightleftharpoons O && K_{p,2} = \frac{X_8 p^{1/2}}{X_4^{1/2}}\\
&\frac{1}{2}H_2 + \frac{1}{2}O_2 \rightleftharpoons OH && K_{p,3} = \frac{X_9}{X_4^{1/2} X_6^{1/2}}\\
&\frac{1}{2}O_2 + \frac{1}{2}N_2 \rightleftharpoons NO && K_{p,4} = \frac{X_{10}}{X_4^{1/2} X_3^{1/2}}\\
&H_2 + \frac{1}{2}O_2 \rightleftharpoons H_2O && K_{p,5} = \frac{X_2}{X_4^{1/2} X_6 p^{1/2}}\\
&CO + \frac{1}{2}O_2 \rightleftharpoons CO_2 && K_{p,6} = \frac{X_1}{X_4^{1/2} X_5 p^{1/2}}
\end{aligned} \tag{2.251}$$

The unit of pressure in the above six equations is in atmospheres (atm).

Olikara and Borman (1975) have curve-fitted the equilibrium constants $K_{p,i}(T)$ to JANAF Table data for the temperature range 600 $<T>$ 4000 K. Their expressions are of the form

$$\log_{10} K_{p,i}(T) = A_i \ln(T/1000) + \frac{B_i}{T} + C_i + D_i T + E_i T^2 \tag{2.252}$$

where T is in Kelvin. The equilibrium constant $K_{p,i}$ curve-fit coefficients are listed in Table 2.11. Given pressure p, temperature T, and equivalence ratio ϕ, Eqs. (2.248), (2.249), and (2.251) will yield 11 equations for the 11 unknowns: the 10 unknown mole fractions X_i and the unknown total product moles n.

Substitution of the six product species equilibrium equations into the atom balance equations results in four equations in four unknowns (X_3, X_4, X_5, X_6). These four equations are solved numerically in the Equilibrium Combustion Solver program `ecp.m` which is listed on the text web page. The program computes the product mole fractions and properties for the six representative fuels given the mixture pressure P, temperature T, and the fuel–air equivalence ratio ϕ.

Table 2.11 Equilibrium Constant $K_{p,i}$ Curve-Fit Coefficients.

$K_{p,i}$	A_i	B_i	C_i	D_i	E_i
$K_{p,1}$	+0.432168E+00	−0.112464E+05	+0.267269E+01	−0.745744E−04	+0.242484E−08
$K_{p,2}$	+0.310805E+00	−0.129540E+05	+0.321779E+01	−0.738336E−04	+0.344645E−08
$K_{p,3}$	−0.141784E+00	−0.213308E+04	+0.853461E+00	+0.355015E−04	−0.310227E−08
$K_{p,4}$	+0.150879E−01	−0.470959E+04	+0.646096E+00	+0.272805E−05	−0.154444E−08
$K_{p,5}$	−0.752364E+00	+0.124210E+05	−0.260286E+01	+0.259556E−03	−0.162687E−07
$K_{p,6}$	−0.415302E−02	+0.148627E+05	−0.475746E+01	+0.124699E−03	−0.900227E−08

The fuels are identified with the following fuel_id, from 1 to 6:

1. methane CH_4
2. gasoline C_7H_{17}
3. diesel $C_{14.4}H_{24.9}$
4. methanol CH_3OH,
5. ethanol C_2H_5OH, and
6. nitromethane CH_3NO_2

With the mixture mole fraction composition known, one can then proceed to compute the thermodynamic properties of interest: enthalpy, entropy, specific volume, internal energy, and specific heat for the given conditions. A reacting mixture of ideal gases has an enthalpy dependent on temperature and pressure, and computing the mixture equilibrium-specific heat c_p requires the change in mole fraction due to a change in temperature, so as discussed earlier in this chapter, the mole fraction partial derivatives $\partial X_i/\partial p$ and $\partial X_i/\partial T$ are also computed. The use of the program is detailed in the following example.

Example 2.14 ***Equilibrium Combustion Mole Fractions***

What are the equilibrium mole fractions and mixture properties resulting from the combustion of a gasoline (fuel_id = 2) mixture at a temperature $T = 2000$ K, pressure $p = 5000$ kPa, and a fuel–air equivalence ratio $\phi = 0.8$?

Solution

The computation is performed using the Equilibrium Combustion Solver input–output program `RunEcp.m` with inputs T, p, ϕ, and fuel_id. As detailed on the web page, the program `RunEcp.m` calls the function `ecp.m`, which in turn calls the functions `fuel.m` for fuel properties and `guess` for initial guess values of mixture properties, then iterates for converged property values, and finally outputs the mole fractions and mixture properties.

The resulting mole fractions and properties are listed below. The equilibrium mole fractions calculated with the equilibrium constant model compare well with the mole fractions of Figure 2.15 computed by the more general Lagrange multiplier method.

```
Equilibrium combustion solver output
Mole fractions
 CO2  = 9.8181e-02
 H2O  = 1.1904e-01
 N2   = 7.4103e-01
 O2   = 3.7728e-02
 CO   = 9.3901e-05
 H2   = 2.5150e-05
 H  = 1.1568e-06
 O  = 1.8367e-05
 OH  = 5.4430e-04
 NO  = 3.3442e-03

Mixture Properties
 h(kJ/kg)  = -190.4
 u(kJ/kg)  = -773.0
 s(kJ/Kg K)  = 8.2124
```

```
v(m3/kg) = 0.1165
cp(kJ/Kg K) = 1.465
Molecular Mass = 28.54
dvdt = 5.8399e-05
dvdp = -2.3301e-05
```

2.17 Thermodynamic Properties of Combustion Products

As shown in the following examples, the Equilibrium Combustion Solver program can be used to compute the thermodynamic properties of a variety of fuels and their dependence on fuel type, temperature, equivalence ratio, and pressure.

The effect of equivalence ratio, pressure, and temperature on the specific heat c_p of the combustion products of a gasoline–air mixture is shown in Figures 2.14 and 2.15. The specific heat increases with temperature due to the activation of additional molecular vibrational modes and decreases with pressure due to the reduction of dissociation. The specific heat also increases as ϕ increases from lean to near stoichiometric, and as the mixture becomes richer than stoichiometric, the specific heat first decreases, then increases. This is due to the changing product concentrations shown in Figure 2.12, primarily governed by the change in the H_2O and CO_2 concentrations. The maximum value of c_p is at stoichiometric where the maximum concentrations of the H_2O and CO_2 species occur.

The effect of temperature on enthalpy of the equilibrium combustion products for three different equivalence ratios is shown in Figure 2.16 for the combustion of gasoline at a pressure of 101.3 kPa. Note that the lowest value of enthalpy occurs at a stoichiometric equivalence ratio, and as the equivalence ratio is made lean or rich, below 2500 K, the enthalpy increases. This behavior is also shown

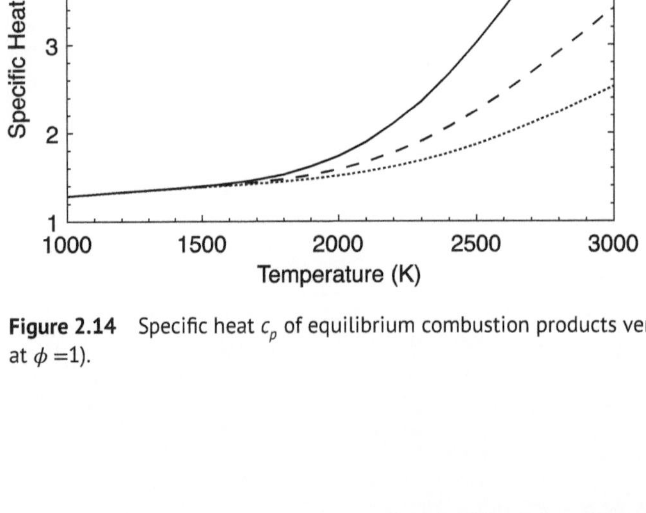

Figure 2.14 Specific heat c_p of equilibrium combustion products versus temperature (gasoline–air mixture at ϕ =1).

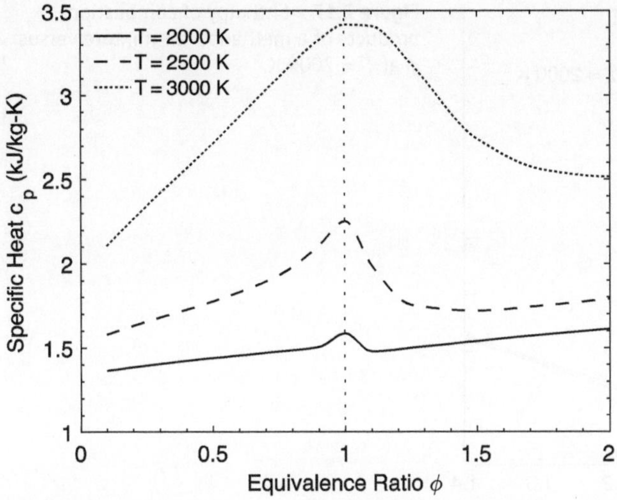

Figure 2.15 Specific heat c_p of equilibrium combustion products versus equivalence ratio (gasoline–air mixture at p = 1000 kPa).

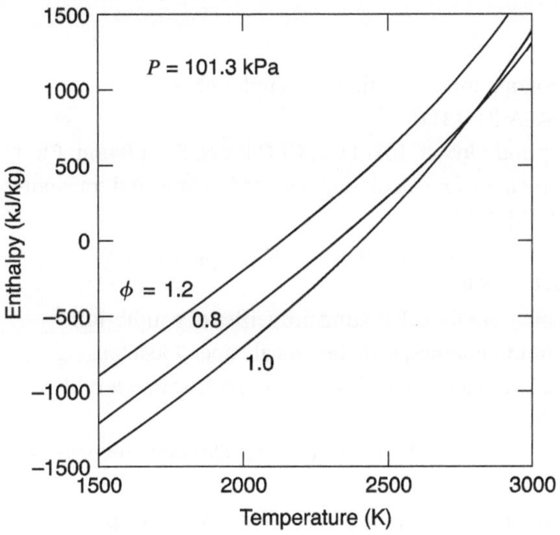

Figure 2.16 Enthalpy of combustion products for a gasoline–air mixture versus temperature at $p = 101.3$ kPa.

in Figure 2.17, a plot of the enthalpy of the combustion products of methanol versus equivalence ratio at pressures of 101 and 2000 kPa. The enthalpy of the combustion products is a minimum at near stoichiometric conditions, since the combustion is complete. Conversely, on either side of stoichiometric, the combustion is incomplete, leading to increased product enthalpy. If the mixture is lean, there is an excess of unburnt oxygen, and if the mixture is rich, there will be unburnt carbon monoxide.

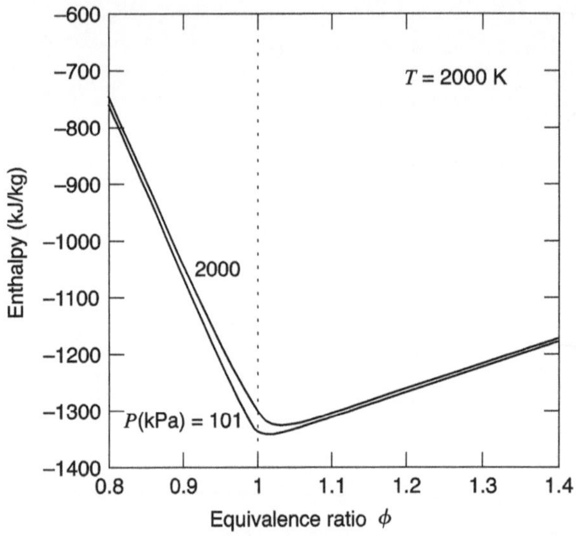

Figure 2.17 Enthalpy of combustion products of a methanol–air mixture versus ϕ at $T = 2000$ K.

References

Gordon, S., and B. McBride (1994), Computer program for calculation of complex chemical equilibrium composition, and applications, NASA RP-1311.

Lide, D., Ed., (2022), *CRC Handbook of Chemistry and Physics*, 103rd ed., CRC Press, Boca Raton, FL.

McBride B., S. Gordon, and M. Reno (1993), Coefficients for calculating thermodynamic and transport properties of individual species, NASA Report TM-4513.

McBride B., M. Zehe, and S. Gordon (2002), NASA Glenn coefficients for calculating thermodynamic properties of individual species, NASA TP-2002-211556.

Olikara, C., and G. Borman (1975), A computer program for calculating properties of equilibrium combustion products with some applications to I.C. engines, SAE Technical Paper 750468.

Rossini, E. (1953), *Selected Values of Physical and Thermodynamic Properties of Hydrocarbons and Related Compounds*, Carnegie Press, Pittsburgh.

Stull, D., E. Westrum, Jr., and G. Singe (1969), *The Chemical Thermodynamics of Organic Compounds*, John Wiley & Sons, New York.

Su, G. (1946), Modified law of corresponding states for real gases, *Ind. Eng. Chem.*, Vol. 38, pp. 803–806.

Vargaftik, N. (1975), *Tables on the Thermophysical Properties of Liquids and Gases*, John Wiley & Sons, New York.

Homework Problems

2.1 What is the molecular weight, enthalpy (kJ/kg), entropy (kJ/kg-K), and Gibbs free energy (kJ/kg) of a combustion product mixture at $p = 1000$ kPa and $T = 1000$ K, if the mixture contains the following species and mole fractions?

Species	X_i
CO_2	0.10
H_2O	0.15
N_2	0.70
CO	0.05

2.2 What is the enthalpy h (kJ/kg) and entropy s (kJ/kg-K) of a mixture of 30% H_2 and 70% CO_2 by volume at a temperature $T = 3000$ K and pressure $p = 2000$ kPa?

2.3 The combustion products from a spark ignition engine have the following composition at $p = 101$ kPa and $T = 298$ K. What is the enthalpy (kJ/kg), specific volume (m^3/kg), and quality χ of the mixture?

Species	X_i
CO_2	0.125
H_2O	0.141
N_2	0.734

2.4 Using bond energies, determine the heat of formation of gaseous normal butane C_4H_{10}.

2.5 By integrating Eq. (2.84) from $p = 0$ to a finite pressure, show that the fugacity **f** of a real gas is related to the pressure p and compressibility factor Z by

$$\ln \frac{\mathbf{f}}{p} = \int_0^{p_r} (Z - 1)\, d(\ln p_r)_T$$

where p_r is the reduced pressure ($p_r = p/p_{\text{critical}}$). Discuss the use of this equation.

2.6 A mixture of 1 mole of gaseous ethene C_2H_4 and 3 moles of oxygen O_2 at 298 K react in a constant-volume combustion chamber. Assume the combustion products are CO_2 and $H_2O_{(g)}$ and there is an insignificant amount of dissociation products. Heat is transferred out of the combustion chamber until the products are cooled to 600 K. Determine the amount of heat transferred in J/mole of the fuel.

2.7 Calculate and compare the enthalpies of vaporization h_{fg} (MJ/kg) of nitromethane, methanol, octane, and tetradecane at 400 K. What is an advantage of a high enthalpy of vaporization for an engine fuel? Why are they different than the corresponding values listed in Table 2.1?

2.8 Calculate the enthalpy of vaporization h_{fg} (MJ/kg) of liquid chlorine at $T = 239.05$ K. The specific volume of chlorine gas at $T = 239.05$ K is 269.1 cm^3/g and that of chlorine liquid is approximately 0.7 cm^3/g. The vapor pressure of liquid chlorine, in centimeters of mercury, can be represented by the expression

$$\log_{10}p = -1414.8T^{-1} + 9.91635 - 1.206 \times 10^{-2}T + 1.34 \times 10^{-5}T^2$$

Note: $\frac{d(\log_{10}p)}{dT} = \frac{1}{2.303}\frac{d(\ln p)}{dT}$

2.9 A stoichiometric mixture of ethanol C_2H_5O, air, and residual gas is burned at constant pressure. If $T_1 = 298$ K, $p_1 = 101.3$ kPa, and residual fraction $f_r = 0.05$, what is the constant pressure adiabatic flame temperature?

2.10 A combustible mixture of air and carbon monoxide that is 10% rich [air/fuel = $\frac{100}{110}$(air/fuel)$_{st}$] is compressed to a pressure of 8.28 bar and a temperature of 555 K. The mixture is ignited, and combustion occurs adiabatically at constant volume. When the maximum temperature is attained, analysis shows 0.228 moles of CO present for each mole of CO supplied. Show that the maximum temperature reached is 2950 K. If the pressure at this temperature is now doubled, calculate the amount of CO present.

2.11 A hydrogen-fueled engine operates with an air mass flowrate $\dot{m}_a = 2.0$ kg/s and produces 1000 kW of power. The exhaust enthalpy is -40,000 kJ/kg$_{fuel}$, and the heat losses from the engine total 25,000 kJ/kg$_{fuel}$. (a) What is the hydrogen fuel flow rate $\dot{m}_f$? (b) At what equivalence ratio ϕ is the engine being operated?

2.12 What are the mole fractions of CO_2, H_2O, CO, N_2 and H_2 produced when methane (CH_4) is burned in rich conditions at $\phi = 1.1$, $T = 1000$ K, and $p = 101$ kPa?

2.13 What is the equilibrium composition for the reaction $H_2 + \frac{1}{2}O_2 \leftrightarrows H_2O$ when the ratio of the number of moles of elemental H to the number of moles of elemental O is equal to four? The temperature $T = 2200$ K and pressure $p = 1$ atm.

3

Chemical Kinetics

3.1 Introduction

In this chapter, we examine chemical reactions and their rates of reaction. Questions such as "How long does it take to complete a reaction?" and "How does a mixture get from an initial state to the final state?" are addressed in this chapter and in Chapter 4. Chemical kinetics is the quantitative study of the rates of chemical reactions and the factors such as temperature, pressure, and species concentrations upon which they depend. It describes the numerous pathways from reactants to products and determines their associated reaction rates. *Global* reactions describe the overall reaction from initial reactants to final products, and *elementary* reactions indicate the sequential processes involving intermediate species formed and consumed during the reaction.

3.2 Rates of Reaction and Their Functional Dependence

All chemical reactions take place at a definite rate, depending on the conditions of the system. Chemical reaction rates are highly nonlinear functions of species concentrations and temperature. The thermal energy that is released during a chemical reaction raises the reactant temperature, resulting in the reaction continuing to proceed at a high rate. Important conditions for reaction rates include: (1) concentrations of the chemical compounds, (2) temperature, (3) pressure, (4) presence of a catalyst or inhibitor, and (5) radiative effects.

As we know, some elementary chemical reactions proceed more slowly in comparison with other elementary reactions occurring in a combustion system. These slow reactions are important to characterize, since they control the reaction rate of the overall combustion process.

The rate of reaction can be expressed in either molar or mass units. The molar reaction rate per unit volume $\dot{\Omega}$ has units (kmol/m^3-s), and the mass reaction rate per unit volume $\dot{\omega}$ has units (kg/m^3-s). For example, if a stoichiometric mixture of fuel (F) and air (A) has a total mixture mass m, where

$$m = m_F + m_A \tag{3.1}$$

then the reactant mass consumption rate $\dot{m}$ is

$$\begin{aligned}\dot{m} &= \dot{m}_F + \dot{m}_A = \dot{m}_F\left(1 + \frac{\dot{m}_A}{\dot{m}_F}\right) \\ &= \dot{m}_F(1 + s_A)\end{aligned} \tag{3.2}$$

Principles of Combustion, Third Edition. Allan T. Kirkpatrick and Kenneth K. Kuo.

Companion website: www.wiley.com/go/kirkpatrick/principlesofcombustion3e

where s_A is the stoichiometric mass air/fuel ratio. The mixture mass reaction rate per unit volume is therefore

$$\dot{\omega} = \dot{\omega}_F(1 + s_A) \tag{3.3}$$

3.2.1 Reaction Order

The reaction order m is the combustion parameter that relates the rate of a reaction to the concentrations of the reacting species. It indicates to what extent the concentration of a species affects the rate of a reaction, as well as which species has the greatest effect on the reaction rate. A forward chemical reaction can be represented by the following stoichiometric, i.e., atom balance equation:

$$\sum_{i=1}^{N} \nu_i A_i \rightarrow \sum_{i=1}^{N} \nu_i' A_i \tag{3.4}$$

where ν_i are the stoichiometric coefficients of the reactants, ν_i' the stoichiometric coefficients of the products, A_i represents the ith chemical species, and N is the sum of the number of species, including both reactants and products, involved in the reaction. If a species represented by A_i is not a reactant then $\nu_i = 0$, and if it is not a product, then $\nu_i' = 0$.

The process represented by Eq. (3.4) is said to be of order ν_i with respect to species A_i. The overall order of the reaction m is

$$m = \sum_{i=1}^{N} \nu_i \tag{3.5}$$

That is, the overall order of the reaction is equal to the sum of the coefficients of the reactants. A zero order indicates that the concentration of that species does not affect the rate of a reaction. In combustion studies, most reactions we shall deal with are either second or third order.

For example, consider the reaction between two hydrogen atoms in the presence of a third hydrogen atom to form a hydrogen molecule, with the third hydrogen atom gaining translational energy during the collision process,

$$3\mathrm{H} \rightarrow \mathrm{H_2} + \mathrm{H} \tag{3.6}$$

Here, the number of species $N = 2$ (H and H_2), and

$$A_1 = \mathrm{H}, \qquad A_2 = \mathrm{H_2}$$
$$\nu_1 = 3, \qquad \nu_2 = 0$$
$$\nu_1' = 1, \qquad \nu_2' = 1$$

Observe that no distinction is made between hydrogen atoms having different energies. The reaction (3.6) is the third order with respect to H, the zeroth order with respect to H_2, and it has an overall order of three.

Let us now consider the overall or global chemical reaction between gaseous hydrogen and oxygen to form water vapor. The reaction can be expressed by

$$2\mathrm{H_2} + \mathrm{O_2} \rightarrow 2\mathrm{H_2O} \tag{3.7}$$

Here, the number of species $N = 3$ (H_2, O_2, and H_2O), and

$$A_1 = \mathrm{H_2}, \quad A_2 = \mathrm{O_2}, \qquad A_3 = \mathrm{H_2O}$$
$$\nu_1 = 2, \qquad \nu_2 = 1, \qquad \nu_3 = 0$$
$$\nu_1' = 0, \qquad \nu_2' = 0, \qquad \nu_3' = 2$$

The chemical reaction (3.7) is balanced with the above sets of stoichiometric coefficients. The hypothetical chemical reaction described by Eq. (3.7) represents a third-order process that is of the second order with respect to H_2, and the first order with respect to O_2. More complicated examples in which the values of ν_i and ν_i' are not necessarily integers can be given without difficulty.

Since Eq. (3.7) is an overall or global chemical reaction, it does not describe the detailed reaction pathways. The overall result of the reaction between 2 moles of H_2 and 1 mole of O_2 is the production of 2 moles of H_2O; but the conversion of H_2 and O_2 occurs by means of a series of successive, interdependent, elementary chemical reactions.

3.2.2 Law of Mass Action

The *law of mass action* which is confirmed by numerous experimental observations, states that the overall rate of reaction $\dot{\Omega}$ is proportional to the products of the concentrations $[A_i]$ of the reacting chemical species, each concentration being raised to a power equal to the corresponding stoichiometric coefficient. Thus, the overall rate of reaction $\dot{\Omega}$ (kmol/m^3-s) is

$$\dot{\Omega} = k\prod_{i=1}^{N}[A_i]^{\nu_i} \tag{3.8}$$

where the coefficient k is called the *specific reaction rate coefficient* or rate constant.

For example, the second-order ($m = 2$) bimolecular reaction of Eq. (3.9)

$$A + B \longrightarrow C \tag{3.9}$$

has a rate of reaction $\dot{\Omega}$

$$\dot{\Omega} = -\frac{d[A]}{dt} = -\frac{d[B]}{dt} = +\frac{d[C]}{dt} = k\,[A][B] \tag{3.10}$$

For a given chemical reaction, the reaction rate coefficient k is independent of the concentrations $[A_i]$ and depends only on the temperature. The SI units of the rate coefficient k depend on the order m of the reaction,

$$\frac{\text{kmol}}{\text{m}^3\cdot\text{s}}\frac{1}{(\text{kmol/m}^3)^m} = \text{kmol}^{1-m}\text{m}^{3m-3}\text{s}^{-1} \tag{3.11}$$

For a first-order reaction ($m = 1$), the units of k are s^{-1}, and for a second-order reaction ($m = 2$), the units of k are m^3/kmol-s in SI units and cm^3/gmol-s in CGS units.

The rate of change of the molar concentration $[A_i]$ of species A_i is

$$\dot{\Omega}_i = \frac{d[A_i]}{dt} = (\nu_i' - \nu_i)\dot{\Omega} \tag{3.12}$$

In Eq. (3.12), the overall reaction rate $\dot{\Omega}$ is multiplied by the difference $(\nu_i' - \nu_i)$ since species A_i may appear on both sides of Eq. (3.4) with different values for ν_i' and ν_i. Applying Eq. (3.12) to the reaction (3.9) is

$$\dot{\Omega}_A = \frac{d[A]}{dt} = (\nu_A' - \nu_A)\dot{\Omega} = (0-1)\dot{\Omega} = -\dot{\Omega}$$
$$\dot{\Omega}_B = \frac{d[B]}{dt} = (\nu_B' - \nu_B)\dot{\Omega} = (0-1)\dot{\Omega} = -\dot{\Omega}$$
$$\dot{\Omega}_C = \frac{d[C]}{dt} = (\nu_C' - \nu_C)\dot{\Omega} = (1-0)\dot{\Omega} = +\dot{\Omega}$$

For reversible reactions, we will use the subscript f to identify a forward reaction, and the subscript b to identify a backward or reverse reaction. Using Eqs. (3.4) and (3.8), the net reaction rate $\dot{\Omega}$ and the net rate of production $\dot{\Omega}_A$ of species A_i are

$$\dot{\Omega} = k_f \prod_{i=1}^{N} [A_i]^{\nu_i} - k_b \prod_{i=1}^{N} [A_i]^{\nu_i'} \tag{3.13}$$

$$\dot{\Omega}_A = \frac{d[A_i]}{dt} = (\nu_i' - \nu_i)\dot{\Omega} \tag{3.14}$$

For example, given the reaction

$$H_2 + O_2 \underset{k_b}{\overset{k_f}{\rightleftharpoons}} HO_2 + H$$

the net rate of production of product HO_2 is given by

$$\frac{d[HO_2]}{dt} = (1-0)k_f[H_2][O_2] - (1-0)k_b[HO_2][H]$$
$$= k_f[H_2][O_2] - k_b[HO_2][H]$$

and for a general reaction of the form

$$aA + bB \underset{k_b}{\overset{k_f}{\rightleftharpoons}} cC + dD$$

the net rate of reaction of A is given by

$$\frac{d[A]}{dt} = -a\,k_f[A]^a[B]^b + a\,k_b[C]^c[D]^d$$

3.2.3 Elementary Reactions and Reaction Mechanisms

An *elementary reaction* is defined as a single-step reaction with a single transition state and no intermediates. In an elementary reaction, the reactant chemical species undergo a direct transformation to create the final products in one step. If no intermediate products are detected during a chemical reaction, that reaction is classified as an elementary reaction. An elementary reaction thus occurs on a molecular level exactly in the way described by the law of mass action given by Eq. (3.13).

A multistep reaction process is modeled with a *reaction mechanism*, a set of elementary reactions which begin with the initial reactants and end with the final products. The law of mass action is applied to the elementary reactions to determine their reaction rates. As we will see in Chapter 4, the detailed reaction mechanisms for many chemical reactions can be very lengthy and complex, so much research has been conducted to develop reduced reaction mechanisms that still capture the essential chemistry of the reacting system.

As an example, two elementary reactions involving the hydroxyl radical OH and the hydrogen atom H are intermediate reactions that are included in the reaction mechanism for the production of H_2O from H_2 and O_2,

$$H + O_2 \rightarrow OH + O$$
$$OH + H_2 \rightarrow H_2O + H$$

3.2.4 Equilibrium Constant K_c

In general, chemical reactions can proceed in both the forward direction, i.e., reactants forming products with a rate coefficient k_f, and the reverse or backward direction, i.e., reaction products reforming back into the initial reactants with a rate coefficient k_b. If the system is in thermodynamic equilibrium, there is no net change in the chemical composition of the mixture. The general set of opposing (forward and backward) chemical reactions is given by

$$\sum_{i=1}^{N} \nu_i A_i \rightleftharpoons \sum_{i=1}^{N} \nu_i' A_i \tag{3.15}$$

Rearranging Eqs. (3.13) and (3.14), the rate of change of concentration ($d[A_i]/dt$) of species A_i can be expressed as

$$\frac{d[A_i]}{dt} = (\nu_i' - \nu_i)k_f \prod_{i=1}^{N} [A_i]^{\nu_i} + (\nu_i - \nu_i')k_b \prod_{i=1}^{N} [A_i]^{\nu_i'} \tag{3.16}$$

At thermodynamic equilibrium, the concentrations of the various species A_i are unchanged,

$$\frac{d[A_i]}{dt} = 0 \quad \text{and} \quad [A_i] = [A_{i,e}] \tag{3.17}$$

where $[A_{i,e}]$ denotes the thermodynamic equilibrium value for species A_i.

The rate coefficients k_f and k_b must therefore be related to an equilibrium constant, K_c, based on concentration, through Eqs. (3.16) and (3.17).

$$K_c \equiv \frac{k_f}{k_b} = \prod_{i=1}^{N} ([A_{i,e}])^{\nu_i' - \nu_i} \tag{3.18}$$

Equation (3.16) can be rewritten in terms of K_c as follows:

$$\frac{d[A_i]}{dt} = (\nu_i' - \nu_i)k_f \prod_{i=1}^{N} ([A_i])^{\nu_i} \left(1 - \frac{1}{K_c} \prod_{i=1}^{N} ([A_i])^{\nu_i' - \nu_i}\right) \tag{3.19}$$

The equilibrium constant K_c can be calculated quite accurately, for example, by quantum-statistical methods from molecular properties. Knowing K_c and the measured value of $d[A_i]/dt$, it is possible to calculate the forward rate coefficient from Eq. (3.19).

In Chapter 2, we defined an equilibrium constant K_p based on partial pressures. The two equilibrium constants K_p and K_c are related, since the molar concentrations of a reaction are related to the partial pressures and the mole fractions through the ideal-gas law,

$$[A_i] = \frac{n_{A_i}}{V} = \frac{p_{A_i}}{R_u T} = X_{A_i} \frac{p}{R_u T} \tag{3.20}$$

and therefore

$$K_c = K_p (R_u T)^{-\Delta n} \tag{3.21}$$

where

$$\Delta n = \sum_{i=1}^{N} (\nu_i' - \nu_i) \tag{3.22}$$

If the number of reactant species $\sum_{i=1}^{N} \nu_i$ is the same as the number of product species $\sum_{i=1}^{N} \nu_i'$ in a reaction, then $K_c = K_p$.

3.2.5 Molecular Collision Analysis

Molecular collision analysis is used to provide a physical understanding of the law of mass action. In the study of molecular collisions, one must first consider the effect of the size of the molecules. The molecular diameter is a measure applied to molecules that express the likelihood that a molecule in a gas will collide with another molecule. It is an indication of the size of the molecule as a target. The molecular diameter is not the same as the atomic diameter which is defined in terms of the size of the atom's electron shell, which is generally a lot smaller. Rather, it is the size of the sphere of influence that can lead to a scattering event. Molecular diameters in Angstroms ($\AA = 10^{-8}$ cm) for various molecules are listed in Table 3.1. Additional molecular diameters are given in the Appendix F.

For simplicity, let us consider the collision of two identical spherical molecules with molecular diameter σ. When they collide, the distance between their centers must be less than or equal to σ. If the distance between the two centers is greater than σ, the two molecules do not collide. This is illustrated in Figure 3.1. In terms of collision cross-sectional area, the larger the value of σ, the greater the probability that two molecules will collide. The collision (b) of the two molecules, each with a diameter of σ, can be considered equivalent to the collision (c) of a molecule with diameter 2σ and another molecule represented as a point. In representations (b) and (c), collision is judged to have occurred if the distance between the two centers of the molecules is less than σ. Calculation of the collision rate or frequency is simpler when based on the (c) representation.

As illustrated in Figure 3.2, the average collision rate for pairs of molecules in a gas is approximately equal to the total number of point molecules in the volume swept out in unit time by a molecule of radius σ moving with average velocity. In the volume swept out by the moving molecule, there are 12 point molecules. This is the number of collisions in the time interval in question according to this simple model.

From the kinetic theory of gases (see Masel 2001), the average molecular velocity $\bar{u}$ is

$$\bar{u} = \left(\frac{8R_uT}{\pi \mathrm{M}}\right)^{1/2} = 1.455 \times 10^4 \left(\frac{T}{\mathrm{M}}\right)^{1/2} \text{ cm/s} \tag{3.23}$$

Table 3.1 Molecular Collision Diameters.

Species	Diameter σ ($\AA$)
H_2O	2.641
H	2.708
H_2	2.827
O	3.050
OH	3.147
O_2	3.467
NO	3.492
CO	3.690
N_2	3.798
CO_2	3.941

Source: Svehla (1995)/NASA/Public domain.

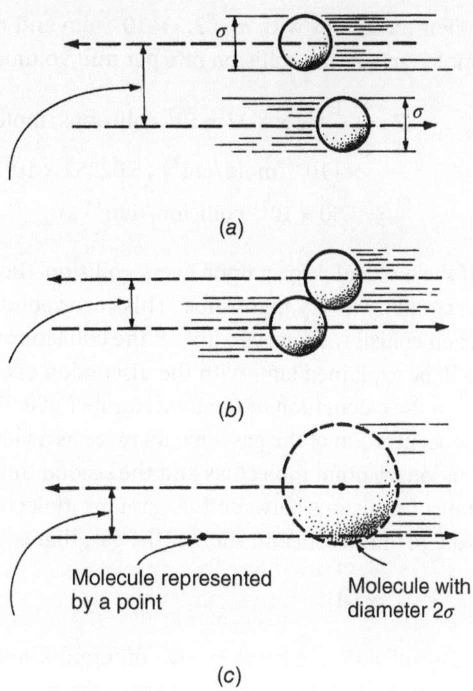

Figure 3.1 Minimum distance between molecules for a collision. (a) Greater than σ; these two molecules do not collide. (b) Less than σ; these two molecules do collide. (c) Less than σ; these two molecules do collide.

We can do a rough analysis to get a sense of the relationship between collision rate and concentration. For a molecule with radius $\sigma = 3.5 \times 10^{-8}$ cm and molecular mass $M = 130$, the volume of the cylinder through which the molecule sweeps in a gas container in one second at 773 K is

$$V = \pi r^2 l = \pi\sigma^2[\bar{u} \cdot 1] = \pi\left(3.5 \times 10^{-8}\right)^2 \left[1.455 \times 10^4 \left(\frac{773}{130}\right)^{1/2}\right]$$
$$= 1.365 \times 10^{-10} \text{cm}^3$$

If the concentration of gas molecules is 10^{-6} mole/cm^3, there are (10^{-6} mole/ cm^3) (1.365×10^{-10} cm^3) (6.02252×10^{23} molecules/mole) = 8.22×10^7 molecules present in this volume. This is the number of collisions that the molecule makes in one second, based on the assumption that all molecules are stationary. If motion of all molecules is taken into account, it can be shown that there is a $\sqrt{2}$-fold higher collision frequency. The total rate of collisions of pairs of molecules in 1 cm^3 of gas is obtained by multiplying the collision rate for one molecule by the number of molecules in 1 cm^3 and then dividing by 2, since each collision of two molecules has been counted twice.

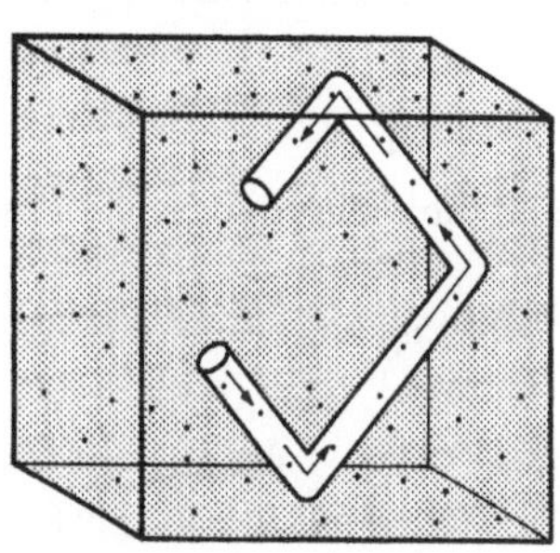

Figure 3.2 Collisions of a moving molecule of radius σ with stationary point molecules.

For molecules with $\sigma = 3.5 \times 10^{-8}$ cm and molecular mass $M = 130$ at a concentration of 10^{-6} mole/cm^3, Z, the collision rate per unit volume (cm^3) is therefore

$$\begin{aligned} Z &= \left(\frac{1}{2}\sqrt{2} \times 8.22 \times 10^7 \text{ collisions/molecules} \cdot \text{s}\right) \\ &\quad \times \left(10^{-6} \text{ mole/cm}^3\right)\left(6.02252 \times 10^{23} \text{ molecules/mole}\right) \\ &= 3.50 \times 10^{25} \text{ collision/(cm}^3 \cdot \text{s)} \end{aligned}$$

If the molecules react upon every collision, the reaction rate will be equal to the collision rate. This is certainly a very high value. This corresponds to the instantaneous rate if a reaction occurs with each collision. However, not all the collisions will have sufficient energy for reactions to occur, as will be explained later with the discussion of activation energy.

In the calculation of the total number of collisions of two molecules in 1 cm^3 in one second, the concentration of the gas appears twice as a factor: the first time in the evaluation of the number of stationary point molecules and the second time in the calculation of the overall collision rate. The rate of collisions between two gaseous molecules of the same kind is therefore proportional to the square of the concentration of the gas, that is

$$Z \propto [A]^2 \tag{3.24}$$

The collision rate between two different kinds of gas molecules (B and C) is also proportional to the product of the concentrations of the two gases:

$$Z \propto [B][C] \tag{3.25}$$

We therefore conclude that a second-order rate law will be associated with a gas-phase reaction whose mechanism is bimolecular collisions. Notice the use of the terms *second-order* and *bimolecular*. The two are not synonymous. The reaction order depends on the number of concentration factors in the experimentally established rate law; the molecularity pertains to the number of species taking part in a postulated reaction step.

From kinetic theory (see Masel 2001), the collision rate (moles/cm^3-s) of molecules B and C can be expressed more precisely as

$$Z_{BC} = \pi\sigma_{BC}^2\left(\frac{8\pi k_B T}{\pi\mu_r}\right)^{1/2} N_A [B][C] \tag{3.26}$$

where k_B is Boltzmann's constant = 1.381×10^{-23} J/K = 1.381×10^{-16} g-cm^2/s^2-K, N_A is Avagrado's number = 6.022×10^{23} molecules/mole, σ_{BC} is the collision radius of molecules B and C, and μ_r is the reduced mass of molecules B and C.

The collision radius σ_{BC} is given by

$$\sigma_{BC} = (\sigma_B + \sigma_C)/2 \tag{3.27}$$

and the reduced mass μ_r of molecules B and C is

$$\begin{aligned} \mu_r &= \frac{m_B m_C}{(m_B + m_C)} \\ &= \frac{M_B M_C}{(M_B + M_C)} \frac{1}{N_A} \end{aligned} \tag{3.28}$$

3.2.6 Activation Energy E_a

Chemists have learned that chemical reactions occur only when collisions between molecules have sufficient energy to cause a rearrangement of their atoms. If the two colliding molecules bring enough energy to the collision, then a rearrangement of atoms to form new molecules occurs. The minimum amount of energy is called the *threshold energy* or the *activation energy*, E_a.

Under favorable conditions, molecular collisions lead to the formation of a transitory chemical species, the *activated complex*. By using the methods of quantum mechanics, it is possible, in principle, to calculate the forces between atoms and molecules. A detailed description of the activated complex will provide insight into the nature of the changes in electronic and nuclear arrangement that characterize the chemical reaction. Consider the reaction between reactants A and B to form the reaction products C and D. The formation of reaction products is preceded by the production of an aggregate, the activated complex, which is designated by the symbol X^*. Thus,

$$A + B \rightleftharpoons X^* \rightleftharpoons C + D \tag{3.29}$$

The energy between two atoms of a diatomic molecule A_2 is a function of the distance between the two atoms. When a free atom B is approaching the molecule A_2, the distance between A and A changes. This is shown in Figure 3.3. Immediately after the collision of B with A_2, the activated complex $BA_2{}^*$ may be formed. The activated complex has a much higher reactivity than normal atoms and may separate in such a way as to give either the components BA and A or the initial components B and A_2.

A *reaction coordinate* may be associated with the path from reactants to reaction products. Depending on the choice of the reaction coordinate, different amounts of energy will be required for the reaction described by Eq. (3.29). The activated complex is located at the point of highest energy on the most favorable reaction path, and the activation energy E^* per molecule is required to form X^*. For the most favorable reaction path, the highest point (X^*) has lower total energy above the reactants (namely, $E_a = N_A E^*$ per mole) than any other reaction path. Figure 3.4 illustrates the activation complex and the required activation energy for exothermic and endothermic reactions in both forward and reverse directions.

An examination of Figure 3.4 reveals that the potential energy of the reactants, E(reactants) > E(product) for an exothermic reaction, while E(product) > E(reactants) for an endothermic reaction. Also, note that the activation energy for forward and reverse reactions is not equal; that is, the forward and reverse reactions have different specific reaction rate coefficients.

In summary, the activation energy E_a is the energy required for the reaction to occur; that is, it is the energy required to move the reactants over an energy barrier in order for the reaction to begin. The activation energy is usually recovered via the thermal energy release during the reaction process.

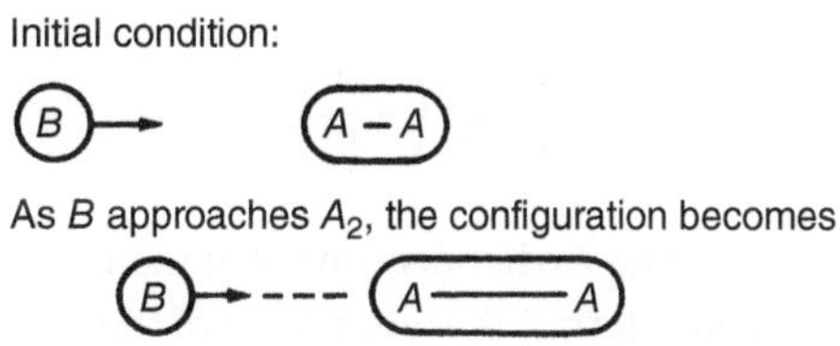

Figure 3.3 Change in the distance between two atoms in the A_2 molecule as a free atom B approaches.

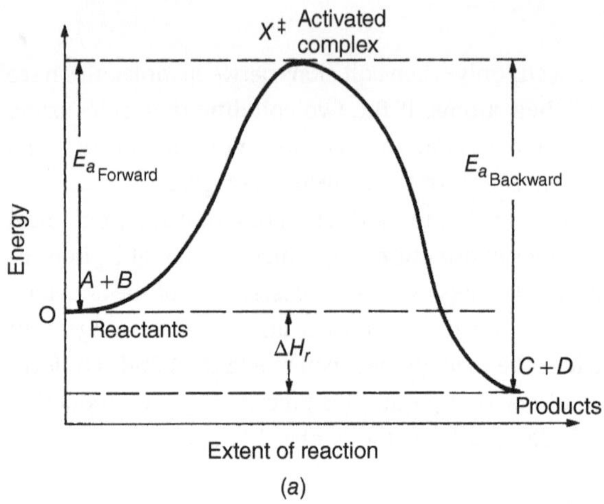

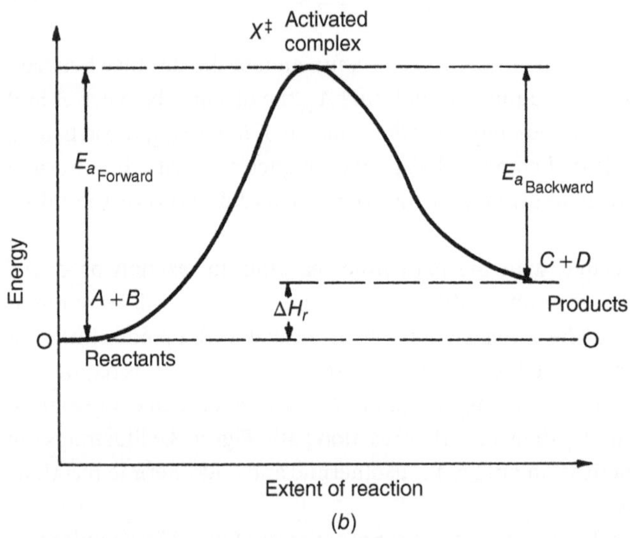

Figure 3.4 Variation of potential energy along reaction coordinate: (*a*) exothermic reaction and (*b*) endothermic reaction.

3.2.7 Arrhenius Equation

The concept that only those molecules that possess energy greater than a certain amount E_a will react during a collision is due to the Swedish scientist Svante Arrhenius (1859–1927). He also introduced the Boltzmann factor, exp $(-E_a/R_uT)$, to calculate chemical reaction rates. Accordingly, the equation

$$k(T) = A_p \exp\left(-\frac{E_a}{R_u T}\right) \tag{3.30}$$

is called the Arrhenius equation; the pre-exponential factor A_p includes the effect of the collision frequency terms, the orientation of the colliding molecules, and the mild temperature dependence of the pre-exponential factor. The parameter A_p can be expanded to a form AT^b in Eq. (3.31), where the exponent b lies between 0 and 1.

$$k(T) = AT^b \exp\left(-\frac{E_a}{R_u T}\right) \tag{3.31}$$

The Boltzmann factor specifies the fraction of collisions that have energy levels greater than the activation energy E_a, and $R_u = 8.314$ kJ/kmol-K is the universal gas constant. The values of A, b, and E_a are based on the nature of the elementary reaction and are not functions of concentration or temperature.

For a bimolecular reaction, the reaction occurs in a collision only if the relative translational energy along the line of the centers of the two molecules at the moment of impact is in excess of E_a. For the second-order reaction

$$B + C \xrightarrow{k} \text{products}$$

the rate law is given by

$$\frac{d[B]}{dt} = -k[B][C] = -A_p[B][C]\exp\left(-\frac{E_a}{R_u T}\right) \tag{3.32}$$

According to Arrhenius,

$$\frac{d[B]}{dt} = -Z_{BC}\mathcal{P}\exp\left(-\frac{E_a}{R_u T}\right) \tag{3.33}$$

where T is the absolute temperature, Z_{BC} is the total collision rate between the B and C molecules, and $\mathcal{P}$ is the steric factor. The steric factor depends on the orientation of the colliding molecules. Its value is typically less than unity due to the orientation dependence of colliding molecules, in addition to the necessary activation energy, for the reaction to occur.

The influence of the steric factor $\mathcal{P}$ can be illustrated by considering the collision of two hydrogen iodide molecules. If the orientation of the colliding molecules is that shown in Figure 3.5a, hydrogen and iodine molecules can be formed; if, however, the orientation is that shown in Figure 3.5b, the colliding molecules will simply bounce off each other. The numerical value of the steric factor for the hydrogen iodide reaction is found to be approximately 0.2. For more complex molecules, the steric factor can be as low as 0.01.

Comparing Eqs. (3.32) and (3.33),

$$A_p\,[B][C] = Z_{BC}\mathcal{P} \tag{3.34}$$

Solving for A_p and substituting the value for Z_{BC} from Eq. (3.26) leads to

$$A_p = \sigma_{BC}^2\left(\frac{8\pi k_B T}{\mu_r}\right)^{1/2} N_A\mathcal{P} \tag{3.35}$$

Example 3.1 ***Arrhenius Equation***

Determine the pre-exponential factor A_p for the reaction

$$O + H_2 \longrightarrow OH + H$$

at $T = 2000$ K. Assume the steric factor $\mathcal{P} = 0.10$.

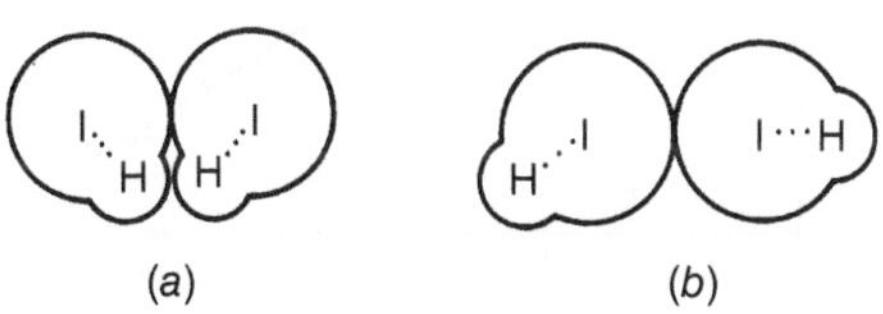

Figure 3.5 Two possible orientations for the collision of two HI molecules.

Solution

The collision radius σ_{BC} is

$$\sigma_{BC} = (\sigma_O + \sigma_{H_2})/2$$
$$= (3.050 + 2.827)/2 = 2.939\,\text{Å} = 2.939 \times 10^{-8}\,\text{cm}$$

The reduced mass μ_r is

$$\mu_r = \frac{m_O\, m_{H_2}}{(m_O + m_{H_2})}$$
$$= \frac{M_O\, M_{H_2}}{(M_O + M_{H_2})}\frac{1}{N_A}$$
$$= \frac{(16.00)(2.008)}{(16.00 + 2.008)}\frac{1}{6.022 \times 10^{23}} = 2.96 \times 10^{-24}\,\text{g}$$

The pre-exponential factor A_p is

$$A_p = \sigma_{BC}^2\left(\frac{8\pi k_B T}{\mu_r}\right)^{1/2} N_A \mathcal{P}$$
$$= (2.939 \times 10^{-8})^2\left(\frac{8\pi(1.381 \times 10^{-16})2000}{2.96 \times 10^{-24}}\right)^{1/2}(6.022 \times 10^{23})(0.10)$$
$$= 7.97 \times 10^{13}\,\text{cm}^3/\text{mole-s}$$

Theoretical treatment indicates that the activation energy E_a obtained from the use of Eq. (3.30) applied to reactions of other orders has significance analogous to that already indicated for bimolecular reactions; hence, Eq. (3.30) can be applied to reactions of all orders.

It should be mentioned that the specific rates of many reactions follow the Arrhenius law. For these reactions, the kinetic data plotted on a graph of $\ln k$ versus T^{-1} follow a straight line. Figure 3.6 shows that in a given chemical reaction, the specific reaction rate coefficient k is independent of the concentrations $[A_i]$ and depends only on temperature. The equation for $\ln k$, as shown in Figure 3.6, can be derived from the natural logarithm of the Arrhenius equation (3.30), which gives

$$\ln k = \ln A_p - \frac{E_a}{R_u T} \tag{3.36}$$

It is important to note that the specific reaction rate coefficient depends on both temperature and the temperature range. The Arrhenius equation generally cannot describe the combustion process over a wide temperature range. For example, a set of reactions that matches with test data at low temperatures may provide erroneous results at high temperature. However, another set of reactions

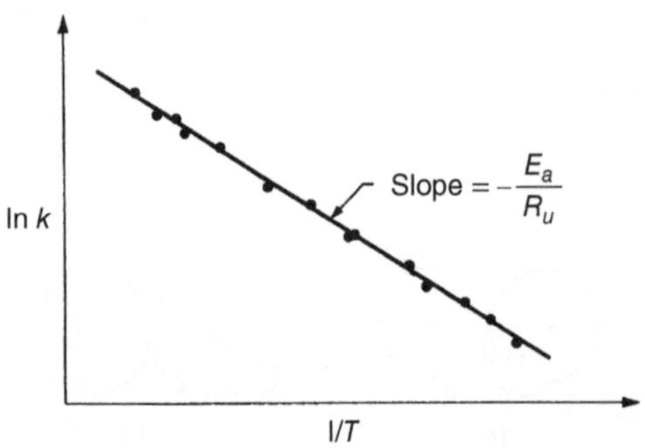

Figure 3.6 Temperature dependence of the specific reaction rate coefficient k.

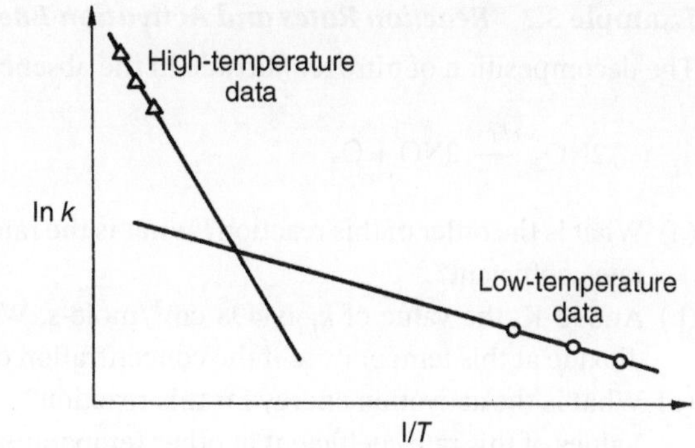

Figure 3.7 The specific reaction rate is dependent upon the temperature range as well as the temperature.

may better match the experimental results, as illustrated in Figure 3.7. Caution must therefore be exercised in extrapolating the specific reaction rate to broader temperature ranges. It is usually very difficult to measure the concentrations of every significant species.

Although many reactions follow the Arrhenius law, there are two classes of reactions for which Eq. (3.30) does not hold. These include the following:

A. *Low activation energy free radical reactions* In these reactions, temperature dependence in the pre-exponential term assumes greater importance, and the so-called absolute theory of reaction appears to provide better correlation of kinetic data with temperature.
B. *Radical recombination* When simple radicals recombine to form a single product, energy must be removed from the product upon its formation in order to stabilize it. A third body is necessary to remove this energy. The pressure dependence of third-body recombination reactions can be quite pronounced. Hence, the specific reaction rate does not follow Eq. (3.30).

In the kinetic study of a given reaction, often there is no simple way to measure the reaction rate directly; normally, the concentration of a reactant or product is determined at various times. As shown in Figure 3.8, a smooth curve is obtained if concentration is plotted against time, where $[A]$ is the initial concentration of the reactant. At any given time, the rate of reaction can be determined from the slope $-d([A]-[x])/dt$ of the instantaneous reactant concentration $([A]-[x])$ curve, or by the rate of generation of product species concentration $(d[x]/dt)$. In particular, the initial slope, at $t = 0$, provides the rate of the reaction corresponding to the concentrations with which the experiment was begun. Another method commonly employed involves the use of expressions relating concentration to time for reactions of various orders.

Figure 3.8 Time variation of the concentrations of reactant and product in terms of $[x]$, the concentration of product.

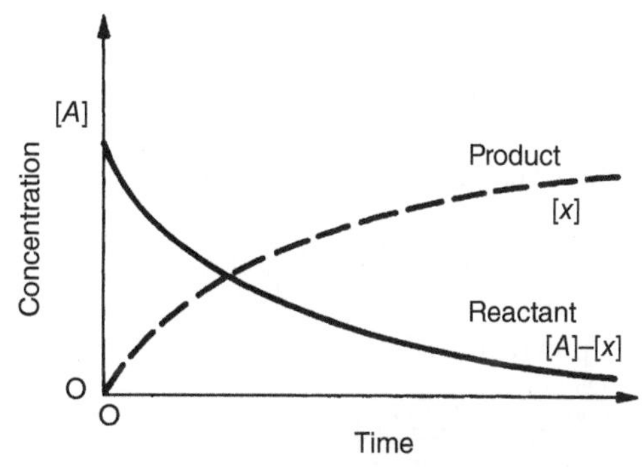

Example 3.2 ***Reaction Rates and Activation Energy***

The decomposition of nitrogen dioxide in the absence of an inert gas can be written as

$$2NO_2 \xrightarrow{k_f} 2NO + O_2$$

(a) What is the order of this reaction? What is the rate law? What is the unit of the specific reaction rate coefficient?

(b) At 592 K, the value of k_f is 498 cm^3/mole-s. What is the rate of decomposition of nitrogen dioxide at this temperature if the concentration of nitrogen dioxide is 3.0×10^{-6} mole/cm^3?

(c) What is the activation energy for this reaction?
Values of this rate coefficient at other temperatures are

T (K)	k_f
603.5	775
627.0	1810
651.5	4110
656.0	4740

Solution

(a) The decomposition of nitrogen dioxide is a second-order reaction, since the sum of the coefficients of the reactants is equal to two. The units of the reaction rate coefficient k can be determined by substituting $m = 2$ into Eq. (3.11),

$$\text{mole}^{1-2}\text{cm}^{6-3}\text{s}^{-1} = \text{cm}^3/(\text{mole} \cdot \text{s})$$

(b) The rate of decomposition is

$$\begin{aligned}\frac{d[NO_2]}{dt} &= -2k_f[NO_2]^2 \\ &= -2 \times 498\frac{\text{cm}^3}{\text{mole} \cdot \text{s}} \times \left(3 \times 10^{-6}\frac{\text{mole}}{\text{cm}^3}\right)^2 \\ &= -8.964 \times 10^{-9}\frac{\text{mole}}{\text{cm}^3\text{s}}\end{aligned}$$

(c) The activation energy for a reaction can be calculated directly from values of rate coefficients measured at two different temperatures. Taking the natural logarithm of both sides of the Arrhenius equation, Eq. (3.30),

$$\ln k_1 = \ln A_p - \frac{E_a}{R_u T_1}$$

$$\ln k_2 = \ln A_p - \frac{E_a}{R_u T_2}$$

Subtracting these two equations yields

$$\ln\frac{k_1}{k_2} = \frac{E_a}{R_u}\left(\frac{1}{T_2} - \frac{1}{T_1}\right) \tag{3.37}$$

Upon rearranging,

$$E_a = \frac{R_u T_2 T_1}{T_1 - T_2}\ln\frac{k_1}{k_2} \tag{3.38}$$

Substituting in any two temperature values and the corresponding specific reaction rate coefficient, we have

$$E_a = \frac{(8.314)(627)(592)}{(627-592)} \ln \frac{1810}{498}$$
$$= 1.138 \times 10^5 \text{ kJ/kmole} = 113.8 \text{ kJ/mole}$$

Comment: Experimentally, E_a is found by plotting measured values of $\ln k$ against $1/T$ and computing the slope of the best-fit straight line through the data points (slope = $-E_a/R_u$). After E_a is obtained, the factor A_p in the Arrhenius equation (3.30) may be calculated, using this equation and the measured values of k.

3.3 One-Step Chemical Reactions of Various Orders

3.3.1 First-Order Reactions

The rate law for a first-order reaction, usually associated with dissociation of a molecule, such as the dissociation of O_2 or H_2, is

$$A_2 \xrightarrow{k_f} 2A \tag{3.39}$$

is

$$\frac{d[A_2]}{dt} = -k_f[A_2] \tag{3.40}$$

After separating the variables and performing integration from time 0 to t, we have

$$\ln\,[A_2]|_0^t = -k_f(t-0) \tag{3.41}$$

or

$$\ln\left(\frac{[A_2(t)]}{[A_2(0)]}\right) = -k_f t \tag{3.42}$$

which gives the concentration of A_2 as a function of time. The rate expression given in Eq. (3.40) can also apply to the reaction in Eq. (3.43) if the concentration of C is much greater than the concentration of A, $[C] \gg [A]$,

$$A + C \xrightarrow{k_f} D \tag{3.43}$$

In this case, the rate expression becomes

$$\frac{d[A]}{dt} = -\frac{d[D]}{dt} = -k_f[A][C] = k'[A] \tag{3.44}$$

where k' is a new specific rate coefficient, which can be defined because the concentration of C is approximately constant (i.e., $[C]$ = constant).

The decomposition of A_2 is a unimolecular reaction that obeys first-order kinetics, and the reaction $A + C \rightarrow D$ is a bimolecular reaction that also obeys first-order kinetics. Thus, all unimolecular reactions are of order one, but not all first-order reactions are unimolecular.

Another example of a first-order reaction is the dissociation of molecule AB

$$AB \xrightarrow{k_f} A + B \tag{3.45}$$

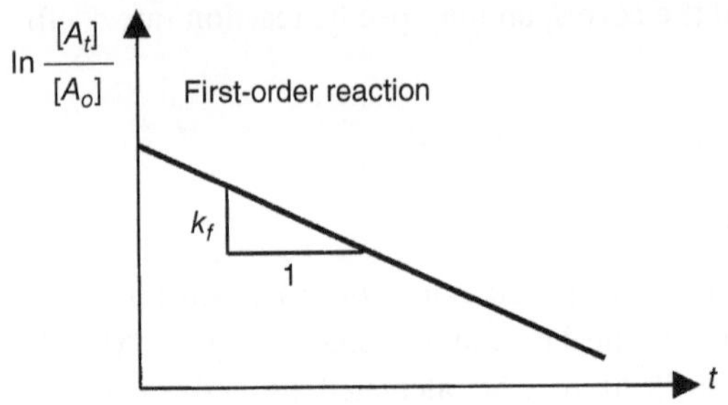

Figure 3.9 Time variation of reactant concentration for first-order reactions.

where the rate law is given as

$$\frac{d[AB]}{dt} = -k_f[AB] \tag{3.46}$$

One can also consider the following reaction as the first order if $m = 1$:

$$mA \xrightarrow{k_f} \text{products} \tag{3.47}$$

The rate of change of the concentration of species A is

$$\frac{d[A]}{dt} = -k_f[A] \tag{3.48}$$

By integration of the above equation, we have

$$\ln\frac{[A_t]}{[A_0]} = -k_f t \tag{3.49}$$

where $[A_t]$ and $[A_0]$ denote the concentrations of species A at time t and 0, respectively. Obviously, a plot of the natural log of their ratio with respect to time will yield a straight line with negative slope (see Figure 3.9). When the time behavior of the concentration ($[A]$) is measured, the rate coefficient, k_f, is determined. From the above reaction rate expression, we can see that when the initial concentration of the first-order reaction is doubled, the reaction rate is also doubled. It is typical for first-order reactions to have this kind of time dependency, which is characteristically different from second-order reactions as we shall discuss next.

Example 3.3 ***First-Order Reaction***

Denoting the concentration of the products at any time by $[P]$ and the concentration of the reactants by $[A_0] - [P]$ (where $[A_0]$ is the initial reactant concentration), develop an expression for the specific reaction rate coefficient. Also, develop an expression for the time varying concentration of $[P]$ (see Figure 3.8).

Solution

From Eq. (3.42)

$$\ln\left(\frac{[A_0] - [P]}{[A_0]}\right) = -k_f t$$

Therefore,

$$k_f = \frac{1}{t}\ln\left(\frac{[A_0]}{[A_0] - [P]}\right)$$

Thus, the concentration of $[P]$ at time t is

$$[P] = [A_0](1 - e^{-k_f t})$$

3.3.2 Second-Order Reactions

Most chemical reactions are bimolecular and proceed as the result of binary collisions. It is therefore not surprising to see that most chemical reactions frequently follow second-order kinetics. If a complex reaction process displays second-order kinetic behavior, then it indicates that the rate-determining (slow) step is a bimolecular process. For the second-order bimolecular reaction

$$A + B \xrightarrow{k_f} AB$$

the rate law becomes

$$\frac{d[A]}{dt} = \frac{d[B]}{dt} = -\frac{d[AB]}{dt} = -k_f[A][B]$$

In this reaction, if the concentration of A is equal to the concentration of B, $[A] = [B]$, the differential equation for this second-order reaction can be solved readily.

For the second-order bimolecular reaction

$$2A \xrightarrow{k_f} C + D$$

the rate law is

$$\frac{d[A]}{dt} = -2\frac{d[C]}{dt} = -2\frac{d[D]}{dt} = -2k_f[A]^2$$

Some representative second-order processes, which have been postulated by Lewis and von Elbe (1987) as elementary reactions occurring in flames, are given below:

$$Cl + H_2 \rightarrow HCl + H$$

$$OH + H_2 \rightarrow H_2O + H$$

$$H + O_2 \rightarrow OH + O$$

$$O + H_2 \rightarrow OH + H$$

$$O_3 + CO \rightarrow CO_2 + 2O$$

$$OH + CH_4 \rightarrow H_2O + CH_3$$

The rate law can also be expressed in terms of the concentration of the consumed portion of the reactant in the reaction. For example, consider the following second-order reaction (sometimes called the atom-transfer reaction):

$$A + B \xrightarrow{k_f} C + D$$

The concentrations of species A and B in time are given as

$$[A] = [A_o] - [x]$$

$$[B] = [B_o] - [x]$$

where $[A_o]$ and $[B_o]$ are the initial concentrations at time $t = 0$, and $[x]$ is the portion of A and B that has been consumed in the reaction. If the rate of change of concentration is

$$\frac{d[A]}{dt} = \frac{d[B]}{dt} = -\frac{d[x]}{dt}$$

then the rate law for this reaction becomes

$$\frac{d[x]}{dt} = k_f[A][B] = k_f([A_o] - [x])([B_o] - [x])$$

If $[B_o] \neq [A_o]$, we can multiply both sides of the equation by

$$\frac{([B_o] - [A_o])dt}{([A_o] - [x])([B_o] - [x])}$$

to obtain

$$\frac{([B_o] - [A_o])d[x]}{([A_o] - [x])([B_o] - [x])} = k_f([B_o] - [A_o])dt$$

Splitting the denominator of the fraction on the left-hand side of the above equation,

$$\left(\frac{1}{([x] - [B_o])} - \frac{1}{([x] - [A_o])}\right) d[x] = k_f([B_o] - [A_o])dt$$

and integrating,

$$\ln\left(\frac{[x] - [B_o]}{[x] - [A_o]}\right) = k_f\left([B_o] - [A_o]\right) t + c_1$$

To solve for k_f, use the fact that at $t = 0$, $[x] = 0$. From the definition of $[x]$, we have

$$c_1 = \ln\left(\frac{[B_o]}{[A_o]}\right)$$

Solving for k_f, the result is

$$k_f = \frac{1}{t([A_o] - [B_o])} \ln\left(\frac{[B_o]([A_o] - [x])}{[A_o]([B_o] - [x])}\right)$$

If $[B_o] = [A_o]$, one can show that

$$k_f = \frac{[x]}{t[A_o]([A_o] - [x])}$$

One can also consider the following reaction as the second order if $m = 2$:

$$mA \xrightarrow{k_f} \text{products}$$

The rate of change of the concentration of species A is

$$\frac{d[A]}{dt} = -k_f[A]^2$$

This rate expression implies that doubling the concentration of each reaction partner quadruples the reaction rate. By integration of the above equation, we have

$$\frac{1}{[A_t]} - \frac{1}{[A_o]} = k_f t \tag{3.50}$$

where $[A_t]$ and $[A_o]$ denote the concentrations of species A at time t and time $t = 0$, respectively.

As shown in Figure 3.10, a plot of the reciprocal of the concentration $[A]$ with respect to time yields a straight line with a positive slope. This linear time dependency of $1/[A]$ is typical for second-order reactions. When the time behavior of the concentration $[A]$ is known, the rate coefficient, k_f, can then be determined from Eq. (3.50).

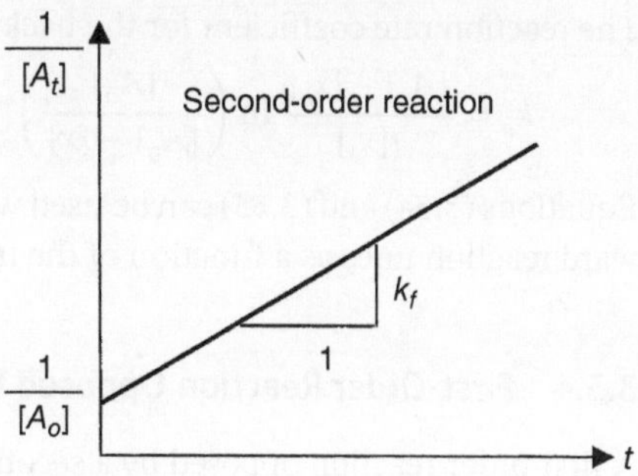

Figure 3.10 Time variation of reactant concentration for second-order reactions.

3.3.3 First-Order Reaction Opposed by a First-Order Reaction

Consider the first-order reaction

$$A \underset{k_b}{\overset{k_f}{\rightleftharpoons}} B$$

with the initial concentration of species B at $t = 0$ equal to zero, i.e., $[B_o] = 0$, where the subscript o denotes initial conditions. If $[x]$ is the portion of A converted to B, Eq. (3.16) becomes

$$\frac{d[x]}{dt} = k_f([A_o] - [x]) - k_b[x]$$

Therefore, at any time,

$$[A] = [A_o] - [x]$$

$$[B] = [x]$$

Also,

$$\frac{k_f}{k_b} = K_c = \frac{[x_e]}{[A_o] - [x_e]} \tag{3.51}$$

where the subscript e identifies the value of $[x]$ at thermodynamic equilibrium. If $[x] = 0$ at $t = 0$, then substituting k_b of Eq. (3.51) into the ordinary differential equation for $[x]$, integrating, and solving for the forward rate coefficient k_f,

$$k_f = \frac{[x_e]}{[A_o]t} \ln\left(\frac{[x_e]}{[x_e] - [x]}\right) \tag{3.52}$$

Hence, if the equilibrium concentration $[x_e] = [B_e]$ is known, we can determine both k_f and k_b from experimental measurements of $[x]$ as a function of time.

Eq. (3.18) can also be written in the form

$$\frac{[x_e]}{[A_o]} = \frac{k_f}{k_f + k_b}$$

from which Eq. (3.52) becomes

$$k_f + k_b = \frac{1}{t} \ln\left(\frac{[x_e]}{[x_e] - [x]}\right) \tag{3.53}$$

Equation (3.53) is formally identical to the first-order rate law for the forward reaction alone (see Example 3.3).

$$k_f = \frac{1}{t} \ln\left(\frac{[A_o]}{[A_o] - [x]}\right) \tag{3.54}$$

The reaction rate coefficient for the backward reaction is

$$k_b = \frac{[A_o] - [x_e]}{t[x_e]} \ln\left(\frac{[A_o]}{[A_o] - [x]}\right) \tag{3.55}$$

Equations (3.54) and (3.55) can be used with experimental data to determine the forward and backward reaction rates as a function of the initial concentration $[A_o]$.

3.3.4 First-Order Reaction Opposed by a Second-Order Reaction

A first-order reaction opposed by a second-order reaction is represented by the relation

$$A \underset{k_b}{\overset{k_f}{\rightleftharpoons}} B + C$$

Using Eq. (3.19), the rate of increase of product species B or C becomes

$$\frac{d[x]}{dt} = k_f([A_o] - [x]) - k_b[x]^2$$

For the case in which the initial concentrations of B and C are equal to zero, the forward rate equation is

$$k_f = \frac{[x_e]}{t(2[A_o] - [x_e])} \ln\left(\frac{[A_o][x_e] + [x]([A_o] - [x_e])}{[A_o]([x_e] - [x])}\right) \tag{3.56}$$

For the case in which the initial concentrations of B and C are a, while that of A is zero, the backward rate equation integrates to

$$k_b = \frac{[x_e]}{t(a^2 - [x_e]^2)} \ln\left(\frac{[x](a^2 - [x][x_e])}{a^2([x_e] - [x])}\right) \tag{3.57}$$

where the equilibrium concentration of A is $[A_e] = [x_e]$, and k_b is the specific rate of formation of A.

3.3.5 Second-Order Reaction Opposed by a Second-Order Reaction

For any second-order opposing reactions, represented symbolically by the following reaction:

$$A + B \underset{k_b}{\overset{k_f}{\rightleftharpoons}} C + D$$

the forward rate coefficient can be expressed as

$$k_f = \frac{[x_e]}{2at(a - [x_e])} \ln\left(\frac{[x](a - 2[x_e]) + a[x_e]}{a([x_e] - [x])}\right) \tag{3.58}$$

where the initial concentration of reactants $a = [A_o] = [B_o]$ and the initial concentration of products $[C_o] = [D_o] = 0$.

3.3.6 Third-Order Reactions

Examples of third-order reactions involving three molecules are

$$3H \rightarrow H_2 + H$$

$$2NO + O_2 \rightarrow 2NO_2$$

Another example representing a recombination reaction is

$$M + 2A \xrightarrow{k_f} A_2 + M^*$$

M is called a third body, which facilitates the recombination reaction of two A atoms to form A_2 during the collision process. After collision, the third body is designated as M^* with a slightly different characteristic from that of M, since the energy level of the third body is changed by the energy transfer process. Some molecules may even emit radiation (light) after collisions. The rate law for the above reaction is given as

$$\frac{d[A_2]}{dt} = k_f[M][A]^2$$

$$\frac{d[A]}{dt} = -2k_f[M][A]^2$$

In case the concentration of M is truly a constant, it can then be combined with k_f to give

$$\frac{d[A_2]}{dt} = k'[A]^2$$

where k' can be regarded as a new rate coefficient. The order of the reaction is then reduced from three to two. If $[M]$ is not truly constant, it is then a function of time, and the order of the reaction remains three.

3.3.7 Consecutive Reactions

In many reaction mechanisms, the products of one reaction undergo further reaction to yield other products. A simple example of this type is

$$A + B \xrightarrow{k_1} AB \xrightarrow{k_2} C + D \tag{3.59}$$

As can be seen from this equation, a consecutive reaction is a series reaction in which k_1 and k_2 are the two specific reaction rate coefficients. Here, the reverse reactions are neglected. The rate laws for the first and second reactions can be determined as follows.

First reaction:

$$\frac{d[AB]}{dt} = k_1[A][B] = -\frac{d[A]}{dt} = -\frac{d[B]}{dt}$$

Second reaction:

$$\frac{d[AB]}{dt} = -k_2[AB] = -\frac{d[C]}{dt} = -\frac{d[D]}{dt}$$

The net rate of change of $[AB]$ is obtained by adding the first and second rate laws:

$$\left(\frac{d[AB]}{dt}\right)_{\text{net}} = k_1[A][B] - k_2[AB]$$

As the reaction proceeds, the concentrations of A and B decrease, and the concentrations of C and D increase. The concentration of AB, therefore, may have a peak at a particular time, as shown in Figure 3.11.

A consecutive reaction in which both reactions are of first order is given by the equation

$$A \xrightarrow{k_1} B \xrightarrow{k_2} C + D$$

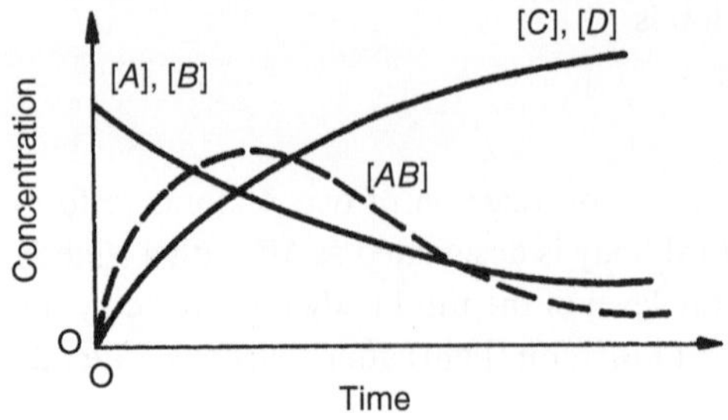

Figure 3.11 Illustration of a concentration peak of the intermediate product AB in a consecutive reaction.

where k_1 and k_2 are the two specific rate coefficients. If the initial conditions at $t = 0$ are $[A] = [A_o]$, $[B] = [C] = [D] = 0$, then the consumption rate of A is given by

$$\frac{d[A]}{dt} = -k_1[A]$$

which integrates to

$$[A] = [A_o]e^{-k_1 t} \tag{3.60}$$

The rate of formation of species C is given by

$$\frac{d[C]}{dt} = k_2[B]$$

The net rate of production of B is equal to the rate of its formation from A minus that of its destruction to produce C and D; i.e.,

$$\frac{d[B]}{dt} = k_1[A] - k_2[B] \tag{3.61}$$

Substituting Eq. (3.60) into Eq. (3.61) gives

$$\frac{d[B]}{dt} = k_1[A_o]e^{-k_1 t} - k_2[B] \tag{3.62}$$

which contains only the variables $[B]$ and t. It integrates to

$$[B] = [A_o]\frac{k_1}{k_2 - k_1}\left(e^{-k_1 t} - e^{-k_2 t}\right) \tag{3.63}$$

Using the above result, the rate of change of $[C]$ can also be obtained as

$$[C] = \frac{[A_o]}{k_2 - k_1}[k_2(1 - e^{-k_1 t}) - k_1(1 - e^{-k_2 t})] \tag{3.64}$$

Once $[C]$ is known, $[D]$ is known, since $[D] = [C]$.

3.3.8 Competitive Reactions

Competitive reactions occur when two or more sets of combustion products are produced from the same set of reactants. This situation is illustrated by

$$A + B \xrightarrow{k_1} AB$$

$$A + B \xrightarrow{k_2} E + F$$

The rate laws for the first and second reactions can be determined as follows:

First reaction:

$$\frac{d[A]}{dt} = -k_1[A][B]$$

Second reaction:

$$\frac{d[A]}{dt} = -k_2[A][B]$$

The net rate of disappearance of species A is found by summing the two equations given above:

$$\frac{d[A]}{dt} = -(k_1 + k_2)[A][B]$$

The extrapolation of the rate law to a higher temperature range can lead to incorrect results because the specific reaction rate coefficients are strongly dependent on temperature. One set of reactions may be dominant at lower temperatures, whereas at higher temperatures, other competitive reactions may also need to be considered.

3.3.9 Time Constant τ_c

The time constant τ_c of a reacting system is defined as the time it takes for the concentration to decrease to $e^{-1} = 0.367$ of the original concentration. For example, consider the following first-order decomposition reaction of species A in collision with molecule M having a constant concentration:

$$A + M \rightarrow \text{products} \tag{3.65}$$

The rate of depletion of A can be written as

$$\frac{d[A]}{dt} = -k[M][A] \tag{3.66}$$

whose solution is

$$[A_t] = [A_o]\exp(-k[M]t) \tag{3.67}$$

where $[A_o]$ is the concentration of A at time $t = 0$. In this case, inspection shows that

$$\tau_c = \frac{1}{k[M]} \tag{3.68}$$

A complex kinetic system is usually composed of many species whose concentrations can decay (or grow) at different rates; thus, a kinetic system usually has disparity in time scales for a very broad range of time constants. During the integration of the governing equations, the numerical solution is dominated by the species that have the smallest time constants. Such a system is termed stiff. Mathematically, the problem of stiffness can be resolved by ensuring the numerical stability of the solution procedure.

Example 3.4 ***Time Constant***

A chamber initially contains 1000 ppm CO and 3% O_2 with the remainder N_2 at a temperature $T = 2000$ K and pressure $p_o = 101.3$ kPa. What is the time constant of the following CO oxidation reaction?

$$CO + O_2 \longrightarrow CO_2 + O$$

The rate coefficient $k = 2.50 \times 10^9 \exp(-24{,}600/T)$ with units m^3/kmol-s.

Solution

The mole fraction of CO is $X_{CO} = 1000/10^6 = 0.001$, and the mole fraction of O_2 is $X_{O_2} = 0.03$. The concentration of CO is

$$[CO] = X_{CO}\frac{p_o}{R_u T} = (0.001)\frac{101.3}{(8.314)(2000)}$$
$$= 6.09 \times 10^{-6}\,\text{kmol/m}^3$$

The concentration of O_2 is

$$[O_2] = X_{O_2}\frac{p_o}{R_u T} = (0.03)\frac{101.3}{(8.314)(2000)}$$
$$= 1.83 \times 10^{-4}\,\text{kmol/m}^3$$

The rate constant k is

$$k = 2.50 \times 10^9 \exp(-24{,}600/2000)$$
$$= 1.138 \times 10^4\,\text{m}^3/\text{kmol-s}$$

Since the concentration of O_2 is much greater than that of CO, the concentration of O_2 can be assumed to be a constant, and from Eq. (3.68), the time constant τ_c is

$$\tau_c = \frac{1}{k[O_2]} = \frac{1}{(1.138 \times 10^4)(1.83 \times 10^{-4})} = 0.48\,\text{second} \tag{3.69}$$

3.4 Chain Reactions

Chain reactions are the most common type of chemical reactions. They consist of a series of consecutive, competitive, and opposing reaction steps with different reaction rate coefficients. These complex chemical reactions occur in numerous combustion processes. Chain reactions can be defined as those reactions in which an intermediate product produced in one step generates a reactive intermediate species in a subsequent step, then that intermediate generates another reactive intermediate, and so on. Combustion processes such as H_2/O_2 reactions, which are understood in considerable detail, will be discussed in terms of their reaction mechanisms after the introduction of the basic chemical kinetics background. In Section 3.5, we shall define free radical species and some initiation processes for their generation, since they play an extremely important role in chain reactions.

3.4.1 Free Radicals

In a reaction process, usually, the highly reactive species are atoms (such as H, O, N, F, and Cl) or radical species (such as CH_3, OH, CH, and C_2H_5). These chemical species, which have unpaired electrons and can react very actively with other molecules, are called *free radicals* or sometimes just *radicals*. In chemical terminology, a free radical is either an atom or a group of atoms, charged or uncharged, that acts as a unit in chemical changes. If one hydrogen atom H · is taken away from CH_4, two free radicals are formed, as shown in Figure 3.12, where the dots symbolize electrons.

For simplicity in notation, we will use either "+" or "−" symbols for positively or negatively charged radicals. For example, after losing its electron, the hydrogen atom becomes a positively

Figure 3.12 Formation of two free radicals from CH_4.

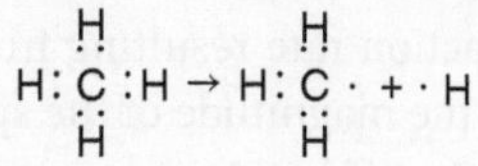

charged free radical, denoted by H^+. Other examples are Cl^-, Na^+, $CH_3{}^+$, etc. (For $CH_3{}^+$, one electron is lost from carbon.) The chlorine atom has an electronic configuration of $1s^2 2s^2 2p^6 3s^2 3p^5$. It takes one electron from another atom to fill the 3p subshell and becomes a chloride ion, Cl^-.

Elementary reactions are called chain-initiating reactions when free radicals are produced from these reactions. Similarly, chain-terminating reactions are those in which free radicals are destroyed. The governing parameter is α, defined as the radical multiplication factor, i.e., the ratio of the number of free radicals in the product to that in the reactant. Elementary reactions are called chain-propagating (or chain-carrying) reactions if α is equal to 1, and chain-branching reactions if α is greater than 1. With chain branching, the concentrations of radical species can increase rapidly, potentially leading to a chemical explosion.

We can illustrate various types of chain reactions using the global reaction

$$A_2 + B_2 \rightarrow 2AB$$

Chain-initiation reactions produce the free radical A through collisions of A_2 with molecule M,

$$A_2 + M \rightarrow A + A + M$$

Chain-propagating reactions involving the free radicals A and B and produce the same number of free radicals as they consume,

$$A + B_2 \rightarrow AB + B$$

$$B + A_2 \rightarrow AB + A$$

Chain-terminating reactions consume free radicals and form stable molecules.

$$A + B + M \rightarrow AB + M$$

Examples of chain-initiating reactions are

$$M + \mathrm{O_2} \rightarrow 2\mathrm{O} + M$$

$$M + Br_2 \rightarrow 2Br + M$$

Examples of chain-propagating reactions ($\alpha = 1$) are

$$\mathrm{CO + OH \rightarrow CO_2 + H}$$

$$\mathrm{H_2 + OH \rightarrow H_2O + H}$$

Examples of chain-branching reactions ($\alpha = 2$) are

$$\mathrm{H + O_2 \rightarrow OH + O}$$

$$\mathrm{H_2 + O \rightarrow OH + H}$$

3.5 Chain-Branching Reactions and Explosions

A net generation in the number of radicals will lead to a rapid increase in the overall rate of reaction, which may eventually culminate into an explosion. Explosions are classified into two distinct categories: *branched-chain explosions*, in which the reaction rate increases without limit because of chain branching, and *thermal explosions*, in which there is an exponential increase in reaction rate resulting from exothermic chemical reaction, heating of reactants, and an increase in the magnitude of the specific reaction rate coefficients.

Consider a 1-cm^3 container that contains initially one chain molecule, that is, one free radical per $centimeter^3$. Assume that the number density in the container is 10^{19} molecules/cm^3 and the average collision rate is 10^8 collisions/s. If the reaction in the volume is a chain-carrying reaction (i.e., one free radical can generate another free radical in the reaction [$\alpha = 1.0$]), then the time required for all of the molecules to react (i.e., 10^{19} collisions) will be

$$t = \frac{10^{19}\ \text{collisions}}{10^{8}\ \text{collisions/s}} = 10^{11}\text{s} \approx 30,000\ \text{years}$$

Such a slow process cannot be called combustion. If the reaction in the volume is a chain-branching reaction (i.e., one free radical or chain particle can generate two chain particles in the reaction [$\alpha = 2.0$]), then the time required for all of the molecules to react can be estimated as follows:

$$1 + 2 + 2^2 + 2^3 + \cdots + 2^N = \frac{2^{N+1} - 1}{2 - 1} = 10^{19}\ \text{molecules}$$

where

$$N = 64\,\text{generations}$$

or

$$t = 64 \times 10^{-8}\text{s} \approx 10^{-6}\text{s} = 1\ \mu\text{s}$$

This is certainly an extremely rapid combustion process. In an actual combustion process, not all reactions are chain-branching reactions. However, the reaction rate can still be very fast even for a very small fraction of chain-branching reactions. For example, in a combustion process in which 1% ($\alpha = 1.01$) of the elementary reactions are of the chain-branching type, the time required for all the molecules in the volume to react would be

$$\frac{\alpha^{N+1} - 1}{\alpha - 1} = \frac{1.01^{N+1}}{0.01} = 10^{19}\text{molecules/cm}^3$$
$$N = 3934$$
$$t = 3934 \times 10^{-8}\text{s} \approx 40\ \mu\text{s}$$

which is still a very fast reaction.

In general, the general behavior of branched-chain reactions and explosions in response to changes in temperature and pressure can be examined by considering the following reaction mechanism beginning with the initial dissociation of a parent molecule M to form the radical species R (Strehlow 1985):

$$nM \xrightarrow{k_1} R \qquad \text{(chain-initiating reaction)}$$
$$R + M \xrightarrow{k_2} \alpha R + M^* \qquad \text{(chain-branching reaction)}$$
$$R + M + M \xrightarrow{k_g} P \qquad \text{(gas chain-terminating reaction)}$$
$$R \xrightarrow{k_w} P \qquad \text{(wall chain-terminating reaction)}$$

The gas chain-terminating reaction is a three-body reaction and the wall chain-terminating reaction is a one-body process. The rate of production of the radical species R is

$$\frac{d[R]}{dt} = k_1[M]^n + (\alpha - 1)k_2[R][M] - k_g[R][M]^2 - k_w[R] \tag{3.70}$$

If we assume a steady-state concentration of $[R]$, then $d[R]/dt = 0$, and upon solving Eq. (3.70) for the concentration $[R]$,

$$[R] = \frac{k_1[M]}{k_g[M]^2 + k_w - k_2(\alpha - 1)[M]} \tag{3.71}$$

The rate of change of the product P concentration can be expressed as

$$\frac{d[P]}{dt} = k_g[R][M]^2 = \frac{k_1 k_g[M]^3}{k_g[M]^2 + k_w - k_2(\alpha - 1)[M]} \tag{3.72}$$

The quantity $k_2(\alpha - 1)[M]$ is positive; as its value increases, it decreases the denominator in Eq. (3.72). The critical value of α where $d[P]/dt \to \infty$ is given as

$$\alpha_c = 1 + \frac{k_g[M]^2 + k_w}{k_2[M]} \tag{3.73}$$

which implies

$$\alpha \geq \alpha_c \Rightarrow \text{chain-branching explosion} \tag{3.74}$$

$$\alpha < \alpha_c \Rightarrow \text{no explosion} \tag{3.75}$$

where α_c is the critical multiplication factor beyond which the mixture becomes explosive.

Equations (3.73) and (3.74) show that explosions can occur for small α_c, which corresponds to fast chain-branching reactions (large k_2) and/or slow change-termination reactions (small k_g and k_w). In addition, since the concentration $[M]$ is proportional to pressure p, then

$$\alpha_c \to 1 + \frac{k_w}{k_2[M]} \to \infty, \quad \text{as}\, p \to 0 \tag{3.76}$$

$$\alpha_c \to 1 + \frac{k_g[M]}{k_2} \to \infty, \quad \text{as}\, p \to \infty \tag{3.77}$$

which indicates that explosion is not possible at very low or high pressures. Physically, at low pressures, the gas density is also low, so the chain-branching reactions become less efficient. At high pressures, the increase in density favors the three-body gas termination reaction compared to the two-body chain-branching reaction.

3.5.1 Explosion Limits of H_2/O_2 Systems

An illustrative example of explosion limits is the H_2/O_2 reaction system. Typical explosion limits for stoichiometric hydrogen–oxygen mixtures measured by Lewis and von Elbe (1987) are plotted schematically in Figure 3.13. The explosion-limit boundary is Z-shaped. At a given temperature, increasing the pressure of a nonexplosive mixture will cause it to become explosive, nonexplosive, and then explosive again. Increasing the temperature will widen the explosive range. In Figure 3.13, the first or lower explosion limit occurs at roughly the same pressure over a relatively large temperature range.

The existence of these explosion limits can be understood from the competition between chain-terminating and chain-branching reactions on wall surfaces and in the gas phase. Intuitively, one would assume that as the pressure is raised, the concentration of free radicals in the gas will increase. The first or lower explosion limit is thus determined by a balance between the removal of chain carriers by the wall surface and the production of chain carriers by gas-phase reactions.

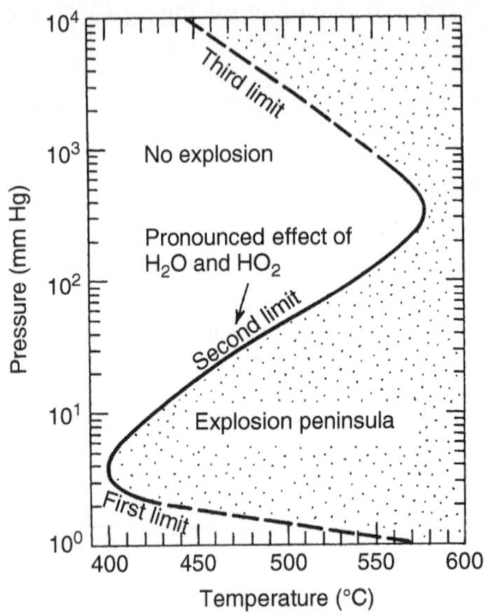

Figure 3.13 Pressure–temperature explosion diagram of a stoichiometric H_2/O_2 mixture.

Postulated chain-initiation reactions for the H_2/O_2 system are the dissociation of H_2, and a reaction between H_2 and O_2,

$$H_2 + M \rightarrow H + H + M \tag{3.78}$$

$$H_2 + O_2 \rightarrow HO_2 + H \tag{3.79}$$

With the production of H from the chain-initiation reactions, the following chain-propagation reactions are initiated,

$$H + O_2 \rightarrow OH + O \tag{3.80}$$

$$O + H_2 \rightarrow OH + H \tag{3.81}$$

$$OH + H_2 \rightarrow H_2O + H \tag{3.82}$$

For low pressures and temperatures, explosion is not possible. The key reaction (3.80) generating the radicals OH and O is not favored, since it is endothermic by 67 J/mole, and the other two reactions depend on the availability of OH and O. In addition, at low pressures, the radicals OH, H, and O rapidly diffuse to the vessel wall where they are consumed.

As the pressure increases, the chain-branching reactions increase relative to the rate of consumption at the wall, leading to an explosion. As the chamber pressure continues to be increased on the explosion peninsula, a second limit is approached, leading to a nonexplosion range.

The second limit is readily explained if the chain-terminating three-body reaction

$$H + O_2 + M \rightarrow HO_2 + M \tag{3.83}$$

is added to the reaction scheme. In the above elementary reaction, the symbol M denotes any third molecule that stabilizes the combination of H and O_2. As the pressure increases, the reaction (3.83) becomes the dominant reaction between H and O_2, since the frequency of ternary collisions $H + O_2 + M$ increases relative to the frequency of binary collisions $H + O_2$.

Because the metastable intermediate hydroperoxy molecule (HO_2) is relatively unreactive, it is able to survive multiple collisions and diffuse to the wall. The destruction of the HO_2 molecule on the wall can be expressed by the reaction

$$HO_2 \rightarrow \frac{1}{2}H_2 + O_2$$

$$HO_2 \rightarrow \frac{1}{2}H_2O + \frac{3}{4}O_2$$

In the nonexplosion region above the peninsula, HO_2 molecules can be assumed to play no part in chain propagation and are destroyed at the wall. At some pressure levels above the second explosion limit, however, there is a rapid increase in the number of radicals, which defines the third explosion limit. The molecule HO_2 collides and reacts with H_2 molecules to form H_2O_2 and H atoms, then dissociates to generate OH radicals,

$$HO_2 + H_2 \rightarrow H_2O_2 + H \rightarrow 2OH + H$$

Another factor worth mentioning is due to the molecular structure similarity between H_2O and HO_2. Based on their molecular structure similarity, H_2O has a bond frequency very close to that of HO_2. Therefore, H_2O is an excellent third body for the trimolecular reaction $H + O_2 + M \rightarrow HO_2 + M$. The region for pronounced effect of H_2O and HO_2 is indicated in the explosion-limit diagram. It is useful to note that for $T > 600$ °C, HO_2 cannot be stabilized and therefore explosion is observed at all pressures.

3.6 Pressure Dependence of Rate Coefficients

A change of overall reaction order with pressure has been observed in many first-order reactions. A model for the pressure dependence of rate coefficients is given by the classic Lindemann model for unimolecular reactions. Lindemann et al. (1922) proposed that a chain-initiating reaction occurs as a result of a two-step reaction sequence, as indicated by Eq. (3.84).

In the Lindemann reaction model, reactant molecules receive energy by collisions with a collision partner molecule M, and at any given time a small fraction of the molecules have sufficient energy to decompose into reaction products. Such molecules are referred to as *energized* molecules, A^*. The concentration of A^* depends on the net rate of generation (rate of energization of A minus the rate of deenergization of A^*) by collision, as well as on the rate of destruction of A^* by decomposition into products.

$$A + M \underset{k_b}{\overset{k_f}{\rightleftharpoons}} A^* + M \tag{3.84}$$

$$A^* \xrightarrow{k_f'} P \text{ (products)} \tag{3.85}$$

Equations (3.84) and (3.85) indicate that an energized molecule A^* may decompose into products P or it can deactivate to A through a collision with molecule M. If these energized molecules are converted into products at a rate that is small compared to the rate at which they are energized by collision or that at which they are deenergized, they will maintain an equilibrium concentration. In addition, since these energized molecules are in equilibrium with the normal molecules, their concentration is proportional to that of the normal molecules. The rate of reaction is proportional to the concentration of energized molecules and is therefore proportional to the concentration of normal molecules, which implies that the reaction is first order.

The Lindemann model predicts that the overall process will obey a first-order rate law as long as the formation of A^* according to Eq. (3.84) is sufficiently rapid to maintain an equilibrium concentration of A^*. Since the frequency of binary collisions decreases as the pressure is reduced, it is reasonable to expect that the process symbolized by Eq. (3.84) will cease to be fast at reduced pressures; therefore, we expect that a first-order reaction should become a second order at sufficiently low pressures.

The differential equations corresponding to the reaction processes of Eqs. (3.84) and (3.85) are

$$\frac{d[A^*]}{dt} = k_f[A][M] - k_b[A^*][M] - k_f'[A^*] \tag{3.86}$$

and

$$\frac{d[A]}{dt} = -k_f[A][M] + k_b[A^*][M] \tag{3.87}$$

The product concentration can be obtained from

$$\frac{d[P]}{dt} = k_f'[A^*] \tag{3.88}$$

Assuming that the concentration of the energized molecule A^* is in a quasi-steady state at an equilibrium concentration, where

$$\frac{d[A^*]}{dt} = 0 \tag{3.89}$$

then Eq. (3.86) can be expressed by

$$[A^*] = \frac{k_f[A][M]}{k_b[M] + k_f'} \tag{3.90}$$

which can be combined with Eq. (3.87) to yield the result

$$\frac{d[A]}{dt} = -\frac{k_f k_f'[A][M]}{k_b[M] + k_f'} \tag{3.91}$$

This equation can be integrated directly after setting $[M] = [A]$ (since the collision partner in the collision process is molecule A).

To understand the effect of pressure on the order of reaction, let us substitute the $[A^*]$ expression in Eq. (3.90) into Eq. (3.88) to obtain

$$\frac{d[P]}{dt} = \frac{k_f k_f'[A][M]}{k_b[M] + k_f'} \tag{3.92}$$

At high pressures, the concentration of the collision partner M is high; hence, $k_b[M] \gg k_f'$ and Eq. (3.92) becomes

$$\frac{d[P]}{dt} = \frac{k_f k_f'}{k_b}[A] = k_\infty[A] \tag{3.93}$$

Reaction (3.93) follows a first-order rate law, where k_∞ represents the equivalent specific reaction rate.

At low pressures, the concentration of the collision partner M is very low; hence, $k_b[M] \ll k_f'$ and Eq. (3.92) becomes

$$\frac{d[P]}{dt} = k_f[A][M] \tag{3.94}$$

Reaction (3.94) follows a second-order rate law. Thus, the Lindemann reaction mechanism (Eqs. (3.84) and (3.85)) indicates that the order of reaction is indeed pressure dependent. The rate

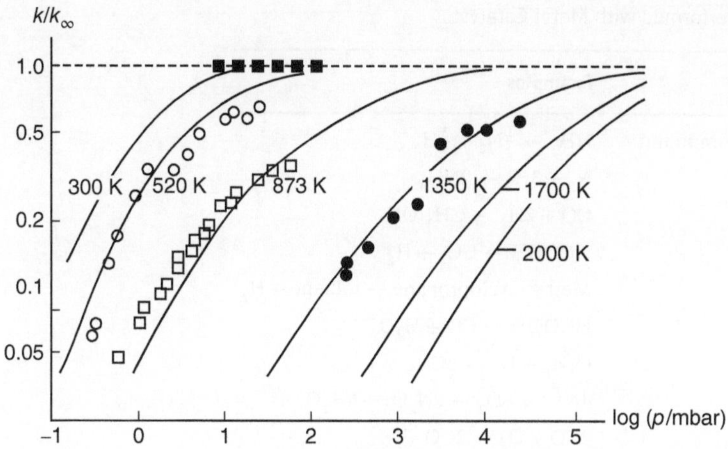

Figure 3.14 Fall-off curves for the unimolecular reaction $C_2H_6 \rightarrow CH_3 + CH_3$.

law of unimolecular reactions can be rewritten from Eq. (3.92) into the following form:

$$\frac{d[P]}{dt} = \frac{k_f k_f'[M]}{k_b[M] + k_f'}[A] \equiv k[A] \tag{3.95}$$

$$\text{where} \quad k \equiv \frac{k_f k_f'[M]}{k_b[M] + k_f'} \rightarrow k_\infty \left(\equiv \frac{k_f k_f'}{k_b}\right) \quad \text{as} \quad p \rightarrow \infty \tag{3.96}$$

At very high pressures, k approaches k_∞ and becomes independent of pressure. At low pressures, k is proportional to the system pressure p, since $[M]$ is proportional to pressure.

Using the above definition of k and k_∞, we have

$$\frac{k}{k_\infty} \equiv \frac{k_b[M]}{k_b[M] + k_f'} \rightarrow 1 \qquad \text{at high pressures} \tag{3.97}$$

Based on this kind of relationship, the theory of unimolecular reaction yields *fall-off* curves. A typical set of *fall-off* curves obtained by Warnatz (1983) for the unimolecular reaction $C_2H_6 \rightarrow CH_3 + CH_3$ is shown in Figure 3.14, which describes the pressure dependence of k for different temperatures. The Lindemann mechanism is a relatively simple model. Hinshelwood (1926) added the contribution of the internal degrees of freedom and the classical probability of the internal energy ranging from E to $E + dE$. Further development of the Lindemann mechanism with inclusion of vibrational modes can be found in Steinfeld et al. (1998).

A steady-state assumption, such as Eq. (3.89), for reaction intermediates is sometimes justified as a first approximation for modeling chemical reactions in flow systems. However, it is necessary to examine the conditions in any given problem in order to at least verify the fact that a steady state is possible. The limitations of the steady-state assumption can be assessed most simply by comparing the complete solution with results derived from a steady-state treatment.

3.7 Surface Reactions

Surface reactions play an important role in many combustion applications, e.g., in catalytic combustion, soot formation and oxidation, wall recombination processes during auto-ignition, coal combustion, and in metal combustion. The concept of *molecularity* is important for a

Table 3.2 Surface Reactions Performed with Metal Catalysts.

Type of Surface Reaction	Examples
Decomposition of Chemical Compound	$NH_3 \rightarrow \frac{1}{2}N_2 + \frac{3}{2}H_2$
Hydrogenation	$N_2 + 3H_2 \rightarrow 2NH_3$
	$CO + 2H_2 \rightarrow CH_3\ OH$
Dehydrogenation	$HCOOH \rightarrow CO_2 + H_2$
	Methyl-cyclopropane → toluene + H_2
Dehydration	$HCOOH \rightarrow CO + H_2O$
Hydrogenolysis	$C_2H_6 + H_2 \rightarrow 2CH_4$
Total Oxidation	$4NH_3 + 8O_2 \rightarrow 2N_2O_5 + 6H_2O$
	$2CO + O_2 \rightarrow 2CO_2$
Partial Oxidation	$2C_2H_4 + O_2 \rightarrow 2C_2H_4O$
Steam Reforming	$C_7H_{16} + 14H_2O \rightarrow 22H_2 + 7CO_2$

basic understanding of surface reactions. The molecularity is defined as the number of reactant molecules taking part in the reaction. The molecularity of a surface reaction is deduced from the kinetics based on the experimental results and theoretical analysis.

The rates of many heterogeneous chemical reactions are influenced by solid surfaces. The rate of a surface reaction can be markedly increased or decreased by the presence of a small amount of some substance, which may be recovered in its original form after the reaction is completed. This phenomenon is referred to as *catalysis*, and the substance the presence of which changes the rate of the surface reaction is called a *catalyst*.

Catalysis is of great importance in the refining of petroleum, the manufacturing of plastics, the oxidation of ammonia, and many industrial processes. For example, two important practical applications of catalysis are low NO_x combustion and generation of low-temperature process heat. In a subsequent chapter, we will examine the catalytic reactions used to reduce pollutant emissions from internal combustion engines, for example, the oxidation of carbon monoxide CO by a platinum Pt catalyst. Table 3.2 shows different types of reactions commonly performed with metal catalysts.

In a pioneering study in surface reaction research, the ability of platinum to facilitate reactions was reported by Henry in 1824. He discovered that adding a platinum powder to the reacting surface in miners' safety lamps enhanced the light emitted from the reaction. The oxidation of ammonia to nitric oxide (NO) is another example of catalysis. When ammonia burns, two independent reactions take place. These are illustrated by the following two reactions:

$$4NH_3 + 3O_2 \rightarrow 2N_2 + 6H_2O$$

$$4NH_3 + 5O_2 \rightarrow 4NO + 6H_2O$$

In the absence of a catalyst, the first of these reactions takes place almost exclusively, but if the combustion takes place on the surface of metallic platinum, the second reaction is so dominant that nearly all of the ammonia is converted to NO rather than to N_2. In this case, the evidence indicates that the reactivity of the gases is greatly increased by their adsorption on the surface of the platinum. Thus, platinum serves as an effective catalyst for the heterogeneous reaction.

When a gas atom or molecule strikes a solid surface it may either rebound, the collision being accompanied by an interchange of energy, or it may be adsorbed. One of the key results associated with surface catalysis involving specific chemical interaction between the surface and the reacting gas molecules is that the molecules must first become adsorbed on the surface before reaction can occur. This view was first proposed in 1825 by Faraday, who studied the nature of adsorption on surfaces and considered that reaction occurred in adsorbed films.

3.7.1 Catalytic Surface Reactions

A surface reaction facilitated by a catalyst such as platinum or palladium is called a catalytic surface reaction. The catalyst itself increases the reaction rate without being consumed in the reaction. The overall process of catalytic surface reactions can be divided into three subprocesses:

- Adsorption of the reactant molecules on the surface.
- Elementary reaction steps, involving various combinations of adsorbed molecules, the surface itself, and gas-phase molecules.
- Desorption of the product molecules from the surface.

For example, the three subprocesses of an H_2 oxidation reaction on a Pt surface are shown in Figure 3.15. First, the H_2 and O_2 are adsorbed on the surface to form H and O atoms, the adsorbed atoms then collide and react with each other to form OH, and then form H_2O. Finally, the H_2O desorbs into the gas phase.

3.7.2 Surface Adsorption Processes

In view of the importance of adsorption forces in connection with surface catalysis, we shall first discuss the adsorption processes before addressing the issue of heterogeneous reaction. The word "adsorption" was coined in 1881 by German physicist Heinrich Kayser. In the early 1900s, Langmuir first investigated the process of adsorption and developed the ideas of sticking and trapping. There are two main types of adsorption, and they are distinctly different. In the first type, called *physisorption*, the forces are of a physical nature and the adsorption is relatively weak. Physisorption is a process in which there is no direct chemical bond between the adsorbate and surface. The adsorbate is held by physical forces such as van der Waals forces. The thermal energy changes found with *van der Waals adsorption* are usually less than 20 kJ/mol.

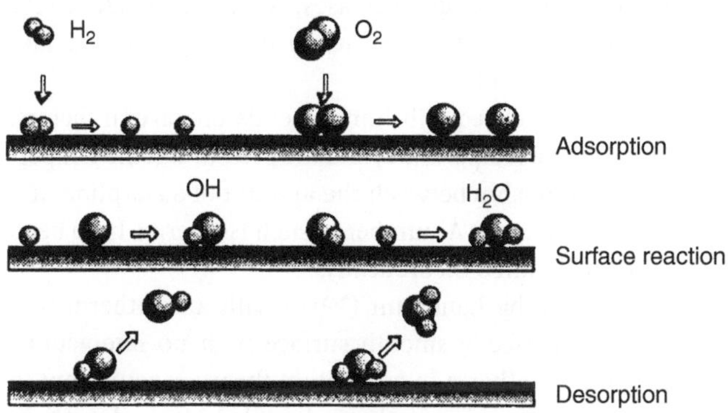

Figure 3.15 Processes of catalytic surface reaction of hydrogen oxidation.

According to Langmuir's original concept (Langmuir 1916), the second type of adsorption is considerably stronger and the adsorbed molecules are held to the surface by valence forces of the same type as those occurring between bound atoms in molecules. The thermal energy evolved in this type of adsorption, known as *chemisorption*, is on the order of 40–400 kJ/mol. Chemisorption occurs when the adsorbate and surface have a direct chemical bond resulting in the sharing of electrons. Usually, the adsorbate first physisorbs and then converts to a chemisorbed state. However, at equilibrium, there are always some physisorbed molecules, and they can contribute to the overall surface reaction process. The rate of adsorption by a surface has been determined to be the product of the total flux of molecules onto the surface and the sticking coefficient S, which is defined as the ratio of molecules that stick to the surface to the number of impinging molecules.

An important consequence of Langmuir's concept of chemisorption is that after a surface has become covered with a single layer of adsorbed molecules, it is essentially saturated. This idea of the *unimolecular layer* was proved experimentally by the existence of a definite adsorption limit for a given surface area; beyond the specific limit additional molecules cannot be readily adsorbed. This result is strong evidence for the theory that chemical forces are involved in chemisorption.

Taylor (1931) suggested that chemisorption is frequently associated with appreciable activation energy and may therefore be a relatively slow process. The energies of activation are often on the order of 80 kJ/mol, and, consequently, the adsorption is extremely slow at low temperatures; under these conditions, *van der Waals adsorption*, which requires very small or no activation energy, will predominate. Surface conditions can influence the chemisorption process. Not all surfaces are smooth, and, in reality, surfaces are never smooth from the molecular viewpoint, and the kind of behavior to be expected on rough surfaces was discussed by Taylor and Thon (1952) and many others. Some surface sites will be more active than others, and chemical processes will occur predominantly on the most active sites, which are called active centers by Taylor. These active centers usually correspond to certain types of lattice defects.

Another complication that exists in connection with chemisorption is that there are interactions, usually of a repulsive nature, between atoms or molecules adsorbed side by side on a surface. The first evidence for this was obtained by Roberts (1935), who measured the heat of adsorption of hydrogen on a smooth tungsten surface. The heat of adsorption was found to decrease from about 190 kJ/mol for a bare surface to 62 kJ/mol for a 70%-covered surface and became close to zero when the surface was fully covered. If a surface is homogeneous, this change in heat of adsorption must be due to repulsive interactions between the adsorbed molecules; as when the surface is sparsely covered, the molecules are sufficiently far apart so as not to interact with each other, but the repulsive interactions will become more important as the coverage increases. Additional evidence for repulsive interactions was obtained by Kummer and Emmett (1951) using an isotope method, and by Weber and Laidler (1951) using microwave spectroscopy.

The amount of gas adsorbed by chemisorption after equilibrium depends on several factors, including the nature of the surface, the adsorbate, the temperature, and the pressure. If the temperature is kept constant for a given system, the relationship between the amount of adsorption and the chamber pressure is known as an *adsorption isotherm*. A number of such isotherms have been proposed, some obtained empirically and others obtained theoretically.

The simplest theoretical equation was developed by Langmuir (1918) with an isotherm that applies to the ideal case of chemisorption on a perfectly smooth surface with no interactions between adsorbed molecules. Thus, the Langmuir isotherm in adsorption theory is equivalent to that of the ideal-gas law. Adsorption that obeys Langmuir's isotherm is called ideal adsorption. The equations for ideal adsorption play an important role in surface kinetics. In general, the

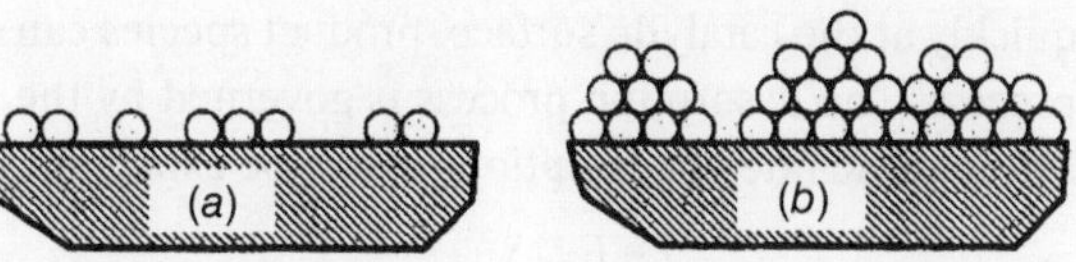

Figure 3.16 Illustration of (*a*) monolayer and (*b*) multilayer adsorption.

quantity of gas that can be adsorbed by a material is proportional to the surface area. Two different types of adsorption are possible, monolayer and multilayer, as illustrated in Figure 3.16.

Monolayer adsorption is more common and dominated by the interaction between the adsorbate and the surface. Multilayer adsorption is dominated by adsorbate–adsorbate interactions and occurs only at temperatures near the boiling point of the adsorbate. In the following derivation, we shall consider the monolayer case. The rate of adsorption has an upper limit; it cannot exceed the rate at which gas-phase molecules collide with the surface. From kinetic theory of gases, the collision frequency (in units of mole/m^2-s) of the ith molecule on the surface is

$$Z_{\max,i} = [C_i]\frac{\bar{u}}{4} \tag{3.98}$$

where $[C_i]$ is the concentration and $\bar{u}$ is the average molecular speed. Upon substitution of Eqs. (3.20) and (3.23),

$$Z_{\max,i} = \frac{p_i}{R_u T}\left(\frac{R_u T}{2\pi M_i}\right)^{1/2} = \frac{p_i}{\sqrt{2\pi M_i R_u T}} \tag{3.99}$$

where M_i is the molecular mass of molecule i. Since the mass of a single molecule $m_i = M_i/N_A$ and Boltzmann's constant $k_B = R_u/N_A$, where N_A is Avagadro's number, the maximum mass adsorption rate $k_{\max,i}$ (kg/m^2/s) is

$$k_{\max,i} = N_A m_i Z_{\max,i} = \frac{p_i m_i N_A}{\sqrt{2\pi M_i R_u T}} = \frac{p_i m_i}{\sqrt{2\pi m_i k_B T}} \tag{3.100}$$

The Boltzmann constant k_B is the proportionality factor that relates the average relative kinetic energy of particles in a gas with the thermodynamic temperature of the gas.

Since only a fraction of molecular collisions on the surface will lead to surface adsorption, we can write the adsorption rate in the following forms:

$$k_{a,i} = S_i k_{\max,i} = S_i p_i \sqrt{\frac{M_i}{2\pi R_u T}} = S_i [C_i] \sqrt{\frac{M_i R_u T}{2\pi}} \tag{3.101}$$

where S_i is defined as the dimensionless sticking coefficient of the ith molecule on the surface. The sticking coefficient S_i is a probability with a value between 0 and 1. For very low pressures, the value of S_i can be close to unity. Under other situations, it can have extremely low values down to 10^{-6}.

A specific reaction rate coefficient for a surface adsorption (a) process can also be expressed in Arrhenius form as:

$$k_a = A_a \exp\left(\frac{-E_a}{R_u T}\right) \tag{3.102}$$

3.7.3 Surface Desorption Processes

The final step in a catalytic reaction process is the desorption of product species. Desorption requires that the molecules have sufficient energy to overcome the bond strength between the adsorbed species and the surface. If the desorption process of product species does not occur quickly at the catalytic surface, product species can become saturated to stop the surface reaction process. The desorption process is governed by the same mechanisms as the adsorption process. The specific rate of desorption (d) can be expressed in the following Arrhenius form:

$$k_d = A_d \exp\left(\frac{-E_{a,d}}{R_u T}\right) \tag{3.103}$$

The pre-exponential factor A_d can be estimated from the vibrational frequencies of the corresponding bond. According to Warnatz et al. (2006), the pre-exponential factors for OH and H_2O desorption processes are on the order of 10^{-13} s^{-1}, which is closely related to the vibrational frequencies of the corresponding bonds. Thus, surface bond energies can be utilized to estimate the activation energy of the desorption process.

The rate of desorption is given by Eq. (3.103) or in terms of Boltzmann's constant as

$$k_{d,i} = \upsilon_{d,i} \cdot \exp\left(\frac{-E_{a,d,i}}{k_B T}\right) \tag{3.104}$$

where $\upsilon_{d,i}$ represents the vibrational frequencies of the corresponding bond to the surface and $E_{a,d,i}$ is the surface bond energy for desorption of species i.

3.8 Surface Reaction Kinetics

3.8.1 The Langmuir Adsorption Isotherm

The simplest situation exists when the gas atoms or molecules occupy single sites on the surface and are not dissociated. A schematic of this simple adsorption and desorption process is shown in Figure 3.17.

Let θ be the fraction of surface that is covered by the adsorbed gas molecules (G) and let $(1 - \theta)$ be the fraction that is bare. The rate of adsorption is then $k_a p(1 - \theta)$, where p is the gas partial pressure and k_a the adsorption rate coefficient. Since the gas concentration is directly proportional to the gas partial pressure, the rate of adsorption can also be expressed in terms of the concentration. The rate of desorption is $k_d\theta$. At equilibrium, the adsorption rate is balanced by the desorption rate, so that

$$\frac{\theta}{1-\theta} = \frac{k_a}{k_d} p \equiv Kp \tag{3.105}$$

where the equilibrium constant K is defined as the ratio k_a/k_d, which is a constant at a given temperature. The above equation can be rewritten as

$$\theta = \frac{Kp}{1 + Kp} \tag{3.106}$$

One can easily plot the above Langmuir's adsorption–desorption isotherm equation to show that the value of θ increases monotonically with pressure. At very low pressures, $\theta \approx Kp$, while at high pressures, $\theta \approx 1$.

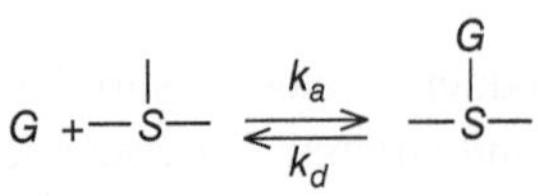

Figure 3.17 Langmuir adsorption schematic.

3.8.2 Unimolecular Surface Reactions

The simplest surface reaction involving one reactant molecule at pressure p is described by the Langmuir adsorption isotherm given in Eq. (3.106). The surface reaction rate $\dot{R}$ is proportional to θ and can therefore be written as

$$\dot{R} = k_a\theta = \frac{k_a Kp}{(1 + Kp)} \tag{3.107}$$

where k_a is the proportionality coefficient. This rate expression is based on the assumption that the adsorption equilibrium is not disturbed by the occurrence of the reaction. This assumption is generally satisfied. At sufficiently high pressures, $Kp \gg 1$ and the rate of reaction is equal to k_a and is independent of pressure, which means that the kinetics are zeroth order. At low pressures, when $Kp \ll 1$, Eq. (3.107) becomes $\dot{R} = k_a Kp$, and the kinetics are first order.

One complication that frequently occurs in surface reactions is when an inhibitor substance I (also called a poison substance) other than the reactant is adsorbed on the surface. The resulting effect is the reduction of catalytic surface area and the corresponding decrease of reaction rate. If the fraction of the surface covered by A is θ, and that covered by I is θ_I, the fraction covered by A can be expressed by

$$\theta = \frac{Kp}{1 + Kp + K_I p_I} \tag{3.108}$$

where p_I is the partial pressure of the inhibitor and K_I its adsorption constant. The rate of reaction is therefore

$$\dot{R} = \frac{k_a Kp}{1 + Kp + K_I p_I} \tag{3.109}$$

When the partial pressure of the reactant is very low, the denominator is reduced to $(1 + K_I p_I)$. The rate of reaction is

$$\dot{R} = \frac{k_a Kp}{1 + K_I p_I} \tag{3.110}$$

If the inhibitor is very strongly adsorbed, $K_I p_I$ is large compared with unity; thus,

$$\dot{R} = \frac{k_a Kp}{K_I p_I} \tag{3.111}$$

The surface reaction is therefore first order. The decomposition of ammonium on platinum can be considered a good example of a first-order reaction at low ammonia partial pressures. The decomposition reaction is partially inhibited by hydrogen produced in the reaction. The rate of reaction follows the following functional dependence:

$$\dot{R} = \frac{k[NH_3]}{[H_2]} \tag{3.112}$$

3.8.3 Adsorption with Dissociation

There is experimental evidence that the adsorption process can be accompanied by the dissociation of the molecule on the surface. For example, hydrogen can be adsorbed on the surfaces of many metals in the form of atoms, each of which occupies a surface site. Similarly, methane adsorbed on metals is usually dissociated into CH_3, CH_2, and hydrogen atoms. The dissociation of a molecule into two species (e.g., H_2 into 2H) is shown schematically in Figure 3.18.

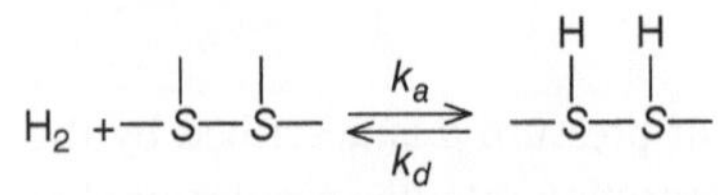

Figure 3.18 Adsorption with dissociation.

In this situation, the process of adsorption can be modeled as a reaction between the gas molecule and two surface sites, and the reaction rate of adsorption may be written as

$$\dot{R}_a = k_a p(1-\theta)^2 \tag{3.113}$$

The reaction rate of desorption, involving two adsorbed atoms, is proportional to the square of the fraction of surface covered by the gas molecule; thus,

$$\dot{R}_d = k_d \theta^2 \tag{3.114}$$

At equilibrium, these two rates are equal; therefore,

$$\frac{\theta}{1-\theta} = \left(\frac{k_a}{k_d}p\right)^{1/2} \equiv K^{1/2}p^{1/2} \tag{3.115}$$

The above equation can be rewritten as

$$\theta = \frac{K^{1/2}\,p^{1/2}}{1+K^{1/2}\,p^{1/2}} \tag{3.116}$$

When the pressure is very low, the denominator approaches unity and θ is proportional to $p^{1/2}$. When the pressure is very high, $(1-\theta)$ is inversely proportional to the square root of pressure.

3.8.4 Competitive Adsorption

The isotherm for two different gases adsorbed on the same surface is of considerable interest. Let the fraction of the surface covered by molecules A be θ, and the fraction covered by B be θ'; the bare fraction is $(1-\theta-\theta')$. The reaction rate of adsorption of A may be written as

$$\dot{R}_a = k_a\, p(1-\theta-\theta') \tag{3.117}$$

The reaction rate of desorption of A is

$$\dot{R}_d = k_d\,\theta \tag{3.118}$$

At equilibrium, the adsorption rate is balanced by the desorption rate, so that

$$\frac{\theta}{1-\theta-\theta'} = \frac{k_a}{k_d}p \equiv Kp \tag{3.119}$$

where p is the partial pressure of A. From a similar step, the equilibrium condition of the B molecules is

$$\frac{\theta'}{1-\theta-\theta'} = \frac{k'_a}{k'_d}p' \equiv K'p' \tag{3.120}$$

where p' is the partial pressure of B and K' is the equilibrium constant for the adsorption–desorption reaction of B. The fractions covered by A and B can be expressed as

$$\theta = \frac{Kp}{1+Kp+K'p'} \quad \text{and} \quad \theta' = \frac{K'p'}{1+Kp+K'p'} \tag{3.121}$$

It can be seen from Eq. (3.121) that the fraction of the surface covered by one gas is reduced if the partial pressure of the other gas is increased; this is caused by the fact that the molecules are competing with one another for a limited number of surface sites in the competitive chemisorption process.

3.8.5 Bimolecular Surface Reactions

Bimolecular surface reactions represent reactions between two adsorbed molecules $(A+B)$ on neighboring surface sites. For the vast majority of surface catalytic reactions, the Langmuir–Hinshelwood (LH) mechanism is used to model the kinetics of a heterogeneous catalytic process. This reaction is shown schematically in Figure 3.19.

The LH mechanism assumes that the reaction rate between two species A and B is proportional to the probability that A and B are adsorbed on neighboring sites and is also proportional to the fractions of the surface, θ and θ', covered by A and B. Therefore, the surface reaction rate can be written as

$$\dot{R} = k\theta\theta' = \frac{kKK'pp'}{(1+Kp+K'p')^2} \tag{3.122}$$

For sparsely covered surfaces, the above equation reduces to

$$\dot{R} \approx kKK'pp' \tag{3.123}$$

If one reactant is weakly adsorbed, Eq. (3.122) reduces to

$$\dot{R} = \frac{kKK'pp'}{(1+K'p')^2} \quad \text{or} \quad RR = \frac{kKK'pp'}{(1+Kp)^2} \tag{3.124}$$

The rate is then proportional to the partial pressure of A in the first expression. As the pressure of B increases, the rate first increases, passes through a maximum, and then decreases. If high concentration of reactant B is strongly adsorbed, $K'p'$ is much greater than unity, and the rate equation becomes

$$\dot{R} = \frac{kKp}{K'p'} \tag{3.125}$$

Another type of mechanism, the Eley and Rideal (1941) mechanism, involves the reaction between a gas phase molecule B and an adsorbed molecule A, shown schematically in Figure 3.20. The fraction of surface covered by A is given by Eq. (3.121), and the rate is proportional to the partial pressure p' of B, namely,

$$\dot{R} = k\theta p' = \frac{kKpp'}{1+Kp+K'p'} \tag{3.126}$$

In this mechanism, B is not at all adsorbed. However, in the above equation with $K'p'$ in the denominator, B molecules do affect the rate by occupying surfaces that might otherwise be occupied by A.

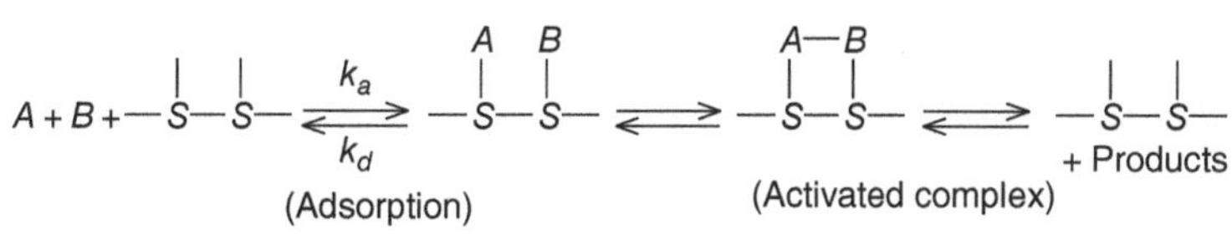

Figure 3.19 Langmuir–Hinshelwood mechanism.

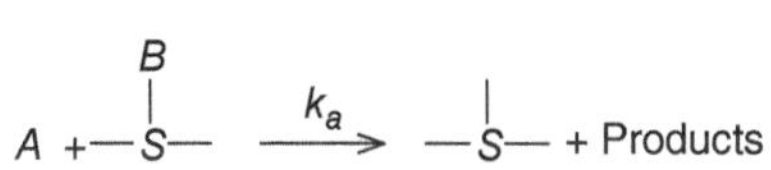

Figure 3.20 Eley–Rideal mechanism.

3.8.6 Kinetic Modeling of Hydrogen–Oxygen Surface Reactions on Platinum

The surface reaction between H_2 and O_2 is a very useful example of a surface reaction since it has a very limited number of elementary reactions. This reaction has been investigated under a variety of conditions, and many types of behavior have been observed. For example, see Ljungstrom et al. (1989) and Warnatz et al. (1994). On a porcelain surface, the rate was found to be proportional to partial pressures of both H_2 and O_2. On the surfaces of various metals, including silver, gold, copper, and nickel, the reaction rate has been found to be the first order in H_2 and the zeroth order in O_2. On a platinum surface, inhibition by O_2 was detected by Langmuir (1922). He also found that H_2O and other product species could inhibit the H_2/O_2 reaction on the surface of silver.

Consider the following elementary reaction steps for H_2/O_2 reaction on Pt surface under high-temperature ($T > 1200$ K) conditions (Ljungstrom et al. 1989):

$$\mathrm{H_2} \underset{k_{d,\mathrm{H_2}}}{\overset{k_{a,\mathrm{H_2}}}{\rightleftharpoons}} 2\mathrm{H}_{(a)} \tag{3.127}$$

$$\mathrm{O_2} \xrightarrow{k_{a,\mathrm{O_2}}} 2\mathrm{O}_{(a)} \tag{3.128}$$

$$\mathrm{H}_{(a)} + \mathrm{O}_{(a)} \xrightarrow{k_{f,\mathrm{OH}}} \mathrm{OH}_{(a)} \tag{3.129}$$

$$\mathrm{H}_{(a)} + \mathrm{OH}_{(a)} \xrightarrow{k_{f,\mathrm{H_2O}}} \mathrm{H_2O}_{(a)} \rightarrow^{k_{d,\mathrm{H_2O}}} \mathrm{H_2O} \tag{3.130}$$

$$\mathrm{OH}_{(a)} \xrightarrow{k_{a,\mathrm{OH}}} \mathrm{OH} \tag{3.131}$$

where subscript a is used for adsorbed species. Similarly, we shall use subscript d for desorption and subscript f for formation reaction. Several reaction steps were excluded from the above reaction mechanism in order to match the high-temperature experimental condition:

1. $\mathrm{OH}_{(a)} + \mathrm{OH}_{(a)} \rightarrow \mathrm{O}_{(a)} + \mathrm{H_2O}_{(a)} \rightarrow \mathrm{O}_{(a)} + \mathrm{H_2O}$ (a low-temperature path).
2. $\mathrm{OH}_{(a)} \rightarrow \mathrm{H}_{(a)} + \mathrm{O}_{(a)}$ (reactions (3.129) and (3.130) are assumed to be much faster).
3. O_2 desorption neglected (due to large desorption energy, which is less likely to occur).
4. H_2O is weakly adsorbed and desorbed rapidly enough, implying no coverage by adsorbed $\mathrm{H_2O}_{(a)}$.
5. There is no coverage of surface by $\mathrm{H}_{2(a)}$ due to low desorption energy.

Equilibrium values of the coverage parameters of species are determined by the steady-state solution of the coupled rate equations given below:

$$0 = \frac{d\theta_\mathrm{H}}{dt} = 2k_{a,\mathrm{H_2}} f_{\mathrm{H_2}}(\theta) - k_{f,\mathrm{OH}}\,\theta_\mathrm{H}\theta_\mathrm{O} - k_{f,\mathrm{H_2O}}\,\theta_\mathrm{H}\theta_\mathrm{OH} - 2k_{d,\mathrm{H_2}}\,{\theta_\mathrm{H}}^2 \tag{3.132}$$

$$0 = \frac{d\theta_\mathrm{O}}{dt} = 2k_{a,\mathrm{O_2}} f_{\mathrm{O_2}}(\theta) - k_{f,\mathrm{OH}}\,\theta_\mathrm{H}\theta_\mathrm{O} \tag{3.133}$$

$$0 = \frac{d\theta_\mathrm{OH}}{dt} = k_{f,\mathrm{OH}}\,\theta_\mathrm{H}\theta_\mathrm{O} - k_{f,\mathrm{H_2O}}\,\theta_\mathrm{H}\theta_\mathrm{OH} - k_{d,\mathrm{OH}}\,\theta_\mathrm{OH} \tag{3.134}$$

where the total surface coverage parameter θ is

$$\theta = \theta_\mathrm{H} + \theta_\mathrm{O} + \theta_\mathrm{OH}$$

Making the assumption that the adsorbed hydrogen $\mathrm{H}_{(a)}$ is more mobile than $\mathrm{O}_{(a)}$ and $\mathrm{OH}_{(a)}$ simplifies the reaction rate equation as follows.

From Eq. (3.104), the formation (f) and desorption (d) rate coefficients are:

$$k_{f,\mathrm{OH}} = \upsilon_{\mathrm{H}} \cdot \exp \frac{-E_{af,\mathrm{OH}}}{k_B T}$$

$$k_{f,\mathrm{H_2O}} = \upsilon_{\mathrm{H}} \cdot \exp \frac{-E_{af,\mathrm{H_2O}}}{k_B T}$$

$$k_{d,\mathrm{H_2}} = \upsilon_{\mathrm{H}} \cdot \exp \frac{-E_{ad,\mathrm{H_2}}}{k_B T}$$

$$\text{where } \upsilon_{\mathrm{H}} = \frac{D_H}{A}\sqrt{\frac{\pi k_B T}{2 m_H}} \approx 10^{13}$$

$$k_{d,\mathrm{OH}} = \upsilon_{d,\mathrm{OH}} \cdot \exp \frac{-E_{ad,\mathrm{OH}}}{k_B T}$$

From Eq. (3.100), the adsorption (a) rate coefficients are:

$$k_{a,\mathrm{H_2}} f_{\mathrm{H_2}}(\theta) = \frac{p_{\mathrm{H_2}} \cdot A}{\sqrt{2\pi\, m_{\mathrm{H_2}} k_B T_{\mathrm{gas}}}} \cdot S_{\mathrm{H_2}}(0) \cdot f_{\mathrm{H_2}}(\theta) \tag{3.135}$$

$$k_{a,\mathrm{O_2}} f_{\mathrm{O_2}}(\theta) = \frac{p_{\mathrm{O_2}} \cdot A}{\sqrt{2\pi\, m_{\mathrm{O_2}} k_B T_{\mathrm{gas}}}} \cdot S_{\mathrm{O_2}}(0) \cdot f_{\mathrm{O_2}}(\theta) \tag{3.136}$$

$$\text{where } f_{\mathrm{H_2}}(\theta) = 1 - \theta,\ f_{\mathrm{O_2}}(\theta) = (1-\theta)^2, \text{ and } A \text{ is the site area.} \tag{3.137}$$

The rate of production of OH species from the surface can be calculated from the product of OH desorption rate $k_{d,\mathrm{OH}}$ and the coverage parameter θ_{OH} given in Eq. (3.134). Similarly, the H_2O production rate can be determined from the formation rate of water $k_{f,\mathrm{H_2O}}$ and the coverage parameters θ_{OH} and θ_{H},

$$\dot{R}_{d,\mathrm{OH}} = k_{d,\mathrm{OH}}\, \theta_{\mathrm{OH}}\, N_o$$

$$\dot{R}_{d,\mathrm{H_2O}} = k_{f,\mathrm{H_2O}}\, \theta_{\mathrm{H}}\, \theta_{\mathrm{OH}}\, N_o \tag{3.138}$$

where N_o is the number of sites per unit area.

The maximum water production rate occurs when the H_2 and O_2 adsorption rates follow a stoichiometric relationship, i.e.,

$$2\mathrm{H_2} + \mathrm{O_2} \rightarrow 2\mathrm{H_2O} \tag{3.139}$$

In addition, OH and H_2 desorption are assumed to be negligible. Under these conditions, we have

$$\frac{k_{a,\mathrm{H_2}} f_{\mathrm{H_2}}(\theta)}{k_{a,\mathrm{O_2}} f_{\mathrm{O_2}}(\theta)} = 2, \qquad \text{for} \quad \theta \leq 1 \tag{3.140}$$

From Eqs. (3.133) and (3.134), we obtain the following relationship using the condition that OH desorption and H_2 desorption are small,

$$k_{f,\mathrm{H_2O}}\, \theta_{\mathrm{H}} \theta_{\mathrm{OH}} = k_{f,\mathrm{OH}}\, \theta_{\mathrm{H}} \theta_{\mathrm{O}} = 2 k_{a,\mathrm{O_2}} f_{\mathrm{O_2}}(\theta) \tag{3.141}$$

Substituting this result into Eq. (3.138),

$$\dot{R}_{\max,\mathrm{H_2O}} = k_{f,\mathrm{H_2O}}\, \theta_{\mathrm{H}}\, \theta_{\mathrm{OH}} = 2 k_{a,\mathrm{O_2}} f_{\mathrm{O_2}}(\theta) \approx 2 k_{a,\mathrm{O_2}} = k_{a,\mathrm{H_2}} \tag{3.142}$$

By using the definition of adsorption rate of H_2 in Eq. (3.135), we can finally get the maximum H_2O production rate:

$$\dot{R}_{\max,H_2O} = k_{a,H_2} = \frac{p_{H_2}}{\sqrt{2\pi A_{H_2} k_B T_{gas}}} \cdot S_{H_2}(0) \tag{3.143}$$

The maximum H_2O production rate is thus determined from either the hydrogen or the oxygen sticking coefficient S at zero surface coverage.

3.9 Experimental Methods for Measurement of Gas-Phase Reaction Rates

Measurement of reaction rate coefficients has always been an important issue for chemists and combustion engineers. However, it is not always easy to measure these reaction rates since some reactions can be very fast (a fraction of a millisecond, for instance) or very slow (several hours, for instance).

In this section, some common methods for the measurement of reaction rate coefficients of gas-phase reactions are discussed. For gas-phase reactions, there are three kinds of general methods: static methods (e.g., reaction rates deduced from measured temperature/pressure- time records in closed vessels or through direct chemical analysis), dynamic methods (e.g., reaction rates measured in isothermal flow systems), and methods for very fast reactions. The degree of accuracy of measurement often depends on the method chosen. In order to conduct rate measurement of very fast reactions, some special techniques have to be utilized. According to Eq. (3.13), the *rate of reaction* is the rate of disappearance of a reactant or the rate of appearance of a product. The purpose of all the rate measurement methods is to determine either directly or indirectly the initial state and the final state. By comparing these two states, one is able to determine rates of reaction. It is also important to note that the theoretical prediction of rate coefficients from first principles using atomic structure kinetics now has an accuracy comparable to experiment (Miller et al. 2021).

3.9.1 Static Methods

The most widely used static method for determining the rate of a reaction involving gaseous reactants is to introduce the reactants into a bulb maintained at the desired temperature and to make determinations of the species concentrations at various time intervals. The measurement of the gas composition can be carried out either directly by chemical analysis or indirectly by measurement of pressure of the system. However, the measure of the pressure variation is only possible if the number of moles changes, which is not always the case. The pressure measurements can be made while the reaction is in progress by connecting a pressure sensor directly to the reaction vessel.

Stopping the reaction after various intervals of time by rapid cooling of the reacting system using cryogenic fluids can also be used to measure the concentration of the species in the intermediate products. The total pressure of the system can be measured. However, in many cases, the total pressure does not provide very reliable information since many intermediate products can form and disappear. Thus, the results of the static method can contain some serious errors. Gas-phase reactions involving certain specific reactants or products are sometimes detected by observing the light absorption at a particular wavelength.

For the reaction of species $A \rightarrow$ Products, the rate of change of the concentration $[A]$ is

$$\frac{d[A]}{dt} = -k_{\text{exp}}[A]^m$$

where k_{exp} is measured value of the specific rate coefficient. The change in time of the concentration of species A can be calculated by integrating this differential equation. For *first-order reactions* ($m = 1$), by integration of the above equation, we obtain the first-order time behavior

$$\ln\frac{[A_t]}{[A_0]} = -k_{\text{exp}}(t - 0) = -k_{\text{exp}}\, t$$

where $[A_0]$ and $[A_t]$ denote the concentrations of species A at time 0 and t, respectively.

For *second-order reactions* ($m = 2$), the time-dependent behavior of species A can be expressed by

$$\frac{1}{[A_t]} - \frac{1}{[A_0]} = k_{\text{exp}}\, t$$

and for *third-order reactions* ($m = 3$), the time-dependent behavior of species A can be expressed by

$$\frac{1}{[A_t]^2} - \frac{1}{[A_0]^2} = 2k_{\text{exp}}\, t$$

There are several static experimental techniques to determine the specific rate coefficient k_{exp}. These are as follows:

1. Flash photolysis resonance fluorescence technique.
2. Relative rate-constant photolysis technique.
3. Laser photolysis/laser-induced fluorescence (LIF) technique.

Brief discussions of these static methods are given below.

3.9.1.1 Flash Photolysis Resonance Fluorescence Technique

Let us first define a *photolysis reaction* as a reaction caused by absorption of a photon; this is different from *thermolysis reactions* caused by vigorous intermolecular collisions. A classic example of this technique is the measurement of the dissociation of Freon refrigerants (CF_2Cl_2 and $CFCl_3$) by absorption of a photon. This reaction is of importance since the dissociation of Freon produces chlorine Cl atoms which then cause the catalytic destruction of ozone (O_3) in the stratosphere. This reaction was first proposed by Molina and Rowland (1974), who subsequently were awarded the 1995 Nobel Prize in Chemistry. Since stratospheric ozone has a strong role in reducing ground-level ultraviolet solar radiation, the manufacture and use of Freons were banned in the 1980s.

The suggested reactions for the chlorine–ozone catalytic destruction cycle are

$$Cl + O_3 \rightarrow ClO + O_2$$

$$ClO + O \rightarrow Cl + O_2$$

Kinetic studies were performed on the first reaction by Kurylo and Braun (1976). Subsequent publications by Orkin et al. (1996) detailed the global warming potential (GWP) of the refrigerants HFC-143a and HFC-245fa. In their experimental setup, a stainless steel cylindrical cell is maintained at a desirable temperature, with an outer tube for circulating fluids between the walls and the cell. The interior of the cell is coated with Teflon to minimize light scattering while maintaining chemical inertness. An absorption cell is used to monitor O_3 concentrations by absorption of radiation from a low-pressure Hg source. A pulsed N_2 flash lamp produces chlorine atoms from ultraviolet flash photolysis in a vacuum environment, and a chlorine atom resonance lamp excites a small fraction of the atoms. Resonantly scattered, the atom signal is detected by a cooled photomultiplier at right angles to both of the lamps. The expression for the rate of chlorine atom removal is given by

$$-\frac{d[\mathrm{Cl}]}{dt} = k_1[\mathrm{Cl}][\mathrm{O_3}] + k_p[\mathrm{Cl}] + k_M[M][Cl]$$

where k_1 is the rate coefficient for the first catalytic cycle discussed earlier, k_p is a rate coefficient associated with both mechanical pumping and diffusion of atoms out of the viewing zone, and k_M is a rate coefficient for reaction with some impurity species M. When the above removal rate equation is integrated, the following relationship is found:

$$\ln\left(\frac{[\mathrm{Cl}_t]}{[\mathrm{Cl}_0]}\right) = (-k_1[\mathrm{O_3}] - k_p - k_M[M]) \cdot t$$

where $[\mathrm{Cl}_0]$ and $[\mathrm{Cl}_t]$ denote the concentrations of chlorine Cl at time 0 and t, respectively. At a fixed total pressure and constant temperature, k_1 can be determined from the slope of the first-order decay rate versus $[\mathrm{O_3}]$. Then, by using a nonlinear least-squares fit of data sets to an Arrhenius type equation, $k_1(T)$ can be determined. Kurylo and Braun obtained the following expression for k_1 in the unit of (cm^3/molecule/s).

$$k_1(T) = (2.94 \pm 0.49) \times 10^{-11} \exp\left[\frac{-(298 \pm 39)}{T}\right]$$

They claimed the measurement of rate coefficients as small as 10^{-16} cm^3/molecule/s can be achieved using this method.

3.9.1.2 Relative Rate-Coefficient Photolysis Technique

There is a need to predict the atmospheric lifetimes of compounds that are greenhouse gases. The relative atmospheric lifetimes of methane and methyl chloroform are inversely proportional to the ratio of the corresponding rate coefficients. According to DeMore (1992), the relative rate coefficients can be measured in a low-speed flow, temperature-controlled photochemical reactor and are based on rates of disappearance of the parent compounds as measured by Fourier transform infrared (FTIR) spectroscopy. The required rate-coefficient ratios can be obtained from tabulations of absolute reaction rate coefficients. However, uncertainties in individual rate coefficients are usually in the range of 10–30%, and the combined error may be substantial. Rate coefficient ratios, on the other hand, can usually be determined more accurately through relative rate techniques. The results permit the relative lifetimes of $\mathrm{CH_3CCl_3}$ and $\mathrm{CH_4}$ (for removal by OH in the troposphere) to be estimated within about 7% at relevant temperatures. The major reaction mechanism considered is

$$\mathrm{O_3} + h\nu_{\text{at } \lambda\, =\, 254\ \text{nm}} \rightarrow \mathrm{O_2} + \mathrm{O(^1D)}$$

$$\mathrm{O(^1D)} + \mathrm{H_2O} \rightarrow \mathrm{OH} + \mathrm{OH}$$

$$\mathrm{OH} + \mathrm{CH_4} \rightarrow \mathrm{H_2O} + \mathrm{CH_3}$$

$$\mathrm{OH} + \mathrm{CH_3CCl_3} \rightarrow \mathrm{H_2O} + \mathrm{CH_2CCl_3}$$

The depletion factor (DF) for a given chemical compound can be defined as

$$\mathrm{DF} = \frac{\text{Initial concentration}}{\text{Final concentration}}$$

It can be determined that the reaction losses are due almost entirely to OH, and the desired rate-coefficient ratio can thus be estimated from

$$\frac{k(CH_3CCl_3)}{k(CH_4)} = \frac{\ln DF(CH_3CCl_3)}{\ln DF(CH_4)}$$

A temperature-controlled quartz photochemical cell was used to contain the test sample by photolysis. The concentrations of CH_4 and CH_3CCl_3 are measured by passing the effluent mixture through a small White cell on a Nicolet spectrometer. Tests based on relative flow rates were conducted to verify that the reactant concentrations measured in the White cell were proportional to those in the photolysis cell under all conditions of the experiment. DeMore (1992) showed that each reactant is depleted by the OH reaction in a manner similar to that of the atmosphere. The rate-coefficient ratio is obtained from the following relationship:

$$\frac{k_{react}}{k_{reference}} = \frac{\ln DF_{react}}{\ln DF_{reference}}$$

3.9.1.3 Laser Photolysis/Laser-Induced Fluorescence (LIF) Technique

Hazardous organic wastes can be disposed through high-temperature incineration methods, which also produce considerable chlorinated hydrocarbons. Hydroxyl radicals (OH) are an important species in hydrocarbon oxidation, and reliable kinetic data for the radical species are necessary for successful modeling of high-temperature systems. Hydrogen atom metathesis rate-coefficient measurements for the reaction of OH with methyl chloride (CH_3Cl), methylene chloride (CH_2Cl_2), and chloroform ($CHCl_3$) over an extended temperature range (295–800 K) were conducted by Kasner et al. (1990). Similar measurements of the following reaction over a temperature range of 294–789 K at a total pressure of 740 ± 10 Torr were also reported:

$$C_2H_5\,Cl + OH \xrightarrow{k_1} C_2H_4\,Cl + H_2O \qquad \Delta H^o_{298} = -35.1\ \text{kcal/mole}$$

A resistively heated quartz optical test cell allowed data to be acquired in the range from ambient temperature to 1000 K. Mutually perpendicular axes of the quartz cell permitted irradiation of the reaction volume by the photolysis and probe lasers and detection of the laser-induced OH fluorescence. Data were taken at a 10-Hz rate coinciding with the firing of the lasers. In Kasner's experiment, hydroxyl radicals were generated by the photolysis of nitrous oxide (N_2O) in the presence of water vapor (H_2O). The photolysis laser was an ArF excimer laser, which operated at 193 nm. This N_2O photolysis produced excited O atoms, which quickly reacted with H_2O to generate OH:

$$N_2O + h\nu_{\text{at }\lambda = 193\text{ nm}} \rightarrow N_2 + O(^1D) \qquad\qquad O(^1D) + H_2O \rightarrow 2OH$$

To ensure complete conversion of O atoms to OH, the ratio of H_2O to N_2O was typically 15:1. Once created, the relative OH population was monitored by a LIF device. A frequency-doubled, Nd:YAG pumped dye laser was used to produce 2–3-mJ pulses at the wavelength of 282.08 nm. This light was used to excite the transition of the OH radical. Hydroxyl fluorescence was collected at 309 nm, corresponding to the 0–0 vibrational transition.

All experiments were performed under pseudo-first-order kinetic conditions with concentration of C_2H_5Cl much greater than that of OH. Over the entire temperature range, OH decay profiles exhibited exponential behavior and were fit by the following nonlinear expression:

$$[OH]_t = [OH]_0 \exp(-k't) + c$$

where $[OH]_0$ and $[OH]_t$ denote the concentrations of OH at time 0 and t, $k' = k \times [C_2H_5\,Cl] + k_d$, and c is an additional term accounting for background light levels. This expression was fitted by using the Levenberg–Marquardt method, an iterative least squares technique. The bimolecular rate coefficient, k, was separated from the diffuse rate coefficient, k_d, by fitting a weighted least-squares line through the data points on the plot of C_2H_5Cl concentration versus k'. Observed pseudo-first-order OH decay rates, in the absence of C_2H_5Cl, were $80 \pm 20\ s^{-1} \times C_{C_2H_5\ Cl}$ at various temperatures.

3.9.2 Dynamic Methods for Reaction Rates in Flow Systems

Kinetic rates can also be determined from isothermal flow reactors. One of the earliest flow reactors is described by Caldin and Trowse (1954). In their setup, two reactant mixtures initially in separate containers flow through a mixing chamber into an observation tube. At various points along the observation tube, the composition of the mixture was determined by optical, thermal, or other methods. By taking measurements at various distances, concentration–time curves were obtained, and from the concentration distribution, the rate constants were determined.

In general, for flow reactor systems, the reactants (gas or liquid) are supplied at known flow rates to a mixing section followed by a long reaction channel of known volume maintained at a desirable temperature. Sampling tubes are connected to multiple stations between the upstream and downstream ends of the reactor, allowing samples to be easily collected for chemical analysis. From the volume of the reactor vessel and the rate of flow, it is possible to calculate the mean time spent by the reaction mixture in the reactor. The difference between the compositions at an upstream station and a downstream station represents the amount of reaction that has occurred in the flow residence time between the two stations. The reaction rate is then deduced from the concentration difference divided by the flow residence time. The order of the reaction can be determined by varying the rate of flow through the reactor and the concentrations of the entering gases or solutions. One can also make measurements at a fixed station by varying the mixture flow velocity. By varying flow rates and the volume of the reaction vessel, the residence time can be varied from 0.025 to 25 seconds. In view of certain minimum time requirements for mixing, flow reactor techniques are not suitable for the study of reactions whose lifetimes are smaller than about 1 ms.

For a plug-flow reactor with a homogeneous reacting mixture, the flow in the reactor is modeled as one-dimensional with the mass fraction Y_i varying only in the x-direction. The streamwise gradient of Y_i is

$$\frac{dY_i}{dx} = \frac{\dot{\Omega}_i M_i}{\rho u_x}$$

where u_x is the average velocity of the gas in the reactor, M_i is the molecular mass, and $\dot{\Omega}_i$ is the production rate of species i in a multiple reaction system with N elementary reactions. The overall production rate $\dot{\Omega}_i$ can then be determined from the measured mass fraction gradient and the known flow rate.

3.9.3 Methods for Measuring Rapid Reaction Rates

There are a number of different methods for measuring rapid reaction rates. These include, for example, the stopped-flow spectrophotometric method, relaxation methods, shock tube methods, flash photolysis methods, reaction in molecular beams, multiple photoionization (MPI) techniques, and femtochemistry techniques.

Table 3.3 Typical Half-Lives of Chemical Species for Different Techniques.

Techniques for Reaction Rate Measurements	Typical Half-Lives (s)
Conventional Static Methods	10^2 to 10^8
Flow Reactors	10^{-3} to 10^2
Rapid Mixing	10^{-4} to 1
Nuclear Magnetic Resonance	10^{-6} to 1
Pulse Radiolysis	10^{-11} to 1
Flash Photolysis	10^{-12} to 1
Relaxation Methods	10^{-10} to 1
Pressure Jump	10^{-6} to 1
Temperature Jump	10^{-8} to 1
Shock Tube Method	10^{-9} to 10^{-3}
Kinetic Spectroscopy Methods	10^{-15} to 10^{-10}
Fluorescence Decay	10^{-10} to 10^{-6}
Photo-stationary Methods	10^{-10} to 1
Electrochemical Methods	10^{-8} to 1
Acoustical Methods	10^{-11} to 10^{-3}
Electrical Pulses and Waves	10^{-10} to 10^{-2}
Femtochemistry Techniques	10^{-14} to 10^{-10}

Source: Adapted from Laidler (1987) and Hammes (1978).

In order to study fast kinetic rates, the reactive atoms (such as O, H, and N) and radicals (such as OH, CH_3, and C_2H_5) have to be produced. Depending on the selected experimental method, generation of highly reactive species is usually achieved by using the following:

- Microwave discharge to disassociate parent molecules (such as O_2 and H_2) to form O, H, N, etc.
- Flash photolysis to supply light flash of very high intensity for very short duration ($\sim 10^{-6}$ second) to generate atoms, free radicals, and excited species.
- Laser photolysis to dissociate molecules by energetic monochromatic waves of laser energy.
- Shock-wave heating to thermally heat reactants to high temperatures by the shock wave in the test section of the shock tube.

Table 3.3 summarizes the range of half-life times of chemical species corresponding to different techniques for reaction rate measurements. For very fast reactions, kinetic spectroscopy methods, resonant multiple photoionization (REMPI), and femtochemistry techniques seem to be the most appropriate.

3.9.3.1 Stopped-Flow Spectrophotometric Technique

The stopped-flow spectrophotometric technique (developed initially for reactions in solution by Chance (1940) possesses some of the best features of both the static and the dynamic flow systems. Johnston (1954) contributed to the initial development of spectrophotometric apparatus for gas-phase reaction measurements in stopped-flow reaction vessels. In the experimental setup, two mixtures are forced through multiple opposing jets into a mixing chamber, in which mixing is extremely rapid and accomplished within 10^{-3} seconds. From the mixing chamber, the mixture passes immediately into a reaction vessel. In some designs, the mixing chamber and reaction vessel are one component. The flow is stopped suddenly, and time-dependent concentration measurements are made by spectrophotometric methods.

The main advantages of the stopped-flow technique over continuous-flow methods are that (1) this method is not affected by the rate and character of the flow, as long as the mixing is satisfactory; (2) the progress of the reaction, covering a wide range of rates, can be measured; and (3) the volumes of reagents used are smaller, which is particularly important when it is desired to work with substances that are difficult to obtain in large quantities. The disadvantages of the method are that it is less sensitive than continuous-flow methods and that there are fewer convenient methods of observation.

Most chemical reactions involve either the liberation or the absorption of heat; thus, it is useful to measure the temperature rise in the system. When a continuous-flow technique is employed, the products may be passed into a solution medium that quenches the reaction; chemical procedures, such as gas-phase titrations by NO_2 injection, can then be used to determine the extent of reaction. This kind of measurement can be carried out at different residence times, thus allowing the rate constant to be obtained. The method of electron-spin resonance has also been applied to reactions occurring in flow systems containing free radicals as intermediate species.

To avoid the difficulties arising from mixing for reactions whose half-lives are smaller than 1 ms, *relaxation methods* have been developed. In this method, the reaction is allowed to compete with a physical process (such as diffusion), the rate of which is known. When the relaxation method is used, the reaction is not started by initially mixing the reactants or raising their temperature. Instead, the reaction is first allowed to go to equilibrium and is then disturbed in some way; its approach to a new equilibrium is then followed using high-speed techniques. From these measurements, the relaxation time of the process can be calculated. Through the relationships between the relaxation time and the rate constants, the reaction rates are determined.

In order for the relaxation methods to be effective, the perturbation from initial equilibrium must be accomplished in a time period that is much shorter than the relaxation time. Usually, the perturbation has to take place over a period of 10^{-6} or 10^{-7} seconds. Various experimental techniques have been employed. These include the pressure-jump and temperature-jump methods, in which pressure and temperature changes are brought about rapidly. Perturbations using, for example, high electric fields or ultrasonic vibrations can produce oscillating temperatures. Various techniques can be used to follow the response of the system to the perturbation and to measure the relaxation time for deducing the reaction rates. This method is particularly suitable for reaction times that are longer than 10^{-5} seconds.

3.9.3.2 Flash Photolysis Method

Another very important method for studying very fast reactions in gas mixtures or liquid solutions is the flash photolysis method developed initially by Basco and Norrish (1961). This method is also called the *kinetic spectroscopy method*. The basic step in the procedure is to produce a light flash of very high intensity and very short duration ($\sim 10^{-6}$ seconds) in the neighborhood of a reaction vessel. The purpose is to produce atoms, free radicals, and excited species in the reaction system, and these active species will undergo further reactions, which can be followed by spectroscopic means.

3.9.3.3 Shock Tube Methods

Fast gas-phase reactions can also be investigated by shock tube methods. The principle of this method is that the chemical reaction system of a gas mixture is contained in a long metal tube, through which a shock wave is allowed to pass. As the shock wave propagates through the gaseous mixture, it raises the mixture temperature, and reaction occurs in and behind the shock front. The species concentration and mixture temperature are determined by various methods and, in this way, knowledge of the kinetic rates can be obtained. The shock wave is usually produced by the rupture of a thin metal disc separating the high-pressure and low-pressure compartments of the shock tube. The reaction mixture is placed in the low-pressure compartment. Helium gas is usually used in the high-pressure compartment to support the displacement of the traveling shock wave into the low-pressure compartment after the mechanical rupture of the diaphragm. As the shock wave passes, the mixture is not at thermal or chemical equilibrium; the perturbation caused by the shock wave is a much stronger one than is usually employed in relaxation methods.

The problem of following the course of a chemical reaction in and behind a shock wave is a very difficult one. Measurement of the change in concentration of at least one reactant or product species is required. This is usually performed by observing its absorption spectrum. Other alternative techniques can also be used. For example, Bradley and Kistiakowsky (1961) used a time-of-flight mass spectrometer to follow concentrations of nitrous oxide (N_2O) decomposition species produced by the shock wave. Even though the time resolution of the time-of-flight mass spectrometer is relatively long (about 50 μs), the concentration of the oxygen atoms can be measured accurately by this method. This method can also follow the concentrations of several species simultaneously. The shock wave technique has also been applied to the study of vibrational and rotational energy transfers between gaseous molecules.

3.9.3.4 Molecular Beam Scattering

Physicists and chemists have also developed molecular beam scattering techniques to produce mono-energetic molecular beams (Steinfeld et al. 1998). The advantage of reactions in molecular beams is the ability to study chemical reactions under conditions in which there is no Maxwellian distribution of velocity (at a given temperature, the Maxwellian distribution of velocity is a probability function based on the number density and the mass of molecules) so that it is easier to interpret the data for determining reaction rates. By allowing two molecular beams to impinge on each other, one can study the kinetic rates of bimolecular reactions.

The molecular beam scattering experiment consists of intersecting two beams of molecules (A and B) and detecting the product molecules C in a suitable reactor. While this may be simple, there are a number of constraints that must be met in order for the experiment to be successful. The product molecules must not undergo collisions with background gas molecules before they reach the detector. This means that the mean free path in the gas must be greater than the distance from the scattering center to the detector, which may be on the order of 10–30 cm. Therefore, the molecular beam scattering chambers are operated at very low pressures on the order of 10^{-6} torr. Differential pumping is frequently employed to reduce the pressure down to 10^{-10} torr or lower. Another constraint is that the beam densities should be very high, typically 10^{11} molecules/cm^3, so that a sizable rate of production of C is obtained. Only a small fraction of this production rate is intercepted by the detector. The detection techniques involve making use of laser-excited fluorescence or multiphoton ionization of the product molecules.

3.9.3.5 Resonant Multiple Photoionization (REMPI)

The REMPI technique utilizes one or more pulsed visible- or ultraviolet-light lasers focused between two biased ion collection plates housed inside a sample cell. The two plates have opposite polarities. If the energy absorbed by the product molecules exceeds their ionization threshold, ions will be generated and detected by collection plates. The detected molecules, ionized using the *MPI process*, provide the resulting signal, which represents the number density of molecules in the cell. The ionization of molecules is achieved by first exciting the product molecule from ground electronic state to an intermediate state and then ionizing the molecule from the intermediate state by one-photon excitation. This technique is referred to as the $n + 1$ MPI process, in which n photons are used to reach the intermediate state and one additional photon is used to ionize the molecule from that excited state.

In general, the $n + 1$ MPI process has high sensitivity. The concentration of product species in a reaction is proportional to the product ion signal; the specific rate parameter k can be determined by fitting the half-life of the ion-count growth to its steady-state value. The photoionization technique is a sensitive laser technique, which is well suited for kinetic measurements. Its main advantage is that the ionization process is highly effective, so that one can ionize and identify transient or stable species at low concentration in the presence of a large excess of other species. This cannot be accomplished using LIF because fluorescence quantum yields are too low.

The *reaction product imaging technique* is a further development of MPI. It utilizes a two-dimensional image of the product angular distribution by capturing the photo ions produced at the scattering center on a position-sensitive detector. Such a detector usually consists of an array of micro-channel plates coupled to a phosphor screen that lights up when struck by an ion; the luminescence is then captured by a charge-coupled device (CCD) camera, which in turn transmits the data to image analysis software. This technique was originally developed to image the angular distributions of the fragmented species following photodissociation.

3.9.3.6 Femtochemistry Technique

The femtochemistry technique is a rather different approach represented by the use of very short-pulse (and therefore wide-bandwidth) lasers to probe real-time reaction dynamics (Steinfeld et al. 1998). This approach is analogous to the laser photolysis method but uses laser pulses as short as 10^{-14} seconds (10 femtoseconds). Using successive delays between initiation (pump) and interrogation (probe) pulses, one can obtain a "stroboscopic" picture of a reaction as it proceeds in time. One example where it has been used is the decomposition of cyclic ketones. The reaction is initiated by absorption of two photons from a femtosecond pulse laser operating at a wavelength of 280–320 nm. This provides enough energy to eliminate CO, leaving the diradical – $CH_2CH_2CH_2CH_2$ –. This diradical is ionized by a delayed femtosecond probe pulse and detected in a time-of-flight mass spectrometer. Buildup and decay occur on a time scale of several hundred femtoseconds or 0.1–1.0 picoseconds. The detected signal is proportional to the product ion signal; the rate coefficient k can be determined by fitting the half-life of the ion-count growth to its steady-state value.

3.10 Experimental Methods to Study Surface Reactions

In general, there are two methods that appear to be the most suitable for studying surface phenomena: spectroscopic methods and temperature-programmed methods. By utilizing spectroscopic methods, it is possible to determine the surface structure, topography, and surface composition of solid materials. In addition, the structure of adsorbed molecules and the binding between solid surface and adsorbents can be evaluated. In general, spectroscopic methods utilize characteristic bands in the infrared (IR) spectra of some fractions of products, which may help identify organic compounds present. Temperature-controlled methods have added greatly to the understanding of chemical kinetics of adsorption, but even more so to the understanding of desorption processes. These methods can provide information on the adsorption energy, kinetic order, activation energy, and the preexponential factors of desorption.

3.10.1 Auger Electron Spectroscopy

Perhaps the most commonly used spectroscopic technique for the determination of surface chemical composition is Auger electron spectroscopy (AES). The basis for Auger electron spectroscopy came from the discovery by P. Auger (1899–1993), a French scientist, who in 1925 recognized that an atom in the excited state could return to the equilibrium state by releasing energy through ejection of an electron from an upper level – hence, the name for the technique. Basically, AES systems input a specified amount of energy through an excitation beam and count the number of electrons that are ejected. Elements can then be identified by the energy required to eject the electrons.

AES is a widely used surface analysis technique that has been successfully applied to many diverse fields ranging from gas-phase chemistry to nanostructure characterization. There are a number of electron microscopes that have been specifically designed for use in Auger spectroscopy; these are termed scanning Auger microscopes (SAMs) and can produce high-resolution, spatially resolved chemical images. SAM images are obtained by stepping a focused electron beam across a sample surface and measuring the intensity of the Auger peak above the background of scattered electrons. One of the major advantages of the SAM system over early AES systems was that it is able to provide a two-dimensional map of the chemical composition of the surface. Additional information about AES is given in Thompson et al. (1985).

3.10.2 Temperature-Controlled Methods

Three common types of temperature-controlled methods are temperature-programmed desorption (TPD), thermal desorption spectroscopy (TDS), and temperature-programmed reaction spectra (TPRS). The basic idea behind the method is that if a gas is adsorbed onto a surface, gradual heating will cause it to desorb. As the temperature is increased, the desorption rate increases, reaches a maximum, and then decreases as the surface is depleted of the adsorbed species. In particular, the identification of "desorption peaks" observed at controlled temperatures makes it possible to distinguish different adsorption states coexisting on a given surface and to evaluate approximately the associated energies of interaction. It is possible for the gas to take more than one adsorbed form on the surface, in which case the number of so-called adsorption peaks will correspond to the number of adsorbed forms on the surface. The desorption peaks provide insight regarding the order in which reactions occur and which steps in the overall reaction could be rate-determining steps.

3.10.3 Combination of Spectroscopic and Temperature-Controlled Methods

Techniques have been developed to use both the spectroscopic and temperature-controlled techniques together. One example of this is scanning kinetic spectroscopy (SKS). This technique is similar to other temperature-controlled methods except that it is carried out in a high vacuum. SKS is generally used in the study of catalysis. The sample used is a catalytic crystal, which is "shot" with a molecular beam of reactant whose temperature is raised according to a computer program. A mass spectrometer is usually utilized to detect both products and reactants leaving the surface of the crystal. By continuously feeding a supply of reactants to the surface, this technique eliminates one of the main drawbacks of temperature-controlled methods: contamination of the surface due to adsorption. Using the SKS technique, one can detect both positive and negative peaks, which has a great significance. Negative peaks indicate that the reactant is being consumed, while positive peaks are an indication of products being formed by chemical reaction or reactants that have left the surface due to desorption. This phenomenon is significant because it can lend insight into the overall surface reaction process. By analyzing the peaks, one can determine what reactions are taking place, or what reactions are significant to the overall reaction.

3.10.4 Surface Reaction Rate Determination

The experimental evaluation of the kinetics of catalytic surface reaction is a complex procedure, involving not only the determination of the overall kinetics but also the determination of the intrinsic kinetics. Application of the law of mass action to elementary reactions on a catalyst surface requires both identification of reactant species on the surface and measurement of their surface concentrations. For many years, equilibrium adsorption isotherms, such as the Langmuir isotherm, the Temkin isotherm, and the Freundlich isotherm, were used to relate surface concentrations of reactant species to gas-phase partial pressures. Eischens et al. (1956) were the first to use an IR spectra technique to identify chemisorbed CO, CO_2, olefins, and acetylene, among others.

Various techniques in addition to IR spectroscopy have been applied to the identification of adsorbed species in the absence of reaction. Among these techniques are FTIR spectroscopy, field-emission microscopy (FEM), field-ion microscopy (FIM), low-energy electron diffraction (LEED), secondary ion mass spectroscopy (SIMS), electron energy loss spectroscopy (EELS), surface-extended X-ray-absorption fine structure (SEXAFS), X-ray photoelectron spectroscopy (XPS), ultraviolet photoelectron spectroscopy (UPS), and electron spin resonance (ESR). Note that many of these techniques require ultra-high vacuum conditions.

LIF uses laser spectroscopic methods to detect intermediate species, which are desorbed from the surface during dynamic reaction conditions, before they react to form new species, a surface technique pioneered by Tevault et al. (1980). LIF is a spectroscopic method in which an atom or a molecule is excited to a higher energy level by the absorption of laser light followed by spontaneous emission of light. It can be carried out at almost any pressure and surface temperature. This method also allows one to simultaneously identify the type of intermediate species, their internal quantum states, and their desorption energies.

An advantage over absorption spectroscopy is that it is possible to get two- and three-dimensional images since fluorescence takes place in all directions, as the fluorescence signal is usually isotropic. Unlike techniques involving high-energy electron impact, LIF does not cause splitting of the radical or radical precursor and, hence, offers a more definitive detection of radical species. For example, Williams et al. (1992) used LIF to measure OH radical desorption over a polycrystalline Pt foil exposed to mixtures of H_2, O_2, and H_2O.

References

Basco, N., and R. Norrish (1961), The production of vibrationally excited hydroxyl radicals under isothermal conditions by flash photolysis, *Proc. R. Soc., Ser. A*, Vol. 260, pp. 293–303.

Bradley, J., and G. Kistiakowsky (1961), Shock wave studies by mass spectrometry. I. Thermal decomposition of nitrous oxide, *J. Chem. Phys.*, Vol. 35, pp. 256–263.

Caldin, E., and F. Trowse (1954), A semi-automatic apparatus for the photometric measurement of the rates of fast reactions, *Discuss. Faraday Soc.*, Vol. 17, pp. 133–137.

Chance, B. (1940), The accelerated flow method for rapid reactions, *J. Franklin Inst.*, Vol. 229, pp. 737–766.

DeMore, W. (1992), Relative rate constants for the reactions of OH with methane and methyl chloroform, *Geophys. Res. Lett.*, Vol. 19, pp. 1367–1370.

Eischens, R., S. Francis, and S. Pliskin (1956), The effect of surface coverage on the spectra of chemisorbed CO, *J. Phys. Chem.*, Vol. 60, pp. 194–201.

Eley, D., and E. Rideal (1941), The catalysis of the parahydrogen conversion by tungsten, *Proc. R. Soc., Ser. A*, Vol. 178, pp. 429–451.

Hammes, G. (1978), *Chemical Kinetics and Dynamics*, Academic Press, New York.

Hinshelwood, C. (1926), On the theory of unimolecular reactions. *Proc. R. Soc., Ser. A*, Vol. 113, pp. 230–233.

Johnston, H. (1954), Photoelectric methods for following fast gas-phase reactions, *Discuss. Faraday Soc.*, Vol. 17, pp. 14–21.

Kasner M., P. Taylor, and B. Dellinger (1990), Laser photolysis/laser induced fluorescence study of OH-C2H5Cl rate constants from 294 - 789 K, *J. Phys. Chem.*, Vol. 94, pp. 3250–3253.

Kummer, J., and P. Emmett (1951), Chemisorption of carbon monoxide and the heterogeneity of the surface of iron catalysts, *J. Am. Chem. Soc.*, Vol. 73, pp. 2886–2889.

Kurylo, M., and W. Braun (1976), Flash photolysis resonance fluorescence study of the reaction $Cl + O_3 \rightarrow ClO + O_2$ over the temperature range 213 - 298 K, *Chem. Phys. Lett.*, Vol. 37, pp. 232–235.

Laidler, K. (1987), *Chemical Kinetics*, 3rd ed., Prentice-Hall, New York.

Langmuir, I. (1916), The constitution and fundamental properties of solids and liquids, *J. Am. Chem. Soc.*, Vol. 38, pp. 2221–2295.

Langmuir, I. (1918), The absorption of gases on plane surfaces of glass, mica and platinum, *J. Am. Chem. Soc.*, Vol. 40, pp. 1361–1403.

Langmuir, I. (1922), *Trans. Faraday Soc.*, Vol. 17, p. 621.

Lewis, B., and G. von Elbe (1987), *Combustion, Flames, and Explosions of Gases*, 2nd ed., Academic Press, New York.

Lindemann, F. A., S. Arrhenius, I. Langmuir, N. R. Dhar, J. Perrin, and W. C. McCLewis (1922), Discussion on "the radiation theory of chemical action", *Trans. Faraday Soc.*, Vol. 17, p. 599.

Ljungstrom S., B. Kasemo, A. Rosen, T. Wahnstrom, and E. Fridell (1989), An experimental study of the kinetics of OH and H_2O formation on Pt in the $H_2 + O_2$ reaction, *Surf. Sci.*, Vol. 216, pp. 63–92.

Masel, R. (2001), *Chemical Kinetics and Catalysis*, Wiley Interscience, New York.

Miller, J., R. Sivaramakrishnan, Y. Tao, C. Goldsmith, M. Burke, A. Jasper, N. Hansen, N. Labbe, P. Glarborg, and J. Zádor (2021), Combustion chemistry in the twenty-first century: developing theory-informed chemical kinetics models, *Prog. Energy Combust. Sci.*, Vol. 83, p. 100886.

Molina M., and S. Rowland (1974), Stratospheric sink for chlorofluoromethanes: chlorine atom catalysed destruction of ozone, *Nature*, Vol. 249, pp. 810–812.

Orkin, V., R. Huie, and M. Kurylo (1996), Atmospheric lifetimes of HFC-143a and HFC-245fa: flash photolysis resonance fluorescence measurements of the OH reaction rate constants, *J. Phys. Chem.*, Vol. 94, pp. 8907–8912.

Roberts, J. (1935), The adsorption of hydrogen on tungsten, *Proc. R. Soc., Ser. A*, Vol. 152, pp. 445–463.

Steinfeld, J., J. Francisco, and W. Hase (1998), *Chemical Kinetics and Dynamics*, 2nd ed., Pearson, New York.

Strehlow, R. (1985), *Combustion Fundamentals*, McGraw-Hill, New York.

Svehla, R. (1995), Transport coefficients for the NASA Lewis chemical equilibrium program, *NASA-TM-4647*.

Taylor H. (1931), The activation energy of adsorption processes, *J. Am. Chem. Soc.*, Vol. 53, pp. 578–597.
Taylor H., and N. Thon (1952), Kinetics of chemisorption, *J. Am. Chem. Soc.*, Vol. 74, pp. 4169–4173.
Tevault, D., L. Talley, and M. Lin (1980), Matrix isolation and laser diagnostic studies of catalytic oxidation of H_2 and D_2 on platinum surface, *J. Chem. Phys.*, Vol. 72, No. 5, pp. 3314–3319.
Thompson, M., M. Baker, A. Christie, and J. Tyson (1985), *Auger Electron Spectroscopy*, John Wiley & Sons, Chichester.
Warnatz, J. (1983), The mechanism of high temperature combustion of propane and butane, *Combust. Sci. Technol.*, Vol. 34, pp. 177–200.
Warnatz, J., M. Allendorf, R. Kee, and M. Coltrin (1994), A model of elementary chemistry and fluid mechanics in the combustion of hydrogen on platinum surfaces, *Combust. Flame*, Vol. 96, pp. 393–406.
Warnatz, J., U. Maas, and R. Dibble (2006), *Combustion*, Springer-Verlag, Berlin, Germany.
Weber, J., and K. Laidler (1951), Microwave spectroscopic investigation of the kinetics of the heterogeneous ammonia-deuterium exchange, *J. Chem. Phys.*, Vol. 19, pp. 1089–1096.
Williams, W., C. Marks, and L. Schmidt (1992), Steps in the reaction $H_2 + O_2 = H_2O$ on Pt: OH desorption at high temperatures, *J. Phys. Chem.*, Vol. 96, pp. 5922–5931.

Homework Problems

3.1 For the rate expression

$$-\frac{d[A]}{dt} = k[A]^2[C]$$

where $[C]$ represents the concentration of all of the constituents of the mixture. (1) What is the order of the reaction? (2) How would the rate of reaction depend on pressure?

3.2 The rate of a given reaction doubles for a 10 °C rise in temperature from T_o to $T_o + 10$. Develop an analytical expression for the activation energy E_a of the reaction.

3.3 Consider the first-order reaction

$$A \underset{k_b}{\overset{k_f}{\rightleftharpoons}} B$$

with $[A_o]$ and $[B_o]$ the initial concentrations of the two chemical species involved. If x represents the consumed portion of A,

(a) Show that

$$k_f + k_b = \frac{1}{t} \ln\left(\frac{[x_e]}{[x_e] - [x]}\right)$$

and

(b) Show that the equilibrium concentration $[x_e]$ of x is

$$[x_e] = \frac{k_f[A_o] - k_b[B_o]}{k_f + k_b}$$

3.4 Develop an expression for the half-life of a first-order reaction when the initial concentration $[C_o]$ decreases to 1/2 of its original value.

3.5 Suppose the reaction order of a chemical reacting system is m, that is,

$$\frac{d[C]}{dt} = -k[C]^m$$

where $[C]$ is the concentration of the reacting species. Show that the rate of change of mole fraction of reacting species, dX/dt, is proportional to $-kX^m\rho^{m-1}$. What is the pressure dependence of dX/dt? List your assumptions. Can we say that

$$\frac{d\epsilon}{dt} \propto -k\epsilon^m \rho^{m-1}$$

where ϵ is the reaction progress variable defined in Section 2.7?

3.6 Using dimensional analysis, express the concentration $[C_i]$ of a species i in a mixture of gases, in terms of the following parameters:
(a) density ρ_i and molecular mass M_i
(b) mixture density ρ, mass fraction Y_i, and molecular mass M_i
(c) mixture concentration $[C]$ and X_i
(d) pressure p, mass fraction Y_i, temperature T, and molecular mass M_i
(e) pressure p, temperature T, and mole fraction X_i

3.7 Express the reaction rate k_f of the following third-order reaction in terms of the consumed portion $[x]$ of the reactant concentration in the reaction

$$B + 2A \xrightarrow{k_f} C + D$$

3.8 An important chain-branching reaction in hydrogen combustion is the elementary reaction of hydrogen peroxide H_2O_2 resulting in two OH radicals,

$$H_2O_2 + M \longrightarrow OH + OH + M$$

If a hydrogen-fueled engine operates at $T = 1000$ K and $p = 4.0$ MPa, what is the time constant τ_c of the hydrogen peroxide reaction? The pre-exponential factor for this reaction is $A_p = 1.2 \times 10^{17}$ cm^3/mole-s, and the activation energy $E_a = 190$ kJ/mole.

3.9 A mixture of high-temperature gases flows over a horizontal graphite slab. Some of the gas phase species $H_2O_{(g)}$, $CO_{2(g)}$, and $H_{2(g)}$ in the mixture can react with graphite surface by heterogeneous reaction and cause thermochemical erosion of the solid (s). (1) Determine the heats of reaction of the three heterogeneous reactions given below at $T = 2000$ K. (2) Are these reactions exothermic or endothermic? Consider the following heterogeneous reactions:

(1) $H_2O_{(g)} + C_{(s)} \rightarrow CO_{(g)} + H_{2(g)}$

(2) $CO_{2(g)} + C_{(s)} \rightarrow 2CO_{(g)}$

(3) $H_{2(g)} + 2C_{(s)} \rightarrow C_2H_{2(g)}$

4

Reaction Mechanisms

4.1 Introduction

In this chapter, we introduce a number of reaction mechanisms for fuels used in engineering applications. We examine the elementary reaction steps involved in hydrogen, ammonia, and hydrocarbon combustion that are used to create reaction mechanisms. Methods for mechanism reduction, including sensitivity analysis, reaction flow analysis, and computational singular perturbation are introduced in this chapter.

For numerical simulations of reacting flows, a chemical reaction mechanism must be specified. This implies that all chemical species involved in all elementary reactions should be known along with their respective Arrhenius parameters before the computation can be performed. In the combustion community, the CHEMKIN format (Kee et al. 1989) has become a practical standard for representing reaction mechanisms in numerical computations.

Despite the increase in computational capability, there are many situations that require rapid and accurate computational methods for the solution of combustion problems. One way of meeting these needs is the reduction of the full chemical reaction mechanism to a simpler form. Often, the full set of equations is not necessary to get a reasonably close estimate, instead, smaller sets of equations that approximate the full solution are used. The reduced reaction mechanisms are validated using experimental measurements of flame speed, ignition delay, and concentration profiles of product species.

4.2 Multiple Reactions

Most practical reactions occur in multiple steps involving many elementary reactions. In this chapter, we will build reaction mechanisms of varying complexity to model these reaction systems. Generalizing Eq. (3.15), the set of the forward and backward chemical reactions for a system of L reactions and N species, we have

$$\sum_{i=1}^{N} \nu_{i,j} A_i \rightleftharpoons \sum_{i=1}^{N} \nu'_{i,j} A_i \tag{4.1}$$

where $i = 1, 2, \ldots N$ and $j = 1, 2, \ldots L$. The forward and backward molar stoichiometric coefficients of species i and reaction j are $\nu_{i,j}$ and $\nu'_{i,j}$, respectively. Generalizing Eq. (3.13), the net rate of reaction Ω_j for reaction j,

$$\dot{\Omega}_j = k_{f,j} \prod_{i=1}^{N} [A_i]^{\nu_{i,j}} - k_{b,j} \prod_{i=1}^{N} [A_i]^{\nu'_{i,j}} \tag{4.2}$$

Principles of Combustion, Third Edition. Allan T. Kirkpatrick and Kenneth K. Kuo.

Companion website: www.wiley.com/go/kirkpatrick/principlesofcombustion3e

and generalizing Eq. (3.14), the net rate of change of $\dot{\Omega}_i$ of species i is the sum of the rate of change of the L reactions,

$$\dot{\Omega}_i = \sum_{j=1}^{L} \frac{d[A_i]}{dt} = \sum_{j=1}^{L} (\nu'_{i,j} - \nu_{i,j})\dot{\Omega}_j \tag{4.3}$$

4.3 Steady-State Approximation

In most chemical reactions, highly reactive intermediate species such as radicals (e.g., OH, H, and O) are formed. Modeling of these reaction systems can be simplified by using a steady-state approximation for the concentration of these radicals. If the radicals are consumed as rapidly as they are formed, then the rates of formation and consumption are equal. This allows us to assume that the concentration of the species is low. The steady-state approximation is very useful for the determination of individual species in a complex reaction mechanism.

For example, consider the reaction system

$$\begin{aligned} A + B &\xrightarrow{k_1} C + D \\ D + E &\xrightarrow{k_2} C + A \end{aligned} \tag{4.4}$$

where D is the radical produced by the first reaction and consumed by the second reaction. The net production of species D is

$$\frac{d[D]}{dt} = k_1[A][B] - k_2[D][E] \tag{4.5}$$

Assuming a steady-state concentration of D,

$$\frac{d[D]}{dt} = 0 \tag{4.6}$$

and therefore

$$k_1[A][B] = k_2[D_e][E] \tag{4.7}$$

where D_e is the equilibrium concentration of species D. Solving Eq. (4.7) for D_e,

$$[D_e] = \frac{k_1[A][B]}{k_2[E]} \tag{4.8}$$

The algebraic equation (4.8) is much simpler to solve for the concentration of species D compared to the integration of the differential equation (4.5).

4.4 Partial Equilibrium

Under certain circumstances, a group of reactions will proceed rapidly and reach a quasi-equilibrium state. Concurrently, one or more reactions may proceed slowly. If the rate or rate coefficient of this slow reaction is to be determined and if the reaction contains a species difficult to measure such as a radical species, it is possible through a partial equilibrium assumption to express the unknown concentrations in terms of measurable concentrations in the reaction mechanism.

The partial equilibrium assumption is very much like the steady-state approximation discussed in Section 4.3. The difference is that the partial equilibrium assumption is concerned with a set of given reactions and steady-state approximation is concerned with a given species.

Both approximations are useful as they reduce the calculation of the concentration of a radical species to an algebraic equation.

Partial equilibrium comes about when the forward and backward rates of a reaction are very large and the contribution that a particular species makes to a given reaction can be accounted for with small differences in the forward and backward rates of those reactions in partial equilibrium. Many combustion mechanisms involve both fast and slow reactions. The fast reactions are chain propagating or chain branching, producing radicals, and the slow reactions are recombination reactions. The fast reactions can be modeled as if they are in equilibrium in the forward and back directions.

If the forward and backward rates of a reaction j are approximately the same and much faster than the net reaction rate $\dot{\Omega}_j$, then from Eq. (4.9),

$$\dot{\Omega}_j = k_{f,j}\prod_{i=1}^{N}[A_i]^{\nu_{i,j}} - k_{b,j}\prod_{i=1}^{N}[A_i]^{\nu'_{i,j}} \approx 0 \tag{4.9}$$

and

$$k_{f,j}\prod_{i=1}^{N}[A_i]^{\nu_{i,j}} \approx k_{b,j}\prod_{i=1}^{N}[A_i]^{\nu'_{i,j}} \tag{4.10}$$

For example, consider the two following reactions:

$$\begin{aligned} A + B_2 &\underset{k_{1b}}{\overset{k_1}{\rightleftharpoons}} AB + B \\ B + A_2 &\underset{k_{2b}}{\overset{k_2}{\rightleftharpoons}} AB + A \end{aligned} \tag{4.11}$$

where the radical [AB] is a difficult measurement. Assuming the forward and backward rates are the same,

$$\begin{aligned} k_1[A][B_2] &= k_{1b}[AB][B] \\ k_2[B][A_2] &= k_{2b}[AB][A] \end{aligned} \tag{4.12}$$

The equilibrium constants K_{c1} and K_{c2} for both reactions are

$$\begin{aligned} K_{c1} &= \frac{k_1}{k_{1b}} = \frac{[AB][B]}{[A][B_2]} \\ K_{c2} &= \frac{k_2}{k_{2b}} = \frac{[AB][A]}{[B][A_2]} \end{aligned} \tag{4.13}$$

Solving for the radical [AB],

$$[\mathrm{AB}] = \left(\mathrm{K}_{c1}\mathrm{K}_{c2}\,[\mathrm{A}_2]\,[\mathrm{B}_2]\right)^{1/2} \tag{4.14}$$

which is now an algebraic relationship between the concentrations.

4.5 Hydrogen H_2 Reaction Mechanisms

Even though there are only two elements involved, the chemical reaction mechanisms of H_2/O_2 systems are quite complex, involving various steps of chain reactions such as initiation, chain-propagating, and chain-branching reactions as well as chain-termination reactions. At room temperatures and ordinary pressures, the reaction is very slow. When the mixture is heated and subjected to a spark or an activating surface, however, the process can be very rapid, even explosive.

Hydrogen fuel combustion systems have many attractive features, including carbon-free combustion products and high specific impulse for rocket propulsion applications. As we shall see in Section 4.6, hydrogen/oxygen reactions also play an important role in the oxidation of hydrocarbons.

The global (g) reaction for H_2 oxidation to H_2O is

$$2H_2 + O_2 \rightarrow 2H_2O \tag{4.15}$$

Numerous H_2/O_2 reaction mechanisms have been evaluated to determine which reactions and which species are necessary to predict the behavior of the system with high accuracy. A reduced mechanism with 21 elementary reactions is given below in Eqs. (4.16)–(4.36) (Glassman and Yetter 2014). The oxidation initiation reactions are primarily hydrogen dissociation and H radical generation steps, since the dissociation energy of H_2 is less than that of O_2.

Also, the maximum flame temperature T_f of an H_2/air reaction is around 2400 K, while the maximum T_f of H_2/O_2 flame is around 3100 K. The difference is due to the presence of nitrogen in air, which also leads to the formation of nitrogen oxides (NO_x) as one of the combustion products.

The initiation reactions providing H radicals are

$$H_2 + M \rightarrow 2H + M \tag{4.16}$$

(requires 435 kJ/mol, more likely at high temperatures)

$$H_2 + O_2 \rightarrow H + HO_2 \tag{4.17}$$

(requires 210 kJ/mol, more likely at low temperatures)

Subsequent chain-branching steps generating radicals O, OH, and additional H radicals are

$$H + O_2 \leftrightarrows OH + O \tag{4.18}$$

$$O + H_2 \leftrightarrows H + OH \tag{4.19}$$

$$H_2 + OH \leftrightarrows H + H_2O \tag{4.20}$$

$$O + H_2O \leftrightarrows 2OH \tag{4.21}$$

Chain-terminating steps involving O, H, and OH radicals are the three-body recombination reactions:

$$H + H + M \rightarrow H_2 + M \tag{4.22}$$

$$O + O + M \rightarrow O_2 + M \tag{4.23}$$

$$H + O + M \rightarrow OH + M \tag{4.24}$$

$$H + OH + M \rightarrow H_2O + M \tag{4.25}$$

When the hydroperoxyl radical HO_2 and hydrogen peroxide H_2O_2 species are considered in this reaction mechanism, many other elementary reactions must be included. Some of these elementary reactions are given below:

$$H + O_2 + M \rightarrow HO_2 + M \tag{4.26}$$

$$HO_2 + H \rightarrow 2OH \tag{4.27}$$

$$HO_2 + H \rightarrow H_2O + O \tag{4.28}$$

$$HO_2 + O \rightarrow O_2 + OH \tag{4.29}$$

$$HO_2 + HO_2 \rightarrow H_2O_2 + O_2 \tag{4.30}$$

$$HO_2 + H_2 \rightarrow H_2O_2 + H \tag{4.31}$$

$$H_2O_2 + OH \rightarrow H_2O + HO_2 \tag{4.32}$$

$$H_2O_2 + H \rightarrow H_2 + HO_2 \tag{4.33}$$

$$H_2O_2 + H \rightarrow H_2O + OH \tag{4.34}$$

$$H_2O_2 + M \rightarrow 2OH + M \tag{4.35}$$

$$H_2O_2 + O_2 \rightarrow 2HO_2 \tag{4.36}$$

Table 4.1 Reduced H_2/O_2 Mechanism with 12 Reactions.

	Reaction		A_p	n	E_a/R_u
1f	$H + O_2 \rightarrow OH + O$		3.52×10^{16}	−0.7	8,590
1b	$OH + O \rightarrow H + O_2$		1.05×10^{14}	−0.313	132
2f	$H_2 + O \rightarrow OH + H$		5.06×10^{4}	2.67	3,165
2b	$OH + H \rightarrow H_2 + O$		2.94×10^{4}	2.64	2,430
3f	$H_2 + OH \rightarrow H_2O + H$		1.17×10^{9}	1.3	1,825
3b	$H_2O + H \rightarrow H_2 + OH$		1.42×10^{10}	1.18	9,379
4f	$H + O_2 + M \rightarrow HO_2 + M$	k_o	5.75×10^{19}	−1.4	0
		k_∞	4.65×10^{12}	0.44	0
5f	$HO_2 + H \rightarrow OH + OH$		7.08×10^{13}	0	148
6f	$HO_2 + H \rightarrow H_2 + O_2$		1.66×10^{13}	0	414
6b	$H_2 + O_2 \rightarrow HO_2 + H$		2.93×10^{12}	0.356	27,929
7f	$HO_2 + OH \rightarrow H_2O + O_2$		2.89×10^{13}	0	−250
			4.50×10^{14}	0	5,500
8f	$H + OH + M \rightarrow H_2O + M$		4.00×10^{22}	−2.0	0
8b	$H_2O + M \rightarrow H + OH + M$		2.02×10^{23}	−1.85	59,682
9f	$H + H + M \rightarrow H_2 + M$		1.30×10^{18}	−1.0	0
9b	$H_2 + M \rightarrow H + H + M$		5.41×10^{17}	−0.727	52,129
10f	$H_2O_2 + M \rightarrow OH + OH + M$	k_o	7.60×10^{30}	−4.20	25,703
		k_∞	2.63×10^{19}	−1.27	25,703
11f	$HO_2 + HO_2 \rightarrow H_2O_2 + O_2$		1.03×10^{14}	0	5,556
			1.94×10^{11}	0	−709
12f	$HO_2 + H_2 \rightarrow H_2O_2 + H$		7.80×10^{10}	0.61	12,045

Source: Sánchez and Williams (2014)/with permission of Elsevier.

Dougherty and Rabitz (1980) modeled H_2/O_2 combustion over a wide range of temperatures and pressures using 9 chemical species (H_2, O_2, H_2O, OH, O, H, HO_2, H_2O_2, and O_3) and 72 elementary reactions. Using sensitivity analysis, their calculations correctly predicted the explosion limit peninsula in the 650–850 K temperature range and the 1–103 Torr pressure range in addition to quantitatively rank ordering the importance of different reactions in the neighborhood of each explosion limit. They also considered a mechanism with 47 elementary reactions and 9 species for studying oscillatory combustion phenomena of H_2/O_2 systems.

Sánchez and Williams (2014) developed an H_2/O_2-reduced mechanism with 12 elementary reactions for autoignition, premixed and non-premixed combustion valid from low to very high pressures over the whole flammability range. Table 4.1 lists the rate coefficients in Arrhenius form $k = A_p T^n \exp(-E_a/R_u T)$, with the rate parameters in units of mol, s, cm^3, kJ, and K. Note that of the 12 reactions listed in Table 4.1, 6 are reversible, using forward (f) and backward (b) notations. For three-body collisions, the term k_o is the low-pressure rate constant, and term k_∞ is the high-pressure rate constant.

4.6 Ammonia NH_3 Reaction Mechanisms

Ammonia oxidation chemistry has been extensively studied due to potential as a carbon-free fuel. In an early study, Miller et al. (1983) proposed a reaction mechanism with 22 species and 98 elementary reactions. The agreement between theory and experiment was generally very good except for very rich flames. The oxidation pathway proposed by Miller et al. (1983) is presented in Figure 4.1. An ammonia oxidation mechanism at high pressure and intermediate temperature with 32 species and 204 reactions is presented in Song et al. (2016). A recent review of ammonia combustion research is contained in Kobayashi et al. (2019).

The combined use of NH_3 with conventional fuels such as H_2 or CH_4 has also been studied in order to improve shortcomings related to low flame speed and reactivity. The low flame speed of NH_3/air flames makes flame enhancement important for the successful application of ammonia as a fuel. Li et al. (2019) developed a reduced mechanism with 28 species and 213 reactions for ammonia/hydrogen combustion and a reduced mechanism with 51 species and 420 reactions for ammonia/hydrogen/methane combustion. An optimized ammonia combustion mechanism with 24 reactions has been developed by Bertolino et al. (2021) using an evolutionary algorithm.

Figure 4.1 shows the principal reactions through which ammonia NH_3 is oxidized to NO and then the NO is converted to molecular nitrogen N_2 in NH_3/O_2 flames. First, ammonia is converted to NH_2 by hydrogen abstraction. The primary abstraction reaction is

$$NH_3 + OH = NH_2 + H_2O \tag{4.37}$$

The NH_3 reaction with an O atom under fuel-lean conditions and the NH_3 reaction with an H atom under fuel-rich conditions are both significant for NH_2 formation:

$$NH_3 + O = NH_2 + OH \qquad \text{(lean)} \tag{4.38}$$

$$NH_3 + H = NH_2 + H_2 \qquad \text{(rich)} \tag{4.39}$$

Successive smaller amine NH radicals are formed by reaction with H atoms and to a lesser extent by reaction with OH,

$$NH_i + H\,(\text{or OH}) = NH_{i-1} + H_2(\text{or } H_2O) \qquad i = 1, 2 \tag{4.40}$$

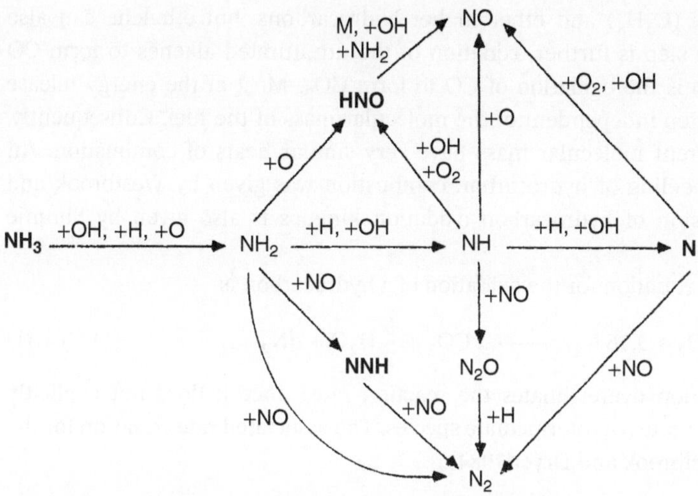

Figure 4.1 Reaction path diagram showing the major routes for the oxidation of NH_3. *Source*: Miller et al. 1983/with permission of Taylor & Francis Group.

Each NH_i free radical can undergo subsequent reaction by one of the two mechanisms:

- Oxidation leading to NO formation.
- Reaction with NO leading to the formation of molecular nitrogen N_2.

Whether N, NH, or NH_2 is the primary radical in determining NO/N_2, product distribution depends on the equivalence ratio (ϕ). For very lean flames, NH_2 is the critical amine radical. As ϕ increases, the greater availability of H atoms results in a shift in the critical radical species from NH_2 to NH to N. Normally, all three radicals play the same role. However, in rich flames, the N atom is more dominant. Amino NH_2 radicals are highly reactive and consequently short-lived, like most radicals; however, they form an important part of nitrogen chemistry. Note that there are reactions other than those shown in Figure 4.1 that form N_2 but do not involve NO. Additional coverage of NO formation mechanisms is contained in Chapter 10.

4.7 Hydrocarbon Reaction Mechanisms

Hydrocarbon oxidation reactions are numerous and complex, as they depend on temperature, pressure, fuel type, and the amount of oxygen introduced into the combustion process. For example, the chain reaction pathways for the oxidation of a hydrocarbon fuel such as paraffin, C_{nH2n+2}, a major component of gasoline, can include at least 10,000 different reactions.

At the molecular level, at low temperatures ($T \approx 600$–1000 K), hydrocarbon fuels are oxidized by O_2 addition in a degenerate chain-branching reaction. Combustion at higher temperatures ($T \geq 1000$ K) involves decomposition into small C_2 and C_1 hydrocarbons which are then oxidized. These higher temperature ($T \geq 1000$ K) hydrocarbon reactions are generally grouped into three distinct phases. The first step in the combustion of a hydrocarbon molecule is breaking up carbon–carbon (dissociation) or carbon–hydrogen (hydrogen abstraction) bonds in the fuel molecule, forming alkenes, i.e., hydrocarbons with double carbon bonds and hydrogen. This step is endothermic, requiring about 400 kJ/mol. In many cases, these intermediate species can be a fuel source themselves. For example, ethylene (C_2H_4) is an important intermediate in the combustion of propane (C_3H_8) and other higher hydrocarbons, but ethylene can also be a primary fuel. The second step is further oxidation of the unsaturated alkenes to form CO and H. The third and last step is the oxidation of CO to form CO_2. Most of the energy release occurs during the last step, a step independent of the molecular mass of the fuel. Consequently, hydrocarbon paraffins of different molecular mass have very similar heats of combustion. An excellent review of kinetic modeling of hydrocarbon combustion was given by Westbrook and Dryer (1984). Detailed discussion of hydrocarbon oxidation kinetics is also given by Simmie (2003).

A single-step global reaction equation for the oxidation of a hydrocarbon is

$$C_aH_bO_c + (a + \frac{b}{4} - \frac{c}{2})(O_2 + 3.76\,N_2) \longrightarrow a\,CO_2 + \frac{b}{2}\,H_2O + dN_2 \tag{4.41}$$

This single-step global reaction overestimates the reaction rate, since it does not explicitly account for the specific reaction rates of intermediate species. The associated rate equation for the above one-step reaction is (Westbrook and Dryer 1984):

$$\frac{d[C_aH_bO_c]}{dt} = -A_p[C_aH_bO_c]^m[O_2]^n \exp(-E_a/R_uT) \tag{4.42}$$

Using experimental data and parametric analysis to match maximum flame speed and the equivalence ratio at which that maximum occurs, Westbrook and Dryer (1984) obtained values for the global parameters A_p, E_a, m, and n for various hydrocarbon fuels, given in Table 4.2.

Detailed reaction mechanisms can be developed systematically, beginning with the simplest C_1 and C_2 species and reactions that are common subelements in the combustion of more complex species. Additional species and reactions can then be incorporated in order of increasing complexity. Hierarchical structures indicating the overall interrelationships between oxidation mechanisms can be developed for simple hydrocarbon fuels (Westbrook and Dryer 1984). The hierarchical structure of the reaction mechanism of aliphatic (nonaromatic) hydrocarbon combustion is shown in Figure 4.2. The oxygen–hydrogen reaction mechanisms contain the chain-branching steps that produce the O, H, and OH radicals for attacking hydrocarbons. (In what follows an equal sign in the reaction equations implies that both forward and backward reactions need to be considered.)

Example 4.1 *Hydrocarbon Reaction Rates*

Compare the one-step stoichiometric reaction rates $\dot{\Omega}_F$ (mol/cm^3-s) of (a) methane CH_4 and (b) iso-octane C_8H_{18} in air at $T = 2000$ K and $p = 1$ atm.

Table 4.2 Pre-exponential Factors, Activation Energies, and Exponents of Some Hydrocarbon Fuels.

Formula	Name	A_p (1-step)	E_a/R_u (K)	m	n
CH_4	Methane	8.3×10^5	15,098	−0.3	1.3
C_3H_8	Propane	8.6×10^{11}	15,098	0.1	1.65
C_7H_{16}	Heptane	5.1×10^{11}	15,098	0.25	1.5
C_8H_{18}	Octane	4.6×10^{11}	15,098	0.25	1.5
CH_4O	Methanol	3.2×10^{12}	15,098	0.25	1.5
C_2H_6O	Ethanol	1.5×10^{12}	15,098	0.25	1.6

Source: Adapted from Westbrook and Dryer (1984). The units of A_p are consistent with a reaction rate of mol/cm^3-s.

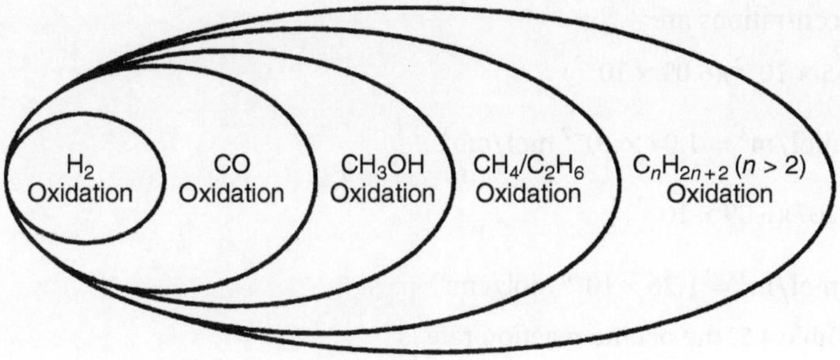

Figure 4.2 Reaction structure of hydrocarbon combustion.

Solution:

The overall mixture concentration is

$$[M] = \frac{P}{R_u T} = \frac{101.3}{(8.314)(2000)} = 6.09 \times 10^{-3}\ \text{kmol/m}^3$$

(a) The methane fuel (F) reaction is

$$1\,CH_4 + 2(O_2 + 3.76\,N_2) \longrightarrow 1\,CO_2 + 2\,H_2O + 9.52\,N_2$$

The mole fractions of reactants methane and oxygen are

$$X_F = \frac{n_F}{n} = \frac{1}{10.52} = 0.095$$
$$X_{O_2} = \frac{n_{O_2}}{n} = \frac{2}{10.52} = 0.190$$

The remaining concentration is nitrogen. The methane and oxygen concentrations are

$$[F] = X_F[M] = (0.095)(6.09 \times 10^{-3}) = 5.788 \times 10^{-4}\ \text{kmol/m}^3 = 5.788 \times 10^{-7}\ \text{mol/cm}^3$$
$$[O_2] = X_{O_2}[M] = (0.190)(6.09 \times 10^{-3}) = 1.158 \times 10^{-3}\ \text{kmol/m}^3 = 1.158 \times 10^{-6}\ \text{mol/cm}^3$$

Using the reaction data in Table 4.2, the methane reaction rate is

$$\begin{aligned}\dot{\Omega}_F = \frac{[CH_4]}{dt} &= -A_p \exp(-E_a/R_u T)[CH_4]^m[O_2]^n \\ &= -(8.3 \times 10^5)\exp(-15098/2000)[5.78 \times 10^{-7}]^{-0.3}[1.158 \times 10^{-6}]^{1.3} \\ &= -6.235 \times 10^{-4}\ \text{mol/cm}^3\text{-s}\end{aligned}$$

(b) The octane fuel (F) reaction is

$$1\,C_8H_{18} + 12.5(O_2 + 3.76\,N_2) \longrightarrow 8\,CO_2 + 9\,H_2O + 47.0\,N_2$$

The mole fractions of reactants methane and oxygen are

$$X_F = \frac{n_F}{n} = \frac{1}{60.5} = 1.65 \times 10^{-2}$$

$$X_{O_2} = \frac{n_{O_2}}{n} = \frac{12.5}{60.5} = 0.207$$

The octane and oxygen concentrations are

$$[\mathrm{F}] = X_F[M] = (1.65\times10^{-2})(6.09\times10^{-3})$$

$$= 1.04\times10^{-4}\,\mathrm{kmol/m^3} = 1.04\times10^{-7}\,\mathrm{mol/cm^3}$$

$$[\mathrm{O_2}] = X_{O_2}[M] = (0.207)(6.09\times10^{-3})$$

$$= 1.26\times10^{-3}\,\mathrm{kmol/m^3} = 1.26\times10^{-6}\,\mathrm{mol/cm^3}$$

Using the reaction data in Table 4.2, the octane reaction rate is

$$\dot{\Omega}_F = \frac{[C_8H_{18}]}{dt} = -A_p\exp(-E_a/R_uT)[[C_8H_{18}]]^m[\mathrm{O_2}]^n$$

$$= -(4.6\times10^{11})\exp(-15098/2000)[1.00\times10^{-7}]^{0.25}[1.26\times10^{-6}]^{1.5}$$

$$= -6.15\times10^{-3}\,\mathrm{mol/cm^3\text{-}s}$$

Comment: The methane combustion rate is an order of magnitude slower when compared to the octane combustion rate since the methane molecule is composed only of C—H bonds, which have greater strength than the C—C bonds in the octane molecule, and thus requires more energy to break. Also, the reaction rate can be quite sensitive to changes in the equivalence ratio due to the resulting changes in flame temperature.

4.7.1 Methane CH_4 Combustion

As discussed previously in Chapter 1, methane CH_4 is a very important practical fuel, constituting approximately 90% of the composition of natural gas. It is also a significant by-product in many industrial processes and is produced during the combustion of most hydrocarbons.

The global reaction for stoichiometric CH_4 oxidation is:

$$CH_4 + 2\,O_2 \rightarrow CO_2 + 2\,H_2O \tag{4.43}$$

Methane chemical kinetics have been widely studied by numerous research groups. A well-known detailed mechanism is GRI-Mech, initially supported by the Gas Research Institute. The current version (Smith et al. 1995), is available online, contains 325 elementary reactions and 53 species and considers both high- and low-temperature reactions. Many of these reactions are also part of the H_2 and CO oxidation mechanisms. An outline of a detailed reaction pathway for methane oxidation is:

Methane $CH_4 \Rightarrow$ *Methyl radical* $CH_3 \Rightarrow$ *Formaldehyde* $CH_2O \Rightarrow$ *Formyl radical* HCO $\Rightarrow$ *Carbon Monoxide* CO $\Rightarrow$ *Carbon Dioxide* CO_2.

The detailed methane reaction mechanism is as follows. At high temperatures (> 1600 K), the thermal decomposition of CH_4 produces the methyl radical CH_3,

$$CH_4 + M = CH_3 + H + M \tag{4.44}$$

An initiation reaction at low temperatures (< 1600 K) under oxidation conditions is

$$CH_4 + O_2 = CH_3 + HO_2 \tag{4.45}$$

which has been shown to be much less important than reaction (4.44). Additional CH_4 reactions with radicals resulting in the production of CH_3 are:

$$CH_4 + H = CH_3 + H_2 \tag{4.46}$$

$$CH_4 + OH = CH_3 + H_2O \tag{4.47}$$

$$CH_4 + O = CH_3 + OH \tag{4.48}$$

$$CH_4 + HO_2 = CH_3 + H_2O_2 \tag{4.49}$$

The methyl radical CH_3 reacts directly with O_2 according to the three reactions listed below:

$$CH_3 + O_2 = CH_3O + O \tag{4.50}$$

$$CH_3 + O_2 = HCO + H_2O \tag{4.51}$$

$$CH_3 + O_2 = CH_2O + OH \tag{4.52}$$

Lately, it has been verified that the major CH_3 oxidation step takes place according to reaction (4.50) as was initially suggested by Brabbs and Brokaw (1974). Bowman (1974) pointed out that reaction (4.52) was probably not an elementary reaction, but rather a convenient model for simulating CH_3 oxidation. Computed model results were found to be very sensitive to variations in the rate of reaction (4.52), with a rather large rate and relatively low activation energy in order to reproduce the observed combustion rate.

During the pyrolysis and oxidation of methane, radical recombination reactions produce significant amounts of larger hydrocarbons containing two or more carbon atoms. These recombination reactions are

$$CH_3 + CH_3 = C_2H_6 \tag{4.53}$$

$$CH_3 + CH_3 = C_2H_5 + H \tag{4.54}$$

$$CH_3 + CH_3 = C_2H_4 + H_2 \tag{4.55}$$

The methyl radical CH_3 reacts with other radicals to form formaldehyde CH_2O and other products,

$$CH_3 + O = CH_2O + H \tag{4.56}$$

$$CH_3 + OH = CH_2O + H_2 \tag{4.57}$$

$$CH_3 + OH = CH_3O + H \tag{4.58}$$

$$CH_3 + CH_2O = CH_4 + HCO \tag{4.59}$$

$$CH_3 + HCO = CH_4 + CO \tag{4.60}$$

$$CH_3 + HO_2 = CH_3O + OH \tag{4.61}$$

At very high temperatures, CH_3 may also dissociate

$$CH_3 + M = CH_2 + H + M \tag{4.62}$$

The methoxy radicals (CH_3O) generated from reactions (4.58) and (4.61) react to form formaldehyde CH_2O

$$CH_3O + M = CH_2O + H + M \tag{4.63}$$

$$CH_3O + O_2 = CH_2O + HO_2 \tag{4.64}$$

$$CH_3O + H = CH_2O + H_2 \tag{4.65}$$

$$CH_3O + O = CH_2O + OH \tag{4.66}$$

$$CH_3O + OH = CH_2O + H_2O \tag{4.67}$$

4.7.2 Formaldehyde CH_2O Combustion

Formaldehyde is an intermediate species in the combustion of most hydrocarbon fuels. The two possibilities for decomposition are

$$CH_2O + M = HCO + H + M \tag{4.68}$$

$$CH_2O + M = CO + H_2 + M \tag{4.69}$$

The chain-branching characteristics of these two paths are vastly different. Reaction (4.68) produces one H atom directly, and, in most environments, the formyl radical HCO rapidly decomposes further into CO and H. Thus, the net difference between reactions (4.68) and (4.69) is H + H as opposed to H_2. Reaction (4.68) is a very active chain-branching sequence in oxidizing environments, while reaction (4.69) is effectively a termination sequence.

Formaldehyde is consumed primarily by reactions with OH, H, and O radicals to produce the formyl radical HCO:

$$CH_2O + OH = HCO + H_2O \tag{4.70}$$

$$CH_2O + H = HCO + H_2 \tag{4.71}$$

$$CH_2O + O = HCO + OH \tag{4.72}$$

The relative importance of reactions (4.70)–(4.72) depends on the particular reaction environment. In fuel-rich mixtures, reaction (4.71) dominates; in lean and stoichiometric condition, reactions (4.70) and (4.72) dominate. Another reaction that is less important but can contribute to CH_2O consumption in some cases is

$$CH_2O + HO_2 = HCO + H_2O_2 \tag{4.73}$$

The formyl radicals (HCO) produced by reaction of CH_2O and other species are in turn consumed in a variety of elementary reactions producing CO,

$$HCO + M = H + CO + M \tag{4.74}$$

$$HCO + O_2 = CO + HO_2 \tag{4.75}$$

This pair of reactions for a hydrocarbon radical illustrates a pattern that will be repeated for many larger radicals; a reaction with O_2 competes with a thermal decomposition reaction, and their relative rates determine much of the behavior of the overall mechanism. Typically, the activation energy of the decomposition reaction (4.74) is substantially higher than that of the reaction with O_2.

The formyl radical also reacts with other radicals, with reaction rates that are close to kinetic collision rates due to negligible activation energies, to produce CO. These reactions include

$$HCO + OH = CO + H_2O \tag{4.76}$$

$$HCO + H = CO + H_2 \tag{4.77}$$

$$HCO + O = CO + OH \tag{4.78}$$

4.7.3 Carbon Monoxide CO Oxidation

The oxidation of CO to CO_2 is the final pathway in hydrocarbon combustion. If water H_2O is present in the reaction as a catalyst, the CO oxidation mechanism proceeds through the following reaction sequence (Glassman and Yetter 2014),

$$CO + O_2 \rightarrow CO_2 + O \tag{4.79}$$

$$O + H_2O \rightarrow OH + OH \tag{4.80}$$

$$CO + OH \rightarrow CO_2 + H \tag{4.81}$$

$$H + O_2 \rightarrow OH + O \tag{4.82}$$

All high-temperature hydrocarbon oxidation eventually involves H_2 and CO oxidation kinetics, and most of the CO_2 that is produced results from reaction (4.81). The oxidizing reaction (4.79) is slow but does not generate significant amounts of CO_2, but its O radical initiates the chain reaction sequence through the chain-branching reaction (4.80). Reaction (4.81) is a chain-propagating step, with the oxidation of CO to CO_2 and the production of an H radical, which in chain-branching reaction (4.82) forms the radicals OH and O.

Because the reaction of CO with the hydroxyl radical OH, reaction (4.81), is the main CO oxidation reaction, the rate of CO oxidation depends very much on the availability of OH radicals. However, the rate of reaction of (4.81) is considerably lower than the rates of reactions between OH radicals and typical hydrocarbon species. As a result, the presence of most hydrocarbon species will effectively inhibit the oxidation reaction of CO. During the oxidation of hydrocarbons, CO is produced in substantial amounts, but the subsequent oxidation of CO to CO_2 is usually retarded until *after* the original hydrocarbon and the intermediate hydrocarbon species have all been consumed. Only then does the OH concentration rise to higher levels, rapidly converting the CO to CO_2.

If hydrogen is the catalyst instead of water, the following reactions should be included:

$$O + H_2 \rightarrow OH + H \tag{4.83}$$

$$OH + H_2 \rightarrow H_2O + H \tag{4.84}$$

If HO_2 is present, the following reactions (4.85) and (4.86) provide an additional route for oxidation of CO to CO_2. Reaction (4.86) is rarely as important as reaction (4.81), although at very high pressures or in the initial stages of hydrocarbon oxidation, the high HO_2 concentrations can make it competitive.

$$H + O_2 + M \rightarrow HO_2 + M \tag{4.85}$$

$$CO + HO_2 \rightarrow CO_2 + OH \tag{4.86}$$

The mole fraction profiles of the major species O_2, H_2O, CO_2, and CO for the stoichiometric combustion of methane in air at $T = 1600$ K at pressures of $p = 1$ atm and $p = 10$ atm using the GRI 3.0 mechanism are shown in Figure 4.3. At 1 atm, the major reactants, CH_4 and O_2, are consumed in about 0.6 ms. H_2O is formed as the result of the consumption of the major species. The concentration of the intermediate species CO peaks around 0.65 ms when the CH_4 is completely consumed. The CO molecules are rapidly oxidized to form CO_2. The oxidation process at 10 atm is similar to that at 1 atm but at a rate about five times faster.

The mole fraction profiles of other intermediate species, namely CH_3, CH_2O, CHO, and the radical species, H, OH, and O, are presented in Figure 4.4. The thermal decomposition of CH_4

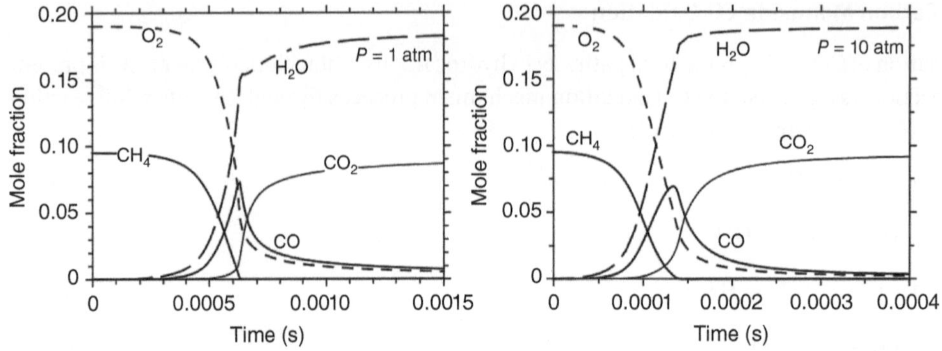

Figure 4.3 Mole fraction profiles of major species versus time for stoichiometric combustion of methane in air. (Adapted from McAllister et al. 2011.)

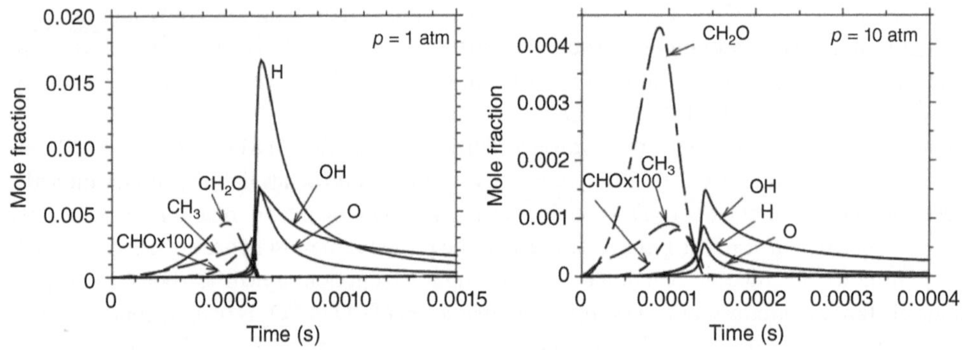

Figure 4.4 Mole fraction profiles of intermediate and radical species versus time for stoichiometric combustion of methane in air. (Adapted from McAllister et al. 2011.)

into CH_3 is accomplished through Eq. (4.44). The CH_3 is consumed by about 0.6 ms. Other intermediate species, CH_2O and CHO, are also formed at this time. The O, H, and OH radicals are formed when the CH_4 has been consumed. When pressure increases to 10 atm, the concentrations of the intermediate and radical species decrease, with the exception of the formaldehyde CH_2O concentration.

4.8 Additional Hydrocarbon Reaction Mechanisms

4.8.1 Ethane C_2H_6 Reactions

The chief use of ethane is as feedstock for ethylene production. Ethane was first synthesized in 1834 by the English scientist Michael Faraday (1791–1867), who made significant contributions to the fields of electrochemistry and electromagnetism. After methane, ethane is the second-largest component of natural gas. It can be used as a refrigerant in cryogenic refrigeration systems. Ethane can be viewed as two methyl CH_3 groups joined together.

The ethane decomposition reaction (4.87) is at neither its high- nor low-pressure limit in most combustion regimes; therefore, some treatment or estimate of its fall-off behavior (see Section 3.6) is usually necessary.

$$C_2H_6 = CH_3 + CH_3 \tag{4.87}$$

The following reaction with oxygen has been considered by chemists for low temperature and in oxidizing environments:

$$C_2H_6 + O_2 = C_2H_5 + HO_2 \tag{4.88}$$

Ethane can also react relatively fast with CH_3 to produce reactive ethyl radicals by

$$C_2H_6 + CH_3 = C_2H_5 + CH_4 \tag{4.89}$$

The other key reactions involving C_2H_6 include

$$C_2H_6 + H = C_2H_5 + H_2 \tag{4.90}$$

$$C_2H_6 + O = C_2H_5 + OH \tag{4.91}$$

$$C_2H_6 + OH = C_2H_5 + H_2O \tag{4.92}$$

The rates of these reactions exhibit substantial non-Arrhenius behavior. Two principal reactions involving ethyl radical (C_2H_5) are

$$C_2H_5 + M = C_2H_4 + H + M \tag{4.93}$$

$$C_2H_5 + O_2 = C_2H_4 + HO_2 \tag{4.94}$$

In shock-tube conditions, reaction (4.93) can be assumed to be second order, while other studies at lower temperature indicate the first-order behavior. In most typical combustion situations, this reaction is in the fall-off region. Other reactions of the ethyl radicals are

$$C_2H_5 + O = CH_3CHO + H \tag{4.95}$$

$$C_2H_5 + O = CH_2O + CH_3 \tag{4.96}$$

$$C_2H_5 + H = CH_3 + CH_3 \tag{4.97}$$

When ethane is the primary fuel, another key feature of the oxidation mechanism is the recombination of C_2 and C_1 radicals to produce C_3 and C_4 species, as shown by the following reactions:

$$C_2H_5 + C_2H_5 = C_4H_{10} \tag{4.98}$$

$$C_2H_5 + CH_3 = C_3H_8 \tag{4.99}$$

$$CH_3 + C_2H_3 = C_3H_6 \tag{4.100}$$

$$C_2H_5 + C_2H_3 = C_4H_8 \tag{4.101}$$

$$C_2H_3 + C_2H_3 = C_4H_6 \tag{4.102}$$

4.8.2 Ethylene C_2H_4 Reactions

Ethylene is a primary fuel and is also produced in large amounts during the combustion of CH_4, C_2H_6, and other higher hydrocarbons. Ethylene is widely used in the chemical industry for the manufacture of polyethylene, a widely used plastic containing polymer chains of ethylene units in various chain lengths.

Three initiation reactions can be important, including

$$C_2H_4 + M = C_2H_2 + H_2 + M \tag{4.103}$$

$$C_2H_4 + M = C_2H_3 + H + M \tag{4.104}$$

$$C_2H_4 + C_2H_4 = C_2H_3 + C_2H_5 \tag{4.105}$$

Reaction (4.103) is the fastest in most cases, while (4.105) rarely competes except in conditions in which C_2H_4 concentrations are very high (Westbrook and Dryer 1984). The two important reactions of the vinyl radical (C_2H_3) are

$$C_2H_3 + M = C_2H_2 + H + M \tag{4.106}$$

$$C_2H_3 + O_2 = C_2H_2 + HO_2 \tag{4.107}$$

Reactions of radicals with ethylene fall into two general classes. One consists of H atom abstraction (removal of an H atom from a molecule):

$$C_2H_4 + H = C_2H_3 + H_2 \tag{4.108}$$

$$C_2H_4 + OH = C_2H_3 + H_2O \tag{4.109}$$

The H atom abstraction reactions at room temperature are often negligibly slow, but under high-temperature conditions, the H atom abstraction reactions can be very important.

The other family of reactions (4.110)–(4.112), involving the formation of an activated complex followed by rearrangement and fragmentation, is given below. They generally have lower activation energies than those of H atom abstraction reactions.

$$C_2H_4 + O = CH_3 + HCO \tag{4.110}$$

$$C_2H_4 + O = CH_2O + CH_2 \tag{4.111}$$

$$C_2H_4 + OH = CH_3 + CH_2O \tag{4.112}$$

Additional reactions of vinyl radicals (C_2H_3) include

$$C_2H_3 + H = C_2H_2 + H_2 \tag{4.113}$$

$$C_2H_3 + O = CH_2CO + H \tag{4.114}$$

$$C_2H_3 + OH = C_2H_2 + H_2O \tag{4.115}$$

A reaction as a possible step leading to soot formation produces 1,3 butadiene (C_4H_6).

$$C_2H_3 + C_2H_4 = C_4H_6 + H \tag{4.116}$$

4.8.3 Acetylene C_2H_2 Reactions

Acetylene is widely used as a fuel and is also an important contributor to the formation and growth of soot. Acetylene was discovered in 1836 by the Irish scientist Edmund Davy (1785–1857). Acetylene pyrolysis and oxidation reactions are part of the reaction mechanisms for most other hydrocarbon fuels, particularly in fuel-rich conditions. As an alkyne, acetylene is unsaturated because its two carbon atoms are bonded together in a triple bond.

Important initiation reactions include

$$C_2H_2 + M = C_2H + H + M \tag{4.117}$$

$$C_2H_2 + C_2H_2 = C_4H_3 + H \tag{4.118}$$

$$C_2H_2 + O_2 = HCCO + OH \tag{4.119}$$

$$C_2H_2 + O_2 = HCO + HCO \tag{4.120}$$

In acetylene pyrolysis, reaction (4.117) dominates under dilute conditions, while reaction (4.118) is more important at high fuel concentrations. Unlike the other hydrocarbons already discussed, the pyrolysis of acetylene is *exothermic*, so decomposition flames and detonation processes are possible even in pure C_2H_2. Under oxidation conditions, the reaction of C_2H_2 with O_2 is extremely important, but there is not yet positive identification of the product distribution.

The following hydrogen atom abstraction reaction competes with the recombination reaction (4.106) only at high temperature:

$$C_2H_2 + H = C_2H + H_2 \tag{4.121}$$

Reactions between C_2H_2 and O atoms include (4.122)–(4.124) with reaction (4.123) being predominant under most conditions.

$$C_2H_2 + O = C_2H + OH \tag{4.122}$$

$$C_2H_2 + O = CH_2 + CO \tag{4.123}$$

$$C_2H_2 + O = HCCO + H \tag{4.124}$$

Acetylene reacts with OH radicals by the following reactions:

$$C_2H_2 + OH = CH_2CO + H \tag{4.125}$$

$$C_2H_2 + OH = C_2H + H_2O \tag{4.126}$$

$$C_2H_2 + OH = CH_3 + CO \tag{4.127}$$

An additional path proposed by Miller et al. (1982) is

$$C_2H_2 + OH = C_2H_2OH \tag{4.128}$$

This reaction is then followed by several reactions involving C_2H_2OH species and other radicals or molecules that produce ketene CH_2CO:

$$C_2H_2OH + H = CH_2CO + H_2 \tag{4.129}$$

$$C_2H_2OH + O = CH_2CO + OH \tag{4.130}$$

$$C_2H_2OH + OH = CH_2CO + H_2O \tag{4.131}$$

$$C_2H_2OH + O_2 = CH_2CO + HO_2 \tag{4.132}$$

$$C_2H_2OH + M = CH_2CO + H + M \tag{4.133}$$

An alternative path for acetylene oxidation under rich conditions has been examined recently, proceeding by means of

$$C_2H_2 + C_2H = C_4H_2 + H \tag{4.134}$$

$$C_4H_2 + OH = C_3H_2 + HCO \tag{4.135}$$

followed by reactions between C_3H_2 and radical species to give a variety of other products. Warnatz et al. (1982) attributed as much as 35% of the fuel consumption in rich acetylene flames to this sequence, which avoids ketene CH_2CO formation entirely.

An additional acetylene pyrolysis sub-mechanism has been proposed as a possible factor in soot formation. The formation processes of polyacetylene species are

$$C_4H_3 + M = C_4H_2 + H + M \tag{4.136}$$

$$C_2H_2 + C_2H + M = C_4H_2 + H + M \tag{4.137}$$

$$C_4H_2 + M = C_4H + H + M \tag{4.138}$$

4.8.4 Ketene CH_2CO Reactions

Due to their cumulated double bonds, ketenes are very reactive and cannot be stored. A summary of ketene reactions, which have been used in model simulations, includes

$$CH_2CO + M = CH_2 + CO + M \tag{4.139}$$

$$CH_2CO + OH = CH_2O + HCO \tag{4.140}$$

$$CH_2CO + OH = HCCO + H_2O \tag{4.141}$$

$$CH_2CO + O = HCCO + OH \tag{4.142}$$

$$CH_2CO + O = HCO + HCO \tag{4.143}$$

$$CH_2CO + H = HCCO + H_2 \tag{4.144}$$

$$CH_2CO + H = CH_3 + CO \tag{4.145}$$

Rate expressions for reactions (4.139) and (4.140) have been obtained from experimental studies, but rate expressions for the other reactions represent either estimates or results of modeling studies.

The ketyl radical HCCO produced from ketene, directly from acetylene by reaction (4.124), or by reaction (4.164) between C_2H and O_2, can be consumed by the following set of reactions suggested by Miller et al. (1983):

$$HCCO + O_2 = CO + CO + OH \tag{4.146}$$

$$HCCO + O = CO + CO + H \tag{4.147}$$

$$HCCO + H = CH_2 + CO \tag{4.148}$$

$$HCCO + OH = HCO + H + CO \tag{4.149}$$

$$HCCO + CH_2 = C_2H_3 + CO \tag{4.150}$$

However, there is very little information available on these reactions at high temperatures near flame temperatures T_f. As a result, most detailed mechanisms avoid modeling the reactions involving the ketyl radical by including only reactions (4.139), (4.143), and (4.145).

Other species produced during the oxidation of ketene and acetylene include C_2H, CH_2, and CH. The reactions of methylene radical (CH_2) near room temperature include

$$CH_2 + O_2 = CO_2 + H_2 \tag{4.151}$$

$$CH_2 + O_2 = CO_2 + H + H \tag{4.152}$$

$$CH_2 + O_2 = CO + H_2O \tag{4.153}$$

$$CH_2 + O_2 = CO + OH + H \tag{4.154}$$

$$CH_2 + O_2 = HCO + OH \tag{4.155}$$

$$CH_2 + O = CH + OH \tag{4.156}$$

$$CH_2 + O = CO + H + H \tag{4.157}$$

$$CH_2 + O = CO + H_2 \tag{4.158}$$

$$CH_2 + OH = CH + H_2O \tag{4.159}$$

$$CH_2 + H = CH + H_2 \tag{4.160}$$

$$CH_2 + CH_2 = C_2H_3 + H \tag{4.161}$$

$$CH_2 + CH_2 = C_2H_2 + H_2 \tag{4.162}$$

$$CH_2 + C_2H_3 = CH_3 + C_2H_2 \tag{4.163}$$

Reaction (4.155) followed by the formyl radical HCO reaction (4.74) results in the same product distribution as reaction (4.154) alone.

Reactions of C_2H with alkanes and with O_2 have been examined at low temperatures. Reactions of C_2H and CH include

$$C_2H + O_2 = HCCO + O \tag{4.164}$$

$$C_2H + O_2 = HCO + CO \tag{4.165}$$

$$C_2H + O = CO + CH \tag{4.166}$$

$$C_2H + C_2H_3 = C_2H_2 + C_2H_2 \tag{4.167}$$

$$CH + O_2 = HCO + O \tag{4.168}$$

$$CH + O_2 = CO + OH \tag{4.169}$$

4.8.5 Methanol CH_3OH Combustion

Methanol consists of a methyl CH_3 group linked to a hydroxyl OH group. The hydroxyl radical increases the hydrocarbon polarity, so that it is miscible in water and has a relatively low vapor pressure. Since oxygen is part of the chemical structure, less air is required from complete combustion. The combustion of alcohols begins with hydrogen H abstraction by H, O, and OH radicals. In general, the oxidation path for methanol proceeds in a sequential manner through

$$CH_3OH \Rightarrow \text{Methyloxy } CH_2OH \Rightarrow \text{Formaldehyde } CH_2O \Rightarrow \text{Formyl } HCO \Rightarrow$$
$$\Rightarrow \text{Carbon monoxide } CO \Rightarrow \text{Carbon dioxide } CO_2$$

Possible initial reactions include

$$CH_3OH + M = CH_3 + OH + M \tag{4.170}$$

$$CH_3OH + M = CH_3O + H + M \tag{4.171}$$

$$CH_3OH + M = CH_2OH + H + M \tag{4.172}$$

$$CH_3OH + O_2 = CH_2OH + HO_2 \tag{4.173}$$

However, based on the relative bond energies, reactions (4.171) and (4.172) can be rejected from the reaction mechanism. The addition of these two reactions to a mechanism already containing reaction (4.170) had no appreciable effect on computed induction times under shock-tube oxidation conditions (Westbrook and Dryer 1984). Reaction (4.173) provides a much different chain initiation function than do the alternate initiation reactions, since both the methoxy CH_3O and hydroperoxyl HO_2 radicals quickly decompose, producing additional H atoms. Reaction (4.170) yields only a relatively nonreactive methyl radical and the chain-propagating radical OH. The lower rates and the short period over which initiation reactions are important make reactions (4.171) and (4.172) negligible. Reaction (4.173) becomes unimportant at high-temperature conditions relative to reaction (4.170).

Abstraction of methanol's H atoms proceeds primarily by breaking C–H bonds rather than the O–H bond:

$$CH_3OH + H = CH_3 + H_2O \tag{4.174}$$

$$CH_3OH + H = CH_2OH + H_2 \tag{4.175}$$

$$CH_3OH + OH = CH_2OH + H_2O \tag{4.176}$$

$$CH_3OH + O = CH_2OH + OH \tag{4.177}$$

$$CH_3OH + CH_3 = CH_2OH + CH_4 \tag{4.178}$$

$$CH_3OH + HO_2 = CH_2OH + H_2O_2 \tag{4.179}$$

For the reactions between CH_3OH and H, reaction (4.174) is faster below 650 K, while the rates reverse at higher temperatures ($T > 1000$ K). If reaction (4.174) or any other faster reaction produced CH_3 radicals in large amounts, then methyl radical recombination reactions (4.176)–(4.178) would result in considerable amounts of C_2 species. However, very low concentrations of C_2 species are observed in CH_3OH–air flames. The CH_2OH radicals decompose by means of reaction (4.180) or react with O_2 and H by reactions (4.181) and (4.182).

$$CH_2OH + M = CH_2O + H + M \tag{4.180}$$

$$CH_2OH + O_2 = CH_2O + HO_2 \tag{4.181}$$

$$CH_2OH + H = CH_2O + H_2 \tag{4.182}$$

4.8.6 Ethanol C_2H_5OH Combustion

Ethanol is an ethyl C_2H_5 group linked to a hydroxyl OH group. It is also called *grain alcohol*. Based in principle on mechanisms for methanol oxidation, the initiation reactions are

$$C_2H_5OH + M = CH_3 + CH_2OH + M \tag{4.183}$$

$$C_2H_5OH + O_2 = CH_3CHOH + HO_2 \tag{4.184}$$

Reactions between radical species and ethanol are

$$C_2H_5OH + OH = CH_3CHOH + H_2O \tag{4.185}$$

$$C_2H_5OH + H = CH_3CHOH + H_2 \tag{4.186}$$

$$C_2H_5OH + O = CH_3CHOH + OH \tag{4.187}$$

$$C_2H_5OH + HO_2 = CH_3CHOH + H_2O_2 \tag{4.188}$$

$$C_2H_5OH + CH_3 = CH_3CHOH + CH_4 \tag{4.189}$$

The hydroxyethyl radical (CH_3CHOH) is consumed by the same paths as CH_2OH, see reactions (4.180) and (4.181), and results in the production of acetaldehyde CH_3CHO:

$$CH_3CHOH + M = CH_3CHO + H + M \tag{4.190}$$

$$CH_3CHOH + O_2 = CH_3CHO + HO_2 \tag{4.191}$$

The production of acetaldehyde as the dominant intermediate in ethanol combustion is analogous to the production of formaldehyde as the major intermediate in methanol combustion. A turbulent flow reactor study of ethanol pyrolysis and oxidation (Westbrook and Dryer 1984) has shown that, similar to the case for methanol CH_3OH, two hydrogen abstraction reactions are important at high temperature, consisting of reaction (4.186) as well as

$$C_2H_5OH + H = C_2H_5 + H_2O \tag{4.192}$$

4.8.7 Acetaldehyde CH_3CHO Reactions

Acetaldehyde is an important intermediate chemical species in the combustion of higher hydrocarbon fuels. The principal initiation reactions are

$$CH_3CHO = CH_3 + HCO \tag{4.193}$$

$$CH_3CHO = CH_3CO + H \tag{4.194}$$

$$CH_3CHO + O_2 = CH_3CO + HO_2 \tag{4.195}$$

At shock-tube temperatures above 1500 K, the reaction (4.193) is more important than reaction (4.194). At lower flow reactor temperatures around 1000 K, reactions (4.195), (4.26), and (4.75) produce high concentrations of hydroperoxyl HO_2 radicals. Much of the subsequent CH_3 radical consumption can be attributed to reaction (4.61) with hydroperoxyl radicals.

The principal H atom abstraction reactions are

$$CH_3CHO + H = CH_3CO + H_2 \tag{4.196}$$

$$CH_3CHO + OH = CH_3CO + H_2O \tag{4.197}$$

$$CH_3CHO + O = CH_3CO + OH \tag{4.198}$$

$$CH_3CHO + CH_3 = CH_3CO + CH_4 \tag{4.199}$$

$$CH_3CHO + HO_2 = CH_3CO + H_2O_2 \tag{4.200}$$

These reactions are followed by the decomposition reaction of the acetic anhydride CH_3CO radical,

$$CH_3CO = CH_3 + CO \tag{4.201}$$

The above mechanism leads to large concentrations of the methyl radical CH_3 and carbon monoxide CO. Some methyl radicals produce methane, and a fraction of methyl radicals recombine to produce ethane C_2H_6, see reaction (4.53). Subsequent reactions of C_2H_6 lead to ethylene C_2H_4, acetylene C_2H_2, and other related species as already discussed.

The chemical kinetics of oxidation and pyrolysis reactions of the higher hydrocarbons (C_4, C_5, etc.) are given in the excellent review papers by Westbrook and Dryer (1984) and Simmie (2003). Due to space limitations, these reactions are not included in this chapter.

4.9 Reduced Mechanism of Hydrocarbon Combustion

In this section, we present an example of a reduced kinetic mechanism of hydrocarbon auto-ignition which incorporates both high-temperature thermal reactions and low-temperature chain reactions. As discussed previously in this Chapter, ignition of a hydrocarbon fuel at high temperatures ($T \geq 1000$ K) involves decomposition into small C_2 and C_1 hydrocarbons which are then oxidized. At lower temperatures ($T < 1000$ K), the fuel is oxidized by O_2 in a degenerate chain-branching reaction.

The kinetic mechanism in this section is a reduced five-step reaction sequence based on Schreiber et al. (1994). The Schreiber model was developed by examining the results of larger kinetic models and associated experiments, and choosing the reactions which best characterized the pressure (0.3 to 4.2 MPa), temperature (600 to 1500 K), and ignition behavior of the actual reactions. Of particular interest was modeling the negative temperature coefficient behavior inherent in alkene combustion. The model uses two reactions for high-temperature oxidation and three reactions for low-temperature oxidation of mixtures of iso-octane C_8H_{18} and *n*-heptane C_7H_{16}. These two fuels are classified as primary reference fuels (PRFs) and are used to characterize the autoignition behavior of spark ignition fuels through the octane number (*ON*). The straight chain n-heptane octane number is defined as $ON = 0$, and the octane number of the branched chain iso-octane is defined as $ON = 100$. Since it has a cetane number of 50, *n*-heptane is also used as a surrogate diesel fuel in experiments and computations.

The five reactions of the mechanism are given in Table 4.3, along with the reaction enthalpies. The enthalpies of the low- and high-temperature paths are equal, in accordance with Hess's Law. As mentioned in Chapter 2, G. Hess (1802–1840) was a Russian chemist who observed that the enthalpy change during a chemical reaction is independent of the path of the reaction. The parameters of the model are:

- F represents the hydrocarbon fuel C_nH_{2n+2}.
- O_2 is oxygen.
- P is the final product.
- x is the high-temperature intermediate species representing the small hydrocarbons CH_2, CH_3, H, etc.
- R is an oxygenated radical representing low-temperature species such as $OC_8H_{15}O_2H$.
- y is a low-temperature chain-propagating partially oxygenated species such as OH. Autoignition at low temperature depends on the continual increase in the concentration of y to produce a branched chain reaction.

The first two reactions (1) and (2) are the high-temperature reactions. They correspond to a two-step scheme with an endothermic fuel decomposition into small hydrocarbons, and their exothermic oxidation into combustion products. The stoichiometric coefficient in the O_2 parentheses is for *n*-heptane. The other three reactions (3), (4), and (5) represent the low-temperature reactions. Reaction (3) is reversible, with the forward reaction (3+) producing an oxygenated

Table 4.3 Five-Step Reduced Kinetic Mechanism.

	Reaction	Enthalpy of Reaction $\bar{q}_{r,i}$ (MJ/kmol)
(1)	$F \longrightarrow x$	+709.9
(2)	$x + 12.5(11)\,O_2 \longrightarrow P$	−4709.9
(3+)	$F + 2\,O_2 \longrightarrow I$	−53.9
(3−)	$I \longrightarrow F + 2\,O_2$	+53.9
(4)	$I \longrightarrow 2\,y$	−60.0
(5)	$y + 0.5\,F + 11.5(10)\,O_2 \longrightarrow P$	−3913.1

radical I. The reverse reaction (3-) has a much larger activation energy, and dominates the forward reaction (3+) at high temperature, thus reducing the relative importance of reactions (4) and (5) at higher temperatures. Reaction (4) is an autocatalytic reaction, in which at least one of the products is a reactant. The low-temperature oxidation path is completed in reaction (5).

The reaction rate equations depend on the concentration of the other molecules with which a molecule may react. The reaction rates $\dot{\Omega}_i$ (kmol/m^3-s) for the five reactions are:

$$\begin{aligned}
\dot{\Omega}_1 &= k_1[\mathrm{F}](p/p_o)^{0.5} \\
\dot{\Omega}_2 &= k_2[\mathrm{x}][\mathrm{O}_2][\mathrm{M}] \\
\dot{\Omega}_{3+} &= k_{3+}[\mathrm{F}][\mathrm{O}_2][M](p/p_o)^{-2.2}C_{3+} \\
\dot{\Omega}_{3-} &= k_{3-}[\mathrm{I}](p/p_o)^{-3.5} \\
\dot{\Omega}_4 &= k_4[\mathrm{I}]C_4 \\
\dot{\Omega}_5 &= k_5[\mathrm{y}][\mathrm{F}][\mathrm{O}_2]
\end{aligned} \tag{4.202}$$

where the concentrations [ ] are in units of (kmol/m^3) and the overall mixture concentration [M] is

$$[\mathrm{M}] = \frac{p}{R_u T} \tag{4.203}$$

The Arrhenius rate coefficients $k_i = A_i \exp(-E_i/R_u T)$ for the five-step reactions of Equation (4.205) are (kmol, m^3, second units)

$$\begin{aligned}
k_1 &= 5.0 \times 10^5 \exp(-18,050/T) \\
k_2 &= 7.0 \times 10^3 \exp(-7200/T) \\
k_{3+} &= 3.5 \times 10^6 (3.85 \times 10^7) \exp(-19,500/T) \\
k_{3-} &= 6.0 \times 10^{24} \exp(-37,500/T) \\
k_4 &= 6.0 \times 10^4 (2.0 \times 10^5) \exp(-5000/T) \\
k_5 &= 1.0 \times 10^6 \exp(-16,500/T)
\end{aligned} \tag{4.204}$$

The pressure term p/p_o has a reference pressure $p_o = 10$ bar, with the exponents chosen so the region of negative temperature dependence moves to higher temperatures as the pressure increases, and the ignition retarding effect of the octane number in the low-temperature reactions R_{3+} and R_4 is represented empirically by the parameters C_{3+} and C_4:

$$\begin{aligned}
C_{3+} &= (110 - ON)/10 \\
C_4 &= [(110 - ON)/10]^{1/2}
\end{aligned} \tag{4.205}$$

From the first law for this constant pressure process with no heat transfer and shaft work, the initial and final enthalpy of the mixture is unchanged. Therefore, as the chemical bond energy is released during combustion, there will be a corresponding increase in the thermal energy, i.e., temperature, of the reactants. If the specific heats of the reactants are assumed to have the constant value $\overline{c}_p$, then for a constant enthalpy process, the rate of temperature change of the mixture is related to the enthalpies or heats of reaction $\overline{q}_{r,i}$ and reaction rates $\dot{\Omega}_i$ as follows:

$$\begin{aligned}\frac{dT}{dt} &= \frac{\sum_{i=1}^{5}\overline{q}_{r,i}\dot{\Omega}_i}{[\mathrm{M}]\overline{c}_p} \\ &= \frac{1}{[\mathrm{M}]\overline{c}_p}(\overline{q}_{r,1}\dot{\Omega}_1 + \overline{q}_{r,2}\dot{\Omega}_2 + \overline{q}_{r,3}\dot{\Omega}_3 + \overline{q}_{r,4}\dot{\Omega}_4 + \overline{q}_{r,5}\dot{\Omega}_5)\end{aligned} \tag{4.206}$$

The net production rate (kmol/m^3-s) of a species is the sum of the elementary rates producing that species minus the sum of the elementary rates consuming that species:

$$\begin{aligned}\frac{d[\mathrm{F}]}{dt} &= -\dot{\Omega}_1 - \dot{\Omega}_3 + \dot{\Omega}_{3-} - 0.5\dot{\Omega}_5 \\ \frac{d[\mathrm{x}]}{dt} &= +\dot{\Omega}_1 - \dot{\Omega}_2 \\ \frac{d[\mathrm{I}]}{dt} &= +\dot{\Omega}_{3+} - \dot{\Omega}_{3-} - \dot{\Omega}_4 \\ \frac{d[\mathrm{y}]}{dt} &= +2\dot{\Omega}_4 - \dot{\Omega}_5 \\ \frac{d[\mathrm{P}]}{dt} &= +\dot{\Omega}_2 + \dot{\Omega}_5 \\ \frac{d[\mathrm{O_2}]}{dt} &= -12.5(11)\dot{\Omega}_2 - 2\dot{\Omega}_{3+} + 2\dot{\Omega}_{3-} - 11.5(10)\dot{\Omega}_5\end{aligned} \tag{4.207}$$

The initial conditions for the above differential equations for T, F, x, I, y, P, and O_2 are $T = T_i$, $[\mathrm{F}] = [\mathrm{F}]_i$, $[\mathrm{O_2}] = [\mathrm{O_2}]_i$, and $[\mathrm{x}]_i = [\mathrm{y}]_i = [\mathrm{I}]_i = 0$. The initial fuel and oxygen concentrations are

$$\begin{aligned}[\mathrm{F}]_i &= X_F\frac{p}{R_u T} \\ [\mathrm{O_2}]_i &= X_{\mathrm{O_2}}\frac{p}{R_u T}\end{aligned} \tag{4.208}$$

With reference to Chapter Two, for the hydrocarbon C_aH_b, the stoichiometric parameter a_{st} is

$$a_{st} = a + b/4 \tag{4.209}$$

and for one mole of fuel, the initial fuel and oxygen mole fractions are

$$\begin{aligned}X_F &= \left(1 + 4.76\frac{a_{st}}{\phi}\right)^{-1} \\ X_{\mathrm{O_2}} &= \frac{a_{st}}{\phi}X_F\end{aligned} \tag{4.210}$$

For a blend of n-heptane (hep) and iso-octane (oct) with octane number ON, the fuel and oxygen mole fractions are

$$\begin{aligned} X_F &= \left(1 - \frac{ON}{100}\right) X_{F,hep} + \left(\frac{ON}{100}\right) X_{F,oct} \\ X_{O_2} &= \left(1 - \frac{ON}{100}\right) X_{O_2,hep} + \left(\frac{ON}{100}\right) X_{O_2,oct} \end{aligned} \tag{4.211}$$

Example 4.2 ***Kinetic Modeling of n-Heptane and iso-Octane Combustion***

(a) A stoichiometric mixture of iso-octane is initially at 600 K and 10 bar (1000 kPa). What are the mixture [M], and initial fuel $[F]_i$ and oxygen $[O_2]_i$ concentrations? Plot the temperature and species concentration profiles during the autoignition process for initial temperatures of 600 K and 1000 K. Describe the differences in the low- (600 K) and the high-temperature (1000 K) combustion processes. Assume the mixture-specific heat $\bar{c}_p$ has a constant value of 36 kJ/kmol.
(b) Plot the ignition delay of mixtures of n-heptane and iso-octane with octane numbers of $ON = 0$, 50, 75, 90, and 100, and initial temperature from 600 to 1200 K. Describe the influence of octane number.

Solution:
The overall mixture concentration is

$$[M] = \frac{P}{R_u T} = \frac{1000}{(8.314)(600)} = 0.20\,\text{kmol/m}^3$$

For C_8H_{18} with $\phi = 1$,

$$\begin{aligned} a_{st} &= a + b/4 = 8 + 18/4 = 12.5 \\ X_F &= \left(1 + 4.76\frac{a_{st}}{\phi}\right)^{-1} = (1 + (4.76)(12.5)/1)^{-1} = 0.0165 \\ X_{O_2} &= \frac{a_{st}}{\phi} X_F = (12.5/1)(0.0165) = 0.207 \\ [F]_i &= X_F[M] = (0.0165)(0.20) = 0.0033\,\text{kmol/m}^3 \\ [O_2]_i &= X_{O_2}[M] = (0.207)(0.20) = 0.0414\,\text{kmol/m}^3 \end{aligned}$$

The above seven differential equations are integrated in time numerically in the Matlab program `Kinetic.m` which is available on the text website. The program computes the instantaneous concentrations of the species F, O_2, x, y, I, P and the mixture temperature T, from given initial conditions to the onset of ignition. The ignition delay is defined as a 400 K increase in the mixture temperature. The inputs to the program are the initial temperature T_i and pressure, P_i, equivalence ratio ϕ, and ON. The program returns the ignition delay, and the species concentration and temperature versus time.

Given the above initial conditions, the autoignition of the gas mixture was calculated using the code `Kinetic.m`. The development of the temperature and concentration profiles are shown in Figures 4.5 and 4.6.

At 600 K, the autoignition is a low-temperature combustion process with an ignition delay of about 500 ms. Reaction (3+) produces the radical intermediate species I which remains at a steady-state concentration. Reaction (4) produces the chain-branching species y which grows

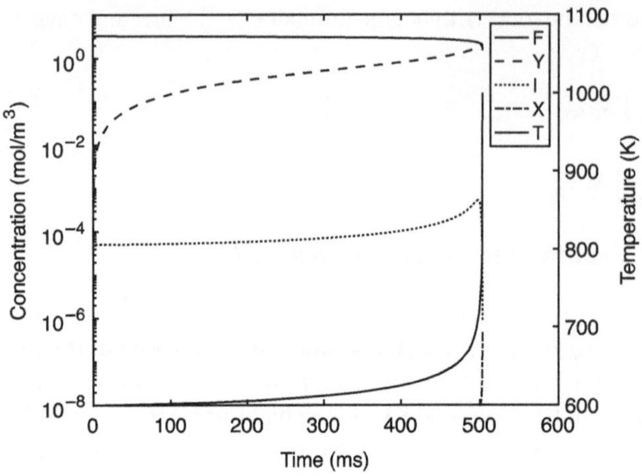

Figure 4.5 Temperature and concentration profiles for T_i = 600 K (iso-octane).

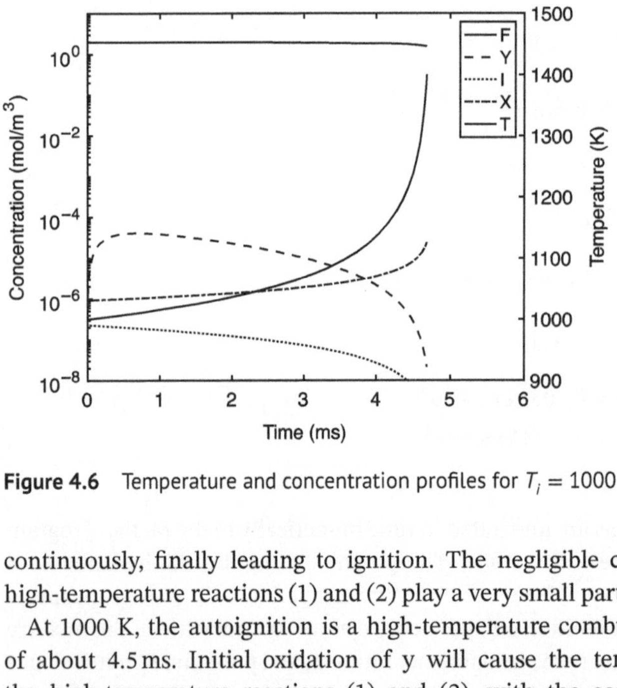

Figure 4.6 Temperature and concentration profiles for T_i = 1000 K (iso-octane).

continuously, finally leading to ignition. The negligible concentration of x indicates that the high-temperature reactions (1) and (2) play a very small part until late in the autoignition process.

At 1000 K, the autoignition is a high-temperature combustion process with an ignition delay of about 4.5 ms. Initial oxidation of y will cause the temperature to rise quickly, activating the high-temperature reactions (1) and (2), with the concentration of the high-temperature intermediate species x exceeding the concentration of the low-temperature intermediate y after 4 ms.

The ignition delays of mixtures of n-heptane and iso-octane are plotted in Figure 4.7. As per convention, the x-axis coordinate is the dimensionless parameter 1000 K/T, so as the initial temperature T_i is increased, 1000 K/T decreases. The ignition delay increase in the region from about 700 to 800 K is known as the negative temperature coefficient region and is a characteristic of the low-temperature combustion of alkenes. Degenerate branching via reactions R_3 and R_4 is responsible for the negative temperature dependence of the reaction rate at intermediate temperatures. As the octane number is increased, the ignition delay increases.

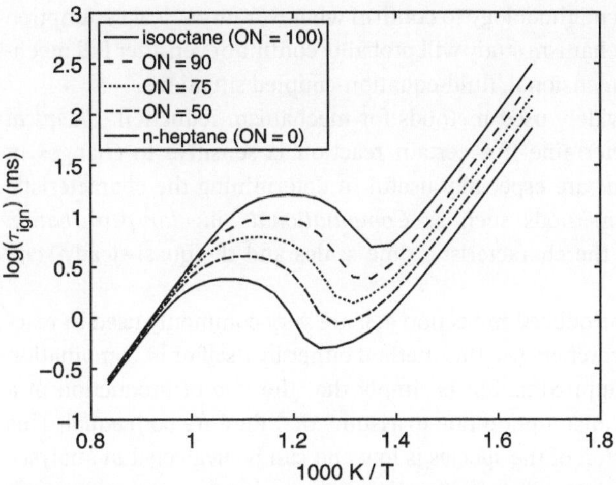

Figure 4.7 Ignition delay as a function of initial temperature T_i and octane number *ON*.

4.10 Chemical Mechanism Reduction

Combustion problems, in general, are quite challenging due to the number of species involved and the necessary reactions needed to obtain an accurate solution. For systems with large numbers of species and reactions, the interactions between species can become quite complex. The size of a chemical mechanism (i.e., the number of reactions) increases approximately linearly with the number of chemical species considered; the scaling factor is approximately a factor of five between the chemical mechanism size and the number of species involved (Lu et al. 2001).

Since large chemical equation sets are difficult to solve, efforts are warranted to reduce these equation sets to manageable proportions. If some chemical reactions can be removed from consideration while retaining accuracy, then determination of the solution space becomes much more tractable. Reduction procedures involve several steps, which include specifying the initial reactants and main reaction product species, determining the temperature range for non-isothermal reactions, and examining the remaining species to determine if the species is necessary or redundant.

From the beginning, it should be stated that chemical mechanism reduction is a field of study by itself. The effort to reduce chemical mechanisms probably started from the simple fact that sufficient computing resources were not available to handle broad-scope, multifaceted combustion problems, and it is challenging to identify the dominant reactions and species through inspection only. Even today, as computing power has increased nearly exponentially, the effort to reduce chemical mechanisms is still just as valid. Another benefit to the study of mechanism reduction is that sometimes greater insight into species interaction and importance of certain mechanisms is achieved when different reduced combustion schemes are compared to each other and to the full equations.

How a method of reduction is typically evaluated is by comparison with experimental results and not necessarily with the results of a full model. Experimental data can be sparse and the conditions of combustion may be difficult to simulate, but the true test of a kinetic mechanism will usually reside with how well the model fits the experimental data and over how large of a parameter range this model is valid. Some might say that it is only a matter of time before the raw computing power necessary to tackle the full mechanisms in combustion is available. But, if one considers the fact that experimentation is the only trusted methodology to confirm whatever numerical assumption or scheme that is produced, reduced mechanism study will probably continue even after full mechanism studies are possible with multidimensional, fluid-equation-coupled situations.

In this section, we introduce some widely used methods for mechanism reduction. *Chemical Sensitivity Analysis* (SA) is used to determine if a certain reaction is sensitive to changes in other parameters. *Reaction-flow methods* are especially useful in determining the characteristic reaction paths. Eigenvector analysis methods such as *Computational Singular perturbation* (CSP) are very helpful in determining the characteristic time scales and the quasi-steady-state species.

Quasi-steady-state approximations, introduced in Section 4.3, are very commonly used in reaction modeling and reduction. Many researchers use this method either by itself or in combination with other similar assumptions. This approximation is simply that the rate of production of a chemical species is close enough to its consumption rate to assume that they are both equal. This allows us to assume that the concentration of the species is low and can be neglected in analysis. A species can be assumed to be in steady state when the reactions consuming the species are much faster than the transport processes within a flame. An indicator for this criterion is the maximum concentration of the species. For example, all species with a weighted mole fraction, X_w, less than 1% can be considered to be in steady state. The steady-state approximation implies that the overall reaction rate of a species is zero. For each steady-state species, one reaction rate can then be expressed as a function of other reaction rates.

As a whole, this approximation allows some of the ordinary differential equations of the full mechanism to be converted to algebraic expressions. With algebraic relations, the number of ordinary differential equations is reduced and the mechanism becomes simpler. The drawback to this type of approximation is that the modeler must have some a priori knowledge regarding what species are likely to exhibit this behavior and are good candidates for this approximation without compromising accuracy excessively. The other drawback is that models based entirely on this type of approximation are usually valid only under a narrow range of parameters, and thus each model tends to fit only a small number of cases.

It is useful to note that the partial equilibrium assumption, introduced in Section 4.4, is not to be confused with the steady-state assumption. Although these two assumptions are both frequently used in combination with each other, it should not be assumed that they are the same thing. The steady-state assumption relates to species, and the partial equilibrium relates to a set of reactions. It is usually more convenient to eliminate species than reactions as this reduces the number of species conservation equations that need to be solved.

At high temperatures, if the forward and backward reaction rates are large, then we can assume partial equilibria for certain reactions, for example, (the equal sign indicates both forward and backward reactions need to be considered)

$$\mathrm{H + O_2 = OH + O} \tag{4.212}$$

$$\mathrm{O + H_2 = OH + H} \tag{4.213}$$

$$\mathrm{OH + H_2 = H_2O + H} \tag{4.214}$$

From these equilibrium reactions, the concentrations of reactive species (such as H, O, and OH), which are difficult to measure, can be expressed in terms of concentrations of stable species such as H_2, O_2, and H_2O, which are easier to measure. Note that the partial equilibrium assumption is typically valid only at very high temperatures (Eggels and de Goey 1995).

4.11 Sensitivity Analysis

The solution of the numerical model of a combustion process depends on a number of input parameters, including, for example, the initial conditions and the rate coefficients. However, the rate coefficients are often not known with great precision, and the experimental data are not detailed enough to estimate accurately the rate coefficient parameters. It is necessary, however, to know how variations of the input parameters affect the solution of the combustion model.

Sensitivity analysis is a way to understand quantitatively how the solution to a model depends on parameters in the model. More precisely, sensitivity analysis can help determine the effects of the uncertainties in rate constant parameters and errors in initial conditions on the model predictions. Sensitivity analyses have been used with several different parameters, the most popular being species concentration or reaction rate coefficients but may also include pressure, temperature, mass fraction, mixture molecular weight, density, mole fraction, and chemical potential. Sensitivity analysis per se can be used to study the effects of any parameter change on a complex system. The main usefulness of sensitivity analysis is based on its unique ability to identify rate-limiting steps (e.g., the slowest step in a series of elementary reactions and the fastest step in a group of parallel elementary reactions).

Sensitivity analyses are used for three main purposes with respect to kinetic mechanisms: (1) to reduce the number of reactions in a given problem, (2) to identify the rate-limiting reaction steps in the system, and (3) to understand the relative importance of the reactions in the system. The concept of sensitivity analysis, although not new to the field of chemical kinetics, has recently received considerable interest in chemical kinetic research due to the development of very efficient methods for obtaining sensitivity analysis information. These newer, less time-consuming methods have enabled the study of larger systems and extended to use of past linear gradients, e.g., higher-order gradients, "derived" gradients, and Green's function gradients (Yetter et al. 1985).

According to Warnatz et al. (2006), for many elementary reactions, a change in the rate coefficients has nearly no effect on the time-dependent solution, showing that quasi-steady states or partial equilibria are in force for those reactions. Even if one chooses to include the reaction explicitly in the reaction mechanism, one does not need a highly accurate rate coefficient. On the other hand, for a few of the elementary reactions, changes in the rate coefficients have very significant effects on the solution. Accordingly, accurate rate coefficients are required. These few important reaction steps are the *rate-determining steps* or *rate-limiting steps*.

An analysis that accounts for simultaneous parameter variations of arbitrary magnitude can be termed a *global* sensitivity analysis. This analysis produces coefficients that have a measure of sensitivity over the entire admissible range of parameter variation. The interpretation of system sensitivities in terms of first-order elementary sensitivity coefficients is called a *local* sensitivity analysis. Here, the first-order elementary sensitivity coefficient is defined as the gradient, $\partial C_i/\partial \alpha_j$, where C_i is the concentration of the ith species at time t and α_j is the jth input parameter, and the gradient is evaluated at a set of nominal parameter values α. Although the sensitivity coefficients $\partial C_i/\partial \alpha_j$ provide direct information on the effect of a small perturbation in each parameter about its nominal value on each concentration, they do not necessarily indicate the effect of simultaneous, large variations in all parameters on each species concentration.

Both local and global analyses are useful in studying the behavior of a system since each has advantages and disadvantages. Global methods can account for simultaneous parameter variations of arbitrary magnitudes and can provide an average effect of all uncertainties examined. However, the computational effort for the global methods is much larger than for the local methods, since global methods require repeated solution of the chemical kinetics equations. Thus, local methods are used more often than global methods. For an excellent review of the different approaches, see Rabitz et al. (1983). Examples of sensitivity analysis applied to steady-state and transient one-dimensional problems are given in Smooke et al. (1988) and Reuven et al. (1987).

Sensitivity Analysis Theory

In sensitivity analysis theory, it is useful to define a generalized Jacobian matrix by the following definition. Given a solution $\mathbf{x}$ that is dependent on a set of parameters $\mathbf{p}$, a Jacobian matrix J can be defined to describe the marginal changes in the solution $\mathbf{x}$ as the parameters change.

$$J = \left(\frac{\partial \mathbf{x}^{\mathrm{T}}}{\partial \mathbf{p}^{\mathrm{T}}} \right)^{\mathrm{T}} \tag{4.215}$$

This Jacobian matrix is called the *first-order sensitivity matrix*. Sensitivity matrices of higher order n are simply the Jacobian matrices of the nth derivatives. Second-order sensitivity matrices have also been used before (Yetter et al. 1985). Second-order sensitivities were shown to yield valuable information on how these linear sensitivities vary when an initial condition or rate constant is varied. A normalized first-order matrix can be established to relate the rate of production of a given species f_j to the change in concentration C_i of another species by defining J as

$$J = \frac{\partial(\ln f_j)}{\partial(\ln C_i)} \tag{4.216}$$

By taking the sum of the squares of the normalized Jacobian, an expression B_i can be obtained for the concentration sensitivity as defined in Eq. (4.217).

$$B_i \equiv \sum_{j=1}^{N} \left(\frac{\partial \ln f_j}{\partial \ln C_i} \right)^2 \tag{4.217}$$

The concentration sensitivity B_i relates a change in ith species concentration to the rate of production of another species. The concentration sensitivity allows the determination of whether individual species are necessary or redundant, as opposed to the reactions.

The rate laws for a reaction mechanism consisting of M reactions involving N chemical species can be written as a system of first-order nonlinear ordinary differential equations

$$\begin{aligned} &\frac{dC_i}{dt} = F_i(C_1, C_2, \ldots, C_N; k_1, k_2, \ldots, k_M) \\ &C_i(t=0) = C_{i,0} \qquad i = 1, 2, \ldots, N \end{aligned} \tag{4.218}$$

where $C_{i,0}$ denotes the initial concentration of the ith species. Here, only the specific rate coefficients shall be considered as parameters of the system, even though other parameters such as pressure and temperature can also be treated as system parameters. The absolute and relative sensitivities can be defined as the dependence of the solution C_i on the rate coefficient parameter k_r by the following definitions:

$$E_{i,r} \equiv \frac{\partial C_i}{\partial k_r} \quad \text{and} \quad E_{i,r}^{(\mathrm{rel})} \equiv \frac{k_r}{C_i}\frac{\partial C_i}{\partial k_r} = \frac{\partial \ln C_i}{\partial \ln k_r} \tag{4.219}$$

Differentiating Eq. (4.218) with respect to k_r, we have

$$\begin{aligned} &\frac{\partial}{\partial k_r}\left(\frac{\partial C_i}{\partial t} \right) = \frac{\partial}{\partial k_r} F_i(C_1, C_2, \ldots, C_N; k_1, k_2, \ldots, k_M) \\ &\frac{\partial}{\partial t}\left(\frac{\partial C_i}{\partial k_r} \right) = \left(\frac{\partial F_i}{\partial k_r} \right)_{C_n, k_{m\neq r}} + \sum_{n=1}^{N} \left\{ \left(\frac{\partial F_i}{\partial C_n} \right)_{C_{l\neq n}, k_m} \left(\frac{\partial C_n}{\partial k_r} \right)_{k\neq r} \right\} \end{aligned} \tag{4.220}$$

$$\frac{\partial}{\partial t}E_{i,r} = \left(\frac{\partial F_i}{\partial k_r}\right)_{C_n,k_{m\neq r}} + \sum_{n=1}^{N}\left\{\left(\frac{\partial F_i}{\partial C_n}\right)_{C_{l\neq n},k_m} E_{n,r}\right\} \tag{4.221}$$

In these equations, C_n in the subscripts of the partial derivatives means that all C_n are held constant during the differentiation, and $C_{l\neq n}$ means that all C_l are held constant, except C_n. Usually, the relative sensitivities of fast reactions decay rapidly with time to very low values, while the product concentrations have high relative sensitivities with respect to slow reactions and vice versa for fast reactions.

A set of rate equations for a complex system of chemical reactions can also be expressed in terms of nonlinear ordinary differential equations given by Eq. (4.222). The set of species mass fractions Y_i can be written as a vector $\vec{Y}$. Therefore, it is possible to rewrite the rate of change of $\vec{Y}$ in terms of the global reaction rates $\tilde{g}$.

$$\frac{d\vec{Y}}{dt} = \tilde{g}(\vec{Y}) \tag{4.222}$$

The matrix quantity of global reaction rates $\tilde{g}$ can be expressed as the sum of the contributions from the stoichiometric coefficient $\nu_{ij} = (\nu_i' - \nu_i)_j$ and the reaction rate $\dot{\Omega}_j$ of the jth reaction in order to determine reaction rate sensitivity.

$$\tilde{g}(\vec{Y}) = \sum_{j=1}^{M} \nu_{ij}\dot{\Omega}_j(\vec{Y}) \tag{4.223}$$

The mathematical goal of sensitivity analysis is to reduce the size of the $\vec{Y}$ vector. There are several computational methods that can be used for this purpose. The most common method is through principal component analysis. For the reaction rate sensitivity, the matrix $\tilde{g}$ can be taken with the transpose of itself and decomposed into a system of eigenvalues and eigenvectors. The eigenvectors thus formed represent sets of coupled reactions. The eigenvalues represent the overall importance of these sets in the overall reaction. By defining certain numerical thresholds, it becomes possible to categorize the eigenvectors and eigenvalues (Tomlin et al. 1992).

In addition to the eigenvalue/eigenvector system, it is possible to define a normalized sensitivity coefficient for a given reaction. An overall rate sensitivity coefficient S_j for the jth reaction can be defined by taking the sum of the squares of the coefficients for that reaction with respect to a change in a given species,

$$S_j \equiv \sum_{i=1}^{N}\left[\frac{\nu_{ij}\dot{\Omega}_j}{f_i}\right]^2 \tag{4.224}$$

where f_i represents the rate of change of the species i.

In order to be necessary, a reaction must have normalized rate sensitivity above a certain threshold value and, in addition, must have an effect on the important species. In most cases, these two conditions represent the same condition, but the sensitivity analysis does have certain flaws that occasionally allow the selection of reactions that are not required. The second condition also allows the addition of other parameters that affect the model, such as temperature or pressure.

The determination of the effect of other parameters is made through a separate form of sensitivity analysis. Conversely, in order for a reaction to be redundant, its normalized rate sensitivity must be below the threshold value, and its removal from the system must not affect the important species. If the reaction fulfills these two conditions, it may be removed from the system of reactions within the model. A similar method is used for the temperature sensitivity analysis. Temperature sensitivities become important when thermodynamic effects play an important role in the chemical reactions. The temperature sensitivity can be defined as

$$TS \equiv \frac{\partial \ln(dT/dt)}{\partial \ln k_j} \tag{4.225}$$

Local methods typically produce the effect of a small change in one parameter by computing the first-order sensitivity coefficient. The first-order local sensitivity coefficients $S_{i,j}$ can be defined by

$$S_{i,j} = \left[\frac{\partial Y_i}{\partial k_j}\right]_{k_{i \neq j}} \qquad j = 1, 2, \dots, M; \quad i = 1, 2, \dots, N \tag{4.226}$$

In the above equation, Y_i is a variable of interest (e.g., a concentration or temperature) and k_j is a rate constant parameter (A, b, and E_a) or an initial condition value. For isothermal reactions, the rate constants are time invariant; there is then no need to consider the individual rate constant parameters.

In a more general notation, if a system is described by a dependent variable $\mathbf{\Phi} = \mathbf{\Phi}(\mathbf{p}, x)$, where $\mathbf{\Phi} \equiv (\Phi_1, \Phi_2, \dots, \Phi_N)$, $\mathbf{x}$ is the independent variable, and $\mathbf{p}$ is a set of parameters, the dependence of $\mathbf{\Phi}$ on the parameters p_j for $j = 1, 2, \dots, N_k$ is described by absolute and relative sensitivities (which are local properties with respect to the independent variable $\mathbf{x}$):

$$S_{\Phi_i, p_j} \equiv \frac{\partial \Phi_i}{\partial p_j} \qquad S^{(\mathrm{rel})}_{\Phi_i, p_j} \equiv \frac{p_j}{\Phi_i}\frac{\partial \Phi_i}{\partial p_j} \tag{4.227}$$

where the term $S^{(\mathrm{rel})} = 0$ implies no dependence, $S^{(\mathrm{rel})} = 1$ indicates linear dependence, and $S^{(\mathrm{rel})} = 2$ represents quadratic dependence on the parameter considered.

Green's Function Method

The Green's function method can be an efficient computational technique for determining local linear sensitivities, and it was adopted by Mishra et al. (1994) for chemical kinetic studies. In the Green's function method, the coupled first-order ordinary differential equations for describing a chemical kinetic system reacting under isothermal conditions without spatial transport [Eq. (4.216)] may be written as

$$\frac{dC_i}{dt} = f_i(\mathbf{C}, \alpha), \qquad C_i(t = 0) \equiv C_{i,0} \tag{4.228}$$

where f_i is a nonlinear first-order, second-order, or third-order function of the species concentrations. The parameter vector α includes all physically definable input parameters of interest, all of which are treated as constant.

Equation (4.228) may be differentiated with respect to the parameter α_j to yield the following set of linear coupled equations in terms of the desired linear sensitivity coefficients, $\partial C_i/\partial \alpha_j$:

$$\frac{\partial^2 C_i}{\partial t\, \partial \alpha_j} = \frac{\partial f_i}{\partial \alpha_j} + \sum_s \frac{\partial f_i}{\partial C_s}\frac{\partial C_s}{\partial \alpha_j} \tag{4.229}$$

This set of equations is solved in closed form to give

$$\frac{\partial \mathbf{C}}{\partial \alpha_j} = G(t, 0)\left[\frac{\partial \mathbf{C}}{\partial \alpha_j}\right]_0 + \int_0^t G(t, t')\left(\frac{\partial \mathbf{f}(t)}{\partial \alpha_j}\right) dt' \tag{4.230}$$

where Green's function satisfies the following linear ODE:

$$\left[\frac{\partial}{\partial t} - J(t)\right] G(t, t') = 0, \qquad t > t' \quad \text{and} \quad G(t', t') = 1 \tag{4.231}$$

where $J(t)$ is the $n \times n$ Jacobian matrix with elements $\partial f_i/\partial C_j$. The governing equations for second-order sensitivities are obtained by taking consecutive derivatives of Eq. (4.228) with respect to the input parameters α_j. The solutions of these equations are again obtained by Green's function techniques. One of the reasons to use Green's function techniques is due to the fact that the original nonlinear differential equations are replaced by a combined linear ODE and an integral equation, which are generally more convenient to solve.

The initial conditions for Eq. (4.229) result from mathematical considerations versus physical considerations for Eq. (4.228). Here, the initial condition $(\partial C_i/\partial \alpha_j)_0$ is the zero vector, unless α_j is the initial concentration of the ith species, in which case the initial condition is a vector whose components are all zero except the ith component, which has the value 1. It is important to recognize that the Green's function matrix itself has an important physical interpretation. In particular, elements of the Green's function matrix may be identified as the gradients

$$G_{ij} = \frac{\partial C_i(t)}{\partial C_j(t')}, \qquad t' \leq t \tag{4.232}$$

These terms may be directly interpreted as "memory functions" in that each term represents the sensitivity of species i at time t with respect to variations of species j at prior time t'. The result is a very detailed probe for the operational mechanistic pathways of a complex chemical system. Perturbations of the chemical species can be introduced at varying times t', and the subsequent response at later times can be studied to deduce mechanistic information. Green's function coefficients are, in principle, the *only* sensitivity coefficients directly measurable in the laboratory, and thus, unexpected behavior as indicated by these response functions can be experimentally investigated.

For example, the response of hydroxyl radicals to a small source of molecular hydrogen was examined by Yetter et al. (1985). Figure 4.8 shows that when H_2 is perturbed between 0 and 10 ms or between 0.5 and 100 s, the OH radical concentration is disturbed at some later time.

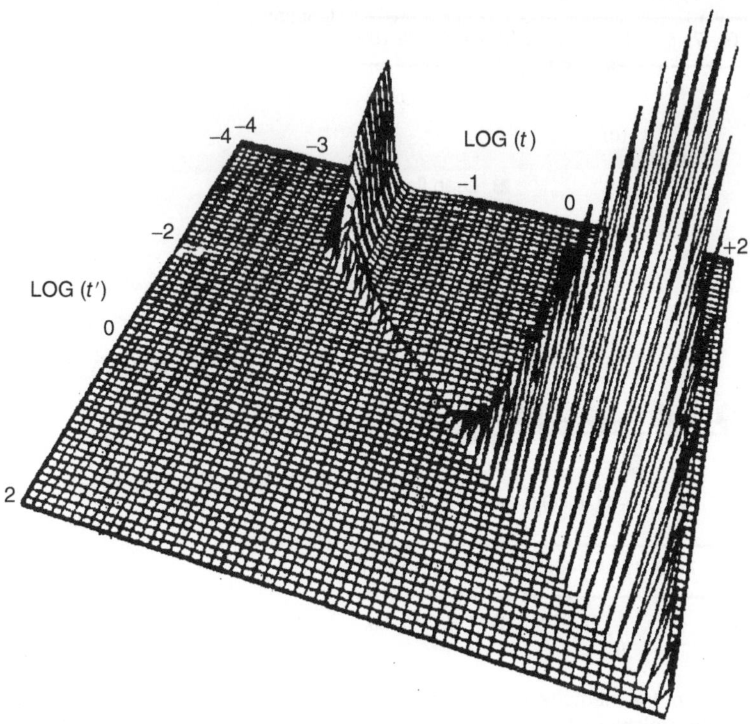

Figure 4.8 Green's function coefficient, $\partial C_{OH}(t)/\partial C_{H_2}(t')$, versus $\log_{10}(t')$ and $\log_{10}(t)$. *Source*: Yetter et al. 1985/with permission of Elsevier.

The minimum and maximum values of $\partial C_{OH}(t)/\partial C_{H_2}(t')$ are −0.03 and 0.7. The later response in OH is significantly different from the earlier response in that it resides for a considerably longer period. That is, the H_2 is not directly converted to H_2O but exists in part as OH for a lengthy period of time prior to the formation of H_2O. This results primarily from the equilibration of reactions $H + H_2O \rightleftarrows OH + H_2$.

Sensitivity Analysis of Laminar Flame Velocity

Nowak and Warnatz (1988) applied a sensitivity analysis to aliphatic hydrocarbon combustion to identify the rate-limiting steps. Figure 4.9*(a)* and 4.9*(b)* shows the results of a sensitivity analysis of the flame velocity of stoichiometric methane CH_4/air and propane C_3H_8/air flames at atmospheric pressure, respectively. The figure shows that the flame velocities are relatively insensitive

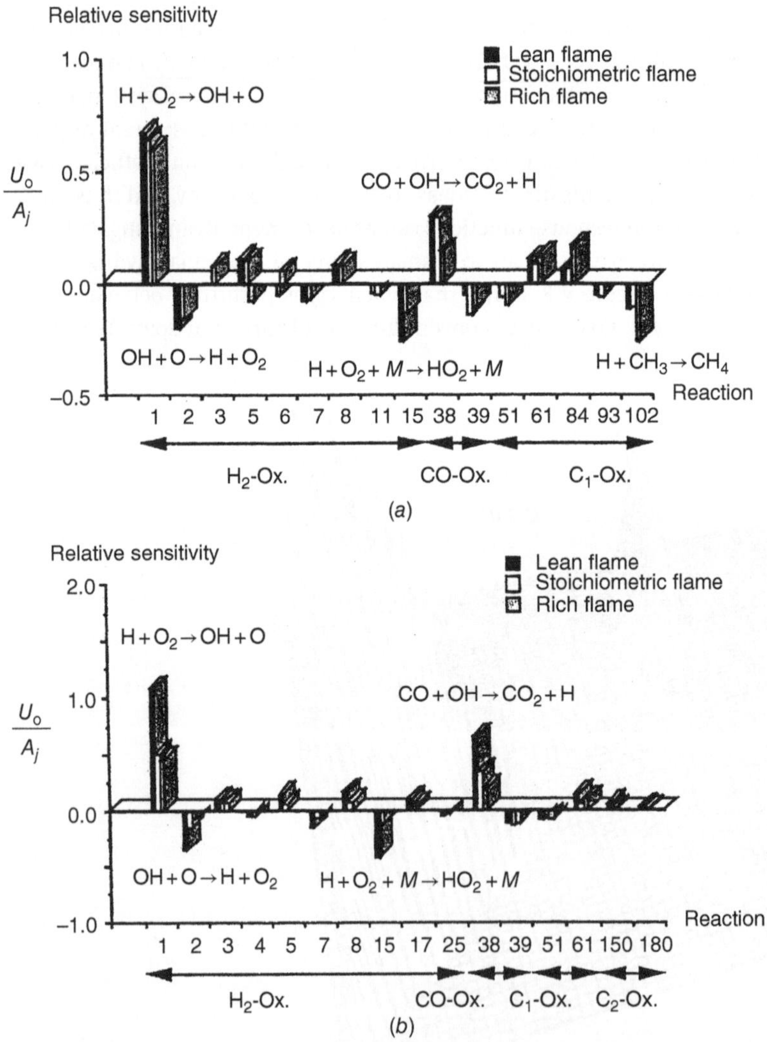

Figure 4.9 *(a)* Sensitivity analysis of the methane/air laminar flame velocity. *(b)* Sensitivity analysis of the propane/air laminar flame velocity (stoichiometric, $p = 1$ bar, and $T_u = 298$ K). *Source*: Adapted from Nowak and Warnatz 1988.

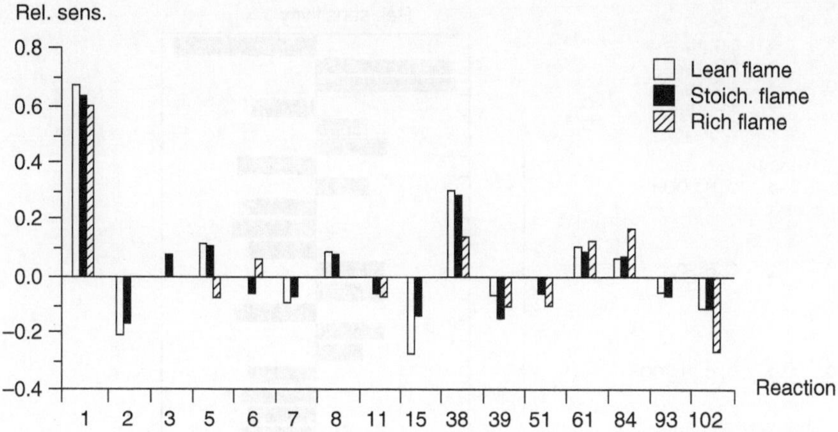

Figure 4.10 Sensitivity analysis of flame velocity U_0 to equivalence ratio for methane/air combustion. *Source*: Nowak and Warnatz 1988/with permission of American Institute of Aeronautics and Astronautics.

to reactions related specifically to methane and propane oxidation. However, there is a strong influence of the several elementary reactions involving oxidation of H atoms and CO reaction with OH radicals, i.e., $H + O_2 \rightarrow OH + O$, $H + O_2 + M \rightarrow HO_2 + M$, and $CO + OH \rightarrow CO_2 + H$.

The similar sensitivity results between CH_4/air and C_3H_8/air laminar flames are significant. For both cases, only a few elementary reactions are sensitive, giving nearly the same qualitative results. This implies that some elementary reactions are always rate limiting, independent of the type of fuels considered. The effect of the stoichiometric ratio on the relative sensitivity of the laminar flame speed for methane CH_4/air combustion is shown in Figure 4.10. Again, the differences between lean and rich flames are not pronounced. Therefore, the sensitivity results are more or less independent of the equivalence ratio of the initial mixture.

In Figure 4.11, the sensitivity analysis of mass fractions of hydrogen Y_H, water vapor Y_{H_2O}, and temperature T to reaction rate constants is presented. The mass fraction of hydrogen atoms is much

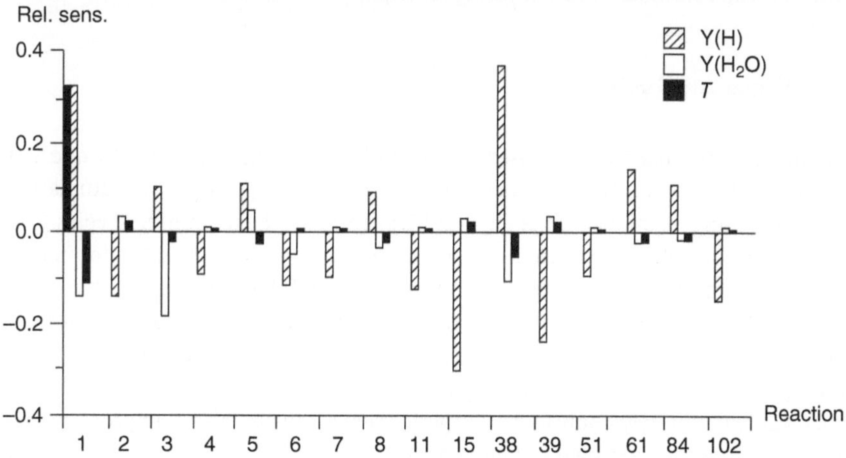

Figure 4.11 Sensitivity of mass fractions Y of hydrogen, water vapor H_2O, and temperature T to reaction rate constants for a methane/air laminar flame at $p = 1$ bar and $T_u = 298$ K. *Source*: Nowak and Warnatz 1988/with permission of American Institute of Aeronautics and Astronautics.

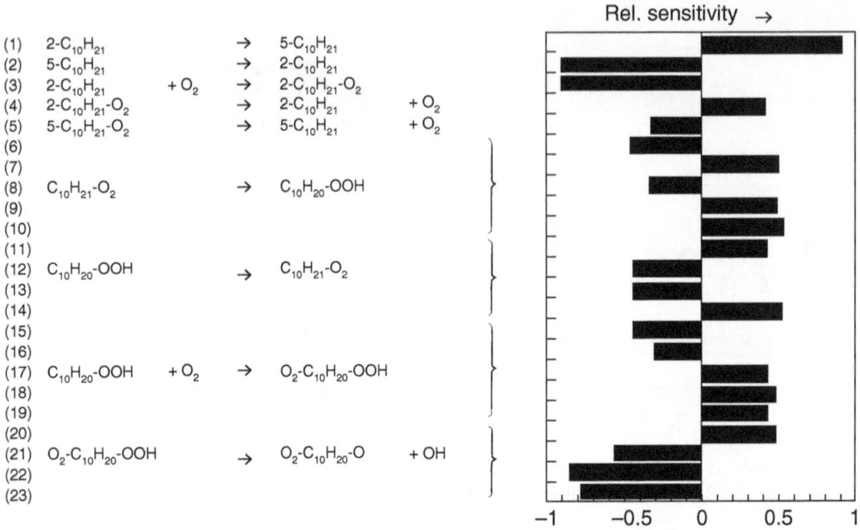

Figure 4.12 Sensitivity analysis for the OH concentration in an igniting stoichiometric $C_{10}H_{22}$ air mixture at $p = 13$ bar and $T_u = 800$ K. *Source*: Nehse et al. 1996/with permission of Elsevier.

more sensitive to the rate coefficients of different reactions than the mass fraction of the stable species H_2O and the temperature T.

Sensitivity Analysis of Ignition Delay

Figure 4.12 shows the results of a sensitivity analysis by Nehse et al. (1996) for the OH concentration in an igniting stoichiometric dodecane ($C_{10}H_{22}$)/air mixture at $p = 13$ bar and a relatively low initial temperature of 800 K. For this ignition delay problem, the sensitivity analysis indicated that the induction time is strongly dependent on many elementary reactions. This dependency is much stronger than the results shown previously in Figures 4.9 and 4.10 for stationary laminar flames, as Figure 4.12 shows a greater number of rate-limiting reactions.

4.12 Reaction Flow Analysis

In numerical simulations of combustion processes with complex reactions, *reaction flow analysis* can often be very helpful. The method of reaction rate analysis assumes a species to be redundant if elimination of all its consuming reactions induces no significant error to the remaining species. The percentage of the contributions of different elementary reactions $r(r = 1, \dots, M)$ to the formation (or consumption) of the chemical species $s(s = 1, \dots, N)$, is shown in Table 4.4.

In this example, 31.0% of the formation of species 1 can be attributed to reaction 1, 0.1% to reaction 2, 2.9% to reaction 3, and 66% to reaction $M - 1$. The percentages in the columns have to add to 100%. After obtaining the percentage output of all species for each reaction, an integral reaction flow analysis and a local reaction flow analysis can be performed for the combustion problem. An example of a *reaction flow diagram* is shown in Figure 4.13.

An *integral reaction flow analysis* considers the overall formation or consumption of chemical species during the combustion process. The results for zero-dimension homogeneously mixed time-dependent systems are integrated over the whole reaction time. The results for 1-D stationary

Table 4.4 Sample Output Percentage of a Reaction Flow Analysis.

Reaction Number r	Chemical Species s 1	2	3	…	$N-1$	N
1	31.0	2.2	0	…	0	0
2	0.1	0	0	…	0	0
3	2.9	5.0	0	…	80.0	5.0
4	0	0	0	…	20.0	92.0
⋮	⋮	⋮	⋮	…	⋮	⋮
$M-1$	66.0	90.4	98.0	…	0	3.0
M	0	2.4	0	…	0	0

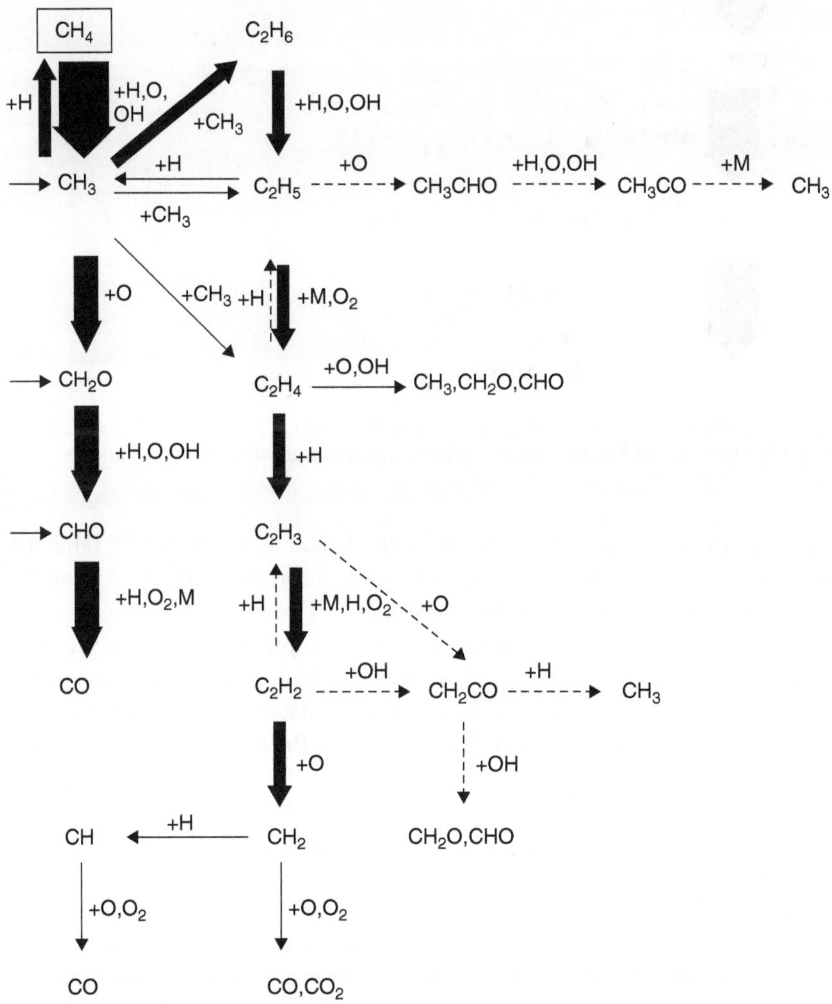

Figure 4.13 Integral reaction flow analysis in a premixed stoichiometric CH_4/air flame at $p = 1$ bar and $T_u = 298$ K. *Source*: Warnatz 1984/with permission of Springer Nature.

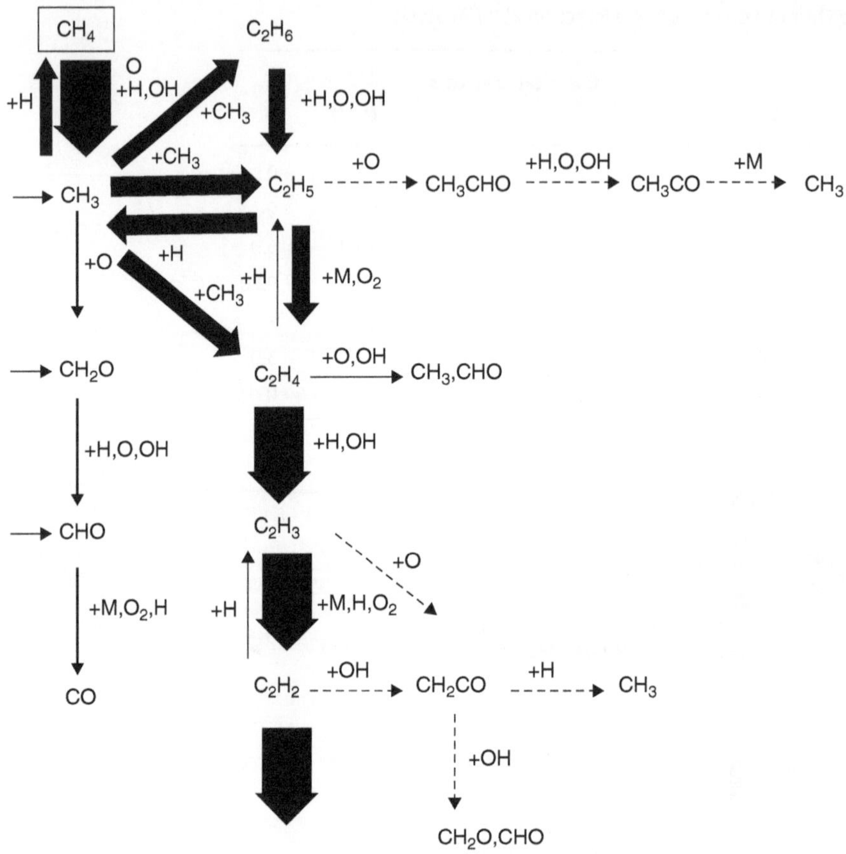

Figure 4.14 Integral reaction flow analysis in a premixed rich CH_4/air flame at $p = 1$ bar and $T_u = 298$ K. *Source*: Warnatz 1984/with permission of Springer Nature. (The analysis of acetylene is left out due to its complexity.)

flames are integrated over the entire reaction zone. A reaction such as reaction 2 in Table 4.4 can be regarded as unimportant if all entries in a row are below a certain limit, e.g., 1% for the formation as well as for the consumption of a chemical species.

A *local reaction flow analysis* considers the formation and consumption of species locally, i.e., at specific times in time-dependent problems (e.g., a homogeneous ignition process) or at specific locations in steady processes (e.g., a flat flame). According to the local analysis, a reaction rate $\dot{\Omega}$ is unimportant for all times t or locations x if the reaction rate satisfies the following condition:

$$|\dot{\Omega}_{t,r,s}| < \epsilon|\text{Max}_{r=1,2,\ldots,M}(\dot{\Omega}_{t,r,s})| \qquad s = 1, 2, \ldots, N, \quad t = 0, \ldots, t_{\text{total}} \tag{4.233}$$

This is a much stricter requirement than in the case of integral analysis. In Eq. (4.233), ϵ is a limit that has to be specified arbitrarily, e.g., $\epsilon = 1\%$.

Figure 4.13 shows an integral reaction flow analysis for a premixed stoichiometric methane CH_4 / air flame. A premixed-rich methane CH_4/air flame reaction flow analysis was also obtained and is shown in Figure 4.14. Note that different reaction paths are followed, depending on the equivalence ratio, even though the chemical mechanism used to model the reactions is the same. In the stoichiometric flame, methane CH_4 is oxidized directly, whereas the methyl radicals (CH_3) in the rich flame recombine to form ethane (C_2H_6), which is then oxidized.

4.13 Computational Singular Perturbation (CSP)

The CSP method (Lam and Goussis 1988) is a systematic method used to identify the quasi-steady state species (QSS) and the time scales of different modes. The method uses the concept of optimal representation of the rate of change of species mass fraction and temperature (represented by a vector **g**) and continuously monitors the contribution to **g** by each of its components, representing linearly independent reaction groups. At any given time, the technique can identify the fast and slow modes, i.e., which terms are nearly constant, which terms are controlling the reactions, and which chemical species are nearly exhausted. No term is neglected a priori, it is simply discarded when it has become numerically too small to make any difference. A similar technique, the method of *intrinsic low-dimensional manifolds in composition space*, developed by Maas and Pope (1992), is based on the same basic methods as CSP.

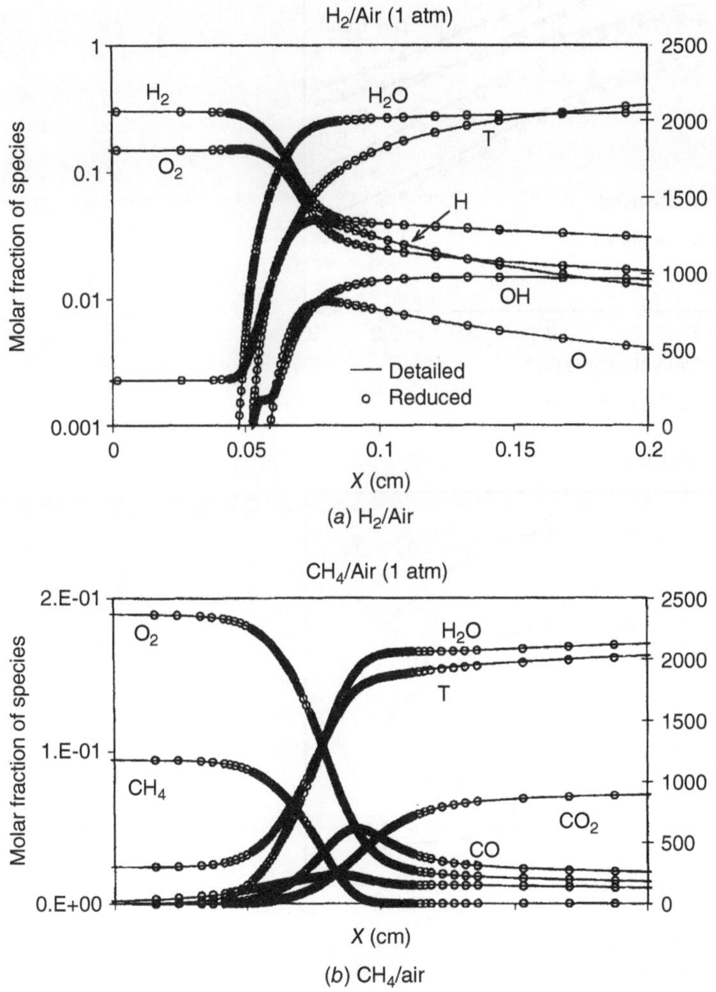

Figure 4.15 Comparison of the flame structures of temperature and species predicted by the reduced and detailed mechanisms for stoichiometric flames at 1 atm. (*a*) H_2/air; (*b*) CH_4/air. *Source*: Lu et al. 2001/with permission of Elsevier.

The CSP method models a given chemical system as a system of ordinary differential equations,

$$\frac{d\mathbf{y}}{dt} \equiv \mathbf{g}(\mathbf{y}) \tag{4.234}$$

where $\mathbf{y}$ is the concentration vector of all the species, and $\mathbf{g}$ is the source term. By taking the time derivative of Eq. (4.234), the following form with the time-dependent Jacobian matrix $\mathbf{J}$ can be derived through the chain rule:

$$\frac{d\mathbf{g}}{dt} = \mathbf{J} \cdot \mathbf{g}(\mathbf{y}), \tag{4.235}$$

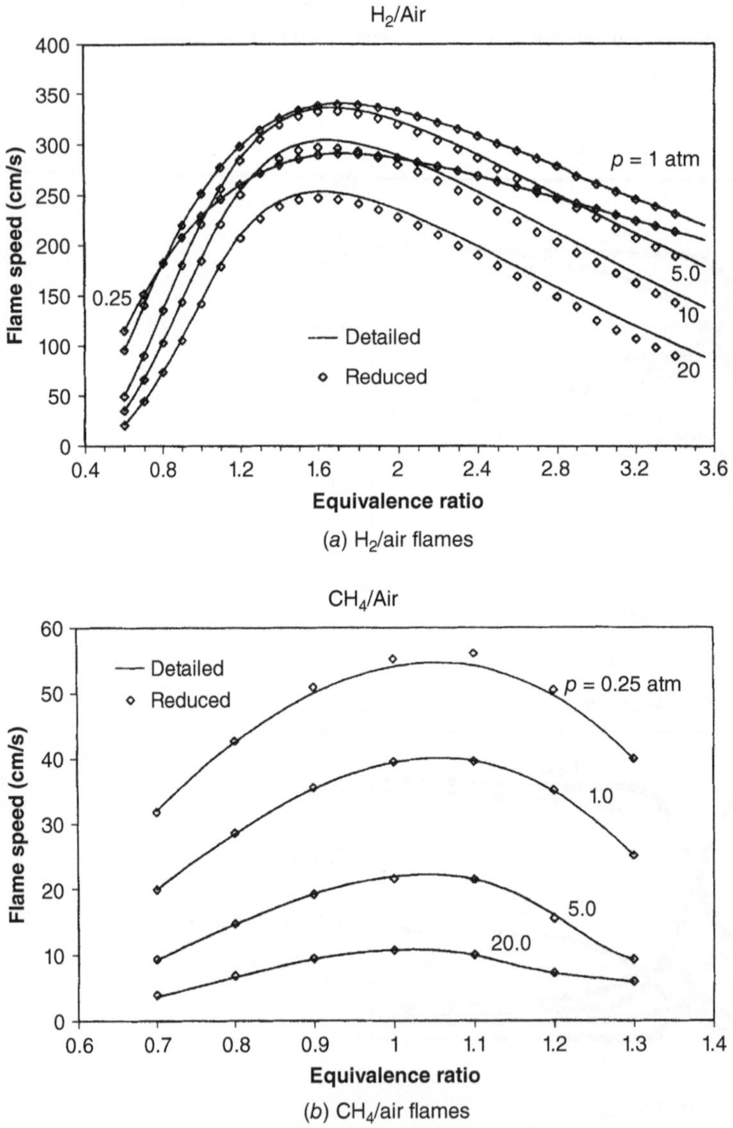

Figure 4.16 Comparison of the dependence of flame speed on the equivalence ratio between reduced and detailed mechanisms for (*a*) H_2/air flames and (*b*) CH_4/air flames at various pressures. *Source*: Lu et al. 2001/with permission of Elsevier.

The Jacobian matrix **J** is

$$\mathbf{J} = \frac{d\mathbf{g}}{d\mathbf{y}} \tag{4.236}$$

By using the eigen-decomposition of the Jacobian matrix $\mathbf{J} = \mathbf{A}\Lambda\mathbf{B}$, Eq. (4.235) can be written as

$$\frac{d\mathbf{f}}{dt} = \Lambda \cdot \mathbf{f}, \qquad \text{where } \mathbf{f} \equiv \mathbf{B} \cdot \mathbf{g} \tag{4.237}$$

and

$$\Lambda = \left(\frac{d\mathbf{B}}{dt} + \mathbf{B} \cdot \mathbf{J}\right) \cdot \mathbf{A}, \qquad \mathbf{A} = \mathbf{B}^{-1} \tag{4.238}$$

The matrices **A** and **B** are matrices consisting of the column and row basis vectors, respectively. If **A** is built up by ideal basis vectors, then matrix Λ reduces to a diagonal matrix and its diagonal elements are the eigenvalues of **J**. The equations for different modes are then decoupled and can be solved independently. The reciprocal of an eigenvalue has the dimension of time and provides the time scale of the corresponding mode. For nonlinear systems, the Jacobian matrix **J** is generally time dependent, the basis vectors rotate in time, and it is not straightforward to obtain a diagonal matrix Λ. In such cases, a CSP refinement procedure (Lam 1993) can be employed to find a pair of **A** and **B** iteratively. For practical purposes in generating reduced mechanisms, CSP is often used together with a constant Jacobian assumption. Recently, an efficient CSP method which does not involve eigen-decomposition has been developed by Zhao and Lam (2019).

Lu et al. (2001) employed the CSP method to generate a four-step and a ten-step reduced mechanism for high-temperature H_2/air and CH_4/air oxidation, respectively. The validity of these reduced mechanisms was evaluated based on the responses of perfectly stirred reactors and one-dimensional planar propagating premixed flames. Comparisons (see Figures 4.15 and 4.16) between the reduced and detailed chemistries over a wide range of pressures and equivalence ratios show very good agreement on the flame speed, flame temperature, and flame structure.

References

Bertolino, A., M. Furst, A. Stagni, A. Frassoldati, M. Pelucchi, C. Cavallotti, T. Faravelli, and A. Parente (2021), An evolutionary, data driven approach for mechanism optimization: theory and application to ammonia combustion, *Combust. Flame*, Vol. 229, p. 111366.

Bowman, C. (1974), Non-equilibrium radical concentrations in shock- initiated methane oxidation, *15th Symposium (Intl.) on Combustion*, The Combustion Institute, pp. 869–882.

Brabbs, T., and R. Brokaw (1974), Shock tube measurements of specific reaction rates in branched chain CH_4-CO-O_2 system, *15th Symposium (Intl.) on Combustion*, The Combustion Institute, pp. 893–901. (also NASA-TM-X-71534).

Dougherty, E., and H. Rabitz (1980), Computational kinetics and sensitivity analysis of hydrogen-oxygen combustion, *J. Chem. Phys.*, Vol. 72, pp. 6571–6586.

Eggels, R., and L. de Goey (1995), Mathematically reduced reaction mechanisms applied to adiabatic flat hydrogen/air flames, *Combust. Flame*, Vol. 100, pp. 559–570.

Glassman, I., and R. Yetter (2014), *Combustion*, 5th ed., Academic Press, New York.

Kee, R., F. Rupley, and J. Miller (1989), Chemkin-II: A Fortran chemical kinetics package for the analysis of gas-phase chemical kinetics, *SAND-89-8009*, Sandia National Lab., Livermore, CA.

Kobayashi, H., A. Hayakawa, K. Kunkuma, A. Somarathne, and E. Okafor (2019), Science and technology of ammonia combustion, *Proc. Combust. Inst.*, Vol. 37, pp. 109–133.

Lam, S., and D. Goussis (1988), Understanding complex chemical kinetics with computational singular perturbation, *Proc. Combust. Inst.*, Vol. 22, pp. 931–941.

Lam, S. (1993), Using CSP to understand complex chemical kinetics, *Comb. Sci. and Tech.*, Vol. 89, pp. 375–404.

Li, R., A. Konnov, G. He, F. Qin, and D. Zhang (2019), Chemical mechanism development and reduction for combustion of $NH_3/H_2/CH_4$ mixtures, *Fuel*, Vol. 257, p. 116059.

Lu, T., Y. Ju, and C. K. Law (2001), Complex CSP for chemistry reduction and analysis, *Combust. Flame*, Vol. 126, pp. 1445–1455.

Maas, U., and S. Pope (1992), Simplifying chemical kinetics: intrinsic low-dimensional manifolds in composition space, *Combust. Flame*, Vol. 88, pp. 239–264.

McAllister, S., J. Chen, and A. Fernandez-Pello (2011), *Fundamentals of Combustion Processes*, Springer, New York.

Miller, J., R. Mitchell, M. Smooke, and R. Kee (1982), Toward a comprehensive chemical kinetic mechanism for the oxidation of acetylene: comparison of model predictions with results from flame and shock tube experiments, *19th Symposium (Intl.) on Combustion*, The Combustion Institute, pp. 181–196.

Miller, J., R. Mitchell, M. Smooke, and R. Kee (1983), Kinetic modeling of the oxidation of ammonia in flames, *Combust. Sci. Technol.*, Vol. 34, pp. 149–176.

Mishra, M., R. Yetter, Y. Reuven, H. Rabitz, and M. Smooke (1994), On the role of transport in the combustion kinetics of a steady-state premixed laminar $CO + H_2 + O_2$ flame, *Int. J. Chem. Kinet.*, Vol. 26, No. 4, pp. 437–453.

Nehse, M., J. Warnatz, and C. Chevalier (1996), Kinetic modeling of the oxidation of large aliphatic hydrocarbons, *Proc. Combust. Inst.*, Vol. 26, pp. 773–780.

Nowak, U., and J. Warnatz (1988), Sensitivity analysis in aliphatic hydrocarbon combustion, in *Dynamics of Reactive Systems*, Vol. 1, Part 1 (A. Kuhl et al. eds.), AIAA, p. 87.

Rabitz, H., A. Kramer, and D. Dacol (1983), Sensitivity analysis in chemical kinetics, *Annu. Rev. Phys. Chem.*, Vol. 34, pp. 419–461.

Reuven, Y., M. Smooke, and H. Rabitz (1987), Sensitivity analysis of one-dimensional mixed initial-boundary value problems: application to freely propagating premixed laminar flames, *J. Sci. Comput.*, Vol. 2, pp. 345–370.

Sánchez, A., and F. Williams (2014), Recent advances in understanding of flammability characteristics of hydrogen, *Prog. Energy Combust. Sci.*, Vol. 41, pp. 1–55.

Schreiber, M., A. Sakak, A. Lingens, and J. Griffiths (1994), A reduced thermokinetic model for the autoignition of fuels with variable octane ratings, *25th Symposium (Intl.) on Combustion*, The Combustion Institute, pp. 933–940.

Simmie, J. (2003), Detailed chemical kinetic models for the combustion of hydrocarbon fuels, *Prog. Energy Combust. Sci.*, Vol. 29, pp. 599–634.

Smith, G., D. Golden, M. Frenklach, N. Moriarity, B. Eiteneer, M. Goldenberg, C. Bowman, R. Hanson, S. Song, W. Gardiner Jr., V. Lissianski, and Z. Qin (1995), GRI MECH 3.0, http://www.me.berkeley.edu/gri_mech/.

Smooke, M., H. Rabitz, Y. Reuven, and F. Dryer (1988), Application of sensitivity analysis to premixed hydrogen-air flames, *Combust. Sci. Technol.*, Vol. 59, pp. 295–319.

Song, Y, H. Hashemi, J. Christensen, C. Zou, P. Marshall, and P. Glarborg (2016), Ammonia oxidation at high pressure and intermediate temperatures, *Fuel*, Vol. 181, pp. 358–365.

Tomlin, S., M. Pilling, T. Turányl, J. Merkin, and J. Brindley (1992), Mechanism reduction for the oscillatory oxidation of hydrogen: sensitivity and steady state analyses, *Combust. Flame*, Vol. 91, pp. 107–130.

Warnatz, J., H. Bockhorn, A. Moser, and H. Wenz (1982), Experimental investigations and computational simulation of acetylene-oxygen flames from near stoichiometric to sooting conditions, 19th Symposium (Intl.) on Combustion, The Combustion Institute, pp. 197–209.

Warnatz, J., U. Maas, and R. Dibble (2006), *Combustion*, Springer-Verlag, Berlin, Germany.

Westbrook, C., and F. Dryer (1984), Chemical kinetic modeling of hydrocarbon combustion, *Prog. Energy Combust. Sci.*, Vol. 10, pp. 1–57.

Yetter, R., F. Dryer, and H. Rabitz (1985), Some interpretive aspects of elementary sensitivity gradients in combustion kinetics modeling, *Combust. Flame*, Vol. 59, pp. 107–133.

Zhao, P., and S. Lam (2019), Toward computational singular perturbation (CSP) without eigen-decomposition, *Combust. Flame*, Vol. 209, pp. 63–73.

Homework Problems

4.1 For the reaction sequence

$$N_5O_5 \xrightarrow{k_1} NO_2 + NO_3$$

$$NO_2 + NO_3 \xrightarrow{k_2} N_2O_5$$

$$NO_2 + O_3 \xrightarrow{k_3} NO_3 + O_2$$

$$2NO_3 \xrightarrow{k_4} 2NO_2 + O_2$$

(a) Write equations for the rates of formation of N_2O_5, NO_2, NO_3, O_2, and O_3.
(b) Assume that the species NO_2 and NO_3 are in steady state. Solve for their concentrations.
(c) What do the results of the steady-state assumption imply with respect to the rate of N_2O_5 formation?

4.2 At low pressure, assuming no wall reaction, the H_2—O_2 reaction may be accounted for by the mechanism

$$H_2 + O_2 \xrightarrow{k_1} 2OH, \qquad k_1 \approx 10^{12}\exp(-39.0/R_uT)\ \text{cm}^3/\text{mol-s}$$

$$OH + H_2 \xrightarrow{k_2} H_2O + H, \qquad k_2 \approx 10^{13.8}\exp(-5.9/R_uT)\ \text{cm}^3/\text{mol-s}$$

$$H + O_2 \xrightarrow{k_3} OH + O, \qquad k_3 \approx 2.2\times10^{14}\exp(-16.5/R_uT)\ \text{cm}^3/\text{mol-s}$$

$$O + H_2 \xrightarrow{k_4} OH + H, \qquad k_4 \approx 1.1\times10^{13}\exp(-9.4/R_uT)\ \text{cm}^3/\text{mol-s}$$

$$H + OH + M \xrightarrow{k_5} H_2O + M, \qquad k_5 \approx 10^{17}\text{cm}^3/\text{mol-s}\,(E_a = 0\ \text{J/mol})$$

Using the steady-state hypothesis, derive the differential equations expressing $d[H_2O]/dt$ in terms of $[H_2]$ and $[O_2]$.

4.3 A stoichiometric mixture of octane and air is ignited and combusts at a temperature of 2100 K and pressure of 10 atm. (a) What is the rate of reaction at that temperature and pressure? (b) If the temperature and volume are held constant during the combustion process, what is the rate of reaction when 1/2 of the fuel has been burned? Assume a single-step global reaction with parameters given in Table 4.2.

4.4 The combustion chamber in a rocket contains H atoms and OH radicals at equal concentrations of 4×10^{-6} mole/cm^3 and 3000 K. The total gas concentration in the combustion chamber is 4×10^{-4} mole/cm^3. Supposing the gases are to be exhausted from the chamber at 1000 K (where H_2O is essentially undissociated) and at a density equal to 1/40 of the density in the chamber, calculate how far in the exhaust stream, moving at 3048 m/s, one will have to probe before finding that the H and OH are 99% recombined. Ignore recombination reactions other than $H + OH + M \rightarrow H_2O + M$, for which k is given as 10^{16}cm^6/mole^2-s, independent of temperature (this is a typical figure for a three-body recombination rate constant).

4.5 One of the mechanisms of atom and radical combinations, called the *energy-transfer mechanism*, is described by the following chemical steps:

$$2R \underset{k_2}{\overset{k_1}{\rightleftharpoons}} R_2^*$$

$$R_2^* + X \xrightarrow{k_3} R_2 + X$$

where R represents a radical or an atom and X is a third body. Use the steady-state treatment for R_2^* to determine the rate of consumption of R. What is the order of the recombination reaction when the concentration of X is sufficiently large such that $k_3C_X \gg k_2$? What is the order of the recombination reaction at very low pressures?

4.6 Another mechanism of atom and radical combination is called the *atom–molecule* or *radical–molecule-complex mechanism*. This mechanism is particularly likely for atom combinations when the third-body X is a molecule that can readily form a complex with one of the atoms. The reaction steps may be represented as

$$R + X \underset{k_2}{\overset{k_1}{\rightleftharpoons}} RX^*$$

The resulting complex retains the energy released in its formation and may be de-energized by collision with another molecule X:

$$RX^* + X \underset{k_4}{\overset{k_3}{\rightleftharpoons}} RX + X + q_{(\text{heat release})}$$

Finally, RX may react with R, giving rise to

$$RX + R \xrightarrow{k_5} R_2 + X$$

Apply the steady-state treatment to determine the rate of consumption of R. What is the order of the combination reaction when the concentration of R is sufficiently small? Does the reaction become second-order at high pressures (high concentrations of X)?

4.7 Consider the following set of elementary reactions between hydrogen and chlorine:

$$Cl_2 + M \xrightarrow{k_1} Cl + Cl + M \qquad (R1)$$

$$Cl + H_2 \xrightarrow{k_2} HCl + H \qquad (R2)$$

$$H + Cl_2 \xrightarrow{k_3} HCl + Cl \qquad (R3)$$

$$Cl + Cl + M \xrightarrow{k_4} Cl_2 + M \qquad (R4)$$

(a) Show that this scheme leads to the rate expression

$$\frac{d[\mathrm{HCl}]}{dt} = 2k_2\sqrt{\frac{k_1}{k_4}}[\mathrm{Cl_2}]^{1/2}[\mathrm{H_2}]$$

(b) Nitrogen trichloride NCl_3 is a potential inhibitor for the reaction, according to

$$\mathrm{NCl_3 + Cl} \xrightarrow{k_5} \mathrm{NCl_2 + Cl_2} \qquad \text{(R5)}$$

Derive the kinetic rate expression for the scheme including reaction (R5). What is the form of the rate expression when $k_5 \gg k_1$?

(c) Oxygen O_2 also inhibits the reaction through

$$\mathrm{H + O_2} + M \xrightarrow{k_6} \mathrm{HO_2} + M \qquad \text{(R6)}$$

Why should NCl_3 be a much stronger inhibitor than O_2?

5

Transport Processes

5.1 Introduction

The transport of mass, momentum, and energy in chemical reactions is an essential aspect of most combustion processes. Combustion systems are modeled by the continuum mechanics equations commonly used in fluid mechanics and heat transfer. Additional terms in the mass, momentum, and energy conservation equations account for the effects of chemical reactions. In addition to these conservation equations, the prediction of combustion processes requires additional relations to track each chemical species, as combustion creates and destroys individual species. In a multicomponent system, the various chemical species can move at different velocities, so considerations such as concentration gradients, mass diffusivity, and molar flux also need to be included in the analysis.

In laminar flows, layers of fluid slide past another in a smooth manner, and the mixing between fluid layers is due to molecular diffusion. As the Reynolds number of the mean flow increases above a critical value, the flow will become turbulent. Turbulent flow can be envisioned as a mean fluid flow upon which are superimposed eddies of different sizes randomly dispersed in the flow. In combustion applications, the fluid mechanics of the turbulent flow field is coupled with species reaction chemistry and energy release.

Study of turbulent combustion is an important and also a very challenging field of engineering, since the flow conditions are turbulent in most practical combustion devices. The transport of momentum and energy by turbulent flow is orders of magnitude larger than that by laminar flow, so the chemical energy release rate and hence the power available from a given combustor or engine are increased to values much greater than those that can be obtained with laminar flow conditions. Spark ignition engines and gas turbine combustors rely on turbulent premixed combustion; diesel engines, rocket engines, and industrial furnaces employ turbulent diffusion combustion.

In this chapter, we introduce three major categories of numerical techniques used in analysis of turbulent combustion transport processes: Reynolds averaged Navier–Stokes (RANS) simulation, large-eddy simulation (LES), and direct numerical simulation (DNS). Examples of these three numerical techniques for a diffusion flame are shown in Figure 5.1. Note the different scales of the flame that are available with each of the techniques. Currently, RANS is most commonly employed in industry, but its range of validity is limited. DNS is the most detailed, but it is too computationally demanding for most realistic engineering problems. LES is a compromise between the two and provides excellent reliability and applicability.

Principles of Combustion, Third Edition. Allan T. Kirkpatrick and Kenneth K. Kuo.

Companion website: www.wiley.com/go/kirkpatrick/principlesofcombustion3e

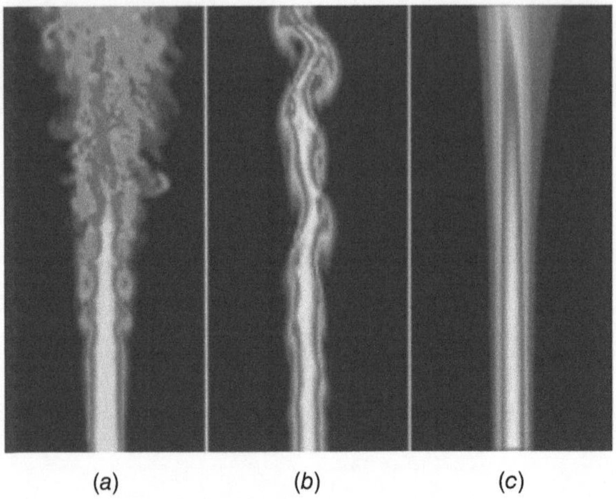

Figure 5.1 Computation of a turbulent diffusion flame using (*a*) DNS, (*b*) LES, and (*c*) RANS. (Courtesy of Dr. Peyman Givi).

5.2 Definitions of Concentrations, Velocities, and Mass Fluxes

The concentration of various species in a multicomponent system may be expressed in numerous ways. We will consider four such ways:

(1) ***Mass concentration or density*** ρ_i, which is the mass of the ith species per unit volume of mixture.
(2) ***Molar concentration*** $[C_i] = \rho_i/\text{M}_i$, which is the number of moles of ith species per unit volume.
(3) ***Mass fraction*** $Y_i = \rho_i/\rho$, which is the mass concentration of ith species divided by the total mass density of the mixture.
(4) ***Mole fraction*** $X_i = n_i/n = [C_i]/[C]$, which is the molar concentration of the ith species divided by the total molar density of the gaseous mixture or liquid solution.

In a multicomponent system, various chemical species will move at different velocities. Let us use $\mathbf{v}_i$ to denote the velocity of the ith species with respect to the stationary coordinate axes.

For a mixture of N species, the local bulk mass-average velocity $\mathbf{v}$ is defined as

$$\mathbf{v} = \frac{\sum_{i=1}^{N} \rho_i \mathbf{v}_i}{\sum_{i=1}^{N} \rho_i} = \frac{\sum_{i=1}^{N} \rho_i \mathbf{v}_i}{\rho} = \sum_{i=1}^{N} Y_i \mathbf{v}_i \tag{5.1}$$

Similarly, a local bulk *molar-average velocity* $\hat{\mathbf{v}}$ is defined as

$$\hat{\mathbf{v}} = \frac{\sum_{i=1}^{N} [C_i]\mathbf{v}_i}{\sum_{i=1}^{N} [C_i]} = \frac{\sum_{i=1}^{N} [C_i]\mathbf{v}_i}{[C]} = \sum_{i=1}^{N} X_i \mathbf{v}_i \tag{5.2}$$

In general, $\hat{\mathbf{v}}$ differs from $\mathbf{v}$ in both magnitude and direction, since their weighting factors are different, as indicated by the above two equations.

In multicomponent reacting flow systems, we are frequently interested in the velocity of a given species with respect to the local bulk mass-average velocity $\mathbf{v}$ or bulk molar-average velocity $\hat{\mathbf{v}}$ of the mixture. This leads to the definition of the *diffusion velocities* (see Figure 5.2).

The mass diffusion velocity $\mathbf{V}_i$ of species i is

$$\mathbf{V}_i = \mathbf{v}_i - \mathbf{v} \quad \text{(mass diffusion)} \tag{5.3}$$

Figure 5.2 Diagram of velocity vectors, where $\mathbf{V}_i = \mathbf{v}_i - \mathbf{v}$ and $\hat{\mathbf{V}}_i = \mathbf{v}_i - \hat{\mathbf{v}}$.

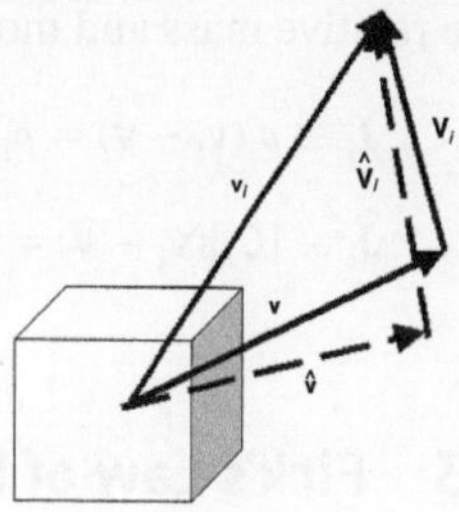

Similarly, the molar diffusion velocity $\hat{\mathbf{V}}_i$ of species i is

$$\hat{\mathbf{V}}_i = \mathbf{v}_i - \hat{\mathbf{v}} \quad \text{(molar diffusion)} \tag{5.4}$$

The diffusion velocities $\mathbf{V}_i$ and $\hat{\mathbf{V}}_i$ indicate the average motion of component i relative to the local bulk motion of the mixture in the fluid stream.

Example 5.1 ***Bulk and Diffusion Velocities***

Consider a simple case of a binary system with species a and b. The mole fraction of species a is $X_a = \frac{1}{3}$, and the mole fraction of b is $X_b = \frac{2}{3}$. Assume the velocity vectors are colinear. The magnitudes of the bulk molar average velocity $\hat{\mathbf{v}} = 10$, and the molar diffusion velocity $\hat{\mathbf{V}}_a = \mathbf{v}_a - \hat{\mathbf{v}} = 2$. The molecular masses are in the ratio $M_a = 3\ \mathrm{M}_b$. What is the species b mass–average velocity $\mathbf{v}_b$, and the bulk mass-average velocity $\mathbf{v}$?

Solution

First, solving for $\mathbf{v}_a$,

$$\mathbf{v}_a = \hat{\mathbf{V}}_a + \hat{\mathbf{v}} = 2 + 10 = 12$$

and from Eq. (5.2),

$$\hat{\mathbf{v}} = X_a\mathbf{v}_a + X_b\mathbf{v}_b$$

$$10 = \frac{1}{3}(12) + \frac{2}{3}(\mathbf{v}_b)$$

$$\mathbf{v}_b = \frac{3}{2}(10 - 4) = 9$$

From Eq. (2.9), the mass fractions Y_i are found from the mole fractions X_i and molecular masses M_i,

$$Y_a = \frac{X_aM_a}{X_aM_a + X_bM_b} = \frac{1/3M_a}{1/3M_a + (2/3)1/3M_a} = 0.6$$

$$Y_b = 1 - Y_a = 0.4$$

and from Eq. (5.1), the mass–average velocity $\mathbf{v}$ is

$$\mathbf{v} = Y_a\mathbf{v}_a + Y_b\mathbf{v}_b = (0.6)12 + (0.4)9 = 10.8$$

Now that concentrations and velocities have been discussed, we are in a position to define mass and molar fluxes. The mass (or molar) flux of species i is a vector quantity denoting the mass (or number of moles) of species i that passes through a unit area per unit time. Thus, the absolute mass $\dot{\mathbf{m}}''$ and molar $\dot{\mathbf{n}}''$ fluxes with respect to stationary coordinates are

$$\text{mass flux:} \quad \dot{\mathbf{m}}_i'' \equiv \rho_i\mathbf{v}_i \quad (\mathrm{kg/m^2\,s}) \tag{5.5}$$

$$\text{molar flux:} \quad \dot{\mathbf{n}}_i'' \equiv [C_i]\mathbf{v}_i \quad (\mathrm{kmol/m^2\,s}) \tag{5.6}$$

The relative mass and molar fluxes are defined as

$$\mathbf{J}_i \equiv \rho_i(\mathbf{v}_i - \mathbf{v}) = \rho_i \mathbf{V}_i \tag{5.7}$$

$$\hat{\mathbf{J}}_i \equiv [C_i](\mathbf{v}_i - \hat{\mathbf{v}}) = [C_i]\hat{\mathbf{V}}_i \tag{5.8}$$

5.3 Fick's Law of Diffusion

In a binary system with overall density ρ with two chemical species A and B, species A always diffuses relative to the mixture in the direction from high concentration of A to low concentration of A. The same behavior is true for species B, diffusing in the direction from high concentration of B to low concentration of B. Therefore, the binary mass diffusivity can be expressed by either $\mathcal{D}_{AB}$ or $\mathcal{D}_{BA}$, with dimensions given in (m^2/s).

Fick's first law of diffusion in terms of the mass diffusion flux $\mathbf{J}_A$ is,

$$\mathbf{J}_A = -\rho \mathcal{D}_{AB} \nabla Y_A \tag{5.9}$$

and in terms of the molar diffusion flux $\hat{\mathbf{J}}_A$ is

$$\hat{\mathbf{J}}_A = -[A]\,\mathcal{D}_{AB} \nabla X_A \tag{5.10}$$

The net mass flux of species A relative to stationary coordinates is determined by superimposing the relative diffusion onto the bulk flow. The mass flux $\dot{\mathbf{m}}''$ of species A relative to stationary coordinates can be expressed as

$$\dot{\mathbf{m}}''_A = \rho_A \mathbf{v} - \rho \mathcal{D}_{AB} \nabla Y_A \tag{5.11}$$

The first term represents the flux of A resulting from the bulk motion of the fluid. The second term with the minus sign represents the relative flux of A resulting from the diffusion of A.

Similar to Eq. (5.11), the molar flux relative to stationary coordinates is the sum of two terms

$$\dot{\mathbf{n}}''_A = [A]\hat{\mathbf{v}} - [A]\,\mathcal{D}_{AB} \nabla X_A \tag{5.12}$$

Again, the first term is the flux of species A due to bulk fluid motion. The second term including the minus sign represents the molar diffusion effect of species A, which is driven by the molar fraction gradient.

It is useful to note that the mathematical form of Fick's law of mass transport for constant-density situation in the transverse direction (y-direction) of a binary system is the same as Newton's law of momentum transport and Fourier's law of conduction heat transfer in the transverse direction. These equations are given here:

$$J_{Ay} = -\mathcal{D}_{AB} \frac{\partial}{\partial y}(\rho_A) \quad \text{(Fick's law for constant } \rho) \tag{5.13}$$

$$\tau_{yx} = -\nu \frac{\partial}{\partial y}(\rho u_x) \quad \text{(Newton's law for constant } \rho) \tag{5.14}$$

$$q_y = -\alpha \frac{\partial}{\partial y}(\rho c_p T) \quad \text{(Fourier's law for constant } \rho c_p) \tag{5.15}$$

Note that the dimensions of mass diffusivity ($\mathcal{D}$), kinematic viscosity (ν), and thermal diffusivity (α) are the same (m^2/s).

5.4 Dimensionless Ratios of Transport Coefficients

In combustion problems, the conservation equations are often written in forms involving dimensionless ratios of the transport coefficients. The dimensionless ratios are named after important figures in engineering.

The *Lewis number* (or Lewis–Semenov number) is defined as

$$\mathrm{Le} \equiv \frac{k}{\rho c_p \mathcal{D}} = \frac{\alpha}{\mathcal{D}} = \frac{\text{rate of thermal energy transport}}{\text{rate of mass transport}} \tag{5.16}$$

The Lewis number is named after Warren Lewis (1882–1975), an American chemical engineering professor who originated the concept of unit operations. Nikolai Semenov (1896–1986) was a Russian engineering professor who won the Nobel Prize in Chemistry in 1956 for his work on chemical kinetics and chain reactions.

The *Prandtl number* is defined as

$$\mathrm{Pr} \equiv \frac{c_p \mu}{k} = \frac{c_p \rho (\mu/\rho)}{k} = \frac{\nu}{\alpha} = \frac{\text{rate of momentum transport}}{\text{rate of thermal energy transport}} \tag{5.17}$$

The Prandtl number is named after Ludwig Prandtl (1875–1953), a German engineering professor who made many significant contributions to fluid mechanics and aerodynamics. The *Schmidt number* is defined as

$$\mathrm{Sc} \equiv \frac{\nu}{\mathcal{D}} = \frac{\text{rate of momentum transport}}{\text{rate of mass transport}} \tag{5.18}$$

The Schmidt number is named after Ernst Schmidt (1892–1975), a German engineering professor who was a pioneer in the field of engineering thermodynamics, especially in heat and mass transfer. The Lewis and Schmidt numbers can be defined for each pair of species in multicomponent mixtures. From the equations above, we see that

$$\mathrm{Le} \equiv \frac{\mathrm{Sc}}{\mathrm{Pr}} \tag{5.19}$$

In many combustion systems, Lewis numbers are close to unity. The approximation $\mathrm{Le} = 1$ is frequently very helpful in theoretical combustion analyses. However, Le can also differ significantly from unity in some situations. Therefore, researchers should examine the combustion conditions before making the assumption of $\mathrm{Le} = 1$ for their analyses and/or modeling work.

5.5 Theory of Mass Diffusion in Gases at Low Density

The mass diffusivity $\mathcal{D}_{AB}$ for binary mixtures of nonpolar gases is predictable within about 5% by kinetic theory. By definition, nonpolar gases contain molecules that have zero permanent electric dipole moment; however, they may temporarily acquire a dipole moment in an electric field as a result of the distortion of their electronic distributions and the locations of their nuclei.

Consider a nonpolar gas containing two molecular species A and A^*, both species having the same mass m_A and the same size and shape. Assume temperature T and molar concentration $[A]$ to be constant. From the kinetic theory of gases, the random molecular velocity relative to the fluid velocity $\mathbf{v}$ has an average magnitude, $\overline{u}$:

$$\overline{u} = \left(\frac{8R_u T}{\pi \mathrm{M}}\right)^{1/2} = \left(\frac{8\mathrm{kBT}}{\pi m_A}\right)^{1/2} \tag{5.20}$$

where $k_B = R_u/N_A$ is Boltzmann's constant = 1.3807×10^{-23} J/K, and N_A is Avogadro's number = 6.0225×10^{23} molecules/mole. The frequency of molecular collisions per unit area (Z) on a stationary surface exposed to the gas is

$$Z = \frac{1}{4}\tilde{n}\overline{u} \tag{5.21}$$

where $\tilde{n}$ represents the molecules per unit volume. The value of $\tilde{n}$ is a constant, since the molar concentration of the mixture $[A]$ is assumed to be constant and $\tilde{n} = [A] \times N_A$. The mean free path l, from kinetic theory, is

$$l = \frac{1}{\sqrt{2}\pi {d_A}^2 \tilde{n}} \tag{5.22}$$

where d_A is the diameter of the molecule A. The molecules reaching any y = constant plane in the gas have, on the average, had their last collision at a distance a from that plane (see Figure 5.3):

$$a = \frac{2}{3}l \tag{5.23}$$

The relative molar flux $\hat{\mathbf{J}}_A$ of species A across any plane of constant y is found by counting the molecules of A that cross a unit area of the plane in the positive y-direction and subtracting the number that crosses in the negative y-direction. Thus, if the concentration profile $X_A(y)$ is assumed to be essentially linear near y = constant plane, we have

$$\hat{\mathbf{J}}_A = \frac{\mathrm{Z}X_A|_{y-a} - \mathrm{Z}X_A|_{y+a}}{N_A}\mathbf{j} \tag{5.24}$$

As a further consequence of this assumption,

$$\frac{dX_A}{dy} = \frac{X_A|_y - X_A|_{y-a}}{a} = \frac{X_A|_{y+a} - X_A|_y}{a}$$

$$X_A|_{y-a} = X_A|_y - \frac{2}{3}l\frac{dX_A}{dy} \tag{5.25}$$

$$X_A|_{y+a} = X_A|_y + \frac{2}{3}l\frac{dX_A}{dy} \tag{5.26}$$

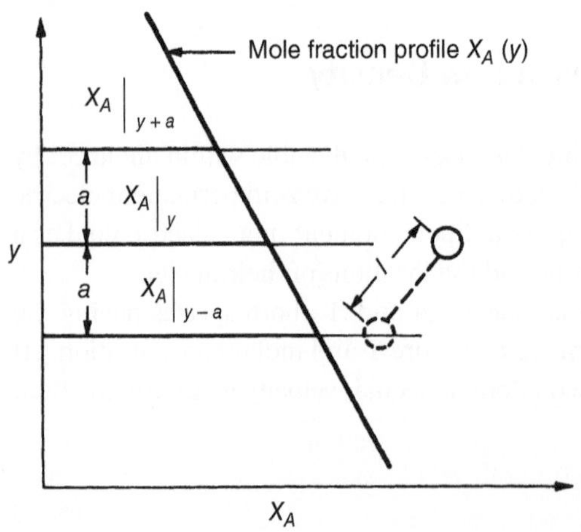

Figure 5.3 Molecular transport of species A from plane at $y - a$ to plane at y.

Substituting Eqs. (5.25) and (5.26) into Eq. (5.24), we have

$$\hat{\mathbf{J}}_{Ay} = \frac{Z}{N_A}\left(-\frac{4}{3}l\frac{dX_A}{dy}\right)\mathbf{j} = -\left(\frac{1}{3}\frac{\tilde{n}}{N_A}\overline{u}l\right)\frac{dX_A}{dy}\mathbf{j}$$

$$\hat{\mathbf{J}}_{Ay} = -[C]\left(\frac{1}{3}\overline{u}l\right)\frac{dX_A}{dy}\mathbf{j} \tag{5.27}$$

This equation corresponds to Fick's law of diffusion in the y-direction, with the following approximate value for $\mathcal{D}_{AA^*}$:

$$\mathcal{D}_{AA^*} = \frac{1}{3}\overline{u}l \tag{5.28}$$

Using Eq. (5.20) for $\overline{u}$ and Eq. (5.22) for l, Eq. (5.28) becomes

$$\mathcal{D}_{AA^*} = \frac{1}{3}\sqrt{\frac{8\text{kBT}}{\pi m_A}}\frac{1}{\sqrt{2}\pi d_A^2\tilde{n}} = \frac{2}{3}\sqrt{\frac{k_B^3}{\pi^3 m_A}\frac{T^{3/2}}{d_A^2}\frac{1}{\tilde{n}\text{kBT}}}$$

$$= \frac{2}{3}\sqrt{\frac{k_B^3}{\pi^3 m_A}\frac{T^{3/2}}{d_A^2}\frac{1}{\tilde{n}\text{kBT}}} \tag{5.29}$$

Using the perfect-gas law $p = [A]R_uT = \tilde{n}k_BT$, an approximate value for $\mathcal{D}_{AA^*}$ can be calculated from

$$\mathcal{D}_{AA^*} = \frac{2}{3}\sqrt{\frac{k_B^3}{\pi^3 m_A}\frac{T^{3/2}}{pd_A^2}} \tag{5.30}$$

The parameter $\mathcal{D}_{AA^*}$ represents the mass diffusivity of a mixture of two species of rigid spheres of identical mass and diameter. The calculation of $\mathcal{D}_{AB}$ for rigid spheres of unequal mass and diameter is slightly more complicated; the corresponding result is

$$\mathcal{D}\text{AB} = \frac{2}{3}\left(\frac{k_B^3}{\pi^3}\right)^{1/2}\left(\frac{1}{2m_A} + \frac{1}{2m_B}\right)^{1/2}\frac{T^{3/2}}{p\left(\frac{d_A+d_B}{2}\right)^2} \tag{5.31}$$

Additional information about the binary gas diffusion coefficient $\mathcal{D}_{AB}$ is given in the Appendix.

5.6 Continuity Equation and Species Mass Conservation Equations

To derive the mass conservation equation of each species in a multicomponent mixture, including chemical reactions that may be taking place, we begin by making a mass balance over an arbitrary differential fluid element in a binary mixture. We apply the law of conservation of mass of species A to a volume element $\Delta x\,\Delta y\,\Delta z$ fixed in space through which a binary mixture of A and B is flowing (see Figure 5.4).

Within this infinitesimal control volume, species A can be produced or consumed by chemical reaction at a mass rate $\dot{\omega}_A$ (kg m^{-3} s^{-1}). If $\dot{\omega}_A < 0$, species A is consumed. Various terms describing their respective contributions to the mass balance are given below.

The rate of accumulation of mass of species A in the control volume is

$$\frac{\partial \rho_A}{\partial t}\Delta x\,\Delta y\,\Delta z \tag{5.32}$$

The rate of mass inflow of A due to the x-direction mass flux at x station is

$$\dot{m}''_{Ax}|_x\,\Delta y\,\Delta z \tag{5.33}$$

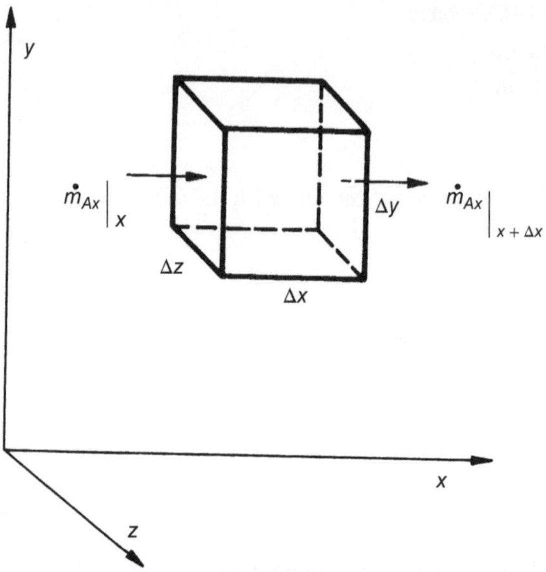

Figure 5.4 Fluid flow through a fixed infinitesimal control volume $\Delta x\,\Delta y\,\Delta z$.

The rate of mass outflow of A due to the x-direction mass flux at $x + \Delta x$ station is

$$\dot{m}''_{Ax}|_{x+\Delta x}\Delta y\,\Delta z = \dot{m}''_{Ax}|_x\Delta y\,\Delta z + \frac{\partial \dot{m}''_{Ax}}{\partial x}\Delta x\,\Delta y\,\Delta z \tag{5.34}$$

The mass rate of production of A by chemical reaction is

$$\dot{\omega}_A \Delta x\,\Delta y\,\Delta z \tag{5.35}$$

There are also inflow and outflow terms in the y and z directions. When the entire mass balance is divided through by $\Delta x\,\Delta y\,\Delta z$, one obtains

$$\frac{\partial \rho_A}{\partial t} + \left(\frac{\partial \dot{m}''_{Ax}}{\partial x} + \frac{\partial \dot{m}''_{Ay}}{\partial y} + \frac{\partial \dot{m}''_{Az}}{\partial z}\right) = \dot{\omega}_A \tag{5.36}$$

This is the *equation of continuity for component A* in a binary mixture. It is also called the mass conservation equation for species A. The quantities $\dot{m}''_{Ax}$, $\dot{m}''_{Ay}$, $\dot{m}''_{Az}$ are the scalar components of the mass flux vector, $\dot{\mathbf{m}}''_A$, in rectangular coordinates. In vector notation, the equation may be written as

$$\frac{\partial \rho_A}{\partial t} + (\nabla \cdot \dot{\mathbf{m}}''_A) = \dot{\omega}_A \tag{5.37}$$

Substituting Eq. (5.11) into Eq. (5.37), we obtain

$$\frac{\partial \rho_A}{\partial t} + \nabla \cdot \rho_A \mathbf{v} = \nabla \cdot \rho \mathcal{D}_{\mathrm{AB}} \nabla Y_A + \dot{\omega}_A \tag{5.38}$$

Similarly, the equation of continuity for component B is

$$\frac{\partial \rho_B}{\partial t} + (\nabla \cdot \dot{\mathbf{m}}''_B) = \dot{\omega}_B \tag{5.39}$$

Adding Eq. (5.37) to Eq. (5.39) gives

$$\frac{\partial \rho}{\partial t} + \nabla \cdot (\rho \mathbf{v}) = 0 \tag{5.40}$$

which is the *mixture continuity equation*.

In arriving at Eq. (5.40), we have made use of both the relation $\dot{\mathbf{m}}_A'' + \dot{\mathbf{m}}_B'' = \rho\mathbf{v}$ and the law of conservation of overall mass in the form $\dot{\omega}_A + \dot{\omega}_B = 0$. For a fluid of constant mass density ρ, Eq. (5.40) becomes

$$\nabla \cdot \mathbf{v} = 0 \tag{5.41}$$

Using the relationships that $\rho_i = Y_i\rho$ and $\mathbf{v}_i = \mathbf{v} + \mathbf{V}_i$ for a multicomponent system, Eq. (5.37) becomes

$$\frac{\partial(\rho Y_i)}{\partial t} + \nabla \cdot [\rho Y_i(\mathbf{v} + \mathbf{V}_i)] = \dot{\omega}_i \tag{5.42}$$

Equation (5.42) in its divergence form can be expressed in an Euler form by the following procedure. Expanding the terms on the left-hand side, we have

$$\rho\frac{\partial Y_i}{\partial t} + Y_i\frac{\partial \rho}{\partial t} + Y_i\nabla \cdot (\rho\mathbf{v}) + \rho\mathbf{v} \cdot \nabla Y_i + \nabla \cdot (\rho Y_i\mathbf{V}_i) = \dot{\omega}_i \tag{5.43}$$

After using the overall continuity equation (5.40), the Euler form of Eq. (5.42) is obtained,

$$\rho\frac{\partial Y_i}{\partial t} + \rho\mathbf{v} \cdot \nabla Y_i + \nabla \cdot (\rho Y_i\mathbf{V}_i) = \dot{\omega}_i, \quad i = 1, 2, \dots, N \tag{5.44}$$

In rectangular coordinates (x, y, z), the mixture continuity equation (5.40) can be written as

$$\frac{\partial \rho}{\partial t} + \frac{\partial}{\partial x}(\rho u_x) + \frac{\partial}{\partial y}(\rho u_y) + \frac{\partial}{\partial z}(\rho u_z) = 0 \tag{5.45}$$

and the mass conservation equation for species i, Eq. (5.44), is expressed as

$$\rho\left(\frac{\partial Y_i}{\partial t} + u_x\frac{\partial Y_i}{\partial x} + u_y\frac{\partial Y_i}{\partial y} + u_z\frac{\partial Y_i}{\partial z}\right) + \frac{\partial}{\partial x}(\rho Y_i V_{ix}) + \frac{\partial}{\partial y}(\rho Y_i V_{iy}) + \frac{\partial}{\partial z}(\rho Y_i V_{iz}) = \dot{\omega}_i \tag{5.46}$$

with mass diffusion velocities

$$V_{ix} = -\frac{\mathcal{D}}{Y_i}\frac{\partial Y_i}{\partial x}, \quad V_{iy} = -\frac{\mathcal{D}}{Y_i}\frac{\partial Y_i}{\partial y}, \quad \text{and} \quad V_{iz} = -\frac{\mathcal{D}}{Y_i}\frac{\partial Y_i}{\partial z}$$

In a general multicomponent system, there are N equations of this kind. In the numerical solution, all the values of Y_i are considered as unknown. Since $\sum_{i=1}^{N} Y_i = 1$, it is not necessary to solve all N partial differential equations for Y_i. One of these N equations can be replaced by a simple algebraic relationship. Usually, $N - 1$ independent equations for Y_i are solved with other conservation equations for the chemically reacting mixture.

For simplicity, the above derivations have been made in rectangular coordinates. However, rectangular coordinates are not always the most convenient for solving various combustion problems. To facilitate the use of the governing conservation equations in cylindrical (r, θ, z) coordinate systems, the continuity equation (5.47) for the gas mixture and the mass conservation equation for species i, Eq. (5.48), are

$$\frac{\partial \rho}{\partial t} + \frac{1}{r}\frac{\partial}{\partial r}(\rho r u_r) + \frac{1}{r}\frac{\partial}{\partial \theta}(\rho u_\theta) + \frac{\partial}{\partial z}(\rho u_z) = 0 \tag{5.47}$$

$$\rho\left(\frac{\partial Y_i}{\partial t}+u_r\frac{\partial Y_i}{\partial r}+\frac{u_\theta}{r}\frac{\partial Y_i}{\partial \theta}+u_z\frac{\partial Y_i}{\partial z}\right)$$
$$+\frac{1}{r}\frac{\partial}{\partial r}(r\rho Y_iV_{ir})+\frac{1}{r}\frac{\partial}{\partial \theta}(r\rho Y_iV_{i\theta})+\frac{\partial}{\partial z}(\rho Y_iV_{iz})=\dot{\omega}_i \tag{5.48}$$

with mass diffusion velocities

$$V_{ir}=-\frac{\mathcal{D}}{Y_i}\frac{\partial Y_i}{\partial r},$$
$$V_{i\theta}=-\frac{\mathcal{D}}{Y_ir}\frac{\partial Y_i}{\partial \theta},$$
$$\text{and}\quad V_{iz}=-\frac{\mathcal{D}}{Y_i}\frac{\partial Y_i}{\partial z}$$

The development given above could have been made equally well in terms of molar units. If $\dot{\Omega}_A$ is the molar rate of production of A per unit volume, then the molar analog of Eq. (5.37) is

$$\frac{\partial[A]}{\partial t}+\nabla\cdot\dot{\mathbf{n}}''_A=\dot{\Omega}_A \tag{5.49}$$

Substituting Eq. (5.12) into Eq. (5.49), we obtain

$$\frac{\partial[A]}{\partial t}+\nabla\cdot[A]\hat{\mathbf{v}}=\nabla\cdot[C]\mathcal{D}_{\mathrm{AB}}\nabla X_A+\dot{\Omega}_A \tag{5.50}$$

If no chemical reactions occur, $\dot{\omega}_A$, $\dot{\omega}_B$, $\dot{\Omega}_A$, and $\dot{\Omega}_B$ are all zero. If, in addition, $\mathbf{v}$ is zero in Eq. (5.38) or $\hat{\mathbf{v}}$ is zero in Eq. (5.50), we get

$$\frac{\partial[A]}{\partial t}=\mathcal{D}_{\mathrm{AB}}\nabla^2[A] \tag{5.51}$$

which is called *Fick's second law of diffusion*. This equation is used for diffusion in solids or stationary liquids and for equimolar counter-diffusion in gases.

5.7 Conservation of Momentum

In this section, we shall derive and discuss the momentum equation in partial differential equation form. The basic assumption is that we are dealing with continuous, isotropic, and homogeneous fluids. We shall consider the special case of a Newtonian fluid, that is, a fluid exhibiting a linear relationship between shear stress and rate of deformation. We derive first the momentum equation in terms of the stress tensor. Then, we consider the constitutive relationships between stress and fluid deformation. We finally substitute the stress–strain relationship to form the Navier–Stokes equations.

Consider a fluid particle of mass dm moving through space with velocity $\mathbf{v}$, as shown in Figure 5.5. If a net force $d\mathbf{F}$ is acting on the particle, the momentum is

$$d\mathbf{F}=\frac{d}{dt}[(dm)\mathbf{v}] \tag{5.52}$$

For the fluid particle of fixed mass, the above equation becomes

$$d\mathbf{F}=dm\frac{d\mathbf{v}}{dt} \tag{5.53}$$

This equation describes the motion of the fluid particle as it moves through space relative to the fixed coordinate system. Thus, $d\mathbf{v}/dt$ represents the acceleration of a particular particle as it moves from point to point through space.

Figure 5.5 An infinitesimal fluid particle moving in three-dimensional space.

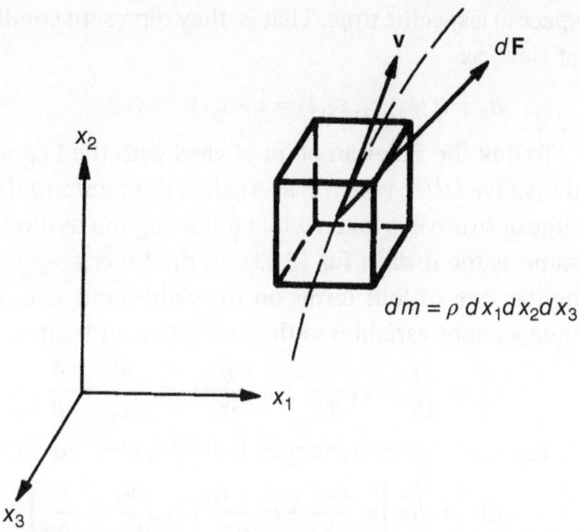

Next, we shall relate the derivative of the velocity to conditions at a fixed point in space. In considering the description of fluid motion, we need to describe the position, velocity, and acceleration throughout the flow for all fluid particles of interest.

Considering the Lagrangian viewpoint in Figure 5.5, we follow a particle and observe changes as it moves in space. Then, the expression for velocity $\mathbf{v}(x_1, x_2, x_3, t)$ does not refer to fixed coordinates but allows for changes in coordinates to locate a distinct particle at various times. The positions are given by

$$x_1 = x_1(t), \quad x_2 = x_2(t), \quad x_3 = x_3(t) \tag{5.54}$$

because the time t is the only independent variable in the Lagrangian approach. Based on the above relationships, we have

$$\mathbf{v} = \mathbf{v}[x_1(t), x_2(t), x_3(t), t] = \mathbf{v}(t) \tag{5.55}$$

where $\mathbf{v}$ refers to the velocity of a distinct particle for different times (as a function of time). Similarly, for acceleration of the particle, we have

$$\mathbf{a} = \mathbf{a}[x_1(t), x_2(t), x_3(t), t] = \mathbf{a}(t) \tag{5.56}$$

or

$$\mathbf{a} = \frac{d\mathbf{v}}{dt} = \frac{\partial \mathbf{v}}{\partial x_1}\frac{dx_1}{dt} + \frac{\partial \mathbf{v}}{\partial x_2}\frac{dx_2}{dt} + \frac{\partial \mathbf{v}}{\partial x_3}\frac{dx_3}{dt} + \frac{\partial \mathbf{v}}{\partial t} \tag{5.57}$$

However,

$$\frac{dx_1}{dt} = u_1, \quad \frac{dx_2}{dt} = u_2, \quad \frac{dx_3}{dt} = u_3 \tag{5.58}$$

where

$$\mathbf{v} = u_1\mathbf{i} + u_2\mathbf{j} + u_3\mathbf{k} = [u_1, u_2, u_3] \tag{5.59}$$

The acceleration then becomes

$$\frac{du_i}{dt} = u_1\frac{\partial u_i}{\partial x_1} + u_2\frac{\partial u_i}{\partial x_2} + u_3\frac{\partial u_i}{\partial x_3} + \frac{\partial u_i}{\partial t} \tag{5.60}$$

The above quantity is the acceleration of a distinct particle moving through space. However, the quantities $u_1, u_2, u_3, \partial u_i/\partial x_1$, etc. represent the flow conditions of the particle at a fixed point in space at a specific time. That is, they represent conditions from a field point of view (Eulerian point of view) as

$$u_i = u_i(x_1, x_2, x_3, t) = u_i(\mathbf{x}, t) \tag{5.61}$$

To link the Eulerian point of view with the Lagrangian parameters, let us consider the material derivative D/Dt, which is also called the *substantial derivative*. In the Lagrangian point of view, this time derivative is taken while following the motion of the fluid particle with a fixed mass. It is the same as the d/dt in Eq. (5.53). In the Eulerian point of view, the D/Dt operator can be expressed by the sum of four terms on the right-hand side of the following equation, since there are four independent variables in the Eulerian coordinates. Thus,

$$\frac{d}{dt} \equiv \frac{D}{Dt} \equiv u_1\frac{\partial}{\partial x_1} + u_2\frac{\partial}{\partial x_2} + u_3\frac{\partial}{\partial x_3} + \frac{\partial}{\partial t} \tag{5.62}$$

In terms of Eulerian independent variables, Eq. (5.53) can now be written as

$$dF_i = dm\left[u_1\frac{\partial u_i}{\partial x_1} + u_2\frac{\partial u_i}{\partial x_2} + u_3\frac{\partial u_i}{\partial x_3} + \frac{\partial u_i}{\partial t}\right] \tag{5.63}$$

or

$$dF_1 = dm\frac{Du_1}{Dt}, \quad dF_2 = dm\frac{Du_2}{Dt}, \quad dF_3 = dm\frac{Du_3}{Dt} \tag{5.64}$$

To write in Cartesian tensor notation, we have

$$dF_i = dm\frac{Du_i}{Dt} \tag{5.65}$$

or

$$dF_i = dm\left[u_j\frac{\partial u_i}{\partial x_j} + \frac{\partial u_i}{\partial t}\right] \tag{5.66}$$

If we consider a particle of mass dm, which is cubical in shape and has sides of length dx_1, dx_2, and dx_3, then

$$dm = \rho\, dx_1\, dx_2\, dx_3 \tag{5.67}$$

The force dF_i acting on the particle is split into a surface force df_i and a body force B_i (force per unit volume). In a mixture of N species, the body forces acting on different chemical species may differ. Thus, for a multicomponent system, we have

$$B_i = \rho\sum_{k=1}^{N} Y_k f_{k,i} \tag{5.68}$$

where $f_{k,i}$ is the force per unit mass acting on the kth species in the ith direction, and

$$dF_i = df_i + B_i\, dx_1\, dx_2\, dx_3 \tag{5.69}$$

Now, we shall express the net surface force in terms of the stress tensor acting on the different faces of the particle. Consider the surface forces acting on the cube of Figure 5.6. In this case, the surface forces for the three directions are

$$df_1 = \left(\frac{\partial\sigma_{11}}{\partial x_1} + \frac{\partial\sigma_{21}}{\partial x_2} + \frac{\partial\sigma_{31}}{\partial x_3}\right) dx_1\, dx_2\, dx_3 \tag{5.70}$$

$$df_2 = \left(\frac{\partial\sigma_{12}}{\partial x_1} + \frac{\partial\sigma_{22}}{\partial x_2} + \frac{\partial\sigma_{32}}{\partial x_3}\right) dx_1\, dx_2\, dx_3 \tag{5.71}$$

$$df_3 = \left(\frac{\partial\sigma_{13}}{\partial x_1} + \frac{\partial\sigma_{23}}{\partial x_2} + \frac{\partial\sigma_{33}}{\partial x_3}\right) dx_1\, dx_2\, dx_3 \tag{5.72}$$

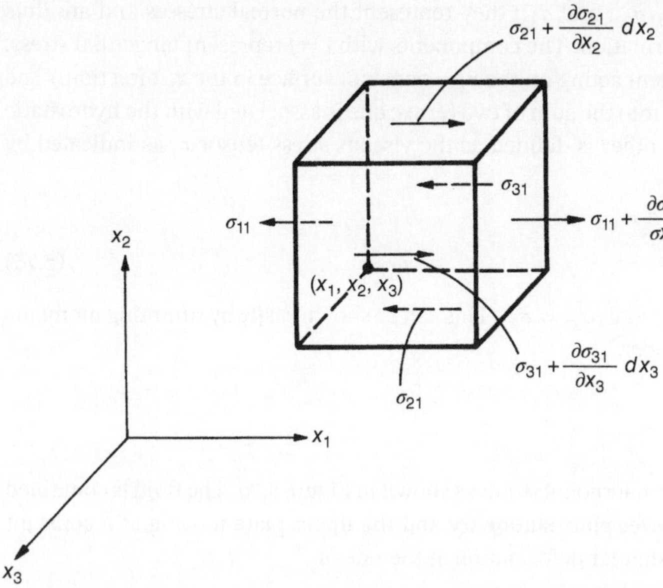

Figure 5.6 Surface-stress components acting on the fluid particle in the x_1-direction.

For example, $\sigma_{21} \times dx_1\, dx_3$ represents the stress force in the x_1-direction and acting on the x_2 = constant plane.

After dividing the terms by $dx_1\, dx_2\, dx_3$, the momentum equation in terms of the stress tensor becomes

$$\underbrace{\rho\left[u_j \frac{\partial u_i}{\partial x_j} + \frac{\partial u_i}{\partial t}\right]}_{\text{Inertia force}} = \underbrace{\frac{\partial \sigma_{ji}}{\partial x_j}}_{\substack{\text{Surface stress}\\ \text{force}}} + \underbrace{\rho \sum_{k=1}^{N} Y_k f_{k,i}}_{\text{Body force}} \tag{5.73}$$

5.7.1 Stress–Strain Rate Relationship (Constitutive Relationship)

Before we develop the constitutive relationship, it is important to point out a fundamental difference between solids and fluids. For solid elastic bodies, shear stress is proportional to the *magnitude* of angular deformation through Hooke's law, whereas for a fluid, shear stress is proportional to the *rate of angular deformation* through Stokes' law.

A fluid deforms when subjected to a shear stress. The rate of deformation varies for different fluids, and it is also dependent on the thermodynamic state for a given fluid. Low resistance to deformation is a natural characteristic property of the fluid.

We will develop the stress–strain relationships for solid elastic bodies and then replace the displacement vector by a time derivative of the displacement. This will relate the stress tensor σ_{ij} to the rate of deformation of a fluid element, i.e., the mean velocity gradients.

The second-order stress tensor σ_{ij} in a three-dimensional space has nine components. It can be written as

$$\sigma_{ij} = \begin{Bmatrix} \sigma_{11} & \sigma_{12} & \sigma_{13} \\ \sigma_{21} & \sigma_{22} & \sigma_{23} \\ \sigma_{31} & \sigma_{32} & \sigma_{33} \end{Bmatrix} = -p\,\delta_{ij} + \tau_{ij} \tag{5.74}$$

The components with $i = j$ are σ_{11}, σ_{22}, and σ_{33}; they represent the normal stresses and are thus related to pressure and linear deformation. The components with $i \neq j$ represent tangential stress. (Note that σ_{21} is the stress component acting on the x_2 = constant surface in the x_1-direction.) The stress tensor σ_{ij} can also be written into the sum of two terms; one is associated with the hydrostatic compressive stress tensor and the other is defined as the viscous stress tensor τ_{ij} as indicated by Eq. (5.74).

The stress tensor is symmetric, i.e.,

$$\sigma_{ij} = \sigma_{ji} \tag{5.75}$$

or specifically, $\sigma_{12} = \sigma_{21}, \sigma_{13} = \sigma_{31}$, and $\sigma_{23} = \sigma_{32}$. This can be shown easily by summing moments and is left as an exercise for the reader.

5.7.2 Strain Rate

Consider the special case of a fluid undergoing strain as shown in Figure 5.7*a*. The fluid is contained between parallel walls with the lower plate stationary and the upper plate moving at a constant velocity U. The fluid experiences angular deformation at the rate of

$$\frac{d\gamma}{dt} = \frac{U}{h} \tag{5.76}$$

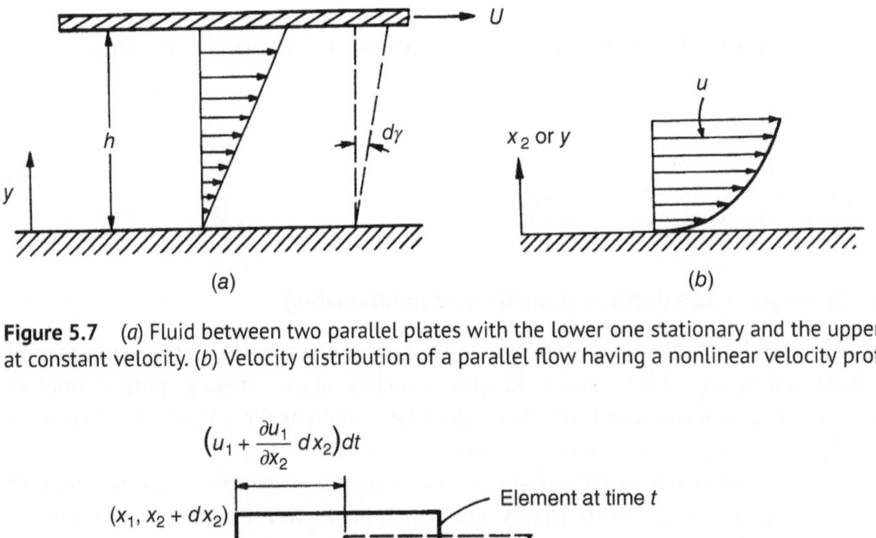

Figure 5.7 (*a*) Fluid between two parallel plates with the lower one stationary and the upper one moving at constant velocity. (*b*) Velocity distribution of a parallel flow having a nonlinear velocity profile.

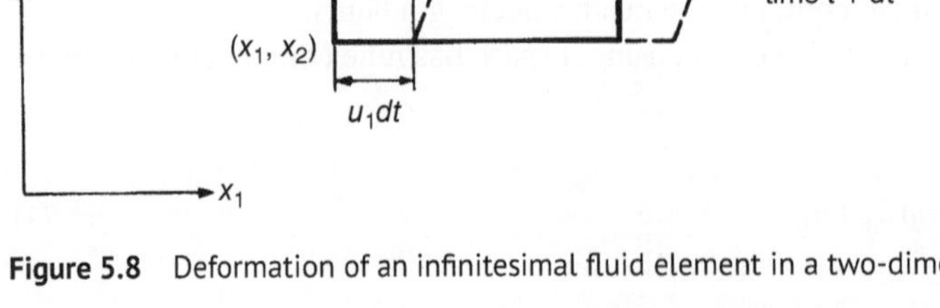

Figure 5.8 Deformation of an infinitesimal fluid element in a two-dimensional parallel flow.

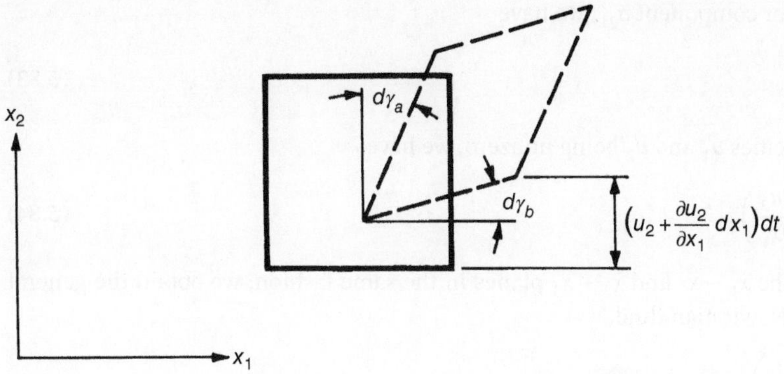

Figure 5.9 Angular deformation of an infinitesimal fluid element in a two-dimensional flow.

In the case of a nonlinear velocity distribution as shown in Figure 5.7*b*, an infinitesimal fluid element experiences a strain rate as shown in Figure 5.8.

The total angular strain in time dt is

$$d\gamma = \frac{\left[u_1 + (\partial u_1/\partial x_2)dx_2\right] dt - u_1\, dt}{dx_2} \tag{5.77}$$

so the strain rate is

$$\frac{d\gamma}{dt} = \frac{\partial u_1}{\partial x_2} \tag{5.78}$$

The above result corresponds to the shear deformation of any small fluid element between two parallel plates allowing angular deformation of this fluid element in one direction. In the case of angular deformation in two directions (i.e., due to the existence of a nonzero velocity in the x_2-direction, u_2), we have angular deformation as shown in Figure 5.9. The total angular deformation is then the sum of $d\gamma_a$ and $d\gamma_b$, i.e.,

$$\begin{aligned} d\gamma = d\gamma_a + d\gamma_b &= \frac{\left[u_1 + (\partial u_1/\partial x_2)dx_2\right] dt - u_1\, dt}{dx_2} \\ &+ \frac{\left[u_2 + (\partial u_2/\partial x_1)dx_1\right] dt - u_2\, dt}{dx_1} \end{aligned} \tag{5.79}$$

so

$$\frac{d\gamma}{dt} = \frac{\partial u_1}{\partial x_2} + \frac{\partial u_2}{\partial x_1} \equiv 2e_{12} = 2e_{21} \tag{5.80}$$

where the general strain-rate tensor is

$$e_{ij} \equiv \frac{1}{2}\left(\frac{\partial u_i}{\partial x_j} + \frac{\partial u_j}{\partial x_i}\right) \tag{5.81}$$

5.7.3 Stress Tensor

Now, let us make the Newtonian fluid assumption, which requires that the shear stress be linearly proportional to the rate of angular deformation. For a parallel-flow system, we have

$$\tau \propto \frac{d\gamma}{dt} = \frac{\partial u_1}{\partial x_2} \tag{5.82}$$

or in terms of stress tensor component σ_{21}, we have

$$\sigma_{21} = \sigma_{12} = \mu \frac{\partial u_1}{\partial x_2} \tag{5.83}$$

For the case of both velocities u_1 and u_2 being nonzero, we have

$$\tau_{12} = \mu \left(\frac{\partial u_1}{\partial x_2} + \frac{\partial u_2}{\partial x_1} \right) \tag{5.84}$$

Further, if we examine the $x_1 - x_3$ and $x_2 - x_3$ planes in the same fashion, we obtain the general shear stress tensor for a Newtonian fluid,

$$\sigma_{ij} = \mu \left(\frac{\partial u_i}{\partial x_j} + \frac{\partial u_j}{\partial x_i} \right) = 2\mu e_{ij}, \quad i \neq j \tag{5.85}$$

The determination of the off-diagonal elements of the general stress tensor σ_{ij} is now complete. This leaves the task of defining the diagonal elements of the stress tensor in terms of pressure and velocity gradient. Let us consider the linear deformation of a fluid particle as shown in Figure 5.10. The linear deformation in time dt is

$$\frac{\left[u_1 + (\partial u_1/\partial x_1)dx_1\right] dt - u_1\, dt}{dx_1} = \frac{\partial u_1}{\partial x_1} dt \tag{5.86}$$

The rate of linear deformation is

$$e_{11} = \frac{\partial u_1}{\partial x_1} \tag{5.87}$$

and for the x_2 and x_3 directions, we have

$$e_{22} = \frac{\partial u_2}{\partial x_2} \quad \text{and} \quad e_{33} = \frac{\partial u_3}{\partial x_3} \tag{5.88}$$

The total rate-of-deformation tensor is

$$\frac{\partial u_i}{\partial x_j} \tag{5.89}$$

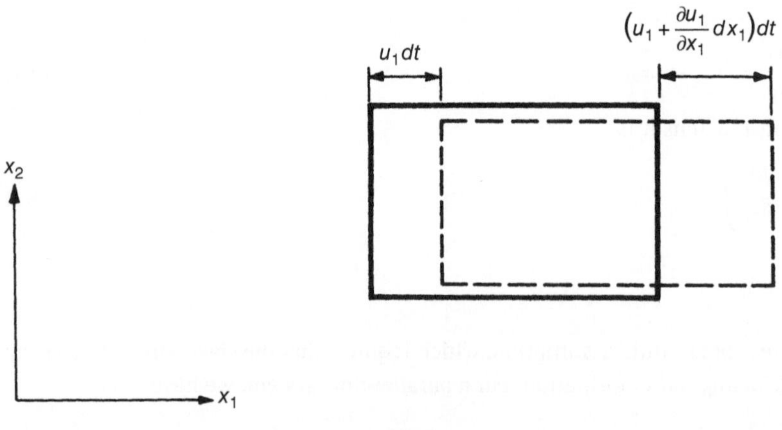

Figure 5.10 Linear deformation of a fluid element.

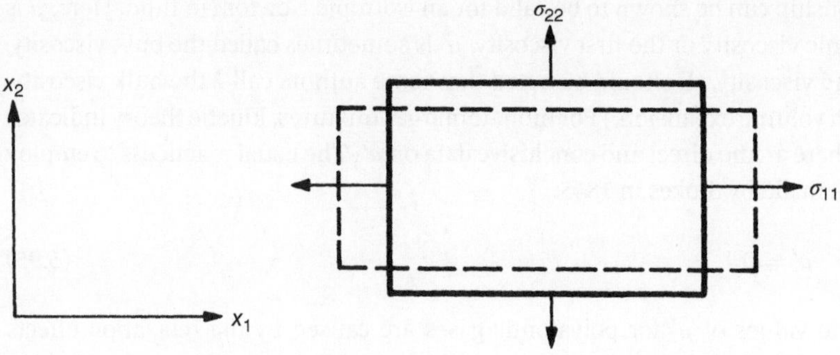

Figure 5.11 Normal stresses acting on a two-dimensional fluid element.

where we have linear deformation for $i = j$ and angular deformation for $i \neq j$. We group the deformation tensor in two parts:

$$\frac{\partial u_i}{\partial x_j} = \frac{1}{2}\left(\frac{\partial u_i}{\partial x_j} + \frac{\partial u_j}{\partial x_i}\right) + \frac{1}{2}\left(\frac{\partial u_i}{\partial x_j} - \frac{\partial u_j}{\partial x_i}\right) \tag{5.90}$$

The first group is a symmetric tensor representing $\frac{1}{2}$ of the angular-deformation tensor. The second group is an antisymmetric tensor representing rotation of a fluid without deformation.

Now, we consider the following relationship between stress and strain for the normal stress components:

$$\begin{aligned}
\sigma_{11} &= -p + ce_{11} + \lambda e_{22} + \lambda e_{33} \\
\sigma_{22} &= -p + \lambda e_{11} + ce_{22} + \lambda e_{33} \\
\sigma_{33} &= -p + \lambda e_{11} + \lambda e_{22} + ce_{33}
\end{aligned} \tag{5.91}$$

These equations state that there is a linear relationship between stress, say σ_{11}, and strain rate in the same direction, e_{11} (see Figure 5.11). Also, the normal stress in the x_1 direction causes strain rates in the other two directions, e_{22} and e_{33}. Since there is no preferred direction, they are related by the same coefficient λ, that is, the fluid is assumed to be isotropic.

In Eq. (5.91), p represents the hydrostatic pressure, which must be included, since when there is no linear deformation ($e_{11} = e_{22} = e_{33} = 0$), we still have a normal stress in all directions and Eq. (5.91) reduces to

$$\sigma_{11} = -p, \quad \sigma_{22} = -p, \quad \text{and} \quad \sigma_{33} = -p \tag{5.92}$$

The normal stress–strain relationships may be written as

$$\begin{aligned}
\sigma_{11} &= -p + \lambda(e_{11} + e_{22} + e_{33}) + (c - \lambda)e_{11} \\
\sigma_{22} &= -p + \lambda(e_{11} + e_{22} + e_{33}) + (c - \lambda)e_{22} \\
\sigma_{33} &= -p + \lambda(e_{11} + e_{22} + e_{33}) + (c - \lambda)e_{33}
\end{aligned} \tag{5.93}$$

The above two coefficients λ and c can be replaced by two other coefficients μ and μ' according to the following relations:

$$\begin{aligned}
c - \lambda &= 2\mu \\
\lambda &= \mu' - \frac{2}{3}\mu
\end{aligned} \tag{5.94}$$

of which the first relationship can be shown to be valid for an isotropic Newtonian fluid. Here, μ is usually called the dynamic viscosity or the first viscosity, μ' is sometimes called the bulk viscosity, and λ is called the second viscosity. (It should be noted that some authors call λ the bulk viscosity, since it is associated with volume expansion.) For monatomic gas mixtures, kinetic theory indicates that $\mu' = 0$. Up to now, there are no direct and conclusive data on μ'. The usual practice is to employ the following hypothesis made by Stokes in 1845:

$$\lambda + \frac{2}{3}\mu = 0, \quad \text{or} \quad \mu' = 0 \tag{5.95}$$

Physically, the nonzero values of μ' for polyatomic gases are caused by the relaxation effects between the translational motion and the various internal degrees of freedom. These effects lead to a positive value of μ'. So far, no theoretical computations or reliable experimental measurements of μ' exist, and in fact μ' is usually negligible in combustion processes.

Substituting Eq. (5.94) into Eq. (5.91) and recognizing $e_{kk} = e_{11} + e_{22} + e_{33}$ from index notation, we have

$$\sigma_{11} = -p + \left(\mu' - \frac{2}{3}\mu\right) e_{kk} + 2\mu e_{11} \tag{5.96}$$

$$\sigma_{22} = -p + \left(\mu' - \frac{2}{3}\mu\right) e_{kk} + 2\mu e_{22} \tag{5.97}$$

$$\sigma_{33} = -p + \left(\mu' - \frac{2}{3}\mu\right) e_{kk} + 2\mu e_{33} \tag{5.98}$$

and the general stress–strain-rate relationship becomes

$$\sigma_{ij} = -p\delta_{ij} + \left(\mu' - \frac{2}{3}\mu\right) e_{kk}\delta_{ij} + 2\mu e_{ij} \tag{5.99}$$

In the above equation, the Kronecker delta function, δ_{ij}, is defined in such way that

$$\delta_{ij} = \begin{cases} 1, & i = j \\ 0, & i \neq j \end{cases} \tag{5.100}$$

The stress tensor can therefore be written as

$$\sigma_{ij} = -p\delta_{ij} + \left(\mu' - \frac{2}{3}\mu\right) \frac{\partial u_k}{\partial x_k}\delta_{ij} + \mu \left(\frac{\partial u_i}{\partial x_j} + \frac{\partial u_j}{\partial x_i}\right) \tag{5.101}$$

5.7.4 Navier–Stokes Equation

When the stress–strain relationship, Eq. (5.101), is substituted into the momentum equation (5.73), the resulting equation is the Navier–Stokes equation in the x_i direction:

$$\rho\left[\frac{\partial u_i}{\partial t} + u_j\frac{\partial u_i}{\partial x_j}\right] = \frac{\partial}{\partial x_j}\left[-p\delta_{ij} + \left(\mu' - \frac{2}{3}\mu\right)\frac{\partial u_k}{\partial x_k}\delta_{ij}\right.$$
$$\left. + \mu\left(\frac{\partial u_i}{\partial x_j} + \frac{\partial u_j}{\partial x_i}\right)\right] + \rho\left(\sum_{k=1}^{N} Y_k f_k\right)_i \tag{5.102}$$

5.8 Conservation of Energy

Before deriving the conservation of energy equation for a multicomponent system, we must be aware of the various effects that contribute to the heat flux. It is known from Fourier's law of heat conduction that the temperature gradient (∇T) in the flow field is the main cause of heat flux ($\mathbf{q}$), as shown by Figure 5.12. There are two additional effects besides the conduction heat flux ($-k\nabla T$), which can contribute to the heat flux in a multicomponent system. When the average velocity $\mathbf{v}_i$ of component i differs from the mass–average velocity of the mixture, there is then a diffusion mass flux $\rho_i \mathbf{V}_i$ of component i relative to the bulk mass flux of the mixture flowing at the mass–average velocity in the local control volume. If the average enthalpy (per unit mass) associated with the ith component is h_i, then the extra enthalpy flux carried by the ith molecules across the control surface per unit time is equal to $h_i \rho_i \mathbf{V}_i$ or $\rho h_i Y_i \mathbf{V}_i$. The overall enthalpy flux carried by all different chemical species per unit time flowing relative to the mass-average motion of the mixture is $\rho \sum_{i=1}^{N} h_i Y_i \mathbf{V}_i$. This enthalpy flux term, caused by interdiffusion processes, constitutes an additional component to the energy flux ($\mathbf{q}_{\text{interdiffusion}}$) in binary and/or multicomponent systems.

Based on Onsager's reciprocal relations for irreversible thermodynamic processes, if temperature gradient ∇T gives rise to diffusion velocities (the "thermal-diffusion effect" or the "Soret effect"), concentration gradients (represented by ∇X_i) must produce a heat flux (see Figure 5.12). This reciprocal effect, known as the *Dufour effect*, provides an additional contribution to the total heat flux $\mathbf{q}$. It is conventional to express the concentration gradients in terms of differences in diffusion velocities, with which the Dufour heat flux is expressed as

$$\mathbf{q}_{\text{Dufour}} = R_u T \sum_{i=1}^{N} \sum_{j=1}^{N} \left(\frac{X_j \alpha_i}{M_i \mathcal{D}_{ij}} \right) (\mathbf{V}_i - \mathbf{V}_j) \tag{5.103}$$

In most cases, the Dufour effect is small. Although it is omitted in most derivations, the term is retained in the general energy equation for completeness.

Conduction and the above two effects yield the heat-flux vector as

$$\begin{aligned} \mathbf{q} &= \mathbf{q}_{\text{conduction}} + \mathbf{q}_{\text{interdiffusion}} + \mathbf{q}_{\text{Dufour}} \\ &= -k\nabla T + \rho \sum_{i=1}^{N} h_i Y_i \mathbf{V}_i + R_u T \sum_{i=1}^{N} \sum_{j=1}^{N} \left(\frac{X_j \alpha_i}{\mathrm{M}_i \mathcal{D}_{ij}} \right) (\mathbf{V}_i - \mathbf{V}_j) \end{aligned} \tag{5.104}$$

We now proceed to derive the conservation of energy equation. The stored energy per unit mass e_t is defined as

$$e_t = e + \frac{u_i u_i}{2} \tag{5.105}$$

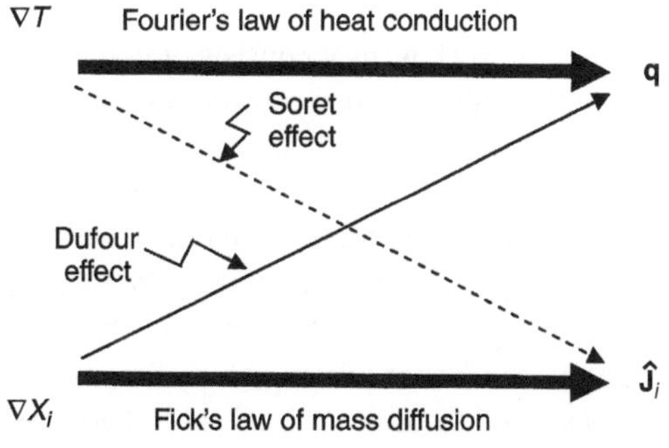

Figure 5.12 Comparison of Dufour and Soret effects with heat conduction and species diffusion.

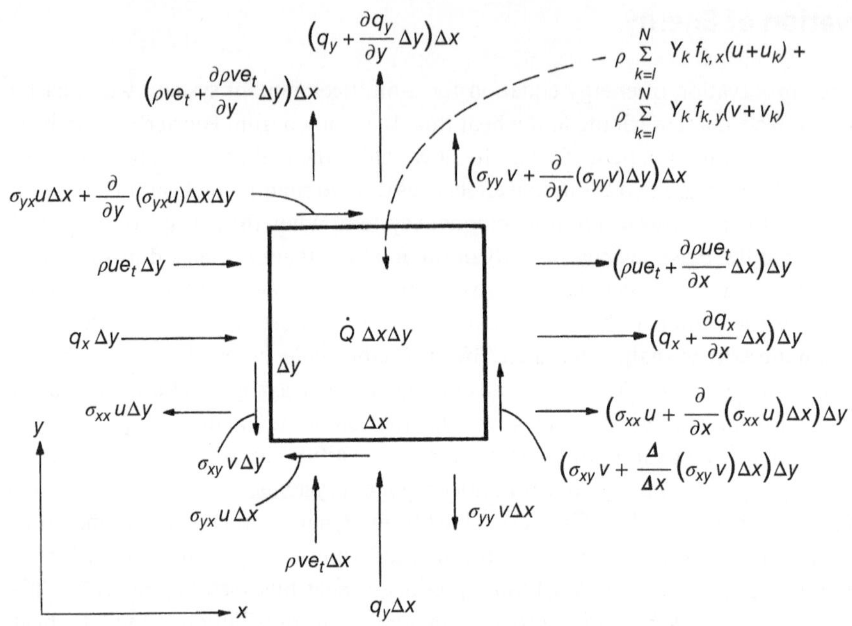

Figure 5.13 Terms in the energy balance equation for a two-dimensional flow.

where e, the specific internal energy, is defined as

$$e = h - \frac{p}{\rho} = \sum_{i=1}^{N} h_i Y_i - \frac{p}{\rho} \tag{5.106}$$

where

$$h_i = h^o_{f,i} + \int_{T^o}^{T} c_{p,i}\, dT \tag{5.107}$$

Beginning with a stationary differential volume element ($\Delta x\, \Delta y \times 1$) through which a pure fluid is flowing, we then write the law of conservation of energy for the fluid contained within the control volume by the balance of the following terms:

(I) Rate of accumulation of internal and kinetic energy in the control volume.
(II) Net rate of influx of internal and kinetic energy by convection.
(III) Net rate of heat addition due to the thermal energy flux **q**, Eq. (5.104).
(IV) Rate of thermal energy added by an external source $\dot{Q}$, such as an electric spark.
(V) Net rate of work done on fluid element by surroundings.

The rate of work term V consists of two parts. One is due to the total surface stress tensor acting on the boundary of the control volume (such as $\sigma_{xx} u \Delta y \cdot 1, \sigma_{yy} v \Delta y \cdot 1, \sigma_{yx} u \Delta x \cdot 1$ as shown in Figure 5.13); the other is due to the body force f. The body force acting on the kth species in the x-direction is $\rho Y_k f_{k,x}$. The rate of work done on the fluid due to this force component is the product of this force component with the x-component of the average velocity of the kth species, that is, $\rho Y_k f_{k,x} u_k$, or $\rho Y_k f_{k,x}(u + U_k)$. Summing over all species in the mixture, we have $\rho \sum_{k=1}^{N} Y_k f_{k,x}(u + U_k)$ as the total contribution of the body force in the x-direction.

The reason for considering a two-dimensional reacting flow in Figure 5.13 is to be able to fit all the terms around the infinitesimal control volume $\Delta x\, \Delta y \cdot 1$ in the same diagram. After obtaining the conservation of energy equation, we shall then generalize it to a three-dimensional space. Based on the energy-balance statement above, the energy equation made of the corresponding terms I through V for a two-dimensional unsteady reacting flow is shown below:

$$\underbrace{\frac{\partial}{\partial t}(\rho e_t)}_{\text{I}} = \underbrace{-\frac{\partial}{\partial x}(\rho u e_t) - \frac{\partial}{\partial y}(\rho v e_t)}_{\text{II}} \underbrace{-\frac{\partial q_x}{\partial x} - \frac{\partial q_y}{\partial y}}_{\text{III}} + \underbrace{\dot{Q}}_{\text{IV}}$$
$$\underbrace{+ \rho\sum_{k=1}^{N} Y_k f_{k,x}(u + U_k) + \rho\sum_{k=1}^{N} Y_k f_{k,y}(v + V_k) + \frac{\partial \sigma_{xx} u}{\partial x} + \frac{\partial \sigma_{yx} u}{\partial y} + \frac{\partial \sigma_{yy} v}{\partial y} + \frac{\partial \sigma_{xy} v}{\partial x}}_{\text{V}} \tag{5.108}$$

where U_k, V_k are relative mass-diffusion velocity components.

To generalize Eq. (5.108) to a three-dimensional space, we have

$$\begin{aligned}
&\frac{\partial}{\partial t}(\rho e_t) + \frac{\partial}{\partial x}(\rho u e_t) + \frac{\partial}{\partial y}(\rho v e_t) + \frac{\partial}{\partial z}(\rho w e_t) \\
&= -\frac{\partial q_x}{\partial x} - \frac{\partial q_y}{\partial y} - \frac{\partial q_z}{\partial z} + \dot{Q} + \frac{\partial \sigma_{xx} u}{\partial x} + \frac{\partial \sigma_{yx} u}{\partial y} + \frac{\partial \sigma_{zx} u}{\partial z} \\
&\quad + \frac{\partial \sigma_{xy} v}{\partial x} + \frac{\partial \sigma_{yy} v}{\partial y} + \frac{\partial \sigma_{zy} v}{\partial z} + \frac{\partial \sigma_{xz} w}{\partial x} + \frac{\partial \sigma_{yz} w}{\partial y} + \frac{\partial \sigma_{zz} w}{\partial z} \\
&\quad + \rho\sum_{k=1}^{N} Y_k f_{k,x}(u + U_k) + \rho\sum_{k=1}^{N} Y_k f_{k,y}(v + V_k) + \rho\sum_{k=1}^{N} Y_k f_{k,z}(w + W_k)
\end{aligned} \tag{5.109}$$

The energy equation in tensor index notation is thus

$$\frac{\partial}{\partial t}(\rho e_t) + \frac{\partial}{\partial x_i}(\rho u_i e_t) = -\frac{\partial q_i}{\partial x_i} + \dot{Q} + \frac{\partial \sigma_{ji} u_i}{\partial x_j} + \rho\sum_{k=1}^{N} Y_k f_{k,i}(u_i + V_{k,i}) \tag{5.110}$$

Using the continuity equation in tensor index form,

$$\frac{\partial \rho}{\partial t} + \frac{\partial \rho u_i}{\partial x_i} = 0 \tag{5.111}$$

the energy equation (5.110) can also be written as

$$\rho\frac{\partial e_t}{\partial t} + \rho u_i \frac{\partial e_t}{\partial x_i} = -\frac{\partial q_i}{\partial x_i} + \dot{Q} + \frac{\partial \sigma_{ji} u_i}{\partial x_j} + \rho\sum_{k=1}^{N} Y_k f_{k,i}(u_i + V_{k,i}) \tag{5.112}$$

Multiplying the momentum equation (5.73) by u_i and making a slight rearrangement, we have

$$\rho\frac{\partial(\frac{1}{2}u_i u_i)}{\partial t} + \rho u_j \frac{\partial(\frac{1}{2}u_i u_i)}{\partial x_j} = u_i \frac{\partial \sigma_{ji}}{\partial x_j} + \rho\sum_{k=1}^{N} Y_k f_{k,i} u_i \tag{5.113}$$

Subtracting Eq. (5.113) from Eq. (5.112), we get

$$\rho\frac{De}{Dt} = -\frac{\partial q_i}{\partial x_i} + \sigma_{ji}\frac{\partial u_i}{\partial x_j} + \dot{Q} + \rho\sum_{k=1}^{N} Y_k f_{k,i} V_{k,i} \tag{5.114}$$

The total rate of deformation tensor, Eq. (5.90), can be expressed in a form with two components, i.e.,

$$\frac{\partial u_i}{\partial x_j} = \frac{1}{2}\left(\frac{\partial u_i}{\partial x_j} + \frac{\partial u_j}{\partial x_i}\right) + \frac{1}{2}\left(\frac{\partial u_i}{\partial x_j} - \frac{\partial u_j}{\partial x_i}\right) = e_{ij} + \hat{\omega}_{ij} \tag{5.115}$$

The first component is the strain-rate tensor (e_{ij}), which is symmetric. The second component represents the angular-velocity tensor ($\hat{\omega}_{ij}$), which is antisymmetric. It is useful to note that

$$(\text{a symmetric tensor}) \times (\text{an antisymmetric tensor}) = 0 \tag{5.116}$$

Therefore,

$$\sigma_{ji}\hat{\omega}_{ij} = 0 \tag{5.117}$$

Separating the stress tensor into pressure p and viscosity μ components, we have

$$\begin{aligned}\sigma_{ji}\frac{\partial u_i}{\partial x_j} &= \sigma_{ji}(e_{ij} + \hat{\omega}_{ij}) = \sigma_{ji}e_{ij} = \frac{1}{2}\left(-p - \frac{2}{3}\mu\frac{\partial u_k}{\partial x_k}\right) \\ &\quad\times \delta_{ij}\left(\frac{\partial u_i}{\partial x_j} + \frac{\partial u_j}{\partial x_i}\right) + 2\mu e_{ij}e_{ij} \\ &= -p\frac{\partial u_j}{\partial x_j} - \frac{2}{3}\mu\left(\frac{\partial u_j}{\partial x_j}\right)^2 + 2\mu e_{ij}e_{ij}\end{aligned} \tag{5.118}$$

In the above operations, we have assumed the bulk viscosity (μ') term can be neglected relative to the other terms. Thus, Eq. (5.114) reduces to

$$\rho\frac{De}{Dt} + p\frac{\partial u_j}{\partial x_j} = -\nabla \cdot \mathbf{q} + \Phi + \dot{Q} + \rho\sum_{k=1}^{N} Y_k\mathbf{f}_k \cdot \mathbf{V}_k \tag{5.119}$$

where Φ, the dissipation by viscous stress, is defined as

$$\Phi = -\frac{2}{3}\mu\left(\frac{\partial u_j}{\partial x_j}\right)^2 + 2\mu e_{ij}e_{ij} \tag{5.120}$$

The internal energy e can be expressed in terms of h, P, and ρ,

$$\begin{aligned}h &= e + \frac{p}{\rho} \\ \frac{De}{Dt} &= \frac{Dh}{Dt} - \frac{1}{\rho}\frac{Dp}{Dt} + \frac{p}{\rho^2}\frac{D\rho}{Dt}\end{aligned} \tag{5.121}$$

From the continuity equation, we have the following relationship between the increase of the mixture density and the volume dilatation:

$$\frac{1}{\rho}\frac{D\rho}{Dt} = -\frac{\partial u_j}{\partial x_j} \tag{5.122}$$

Using Eqs. (5.121) and (5.122), the energy equation (5.119) can be written as

$$\rho\frac{Dh}{Dt} - \frac{Dp}{Dt} = -\nabla \cdot \mathbf{q} + \Phi + \dot{Q} + \rho\sum_{k=1}^{N} Y_k\mathbf{f}_k \cdot \mathbf{V}_k \tag{5.123}$$

5.9 Physical Derivation of the Multicomponent Diffusion Equation

In order to render the conservation equations soluble for a multicomponent system, the diffusion velocities $\mathbf{V}_i$ appearing in these equations must be determined. In general, there are two ways to evaluate the diffusion velocities. The approximate way is to employ Fick's law of mass diffusion and to replace the diffusion velocities with the concentration gradient times some appropriate coefficient (see Eqs. (5.5)–(5.8) and Eqs. (5.10)–(5.11)). The more exact method is to find $\mathbf{V}_i$ from the multicomponent diffusion equation, which is derived in the following through physical arguments.

The binary-collision problem has been studied extensively in classical mechanics. The three-dimensional dynamical problem of binary collision between two particles with masses m_i and m_j is found to be equivalent to a one-body problem in a plane, with the reduced mass

$$\mu_{ij} \equiv \frac{m_i m_j}{m_i + m_j} \tag{5.124}$$

The average momentum transferred from a molecule of species i to a molecule of species j must be equal to the negative of that transferred from a j to an i molecule. It is expected that on the average, the momentum transferred to an i molecule in a collision between i and j molecules is approximately

$$\mu_{ij}(\mathbf{V}_j - \mathbf{V}_i) \tag{5.125}$$

where μ_{ij} is the reduced mass and $\mathbf{V}_j - \mathbf{V}_i$ is the average relative velocity between molecules of species i and j. If we define $\mathbf{f}_i$ as the body force per unit mass on molecules of species i, then the body force per unit volume acting on molecules of species i is

$$\rho Y_i \mathbf{f}_i = \rho Y_i \mathbf{f}_i \sum_{j=1}^{N} Y_j \tag{5.126}$$

since $\sum_{i=1}^{N} Y_j = 1$. Hence, the net rate of change of momentum of molecules of species i per unit volume is

$$\Gamma_i = \sum_{j=1}^{N} \mu_{ij} Z_{ij}(\mathbf{V}_j - \mathbf{V}_i) + \sum_{j=1}^{N} \rho Y_i Y_j \mathbf{f}_i \quad \text{for} \quad i = 1, 2, \dots, N \tag{5.127}$$

where Z_{ij} is the total number of collisions per unit volume per second between the molecules of species i and j,

This change of momentum of species i is evident in changes in both the random velocity and ordered velocity of the molecules. The rate of change of the ordered momentum is $\rho Y_i \, D\mathbf{v}/Dt$. The partial pressure p_i of species i represents physically the momentum of molecules of type i transported per second across a surface of unit area, traveling with the mass–average velocity of the fluid. Thus, ∇p_i is the rate of change of momentum associated with the random motion of molecules of type i per unit volume. The quantity Γ_i is therefore given by the relation

$$\Gamma_i = \rho Y_i \frac{D\mathbf{v}}{Dt} + \nabla p_i, \qquad i = 1, 2, \dots, N \tag{5.128}$$

According to Dalton's law of partial pressures,

$$p_i = X_i p, \qquad i = 1, 2, \dots, N \tag{5.129}$$

Hence,

$$\nabla p_i = p \, \nabla X_i + X_i \, \nabla p, \qquad i = 1, 2, \dots, N \tag{5.130}$$

Neglecting the viscous forces, the momentum equation can be used to show that

$$\rho \frac{D\mathbf{v}}{Dt} = -\nabla p + \rho \sum_{j=1}^{N} Y_j \mathbf{f}_j \tag{5.131}$$

Substituting Eqs. (5.128), (5.130), and (5.131) into Eq. (5.127) yields

$$\nabla X_i = \sum_{j=1}^{N} \frac{\mu_{ij} Z_{ij}}{p} (\mathbf{V}_j - \mathbf{V}_i) + (Y_i - X_i) \frac{\nabla p}{p} + \frac{\rho}{p} \sum_{j=1}^{N} Y_i Y_j (\mathbf{f}_i - \mathbf{f}_j) \tag{5.132}$$

The product $\mu_{ij} Z_{ij}$ may be related to the binary diffusion coefficient by considering the limiting case of a constant-pressure process in a two-component system with no body forces, for which, Eq. (5.132) reduces to

$$\nabla X_A = \frac{\mu_{AB} Z_{AB}}{p} (\mathbf{V}_B - \mathbf{V}_A) \tag{5.133}$$

The relative molar flux $\hat{\mathbf{J}}_A$ can be expressed in terms of the mole fractions of species A and B, and the velocity–vector difference between B and A as

$$\hat{\mathbf{J}}_A = -[C] X_A X_B (\mathbf{V}_B - \mathbf{V}_A) \tag{5.134}$$

Substituting $\mathbf{V}_B - \mathbf{V}_A$ of Eq. (5.134) into Eq. (5.133) and comparing it with Eq. (5.10), we have $\mathcal{D}_{AB} = (X_A X_B p)/(\mu_{AB} Z_{AB})$, which implies that, in general,

$$\mathcal{D}_{ij} = \frac{X_i X_j p}{\mu_{ij} Z_{ij}} \tag{5.135}$$

Utilizing Eq. (5.135), Eq. (5.132) becomes

$$\nabla X_i = \sum_{j=1}^{N} \frac{X_i X_j}{\mathcal{D}_{ij}} (\mathbf{V}_j - \mathbf{V}_i) + (Y_i - X_i) \frac{\nabla p}{p} + \frac{\rho}{p} \sum_{j=1}^{N} Y_i Y_j (\mathbf{f}_i - \mathbf{f}_j), \quad i = 1, 2, \ldots, N \tag{5.136}$$

The multicomponent diffusion equation obtained through more rigorous derivation from the kinetic theory is

$$\begin{aligned} \nabla X_i = {} & \sum_{j=1}^{N} \frac{X_i X_j}{\mathcal{D}_{ij}} (\mathbf{V}_j - \mathbf{V}_i) + (Y_i - X_i) \frac{\nabla p}{p} \\ & + \frac{\rho}{p} \sum_{j=1}^{N} Y_i Y_j (\mathbf{f}_i - \mathbf{f}_j) + \sum_{j=1}^{N} \frac{X_i X_j}{\rho \mathcal{D}_{ij}} \left(\frac{\alpha_j}{Y_j} - \frac{\alpha_i}{Y_i} \right) \frac{\nabla T}{T} \\ & i = 1, 2, \ldots, N \end{aligned} \tag{5.137}$$

where α_j is the thermal diffusivity of species j. Physically, Eq. (5.137) states that concentration gradients result from diffusion velocities, pressure gradients, the differences in the body force per unit mass on molecules of different species, and thermal-diffusion effects. For a given chemical species in a three-dimensional space, Eq. (5.137) can be expressed as three individual equations for vector component balance in all three directions. It may be noted that the last summation of the above equation was absent in Eq. (5.136), since we did not consider the thermal-diffusion (Soret) effect in the derivation. Opposite to the Dufour effect, see Figure 5.12, the Soret effect describes the mass diffusion resulting from temperature gradients.

The above diffusion equation (Eq. (5.137)) reduces to Fick's law of diffusion for binary mixtures when (1) thermal diffusion is negligible, (2) the body force per unit mass is the same for each species ($\mathbf{f}_A = \mathbf{f}_B$), and (3) either the pressure is a constant or the molecular mass of both species are the same.

For a multicomponent system of N species, the diffusion equation simplifies to

$$\nabla X_i = \sum_{j=1}^{N} \frac{X_i X_j}{\mathcal{D}_{ij}} (\mathbf{V}_j - \mathbf{V}_i) \tag{5.138}$$

and for a two-component ($N = 2$) mixture,

$$\nabla X_1 = \frac{X_1 X_2}{\mathcal{D}_{ij}} (\mathbf{V}_2 - \mathbf{V}_1) \tag{5.139}$$

By using Eqs. (5.3) and (5.6), it can be shown that Eq. (5.138) can be written as

$$\nabla X_i = \sum_{j=1}^{N} \frac{X_i X_j}{[C]\mathcal{D}_{ij}} (X_i \dot{\mathbf{n}}_j'' - X_j \dot{\mathbf{n}}_i'') \tag{5.140}$$

In performing calculations for multicomponent systems, it is often convenient to define an effective binary diffusion $\mathcal{D}_{\text{eff}}$ for the diffusion of species i in a mixture m. Equation (5.12) then becomes

$$\dot{\mathbf{n}}_i'' = [C_i]\hat{\mathbf{v}} - [C]\mathcal{D}_{\text{eff}} \nabla X_i \tag{5.141}$$

This equation, simpler than Eq. (5.140), also relates $\dot{\mathbf{n}}_i''$ to ∇X_i.

Useful relationships between the mole and mass fractions X_i and Y_i in multicomponent systems are

$$X_i = \frac{Y_i/\text{M}_i}{\sum_{j=1}^{N}(Y_j/\text{M}_j)} \quad \text{or} \quad Y_i = \frac{X_i \text{M}_i}{\sum_{j=1}^{N} X_j \text{M}_j}, \quad i = 1, 2, \ldots, N \tag{5.142}$$

Finally, the ideal-gas equation of state for a multicomponent system can be written as

$$p = \rho R_u T \sum_{i=1}^{N} \frac{Y_i}{\text{M}_i} \tag{5.143}$$

If $\mathbf{V}_i$ can be substituted by Fick's law in the species and energy equations, then in a system with N species, there are $N + 6$ unknowns. They are

$$Y_1, Y_2, Y_3, \ldots, Y_N, \rho, T, p, u, v, w \tag{5.144}$$

The $N + 6$ equations to be solved are

- The overall mass continuity (Eq. (5.40))
- Three (x, y, z) momentum equations (Eq. (5.102))
- The energy equation (Eq. (5.119))
- $N - 1$ species equations (Eq. (5.44))
- The equation of state (Eq. (5.143))
- Summation equation relating all Y_i $[Y_1 + Y_2 + \cdots + Y_N = 1]$

If $\mathbf{V}_i$ is solved from the diffusion equation for a multicomponent system, there will be $5N + 6$ unknowns. Equations (5.136) and (5.137) constitute a system of differential equations of size N^2, which is computationally intensive. Therefore, Fick's law of mass diffusion is typically adopted for a broad range of combustion problems.

5.10 Shvab–Zel'dovich Formulation

A very useful technique was developed by Shvab and Zel'dovich (Zel'dovich 1951) to express the equations for conservation of energy and of chemical species in a common form. Their formulation enables the chemical-source terms to be removed from many equations by considering certain linear combinations of dependent variables. Their general procedure is called the Shvab–Zel'dovich formulation. The formulation assumes that evaporation and combustion processes are quasi-steady, so that at any instant in time, the processes can be modeled as if they were steady state. In addition, the Lewis number Le is assumed to be equal to one, which implies that the mass diffusivity $\mathcal{D}$ and the thermal diffusivity α are equal.

A number of effects in the governing equations are unimportant for a large class of combustion problems and can therefore be omitted in many applications. These include

(a) Body forces ($\mathbf{f}_i$ terms)
(b) Soret and Dufour effects (terms involving α_i)
(c) Pressure gradient diffusion $[(Y_i - X_i)\nabla p/\rho]$ (term in Eq. (5.137))
(d) Bulk viscosity (μ')

For *quasi-steady flow*, the overall continuity equation (Eq. (5.40)) reduces to

$$\nabla \cdot \rho\mathbf{v} = 0 \tag{5.145}$$

and for *low-speed* steady-flow problems, viscous effects are often negligible, and the momentum equation (5.102) reduces to

$$\frac{\partial p}{\partial x_i} \approx 0 \tag{5.146}$$

In other words, the pressure can be taken to be constant for the combustion zone, i.e.,

$$p = \text{constant} \tag{5.147}$$

With the above assumptions, the energy equation (5.123) reduces to

$$\rho\mathbf{v} \cdot \nabla h = \nabla \cdot (k\nabla T) - \nabla \cdot \left(\rho \sum_{i=1}^{N} h_i Y_i \mathbf{V}_i\right) + \dot{Q} \tag{5.148}$$

since $\Phi = 0$ and the pressure gradient is zero.

Adding $\sum_{i=1}^{N} h_i Y_i \nabla \cdot \rho\mathbf{v} = 0$ on the left-hand side of Eq. (5.148), noting that $h = \sum_{i=1}^{N} h_i Y_i$, and setting the external heat addition $\dot{Q} = 0$, one obtains

$$\nabla \cdot \left[\rho \sum_{i=1}^{N} h_i Y_i (\mathbf{v} + \mathbf{V}_i) - k\nabla T\right] = 0 \tag{5.149}$$

The conservation of species equation (kg/m^3 s) (5.42) for steady flow reduces to

$$\nabla \cdot [\rho Y_i (\mathbf{v} + \mathbf{V}_i)] = \dot{\omega}_i, \qquad i = 1, 2, \ldots, N \tag{5.150}$$

When the *mass-diffusion coefficients of all pairs of species are equal*, the diffusion equation (5.138) reduces to

$$\mathcal{D}\nabla X_i = X_i \sum_{j=1}^{N} X_j \mathbf{V}_j - X_i \mathbf{V}_i, \qquad i = 1, 2, \ldots, N \tag{5.151}$$

Multiplying the above equation by Y_i/X_i and summing over i, we obtain

$$\sum_{i=1}^{N} \frac{Y_i}{X_i} \mathcal{D} \nabla X_i = \sum_{i=1}^{N} Y_i \sum_{j=1}^{N} X_j \mathbf{V}_j - \sum_{i=1}^{N} Y_i \mathbf{V}_i \tag{5.152}$$

Using the fact that

$$\sum_{i=1}^{N} Y_i = 1 \quad \text{and} \quad \sum_{i=1}^{N} Y_i \mathbf{V}_i = 0 \tag{5.153}$$

we obtain

$$\sum_{j=1}^{N} X_j \mathbf{V}_j = \sum_{i=1}^{N} Y_i \mathcal{D} \nabla \ln X_i = \sum_{j=1}^{N} Y_j \mathcal{D} \nabla \ln X_j \tag{5.154}$$

Substituting Eq. (5.154) into Eq. (5.151) and dividing by X_i give

$$\mathcal{D} \left[\nabla \ln X_i - \sum_{j=1}^{N} Y_j \nabla \ln X_j \right] = -\mathbf{V}_i, \qquad i = 1, 2, \ldots, N \tag{5.155}$$

Substituting Eq. (5.142) for X_i in the above equation, we obtain Fick's law

$$\mathbf{V}_i = -\mathcal{D} \nabla \ln Y_i, \qquad i = 1, 2, \ldots, N \tag{5.156}$$

Substituting Eqs. (5.150), (5.156), and (5.107) into Eq. (5.149), we obtain

$$\begin{aligned} \nabla \cdot & \left[\rho \mathbf{v} \int_{T^o}^{T} \left(\sum_{i=1}^{N} Y_i c_{p,i} \right) dT - \rho \mathcal{D} \sum_{i=1}^{N} \nabla Y_i \int_{T^o}^{T} c_{p,i}\, dT - \rho \mathcal{D} \frac{k}{\rho c_p \mathcal{D}} c_p \nabla T \right] \\ &= -\sum_{i=1}^{N} \dot{\omega}_i h^o_{f,i} \end{aligned} \tag{5.157}$$

Note that the species enthalpy h_i has been split into the formation $h^o_{f,i}$ and the specific heat $c_{p,i}$ terms. Assuming that the Lewis number is unity, that is,

$$\mathrm{Le} = \frac{k}{\rho c_p \mathcal{D}} = 1 \tag{5.158}$$

and using the fact that

$$c_p = \sum_{i=1}^{N} Y_i c_{p,i} \tag{5.159}$$

and also noting that

$$\begin{aligned} \nabla \int_{T^o}^{T} c_p\, dT &= \nabla \sum_{i=1}^{N} Y_i \int_{T^o}^{T} c_{p,i}\, dT \\ &= \sum_{i=1}^{N} (\nabla Y_i) \int_{T^o}^{T} c_{p,i}(T^*) dT^* + \sum_{i=1}^{N} Y_i \nabla \int_{T^o}^{T} c_{p,i}(T^*) dT^* \\ &= \sum_{i=1}^{N} (\nabla Y_i) \int_{T^o}^{T} c_{p,i}\, dT + \sum_{i=1}^{N} Y_i c_{p,i}(T) \nabla T \\ &= \sum_{i=1}^{N} (\nabla Y_i) \int_{T^o}^{T} c_{p,i}\, dT + c_p \nabla T \end{aligned} \tag{5.160}$$

Equation (5.157) then becomes

$$\nabla \cdot \left\{ \rho \mathbf{v} \int_{T^o}^{T} c_p \, dT - \rho \mathcal{D} \overbrace{\left[\sum_{i=1}^{N} \nabla Y_i \int_{T^o}^{T} c_{p,i} \, dT + c_p \nabla T \right]}^{\nabla \int_{T^o}^{T} c_p \, dT} \right\} = - \sum_{i=1}^{N} \dot{\omega}_i h_{f,i}^o \tag{5.161}$$

Substituting Eq. (5.160) into Eq. (5.161), we obtain the Shvab–Zel'dovich energy equation (W/m^3):

$$\nabla \cdot \left\{ \rho \mathbf{v} \int_{T^o}^{T} c_p \, dT - \rho \mathcal{D} \nabla \int_{T^o}^{T} c_p \, dT \right\} = - \sum_{i=1}^{N} \dot{\omega}_i h_{f,i}^o \tag{5.162}$$

The physical interpretation of Eq. (5.162) is that the combined rates of convection and diffusion of sensible enthalpy is equal to the rate of production of enthalpy by chemical reaction.

Substituting Eq. (5.156) in Eq. (5.150), we obtain the Shvab–Zel'dovich species equation:

$$\nabla \cdot \{ \rho \mathbf{v} Y_i - \rho \mathcal{D} \nabla Y_i \} = \dot{\omega}_i \tag{5.163}$$

Note the similarity in Eqs. (5.162) and (5.163). In deriving these equations, we have not assumed that the specific heat (or any transport coefficient) of the mixture is a constant or that the specific heats of all the species are equal.

The Shvab–Zel'dovich form of the conservation equations is particularly useful when one may assume that chemical changes occur by a single-step forward reaction involving species A_i (see Chapter 3),

$$\sum_{i=1}^{N} \nu_i A_i \rightarrow \sum_{i=1}^{N} \nu_i' A_i \tag{5.164}$$

In this case, converting the species reaction rate from mass to molar units, we define

$$\dot{\Omega} \equiv \frac{\dot{\omega}_i}{M_i(\nu_i' - \nu_i)}, \qquad i = 1, 2, \ldots, N \tag{5.165}$$

where M_i is the molecular mass, ν_i a stoichiometric coefficient, and $\dot{\omega}_i$ the reaction rate of species i. Defining two new parameters η_T and η_i such that

$$\eta_T \equiv \frac{\int_{T^o}^{T} c_p \, dT}{\sum_{i=1}^{N} h_{f,i}^o M_i(\nu_i - \nu_i')}$$
$$\eta_i \equiv \frac{Y_i}{M_i(\nu_i' - \nu_i)}, \qquad i = 1, 2, \ldots, N \tag{5.166}$$

then the energy equation (5.162) becomes

$$\nabla \cdot [\rho \mathbf{v} \eta_T - \rho \mathcal{D} \nabla \eta_T] = \dot{\Omega} \tag{5.167}$$

and the species equation (5.163) becomes

$$\nabla \cdot [\rho \mathbf{v} \eta_i - \rho \mathcal{D} \nabla \eta_i] = \dot{\Omega} \tag{5.168}$$

The nonlinear rate term $\dot{\Omega}$ may be eliminated from N of the $N + 1$ equations corresponding to

$$L(\eta) = \dot{\Omega} \tag{5.169}$$

where L is a linear operator if $\rho\mathcal{D}$ is independent of η

$$L(\eta) \equiv \nabla \cdot [\rho\mathbf{v}\eta - \rho\mathcal{D}\nabla\eta] \tag{5.170}$$

where η can be $\eta_T, \eta_1, \eta_2, \ldots,$ or η_N. If we select $\eta = \eta_1$ to be solved from the inhomogeneous differential equation, then

$$L(\eta_1) = \dot{\Omega} \tag{5.171}$$

The other N variables are then determined by the linear homogeneous differential equation

$$L(\beta) = 0 \tag{5.172}$$

where β can be defined as $\beta_T, \beta_2, \beta_3, \ldots, \beta_N$:

$$\begin{aligned} \beta_T &\equiv \eta_T - \eta_1 \\ \beta_2 &\equiv \eta_2 - \eta_1 \\ &\vdots \\ \beta_N &\equiv \eta_N - \eta_1 \end{aligned} \tag{5.173}$$

It should be noted that $\rho\mathcal{D}$ may depend on β_T or β_i. Under such conditions, the operator L depends implicitly on β and Eq. (5.172) becomes nonlinear. As we will see in laminar premixed flames in Chapter 7 and laminar diffusion flames in Chapter 8, through the use of the Shvab–Zel'dovich formulation, we can obtain many important relationships. For systems in which the reactants are initially unmixed, combustion rates can even be determined by solving a homogeneous differential equation like (5.172), bypassing the handling of an inhomogeneous differential equation like (5.171).

Example 5.2 ***Shvab–Zel'dovich Species and Energy Equations for a Fuel Droplet***

An evaporating and combusting fuel droplet can be modeled as a sphere with spherically symmetric liquid and gaseous layers. Assuming a quasi-steady gaseous flow from the droplet, what are the Shvab–Zel'dovich fuel species and energy differential equations in spherical coordinates?

Solution

Using a quasi-steady assumption, the gaseous mass flowrate from a fuel droplet is

$$\dot{m} = 4\pi r^2 \rho u_r = \text{constant}$$

where u_r is the bulk flow radial velocity. Note that a constant flowrate implies that

$$\frac{d}{dr}\left(r^2 \rho u_r\right) = 0$$

The vector form of the Shvab–Zel'dovich species equation is given by Eq. (5.163),

$$\nabla \cdot \{\rho\mathbf{v}Y_i - \rho\mathcal{D}\nabla Y_i\} = \dot{\omega}_i$$

For a spherically symmetric fuel (F) droplet, the differential form of the fuel species equation is

$$\frac{1}{r^2}\frac{d}{dr}\left[r^2\left(\rho u_r Y_F - \rho\mathcal{D}\frac{dY_F}{dr}\right)\right] = \dot{\omega}_F$$

Therefore,

$$\frac{\rho u_r r^2}{r^2}\frac{dY_F}{dr} - \frac{1}{r^2}\frac{d}{dr}\left(r^2\rho\mathcal{D}\frac{dY_F}{dr}\right) = \dot{\omega}_F$$

and

$$\rho u_r r^2 \frac{dY_F}{dr} - \frac{d}{dr}\left(r^2 \rho D \frac{dY_F}{dr}\right) = r^2 \dot{\omega}_F$$

The vector form of the Shvab–Zel'dovich energy equation is given by Eq. (5.162),

$$\nabla \cdot \left\{ \rho \mathbf{v} \int_{T^o}^{T} c_p \, dT - \rho D \nabla \int_{T^o}^{T} c_p \, dT \right\} = -\sum_{i=1}^{N} \dot{\omega}_i h_{f,i}^o$$

To simplify the above energy equation, assume that a reference temperature is chosen so that

$$c_p T = \int_{T^o}^{T} c_p \, dT \quad \text{and} \quad \dot{Q} = -\sum_{i=1}^{N} \dot{\omega}_i h_{f,i}^o$$

Since $\rho D = k/c_p$, the differential form of the energy equation can be written as

$$\frac{1}{r^2} \frac{d}{dr} \left[r^2 \left(\rho u_r c_p T - \frac{k}{c_p} \frac{d}{dr} c_p T \right) \right] = \dot{Q}$$

Therefore,

$$\frac{\rho u_r r^2}{r^2} \frac{d}{dr} c_p T - \frac{1}{r^2} \frac{d}{dr} \left(r^2 \frac{k}{c_p} \frac{d}{dr} c_p T \right) = \dot{Q}$$

and

$$\rho u_r r^2 \frac{d}{dr} c_p T - \frac{d}{dr} \left(r^2 \frac{k}{c_p} \frac{d}{dr} c_p T \right) = r^2 \dot{Q}$$

5.11 Boundary Conditions at an Interface

In the formulation of a combustion problem, in addition to applying the governing differential equations, one also has to formulate the boundary and initial conditions. The specification of boundary conditions is not always trivial. The number of boundary conditions required for a set of governing differential equations depends quite often on the flow conditions of the reactive mixture, in other words, on the type (based on the characteristic values or eigenvalues) of the governing equations. Either under- or over-specifying the boundary conditions will make the overall formulation ill posed.

The boundary condition specifications for various types of combustion problems are discussed in later chapters. The purpose of describing boundary conditions in this section is to familiarize readers with a formal way of specifying them through the use of conservation conditions at material interfaces, where the boundary conditions are usually more complex than those in the far field.

Relations expressing the conservation of mass, momentum, and energy at an interface may be derived by converting the conservation equations in differential form to integral form and then taking the limit in which the volume of integration approaches a surface (Williams 1985). Reynolds' Transport Theorem is used for this derivation procedure. This theorem states that the time rate of change of the integral of an arbitrary function $f(x, t)$ over a given volume V can be written as the sum of two parts:

$$\frac{d}{dt}\left(\iiint_V f \, dV \right) = \iiint_V \frac{\partial f}{\partial t} dV + \iint_S \mathbf{v}_B \cdot \mathbf{n} f \, dS \tag{5.174}$$

The first term in Eq. (5.174) represents the contribution due to the time rate of change of the integrand. The second term represents the contribution due to the motion of the control surface S bounding the volume V. Here, $\mathbf{n}$ is the local outward normal unit vector to the surface S, and $\mathbf{v}_B$ is the velocity of the boundary of the control volume. In general, both $\mathbf{n}$ and $\mathbf{v}_B$ can vary from point to point on the control surface S.

5.11.1 Integral Forms of Conservation Equations

The mass conservation condition at an interface is derived by first considering the continuity equation:

$$\frac{\partial \rho}{\partial t} + \nabla \cdot (\rho \mathbf{v}) = 0 \tag{5.175}$$

Upon integration of this equation over a volume V, we have

$$\iiint\limits_V \left[\frac{\partial \rho}{\partial t} + \nabla \cdot (\rho \mathbf{v})\right] dV = 0 \tag{5.176}$$

Using Eq. (5.174) and the divergence theorem,

$$\iiint\limits_V \nabla \cdot (\rho \mathbf{v}) dV = \iint\limits_S \rho \mathbf{v} \cdot \mathbf{n}\, dS \tag{5.177}$$

we obtain

$$\frac{d}{dt}\iiint\limits_V \rho\, dV - \iint\limits_S \mathbf{v}_B \cdot \mathbf{n} \rho\, dS + \iint\limits_S \rho \mathbf{v} \cdot \mathbf{n}\, dS = 0 \tag{5.178}$$

or

$$\frac{d}{dt}\left(\iiint\limits_V \rho dV\right) + \iint\limits_S \rho(\mathbf{v} - \mathbf{v}_B) \cdot \mathbf{n}\, dS = 0 \tag{5.179}$$

Equation (5.179) represents the integral form of the continuity equation.

The momentum equation (5.73) can be expressed in the vector form as

$$\rho\frac{\partial \mathbf{v}}{\partial t} + \rho \mathbf{v} \cdot \nabla \mathbf{v} = \nabla \cdot \tilde{\sigma} + \rho \sum_{i=1}^{N} Y_i \mathbf{f}_i \tag{5.180}$$

Multiplying the continuity equation (5.40) by $\mathbf{v}$, we obtain

$$\mathbf{v}\frac{\partial \rho}{\partial t} + \mathbf{v}\nabla \cdot (\rho \mathbf{v}) = 0 \tag{5.181}$$

Adding Eqs. (5.180) and (5.181) gives

$$\frac{\partial}{\partial t}(\rho \mathbf{v}) + \nabla \cdot (\rho \mathbf{v}\mathbf{v}) = \nabla \cdot \tilde{\sigma} + \rho \sum_{i=1}^{N} Y_i \mathbf{f}_i \tag{5.182}$$

We may integrate this over S to express the integral form of the momentum equation:

$$\frac{d}{dt}\iiint\limits_V \rho \mathbf{v}\, dV + \iint\limits_S \rho \mathbf{v}[(\mathbf{v} - \mathbf{v}_B) \cdot \mathbf{n}] dS = \iint\limits_S \tilde{\sigma} \cdot \mathbf{n}\, dS + \iiint\limits_V \rho \sum_{i=1}^{N} Y_i \mathbf{f}_i\, dV \tag{5.183}$$

The energy equation (5.110) can be written in vector notation as

$$\frac{\partial}{\partial t}(\rho e_t) + \nabla \cdot (\rho e_t \mathbf{v}) = -\nabla \cdot \mathbf{q} + \dot{Q} + \nabla \cdot (\tilde{\sigma} \cdot \mathbf{v}) + \rho \sum_{i=1}^{N} Y_i \mathbf{f}_i \cdot (\mathbf{v} + \mathbf{V}_i) \tag{5.184}$$

After integration, we obtain the integral form of the energy equation,

$$\frac{d}{dt}\iiint_V \rho e_t\, dV + \iint_S \rho e_t[(\mathbf{v}-\mathbf{v}_B)\cdot\mathbf{n}]dS$$
$$= \iiint_V \dot{Q}\, dV - \iint_S \mathbf{q}\cdot\mathbf{n}\, dS + \iint_S \mathbf{v}\cdot\tilde{\sigma}\cdot\mathbf{n}\, dS$$
$$+ \iiint_V \rho\sum_{i=1}^{N} Y_i\mathbf{f}_i\cdot(\mathbf{v}+\mathbf{V}_i)dV \tag{5.185}$$

The vector form of the conservation of chemical species equation (5.42) is

$$\frac{\partial}{\partial t}(\rho Y_i) + \nabla\cdot[\rho Y_i(\mathbf{v}+\mathbf{V}_i)] = \dot{\omega}_i \tag{5.186}$$

Upon integration, we obtain

$$\frac{d}{dt}\left(\iiint_V \rho Y_i\, dV\right) + \iint_S \rho Y_i(\mathbf{v}+\mathbf{V}_i-\mathbf{v}_B)\cdot\mathbf{n}\, dS = \iiint_V \dot{\omega}_i\, dV \tag{5.187}$$

5.11.2 Conservation Equations at an Interface

To form a surface interface, we now consider the control volume V to be a thin slab, the thickness of which is made to approach zero (see Figure 5.14). In the limit, the integrals over the surface S are composed of two parts, namely, integrals over each face of the slab.

Let us arbitrarily choose one face as the "positive" side of the slab, identifying quantities on this side by the subscript + and those on the other side by the subscript −. It is clear that $\mathbf{n}_-$ is equal to $-\mathbf{n}_+$ at corresponding points on the two faces; in the limit, the integral over the surface of V may be replaced by an integral over the interface area S_I, and, hence, Eq. (5.179) becomes

$$\iint_S \left[\rho_+(\mathbf{v}_+-\mathbf{v}_{B+}) - \rho_-(\mathbf{v}_- - \mathbf{v}_{B-})\right]\cdot\mathbf{n}_+ dS = -\lim_{V\to 0}\left[\frac{d}{dt}\iiint_V \rho\, dV\right] \tag{5.188}$$

Integrating over the same interfacial volume, Eq. (5.183) becomes

$$\iint_S \{\rho_+\mathbf{v}_+[(\mathbf{v}_+-\mathbf{v}_{B+})\cdot\mathbf{n}_+] - \rho_-\mathbf{v}_-[(\mathbf{v}_- - \mathbf{v}_{B-})\cdot\mathbf{n}_+] - (\tilde{\sigma}_+ - \tilde{\sigma}_-)\cdot\mathbf{n}_+\}dS$$
$$= \lim_{V\to 0}\left[\iiint_V \rho\sum_{i=1}^{N} Y_i\mathbf{f}_i\, dV - \frac{d}{dt}\iiint_V \rho\mathbf{v}\, dV\right] \tag{5.189}$$

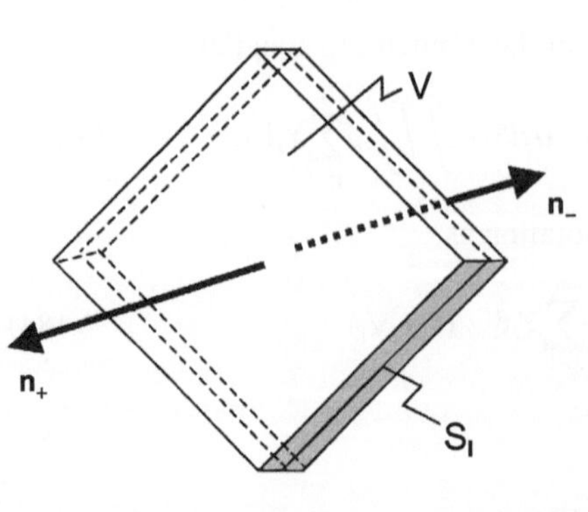

Figure 5.14 Control volume and interface S_I for the derivation of interface conditions.

Using the same integration procedure, the energy equation (5.185) becomes

$$\iint_S \left[\rho_+ e_{t+}(\mathbf{v}_+ - \mathbf{v}_{B+}) - \rho_- e_{t-}(\mathbf{v}_- - \mathbf{v}_{B-}) + \mathbf{q}_+ - \mathbf{q}_- - \mathbf{v}_+ \cdot \tilde{\sigma}_+ + \mathbf{v}_- \cdot \tilde{\sigma}_-\right] \cdot \mathbf{n}_+ \, dS$$

$$= \lim_{V\to 0}\left\{\iiint_V \dot{Q}\, dV + \iiint_V \rho \sum_{i=1}^{N} Y_i \mathbf{f}_i \cdot (\mathbf{v} + \mathbf{V}_i)\, dV - \frac{d}{dt}\iiint_V \rho e_t \, dV\right\} \tag{5.190}$$

Similarly, the species equation (5.187) becomes

$$\iint_S \left[\rho_+ Y_{i+}(\mathbf{v}_+ + \mathbf{V}_{i+} - \mathbf{v}_{B+}) - \rho_- Y_{i-}(\mathbf{v}_- + \mathbf{V}_{i-} - \mathbf{v}_{B-})\right] \cdot \mathbf{n}_+ \, dS$$

$$= \lim_{V\to 0}\left(\iiint_V \dot{\omega}_i \, dV - \frac{d}{dt}\iiint_V \rho Y_i \, dV\right), \quad i = 1, 2, \ldots, N \tag{5.191}$$

The last term in the equation represents surface reactions as well as the nonzero time rate of accumulation of the ith chemical species at the interface.

If the interface is taken to be a solid surface at which chemical reactions are proceeding at a finite rate, then

$$\dot{\omega}_i \to \dot{\omega}_i' \delta(y - y_I)$$

The parameter $\dot{\omega}_i'$ represents the surface reaction rate for producing ith species with units of kg/m^2 s, while $\dot{\omega}_i$ has units of kg/m^3 s. The term $\delta(y - y_I)$ is the Dirac delta function and is nonzero only when the y coordinate normal to the surface is equal to y_I on the surface. The Dirac delta function is named after P. Dirac (1902–1984), an English physicist who was one of the founders of quantum mechanics.

Replacing the volume integral for the species production by a surface integral,

$$\lim_{V\to 0}\iiint_V \dot{\omega}_i \, dV = \iint_S \dot{\omega}_i' \, dS$$

If ρ' represents the mass of species i per unit area, then

$$\rho Y_i \to \rho_i' \delta(y - y_I)$$

The last term of Eq. (5.191) can be written as

$$\lim_{V\to 0}\frac{d}{dt}\iiint_V \rho Y_i \, dV = \frac{d}{dt}\lim_{V\to 0}\iiint_V \rho_i' \delta(y - y_I)\, dV = \frac{d}{dt}\iint_S \rho_i' \, dS \tag{5.192}$$

Thus, the physical meaning of Eq. (5.191) can be expressed by the following species mass-flux balance statement:

The mass flux of the i*th species leaving the surface minus the mass flux of the* i*th species entering the surface is equal to the production rate of the* i*th species at the surface minus the rate of accumulation of the* i*th species on the control surface.*

Similar interpretations can be given for the continuity equation of a mixture's mass flux, Eq. (5.188), the momentum flux-balance equation, Eq. (5.189), and the energy flux-balance equation, Eq. (5.190). In many combustion problems, the body force and accumulation terms in

Eqs. (5.188) through (5.190) involving $\lim_{V \to 0}$ are zero. If it is further assumed that the interface is stationary, i.e.,

$$\mathbf{v}_B = 0 \tag{5.193}$$

and that viscosity is negligible, then the continuity, momentum, and energy equations can be simplified considerably to give

$$\begin{aligned} &\rho_+ \mathbf{v}_+ \cdot \mathbf{n}_+ = \rho_- \mathbf{v}_- \cdot \mathbf{n}_+ \\ &\rho_+ \mathbf{v}_+ (\mathbf{v}_+ \cdot \mathbf{n}_+) + p_+ \mathbf{n}_+ = \rho_- \mathbf{v}_- (\mathbf{v}_- \cdot \mathbf{n}_+) + p_- \mathbf{n}_+ \end{aligned} \tag{5.194}$$

$$\begin{aligned} &\left\{ \rho_+ \left[h_{t+} \mathbf{v}_+ + \sum_{i=1}^{N} h_{i+} Y_{i+} \mathbf{V}_{i+} \right] - k_+ (\nabla T)_+ + \mathbf{q}_{R+} \right\} \cdot \mathbf{n}_+ \\ &= \left\{ \rho_- \left[h_{t-} \mathbf{v}_- + \sum_{i=1}^{N} h_{i-} Y_{i-} \mathbf{V}_{i-} \right] - k_- (\nabla T)_- + \mathbf{q}_{R-} \right\} \cdot \mathbf{n}_+ \end{aligned} \tag{5.195}$$

where $\mathbf{q}_{R+}$ and $\mathbf{q}_{R-}$ represent the radiative heat-flux contribution. These terms have been included because radiative losses from surfaces are often substantial. If there is no surface reaction to generate species i, then the terms on the right-hand side of Eq. (5.191) are also zero and the species mass-flux equation at the interface gives

$$\rho_+ Y_{i+} (\mathbf{v}_+ + \mathbf{V}_{i+}) \cdot \mathbf{n}_+ = \rho_- Y_{i-} (\mathbf{v}_- + \mathbf{V}_{i-}) \cdot \mathbf{n}_+ \tag{5.196}$$

5.11.3 Continuity at a Combusting Interface

As shown in Figure 5.15*a*, the right-hand surface of a solid material is burning toward the left at a rate of v_b, where A_b is the burning surface area and ρ_s is the density of the solid.

Let the solid density ρ_s be equal to the arbitrary function f given in *Reynolds' Transport Theorem* given by Eq. (5.174). The density ρ_s is assumed to be constant. As shown in Figure 5.15*a*, the control volume (*a*) decreases in size as it moves with the burning right-hand side surface. Note that the direction of the burning velocity v_b is opposite in sign to the outward normal $\mathbf{n}$.

For control volume (*a*), the Reynolds Transport Theorem takes the form

$$\begin{aligned} \frac{d}{dt}\left(\iiint_V \rho_s dV \right) &= \iiint_V \frac{\partial \rho_s}{\partial t} dV + \iint_S \mathbf{v}_B \cdot \mathbf{n} \rho_s dS \\ &= \iint_S \mathbf{v}_B \cdot \mathbf{n} \rho_s dS \end{aligned} \tag{5.197}$$

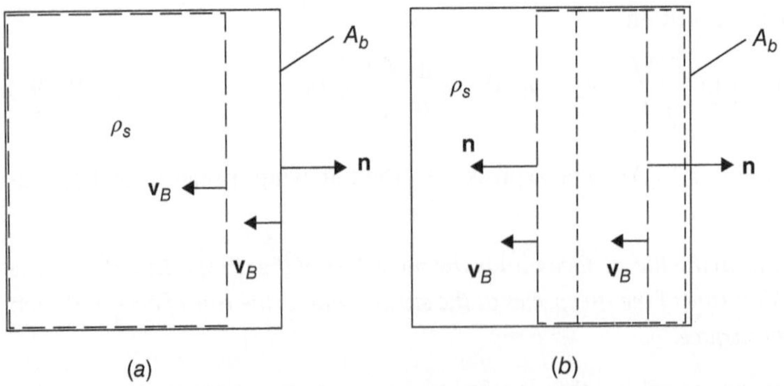

Figure 5.15 Regressing surface of a burning solid material. (*a*) Control volume decreases with time. (*b*) Moving control volume with a fixed volume.

The mass burning rate can be expressed as

$$\frac{dm}{dt} = -v_b \rho_s A_b \tag{5.198}$$

If the solid material is burning away, but we keep the size of the control volume (b) constant by moving both boundaries of the control volume at the same rate as the burning rate. According to Figure 5.15b,

$$\frac{d}{dt}\left(\iiint_V \rho_s \, dV\right) = \iiint_V \frac{\partial \rho_s}{\partial t} dV + \iint_S \mathbf{v}_B \cdot \mathbf{n} \rho_s \, dS$$

$$= \iint_S \mathbf{v}_B \cdot \mathbf{n} \rho_s \, dS \tag{5.199}$$

$$\frac{dm}{dt} = +v_{b_{\text{left}}} \rho_s A_b - v_{b_{\text{right}}} \rho_s A_b$$

$$= +v_b \rho_s A_b - v_b \rho_s A_b = 0 \tag{5.200}$$

The mass m is constant in control volume (b).

5.11.4 Momentum Flux at a Combusting Gas–Solid Interface

The momentum flux at a gas–solid interface depends on the stress tensors on either side of the interface. If the viscous term is negligible, the total stress tensor $\tilde{\sigma}$ in the gas phase is

$$\sigma_{ij} = -p\delta_{ij} + \left(\mu' - \frac{2}{3}\mu\right)\frac{\partial u_k}{\partial x_k}\delta_{ij} + \mu\left(\frac{\partial u_i}{\partial x_j} + \frac{\partial u_j}{\partial x_i}\right)$$

$$= -p\delta_{ij} \tag{5.201}$$

The stress tensor in a linear viscoelastic solid material can be written as

$$\sigma_{ij}(t) = \int_{-\infty}^{t} G_1(t-t')\frac{d}{dt'}\varepsilon_{ij}(t')dt' - \frac{1}{3}\delta_{ij}\int_{-\infty}^{t} G_1(t-t')\frac{d\varepsilon_{kk}}{dt'}dt' + \delta_{ij}K\varepsilon_{kk} \tag{5.202}$$

where $G_1(t-t')$ is called the relaxation modulus in shear, ε_{ij} and ε_{kk} are the strain-rate tensor components, and K is the bulk modulus. Note that if the solid material is either a solid fuel or a solid propellant, such material can usually be assumed to be viscoelastic in shear and elastic in bulk. For most viscoelastic materials, we can express G_1 as a function of time t according to the following exponential series:

$$G_1 = g_0 + \sum_{i=1}^{N} g_i e^{-\beta_i t} \tag{5.203}$$

For simplicity, let us denote the normal stress in the solid phase by $\sigma_p \mathbf{1}$. Note that σ_p is positive for tension. Then, the normal-stress tensor in the solid phase becomes

$$\tilde{\sigma} = \sigma_p \mathbf{1}$$

Now, we shall assume the interface is stationary, i.e., fixed in space by feeding the solid material upward at its burning rate r_b, and the body-force and viscous effects are negligible. Applying the integral momentum equation (5.183), we have

$$A\rho_{g+}\mathbf{v}_{g+}(\mathbf{v}_{g+} \cdot \mathbf{n}_+) - A\rho_{s-}\mathbf{r}_{b-}(\mathbf{r}_{b-} \cdot \mathbf{n}_+) - (-p_+\mathbf{1} - \sigma_{p-}\mathbf{1}) \cdot \mathbf{n}_+ A$$

$$= \lim_{V\to 0}\left[\iiint_V \rho \sum_{i=1}^{N} Y_i \mathbf{f}_i \, dV - \frac{d}{dt}\iiint_V \rho \mathbf{v} \, dV\right] \to 0 \tag{5.204}$$

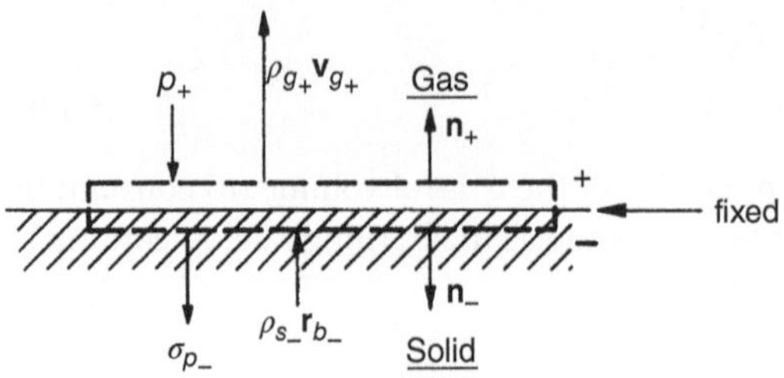

Figure 5.16 An infinitely thin control volume at the gas–solid interface of a burning solid material.

Figure 5.16 indicates the four-term balance between the burned gas momentum flux leaving the control surface $\rho_{g+}v_{g+}$, the momentum flux of the burning solid entering the control surface $\rho_{s-}r_{b-}$, the pressure force exerted on the top surface of the control volume $p_+\mathbf{n}_+$, and the normal stress exerted on the bottom surface of the control volume σ_{p-}.

The momentum equation takes the form

$$A\rho_{g+}\mathbf{v}_{g+}v_{g+} - A\rho_{s-}\mathbf{r}_{b-} + Ap_+\mathbf{n}_+ + A\sigma_{p-}\mathbf{n}_+ = 0 \tag{5.205}$$

$\mathbf{n}_+$ can be eliminated from the above equation because

$$\mathbf{v}_g = v_g\mathbf{n}_+, \quad \mathbf{r}_{b-} = r_b\mathbf{n}_+$$

Thus, we have

$$\rho_g v_{g+}^2 + p_+ = \rho_{s-}r_{b-}^2 - \sigma_{p-}$$

Because $\rho_{g+}v_{g+} = \rho_{s-}r_{b-}$ and usually $u_{g+} \gg r_{b-}$, we can accept the approximation that $\rho_g v_{g+}^2 \gg \rho_{s-}r_{b-}^2$, then,

$$-\sigma_{p-} = p_+ + \rho_{g+}v_{g+}^2 \tag{5.206}$$

Therefore, the normal compressive stress $(-\sigma_{p-})$ in the solid phase immediately below the interface is approximately equal to the sum of the hydrostatic pressure and the momentum of the gas leaving the burning surface.

5.11.5 Species Mass Flux at a Gas–Solid Interface

Figure 5.17 indicates a five-term species i mass flux balance at a gas–solid interface. The mass flux convected and diffused out equals the mass flux supplied from the bottom and generated or accumulated at the surface. Assume:

1. The interface location is fixed in space.
2. Fick's law of mass diffusion is valid.
3. There is no subsurface diffusion.
4. The mass accumulation at the surface is negligible.

From the species equations (5.191) and (5.192), we have

$$A\rho_{g+}Y_{i+}(\mathbf{v}_{g+} + \mathbf{V}_{i+}) \cdot \mathbf{n}_+ - A\rho_{s-}Y_{i-}\mathbf{r}_{b-} \cdot \mathbf{n}_+ = \iint_S \dot{\omega}_i' \, dS - \frac{d}{dt}\iint_S \rho_i' \, dS \tag{5.207}$$

where

$$\mathbf{V}_{i+} = -\frac{\mathcal{D}}{Y_{i+}}\frac{\partial Y_i}{\partial y}\bigg|_+ \mathbf{n}_+ \tag{5.208}$$

Figure 5.17 Mass flux balance of ith species at the gas–solid interface.

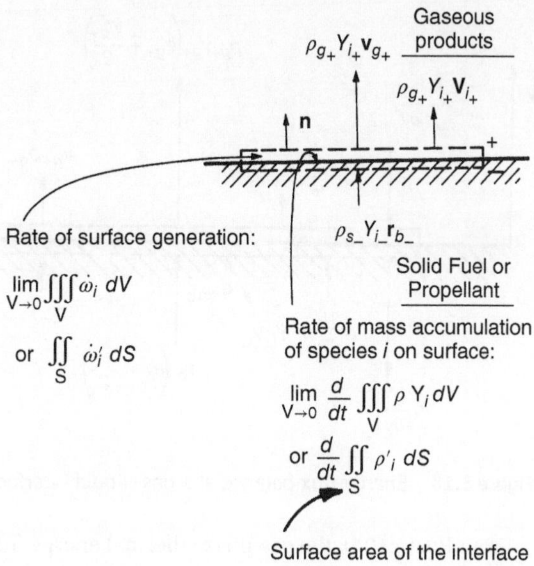

so that

$$\rho_{g+} Y_{i+} v_{g+} - \rho_{g+} \mathcal{D} \left. \frac{\partial Y_i}{\partial y} \right|_+ = \rho_{s-} Y_{i-} r_{b-} + \dot{\omega}_i' - \frac{1}{A} \frac{d}{dt} \iint_S \rho_i' dS \tag{5.209}$$

If we further assume that the system is at steady state with no mass accumulation, then

$$\rho_{g+} Y_{i+} v_{g+} - \rho_{g+} \mathcal{D} \left. \frac{\partial Y_i}{\partial y} \right|_+ = \rho_{s-} Y_{i-} r_{b-} + \dot{\omega}_i' \tag{5.210}$$

In general, a gaseous species i in the product does not exist in the same form in the solid; hence, $Y_{i-} = 0$. Therefore, the mass flux convected and diffused out of the surface is balanced by the mass flux generated on the surface,

$$\rho_{g+} Y_{i+} v_{g+} - \rho_{g+} \mathcal{D} \left. \frac{\partial Y_i}{\partial y} \right|_+ = \dot{\omega}_i' \tag{5.211}$$

5.11.6 Energy Flux at a Gas–Liquid Interface

If the body-force work is zero, and the system is at steady state with no net accumulation of total stored energy inside the interfacial control volume, then the energy equation (5.190) can be expressed as

$$\begin{aligned}
&\iint_{S_l} \rho_{g+} \left(e_{g+} + \frac{v_{g+}^2}{2} \right) v_{g+} \, dS - \iint_S \rho_{l-} \left(e_{l-} + \frac{r_{b-}^2}{2} \right) r_{b-} \, dS \\
&\quad + \iint_S \mathbf{q}_+ \cdot \mathbf{n}_+ \, dS - \iint_S \mathbf{q}_- \cdot \mathbf{n}_+ \, dS \\
&\quad - \iint_S v_{g+}(-p_{g+} + \tau_{ii+}) dS - \iint_S r_{b-} p_{l-} \, dS = \lim_{V\to 0} \iiint_V \dot{Q} \, dV
\end{aligned} \tag{5.212}$$

Figure 5.18 illustrates the various energy flux terms in and out of the interfacial control volume.

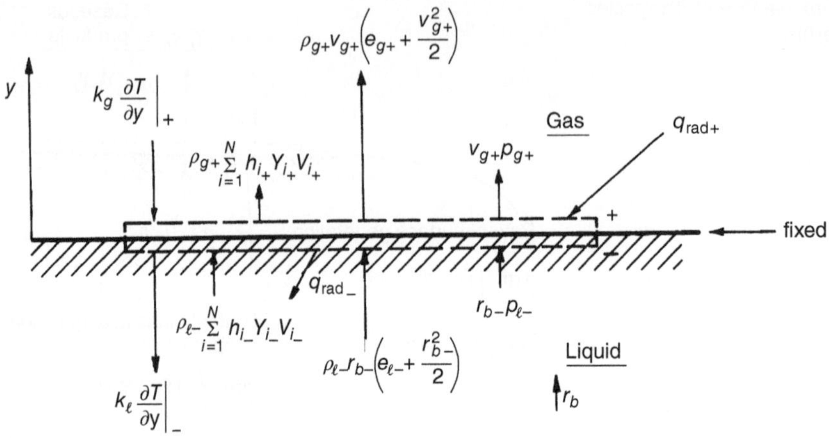

Figure 5.18 Energy-flux balance at a gas–liquid interface.

From Eq. (5.104), the gas-phase thermal energy flux $\mathbf{q}_+$ can be written as the sum of three terms:

$$\mathbf{q}_+ = -k\nabla T + \rho \sum_{i=1}^{N} h_i Y_i \mathbf{V}_i + \mathbf{q}_{\text{Dufour}} \tag{5.213}$$

The term $\mathbf{q}_{\text{Dufour}}$ can be neglected; also, the non-hydrostatic normal stresses, τ_{ii+} ($\approx \frac{4}{3}\mu\nabla \cdot \mathbf{v}$), can be neglected in comparison with p_{g+}.

Now assume the net rate of energy absorption at the surface interface is due to radiation, so

$$\lim_{V\to 0} \iiint_V \dot{Q}\, dV = \iint_S (\mathbf{q}_{\text{rad}-} - \mathbf{q}_{\text{rad}+}) \cdot \mathbf{n}_+\, dS \tag{5.214}$$

The energy-flux balance becomes

$$\begin{aligned} &\rho_{g+} v_{g+} \left(e_{g+} + \frac{p_{g+}}{\rho_{g+}} + \frac{v_{g+}^2}{2} \right) + \rho_{g+} \sum_{i=1}^{N} h_{i+} Y_{i+} V_{i+} - k_g \left.\frac{\partial T}{\partial y}\right|_+ - q_{\text{rad}_+} \\ &= \rho_{l-} r_{b-} \left(e_{l-} + \frac{p_{l-}}{\rho_{l-}} + \frac{r_{b-}^2}{2} \right) + \rho_{l-} \sum_{i=1}^{N} h_{i-} Y_{i-} V_{i-} - k_l \left.\frac{\partial T}{\partial y}\right|_- - q_{\text{rad}-} \end{aligned} \tag{5.215}$$

The energy e can be expressed as

$$e \equiv h - \frac{p}{\rho} = \sum_{i=1}^{N} h_i Y_i - \frac{p}{\rho} \tag{5.216}$$

Upon substitution, we have

$$\begin{aligned} &\rho_{g+} v_{g+} \left(h_{g+} + \frac{v_{g+}^2}{2} \right) + \rho_{g+} \sum_{i=1}^{N} h_{i+} Y_{i+} V_{i+} - k_g \left.\frac{\partial T}{\partial y}\right|_+ - q_{\text{rad}_+} \\ &= \rho_{l-} r_{b-} \left(h_{l-} + \frac{r_{b-}^2}{2} \right) + \rho_{l-} \sum_{i=1}^{N} h_{i-} Y_{i-} V_{i-} - k_l \left.\frac{\partial T}{\partial y}\right|_- - q_{\text{rad}-} \end{aligned} \tag{5.217}$$

The species mass-flux balance at a gas–liquid interface is

$$\rho_{g+} Y_{i+} v_{g+} - \rho_{g+} \mathcal{D} \left.\frac{\partial Y_i}{\partial y}\right|_+ = \rho_{l-} Y_{i-} r_{b-} - \rho_{l-} \mathcal{D} \left.\frac{\partial Y_i}{\partial y}\right|_- + \dot{\omega}_i' \tag{5.218}$$

or using Fick's law,

$$\rho_{g+}[Y_{i+}v_{g+} + Y_{i+}V_{i+}] = \rho_{l-}[Y_{i-}r_{b-} + Y_{i-}V_{i-}] + \dot{\omega}_i' \tag{5.219}$$

Since usually $h_{g+} \gg v_{g+}^2/2$ and $h_{l-} \gg r_{b-}^2/2$, we can drop the kinetic-energy terms from the energy equation,

$$\begin{aligned} &\rho_{g+}\sum_{i=1}^{N} h_{i+}Y_{i+}(v_{g+} + V_{i+}) - k_g \left.\frac{\partial T}{\partial y}\right|_{+} - q_{\text{rad}_+} \\ &= \rho_{l-}\sum_{i=1}^{N} h_{i-}Y_{i-}(r_{b-} + V_{i-}) - k_l \left.\frac{\partial T}{\partial y}\right|_{-} - q_{\text{rad}\_} \end{aligned} \tag{5.220}$$

It should be noted that the enthalpy contains not only the sensible enthalpy but also the chemical enthalpy, that is,

$$h_i = h_{f,i}^o + \int_{T_o}^{T} c_{p_i}\, dT \tag{5.221}$$

Upon substitution,

$$\begin{aligned} &\sum_{i=1}^{N}\left[h_{f,i}^o + \int_{T_o}^{T} c_{p_i}\, dT\right]_{+} \rho_{g_+}Y_{i+}(v_{g_+} + V_{i+}) - k_g \left.\frac{\partial T}{\partial y}\right|_{+} - q_{\text{rad}_+} \\ &= \sum_{i=1}^{N}\left[h_{f,i}^o + \int_{T_o}^{T} c_{p_i}\, dT\right]_{-} \rho_{l-}Y_{i-}(r_{b-} + V_{i-}) - k_l \left.\frac{\partial T}{\partial y}\right|_{-} - q_{\text{rad}\_} \end{aligned} \tag{5.222}$$

Rearranging terms,

$$\begin{aligned} k_l \left.\frac{\partial T}{\partial y}\right|_{-} &= k_g \left.\frac{\partial T}{\partial y}\right|_{+} + (q_{\text{rad}_+} - q_{\text{rad}\_}) \\ &+ \sum_{i=1}^{N} \rho_{l-}Y_{i-}(r_{b-} + V_{i-})h_{f,i-}^o - \sum_{i=1}^{N} \rho_{g+}Y_{i+}(v_{g_+} + V_{i+})h_{f,i+}^o \\ &+ \sum_{i=1}^{N} \rho_{l-}Y_{i-}(r_{b-} + V_{i-}) \int_{T_o}^{T} c_{p_{i-}}\, dT - \sum_{i=1}^{N} \rho_{g+}Y_{i+}(v_{g_+} + V_{i+}) \int_{T_o}^{T} c_{p_{i+}}\, dT \end{aligned} \tag{5.223}$$

The last two terms for the enthalpy change across the interface are equal to

$$\sum_{i=1}^{N} \rho_{l-}Y_{i-}(r_{b-} + V_{i-})(c_{p_{i-}} - c_{p_i+})(T_s - T_o)$$

if $c_{p_i}(T_s - T_o) = \int_{T_o}^{T} c_{p_i}\, dT$ for all i species and if $V_i = 0$ for all i; this term also can be further simplified to

$$\rho_{l-}r_{b-}(c_{pl} - c_{pg})(T_s - T_o)$$

5.12 Turbulent Flow Fields

In laminar flows, the fluid layers slide past another in a smooth manner, and the mixing between the fluid layers is due to molecular diffusion. As the Reynolds number ($\text{Re} = \overline{u}L/\nu$) of the mean flow increases above a critical value, above about 2300 for pipe flow, the flow will become turbulent. The transition to turbulence occurs due to the growth of instability in laminar flows when the Reynolds

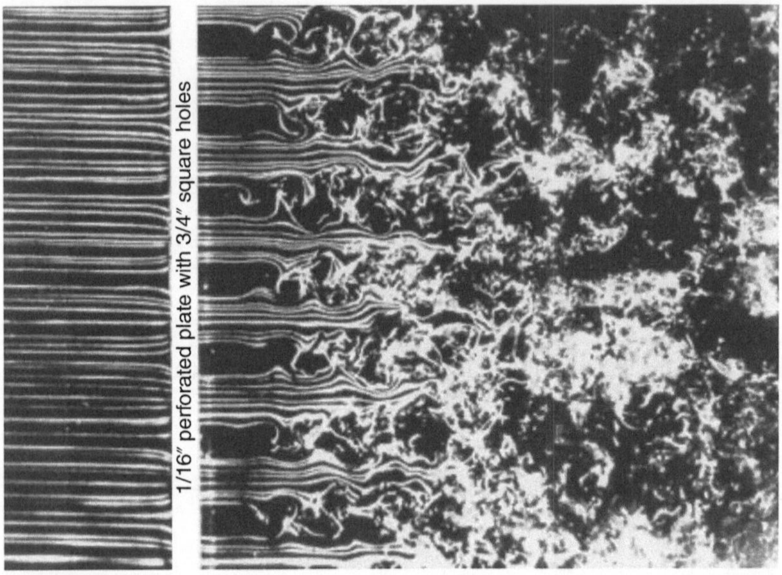

Figure 5.19 Generation of turbulence by grid at $Re_D = 2000$. (Adapted from Van Dyke 1982).

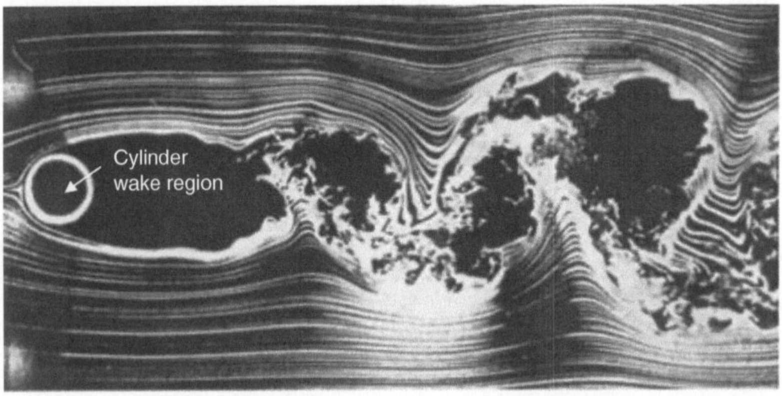

Figure 5.20 Visualization of turbulent wake of a cylinder at $Re_D = 10{,}000$. (Adapted from Van Dyke 1982).

number becomes sufficiently large. The instability is related to the interaction of viscous terms and nonlinear inertia terms in the equation of motion. Turbulent flow can be envisioned as a mean fluid flow upon which are superimposed eddies of different sizes randomly dispersed in the flow, as shown in Figures 5.19 and 5.20.

Turbulence is inherently dissipative, three dimensional, and time dependent. A turbulent flow is composed of numerous vortices or eddies which have finite lifetimes and appear to be born at random times. The axes of the vortices also assume random orientations. There are even vortices within vortices. As we will see, it is not until the flow is analyzed statistically that any regularity in the flow field begins to appear. The statistical properties such as time and space averages, correlation functions, and probability density functions (PDFs) show regular behavior. In turbulent reacting flows, the fluid mechanics of the turbulent flow field is also coupled with species combustion chemistry and energy release.

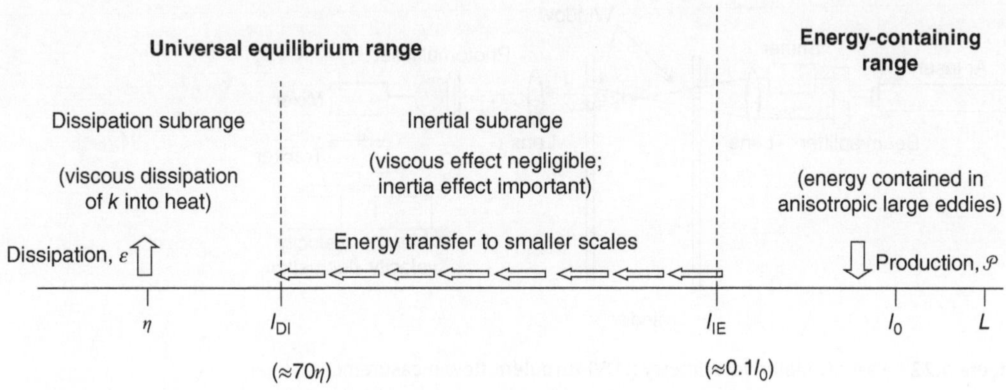

Figure 5.21 Turbulence energy cascade. (Adapted from Pope (2000).)

Turbulence decays without a production mechanism. The largest eddies are generated by shear in the mean flow and account for most of the transport of momentum and energy. Most of the kinetic energy of the turbulence is contained in the largest eddies. Since the eddies have a finite lifetime, observations have indicated that the eddy kinetic energy is transferred to smaller and smaller eddies and are eventually dissipated into heat energy by viscosity, as indicated schematically in Figure 5.21. Inertial effects spread the turbulent energy from the large eddies to smaller and smaller eddies until the viscous stresses are comparable to the inertial forces, a process called an energy cascade. The smallest turbulent eddies are still much larger ($\approx 10^{-6}$ m) than molecular length scales ($\approx 10^{-9}$ m).

The energy cascade concept was first proposed by A. Kolmogorov (1903–1987), a Russian mathematician, who proposed that the smallest scales of turbulence are self-similar. Kolmogorov's first similarity hypothesis was that the properties of the smallest scales depend only on the dissipation rate ϵ and kinematic viscosity ν. Kolmogorov's second similarity hypothesis was that there is an upper subrange (inertial subrange) of small eddies that receive energy at a rate proportional to the dissipation rate ϵ, independent of the kinematic viscosity ν. In the inertial subrange, inertia forces play a dominant role, with insignificant viscous forces. The Kolmogorov postulates imply that the smallest eddies are in a quasi-steady state where the rate of receiving energy from the larger eddies is approximately equal to the rate at which their kinetic energy is dissipated to thermal energy by viscosity.

Using lasers, it has become possible to measure local instantaneous turbulent velocities, gas temperatures, and some species concentrations within the cylinder without insertion of intrusive probes. Figure 5.22 shows an arrangement for measuring velocity using a laser Doppler velocimetry (LDV) measurement technique. The arrangement shown is a steady flow test rig. The beam from an argon ion laser is split into two beams that are then focused to a small volume within the flow. As flow particles pass through the probe volume made by the intersecting laser beams, they scatter radiation in all directions. The Doppler effect shifts the frequency of the scattered light. The frequency shift is proportional to the particle velocity. The electronics of the LDV system filter and process the signal to detect the frequency shifts. Both the mean and turbulent velocity components are measured. By moving the laser beams and thus the probe volume, the velocity can be measured at different points within the cylinder.

Particle image velocimetry (PIV) systems measure turbulent velocity by determining particle displacement over time using a double-pulsed laser technique. A pulsed laser light sheet illuminates a plane in the flow, and the positions of particles in that plane are recorded by a video camera.

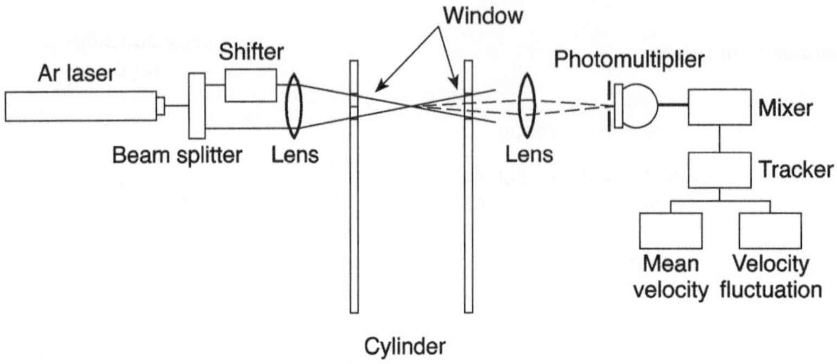

Figure 5.22 Laser Doppler velocimetry (LDV) turbulent flow measurement.

A fraction of a second later, another laser pulse illuminates the same plane, creating a second image. Images on the two planes are analyzed using cross-correlation techniques to compute the turbulent velocity field. Additional information about PIV and other laser-based turbulent velocity measurement techniques is given in Romano et al. (2007).

5.13 Turbulent Length, Velocity, and Time Scales

Turbulent combustion involves various lengths, velocity, and time scales which are used to characterize the turbulent flow field and associated chemical reactions. Turbulence is generally modeled by four length scales: the characteristic length L of the device in which the flow takes place, the integral length scale l, the Taylor length scale λ, and the Kolmogorov length scale η. The characteristic length L of the enclosure represents the largest possible eddy size that the confining geometry of the walls of the combustion device will allow, such as the cylinder bore or clearance height in an engine. In addition, there will be a representative density ρ, temperature T, and kinematic viscosity ν. The length scales determine turbulence parameters such as turbulent dissipation ϵ and eddy viscosity ν_t.

To quantify the magnitude of turbulent fluctuations, a characteristic turbulent velocity, the root mean square turbulent velocity $u_t(x)$, is defined in Eq. (5.224):

$$u_t(x) = \left[\frac{1}{n}\sum_{j=1}^{n} u'^2(x,j)\right]^{1/2} \tag{5.224}$$

where u' is a velocity fluctuation about a mean value $\overline{u}$. The turbulent kinetic energy k per unit mass (m^2/s^2), assuming isotropic turbulence ($u'^2 \approx v'^2 \approx w'^2$), is

$$k = \frac{1}{2}\overline{u_i'u_i'} \approx \frac{3}{2}u_t^2 \tag{5.225}$$

Measurements of turbulent velocities will be correlated if taken inside an eddy. Therefore, the integral scale l is defined mathematically as the distance r between two points where the two-point correlation R_{11} of the fluctuating velocity at the two points decreases to a value of zero,

$$\begin{aligned} R_{11} &= \frac{\overline{u_1'(x)u_1'(x+\Delta x)}}{\overline{\acute{u}_1^2}} \\ l &= \int_0^\infty R_{11}(r)dr \end{aligned} \tag{5.226}$$

Since most of the kinetic energy in a turbulent flow is contained in the larger eddies, the integral length scale l represents the size of the relatively larger and thus most energetic eddies in the turbulent flow field which move together as a unit. When l is very large, there is small influence of turbulence on a flame; only the small-scale eddies affect the structure of the flame front. The integral length scale can also be regarded as the distance an eddy moves through before dissipating. As a first approximation, the integral scale is taken as approximately equal to the characteristic length, $l \approx L$. For an engine cylinder, it is assumed to be equal to the instantaneous combustion chamber height, and for a boundary layer, it is assumed to be the boundary layer thickness.

A number of significant turbulence parameters are based on the integral length scale l. The turbulent Reynolds number Re_t, a measure of the overall turbulence intensity, is based on the integral length scale l, the turbulent velocity u_t, and the fluid kinematic viscosity ν, as given by Equation (5.227).

$$Re_t = \frac{u_t l}{\nu} \tag{5.227}$$

Following Kolmogorov, the turbulent energy transfer from the large eddies of size l is equal to the dissipation of energy at the Kolmogorov scale η. The rate of turbulent kinetic energy dissipation per unit mass ϵ can be related to the integral length scale l using a drag coefficient analogy. Assuming the energy E dissipated per unit time is the product of the large eddy drag force F and velocity u_t,

$$\frac{dE}{dt} = Fu_t \approx (\rho l^2 u_t^2)(u_t) = \rho l^2 u_t^3 \tag{5.228}$$

Since $\rho \approx m/l^3$, the rate of energy dissipated per unit mass (W/kg) is

$$\epsilon = \frac{dE/dt}{m} = \frac{\rho l^2 u_t^3}{\rho l^3} = \frac{u_t^3}{l} = \frac{1}{l}\left(\frac{2}{3}k\right)^{3/2} \tag{5.229}$$

The turbulent eddy viscosity ν_t is the product of the integral scale and turbulent velocity:

$$\nu_t = u_t l \approx \left(\frac{2}{3}\right)^2 \frac{k^2}{\epsilon} = C_\mu \frac{k^2}{\epsilon} \tag{5.230}$$

upon substitution of Eqs. (5.225) and (5.229), and incorporation of a general parameter C_μ, which is generally taken to be equal to 0.09.

The wavenumber κ of a turbulent eddy is defined as its spatial frequency with units of radians/length (m^{-1}), since the velocity reverses across the eddy diameter d. The wavenumber is therefore the inverse of the eddy size, so large eddies have a small wavenumber, and vice versa:

$$\kappa = \frac{2\pi}{\lambda} = \frac{\pi}{d} \tag{5.231}$$

Eddies of different sizes have different amounts of turbulent kinetic energy depending on the velocity fluctuations at each scale of the eddies. The dependence of the turbulent kinetic energy on eddy size is given by the turbulent kinetic energy spectrum $E(\kappa)$ (m^3/s^2) in wavenumber κ space, as shown in Figure 5.23.

The fraction dk of turbulent kinetic energy contained in eddies with wave number from κ to $\kappa + d\kappa$ is

$$dk = E(\kappa)d\kappa \tag{5.232}$$

The overall turbulent kinetic energy k is the integral of the energy spectrum $E(\kappa)$ over all wave numbers.

$$k = \int_0^\infty E(\kappa)d\kappa \tag{5.233}$$

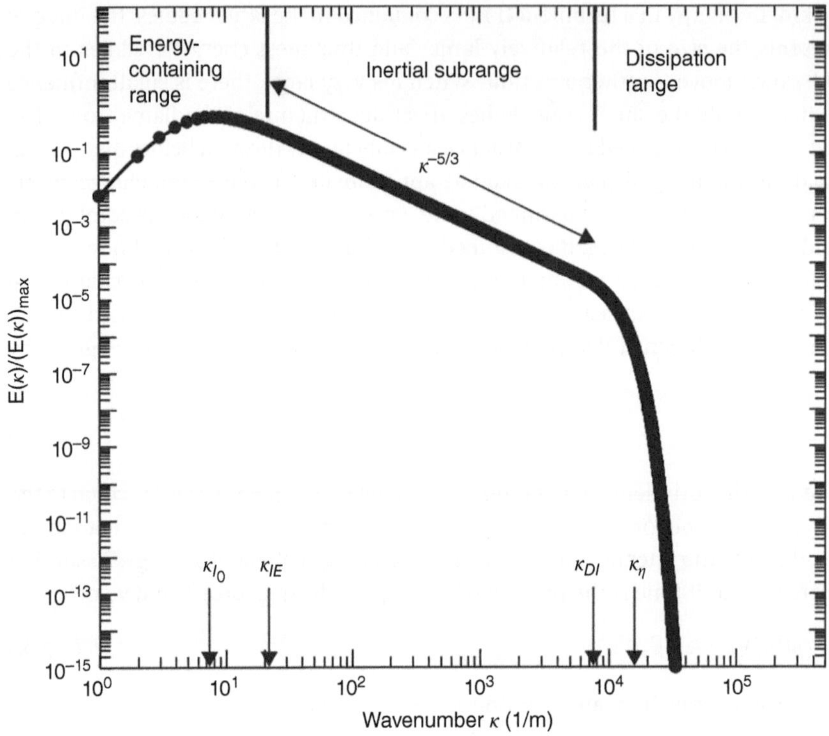

Figure 5.23 Dimensionless turbulent kinetic energy spectrum as a function of wavenumber κ.

In the inertial subrange, we assume

$$E(\kappa) = \frac{dk}{d\kappa} \sim \frac{k}{\kappa} \tag{5.234}$$

From Eq. (5.229)

$$k = \frac{3}{2}(\epsilon l)^{2/3} \sim (\epsilon l)^{2/3} \tag{5.235}$$

and from Eq. (5.231)

$$\kappa \sim d^{-1} \sim l^{-1} \tag{5.236}$$

Upon substitution of Eqs. (5.235) and (5.236) into (5.234),

$$E(\kappa) \sim \epsilon^{2/3}\kappa^{-5/3} \tag{5.237}$$

Equation (5.237) is known as the "$-5/3$ law" for the decay rate of the turbulent kinetic energy in the inertial subrange. The spectrum attains a maximum at a wavenumber that corresponds to the integral scale, since eddies of the integral length scale contain most of the turbulent kinetic energy. There is a cutoff due to viscous effects at the Kolmogorov length scale η. Beyond this cutoff length scale (in the range called the dissipation range), the turbulent kinetic energy per unit wavenumber decreases exponentially due to viscous effects.

The dissipation ϵ is also be defined in terms of the turbulent kinetic energy spectrum $E(\kappa)$,

$$\epsilon = 2\nu \int_0^\infty \kappa^2 E(\kappa) d\kappa \tag{5.238}$$

Note that the dissipation increases with the square of the wave number, i.e., the greatest dissipation occurs at the smallest eddies.

The integral time scale, τ_i, represents the lifetime of a large-scale turbulent eddy, assuming that the large eddies lose most of their energy in one turnover time. For a combustion situation, it represents an estimate of the mixing time of the fuel and air:

$$\tau_i = l/u_t = \left(\frac{l^2}{\epsilon}\right)^{1/3} \tag{5.239}$$

The Taylor length scale λ is useful in estimating the mean strain rate of the turbulence. It is defined in relation to the dissipation ϵ as

$$\epsilon = 15\nu\left(\frac{\partial u'}{\partial x}\right)^2 = 15\nu\frac{u_t^2}{\lambda^2} \tag{5.240}$$

The Kolmogorov length scale η is the smallest eddy size viscous damping that will allow to exist. From dimensional analysis, since it is assumed that the smallest eddies depend only on the dissipation rate ϵ and the kinematic viscosity ν, the Kolmogorov length scale is

$$\eta = \left(\frac{\nu^3}{\epsilon}\right)^{1/4} \tag{5.241}$$

The corresponding Kolmogorov time, velocity, and wavenumber scales are

$$\begin{aligned} \tau_\eta &= \left(\frac{\nu}{\epsilon}\right)^{1/2} \\ u_\eta &= (\nu\epsilon)^{1/4} \\ \kappa_\eta &= \frac{\pi}{\eta} = \pi\epsilon^{1/4}\nu^{-3/4} \end{aligned} \tag{5.242}$$

As the turbulence level u_t increases, from Eq. (5.229), the dissipation rate ϵ increases, and from Eq. (5.241), the smallest turbulent eddies become even smaller. Likewise, as the turbulent Reynolds number increases, the smaller eddies also decrease in size according to Eqs. (5.244) and (5.245).

Dimensional analysis of simple isotropic turbulent flows leads to the following relationships between the four length scales:

$$l = C_l\, L \tag{5.243}$$

$$\frac{\lambda}{l} = \left(\frac{15}{C_\lambda}\right)^{1/2} Re_t^{-1/2} \tag{5.244}$$

$$\frac{\eta}{l} = \left(\frac{15}{C_\eta}\right)^{-1/4} Re_t^{-3/4} \tag{5.245}$$

The constants C_l, C_λ, and C_η are numbers unique to the flow of interest and whose order of magnitude is unity.

The ratio of the integral time scale τ_i to the Kolmogorov time scale τ_η is

$$\frac{\tau_i}{\tau_\eta} = \frac{l/u_t}{(\nu/\epsilon)^{1/2}} = Re_t^{1/2} \tag{5.246}$$

The inertial range is the ratio of the integral length scale to the Kolmogorov length scale and is proportional to Re_t number raised to the 3/4 power. If the turbulent Reynolds number of a flow is 10^4, then the integral length scale is 1000 times larger than the Kolmogorov length scale, and the integral time scale is 100 times larger than the Kolmogorov time scale. By way of comparison, the velocity and species gradients in a laminar flame occur over a small range of scales, corresponding to just one order of magnitude.

The Damköhler number is the ratio of the integral time scale τ_i to the chemical reaction time scale τ_r,

$$Da = \frac{\tau_i}{\tau_r} \tag{5.247}$$

At large Damköhler numbers ($Da \gg 1$), the turbulent eddy time becomes longer than the characteristic chemical reaction time, i.e., the chemical reactions are more rapid than the fluid mixing, corresponding to a thin reaction zone distorted and convected by the flow field. The turbulent eddies wrinkle and strain the flame surface. The internal structure of the flame is not strongly affected by the turbulence and may be described as a laminar flame element called a "flamelet." Combustion in most engineering devices occurs with $Da \gg 1$.

A low Damköhler number ($Da \ll 1$) corresponds to a relatively slow chemical reaction. In this case, reactants and products are mixed by the turbulent eddies before reacting. The formulation of the process becomes that of a system described as a *perfectly stirred reactor*.

Example 5.3 ***Turbulence Length, Velocity, and Time Scales***

An internal combustion engine has a mean piston speed $\overline{U}_p$ of 5.0 m/s and a clearance volume height h of 2.0 mm. At the end of the compression stroke, what are the characteristic, integral, and Kolmogorov turbulence parameters? Assume the fluid kinematic viscosity $\nu = 100 \times 10^{-7}\,\text{m}^2/\text{s}$, and $C_\eta = 1$. Experimentally, it has been found that the turbulent velocity u_t is approximately 1/2 the mean piston speed.

Solution

Since the eddies are constrained by the clearance volume geometry at top dead center, the integral scale l and characteristic length L are

$$l \approx L = h = 2.0\,\text{mm}$$

The turbulence parameters are:

Turbulent velocity: $u_t \approx \overline{U}_p/2 = 2.5\ \text{m/s}$

Turbulent Reynolds number: $Re_t = \dfrac{u_t l}{\nu} = \dfrac{(2.5)(2 \times 10^{-3})}{(100 \times 10^{-7})} = 500$

Turbulent kinetic energy: $k = \dfrac{3}{2}u_t^2 = \dfrac{3}{2}(2.5)^2 = 9.375\ \text{J/kg}$

Integral time scale: $\tau_i = \dfrac{l}{u_t} = \dfrac{2.0 \times 10^{-3}}{2.5} = 0.80\ \text{ms}$

Dissipation: $\epsilon = \dfrac{1}{l}\left(\dfrac{2}{3}k\right)^{3/2} = \dfrac{1}{2 \times 10^{-3}}\left(\dfrac{2}{3}(9.375)\right)^{3/2} = 7812\ \text{W/kg}$

Eddy viscosity: $\nu_t = C_\mu \dfrac{k^2}{\epsilon} = (0.09)\dfrac{9.375^2}{7812} = 1.01 \times 10^{-3}\ \text{m}^2/\text{s}$

Eddy wave number: $\kappa_l = \dfrac{\pi}{l} = \dfrac{\pi}{2 \times 10^{-3}} = 1.57 \times 10^3\ \text{m}^{-1}$

Kolmgorov length: $\eta = \left(\dfrac{\nu^3}{\epsilon}\right)^{1/4} = \left(\dfrac{(100 \times 10^{-7})^3}{7812}\right)^{1/4} = 18.9\ \mu\text{m}$

Kolmgorov velocity: $u_\eta = (\nu\epsilon)^{1/4} = [(100 \times 10^{-7})(7812)]^{1/4} = 0.53\ \text{m/s}$

Note that for this example, $\nu_t/\nu \approx 100$, indicating that in this case, the turbulent transport of momentum is 100 times larger than the laminar transport of momentum.

5.14 Reynolds and Favre Averaging

There are two different statistical averaging procedures commonly used in the analysis of turbulent reacting flows:

(1) Time averaging, also called Reynolds averaging.
(2) Density averaging, also called Favre averaging.

5.14.1 Reynolds Averaging

With time or Reynolds averaging, a variable q is decomposed into mean $\overline{q}$ and fluctuating q' parts. The mean part is determined by taking a time average over a time interval, $\Delta t = t_2 - t_1$, long enough so that $\overline{q}$ is independent of time, i.e., statistically stationary.

$$\overline{q} = \frac{1}{\Delta t}\int_{t_1}^{t_2} q(t)dt \tag{5.248}$$

The decomposition of variables such as velocity, pressure, density, and temperature into a mean $\overline{q}$ and a deviation q' from the mean is represented as

$$\begin{aligned} u_i(t) &= \overline{u}_i + u_i'(t) \\ p(t) &= \overline{p} + p'(t) \\ \rho(t) &= \overline{\rho} + \rho'(t) \\ T(t) &= \overline{T} + T'(t) \end{aligned} \tag{5.249}$$

The time average of a fluctuating variable is zero, so

$$\overline{u_i'} = \overline{p'} = \overline{\rho'} = \overline{T'} = 0 \tag{5.250}$$

In addition, for two variables q and r, Reynolds averaging implies

$$\begin{aligned} \overline{\overline{q}} &= \overline{q} \\ \overline{q + r} &= \overline{q} + \overline{r} \\ \overline{qr} &= \overline{q}\,\overline{r} + \overline{q'r'} \\ \overline{\frac{\partial q}{\partial x}} &= \frac{\partial \overline{q}}{\partial x} \end{aligned} \tag{5.251}$$

The incompressible Reynolds averaged mean momentum equation (in the index form) is

$$\rho\left[\frac{\partial \overline{u}_i}{\partial t} + \overline{u}_j\frac{\partial \overline{u}_i}{\partial x_j}\right] = -\frac{\partial \overline{p}}{\partial x_i} + \frac{\partial}{\partial x_j}\left(\mu\frac{\partial \overline{u}_i}{\partial x_j}\right) - \rho\frac{\partial \overline{u_i'u_j'}}{\partial x_j} + \overline{F}_i \tag{5.252}$$

However, combustion applications have density and temperature fluctuations in addition to velocity and pressure fluctuations, which give rise to additional correlations of fluctuating variables such as $\overline{\rho' u_i'}$ that are unknown and must be measured or modeled.

For example, the continuity equation for a compressible flow is

$$\frac{\partial \rho}{\partial t} + \frac{\partial}{\partial x_i}(\rho u_i) = 0 \tag{5.253}$$

Upon substitution of the Reynolds decomposition,

$$\frac{\partial(\overline{\rho}+\rho')}{\partial t}+\frac{\partial}{\partial x_i}(\overline{\rho}u_i+\rho' u_i+\overline{\rho}u_i'+\rho' u_i')=0 \tag{5.254}$$

Time averaging of Eq. (5.254) results in

$$\frac{\partial(\overline{\rho})}{\partial t}+\frac{\partial}{\partial x_i}(\overline{\rho}\overline{u}_i+\overline{\rho' u_i'})=0 \tag{5.255}$$

In order to achieve closure, we need to measure or model the correlation $\overline{\rho' u_i'}$ between the fluctuation density and the fluctuation velocity. Similarly, triple correlations involving density fluctuations appear in the compressible momentum and energy equations with Reynolds averaging.

5.14.2 Favre Averaging

Density averaging was introduced by the French scientist A. Favre in 1965 for use in compressible or chemically reacting turbulent flows which commonly have large density variations. With Favre averaging, a variable q is also expressed as a mean value $\tilde{q}$ and fluctuation q'',

$$q(t)=\tilde{q}+q''(t) \tag{5.256}$$

To form a Favre average, a variable q is multiplied by the density and a time average is performed,

$$\tilde{q}=\frac{1}{\overline{\rho}\,\Delta t}\int_{t_1}^{t_2}\rho(t)q(t)dt=\frac{\overline{\rho q}}{\overline{\rho}} \tag{5.257}$$

since the time average of a Favre-averaged fluctuation q'' is $\overline{\rho q''}=0$. The velocity, temperature, and enthalpy are represented in Favre averaging as

$$\begin{aligned}
u_i(t) &= \tilde{u}_i+u_i''(t)\\
T(t) &= \tilde{T}+T''(t)\\
h(t) &= \tilde{h}+h''(t)
\end{aligned} \tag{5.258}$$

It should be noted that the density ρ and pressure p are Reynolds averaged, not Favre averaged.
It can be shown that with Favre averaging of variables q and r,

$$\begin{aligned}
\overline{\tilde{q}} &= \tilde{q}\\
\overline{\rho\tilde{q}} &= \overline{\rho}\tilde{q}\\
\overline{q''} &= -\frac{\overline{\rho' q'}}{\overline{\rho}}\\
\overline{\rho q r} &= \overline{\rho}\tilde{q}\tilde{r}+\overline{\rho q'' r''}\\
\widetilde{\rho q} &= \rho\tilde{q}
\end{aligned} \tag{5.259}$$

A Favre-averaged variable is more difficult to measure experimentally than a Reynolds-averaged one. The relationship between these two variables can be found if the correlation between density and the fluctuating variable is known, for example, from Eqs. (5.257) and (5.251),

$$\tilde{u}_i=\overline{u}_i+\frac{\overline{\rho' u_i'}}{\overline{\rho}} \tag{5.260}$$

The advantage of Favre-averaged variables is seen by comparing the averaging of the turbulent stress term $\rho u_i u_j$ that appears in the convective term of the Navier–Stokes equations,

$$\begin{aligned}\overline{\rho u_i u_j} &= \overline{\rho}\,\overline{u}_i\overline{u}_j + \overline{\rho}\,\overline{u'_i u'_j} + \overline{u}_i\overline{\rho' u'_j} + \overline{u}_j\overline{\rho' u'_i} + \overline{\rho' u'_i u'_j} \\ &= \overline{\rho}\,\widetilde{u}_i\widetilde{u}_j + \overline{\rho}\,\widetilde{u''_i u''_j}\end{aligned} \tag{5.261}$$

There are five terms in the averaging process when expressed in terms of Reynolds-averaged variables, and only two terms in the averaging process when expressed in terms of Favre-averaged variables.

5.14.3 Favre-Averaged Transport Equations

If we substitute Eq. (5.260) into the Reynolds-averaged continuity equation (5.255), we obtain the Favre-averaged compressible continuity equation (5.262),

$$\frac{\partial\overline{\rho}}{\partial t} + \frac{\partial(\overline{\rho}\widetilde{u}_j)}{\partial x_j} = 0 \tag{5.262}$$

This is very similar in structure to the incompressible Reynolds-averaged continuity equation.

The Favre-averaged compressible mean momentum ($i = 1, 2, 3$), total enthalpy, and species conservation equations ($k = 1, \dots, N$) are (Veynante and Vervisch 2002)

$$\frac{\partial(\overline{\rho}\widetilde{u}_i)}{\partial t} + \frac{\partial(\overline{\rho}\widetilde{u}_j\widetilde{u}_i)}{\partial x_j} = -\frac{\partial\overline{p}}{\partial x_i} - \frac{\partial(\overline{\rho}\widetilde{u''_i u''_j})}{\partial x_j} + \frac{\partial\overline{\tau}_{ij}}{\partial x_j} + \overline{F}_i \tag{5.263}$$

$$\frac{\partial}{\partial t}(\overline{\rho}\widetilde{h}_t) + \frac{\partial\overline{\rho}\widetilde{h}_t\widetilde{u}_j}{\partial x_j} = \frac{\partial\overline{p}}{\partial t} - \frac{\partial\overline{\rho}\widetilde{h''_t u''_j}}{\partial x_j} + \frac{\partial}{\partial x_j}\left(\overline{\mathcal{J}^h_j} + \overline{u_i\tau_{ij}}\right) + \overline{u_j F_j} \tag{5.264}$$

$$\frac{\partial(\overline{\rho}\widetilde{Y}_k)}{\partial t} + \frac{\partial(\overline{\rho}\widetilde{Y}_k\widetilde{u}_j)}{\partial x_j} = -\frac{\partial}{\partial x_j}\left[\overline{\mathcal{J}^k_j} + \overline{\rho}\widetilde{Y''_k u''_j}\right] + \overline{\dot{\omega}}_k \tag{5.265}$$

where $\overline{\mathcal{J}^h_j}$ and $\overline{\mathcal{J}^k_j}$ are the laminar diffusive fluxes, usually small compared to the turbulent transport, and $\overline{\dot{\omega}}_k$ are the species chemical reaction rates. Note that with Favre averaging, the density and pressure are not mass averaged, and the density fluctuations are removed from the averaged equations.

5.15 Turbulence Modeling

5.15.1 Averaged Navier–Stokes Models

The spatial distribution as well as the temporal change in turbulent flow fields have been modeled for many years with RANS and Favre Averaged Navier–Stokes (FANS) full-field turbulence models. However, the averaged turbulence equations are not closed until the Reynolds stress terms $\overline{\rho}\widetilde{u''_i u''_j}$, the turbulent enthalpy correlation terms $\overline{\rho}\widetilde{h''_t u''_j}$, and species flux correlation terms $\overline{\rho}\widetilde{Y''_k u''_j}$ are also modeled.

The Reynolds stress term can be modeled with a gradient diffusion model, which is the product of a turbulent eddy viscosity μ_t and a gradient of the mean velocity,

$$\overline{\rho}\widetilde{u''_i u''_j} = -\mu_t\frac{\partial\widetilde{u}_i}{\partial x_j} \tag{5.266}$$

Similarly, the species ($\widetilde{u_j''Y_k''}$) and temperature ($\widetilde{u_j''T}$) turbulent fluxes can be modeled with a gradient diffusion model, for example,

$$\bar{\rho}\widetilde{u_j''Y_k''} = -\frac{\mu_t}{Sc_t}\frac{\partial \widetilde{Y}_k}{\partial x_j} \tag{5.267}$$

where Sc_t is a turbulent Schmidt number. In some turbulent premixed flames, there can be counter-gradient transport where the fluxes have the same sign as the mean gradient (Bray and Moss 1977).

There are a variety of turbulent eddy viscosity models for the Reynolds stress currently being used by combustion modelers. One-equation models relate a turbulence length scale to a typical flow dimension. Two-equation models use a differential equation for the turbulence length scale or equivalent.

Widely used two-equation models are the family of $k - \epsilon$ models. The basic $k - \epsilon$ model is a model based on both a partial differential transport equation (5.269), for the turbulent kinetic energy $k = \widetilde{u_i''u_i''}/2$, and a partial differential transport equation (5.270), for the turbulent kinetic energy dissipation ϵ.

The $k - \epsilon$ model is based on a scalar eddy viscosity, so it does not take into account non-isotropic effects on the turbulence field such as streamline curvature and strongly swirling flows. It also assumes a fully developed turbulent flow field and does not compute small scale structures of the flow. In fully developed turbulence, the Reynolds stress tensor can be as much as two orders of magnitude higher than the mean viscous stress tensor.

The $k - \epsilon$ model assumes that the turbulent eddy viscosity depends on k and ϵ as given by Eq. (5.268):

$$\mu_t = C_\mu \bar{\rho} k^2/\epsilon \tag{5.268}$$

The k and ϵ differential equations in terms of Favre averages are given in Eqs. (5.269) and (5.270) (Haworth 2010),

$$\frac{\partial}{\partial t}(\bar{\rho}k) + \frac{\partial \bar{\rho} k \widetilde{u}_j}{\partial x_j} = \frac{\partial}{\partial x_j}\left[\left(\frac{\mu_t}{\sigma_k} + \mu\right)\frac{\partial k}{\partial x_j}\right] - \tau_{ij}\frac{\partial \widetilde{u}_j}{\partial x_i} - \bar{\rho}\epsilon \tag{5.269}$$

$$\frac{\partial}{\partial t}(\bar{\rho}\epsilon) + \frac{\partial \bar{\rho} \epsilon \widetilde{u}_j}{\partial x_j} = \frac{\partial}{\partial x_j}\left[\left(\frac{\mu_t}{\sigma_\epsilon} + \mu\right)\frac{\partial \epsilon}{\partial x_j}\right] + C_1\frac{\epsilon}{k}\tau_{ij}\frac{\partial \widetilde{u}_j}{\partial x_i} - C_2\bar{\rho}\frac{\epsilon^2}{k} \tag{5.270}$$

The three terms on the right-hand side of Eqns. (5.269) and (5.270) represent turbulent transport, production, and dissipation, and τ_{ij} is the apparent turbulent stress tensor given by

$$\tau_{ij} = -\bar{\rho}\widetilde{u_i''u_j''} = \mu_t\left(\frac{\partial \widetilde{u}_i}{\partial x_j} + \frac{\partial \widetilde{u}_j}{\partial x_i} - \frac{2}{3}\frac{\partial \widetilde{u}_l}{\partial x_l}\delta_{ij}\right) - \frac{2}{3}\bar{\rho}k\delta_{ij} \tag{5.271}$$

Equations (5.268)–(5.271) are combined with the continuity, momentum, and energy equations to form a complete system for numerical analysis. The constants $\sigma_k = 1.0, \sigma_\epsilon = 1.3, C_\mu = 0.09$, $C_1 = 1.44$, and $C_2 = 1.92$ are empirical constants used in many combustion applications. One problem with two-equation models is the excessive generation of turbulent kinetic energy in the vicinity of stagnation points, which requires incorporation of limiter terms in the turbulence equations.

5.15.2 Large Eddy Simulation (LES)

Large eddy simulation (LES) is a turbulence modeling procedure in which the large eddies are computed, and the smallest eddies are modeled. With LES modeling, the time-dependent Navier–Stokes equations are spatially filtered over the computational grid, reducing the range of scales in a simulation. Primary flow features that are larger than the filter size are computed and the smaller features smaller than the filter size are modeled. The smallest eddies are more amenable to modeling as they have greater isotropic turbulence characteristics than the larger eddies. Eddies of size less than the grid size are removed from the dynamics. However, since chemical reactions occur on the unresolved molecular scales of the flow, modeling of the combustion process is still required. Because of the substantial computational requirements of LES, there has been a tendency to use quite simple RANS turbulent combustion models.

Both RANS and LES models are similar in that they both average over the small scales, with the LES mesh size much smaller than the RANS mesh size. Accordingly, LES modeling produces flow structures and vortices not seen with RANS models. The time-dependent zones of fresh and burnt gases in a combustor can be clearly visualized, as they are computed, not averaged. LES models are available in many industrial CFD computer programs. It should be noted that LES turbulence models are more computationally intensive than RANS models.

In LES, the conservation equations are filtered in the physical (x_i) space with a filter of width Δ equivalent to the grid size, and in the wavenumber (κ) space. As indicated in Figure 5.24, a low-pass filter passes over turbulent fluctuations with a wavenumber (i.e., frequency) lower than a cutoff wavenumber and smooths over the fluctuations higher than the cutoff value. All of the frequency components greater than the cutoff wavenumber $\kappa = \pi/\Delta$ vanish.

The Favre filtering operation for a variable q is given by Eq. (5.272) (Veynante and Vervisch 2002),

$$\bar{\rho}\tilde{q}(\mathbf{x}, t) = \int_D \rho q(\hat{\mathbf{x}}, t) G(\mathbf{x} - \hat{\mathbf{x}}) d\hat{\mathbf{x}} \tag{5.272}$$

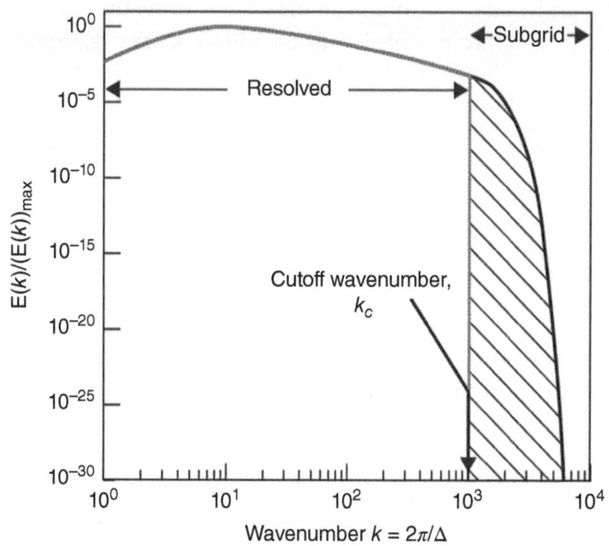

Figure 5.24 Spectral schematic of large eddy simulation methodology. (Adapted from Kuo and Acharya 2012).

where G is the LES filter, a convolution kernel, and the integration is over the entire flow domain D. The LES filter in the physical space is the product of the filters in each of the n space dimensions, For example, a Gaussian filter is

$$G(\mathbf{x}-\hat{\mathbf{x}}) = \prod_{i=1}^{n} \frac{\sqrt{6}}{\sqrt{\pi}\Delta_i} \exp\left[-6(x_i-\hat{x}_i)^2/\Delta_i^2\right], \quad n = 1, 2, \text{or } 3 \tag{5.273}$$

and a box filter is

$$G(\mathbf{x}) = \begin{cases} 1/\Delta^3, & \text{if } |x_i| \leq \Delta/2, \quad i = 1, 2, 3 \\ 0, & \text{otherwise} \end{cases} \tag{5.274}$$

where (x_1, x_2, x_3) are the physical coordinates of the location $\mathbf{x}$. A cutoff filter in wavenumber space is

$$G(\kappa) = \begin{cases} 1, & \text{if } \kappa \leq \pi/\Delta \\ 0, & \text{otherwise} \end{cases} \tag{5.275}$$

The LES filters are normalized over the combustion domain,

$$\int_D G(\mathbf{x})d\mathbf{x} = 1 \tag{5.276}$$

Application of the LES filter, Eq. (5.272), to the momentum equation (5.263), results in the filtered momentum equation, given by

$$\frac{\partial(\overline{\rho}\tilde{u}_i)}{\partial t} + \frac{\partial(\overline{\rho}\tilde{u}_j\tilde{u}_i)}{\partial x_j} = -\frac{\partial\overline{p}}{\partial x_i} - \frac{\partial}{\partial x_j}\left[\overline{\rho}(\widetilde{u_iu_j} - \tilde{u}_i\tilde{u}_j)\right] + \frac{\partial\overline{\tau}_{ij}}{\partial x_j} + \overline{F}_i \tag{5.277}$$

which is somewhat similar to the RANS momentum equation. The unresolved subgrid-scale Reynolds stress tensor τ_{ij} is

$$\tau_{ij} = \overline{\rho}(\widetilde{u_iu_j} - \tilde{u}_i\tilde{u}_j) \tag{5.278}$$

Since the correlation term $\widetilde{u_iu_j}$ is of subgrid scale, τ_{ij} cannot be calculated from the filtered equations. Therefore, the anisotropic terms of τ_{ij} are modeled with an eddy viscosity transport model,

$$\tau_{ij} - \frac{1}{3}\delta_{ij}\tau_{kk} = \overline{\rho}\nu_t\left(\frac{\partial\tilde{u}_i}{\partial x_j} + \frac{\partial\tilde{u}_j}{\partial x_i}\right) = -2\overline{\rho}\nu_t\tilde{S}_{ij} \tag{5.279}$$

where ν_t is the subgrid eddy viscosity and $\tilde{S}_{ij}$ is the resolved rate of strain tensor. Using turbulent mixing length arguments, Smagorinsky (1963) modeled the subgrid eddy viscosity as

$$\nu_t = (C_s\Delta)^2\,|\tilde{S}| \tag{5.280}$$

The term C_s is the Smagorinsky eddy viscosity model coefficient. Values of C_s from 0.17 to 0.3 give satisfactory results in simulations of decaying homogeneous isotropic turbulence, depending on the flow configuration.

However, the Smagorinsky eddy viscosity ν_t has nonzero values for laminar flows and is too large for simulations for turbulent channel flow and mixing layers. The applicability of the Smagorinsky model for general flows can be improved by the application of a dynamic procedure (Germano et al. 1991) which compares the eddy dissipation at different scales, i.e., two filter levels, the original grid size, and a "test filter" which is a multiple, typically twice, of the original grid size. The procedure assumes scale invariance of the model coefficients in the inertial subrange of the turbulence spectrum. Application of the test filter to the momentum equation results in the unresolved stress term T_{ij},

$$T_{ij} = \bar{\rho}(\widehat{\tilde{u}_i \tilde{u}_j} - \hat{\tilde{u}}_i \hat{\tilde{u}}_j) \tag{5.281}$$

Applying the test filter to Eq. (5.278), and then equating the term $\widehat{\tilde{u}_i \tilde{u}_j}$ from Eqs. (5.278) and (5.281), results in the Germano identity,

$$\widehat{\tilde{u}_i \tilde{u}_j} - \hat{\tilde{u}}_i \hat{\tilde{u}}_j = T_{ij} - \hat{\tau}_{ij} \tag{5.282}$$

With this dynamic procedure, the Smagorinsky coefficient C_s is found as a function only of u_i and its gradients by substituting the Smagorinsky equations, (5.279) and (5.280), into the Germano identity, Eq. (5.281). Accordingly, C_s is now a function of position and time. Similar dynamic procedures can be derived for the scalar variables involved in modeling subgrid eddies. Minimum dissipation models have also been developed as an alternative approach for the determination of the sub-grid eddy viscosity (Rozema et al. 2015).

5.15.3 Direct Numerical Simulation (DNS)

In DNS, the Navier–Stokes equations are numerically solved without any turbulence models by resolving the whole range of spatial and temporal scales of the turbulence. Therefore, DNS provides the first-principle flow solutions free from turbulence modeling errors. It is a complete time-dependent solution of the momentum, energy, species, and continuity equations. Since no turbulence model is used at any length scale, the grid must be small enough to resolve the smallest turbulent eddy whose size is of the order of the Kolmogorov length scale η. All the spatial scales of the turbulence must be resolved, up to the integral scale l, which are associated with the motions containing most of the kinetic energy.

However, even for nonreacting flows, the number of grid points required to resolve all the length scales in a 3D computation is proportional to $Re_t^{9/4}$ which means that even for a moderate turbulent Reynolds number of $Re_t = 10^4$, the grid points needed for a DNS computation is $N = 10^9$. With the introduction of massively parallel processors and the corresponding parallelization of DNS codes, calculations with a few billion grid points and realistic chemistry are now possible.

However, many issues remain. In flows of practical interest, the flame structure can be very thin, e.g., in the premixed flamelet regime, the flame thickness $\pm\delta_L$ can be orders of magnitude smaller than the smallest turbulent scale η. Thus, even when the Kolmogorov scale is resolved, resolving the thin flame structure is very difficult. In addition, there are reactions whose rates are much faster than the Kolmogorov time scale τ_η. Two approaches have been attempted to circumvent this limitation: the use of modified chemistry to artificially thicken the flame in order to resolve it (Baum et al. 1994) and the use of a thin flame model where the flame front is tracked without resolving it (Kerstein et al. 1988).

Due to the constraining computational resources (processing time and memory size), DNS has been applied mainly for understanding basic combustion phenomena in very simple geometric configurations. To ensure accurate solutions in space and time, high-order spatial discretization, time integration schemes, and adaptive mesh refinement are used. The results from DNS calculations supply useful information for turbulence modeling required by other numerical approaches. The information from DNS can be used to verify an LES code. When reduced to the low Reynolds number flow in simpler geometries, an accurate LES code should give the same results as DNS code. DNS has also been used as a tool that allows numerical experiments to be carried out in situations where experimental studies cannot be conducted easily and accurately.

5.16 Probability Density Function (PDF) Models

PDF methods address turbulence through use of probabilistic mathematical models. Traditional turbulence models, such as the two-equation models discussed in Section 5.16, are based on Reynolds averaging techniques and yield modeled equations for statistical moments. In contrast, the PDF characterizes all of the parameters of the turbulent flow and computes the probability rather than exact values of the turbulent flow parameters at given positions and times. Since the PDF changes at every point in the flow field, it contains a large amount of statistical information and thus provides a fuller description of unsteady reacting flow fields than two-equation turbulence models. As we will see, the PDF is used as an averaging tool.

There are two types of PDFs: presumed and transport. If a presumed PDF is used, the first and second moments of the PDF are determined from local conditions. If a transport PDF is used, the PDF is computed by solving a transport or balance equation, which is derived from the Navier–Stokes equations. In the transport PDF equation, the convection, mean pressure gradient, and chemical reaction source terms do not require a modeling assumption, and they can be solved exactly.

5.16.1 Presumed PDF

We start this section with the distribution function $F_u(U)$ of a single variable, such as the velocity u, or scalar variable ϕ, such as mixture fraction or temperature. The cumulative distribution function $F_u(U)$ is defined as the probability p of finding a value of $u < U$:

$$F_u(U) = p(u < U) \tag{5.283}$$

where U is the sample space variable associated with the variable u. The probability of finding a value of u in a certain interval $U_- < u < U_+$ is given by:

$$p(U_- < u < U_+) = F_u(U_+) - F_u(U_-) \tag{5.284}$$

The PDF of u is now formally defined as

$$P_u(U) \equiv \frac{dF_u(U)}{dU} \quad \text{or} \quad dF_u(U) = P_u(U)dU \tag{5.285}$$

The function $P_u(U)$ is the probability per unit U of the event $u = U$, i.e., $P_u(U)dU$ is the probability of finding u in the range $U \leq u < U + dU$. If the possible values of u range from $-\infty$ to $+\infty$, it follows that

$$\int_{-\infty}^{+\infty} P_u(U)dU = 1 \tag{5.286}$$

In other words, the probability of finding the value of u between $-\infty$ to $+\infty$ is one. Also, the probability is always greater than zero,

$$P_u(U) \geq 0 \tag{5.287}$$

The PDF represents the weighting factor for a given value of a random variable. In turbulent flows, the PDF of any given variable depends on the variable's position $\mathbf{x}$ and time t. These dependencies are written in the form

$$P_u(U; \mathbf{x}, t) \tag{5.288}$$

The semicolon indicates that P_u is the probability density in U-space and a function of $\mathbf{x}$ and t. In stationary turbulent flows, it is independent of t; and for homogeneous turbulent flow fields, it is independent of **x**. Since a turbulent flow field is a random process, we will not distinguish between the random variable u and the sample space variable U, and in what follows, write the PDF as $P(u; \mathbf{x}, t)$.

A familiar example of a presumed PDF is the normal or Gaussian distribution of a variable u,

$$P(u) = \frac{1}{\sigma\sqrt{2\pi}} \exp\left(-\frac{(u-\overline{u})^2}{2\sigma^2}\right) \tag{5.289}$$

where $\overline{u}$ is the mean value and σ is the standard deviation of u.

The mean (first moment) $\overline{u}$ a of a variable u as function of **x** and t can be determined from the variable's PDF by ensemble averaging,

$$\overline{u} = \int_{-\infty}^{+\infty} uP(u; \mathbf{x}, t)du \tag{5.290}$$

and the mean value of a function $g(u)$ is

$$\overline{g}(\mathbf{x}, t) = \int_{-\infty}^{+\infty} g(u)P(u; \mathbf{x}, t)du \tag{5.291}$$

The variance $\overline{u'^2}$ is the sum of the squared distances from the mean times the probability of being at that distance. It is a measure of the width of a PDF and is the average of the fluctuations squared. The variance (a second moment) is determined from the PDF and is given by

$$\sigma^2 = \overline{u'^2} = \int_{-\infty}^{+\infty} (u-\overline{u})^2 P(u; \mathbf{x}, t)du \tag{5.292}$$

where the variable u has been split into its mean and fluctuating components,

$$u(\mathbf{x}, t) = \overline{u}(\mathbf{x}, t) + u'(\mathbf{x}, t) \tag{5.293}$$

Higher-order moments, skewness (asymmetry), and kurtosis (peakedness) are similarly defined, with the distances, raised to the third and the fourth power, respectively. In general, the n moments of the PDF of a variable u are of the form

$$\overline{u(\mathbf{x}, t)^n} = \int_{-\infty}^{+\infty} u^n P(u; \mathbf{x}, t)du \tag{5.294}$$

where the overbar represents the mean or expectation value of u^n.

Similarly, the first moment of a variable u using a Favre-weighted PDF is

$$\widetilde{u} = \int u\widetilde{P}(u)du \tag{5.295}$$

and more generally,

$$\widetilde{q}(\mathbf{x}) = \int_0^1 q(f)\widetilde{P}(f; \mathbf{x})df \tag{5.296}$$

For reactive scalars such as the species mass fractions Y_i and temperature T, the Favre-averaged values are expressed as

$$\widetilde{Y}_i(\mathbf{x}) = \int_0^1 Y_i(f)\widetilde{P}(f; \mathbf{x})df \tag{5.297}$$

and

$$\widetilde{T}(\mathbf{x}) = \int_0^1 T(f)\widetilde{P}(f; \mathbf{x})df \tag{5.298}$$

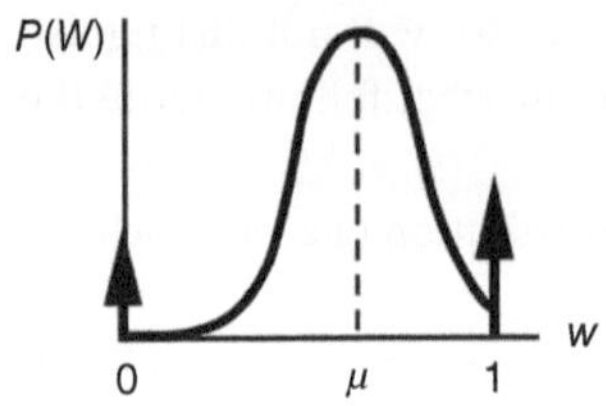

Figure 5.25 PDF of a truncated Gaussian function with two Dirac δ-functions at the boundaries. (Warnatz et al. 2006)/with permission of Springer Nature.

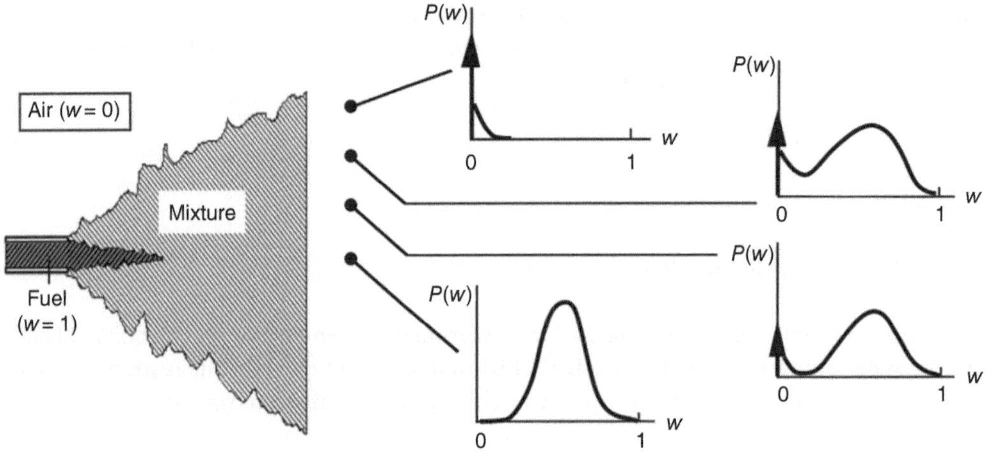

Figure 5.26 PDF distribution of a turbulent fuel jet mixing with surrounding air. (Warnatz et al. 2006)/with permission of Springer Nature.

Truncated Gaussian functions or β-functions are used for approximations to the one-dimensional PDFs. As shown in Figure 5.25, a β function for a variable w consists of a Gaussian distribution and two Dirac δ-functions at each boundary to account for intermittency peaks.

Example 5.4 ***PDF Distribution of a Turbulent Fuel Jet***

A turbulent fuel jet mixing with surrounding air is shown in Figure 5.26. Sketch four equally spaced PDFs across the width of the jet.

Solution

The fuel mixture fraction $w = 1$ in the fuel jet, and $w = 0$ in the surrounding air. At the centerline of the jet, the PDF has a Gaussian shape. As one moves outward from the centerline, the Gaussian component of the PDF decreases, and the delta function component increases. At the edge of the jet, the PDF has the shape of a delta function with a minimal Gaussian component.

The Favre- or density-weighted β function PDF for a single variable ϕ is given by

$$\widetilde{P}(\phi) = \frac{1}{B(a,b)}\phi^{a-1}(1-\phi)^{b-1} = \frac{\Gamma(a+b)}{\Gamma(a)\Gamma(b)}\phi^{a-1}(1-\phi)^{b-1} \tag{5.299}$$

where $B(a, b)$ is a normalization factor,

$$B(a,b) = \int_0^1 \phi^{a-1}(1-\phi)^{b-1}\, d\phi \tag{5.300}$$

The Γ function is defined as

$$\Gamma(x) = \int_0^\infty e^{-t}\, t^{x-1}\, dt \tag{5.301}$$

Upon substitution of Eq. (5.299) into the first and second moments of ϕ,

$$\begin{aligned}\tilde{\phi} &= \int_0^1 \phi\tilde{P}(\phi)d\phi \\ \widetilde{\phi''^2} &= \int_0^1 (\phi - \tilde{\phi})^2\tilde{P}(\phi)d\phi\end{aligned} \tag{5.302}$$

and solving for a and b in terms of $\tilde{\phi}$ and $\widetilde{\phi''^2}$,

$$\begin{aligned}a &= \tilde{\phi}\left[\frac{\tilde{\phi}(1-\tilde{\phi})}{\widetilde{\phi''^2}} - 1\right] \\ b &= \frac{a}{\tilde{\phi}} - a\end{aligned} \tag{5.303}$$

Conversely, given a and b, one can solve for $\tilde{\phi}$ and $\widetilde{\phi''^2}$,

$$\begin{aligned}\tilde{\phi} &= \frac{a}{a+b} \\ \widetilde{\phi''^2} &= \frac{ab}{(a+b)^2(a+b+1)}\end{aligned} \tag{5.304}$$

A joint or multivariate PDF, $P(u, v; \mathbf{x}, t)$, is used to quantify the correlation of variables such as velocity, pressure, temperature, and species variables in a reacting flow. A joint PDF, $P(u, v; \mathbf{x}, t)$, is the probability density of the simultaneous events $u(x, t) = U$ and $v(x, t) = V$. At each point in space and time, the joint PDF function contains a complete statistical description of u and v. Note that the joint PDF function does not contain two-point information, such as the frequency or length scale of the turbulent fluctuations.

The PDF of u is found from a joint PDF by integrating over all values of v and is called the marginal PDF of u,

$$P(u) = \int_{-\infty}^{+\infty} P(u, v)dv \tag{5.305}$$

Some joint PDF statistics follow:

$$\overline{u} = \int\int u\,P(u, v)du\,dv \tag{5.306}$$

$$\overline{v} = \int\int v\,P(u, v)du\,dv \tag{5.307}$$

$$\overline{uv} = \int\int uv\,P(u, v)du\,dv \tag{5.308}$$

The correlation or covariance between u and v is given by

$$\begin{aligned}\overline{u'v'} &= \int\int (u' - \overline{u})(v' - \overline{v})P(u, v)du\,dv \\ &= \overline{uv} - \overline{u}\,\overline{v} - \overline{u}\,\overline{v} + \overline{u}\,\overline{v} \\ &= \overline{uv} - \overline{u}\,\overline{v}\end{aligned} \tag{5.309}$$

If u and v are statistically independent, $\overline{u'v'} = 0$. However, in turbulent flows, $\overline{u'v'}$ represents one of the Reynolds stress components, which is nonzero in turbulent shear flows.

Conditional moments are averages, variances, etc., taken for only those members of a whole ensemble that comply with a specified condition. The prescribed condition may be that a particular quantity be within an infinitesimal range of a set value. For example, the conditional average and variance of a velocity component could be conditioned on the fluid having a certain temperature. The conditional averages will then be functions of not only position in space and time $(\mathbf{x}, t)$ but also the chosen temperature T.

The notation for the conditional PDF is $P(u|v)$, which describes the probability density of u for a fixed value of v. The notation for the conditional average or mean of u, for a fixed value of v, is given by angular brackets $\langle u|v\rangle$, i.e.,

$$\langle u|v\rangle = \int_{-\infty}^{+\infty} uP(u|v)du \tag{5.310}$$

One-dimensional PDF's can be obtained from experiments or DNSs. For example, in a turbulent reactor, the PDF for the mass fraction of fuel (f or Z) corresponds approximately to a Gaussian distribution. If both random variables in a joint PDF and any linear combination of these two random variables are also normal distributions, the distribution is called a joint normal distribution. In nearly homogeneous turbulence, experiments have shown that the joint PDF of velocity and a scalar variable such as mixture fraction is a joint normal distribution. However, in inhomogeneous flows, especially turbulent reactive flows, PDFs can be far from normal.

5.16.2 PDF Transport

The transport equation for the PDF, $P(\phi)$, of a single scalar ϕ is expressed as (Cant and Mastorakos 2008)

$$\frac{\partial}{\partial t}[\rho P(\phi)] = -\frac{\partial}{\partial x_k}[\langle u_k|\phi\rangle \rho P(\phi)] - \frac{\partial}{\partial \phi}[\dot{\omega}(\phi)P(\phi)] - \frac{\partial^2}{\partial \phi^2}[\rho P(\phi)\langle N_\phi|\phi\rangle] \tag{5.311}$$

with $N_\phi = \mathcal{D}(\partial\phi/\partial x_k)^2$ being the dissipation rate of the scalar ϕ, and $\dot{\omega}$ is the reaction rate. The first term on the right-hand side is the turbulent convection, the second term is the chemical reaction, and the last term is the molecular diffusion. The conditional velocity is represented by a gradient diffusion model,

$$\langle u|\phi\rangle P(\phi) = \overline{u}_k P(\phi) - \mathcal{D}_t \frac{\partial P(\phi)}{\partial x_k} \tag{5.312}$$

The transport equation for a multispecies mass-weighted PDF $\widetilde{P}(\phi_k, k = 1, N)$ is written as,

$$\overline{\rho}\widetilde{P}(\phi_k, k = 1, N) = \overline{\rho(\phi_k, k = 1, N)\prod_{k=1}^{N}\delta(Y_k - \phi_k)} \tag{5.313}$$

where the term $\delta(Y_k - \phi_k)$ is the Dirac delta function for species k.

In a reacting flow, solution of the PDF transport equation for the probability $P(\mathbf{u}; \mathbf{x}, t)$ will allow the Reynolds stresses $\widetilde{u_i''u_j''}$ to be determined. Similarly, knowing the joint PDF of the velocity field $\mathbf{u}$ and a scalar ϕ will allow the Reynolds flux for ϕ, $\widetilde{u_i''\phi''}$ to be computed.

The PDF transport equation is usually solved using Monte Carlo/Lagrangian particle methods. The computational cost is proportional to the total dimension number, i.e., the total number of individual parameters considered in the simulations. Even when only 15–20 particles per cell are considered, its total computational cost is still relatively high compared to an RANS computation. Thus, reduced chemical mechanisms are typically used.

5.17 Further Reading

For more detailed information about turbulent combustion modeling, the reader is referred to texts by Peters (2000), Cant and Mastorakos (2008), Kuo and Acharya (2012), and Poinsot and Veynante (2022).

References

Baum, M., T. Poinsot, D. Haworth, and N. Darabiha (1994), Direct numerical simulation of $H_2/O_2/N_2$ flames with complex chemistry in two-dimensional turbulent flows, *J. Fluid Mech.*, Vol. 281, pp. 1–32.

Bray, K., and J. Moss (1977), A unified statistical model of the premixed turbulent flame, *Acta Astronaut.*, Vol. 4, No. 3–4, pp. 291–319.

Cant, R., and E. Mastorakos (2008), *An Introduction to Turbulent Reacting Flows*, Imperial College Press, Cambridge, UK.

Germano, M., U. Piomelli, P. Moin, and W. Cabot (1991), A dynamic subgrid-scale eddy viscosity model, *Phys. Fluids A: Fluid Dyn.*, Vol. 3, No. 7, pp. 1760–1765.

Haworth, D. (2010), Progress in probability density function methods for turbulent reacting flows, *Prog. Energy Combust. Sci.*, Vol. 36, No. 2, pp. 168–259.

Kerstein, A. (1988), A linear-eddy model of turbulent scalar transport and mixing, *Combust. Sci. Technol.*, Vol. 60, No. 4–6, pp. 391–421.

Kuo, K., and R. Acharya (2012), *Fundamentals of Turbulent and Multiphase Combustion*, John Wiley & Sons, New York.

Peters, N. (2000), *Turbulent Combustion*, Cambridge University Press, Cambridge, UK.

Poinsot, T., and D. Veynante (2022), *Theoretical and Numerical Combustion*, 3rd ed., CNRS.

Pope, S. (2000), *Turbulent Flows*, Cambridge University Press, Cambridge, UK.

Romano, G., N. Ouellette, H. Xu, E. Bodenschatz, E. Steinberg, C. Meneveau, and J. Katz (2007), Measurements of turbulent flows, in *Handbook of Experimental Fluid Mechanics*, Springer, New York.

Rozema, W., H. Bae, P. Moin, and R. Verstappen (2015), Minimum-dissipation models for large-eddy simulation, *Phys. Fluids*, Vol. 27, No. 8, p. 085107.

Smagorinsky, J. (1963), General circulation experiments with the primitive equations: I. The basic experiment, *Mon. Weather Rev.*, Vol. 91, No. 3, pp. 99–164.

Van Dyke, M. (1982), *An Album of Fluid Motion*, Parabolic Press, Stanford, CA.

Veynante, D., and L. Vervisch (2002), Turbulent combustion modeling, *Prog. Energy Combust. Sci.*, Vol. 28, No. 3, pp. 193–266.

Warnatz, J., U. Maas, and R. Dibble (2006), *Combustion*, Springer-Verlag, Berlin, Germany.

Williams, F. (1985), *Combustion Theory*, 2nd edition, CRC Press, Boca Raton, FL.

Zel'dovich Y.,(1951), On the theory of combustion of initially unmixed gases, *Sov. Tech. Phys.*, 1949, Vol. 19; translated as NACA Tech. Memo 1296.

Homework Problems

5.1 A combustion system contains a flowing mixture of H_2 and O_2. The mole fraction of H_2 is $X_{H_2} = \frac{1}{4}$, and the mole fraction of O_2 is $X_{O_2} = \frac{3}{4}$. Assume the velocity vectors are colinear.

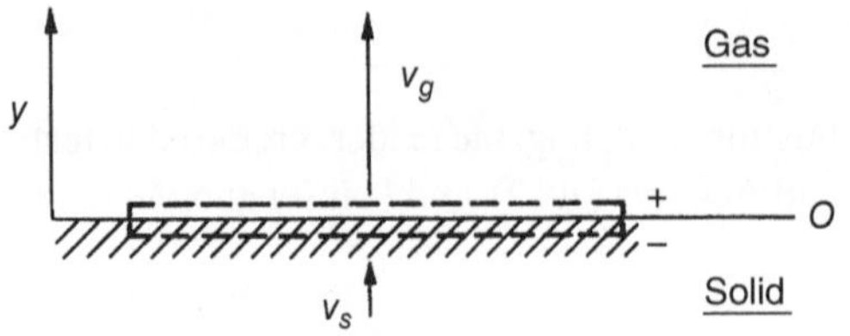

Figure 5.27 Diagram of solid fuel gasification for Problem 5.3.

The bulk molar average velocity $\hat{\mathbf{v}} = 25$ m/s, and the oxygen molar diffusion velocity $\hat{\mathbf{V}}_{O_2} = 3$ m/s. What is the hydrogen mass-average velocity $\mathbf{v}_{H_2}$ and the bulk mass-average velocity $\mathbf{v}$?

5.2 Express the quantity $\dot{Q}$ in the energy equation in terms of the heat of formation and the rate of local production of all the species considered in reacting flow systems, if the internal energy e in the following equation contains *no* chemical energy:

$$\rho\frac{\partial e}{\partial t} + \rho\mathbf{v}\cdot\nabla e = -\nabla\cdot\mathbf{q} + \tilde{\sigma} : \nabla\mathbf{v} + \dot{Q} + \rho\sum_{k=1}^{N} Y_k\mathbf{f}_k\cdot\mathbf{V}_k$$

5.3 If a solid fuel is gasifying at a fixed surface by feeding it in the upward direction at its regression rate as shown in Figure 5.27, (1) write down the boundary conditions for the balance of mass flux at the interface. (2) If the interface control volume moves at the regression rate of the solid in the above problem, what is the mass-flux balance equation across this interface?

5.4 Derive a continuity equation to describe the local balance of the element mass fraction Z_i in a chemically reacting flow where Z_i is defined as

$$Z_i = \frac{\text{mass of atomic element } i \text{ in all chemical species}}{\text{total mass of the mixture}}$$

Note that $Z_i = \sum_{j=1}^{N} a_{ij}Y_j$ where Y_j represents the mass fraction of the jth molecule and a_{ij} represents the mass fraction of the ith atom in the jth molecule.

5.5 Write the energy flux balance at the interface of a burning solid propellant that contains both oxidizer and fuel ingredients. The propellant can be treated as a homogeneous material.

5.6 Using dimensional analysis, derive the Kolmogorov spectrum equation, $E(\kappa) \sim \epsilon^{2/3}\kappa^{-5/3}$. Following Kolmogorov, assume that the rate of energy dissipation ϵ is independent of wavenumber κ, and the energy spectrum $E(\kappa)$ is a function only of ϵ and κ.

5.7 A fuel–air mixture is flowing in a 0.15-m diameter pipe at a Reynolds number ($Re = \overline{u}D/\nu$) of 18,000. Assume the fluid kinematic viscosity ν is 100×10^{-7} m^2/s and the turbulent intensity $u_t/\overline{u} = 0.04$. What is the turbulent Reynolds number Re_t, turbulent kinetic energy k, dissipation ϵ, and Kolmogorov microscale η?

5.8 An internal combustion engine has a mean piston speed $\overline{U}_p$ of 10.0 m/s and a clearance volume height h of 5 mm at the end of the compression stroke. What is the characteristic length L, integral scale l, Taylor microscale λ, and Kolmogorov microscale η at the end of compression? Compare your calculations with Example 5.3, and explain the differences.

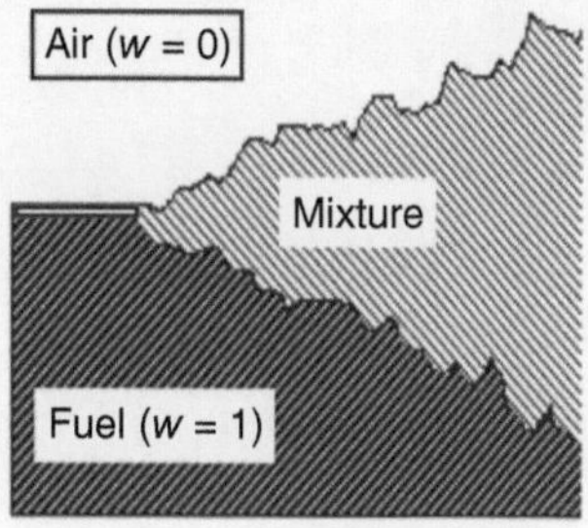

Figure 5.28 Turbulent fuel–air mixing layer schematic. (Warnatz et al. 2006)/with permission of Springer Nature.

Assume the fluid kinematic viscosity ν at the end of compression is $100 \times 10^{-7}\text{m}^2/\text{s}$ and $C_\eta = C_\lambda = 1$, $C_l = 0.2$.

5.9 Starting with the compressible momentum equation (5.314), shown below, derive the Favre-averaged momentum equation (5.315).

$$\frac{\partial \rho u_i}{\partial t} + \frac{\partial \rho u_i u_j}{\partial x_j} = -\frac{\partial p}{\partial x_i} + \frac{\partial \tau_{ij}}{\partial x_j} + F_i \tag{5.314}$$

$$\frac{\partial(\bar{\rho}\tilde{u}_i)}{\partial t} + \frac{\partial(\bar{\rho}\tilde{u}_j\tilde{u}_i)}{\partial x_j} = -\frac{\partial \bar{p}}{\partial x_i} - \frac{\partial(\bar{\rho}\widetilde{u_i''u_j''})}{\partial x_j} + \frac{\partial \bar{\tau}_{ij}}{\partial x_j} + \bar{F}_i \tag{5.315}$$

5.10 A fuel–air turbulent mixing layer is shown in Figure 5.28. The fuel mixture fraction $w = 0$ in the air stream, and $w = 1$ in the fuel stream. In a similar manner as shown in Figure 5.26, sketch qualitatively the fuel mixture fraction PDF at four evenly spaced locations across the mixing layer. Justify the shape of the PDF profiles at each location.

6
Premixed Flames

6.1 Introduction

In this chapter, we will examine the characteristics of premixed flames. A premixed flame, also termed a deflagration, is a finite thickness combustion wave that propagates into a fuel/oxidizer mixture and leaves combustion products in its wake. Depending on the flow conditions, the propagation of the flame can occur through a laminar or a turbulent fluid flow field, as well as through partially premixed reactants.

Premixed flames are used in many applications, for example, in spark ignition internal combustion engines, heating appliances, gas stoves, and industrial processes. Premixed flames are also of scientific interest as they are used as a basis for comparison of experiments, theory, and computations; validation of chemical models; and development of turbulent combustion models.

One of the major tasks in premixed flame studies is to determine the laminar flame speed S_L for a given set of flow conditions. Depending on the pressure and equivalence ratio, the values of laminar flame speeds for hydrocarbon fuels in air are usually less than 50 cm/s. Hydrogen is unique as its flame speed is in the range of 100–300 cm/s. The flame speed governs the resulting flame thickness, shape, and stability. The detailed flame structures, i.e., the species concentration and temperature profiles through the flame, are also of interest.

This chapter also discusses flame ignition, flammability limits, and flame quenching, three limiting parameters which depend on flame speed and flame heat transfer. Proper consideration of these parameters is necessary to reduce flashback and blow-off in the design and operation of combustion devices.

In general, a mixture at a maximum flame temperature is also a mixture with a maximum flame speed. At high temperatures, dissociation reactions are favored. These reactions introduce numerous free radicals into the flame, and these free radicals then act as chain carriers to promote the reaction and, hence, increase the propagation speed of the flame. The lighter these free radicals are, the more easily they are able to diffuse to the preheat zone. This is why the presence of lightweight hydrogen atoms in a reaction can significantly enhance flame speeds.

It has been found that flame stretching due to strain and curvature also has significant effects on the structure and propagation speed of a laminar flame. For example, the flame speeds of lean propane/air spherical flames increase with flame radius, while for rich flames, they decrease. The flame stretching is characterized by the nondimensional Karlovitz and Markstein numbers.

As the nondimensional Reynolds number of the mean flow field increases above a critical value, the flow field will become turbulent. Turbulent flow can be envisioned as a mean fluid flow upon which are superimposed eddies of different sizes randomly dispersed in the flow. The turbulence can be already present in the flow or generated by the combustion process. Due to the presence

Principles of Combustion, Third Edition. Allan T. Kirkpatrick and Kenneth K. Kuo.

Companion website: www.wiley.com/go/kirkpatrick/principlesofcombustion3e

of the turbulent eddies, the propagation speed of a flame through a turbulent flow field is much greater than the laminar flame speed and increases as the Reynolds number increases.

Flame propagation in a turbulent flow field is generally classified into three combustion regimes, depending on the turbulence intensity and related parameters, and characterized by the Damköhler number. The regimes are: wrinkled laminar flame, flamelets, and distributed reaction. The turbulent flame surface is very complex with rapid fluctuations of the flame envelope, and the turbulent reaction zone is much thicker than the laminar reaction zone.

6.2 Laminar Flame Speed Modeling

A basic model of a laminar premixed flame divides the flame into two major zones, as shown in Figure 6.1. Zone I is the preheat (*ph*) zone, with thickness δ_{ph}, in which the gases are heated from T_o to the ignition temperature T_i and ignite at the boundary between the zones. Zone II is a narrower reaction (*r*) zone, with thickness δ_r, in which the reactants combust, release chemical energy, and increase in temperature from the ignition temperature T_i to the flame temperature T_f. The chemical reaction is assumed to be negligible in the preheat zone I. As the reactants' temperatures rise in the reaction Zone II, the chemical reactions, which depend exponentially on temperature, increase until the reactants are consumed and their concentration then decreases, forming the downstream side of the flame front.

The flame is assumed to be one dimensional, with the unburned gas entering the preheat Zone I normal to the zone boundary. For the coordinate system moving with the flame from right to left, the laminar flame speed S_L is defined as the velocity of the unburned gases through the combustion zone in the direction normal to the wave surface. Depending on the situation, three other laminar flame speed definitions are also used in combustion studies and will be covered more fully in Section 6.7. The *displacement flame speed* is a measure of the flame front speed relative to the surrounding mixture. The *absolute speed* is the flame speed relative to a fixed reference frame. The *consumption flame speed* is the speed at which reactants are consumed.

The continuity equation (kg/m^2-s) for the 1-D flow shown in Figure 6.1 is

$$\dot{m}'' = \rho_u S_L = \rho_b u_b \tag{6.1}$$

where the subscripts u and b represent the unburned and burned gases, respectively. In the coordinate system moving with the flame, the gas flow accelerates as it traverses across the flame. For example, with premixed hydrocarbon combustion at atmospheric pressure, the burned gas density is less ($\rho_b \approx 0.2\,\rho$) than the unburned gas density, so the burned gas velocity is greater ($u_b \approx 5\,S_L$) than the unburned gas velocity. The static pressure also decreases slightly ($p_b \approx 0.98\,p$).

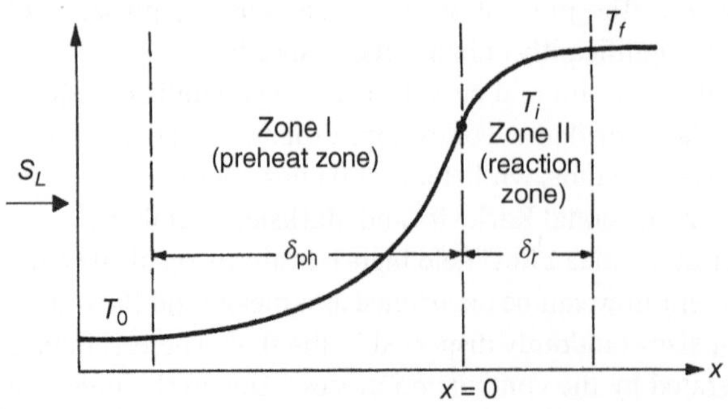

Figure 6.1 Schematic diagram of the temperature variation across a laminar flame.

An early thermal analysis of laminar flames was performed by Mallard and LeChatelier (1883) using the control volume energy equation to determine the laminar-flame speed S_L. Mallard and LeChatelier made the assumption that the combustion process was primarily governed by the conduction of thermal energy from the flame to the upstream unburned gases. The thermal energy conducted from zone II into zone I is assumed to raise the unburned gas temperature from T_o to the ignition temperature T_i.

The open system control volume for zone I is

$$\dot{Q}/A = \dot{m}'' c_p (T_i - T_o) = S_L \rho_u c_p (T_i - T_o) \tag{6.2}$$

where $\dot{Q}/A$ is the thermal energy conducted into zone I, $\dot{m}''$ is the gas mass flux (kg/m^2-s), and A is the gas cross-sectional area. Assuming a linear conduction temperature profile in zone II, the conduction heat flux to zone I from the flame at temperature T_f is

$$\dot{Q}/A = k \frac{T_f - T_i}{\delta_r} \tag{6.3}$$

Equating Eqs. (6.2) and (6.3), and solving for S_L,

$$S_L = \alpha \frac{(T_f - T_i)}{(T_i - T_o)} \frac{1}{\delta_r} \tag{6.4}$$

where $\alpha = k/\rho_u c_p$ is the thermal diffusivity of the gas. The value of S_L can be determined from Eq. (6.4), if the reaction zone thickness δ_r is known.

If the unburned gas flowing into zone II completely reacts in stoichiometric combustion, then

$$\dot{m}'' = \rho_u S_L = \dot{\omega} \delta_r \tag{6.5}$$

where $\dot{\omega}$ is the reactant (fuel and oxidizer) mass consumption rate per unit volume (kg/m^3-s).

Substituting Eq. (6.5) into Eq. (6.4), the laminar flame speed S_L is

$$S_L = \sqrt{\alpha \frac{(T_f - T_i)}{(T_i - T_o)} \frac{\dot{\omega}}{\rho_u}} \sim \left(\frac{\alpha \dot{\omega}}{\rho_u} \right)^{1/2} \tag{6.6}$$

and the laminar flame thickness δ_r can be expressed as

$$\delta_r = \frac{\rho_u S_L}{\dot{\omega}} = \frac{\alpha \dfrac{(T_f - T_i)}{(T_i - T_o)}}{S_L} \tag{6.7}$$

Mallard and LeChatelier did not specify the specific temperature at which the reaction rate $\dot{\omega}$ should be determined. However, their analysis suggests that the flame speed is proportional to the square root of the product of thermal diffusivity and mass reaction rate. This result provides one of the most important relationships in laminar-flame theory, as it predicts that the flame propagation speed results from the diffusion of thermal energy from a reaction zone to a preheat zone.

Based on Eq. (6.6) and the Arrhenius law for the reaction rate $\dot{\omega}$, the temperature dependence of the laminar-flame speed is:

$$S_L \sim \sqrt{e^{-E_a/R_u T}} = e^{-E_a/2R_u T} \tag{6.8}$$

Assuming the mass reaction rate term $\dot{\omega}$ follows the second-order kinetics, then the concentration terms in $\dot{\omega}$ would be proportional to p^2. However, the density term in α and the other density term in Eq. (6.6) also have a p^2 dependence. Thus, for a second-order reaction system, the flame speed is independent of pressure. In general, the pressure dependence of the mass reaction rate $\dot{\omega} \sim p^m$, where m is the overall order of the reaction. Thus, for the pressure dependence, we have

$$S_L \sim p^{(m-2)/2} \tag{6.9}$$

A correlation of experimental data from a number of sources for the stoichiometric combustion of methane–air flames with $p > 5$ atm and T_u from 290 to 300 K by Andrews and Bradley (1972b) is

$$S_L = \frac{43}{p^{1/2}} \tag{6.10}$$

where S_L is in cm/s and p in atm.

Example 6.1 ***Laminar Flame Speed***

Estimate the laminar flame speed S_L and thickness δ_r of a stoichiometric mixture of iso-octane in air at 1 atm and $T_o = 300$ K. The molecular mass M_F of iso-octane is 114.22 kg/kmol, and the mass stoichiometric air/fuel ratio s_A is 15.03. Assume the average concentration of the reactants in the flame is 1/2 of the unreacted concentration.

Solution

The iso-octane fuel (F) stoichiometric reaction is

$$1C_8H_{18} + 12.5(O_2 + 3.76N_2) \longrightarrow 8CO_2 + 9H_2O + 47.0N_2$$

The auto-ignition temperature of iso-octane is $T_i = 650$ K, and the adiabatic flame temperature of iso-octane is $T_f = 2260$ K. With reference to Figure 6.1, the average temperature in the preheat zone (ph) is assumed to be

$$T_{ph} = \frac{T_o + T_i}{2} = \frac{300 + 650}{2} = 475\ \text{K}$$

The average temperature in the reaction zone (r) is assumed to be

$$T_r = \frac{T_i + T_f}{2} = \frac{650 + 2260}{2} = 1455\ \text{K}$$

$$\frac{(T_f - T_i)}{(T_i - T_o)} = \frac{(2260 - 650)}{(650 - 300)} = 4.60$$

In the reaction zone, the average mole fractions and concentrations of reactants octane and oxygen will decrease to zero as the combustion proceeds, so taking the average,

$$\overline{X}_F = \frac{1}{2}X_F = \frac{1}{2}\frac{n_F}{n} = \left(\frac{1}{2}\right)\frac{1}{60.5} = 8.26 \times 10^{-3}$$

$$\overline{X}_{O_2} = \frac{1}{2}X_{O_2} = \frac{1}{2}\frac{n_{O_2}}{n} = \left(\frac{1}{2}\right)\frac{12.5}{60.5} = 0.103$$

The overall mixture concentration in the reaction zone is

$$[M] = \frac{p}{R_u T_r} = \frac{101.3}{(8.314)(1455)} = 8.37 \times 10^{-3}\ \text{kmol/m}^3$$

and the average octane and oxygen concentrations are

$$\begin{aligned}[F] &= \overline{X}_F[M] = (8.26 \times 10^{-3})(8.37 \times 10^{-3}) \\ &= 6.92 \times 10^{-5}\ \text{kmol/m}^3 = 6.92 \times 10^{-8}\ \text{mol/cm}^3 \\ [O_2] &= \overline{X}_{O_2}[M] = (0.103)(8.37 \times 10^{-3}) \\ &= 8.62 \times 10^{-4}\ \text{kmol/m}^3 = 8.62 \times 10^{-7}\ \text{mol/cm}^3\end{aligned}$$

Using the reaction coefficients in Table 4.2, the octane molar consumption rate is

$$\dot{\Omega}_F = \frac{d[C_8H_{18}]}{dt} = A_p \exp(-E_a/R_uT_r)[C_8H_{18}]^m[O_2]^n$$
$$= (4.6 \times 10^{11})\exp(-15098/1455)[6.92 \times 10^{-8}]^{0.25}[8.62 \times 10^{-7}]^{1.5}$$
$$= 1.86 \times 10^{-4}\,\text{mol/cm}^3\text{s} = 0.186\,\text{kmol/m}^3\text{s}$$

and the octane mass reaction consumption rate is

$$\dot{\omega}_F = \dot{\Omega}_F \cdot M_F = (0.186)(114.22) = 21.2\,\text{kg/m}^3\text{s}$$

The overall mass reactant consumption rate is therefore

$$\dot{\omega} = (1 + s_A)\dot{\omega}_F = (1 + 15.03)(21.2) = 340.6\,\text{kg/m}^3\text{s}$$

Using the properties of air from the Appendix for the mixture properties, $k(1455\,\text{K}) \sim 0.10\,\text{W/mK}$, and $c_p(475\,\text{K}) \sim 1020\,\text{J/kgK}$. The unburned air density is

$$\rho_u = \frac{P}{RT_u} = \frac{101.3}{(0.287)(300)} = 1.18\,\text{kg/m}^3$$

The effective thermal diffusivity α is

$$\alpha = \frac{k(1455)}{\rho_u c_p(475)} = \frac{0.10}{(1.18)(1020)} = 8.31 \times 10^{-5}\,\text{m}^2/\text{s}$$

The laminar flame speed S_L is

$$S_L = \left(\frac{\alpha \frac{(T_f - T_i)}{(T_i - T_o)}\dot{\omega}}{\rho_u}\right)^{1/2} = \left(\frac{(8.3 \times 10^{-5})(4.60)(340.6)}{1.18}\right)^{1/2}$$
$$= 0.33\,\text{m/s} = 33\,\text{cm/s}$$

From Eq. (6.5), the flame thickness δ_r is

$$\delta_r = \frac{\rho_u S_L}{\dot{\omega}} = \frac{(1.18)(0.33)}{340.6} = 1.14 \times 10^{-3}\,\text{m} = 1.14\,\text{mm}$$

Comment: The experimental correlation from Metghalchi and Keck (1982) for the iso-octane laminar flame speed at 300 K is $S_L = 26\,\text{cm/s}$, indicating the approximate nature of Eq. (6.6). Note that the effective thermal diffusivity α is a hybrid term as it is constructed from parameters at different temperatures in the flame zones.

6.3 Zeldovich, Frank-Kamenetsky, and Semenov Flame Model

Zeldovich, Frank-Kamenetsky, and Semenov (Zeldovich 1948) adopted Mallard and LeChatelier's concept of dividing the flame into two zones (preheat and reaction zones). However, instead of using a control volume energy equation, they used the differential energy equation and the differential species conservation equation in the two zones. They also assumed that the ignition temperature was very close to the adiabatic flame temperature and consequently replaced T_i with T_f in their estimation of mass reaction rates. Their basic assumptions are as follows:

1. The flame is one-dimensional and steady-state.
2. The pressure is constant.

3. The number of moles does not vary during the reaction. This restriction is later removed to allow a stoichiometric reaction to change the product–reactant mole ratio.
4. The thermodynamic properties c_p and k are constant.
5. The Lewis number $Le = 1$, implying that $k/c_p = \mathcal{D}\rho$. This has the effect of reducing the number of differential equations from two to one by replacing one of the differential equations with an algebraic equation. Later, this restriction is modified to allow Le = constant.

The species continuity and energy conservation equations are first written in their most general forms and then simplified according to the above assumptions.

The general species mass conservation equation (kg/m^3-s) for the reacting species (r) derived in Chapter 5 is

$$\frac{\partial \rho_r}{\partial t} + \nabla \cdot \rho_r \mathbf{v} = \nabla \cdot (\rho \mathcal{D} \nabla Y_r) - \dot{\omega}_r \tag{6.11}$$

where ρ is the gas mixture density (kg/m^3), ρ_r is the reacting (r) species density, and $\dot{\omega}_r$ is the mass rate of reaction, which is positive for the consumption of reactants. For a 1-D steady flame, Eq. (6.11) reduces to

$$\frac{d}{dx}\left(\frac{\rho_r}{\rho} \cdot \rho u\right) = \frac{d}{dx}\left[\rho \mathcal{D} \frac{d}{dx}\left(\frac{\rho_r}{\rho}\right)\right] - \dot{\omega}_r \tag{6.12}$$

or

$$\frac{\rho_r}{\rho}\frac{d(\rho u)}{dx} + \rho u \frac{d(\rho_r/\rho)}{dx} = \rho \mathcal{D} \frac{d^2}{dx^2}\left(\frac{\rho_r}{\rho}\right) - \dot{\omega}_r \tag{6.13}$$

From the continuity equation, $d(\rho u)/dx = 0$, and the species equation (6.13) becomes

$$\rho u \frac{d(\rho_r/\rho)}{dx} = \rho \mathcal{D} \frac{d^2}{dx^2}\left(\frac{\rho_r}{\rho}\right) - \dot{\omega}_r \tag{6.14}$$

The number density $\tilde{n}$ of reactant molecules (molecules/m^3) is

$$\tilde{n} = \rho_r \frac{N_A}{\mathrm{M}_r} \tag{6.15}$$

where N_A is Avogadro's number of molecules per mole of reactant, and M_r is the reactant molecular mass. Equation (6.14) then becomes

$$\rho u \frac{d(\tilde{n}\mathrm{M}_r/\rho N_A)}{dx} = \rho \mathcal{D} \frac{d^2}{dx^2}(\tilde{n}\mathrm{M}_r/\rho N_A) - \dot{\omega}_r \tag{6.16}$$

Multiplying through by N_A/M_r and rearranging the terms, we have

$$\rho \mathcal{D} \frac{d^2(\tilde{n}/\rho)}{dx^2} - \rho u \frac{d(\tilde{n}/\rho)}{dx} - \tilde{\omega} = 0 \tag{6.17}$$

where

$$\tilde{\omega} \equiv \dot{\omega}_r \frac{N_A}{\mathrm{M}_r}$$

The first term in Eq. (6.17) represents the rate of diffusion of reactant species and the second term the rate of convection of reactant species. The inhomogeneous term $\tilde{\omega}$ represents the rate of change of the number of molecules of reactant per unit volume by chemical reaction.

From Chapter 5, the general form of the energy equation (kW/m^3) is

$$\rho \frac{Dh}{Dt} - \frac{Dp}{Dt} = -\nabla \cdot \mathbf{q} + \Phi + \dot{Q} + \rho \sum_{K=1}^{N} Y_K \mathbf{f}_K \cdot \mathbf{v}_K \tag{6.18}$$

Neglecting dissipation, body-force work, and unsteady terms, using the heat conduction equation, and assuming a constant specific heat,

$$q = -k\frac{dT}{dx}, \qquad h = c_pT \tag{6.19}$$

the energy equation in one-dimensional form is:

$$\rho c_p u\frac{dT}{dx} = k\frac{d^2T}{dx^2} + \dot{Q}$$

or

$$\frac{k}{\rho c_p}\frac{d^2T}{dx^2} = u\frac{dT}{dx} - \frac{\dot{Q}}{\rho c_p} \tag{6.20}$$

In this form of the energy equation, the convection and the diffusion fluxes balance the chemical heat production. The combustion term $\dot{Q}$ can be expressed as

$$\dot{Q} = \tilde{\omega}\tilde{q} \tag{6.21}$$

where $\tilde{q}$ represents the heat of combustion per molecule of reactant.

Equation (6.20) can be written as

$$\frac{k}{c_p}\frac{d^2T}{dx^2} - \rho u\frac{dT}{dx} + \frac{\tilde{\omega}\tilde{q}}{c_p} = 0 \tag{6.22}$$

where the product ρu is the eigenvalue of the problem.

Following the Shvab–Zeldovich formulation given in Chapter 5, two new variables θ and ζ are defined as

$$\theta \equiv c_p\frac{T - T_o}{\tilde{q}} \tag{6.23}$$

$$\zeta \equiv \frac{\tilde{n}_o}{\rho_o} - \frac{\tilde{n}}{\rho} \tag{6.24}$$

where the subscript *o* designates initial values at an undisturbed location. Substituting Eqs. (6.23) and (6.24) into Eqs. (6.22) and (6.14), we have

$$\frac{k}{c_p}\frac{d^2\theta}{dx^2} - \rho u\frac{d\theta}{dx} + \tilde{\omega} = 0 \tag{6.25}$$

$$\rho D\frac{d^2\zeta}{dx^2} - \rho u\frac{d\zeta}{dx} + \tilde{\omega} = 0 \tag{6.26}$$

These two equations are identical if $Le = 1$, which implies $\rho D = k/c_p$.

The boundary conditions for (6.25) and (6.26) are prescribed as follows: for $x = -\infty$ (cold boundary),

$$\zeta = 0, \qquad \theta = 0 \tag{6.27}$$

For $x = +\infty$ (hot boundary),

$$\zeta = \frac{\tilde{n}_o}{\rho_o}, \qquad \theta = c_p\frac{T_f - T_o}{\tilde{q}} \tag{6.28}$$

It is interesting to note that the Solutions of Eqs. (6.25) and (6.26) are identical (i.e., $\zeta = \theta$) over the entire combustion wave zone if

$$\frac{\tilde{n}_o}{\rho_o} = c_p\frac{T_f - T_o}{\tilde{q}} \tag{6.29}$$

The above relationship can be satisfied if the flame is adiabatic. Under this assumption, the following algebraic equation can be introduced to replace one of the two differential equations [say, Eq. (6.26)]:

$$c_p T + \frac{\tilde{n}\tilde{q}}{\rho} = c_p T_o + \frac{\tilde{n}_o \tilde{q}}{\rho_o} = c_p T_f \tag{6.30}$$

The meaning of this equation is that the sum of the thermal (sensible) and chemical energies per unit mass of mixture is constant throughout the combustion zone. This implies that only one ordinary differential equation, the energy equation (6.22) or the species equation (6.17), needs to be solved. Using the energy equation, in the preheat zone (region I) where $\tilde{\omega} = 0$, the energy equation is

$$\frac{d^2T}{dx^2} - \frac{(\rho u)c_p}{k}\frac{dT}{dx} = 0 \tag{6.31}$$

with the boundary conditions

$$x = -\infty, \qquad T = T_o \tag{6.32}$$

$$x = 0^-, \qquad T = T_i \tag{6.33}$$

In the reaction zone (region II), the convective term was considered by these researchers to be negligible in comparison with the heat generation term, since T_i is usually very close to T_f. Then, the energy equation can be written as

$$\frac{d^2T}{dx^2} + \frac{\tilde{\omega}\tilde{q}}{k} = 0 \tag{6.34}$$

with boundary conditions

$$x = \infty, \qquad T = T_f, \qquad \frac{dT}{dx} = 0 \tag{6.35}$$

$$x = 0^+, \qquad T = T_i \tag{6.36}$$

In order to have heat-flux balance at the interface of the two zones, it is required that

$$\left(\frac{dT}{dx}\right)_{x=0^-} = \left(\frac{dT}{dx}\right)_{x=0^+} \tag{6.37}$$

By integrating Eq. (6.31) with respect to x, we have

$$\frac{dT}{dx} = \frac{(\rho u)c_p}{k} T + \text{constant} \tag{6.38}$$

Since at $x = -\infty$, we have $T = T_o$, $dT/dx = 0$, it follows that

$$\frac{dT}{dx} = \frac{(\rho u)c_p}{k}(T - T_o) \tag{6.39}$$

This gives

$$\left(\frac{dT}{dx}\right)_{x=0^-} = \frac{(\rho u)c_p(T_i - T_o)}{k} \approx \frac{(\rho u)c_p(T_f - T_o)}{k} \tag{6.40}$$

if T_i is very close to T_f.

In order to obtain the temperature gradient on the 0^+ side, we need to integrate Eq. (6.34) by first multiplying it by $2\,dT/dx$ and rearrange it to give

$$\frac{d}{dx}\left(\frac{dT}{dx}\right)^2 = -2\left(\frac{dT}{dx}\right)\frac{\tilde{q}\tilde{\omega}}{k} \tag{6.41}$$

Integrating from $x = 0^+$ to $x = \infty$, and applying the region II boundary conditions, we have

$$0 - \left(\frac{dT}{dx}\right)^2_{x=0^+} = -\int_{T_i}^{T_f} \frac{2\tilde{q}\tilde{\omega}}{k} dT \tag{6.42}$$

or

$$\left(\frac{dT}{dx}\right)^2_{x=0^+} = \int_{T_i}^{T_f} \frac{2\tilde{q}\tilde{\omega}}{k} dT \tag{6.43}$$

Substituting both (6.40) and (6.43) into (6.37), we have

$$\frac{(\rho u)^2 c_p^2 (T_f - T_o)^2}{k^2} = \frac{2\tilde{q}}{k} \int_{T_i}^{T_f} \tilde{\omega}\, dT \tag{6.44}$$

or

$$(\rho u)^2 = \frac{2k\tilde{q}}{c_p^2 (T_f - T_o)^2} \int_{T_i}^{T_f} \tilde{\omega}\, dT \tag{6.45}$$

Also, we know that $\tilde{n}_o \tilde{q}/\rho_o = c_p(T_f - T_o)$, then

$$(\rho u)^2 = \rho_o^2 S_L^2 = \rho_o^2 u_o^2 = \frac{2k}{c_p(T_f - T_o)} \frac{\rho_o}{\tilde{n}_o} \int_{T_i}^{T_f} \tilde{\omega}\, dT \tag{6.46}$$

Therefore,

$$S_L = u_o = \sqrt{\left(\frac{k}{\rho_o c_p}\right) \frac{2}{(T_f - T_o)} \left(\frac{1}{\tilde{n}_o} \int_{T_i}^{T_f} \tilde{\omega}\, dT\right)} = \sqrt{\left(\frac{k}{\rho_o c_p}\right) \frac{2}{(T_f - T_o)} I} \tag{6.47}$$

where

$$I \equiv \frac{1}{\tilde{n}_o} \int_{T_i}^{T_f} \tilde{\omega}\, dT = \frac{1}{\tilde{n}_o} \int_{T_o}^{T_f} \tilde{\omega}\, dT \tag{6.48}$$

The two integral quantities in the above equation are equal because there is no reaction between T_o and T_i. Now, let the reaction rate $\tilde{\omega}$ be a function of T, using an expression for a zero-order chemical reaction for simplicity:

$$\tilde{\omega} = Ae^{-E_a/R_u T} \tag{6.49}$$

where A, an Arrhenius-type factor, is considered to be a constant. For most hydrocarbon fuels, $E_a/R_u T \gg 1$. Since $E_a \approx 120$–160 kJ/mol, $T_f \approx 1500$–2000 K, and $R_u = 8.3145$ kJ/kmol K, then

$$\frac{E_a}{R_u T_f} \sim 10 \gg 1 \tag{6.50}$$

Set $\sigma \equiv T_f - T$; its values range from $\sigma_1 \equiv T_f - T_i$ down to 0 for zone II. Then,

$$T = T_f - \sigma = T_f \left(1 - \frac{\sigma}{T_f}\right) \tag{6.51}$$

Since σ/T_f is generally small, we have

$$\begin{aligned} e^{-E_a/R_u T} &= \exp\left[-\frac{E_a}{R_u T_f (1 - \sigma/T_f)}\right] \approx \exp\left[-\frac{E_a}{R_u T_f}\left(1 + \frac{\sigma}{T_f}\right)\right] \\ &= e^{-E_a/R_u T_f} e^{-E_a \sigma/R_u T_f^2} \end{aligned} \tag{6.52}$$

The expression for $\tilde{\omega}$ is now introduced into (6.48) to give

$$I = \frac{Ae^{-E_a/R_uT_f}}{\tilde{n}_o}\int_{T_i}^{T_f} e^{-E_a\sigma/R_uT_f^2}\,dT = -\frac{Ae^{-E_a/R_uT_f}}{\tilde{n}_o}\int_{\sigma_1}^{o} e^{-E_a\sigma/R_uT_f^2}\,d\sigma \tag{6.53}$$

Let

$$\beta \equiv \frac{E_a\sigma}{R_uT_f^2} \quad \text{and} \quad \beta_1 \equiv \frac{E_a\sigma_1}{R_uT_f^2} \tag{6.54}$$

Substituting (6.54) into (6.53):

$$I = \frac{Ae^{-E_a/R_uT_f}}{\tilde{n}_o}\frac{R_uT_f^2}{E_a}\int_o^{\beta_1} e^{-\beta}\,d\beta \approx \frac{Ae^{-E_a/R_uT_f}}{\tilde{n}_o}\frac{R_uT_f^2}{E_a} \tag{6.55}$$

since $\int_o^{\beta_1} e^{-\beta}\,d\beta = 1 - e^{-\beta_1} \approx 1$. Therefore,

$$I \approx \frac{Ae^{-E_a/R_uT_f}}{\tilde{n}_o}\left(\frac{R_uT_f^2}{E_a}\right) \tag{6.56}$$

Upon substitution into Eq. (6.47), S_L for a zero-order reaction can be written as

$$S_L = u_o = \sqrt{\frac{2k}{c_p\rho_o}\frac{1}{\tilde{n}_o}\left(\frac{Ae^{-E_a/R_uT_f}}{T_f - T_o}\right)\frac{R_uT_f^2}{E_a}} \tag{6.57}$$

Since $\tilde{n}_o \sim p$ and $\rho_o \sim p$, then $S_L \sim \sqrt{p^{-2}}$ for a zero-order reaction. This agrees with the result obtained before, Eq. (6.9), that the flame speed $S_L \sim \left(p^{m-2}\right)^{1/2}$. Note that this analysis is valid for $E_a/R_uT_f \gg 1$, in other words, for high-activation-energy reactions.

Reactions in which the reaction rate $\tilde{\omega}$ is a function of the reactant concentration as well as temperature can also be derived. For the first-order (unimolecular) reactions, set

$$\tilde{\omega} \sim [C]Ae^{-E_a/R_uT} \tag{6.58}$$

where $[C]$ is the reactant concentration. For the second-order (bimolecular) reactions, the reaction rate scales as

$$\tilde{\omega} \sim [C]^2Ae^{-E_a/R_uT} \tag{6.59}$$

Some of the restricting assumptions may now be removed. If we allow the number of molecules to change in the stoichiometric ratio n_r/n_p, assign k and c_p the mean values $\bar{k}$ and $\bar{c}_p$ in region I and the value for the burned gases k_f and c_{pf} in region II, and set $(k/c_p)/\rho D = Le$, then it can be shown that the equation for a first-order reaction is

$$S_L = u_o = \sqrt{\frac{2k_fc_{pf}A}{\rho_o\bar{c}_p^2}\left(\frac{T_o}{T_f}\right)\left(\frac{n_r}{n_p}\right)(Le)\left(\frac{R_uT_f^2}{E_a}\right)^2\frac{e^{-E_a/R_uT_f}}{(T_f - T_o)^2}} \tag{6.60}$$

and for a second-order reaction is

$$S_L = u_o = \sqrt{\frac{2k_f(c_{pf})^2A\tilde{n}_o}{\rho_o\bar{c}_p^3}\left(\frac{T_o}{T_f}\right)\left(\frac{n_r}{n_p}\right)^2(Le)^2\left(\frac{R_uT_f^2}{E_a}\right)^3\frac{e^{-E_a/R_uT_f}}{(T_f - T_o)^3}} \tag{6.61}$$

The laminar flame speed S_L is now based on the flame temperature T_f. The preceding analysis is still approximate but predicts flame speed trends. Note that we still have the basic relationship between flame speed, thermal diffusivity, and reaction rate:

$$S_L \sim \left[\alpha\,\dot{\omega}(T_f)\right]^{1/2} \tag{6.62}$$

6.4 Diagnostic Methods for Flame Speed and Structure

In the last 20 years, there have been substantial developments in nonintrusive combustion diagnostic techniques. Many technical papers and books are available in the laser diagnostic technique area. For example, see Soid and Zainal (2011) and Gupta (2022). The advances in diagnostics have been due largely to the rapid development of laser-based spectroscopic techniques, computer-coupled data acquisition systems, and digital image capture and analysis. The nonintrusive character of these methods prevents perturbation of the flow systems under study and thus provides more realistic and accurate measurements than those provided by intrusive techniques. The diagnostic techniques are used for a variety of purposes, including achieving a better understanding of flame and flow field structures in order to improve the design of existing combustion systems, validating theoretical models and numerical codes, studying the formation mechanism of undesirable species in emission control, and conducting performance evaluations of combustion systems.

The diagnostic techniques have been applied to various flame studies for measurements of the velocity field of chemically reacting flows, concentrations of chemical species, and temperatures of gaseous mixtures and particles. Various measurement techniques for flame speed have been developed, including the Bunsen burner method, flat-flame burner method, the spherical bomb method, and the soap bubble method (Andrews and Bradley 1972a). These techniques are briefly discussed in the following section.

6.4.1 Bunsen Burner Method

The ubiquitous laboratory premixed flame burner was invented by R. Bunsen (1811–1899), a German scientist, to analyze the emission spectra from chemical compounds. Figure 6.2 shows a schematic diagram of the Bunsen burner. The design principle of the Bunsen burner has been adopted in many gas appliances such as cooking stoves and gas burners since it can be used with a variety of gaseous fuels. In comparison with purely diffusion flames, the premixed flames from Bunsen burners are relatively clean and provide higher-temperature combustion.

In a Bunsen-type burner, the fuel gas enters the burner through a feed line near the bottom. The fuel jet induces the entrainment of primary air through a number of ports controlled by a needle valve. As fuel and airflow up the barrel tube, they mix, and before reaching the top of the burner, the gases are well mixed and the mixture can be considered to be homogeneous. The mixture of gas and primary air passes up the burner tube at a speed that is sufficient to prevent the flame flashback down the tube. At the port, the flame is anchored near the top of the burner.

As long as the fuel feeding rate remains constant, the premixed flame in a quiescent atmosphere will remain stationary and steady. A few streamlines are also sketched in Figure 6.2 to show the general flow characteristics near a steady Bunsen burner flame. The gas flow in the tube is laminar with a parabolic velocity profile. As we will see in a later section, the low flow velocity at the tube wall coupled with the heat conduction to the wall contribute to the stabilization of the flame.

Since the amount of air entrained is usually below that needed for complete combustion of the gas, there is some diffusion combustion of product species such as H and CO, shown as a diffusion envelope in Figure 6.2.

A shadowgraph of a conical premixed laminar flame from Weinberg (1963) is shown in Figure 6.3*a*. A simultaneous photograph of visible (outer cone) and schlieren (inner cone) cones of a premixed flame from Weinberg (1963) is shown in Figure 6.3*b*.

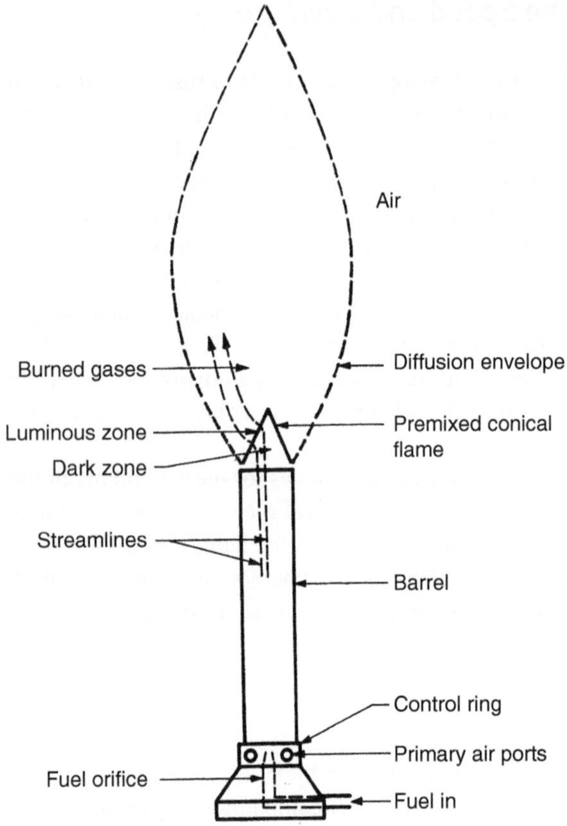

Figure 6.2 Schematic of a premixed laminar flame shown above a Bunsen burner.

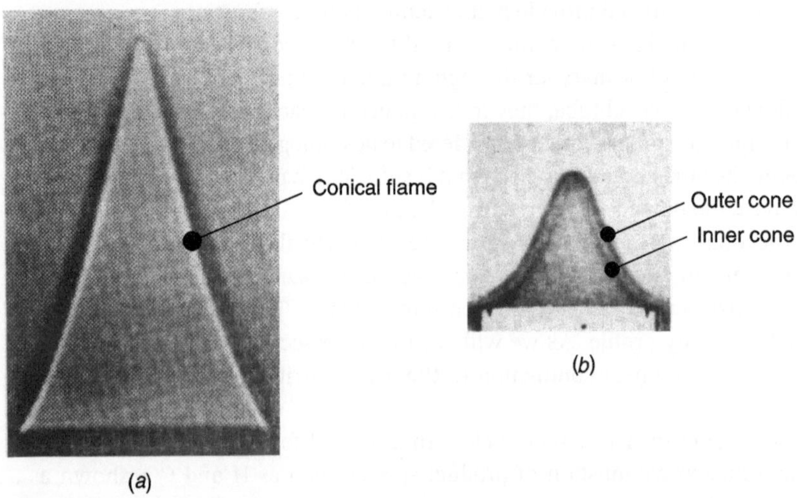

Figure 6.3 (*a*) Shadowgraph of conical premixed ethylene–air laminar flame. (*b*) Simultaneous photographs of visible (outer) and schlieren (inner) cones. *Source*: Adapted from Weinberg 1963/Butterworth & Company. Ltd.

These photographs indicate clearly that there is a luminous conical region within which reaction and heat release are taking place. Underneath the luminous cone, there is a dark zone (also marked in Figure 6.2) within which the unburned gas molecules change their flow direction from the initially vertical direction to outward directions. Immediately beneath the luminous cone, the unburned gases are heated to a critical temperature at which rapid chemical reaction starts. The burned gases expand as they leave the flame zone and are diluted and cooled by the surrounding air.

The luminous zone is usually no more than 1 mm thick. The color of the luminous zone changes with the fuel–air ratio. When the mixture is fuel lean, the color of the cone surface is deep violet due to large concentrations of excited CH radicals. When the mixture is fuel rich, the color of the cone surface is green due to large concentrations of C_2 molecules. The high-temperature burned gases usually show a reddish glow that arises from radiation from CO_2 and water vapor. When the mixture is highly fuel rich, carbon particles (soot) form and an intense yellow radiation appears. The radiation is yellow because the blackbody radiation curve peaks in the yellow for the temperatures that normally exist in these premixed flames.

A sketch of the flow configuration through the conical laminar flame is shown in Figure 6.4. Let us call the conical surface area of the innermost cone A_f, the tube cross-sectional area A_{tube}, and the average flow velocity in the tube V_{tube}. Then, the laminar flame speed S_L, defined as the magnitude of the velocity of the unheated premixed gas just before entering the flame zone, can be determined from the continuity equation:

$$S_L = V_{\text{tube}} \frac{A_{\text{tube}}}{A_f} \tag{6.63}$$

Since the conical surface area A_f is greater than A_{tube}, S_L must be less than V_{tube}.

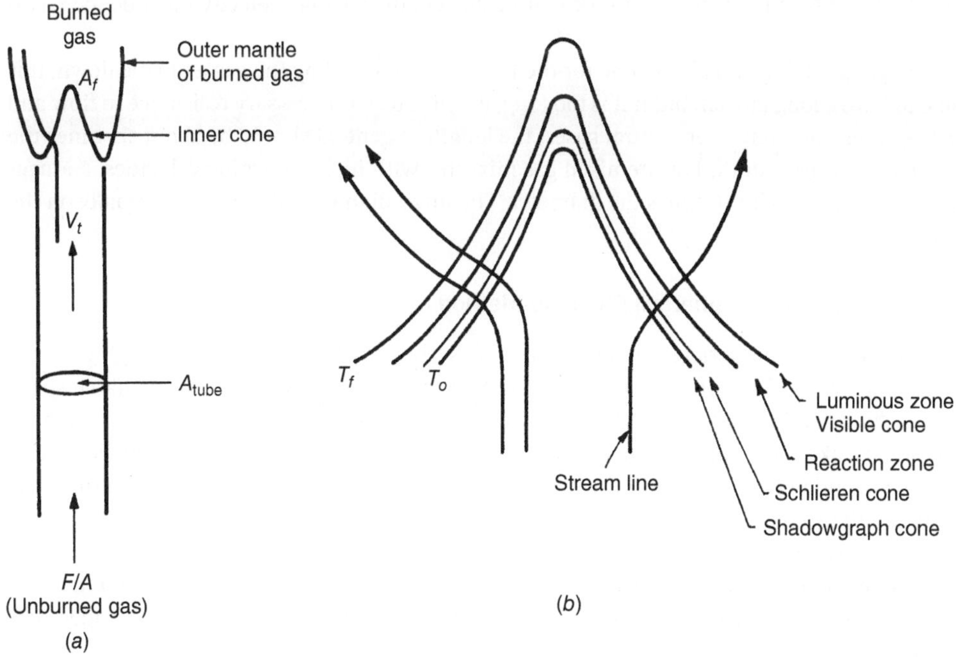

Figure 6.4 (*a*) Flow configuration near the mouth of a Bunsen burner. (*b*) Relative locations of different conical surfaces.

Although Eq. (6.63) is a very simple equation, there are some difficulties in determining S_L. This is due to the uncertainty of the surface area A_f. Depending on the optical method used, the conical area varies in the manner shown in Figure 6.4*b*. Many experimentalists have used the shadow cone because it is simpler than the schlieren techniques. The shadow cone, being cooler, gives a more correct result than the visible cone. The conical surface area A_f can also be determined with the use of OH chemiluminescence imaging (Wu et al. 2016).

The major advantage of the Bunsen burner method for premixed laminar flame study is that the equipment is simple, flexible, and easily adapted for measurements at varying initial temperature and pressure conditions. However, there are several disadvantages, including the following:

1. Since the flame is conical in shape and not one-dimensional (1-D), the flame stretch effect (see Section 6.7) has to be included in the flame speed determination.
2. The diffusional interchange of chemical species with those in the surrounding atmosphere, through the circumferential edge near the base of the conical flame, alters the fuel–oxidant ratio, so that the flame velocity observed may not represent the measured fuel–air ratio.
3. It is difficult to completely eliminate wall-quenching effects on flame speed.
4. The flame cone can act as a lens in a shadow measurement; this causes uncertainty in the proper cone size determination.
5. There may be inadequate entrainment of air and a resultant tendency toward flashback with large diameter burners.
6. A steady supply of gas is required, which may be difficult for pure gases.

There are geometric limits to the cylindrical diameter of Bunsen burners. The lower limit of tube diameter is set by the quenching distance (a term to be discussed later) of the gas–air mixture. The upper limit is set by the increasing tendency toward flashback with large burners. To maintain the flame at the port, there is a maximum gas velocity for a given supply pressure. To obtain sufficient velocity at all points in the tube and to prevent flashback, the average velocity must be increased for larger burners.

To obtain good mixing and allow time for the turbulence created by the gas jet to die down, it is desirable to have a long burner, but if it is too long, it will cause unnecessary resistance to flow and reduce the air entrainment. For natural gases, the length in general should be about six times the inner diameter of the mouth. For premixed gas mixtures with higher burning velocities, a longer burner is required, usually 10 times the diameter. The inner diameter of the tube is usually on the order of 1 cm.

6.4.2 Constant-Volume Spherical Chamber Method

The advantage of the spherical chamber method is the capability to perform high-pressure measurements; meanwhile, one of its main drawbacks is the difficulty in measuring flame speed at relatively high combustion temperatures. It should be noted that a laminar flame does not have to be stationary like the Bunsen burner flame as shown in Figure 6.4; it can be a traveling flame in a spherical windowed chamber. In the closed constant-volume spherical chamber method, the combustible mixture is ignited at the center of a rigid spherical vessel, which is usually on the order of 30 cm in diameter.

As the flame progresses, the expansion of the burned gas causes both the pressure and the temperature of the unburned gas to increase because of the adiabatic compression. The temperature increase causes the flame velocity to increase continuously from the center toward the wall. In this method, simultaneous records of the size of the spherical zone of the burned gas and of the pressure in the vessel must be made to determine S_L. It can be shown from the consideration of the time variation of the instantaneous mass of the unburned gas that laminar flame speed of the spherical deflagration wave is related to dp/dt and dr/dt according to

$$S_L = \left(1 - \frac{R^3 - r^3}{3p\gamma_u r^2}\frac{dp}{dr}\right)\frac{dr}{dt} = \frac{dr}{dt} - \frac{R^3 - r^3}{3p\gamma_u r^2}\frac{dp}{dt} \tag{6.64}$$

where p and r are the instantaneous chamber pressure and spherical flame radius, respectively, R is the inner radius of the spherical vessel, and γ_u is the specific heat ratio of the unburned gas. In Eq. (6.64), S_L is obtained as a difference of two time derivatives, which are of comparable magnitude; therefore, errors in these derivatives can be magnified in the S_L calculation. As noted in Section 6.7, a correction for the flame curvature is also required.

An alternative method of evaluating the burning velocity is to determine the rate of change of the mass fraction Y of burned gas. That is,

$$S_L = \frac{1}{3}\frac{R^3}{r^2}\left(\frac{p_i}{p}\right)^{1/\gamma_u}\frac{dY}{dt} \tag{6.65}$$

where p_i is the initial pressure of the unreacted gas mixture. For small values of Y, the value of Y can be calculated from

$$Y \equiv \frac{p - p_i}{p_e - p_i} \tag{6.66}$$

where p_e is the pressure corresponding to equilibrium chamber pressure based on a constant-volume combustion condition, which can be computed using thermodynamic considerations. The flame speed deduced according to the constant-volume spherical bomb method assumes complete equilibrium behind the flame front and no heat losses. Any delay in attaining equilibrium in a rather large volume immediately behind the flame front can cause error; the value of S_L calculated using the above expression is usually less than the actual value.

6.4.3 Flat-Flame Burner Method

The flat-flame burner or heat flux method is usually attributed to Powling (1949). It is considered the most accurate method, because it offers the simplest flame front and one in which the areas of the shadow, schlieren, and visible fronts are all the same (Bosschaart and de Goey 2003). With this method, the flame is quasi-adiabatic and flat, thus a direct determination of the laminar flame speed is possible without correcting for stretch. It is achieved by placing a porous metal disk or a series of small tubes of 1-mm diameter or less at the exit of a larger flow tube. Botha and Spalding (1954) extended the flat-flame burner method to higher flame speeds by cooling the porous disk. The cooling brings the flame front closer to the porous disk and stabilizes it. In order to obtain the adiabatic laminar flame speed, several tests at different cooling rates are made. The values of S_L are plotted against the cooling rates. The curve is then extrapolated to zero cooling rate to obtain adiabatic flame speed S_L.

A modern flat-flame burner (Nilsson et al. 2013) is shown in Figure 6.5. The device typically consists of a sintered, porous bronze (or 316 stainless steel) water-cooled burner surrounded by a sintered, porous shroud ring for the introduction of a shielding gas (usually nitrogen), both enclosed in a housing with connections for water cooling, combustion gas, and shielding gas.

The gaseous mixture is ignited at a high flow rate and adjusted until the flame is flat. By controlling the rate of efflux of burned gases with a grid above the flame, a stable flame can be obtained. This method is applicable only to mixtures having low burning velocities, on the order of 40–70 cm/s or less. At higher S_L, the flame front positions itself far from the burner and forms conical shapes.

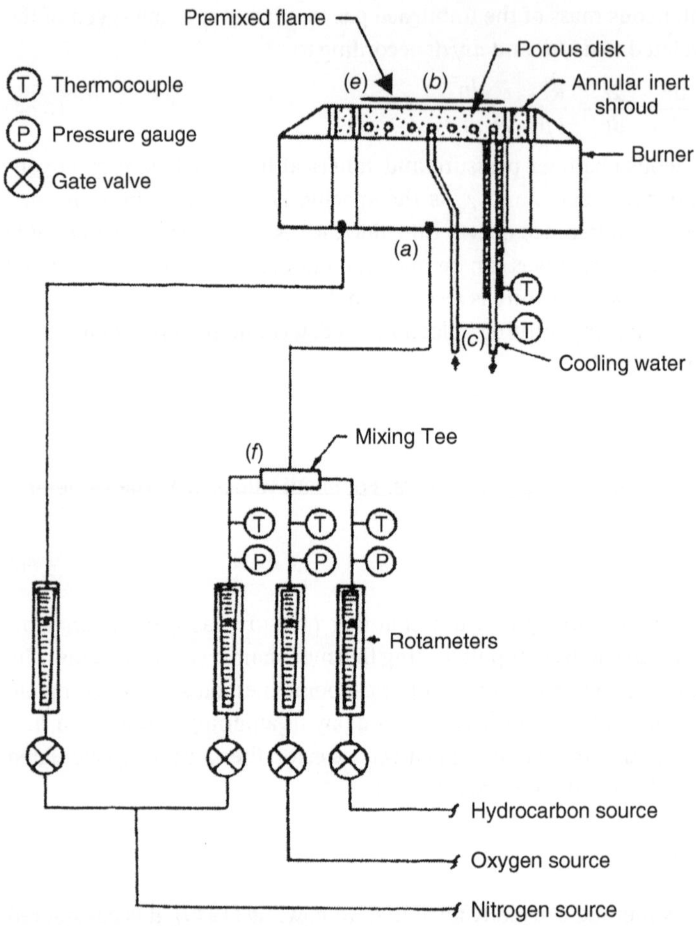

Figure 6.5 A flat-flame burner setup with (*a*) reactant gas supply line, (*b*) a flat premixed flame, and (*c*) a cooling water passage through (*e*) a sintered metal disk. The gas mixture is controlled by (*f*) a mixing tee valve.

6.4.4 Soap-Bubble Method

Flame velocity determinations using the bubble method were first made by F. Stevens in 1923. He devised the procedure of using a soap bubble as a constant pressure volume for a combusting mixture. In the soap-bubble method, a homogeneous combustible mixture is used to blow a soap bubble around a pair of spark electrodes. At time zero, the gas mixture contained in the spherical soap bubble is ignited by the spark (see Figure 6.6). An approximate way to calculate the flame velocity is to use

$$S_L = \frac{V_{\text{flame}} r_i^3}{r_f^3} \tag{6.67}$$

where V_{flame} is the average spatial velocity of the spherical flame front, r_i is the initial radius of the soap bubble, and r_f is the final radius of the sphere of burned gas. In general, we can assume that (1) the spherical flame spreads uniformly in radial direction in gaseous mixture;

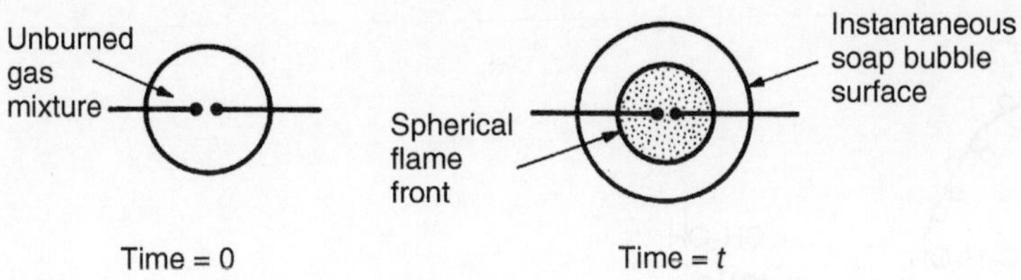

Figure 6.6 Experimental setup of the soap-bubble method.

(2) the pressure remains constant; and (3) the growth of the flame front can be followed by some photographic means.

The instantaneous flame speed can be deduced by equating the mass flux in front of the flame front to that behind the flame front, i.e.,

$$S_L A \rho_u = u_r A \rho_b \tag{6.68}$$

or

$$S_L = u_r \frac{\rho_b}{\rho_u} = u_r \frac{T_u}{T_b} \tag{6.69}$$

where u_r represents the recorded flame velocity.

One obvious disadvantage of this method is the large uncertainty in the temperature ratio T_u/T_b that is needed for obtaining the density ratio ρ_b/ρ_u. Although one can assume that the gases reach the theoretical flame temperature, comparison of measured and calculated expansion ratios often reveals serious discrepancies. Also, for use in Eq. (6.67), the initial and final sizes of the bubble must be known very accurately because they are raised to the third power in the calculation. In fact, however, the final size is difficult to measure. There are several other problems in using the soap-bubble method, namely the following:

1. The method is not suitable for studying the flame spread of dry mixtures, since they can be moistened by evaporation from the soap Solution.
2. There is inevitably some heat loss to the electrodes.
3. For slow flames, the flame front may not remain spherical; also, the thickness of the reaction zone becomes large.
4. For very fast flames, the flame front is no longer smooth due to the formation of a cellular flame structure.

6.5 Laminar Flame Speeds of Hydrocarbon/Air Mixtures

Some typical results for laminar flame speeds in hydrocarbon/air mixtures as a function of fuel volume percentage in air are shown in Figure 6.7 based on the data of Warnatz (1992). For hydrocarbon fuels, the peak of the flame speed occurs near stoichiometric or slightly fuel-rich mixtures.

Clingman et al. (1953) measured the flame propagation speed of methane CH_4 in various oxygen/inert-gas mixtures with a mantled Bunsen burner to elucidate the dominant effects of the thermal diffusivity α and the reaction rate $\dot{\omega}$ on the laminar flame speed S_L. The volumetric ratio of oxygen to inert gas was set at 0.21:0.79 for all cases studied. The inert gases chosen were nitrogen (N_2), helium (He), and argon (Ar). The flame speed in He mixtures was higher than that in Ar mixtures. This is due to the fact that the thermal diffusivity α of helium is much larger than

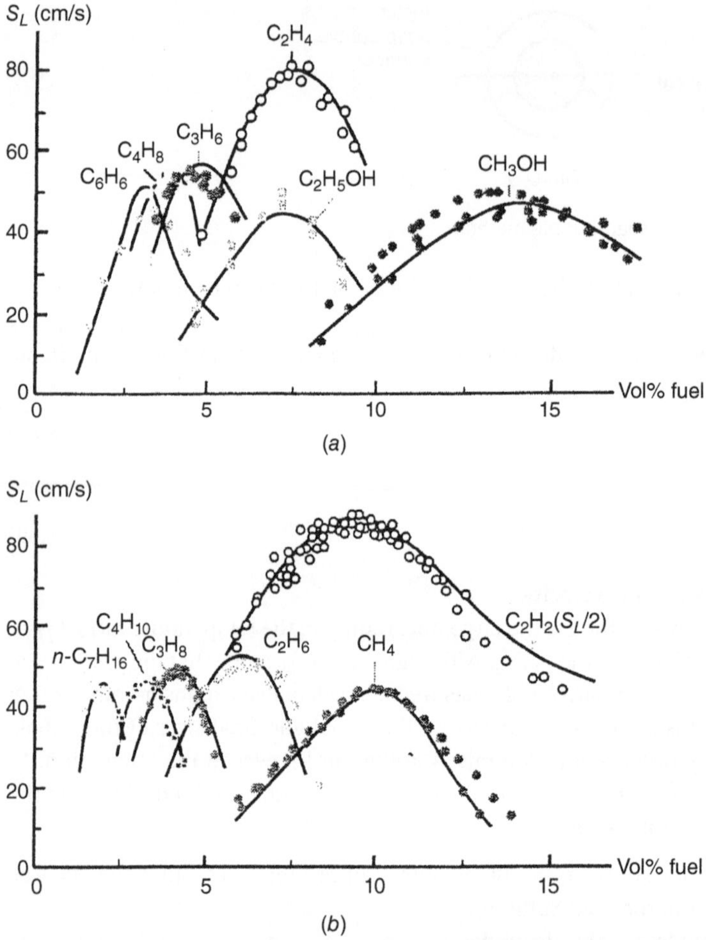

Figure 6.7 Dependence of laminar-flame speed on mixture composition for different fuel–air mixtures. (a) Hydrocarbon alkenes and alcohols. (b) Hydrocarbon paraffins and acetylene. *Source*: Adapted from Warnatz 1992/with permission of Elsevier.

that of argon since the molecular mass of helium is much smaller than that of argon. Recall the relationship given by Eq. (6.62), restated here as Eq. (6.70),

$$S_L \sim [\alpha \dot{\omega}]^{1/2} \tag{6.70}$$

Also, the flame speed in the argon mixtures was higher than that in the nitrogen mixtures (regular air). This is due to the fact that while Ar and N_2 have nearly the same values of thermal diffusivity, Ar is a monatomic gas, which has a lower specific heat ($\frac{5}{2}R$) than the diatomic gas N_2 ($\frac{9}{2}R$). Since the heat release is the same, the flame temperature, and thus the flame speed, will be higher in the argon mixture than that in the nitrogen mixture.

Metghalchi and Keck (1982) developed the following frequently referenced correlation for the laminar flame speed, Equation (6.71), as a function of mixture equivalence ratio ϕ, unburned gas temperature $T_u > 350$ K, and pressure p. They used a spherical combustion chamber, and the fuels used were methanol, propane, iso-octane, and indolene. The measurement emphasis was on the high-temperature and pressure region important for applications to internal combustion engines and gas turbine combustors. The effect of adding residual or recycled exhaust gases (15% carbon dioxide and 85% nitrogen by volume) to stoichiometric iso-octane–air mixtures was also studied for diluent mass fractions f ranging from 0 to 0.20.

$$S_L = S_{L,o}\left(\frac{T_u}{T_{u,0}}\right)^a \left(\frac{P}{P_o}\right)^b (1 - 2.1f) \tag{6.71}$$

where $T_{uo} = 298$ K and $P_o = 1$ atm, and f is the diluent mass fraction.

The reference flame speed $S_{L,o}$ (m/s) is given by Eq. (6.72):

$$S_{L,o} = c_1 + c_2(\phi - \phi_o)^2 \tag{6.72}$$

where c_1, c_2, and ϕ_o are given in Table 6.1.

The parameters a and b depend on the equivalence ratio and for the fuels tested by Metghalchi and Keck are given by

$$a = 2.18 - 0.8(\phi - 1) \tag{6.73}$$

$$b = -0.16 + 0.22(\phi - 1) \tag{6.74}$$

Using Eq. (6.1), the dependence of the laminar flame speed on the mixture equivalence ratio, unburned gas temperature, and pressure is illustrated in Figures 6.8 and 6.9. The laminar flame speed shows a maximum for slightly rich mixtures, is a strong function of unburned gas temperature T_u, and is a weak function of pressure P. The strong dependence of the laminar flame speed on unburned gas temperature is due to the exponential relation between the reaction kinetics and

Table 6.1 Curve-Fit Parameters for Laminar Flame Speed Correlation.

Fuel	c_1	c_2	ϕ_o
Methanol	0.3692	−1.4051	1.11
Propane	0.3422	−1.3865	1.08
Iso-octane	0.2632	−0.8472	1.13

Source: Metghalchi and Keck 1982/with permission of Elsevier.

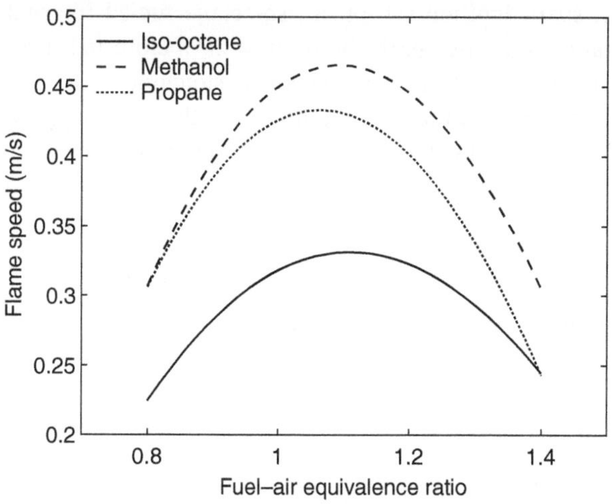

Figure 6.8 Dependence of laminar flame speed S_L on equivalence ratio ϕ ($T_u = 350$ K, $p = 1$ atm, $f = 0.05$).

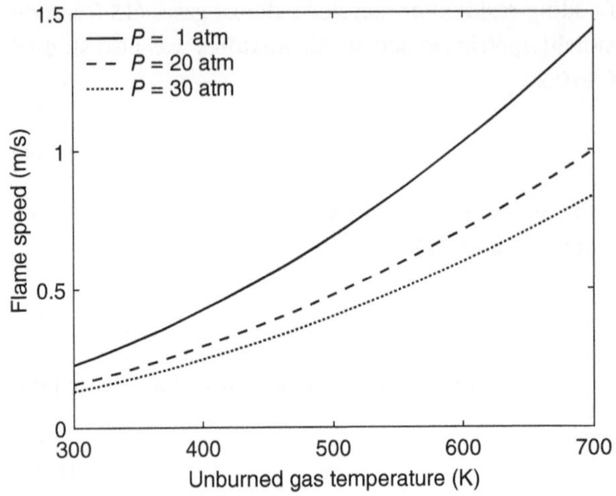

Figure 6.9 Dependence of laminar flame speed S_L on unburned gas temperature T_u (iso-octane, $\phi = 1.0$, $f = 0.05$).

temperature. Also, inspection of Eq. (6.1) indicates that the laminar flame speed decreases linearly with increasing diluent mass fraction.

Based on the theories discussed earlier in this chapter, we know that $S_L \sim p^{(m-2)/2}$, where m is the order of chemical reaction. The pressure dependence observed by Metghalchi and Keck implies that the overall order of the reaction is less than 2 for flames with $S_L < 50\,\text{cm/s}$, equal to 2 for those with $50 < S_L < 100\,\text{cm/s}$, and greater than 2 for those with $S_L > 100\,\text{cm/s}$. The relationship $S_L \sim p^{(m-2)/2}$ is supported by much experimental evidence. For example, the first-order ($m = 1$) decomposition reaction for ethylene oxide monopropellant (C_2H_4O) is shown to follow $S_L \sim \sqrt{p^{-1}}$. For many hydrocarbon fuels undergoing second-order reactions, the laminar-flame speeds have been found to be independent of pressure.

Example 6.2 ***Iso-octane Laminar Flame Speed***

Compute the laminar flame speed $S_{L,s}$ at spark ignition (s) for an iso-octane fueled internal combustion engine with a compression ratio $r = 8$ and spark timing $\theta_s = 25°$ before top dead center. The residual fraction is given by $f = 0.10\,(8/r)$, the combustion is stoichiometric with $\phi = 1$, and $\gamma = 1.3$. The pressure p and unburned mixture temperature T_u at the time of spark ignition relative to bottom dead center (bdc) are given by the ideal-gas $p - V$ equations,

$$T_{u,s} = 350(V_{\text{bdc}}/V_s)^{(\gamma-1)/\gamma}$$

$$p_s = 0.5(V_{\text{bdc}}/V_s)^{\gamma}$$

The dimensionless cylinder volume is given by the following relationship,

$$\frac{V}{V_{\text{bdc}}} = \frac{1}{r} + \left(\frac{r-1}{2r}\right)(1 - \cos\theta)$$

Solution

The reference flame speed $S_{L,o}$ from Eq. (6.72) is

$$S_{L,o} = c_1 + c_2(\phi - \phi_0)^2 = 0.2632 - 0.8472(1 - 1.13)^2 = 0.249\,\text{m/s}$$

The dimensionless volume V_s/V_{bdc} at spark is

$$\frac{V_s}{V_{\text{bdc}}} = \frac{1}{r} + \left(\frac{r-1}{2r}\right)(1 - \cos\theta_s)$$
$$= \frac{1}{8} + \left(\frac{8-1}{2\cdot 8}\right)(1 - \cos(-25)) = 0.166$$

The pressure P_s and temperature T_s at spark are

$$P_s = 0.5\left(\frac{V_{\text{bdc}}}{V_s}\right)^{\gamma} = 0.5(1/0.166)^{1.3} = 5.16\,\text{atm}$$

$$T_{u,s} = 350\left(\frac{V_{\text{bdc}}}{V_s}\right)^{\frac{\gamma-1}{\gamma}} = 350(1/0.166)^{0.3/1.3} = 529.7\,\text{K}$$

From Eqs. (6.73) and (6.74), for $\phi = 1$, $a = 2.18$ and $b = -0.16$. Therefore, the laminar flame speed $S_{L,s}$ at spark ignition is

$$S_{L,s} = S_{L,o}\left(\frac{T_{u,s}}{298}\right)^{a}(P_s)^{b}\ (1 - 2.1f)$$
$$= 0.249\left(\frac{529.7}{298}\right)^{2.18}(5.16)^{-0.16}\ (1 - 2.1\cdot 0.1)$$
$$= 0.53\,\text{m/s}$$

6.6 Numerical Analysis of Laminar-Flame Problems

With the advent of high-speed digital computers, the Solution of laminar-flame problems has changed significantly from the methods discussed in the previous sections. Modern numerical techniques now permit not only the calculation of flame speed but also a determination of the temperature profile and species composition across the flame. These numerical techniques use both the detailed and reduced kinetic mechanisms outlined previously in Chapter 3.

A very useful open-source package is Cantera, which is a suite of software tools for problems involving chemical kinetics, thermodynamics, and transport processes. Since Cantera is open source, users can incorporate detailed chemical kinetics and transport models into their calculations. Cantera can be used from Python and Matlab, or in applications written in C/C++ and Fortran 90. The most popular commercial general-purpose combustion software currently is Chemkin-Pro. Many researchers in the laminar flame area use combustion mechanisms in Chemkin format, which has become a de facto standard in the combustion community. The Chemkin software was originally developed by Kee et al. (1980) at the Sandia National Laboratories and has since become a standard tool in combustion engineering as well as in other fields such as fluid dynamics and chemistry.

6.6.1 Example: Premixed O_3/O_2 Laminar Flames

We shall use Heimerl and Coffee (1980) as a classic example of numerical analysis methods. As shown in Figure 6.10, a one-dimensional, premixed, laminar, steady-state ozone/oxygen flame was considered in their theoretical model. One reason for choosing this mixture was its simplicity; it involves only three species: O, O_2, and O_3. The chemical reactions between these species are

$$O_3 + M \underset{k_{b_1}}{\overset{k_1}{\rightleftharpoons}} O + O_2 + M \tag{6.75}$$

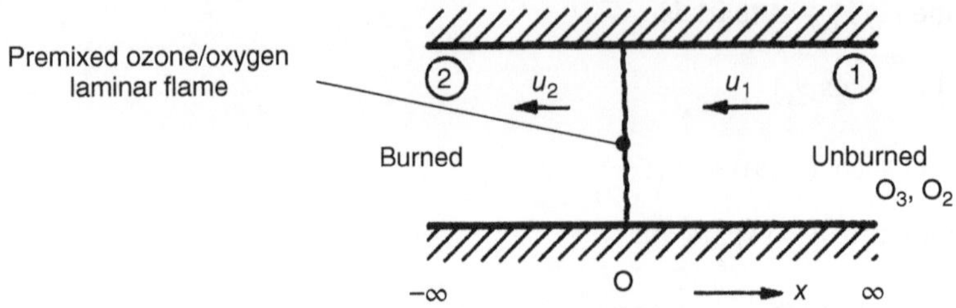

Figure 6.10 Premixed ozone/oxygen laminar flame.

Table 6.2 Kinetic Coefficients.

Coefficient	Expression	Units	Remarks
k_1	$4.31\times10^{14}\exp\left(-\frac{11{,}161}{T}\right)$	cm^3/mole-s	$M = O_3$
k_{b1}	$1.2\times10^{13}\exp\left(\frac{+976}{T}\right)$	cm^6/mole^2-s	
k_2	$1.14\times10^{13}\exp\left(-\frac{2300}{T}\right)$	cm^3/mole-s	
k_{b2}	$1.19\times10^{13}\exp\left(-\frac{50{,}600}{T}\right)$	cm^3/mole-s	
k_3	$1.38\times10^{18}T^{-1}\exp\left(-\frac{171}{T}\right)$	cm^6/mole^2-s	$M = O_2$
k_{b3}	$2.75\times10^{19}T^{-1}\exp\left(-\frac{59{,}732}{T}\right)$	cm^3/mole-s	$M = O_2$

Source: Adapted from Heimerl and Coffee (1980)/with permission of Elsevier.

$$\mathrm{O} + \mathrm{O}_3 \underset{k_{b_2}}{\overset{k_2}{\rightleftharpoons}} 2\mathrm{O}_2 \tag{6.76}$$

$$2\mathrm{O} + M \underset{k_{b3}}{\overset{k_3}{\rightleftharpoons}} \mathrm{O}_2 + M \tag{6.77}$$

where M represents the third body, which could be either O, O_2, or O_3. The kinetic coefficients for these reactions are given as a function of temperature in Table 6.2.

The rate coefficient may vary with the identity of the chaperon third body M. The chaperon relationships used are

$$k_1(M = \mathrm{O}) = k_1(M = \mathrm{O}_2) = 0.44k_1(M = \mathrm{O}_3) \tag{6.78}$$

$$k_3(M = \mathrm{O}) = 3.6k_3(M = \mathrm{O}_2) = 3.6k_3(M = \mathrm{O}_3) \tag{6.79}$$

The mass rate of production of species i for Eqs. (6.75)–(6.77) can be given as

$$\dot{\omega}_i = \mathrm{M}_i \sum_{k=1}^{6} (\nu'_{i,k} - \nu_{i,k}) \underbrace{B_k T^{\alpha_k} e^{-E_a/R_u T}}_{k_k} \prod_{j=1}^{3} \left(\frac{X_j}{R_u} \frac{p}{T} \right)^{\nu_{j,k}} \tag{6.80}$$

which considers all six chemical reactions [Eqs. (6.75)–(6.77)] and all three chemical species.

In addition to the kinetic coefficients, various thermodynamic and transport properties are needed. The sensible enthalpy and constant-pressure specific heat are related to temperature by the following equations:

$$h = R_u\left(a_o + \sum_{n=1}^{5} a_n \frac{T^n}{n}\right) \tag{6.81}$$

$$c_p = R_u \sum_{n=1}^{5} a_n T^{n-1} \tag{6.82}$$

The governing equations used in the model are given in one-dimensional time-dependent forms as follows:

Overall continuity equation

$$\frac{\partial \rho}{\partial t} + \frac{\partial(\rho u)}{\partial x} = 0 \tag{6.83}$$

Species mass continuity equation

$$\rho\frac{\partial Y_k}{\partial t} + \rho u \frac{\partial Y_k}{\partial x} = -\frac{\partial}{\partial x}(\rho Y_k V_k) + \dot{\omega}_k \tag{6.84}$$

where $k = 1, 2,$ or 3 for O, O_2, or O_3, respectively.

Energy equation

$$\rho c_p \frac{\partial T}{\partial t} + \rho u c_p \frac{\partial T}{\partial x} = \frac{\partial}{\partial x}\left(k\frac{\partial T}{\partial x}\right) - \sum_{k=1}^{N} \dot{\omega}_k h_{f,k}^o - \rho \sum_{k=1}^{N} c_{p,k} Y_k V_k \frac{\partial T}{\partial x} \tag{6.85}$$

This energy equation was arrived at under the assumption that the pressure is constant in the reaction zone, the viscous dissipation is negligible, and there is no body-force work.

Diffusion equation:

$$\frac{\partial X_k}{\partial x} = \sum_{j=1}^{N} \frac{X_k X_j}{\mathcal{D}_{kj}}(V_j - V_k) \tag{6.86}$$

The predicted flame velocities are in reasonable agreement with the measurements of Streng and Grosse (1957) as shown in Figure 6.11 and Table 6.3. The calculated results of Heimerl and

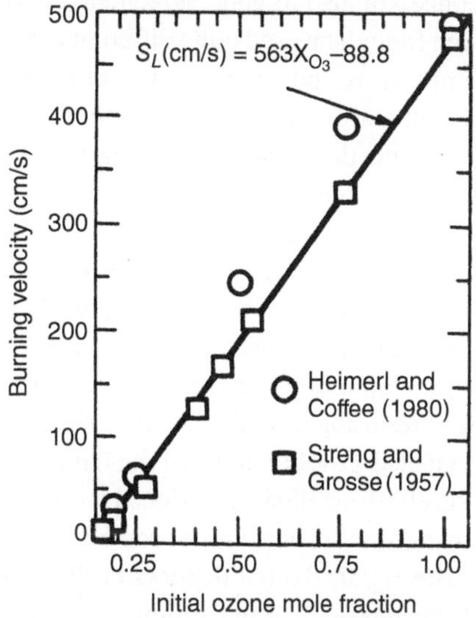

Figure 6.11 Comparison of theoretical results of Heimerl and Coffee (1980) with experimental data of Streng and Grosse (1957).

Table 6.3 Comparison of Flame Speeds.

Initial Ozone Mole Fraction	Heimerl (cm/s)	Flame Speed Streng (cm/s)	Warnatz (cm/s)
1.00	497	474	445
0.75	396	333	350
0.50	248	193	225
0.25	64	52	65
0.20	33	24	37

Source: Heimerl and Coffee (1980), Streng and Grosse (1957), and Warnatz (1977).

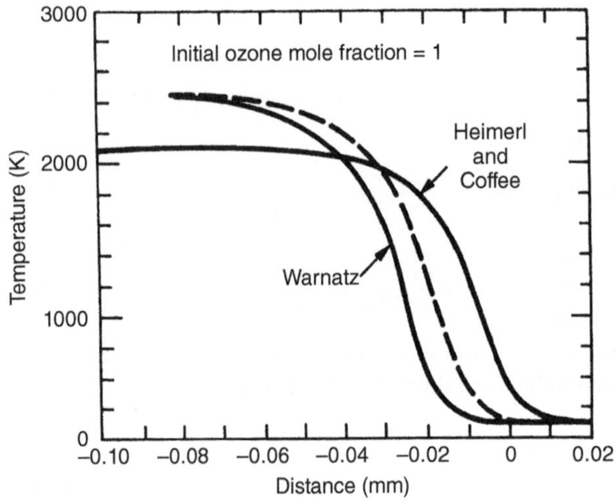

Figure 6.12 Comparison of calculated temperature profiles by Warnatz (1977) and Heimerl and Coffee (1980).

Coffee are also in good agreement (within ±12%) with the computed results of Warnatz (1977) even though he used quite different expressions for some of the kinetic coefficients. The calculated temperature and atomic oxygen profiles are quite different from those of Warnatz. The comparison of temperature profiles across the reaction zone is shown in Figure 6.12, where the dashed profile represents the substitution of Warnatz's expressions for the kinetic coefficients k_{f1} and k_{f2} into the Heimerl–Coffee model.

The computed profiles of the mole fractions in a stoichiometric H_2/O_2 flame at $p = 1$ atm and $T_u = 298$ K are shown in Figure 6.13 (Warnatz 1981). The numerical simulations used a reduced chemical mechanism consisting of 19 elementary reactions. Due to the large diffusion coefficient and thermal conductivity of hydrogen, the mole fraction of H_2 is broadened, leading to low values at the flame front, and a maximum of the O_2 profile at the same location. Note the narrow HO_2 peak resulting from the fact that HO_2 cannot coexist with H, O, and OH.

Lu et al. (2001) compared the flame structure and speed for H_2/air and CH_4/air premixed flames using both detailed and reduced mechanisms as shown in Figures 4.15 and 4.16. The detailed H_2/air mechanism involved nine species: H_2, O_2, N_2, H_2O, O, OH, H, HO_2, and H_2O_2.

The H_2/air flame thickness in Figure 4.15 is about 0.5 mm, with a flame temperature rising to above 2000 K. The maximum H_2 flame speed at 1 atm was about 350 cm/s at a rich equivalence

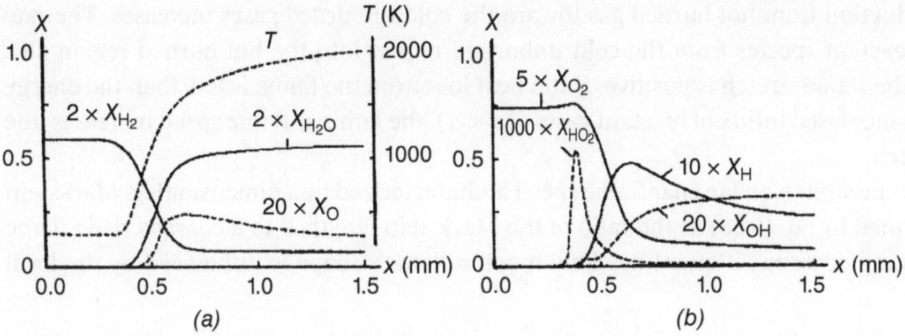

Figure 6.13 Computed mole fractions in a stoichiometric H_2/O_2 flame at $p = 1$ atm and $T_u = 298$ K. (a) H_2, H_2O, and O mole fractions. (b) O_2, H, OH, and HO_2 mole fractions. *Source*: Adapted from Warnatz 1981/with permission of Elsevier.

ratio of 1.6. For CH_4/air flames, GRI MECH 1.2 was used as the detailed mechanism. The CH_4/air flame thickness in Figure 4.16 is also about 0.5 mm, with a flame temperature of about 2000 K. The maximum CH_4 flame speed at 1 atm was about 40 cm/s at an equivalence ratio of 1.05.

6.7 Stretched Laminar Premixed Flames

6.7.1 Flame Stretch Factor, Karlovitz, and Markstein Numbers

It has been found that flame stretching has significant effects on laminar and turbulent flames. The effect of stretch on the flame is to reduce the thickness of the flame front, and hence the flame speed and the flame structure through its coupled effect with mass and heat diffusion. In addition, flame stretch can result in opposite chemical kinetic effects as the fuel–air mixture changes from lean to rich, leading to misinterpretation of experimental data. Accordingly, flame stretch considerations have been applied to (1) flame stability analysis, (2) laminar-flame speed determination, (3) flammability limits, and (4) modeling of turbulent flames.

Karlovitz et al. (1953) and Markstein (1964) initiated the study of stretched premixed flames and demonstrated the importance of flame stretch and preferential diffusion in terms of flame-front hydrodynamic instabilities. Chung and Law (1988), Candel and Poinsot (1990), and Buckmaster (1993) are among those who have made important contributions to our understanding of the mathematical relationships and physical processes associated with flame stretch behavior.

Three related nondimensional parameters are used to characterize the stretch effect: the flame stretch factor κ, the Markstein number Mn, and the Karlovitz number Ka. A flame surface propagating in a nonuniform or turbulent flow field undergoes strain and curvature effects leading to changes in the frontal area. The flame stretch factor κ is defined as "the relative rate of change of flame surface area," which is also the reciprocal of a characteristic time for flame stretching,

$$\kappa \equiv \frac{1}{\delta A}\frac{d(\delta A)}{dt} = \frac{1}{A}\frac{dA}{dt} \tag{6.87}$$

Positive flame stretch implies that the flame surface area increases ($\kappa > 0$); negative flame stretch means that the flame surface area decreases ($\kappa < 0$). Two major effects, heat loss and diffusion of reactants, influence the flame speed due to flame stretching. For example, let us consider the case of a rich mixture with positive flame stretching. When the surface area of the flame increases, the rate of heat conduction from hot burned gas toward the cold unburned gases increases. The rate of diffusion of reactant species from the cold unburned region into the hot burned region also increases when the flame stretch is positive. If the heat loss from the flame is less than the energy generated due to increased influx of reactant gases ($Le < 1$), the laminar flame speed increases due to the flame stretch.

The effect of flame stretch on laminar flame speed is characterized by a dimensionless Markstein number Mn, defined in Eq. (6.88) as the ratio of the Markstein length $\mathcal{L}$ to a characteristic flame thickness δ. In simpler terms, it quantifies how much a flame's shape is influenced by the local flow conditions:

$$\text{Mn} \equiv \frac{\mathcal{L}}{\delta} \tag{6.88}$$

where $\mathcal{L}$ is the Markstein length. The Markstein length is related to the characteristic flame thickness, δ, because both are representative of the scale of distances over which the diffusion of mass and heat occurs in the flame (Kwon and Faeth 1992).

The Markstein number is named after George H. Markstein (1911–2011), an American engineering professor, who showed that thermal diffusion stabilized the curved flame front and proposed a relation between the critical wavelength for hydrodynamic stability of the flame front, called the Markstein length $\mathcal{L}$, and the thermal thickness of the flame.

Note that Markstein lengths can be positive or negative and are generally positive for most fuels. The larger the Markstein length $\mathcal{L}$ and the Markstein number, the greater the effect of curvature on the flame speed and stability. Negative Markstein lengths indicate flame speed increases with positive stretch and can result in an instability leading to cell formation in the wave front.

In terms of the effect of flame stretch factor κ and Markstein length $\mathcal{L}$ on laminar-flame speed, Markstein proposed the relationship given in Eq. (6.89),

$$S_d = S_L^o - \mathcal{L}\kappa \tag{6.89}$$

where S_d is the displacement speed of the stretched flame with respect to the unburnt reactants, and S_L^o is the laminar flame speed for the un-stretched flame. The displacement speed is a kinematic parameter and depends on which surface is chosen to represent the flame, and the laminar flame speed is a thermodynamic parameter. For moderate values of strain and curvature, S_d measured close to the reactant side of the flame remains close to the laminar flame speed.

As shown in Figure 6.14 and in Eq. (6.90), the flame surface velocity $\mathbf{w}$ can be decomposed into two velocity vector components: one in the normal direction $\mathbf{n}$ with a magnitude of S_d, and the other vector the local fluid velocity $\mathbf{v}$ at the flame front,

$$\mathbf{w} = \mathbf{v} + S_d\,\mathbf{n} \tag{6.90}$$

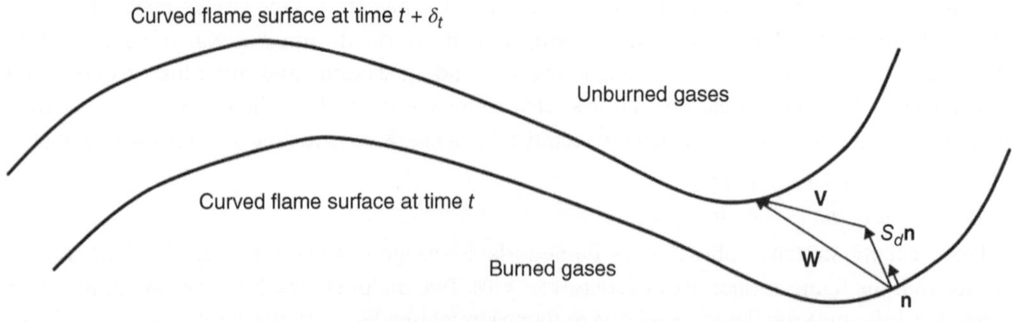

Figure 6.14 Velocity vectors associated with a propagating curved laminar flame front.

Accordingly, the displacement speed S_d can be expressed as

$$S_d = (\mathbf{w} - \mathbf{v}) \cdot \mathbf{n} \tag{6.91}$$

The Karlovitz number Ka is a nondimensional stretch rate and is defined as the dimensionless ratio of a characteristic flame residence time δ/S_L to a characteristic flame stretch time κ^{-1},

$$\mathrm{Ka} \equiv \kappa \frac{\delta}{S_L} \tag{6.92}$$

where δ is a characteristic flame thickness. The Karlovitz number increases with increasing stretch rate κ. The Karlovitz number is named after Bela Karlovitz (1994–2004), a Hungarian/American engineer who made a number of important contributions to combustion engineering. Depending on the combustion situation, Karlovitz numbers based individually on strain or curvature can also be defined.

Substituting Eqs. (6.88) and (6.92) into (6.89) gives the dimensionless relationship,

$$\frac{S_L^o}{S_d} = 1 + \mathrm{Mn}\,\mathrm{Ka} \tag{6.93}$$

Eq. (6.93) indicates the relationship between the unstretched flame speed S_L^o and the Karlovitz and Markstein numbers.

We now derive the basic relationships between the flame stretch rate and the strain rate, dilatation of the fluid element, and the curvature of the flame surface. The three perpendicular coordinates on a curved flame surface are shown in Figure 6.15, which has two unit vectors (ν and η) tangent to the flame surface and an outward normal unit vector $\mathbf{n}$ at the location $\mathbf{r}(\nu, \eta, \mathrm{n})$ as a function of the three independent coordinates.

The elemental arc $(d\mathbf{s})_\nu$ in the $\boldsymbol{\nu}$ direction and the elemental arc $(d\mathbf{s})_\eta$ in the $\boldsymbol{\eta}$ direction are given by

$$(d\mathbf{s})_\nu = \left(\frac{\partial \mathbf{r}}{\partial \nu}\right) d\nu, \qquad (d\mathbf{s})_\eta = \left(\frac{\partial \mathbf{r}}{\partial \eta}\right) d\eta \tag{6.94}$$

The incremental flame surface area can be approximated by

$$dA(t) = \left(\frac{\partial \mathbf{r}}{\partial \nu} \times \frac{\partial \mathbf{r}}{\partial \eta}\right) \cdot \mathbf{n}(d\nu)(d\eta) \tag{6.95}$$

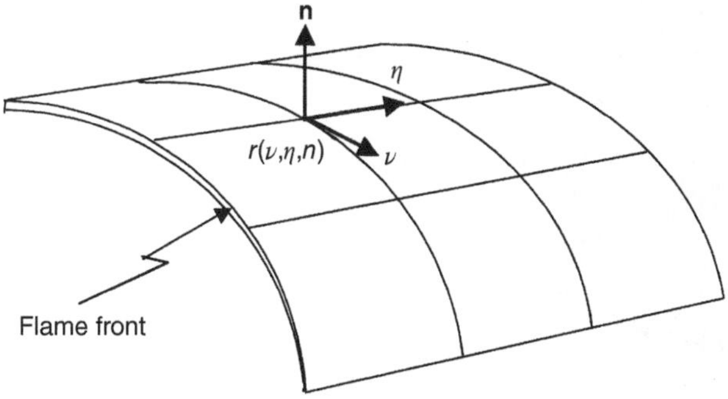

Figure 6.15 Three perpendicular curvilinear coordinates for a curved laminar-flame front. (Adapted from Candel and Poinsot 1990).

In the orthogonal curvilinear coordinate system, the two unit vectors ($\mathbf{e}_\nu$ in the ν direction and $\mathbf{e}_\eta$ in the η direction) can be given by

$$\mathbf{e}_\nu \equiv \frac{\partial \mathbf{r}/\partial \nu}{|\partial \mathbf{r}/\partial \nu|}, \qquad \mathbf{e}_\eta \equiv \frac{\partial \mathbf{r}/\partial \eta}{|\partial \mathbf{r}/\partial \eta|}, \qquad \text{and} \qquad \mathbf{e}_\nu \times \mathbf{e}_\eta = \mathbf{n} \tag{6.96}$$

Thus,

$$dA(t) = \left|\frac{\partial \mathbf{r}}{\partial \nu}\right| \left|\frac{\partial \mathbf{r}}{\partial \eta}\right| (d\nu)(d\eta) \tag{6.97}$$

Now consider the curved flame surface $A(t)$ in Figure 6.16 moving in space with a flame velocity $\mathbf{w}$. The rate of change of the flux of a given vector $\mathbf{H}$ across the flame surface is given by the following equation based on Reynolds' transport theorem.

$$\frac{d}{dt}\int_{A(t)} \mathbf{H}\cdot \mathbf{n}dA = \int_{A(t)} \left[\frac{\partial \mathbf{H}}{\partial t} + \mathbf{w}\cdot\nabla\mathbf{H} - \mathbf{H}\cdot\nabla\mathbf{w} + \mathbf{H}\nabla\cdot\mathbf{w}\right]\cdot \mathbf{n}dA \tag{6.98}$$

By specifying $\mathbf{H} = \mathbf{n}$, Eq. (6.98) yields

$$\frac{d}{dt}\int_{A(t)} dA = \int_{A(t)} \left[\frac{\partial \mathbf{n}}{\partial t} + \mathbf{w}\cdot\nabla\mathbf{n} - \mathbf{n}\cdot\nabla\mathbf{w} + \mathbf{n}\nabla\cdot\mathbf{w}\right]\cdot \mathbf{n}dA \tag{6.99}$$

Since $\mathbf{n}$ is the unit normal vector, we have

$$\frac{\partial \mathbf{n}}{\partial t}\cdot\mathbf{n} = \frac{1}{2}\frac{\partial n^2}{\partial t} = 0 \tag{6.100}$$

$$(\mathbf{w}\cdot\nabla\mathbf{n})\cdot\mathbf{n} = \mathbf{w}\cdot\nabla\left(\frac{n^2}{2}\right) = 0 \tag{6.101}$$

Therefore, the transport Eq. (6.99) becomes

$$\frac{d}{dt}\int_{A(t)} dA = \int_{A(t)} (-\mathbf{nn} : \nabla\mathbf{w} + \nabla\cdot\mathbf{w})dA \tag{6.102}$$

The first term on the right-hand side of Eq. (6.102) is the gradient operator normal to the flame surface, and in tensor or index notation can be written as

$$-\mathbf{nn} : \nabla\mathbf{w} = -n_i n_j \frac{\partial w_i}{\partial x_j} \tag{6.103}$$

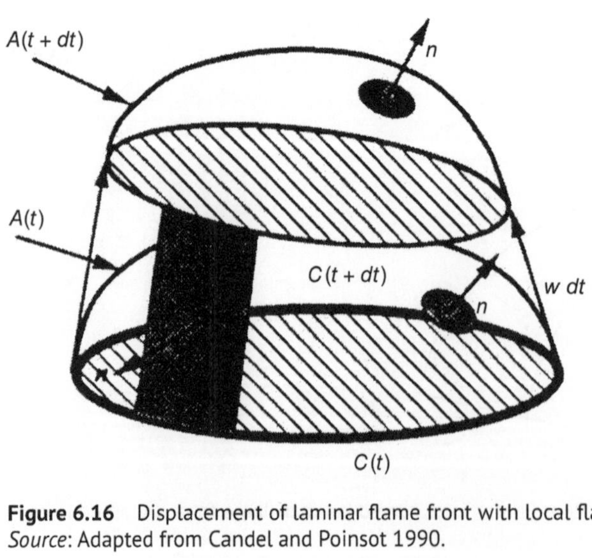

Figure 6.16 Displacement of laminar flame front with local flame velocity $\mathbf{w}$ from time t to $t + dt$. *Source*: Adapted from Candel and Poinsot 1990.

If we consider a differential surface element, then $A(t)$ can be replaced by δA. The flame stretch rate κ can then be expressed as

$$\kappa \equiv \frac{1}{\delta A}\frac{d(\delta A)}{dt} = -\mathbf{nn} : \nabla\mathbf{w} + \nabla\cdot\mathbf{w} \tag{6.104}$$

Several vector operations can be applied to Eq. (6.90) to give

$$\nabla\mathbf{w} = \nabla\mathbf{v} + S_d\,\nabla\mathbf{n} + \mathbf{n}\nabla S_d \tag{6.105}$$

$$\mathbf{nn} : \nabla\mathbf{w} = \mathbf{nn} : \nabla\mathbf{v} + \mathbf{n}\cdot\nabla S_d \tag{6.106}$$

$$\nabla\cdot\mathbf{w} = \nabla\cdot\mathbf{v} + S_d\,(\nabla\cdot\mathbf{n}) + \mathbf{n}\cdot\nabla S_d \tag{6.107}$$

After substituting these equations into Eq. (6.104), we have

$$\begin{aligned}\kappa \equiv \frac{1}{\delta A}\frac{d(\delta A)}{dt} &= -\mathbf{nn} : \nabla\mathbf{v} + \nabla\cdot\mathbf{v} + S_d(\nabla\cdot\mathbf{n}) \\ &= -n_i n_j\frac{\partial v_i}{\partial x_j} - \frac{1}{\rho}\frac{\partial\rho}{\partial t} + \frac{S_d}{R}\end{aligned} \tag{6.108}$$

where R is the local radius of the flame curvature. The three terms on the right-hand side of Eq. (6.108) represent the flame stretch by the strain rate tensor, the volume expansion of the fluid (dilatation), and the curvature of the flame, respectively. These terms arise from the nonuniformities of the flow and the normal propagation of the front. In addition, from geometric considerations, using Eq. (6.96), we have

$$\mathbf{n} \equiv \mathbf{e}_\nu \times \mathbf{e}_\eta \tag{6.109}$$

and in tensor notation,

$$\nu_j\nu_i + \eta_j\eta_i = \delta_{ji} - n_j n_i, \tag{6.110}$$

Therefore, Equation (6.108) can also be expressed as

$$\kappa \equiv \frac{1}{\delta A}\frac{d(\delta A)}{dt} = (\boldsymbol{\nu\nu} + \boldsymbol{\eta\eta}) : \nabla\mathbf{v} + S_d(\nabla\cdot\mathbf{n}) \tag{6.111}$$

6.7.2 Balance Equation for Premixed Laminar-Flame Area

Following Candel and Poinsot (1990), we utilize Reynolds' transport theorem and several aforementioned mathematical relationships to derive various forms of a balance equation for the premixed laminar-flame area. This balance equation describes the rate of change of the flame surface density under transient conditions. It not only is useful for laminar flames but also can be applied to turbulent flames, using the flamelet concept.

The flame surface density Σ is defined as

$$\Sigma \equiv \delta A/\delta V \tag{6.112}$$

From the Reynolds transport theorem for a moving volume $V(t)$, we have

$$\frac{d}{dt}\int_{V(t)} f\,dV = \int_{V(t)}\frac{\partial f}{\partial t}dV + \int_{S(t)} f\mathbf{w}\cdot\mathbf{n}\,dA \tag{6.113}$$

Substituting $f = 1$ in the above equation and using the divergence theorem yields

$$\frac{d}{dt}\int_{V(t)} dV = \int_{V(t)} \nabla\cdot\mathbf{w}dV \tag{6.114}$$

For a volume element δV, the above equation becomes

$$\frac{1}{\delta V}\frac{d(\delta V)}{dt} = \nabla \cdot \mathbf{w} \tag{6.115}$$

From Eqs. (6.112), (6.104), (6.115), and (6.106), we have

$$\begin{aligned}\frac{1}{\Sigma}\frac{d\Sigma}{dt} &= \frac{1}{A}\frac{dA}{dt} - \frac{1}{V}\frac{dV}{dt} = [-\mathbf{nn} : \nabla\mathbf{w} + \nabla \cdot \mathbf{w}] - [\nabla \cdot \mathbf{w}] \\ &= -\mathbf{nn} : \nabla\mathbf{w} = -\mathbf{nn} : \nabla\mathbf{v} - \mathbf{n} \cdot \nabla S_d \end{aligned} \tag{6.116}$$

Since we can write the following relationship based on Eq. (6.90),

$$\frac{d\Sigma}{dt} = \frac{\partial \Sigma}{\partial t} + \mathbf{w} \cdot \nabla\Sigma = \frac{\partial \Sigma}{\partial t} + \mathbf{v} \cdot \nabla\Sigma + S_d \mathbf{n} \cdot \nabla\Sigma \tag{6.117}$$

Eq. (6.116) can be rewritten in the equivalent form

$$\frac{\partial \Sigma}{\partial t} + \nabla \cdot (\mathbf{v}\Sigma) = -(\mathbf{nn} : \nabla\mathbf{v} - \nabla \cdot \mathbf{v})\Sigma - \mathbf{n} \cdot \nabla(S_d \Sigma) \tag{6.118}$$

Based on the relationships between Eqs. (6.108) and (6.111), the above equation can also be written as

$$\frac{\partial \Sigma}{\partial t} + \nabla \cdot (\mathbf{v}\Sigma) = [(\boldsymbol{\nu\nu} + \boldsymbol{\eta\eta}) : \nabla\mathbf{v}]\Sigma - \mathbf{n} \cdot \nabla(S_d \Sigma) \tag{6.119}$$

The flame area balance equation can also be written in terms of flame area per unit mass, a_f, which is defined as

$$a_f \equiv \frac{\Sigma}{\rho} \tag{6.120}$$

We then have

$$\frac{\partial(\rho a_f)}{\partial t} + \nabla \cdot (\rho \mathbf{v} a_f) = \rho[(\boldsymbol{\nu\nu} + \boldsymbol{\eta\eta}) : \nabla\mathbf{v}]a_f - \mathbf{n} \cdot \nabla(\rho S_d a_f) \tag{6.121}$$

The effect of flame curvature can be seen from the following balance equation, which can be shown to be consistent with Eq. (6.121):

$$\frac{\partial(\rho a_f)}{\partial t} + \nabla \cdot [\rho(\mathbf{v} + S_d \mathbf{n})a_f] = \rho[(\boldsymbol{\nu\nu} + \boldsymbol{\eta\eta}) : \nabla\mathbf{v}]a_f + \rho a_f S_d \nabla \cdot \mathbf{n} \tag{6.122}$$

Using the expression of flame stretch factor given by Eq. (6.111), the above equation becomes

$$\frac{\partial(\rho a_f)}{\partial t} + \nabla \cdot [\rho(\mathbf{v} + S_d \mathbf{n})a_f] = \rho \kappa a_f \tag{6.123}$$

6.7.3 Effect of Stretch on Hydrogen–Air Spherical Flames

Various methods to measure the stretch effect on the flame speed have been devised. However, it is a challenging process to generate accurate experimental data that can be utilized conveniently to deduce the flame speed with reference to a 1-D standard premixed flame, so that the stretch effect can be compared with data from other sources. Determining the Markstein number in an experimental or a numerical investigation requires (1) defining an "upstream reference location" where the flame stretch is computed and (2) defining a "flame reference location" where the flame displacement is estimated based on velocity extrapolation from the "upstream reference location."

Dowdy et al. (1990) developed an innovative technique for determining flame speed and the influence of flame stretch in laminar flames. Because of their simple symmetric geometry, Dowdy used expanding spherical flames to determine flame speed and stretch effects and quantified the stretch effect in terms of the Markstein length $\mathcal{L}$. Burning velocities for hydrogen/air mixtures were obtained over a very broad stoichiometric range.

In the Dowdy analysis, an expression for the time variation of the radius of an expanding spherical flame at constant pressure is derived including a flame stretch effect. Since the flame structure is assumed to be one-dimensional, the analysis generally holds for large radii but is not valid for very small radii, where a quasi-1D assumption fails. For an expanding spherical flame with instantaneous surface area $A = 4\pi r^2$, the flame stretch is solely due to the change in curvature with time, since symmetry considerations imply zero velocity gradients in the planes tangent to the flame.

The stretch factor can be evaluated from

$$\kappa = \frac{1}{A}\frac{dA}{dt} = \frac{2}{r}\frac{dr}{dt} \tag{6.124}$$

where r is the instantaneous radius of the flame. Restating Eq. (6.89),

$$S_{u,r} = S_u^o - \mathcal{L}\kappa \tag{6.125}$$

where $S_{u,r}$ is the burning velocity associated with the reference flame surface, S_u^o is the 1-D burning velocity that is to be determined, and $\mathcal{L}$ is the Markstein length. Since both the flame and the flow are moving, the reference surface must be well defined when applying the conventional 1-D definition of burning velocity to flames in divergent flows. The reference surface was selected by comparing the 1-D planar flame to a stationary spherical flame $S_{u,r}$ that, although in a divergent flow, is unstretched. Its position is considered to be the place where the local mass flux in the spherical flame equals the constant value of the 1-D planar flame. The term $S_{u,r}$ also represents the mass flux at this surface divided by the unburned gas density ρ_u.

The mass inside the expanding flame is given by

$$m = \frac{4}{3}\pi r^3 \overline{\rho}_b \tag{6.126}$$

where $\overline{\rho}_b$ is the mean burnt gas density. The mass flux crossing this surface is

$$\dot{m}'' = \left(\frac{dm}{dt}\right)\frac{1}{4\pi r^2} = \left[\overline{\rho}_b + \frac{r}{3}\frac{d\overline{\rho}_b}{dr}\right]\frac{dr}{dt} = \rho_b^o f(r)\frac{dr}{dt} \tag{6.127}$$

where ρ_b^o is the 1-D burnt gas density and

$$f(r) \equiv \left[\overline{\rho}_b + \frac{r}{3}\frac{d\overline{\rho}_b}{dr}\right]\frac{1}{\rho_b^o} \tag{6.128}$$

$f(r)$ can be regarded as the density correction factor. From geometrical considerations, this function can be assumed to have the following form:

$$f(r) = 1 + \frac{c}{r} \tag{6.129}$$

where c is a constant determined by modeling the flame.

The mass flux at the reference surface differs from its 1-D value because of stretch. Using Eqs. (6.124) and (6.125) and the conservation of mass across the flame ($\rho_b^o S_b^o = \rho_u S_u^o$), the following relationship is obtained:

$$\dot{m}'' = \rho_b^o S_b^o - \left(\frac{2\rho_u \mathcal{L}}{r}\right)\frac{dr}{dt} \tag{6.130}$$

where S_b^o is the 1-D burnt gas velocity. Equating (6.127) and (6.130) and after some algebra, we obtain

$$\frac{dr}{dt} = \frac{S_b^o}{f(r) + \ell/r} \tag{6.131}$$

where $\ell = 2\mathcal{L}\rho_u/\rho_b^o$. Substituting (6.129) into (6.131) gives

$$\frac{dr}{dt} = \frac{S_b^o}{1 + b/r} \tag{6.132}$$

where $b \equiv c + \ell$.

Equation (6.132) does not hold for $r \leq -b$; it can be applied only for large radii. Integrating Eq. (6.132) with respect to time, we have

$$r + b \ln r = S_b^o t + \text{constant} \tag{6.133}$$

The value of S_b^o from the experimental data is obtained by fitting a curve in the form of Eq. (6.133) to the data, where S_b^o appears as a fitting parameter. The value of this parameter obtained experimentally was 18.31 m/s, compared with the traditional best-fit straight-line value of 16.36 m/s. The burning velocity S_u^o is then obtained by multiplying S_b^o by ρ_b^o/ρ_u. The Markstein length $\mathcal{L}$ is obtained from (6.134),

$$\mathcal{L} = \frac{(b-c)}{2}\frac{\rho_b^o}{\rho_u} \tag{6.134}$$

where b and c are obtained from computer modeling for each flame.

In the experimental work, flames were propagated in a spherical bomb of radius 300 mm with central ignition, using a spark gap of about 0.7 mm. Data were obtained in the pre-pressure period, with flame radius limited to 35 mm. Flame propagation was recorded by high-speed photography of the schlieren image, from which the radius as a function of time was obtained. Hydrogen (99.95% purity) and compressed air were used to form combustible mixtures. All tests were performed at 1 atm and at an initial temperature of 296 ± 1 K.

Dowdy et al. (1990) also conducted modeling work for three types of flames, including planar 1-D, stationary spherical, and expanding spherical flames. The first two types were carried out by using the Sandia Laboratories' CHEMKIN program (Kee et al. 1985) slightly modified to include thermal diffusion which is important in hydrogen combustion. In all the modeling work, the following important reactions were included: $H + O_2 = OH + O$, $O + H_2 = OH + H$, $OH + H_2 = H_2O + H$, and $H + O_2 + M = HO_2 + M$.

Figure 6.17 shows flame speed versus the equivalence ratio. The maximum flame speed obtained was 2.85 m/s at an equivalence ratio of 1.4 (41% H_2). At an equivalence ratio of 1.0, the value was 2.13 m/s. This figure also shows the burning velocities predicted by planar 1-D modeling.

The Markstein lengths for a hydrogen/air mixture are plotted as a function of the equivalence ratio in Figure 6.18. As discussed earlier, Marstein lengths can be positive or negative. The Markstein lengths are negative (i.e., burning velocity increases with stretch) on the lean side and positive on the rich side. The authors also observed similar behavior in methane/air and the opposite behavior in propane/air mixtures.

6.7.4 Effect of Stretch on Spherical Hydrocarbon–Air Flames

Tseng et al. (1993) measured the effect of flame stretch on laminar flame speeds of hydrocarbon–air mixtures, using outward-propagating spherical flames. The hydrocarbon fuels were methane,

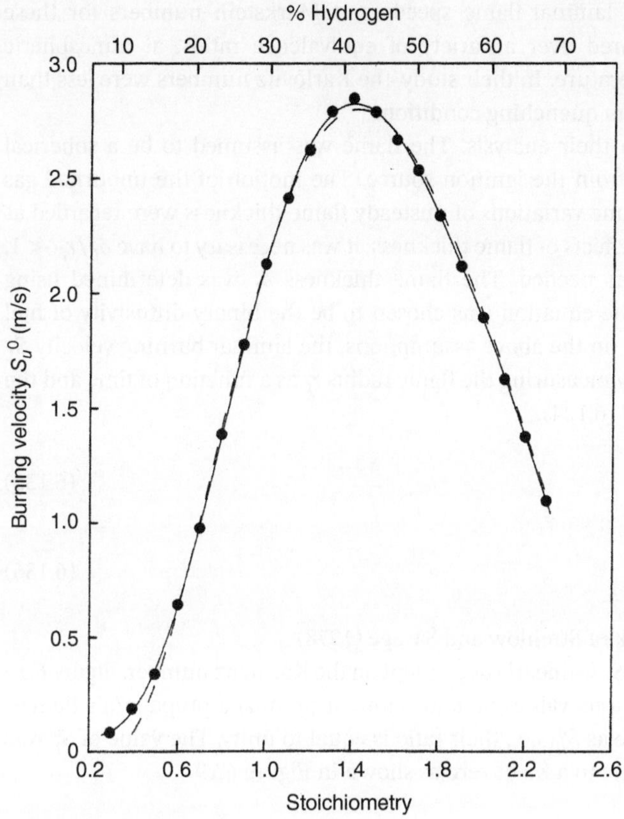

Figure 6.17 Burning velocity as a function of stoichiometry. Points: experimental result of Dowdy et al. (1990). Solid line: best-fit polynomial through experimental data. Broken line: best-fit polynomial through planar modeling results. (Adapted from Dowdy et al. 1990).

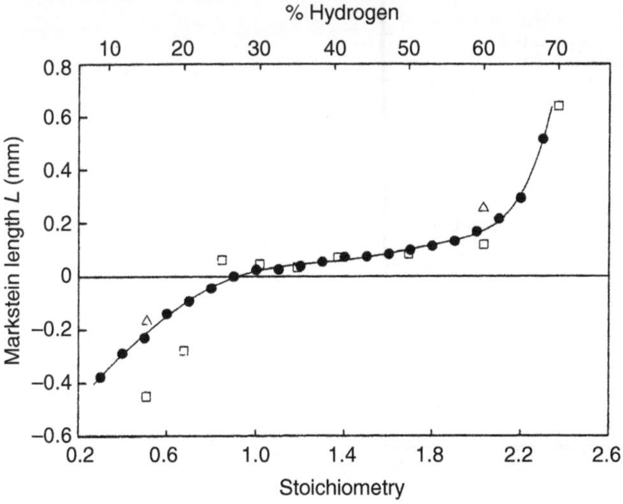

Figure 6.18 Markstein lengths for hydrogen/air as a function of stoichiometry. Black circle: the experimental data of Dowdy et al. (1990). The unfilled symbols are from modeling work by Dixon-Lewis (1983) and Warnatz and Peters (1984). (Adapted from Dowdy et al. 1990).

ethane, ethylene, and propane. The laminar-flame speeds and Markstein numbers for these hydrocarbon/air flames were measured over a variety of equivalence ratios, at atmospheric pressure, and at ambient initial temperature. In their study, the Karlovitz numbers were less than 0.3 so that the flames were remote from quenching conditions.

Several assumptions were made in their analysis. The flame was assumed to be a spherical deflagration wave propagating away from the ignition source. The motion of the unburned gas was considered negligible. Also, the time variations of unsteady flame thickness were regarded as negligible. To minimize the transient effects of flame thickness, it was necessary to have $\delta_L/r_f \ll 1$. This implies that an estimate of δ_L is needed. The flame thickness δ_L was determined using Eq. (6.135). The mass diffusivity in the equation was chosen to be the binary diffusivity of fuel into the diluent (i.e., nitrogen). Based on the above assumptions, the laminar burning velocity S_L and stretch factor κ were calculated by measuring the flame radius r_f as a function of time and the derivative of r_f, using Eqs. (6.136) and (6.124),

$$\delta_L = \frac{\mathcal{D}_u}{S_L} \tag{6.135}$$

$$S_L = \frac{\rho_b}{\rho_u}\frac{dr_f}{dt} \tag{6.136}$$

where Eq. (6.136) is based on the work of Strehlow and Savage (1978).

Based on Eq. (6.135), the ratio of S_L^o/S_L is linearly dependent on the Karlovitz number. Figure 6.19 shows this linear dependency for various values of ϕ for laminar premixed propane/air flames. When Ka is zero, S_L must be the same as S_L^o, i.e., their ratio is equal to unity. The value of S_L^o was determined by extrapolating the S_L data to a Ka of zero as shown in Figure 6.19.

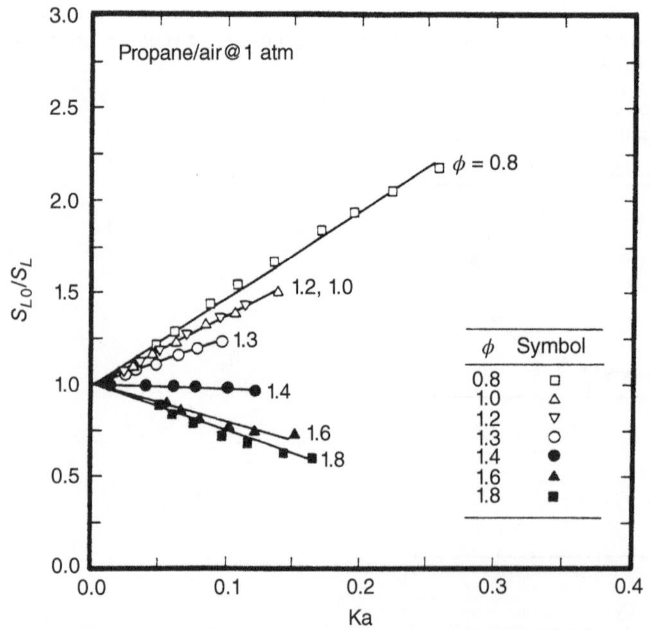

Figure 6.19 Laminar burning velocity as a function of Karlovitz number and fuel-equivalence ratio for propane–air mixtures. *Source*: Adapted from Tseng et al. 1993/with permission of Elsevier.

In view of the straight-line nature of the data for different values of the equivalence ratio ϕ, one can assume that the Markstein number is independent of the Karlovitz number. This implies, from Eq. (6.93), that Mn governs the slopes of the data lines in Figure 6.19. A correlation of the form of Eq. (6.137) is based on these data. Due to the large scatter of the data, it is difficult to justify more than a linear correlation.

$$\mathrm{Mn} = S(\phi - \phi_n) \tag{6.137}$$

The slope S can also be regarded as the sensitivity of Mn to variations of ϕ, and ϕ_n is the neutral preferential diffusion condition where Mn = 0. The slope S can have positive or negative values depending on whether the reactant mixture exhibits unstable preferential diffusion behavior at a fuel equivalence ratio lower or higher than ϕ_n. A least-square fit using Eq. (6.137), based on the data of Tseng et al. (1993), Palm-Leis and Strehlow (1969), Taylor (1991), and Fristrom (1965) is also illustrated in Figure 6.20. The slope S is -8.8 ± 0.8 and the intercept (ϕ_n) is 1.44 ± 0.11. The maximum value of the Karlovitz number Ka = 0.25.

For laminar premixed CH_4/air, C_2H_6/air, and C_2H_4/air flames, similar plots to Figures 6.19 and 6.20 were developed by Tseng et al. (1993). Their summary of Markstein number correlations for four different hydrocarbon fuels burning with air, Taylor's results (1991) and data of Kwon et al. (1992) for $H_2/O_2/N_2$ flames are given in Table 6.4. A linear correlation between Mn and ϕ was assumed. It is noted that the slope S is positive for both CH_4/air flames and $H_2/O_2/N_2$ flames, since these fuel/air mixtures exhibit unstable preferential diffusion at equivalence ratios greater than ϕ_n.

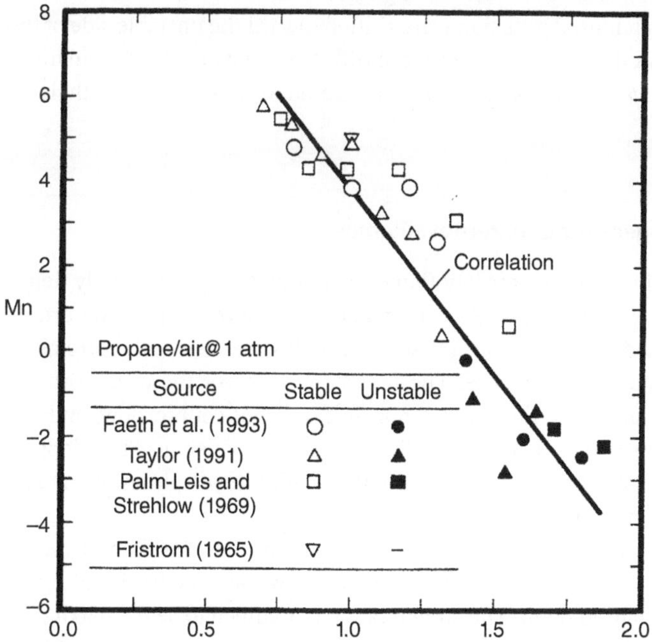

Figure 6.20 Markstein number (Mn) as a function of equivalence ratio for propane–air mixtures. *Source*: Adapted from Tseng et al. 1993/with permission of Elsevier.

Table 6.4 Summary of Markstein Number Correlations, Eq. 6.137, for Four Different Hydrocarbon Fuels Burning with Air and for $H_2/O_2/N_2$ Flames.

Reactants	Slope S	Intercept ϕ_n	Maximum Ka	ϕ_n Range	Std. Dev. of S	Std. Dev. of ϕ_n	Correl. Coeff.
C_3H_8/Air	−8.8	1.44	0.25	0.8–1.8	0.8	0.11	0.95
CH_4/Air	10.2	0.74	0.30	0.6–1.35	1.1	0.09	0.93
C_2H_6/Air	−4.0	1.68	0.25	0.8–1.6	0.7	0.22	0.83
C_2H_4/Air	−2.9	1.95	0.24	0.8–1.8	0.4	0.22	0.86
$H_2/O_2/N_2$	3.9	1.50	0.21	1.00–4.83	0.2	0.22	0.99

Source: Tseng et al. (1993), Taylor (1991), and Kwon et al. (1992).

Based on the results of Tseng et al. (1993) on outwardly propagating spherical flames for the hydrocarbons methane, ethane, ethylene, and propane in air at atmospheric pressure and initially at room temperature, the following major findings are noted:

1. Effects of preferential diffusion/stretch interactions could be correlated according to Eq. (6.93).
2. Effects of flame stretch on laminar burning velocities were substantial, yielding Markstein numbers in the range −2.5 to +7.2 and corresponding variations of S_L^o/S_L in the range of 0.4–2.7, implying significant effects of flame stretch on laboratory measurements of laminar burning velocities.
3. Markstein numbers varied linearly with equivalence ratio from the neutral preferential diffusion condition, according to Eq. (6.137), although there was significant scatter in the data.
4. Equivalence ratios for neutral preferential diffusion were shifted toward the unstable side of the equivalence ratio. The ϕ ranges for the unstable preferential-diffusion (negative Markstein numbers) behavior include propane, $\phi > 1.44$; methane, $\phi < 0.74$; ethane, $\phi > 1.68$; and ethylene, $\phi > 1.95$.

6.7.5 Effect of Stretch on Hydrogen–Air Counterflow Flames

Egolfopoulos and Law (1990) employed a counterflow flame technique to experimentally determine the effect of stretch on laminar-flame speeds of hydrogen/air mixtures. The counterflow technique was originally established by Law et al. (1986). With this geometry, the stretch is equal to the strain since there is no flow curvature. The technique involves the establishment of two symmetrical, planar, nearly adiabatic flames in a nozzle-generated counterflow, and the determination of the axial velocity profile along the centerline of the flow by using a laser doppler velocimeter (LDV). A sketch of the laminar flame in the stagnation-point flow regimes of 2-D and axisymmetric flow situations are shown in Figure 6.21.

In this configuration, the velocity far upstream decreases as it approaches the flame front. It increases as the reactants are preheated and divergence is created in the flow-field and again decreases to become zero at the stagnation point. The stretch rate κ is assumed to be the maximum gradient in the upstream velocity field. By identifying the minimum point of the velocity profile as a reference upstream flame speed S_u, and the velocity gradient ahead of it as the imposed stretch rate κ, the upstream flame speed S_u can be determined as a function of κ for a given mixture concentration and pressure.

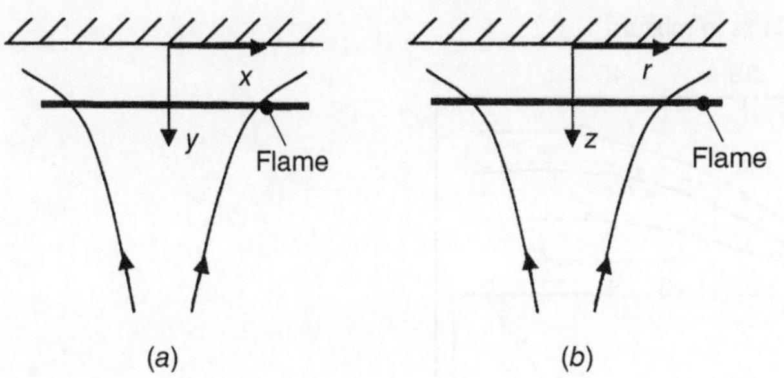

Figure 6.21 Laminar premixed flames in the vicinity of stagnation-point flows. (*a*) Two-dimensional flow and (*b*) axisymmetric flow.

By considering the counterflow conditions, the velocity vector for potential flow in the vicinity of a stagnation-point can be written for a 2-D planar case as

$$\mathbf{v} = [v_x, v_y, v_z] = [ax, -ay, 0] \tag{6.138}$$

and for an axisymmetric case as

$$\mathbf{v} = [v_r, v_z, v_\theta] = \left[\frac{a}{2}r, -az, 0\right] \tag{6.139}$$

The stretch factor κ for a premixed laminar flame in the vicinity of stagnation-point flow is

$$\kappa = a \tag{6.140}$$

where a is the velocity field constant in Eqs. (6.138) and (6.139). For this type of flame, at least three diffusivities must be considered: (1) thermal diffusivity α, (2) mass diffusivity $\mathcal{D}_i$ of the deficient reactant species i (e.g., oxidizer species for a fuel-rich mixture), and (3) mass diffusivity $\mathcal{D}_j$ of the excess reactant species j (e.g., excess fuel for a fuel-rich mixture). Accordingly, two effects associated with these three diffusion coefficients are: a non-unity Lewis number ($Le \equiv \alpha/\mathcal{D}_i$) effect and the preferential diffusion ($\mathcal{D}_i/\mathcal{D}_j$) effect.

The control volume around the flame loses thermal energy by conduction to the incoming reactants in the positive y-direction but can gain chemical energy from an increase in the deficient reactant concentration. If $Le = 1$, the flame temperature T_f of a stretched flame can stay the same as the adiabatic flame temperature. If $Le > 1$, the heat loss effect can exceed the reactant mass species diffusion effect and $T_f < T_{f,\text{ad}}$; conversely, $T_f > T_{f,\text{ad}}$ if $Le < 1$. Due to these diffusion effects, for flames with $Le > 1$, increasing the stretch rate will lower the flame temperature T_f until extinction occurs due to incomplete reaction. However, for $Le < 1$ flames, increasing stretch rate increases the flame temperature T_f, and extinction will occur only when the reaction zone is quenched by the stagnation surface.

In Egolfopoulos and Law's experimental measurements using the counterflow premixed flame setup, the stretch-free flame speed S_u^o was determined by linearly extrapolating $S_u(\kappa)$ to $\kappa = 0$. It was also assumed that the upstream and downstream heat losses were minimized due to the symmetrical nozzle geometry. In the experiments the fuel equivalence ratio was varied from 0.153 to 2.2, and the system pressure was varied from 0.2 to 2.25 atm. The visualization of the nonluminous hydrogen flame was achieved first by "doping" the mixture with trace amounts of methane, which was removed during the data-taking period.

For the numerical simulation, Egolfopoulos and Law adopted a flame code developed by Kee et al. (1985), which employs a hybrid time-integration/Newtonian-iteration technique to solve the steady-state equations of mass and energy conservation, and outputs the complete flame structure

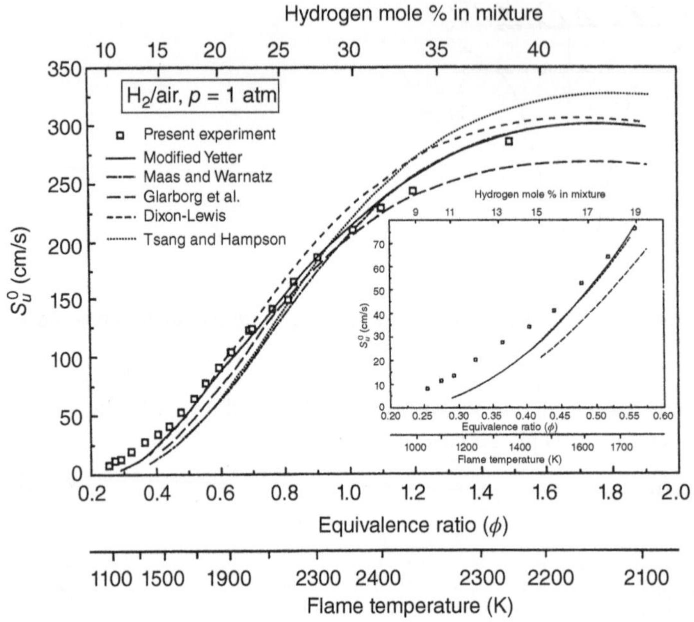

Figure 6.22 Comparison of experimental and numerical $S_u^o(\phi)$ for atmospheric H_2/air mixtures using various kinetic schemes. *Source*: Adapted from Egolfopoulos and Law 1990/with permission of Elsevier.

together with the sensitivities of the local concentrations of species and the mass burning rates of all reactions involved. In performing the comparison between the numerical results with experimental data, Egolfopoulos and Law used five kinetic schemes: Maas and Warnatz (1988), Dixon-Lewis (1972), Glarborg et al. (1986), Tsang and Hampson (1986), and Yetter et al. (1991). The last scheme was modified with the input of Yetter and is called the modified Yetter scheme. For atmospheric H_2/air flames, Figure 6.22 provides the comparison of the experimental data with the calculated results using these various kinetic schemes.

Figure 6.22 indicates that all of the kinetic schemes underpredict the flame speed as ϕ becomes increasingly leaner. Egolfopoulos and Law also studied the effects of nitrogen-diluted flames to determine the effects of temperature, oxygen concentration, and stoichiometry on flame speed. Nitrogen dilution was used in order to maintain low flame temperatures even as the equivalence ratio was increased. Again, the kinetic schemes underpredicted the flame speeds for very lean conditions. However, for fuel-rich conditions, the kinetic schemes predicted the flame speed quite well, even though the flame temperatures were kept in the same range as the lean cases by nitrogen dilution. This result clearly indicates the influence of both the temperature and the oxygen concentration on the kinetic mechanism in lean combustion.

6.8 Stabilization of Combustion Flames in Laminar Flows

In order to produce a stable premixed flame in combustion devices, there is a need to balance the velocity of the unburned gas mixture flowing into the device with the speed of the flame front consuming the gas mixture. If the flame speed and the mixture velocity are not balanced, then flame blowoff or flashback can occur.

Most commercial gas–air premixed burners are basically laminar-flow Bunsen burners and operate at atmospheric pressure, i.e., the primary air is induced from the atmosphere by the fuel flow, followed by mixing in the burner passage and ignition in the burner ports where the flame is stabilized. The induced air flowrate is determined by the position of a throttle at the air inlet. The burning mixture issues from many small and closely spaced drilled holes, typically 1–2 mm in diameter.

A classic stability diagram for a premixed laminar flame at the outlet of a cylindrical tube is given in Figure 6.23 (Wohl et al. 1949). In this plot, the stability boundaries are given as functions of fuel concentration in the mixture and the premixed gas supply velocity. When the supply velocity to a stable flame, represented by a point in the dark shaded region, is decreased until the flame velocity exceeds the approach velocity over some portion of the burner port, the flame flashes back into the burner, shown as the unshaded area of Figure 6.23.

Alternatively, if the supply velocity is increased until it exceeds that of the downward flame velocity at every point, the flame will either be entirely extinguished beyond the blowoff limit or be lifted until a new stable position in the gas stream above the burner is reached as a result of dilution with secondary air from the surroundings. The lift curve in Figure 6.23 is a continuation of the blowoff curve beyond point A. The blowout curve corresponds to the velocity required to extinguish a lifted flame. Once the flame has lifted, the approach velocity must be decreased well below the original lift velocity before the flame will drop back and be reseated on the burner rim. Between fuel concentrations *A* and *B*, the blowout of a lifted flame occurs at a lower velocity than the flame blowoff from the port.

Other common ways to stabilize flames in laminar flows are to use flame holders with sharp corners such as bluff bodies or perforated plates. The flame holder stabilizes the flame against the flow of unburned mixture and fixes the position of the flame. At low flow Reynolds numbers, the mixture flows uniformly without recirculation eddies. As the flow Reynolds number is increased, a recirculation wake will be formed downstream of the bluff body.

In a simple Bunsen flame from a circular tube, the stabilization depends on the velocity gradients in the flow emerging from the tube. As a result, the Bunsen flame is able to remain over the burner exit over a range of flow velocities and mixture concentrations. Figure 6.24 shows four different

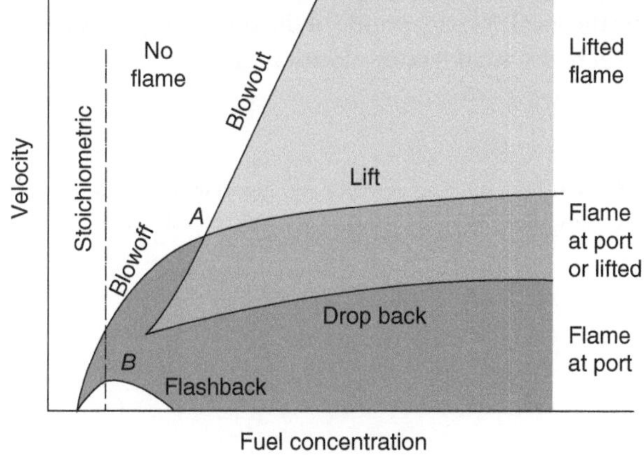

Figure 6.23 Characteristic stability diagram for a premixed open burner flame. (Adapted from Wohl et al. 1949).

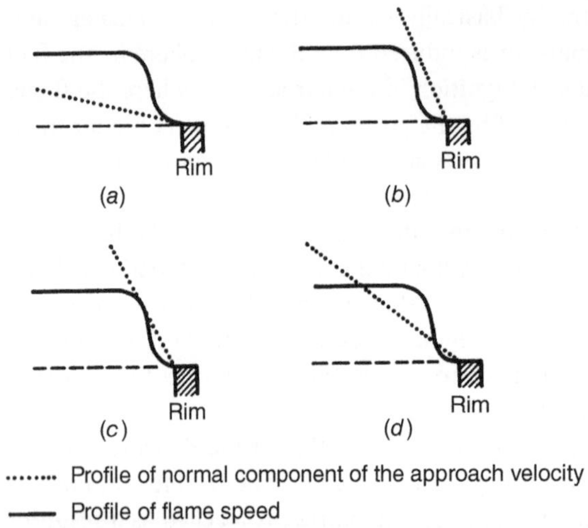

Figure 6.24 Four different radial distributions of velocity profiles in the consideration of laminar-flame stability near the rim of a Bunsen burner. (*a*) Flashback (*b*) blowoff, (*c*) critical, and (*d*) stable. (Adapted from Kanury 1975).

situations of flame stabilization near the exit rim of a Bunsen burner. Near the vicinity of the burner rim, the outward flowing gas mixture velocity profile is assumed to be linear.

The velocity profile of the gas mixture controls the flame stability as shown by comparing the radial profiles of the near-surface region of the approach velocity of the unburned gas mixture (dotted line) with the flame velocity (solid line) adjacent to the burner rim.

When the premixed-gas flow is weak [case (*a*)], $S_L > u_{\text{supply}}$ not only in the near-surface region but also over almost the entire cross section of the barrel; hence, the flame propagates into the barrel to produce a *flashback* situation.

On the other hand, when the premixed-gas flow is very strong [case (*b*)], $S_L < u_{\text{supply}}$ over the entire cross section, producing a *blowoff* situation. The critical condition for blowoff arises when the S_L and u_{supply} profiles are tangent at a given radial location [case (*c*)]. Stable combustion is shown as case (*d*) in Figure 6.24.

Consider a velocity profile slightly inside the barrel and assume that the flame has just entered the tube (see Figure 6.25). The distance d_p from the tube wall is called the penetration distance, which is half of the quenching distance, d_q. Below d_q, no flame can propagate into the tube supplying the premixed combustible gases. If $\bar{u}_1$ represents the local velocity profile of the premixed gas near the tube wall, then there is no radial location where the local flame velocity is greater than the local gas velocity.

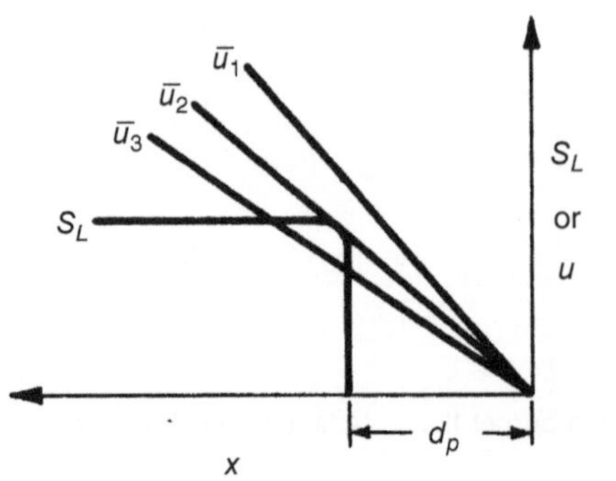

Figure 6.25 Burning velocity and gas velocity inside a Bunsen tube. (Adapted from Lewis and von Elbe 1987).

Figure 6.26 Effect of critical boundary velocity gradient and fuel percentage on flame stability. (Adapted from Kanury 1975).

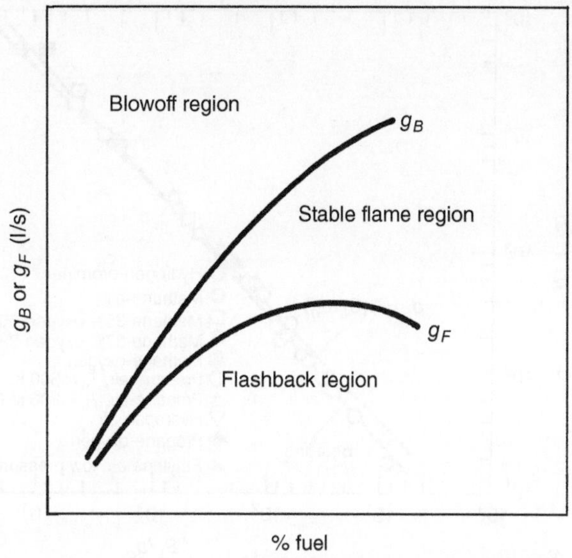

Therefore, any flame inside the tube will then be blown out of the tube. Then, the critical flashback condition is reached if u_{supply} is represented by $\bar{u}_2$. The profile for the mean velocity $\bar{u}_3$ corresponds to the case for which the flame undergoes flashback. The critical value of the boundary velocity gradient g at which the *flashback* condition is first realized is denoted by g_F.

A qualitative sketch of critical boundary velocity gradients versus fuel–oxidant ratio is shown in Figure 6.26. For rich mixtures, the blowoff curve continues to rise instead of decreasing after the stoichiometric value is reached. The cause for this trend is that the experiments are performed in air, and the diffusion of air into the premixed gaseous mixture as the flame lifts off the burner rim increases the local flame speed of the initially fuel-rich mixture.

The flashback velocity gradient g_F is also related to the ratio of flame speed and quenching distance ratio S_L/d_q. Lewis and von Elbe (1987) suggest that twice the depth of penetration of quenching ($2d_p$) should be equal to the quenching distance between parallel plates. For cylindrical tubes d_p should be about $\frac{1}{3}$ the quenching distance (Berlad and Potter 1957). One can thus express the flashback velocity gradient g_F for cylindrical tubes as

$$g_F = 3\frac{S_L}{d_q} \tag{6.141}$$

The flashback velocity gradient g_F results of Berlad and Potter (1957) using different fuel–oxidant mixtures are shown in Figure 6.27. The best line through the data has the equation

$$g_F = 14.125\left(\frac{S_L}{d_q}\right)^{1.168} \tag{6.142}$$

Equation (6.141) is also sketched in Figure 6.27. Considering the variety of mixtures included in the correlation, the agreement with Eq. (6.141) is acceptable.

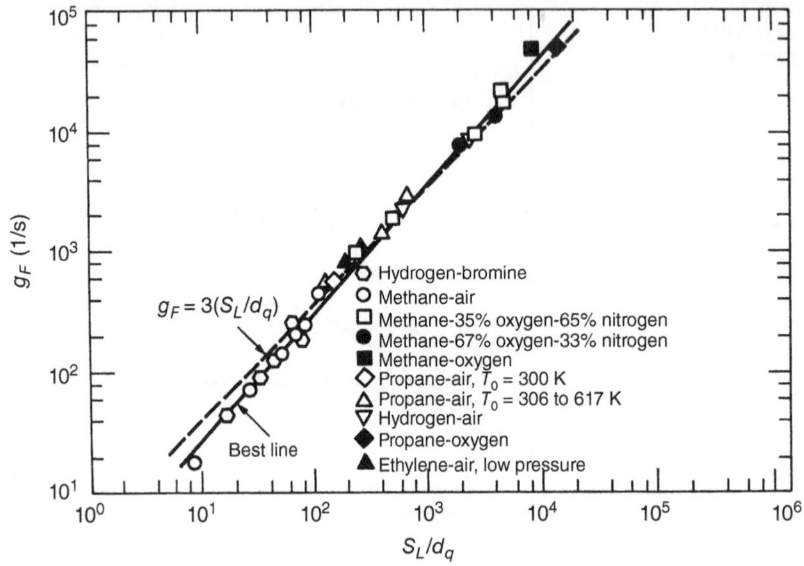

Figure 6.27 Correlation of the flashback velocity gradient g_F with S_L/d. (Adapted from Berlad and Potter 1957).

6.9 Ignition

An energy source is needed to ignite a fuel–air mixture and generate a flame. When the combustion process is initiated by an external energy source which locally heats the mixture to a high temperature, the process is termed piloted ignition. Combustion is then sustained when the ignition source is removed. Piloted ignition can be achieved using a spark, pilot flame, electrical resistance (glow plug), laser, friction, or any sufficiently hot source.

The minimum ignition energy (MIE) is defined as the minimum energy that can ignite a specified fuel–air mixture. If enough energy is deposited in a small volume of the fuel–air mixture, the mixture temperature will increase to a value high enough so that the chemical energy release rate in the small volume is greater than the heat loss to the surroundings, leading to sustained flame propagation through the mixture. Typical values of the MIE are MIE = 0.80 mJ for gasoline, 0.28 mJ for methane, 0.25 for propane, and 0.017 mJ for hydrogen.

An electric spark generated between two electrodes is commonly used to provide the ignition energy to a fuel–air mixture in stationary combustion chamber, an internal combustion, or gas turbine engine. If the electrical energy added is above theMIE, thermal energy is generated by chemical reaction inside the mixture at a greater rate than is lost by conduction to the electrodes or surrounding gas, leading to a glowing plasma, and then sustainable flame propagation through the mixture. The flame kernel produced by an electric spark has a toroidal or spherical shape with internal recirculation that entrains the surrounding unburned mixture. There are two countering effects. Entrainment of fresh unburned mixture will increase the surface area available for ignition, but since the fresh mixture is at ambient temperature, it also will have a cooling effect on the flame. If insufficient electrical energy is deposited, the produced radicals recombine more quickly than they are generated by chain branching reactions, and thermal energy is conducted away from the kernel, leading to flame quenching.

The ignition energy required to heat the flame kernel to a sustained flame temperature decreases with increasing mixture temperature and pressure and is minimum at stoichiometric conditions. As the mixture becomes more lean, the laminar flame speed decreases, increasing the proportion of thermal energy that is conducted away from the plasma kernel, and consequently increasing the MIE required. The minimum energy also increases with increasing flow velocity and turbulence intensity. While increasing turbulence intensity increases flame speed, it also increases the heat transfer loss to the surroundings. Conventional ignition systems provide about 40 mJ to the electrodes to account for the flow velocity and heat transfer loss to the electrodes and the surroundings.

To estimate the MIE from an electric spark, we start by defining a critical combustion radius r_c. At this critical radius, there is a balance between the rate of energy generation by chemical reaction and the conduction heat loss. If the actual radius is greater than the critical radius, the reacting gas will continue to expand, and vice versa.

$$\dot{Q}_r = \dot{Q}_{\text{loss}}$$

$$\dot{\omega}_F V(r) q_c = -k A(r) \left.\frac{dT}{dr}\right|_{r_c} \tag{6.143}$$

where $\dot{\omega}_F$ is the fuel reaction rate, $V(r)$ is the combustion volume, and $A(r)$ the combustion surface area. Therefore,

$$\dot{\omega}_F q_c \frac{4}{3}\pi r_c^3 = -k 4\pi r_c^2 \left.\frac{dT}{dr}\right|_{r_c} \tag{6.144}$$

We approximate the temperature gradient at the edge of the expanding gas as

$$\left.\frac{dT}{dr}\right|_{r_c} \sim \frac{T_b - T_u}{r_c} \tag{6.145}$$

Upon substitution of Eq. (6.145) into Eq. (6.144), and solving for r_c,

$$r_c = \left[\frac{3k(T_b - T_u)}{\dot{\omega}_F q_c}\right]^{1/2} \tag{6.146}$$

and from Chapters 2 and 3, where

$$q_c = (s_A + 1)c_p(T_b - T_u)$$

$$\dot{\omega} = \dot{\omega}_F(s_A + 1)$$

the critical radius becomes

$$r_c = \left(\frac{3k}{\dot{\omega} c_p}\right)^{1/2} \tag{6.147}$$

From Eq. (6.6), the laminar flame speed S_L is approximately

$$S_L \sim \left(\frac{\alpha\,\dot{\omega}}{\rho_u}\right)^{1/2} \tag{6.148}$$

and therefore

$$r_c \sim \sqrt{3}\frac{\alpha}{S_L} \tag{6.149}$$

Next, we assume that the energy added by the spark heats the gas to the flame temperature,

$$\text{MIE} = m_c c_p (T_b - T_u) \tag{6.150}$$

The gas mass at the critical radius is

$$m_c = \rho_b V(r_c) = \rho_b \frac{4}{3}\pi r_c^3 \tag{6.151}$$

so

$$\text{MIE} = \rho_b \frac{4}{3}\pi\, c_p (T_b - T_u)\, r_c^3 \tag{6.152}$$

Upon substitution of Eq. (6.149),

$$\begin{aligned} \text{MIE} &= \frac{4}{3}\pi\, 3^{3/2}\, \rho_b\, c_p (T_b - T_u)\left(\frac{\alpha}{S_L}\right)^3 \\ &\sim 20\, \rho_b\, c_p (T_b - T_u)\left(\frac{\alpha}{S_L}\right)^3 \\ &\sim 20\, \frac{p\, c_p}{R_b}\left(1 - \frac{T_u}{T_b}\right)\left(\frac{\alpha}{S_L}\right)^3 \end{aligned} \tag{6.153}$$

using the ideal-gas equation $\rho_b = p/R_b T_b$. Equation (6.153) indicates that the MIE scales as S_L^{-3} and p^{-2}, since $\alpha \sim 1/p$, and assuming a second-order ($m = 2$) reaction.

Laser ignition can also be used to ignite a fuel–air mixture. Laser ignition has advantages and disadvantages when compared with conventional spark ignition. Laser ignition is nonintrusive and provides greater control over the location and timing of the ignition event. Problems with spark plug fouling, erosion, and flame quenching due to heat loss through the electrodes are eliminated. The lean limit of laser-ignited methane–air mixtures is also lower than that using conventional spark ignition.

The main method for plasma formation in laser ignition presently used is nonresonant breakdown in which a laser beam is focused by a converging lens to an intensity high enough to ionize the fuel and air molecules. The laser pulse forming the ellipsoidal plasma has about a 10-ns duration. The initial ionization leads to the generation of free electrons that have energies high enough to ionize other gas molecules on impact, an electron avalanche process called electron–ion inverse bremsstrahlung, producing a high temperature and pressure plasma.

A sequence of OH* chemiluminescence images in a rapid compression machine is presented in Figure 6.28. The high temperatures in the plasma initiate chemical reactions that evolve into a

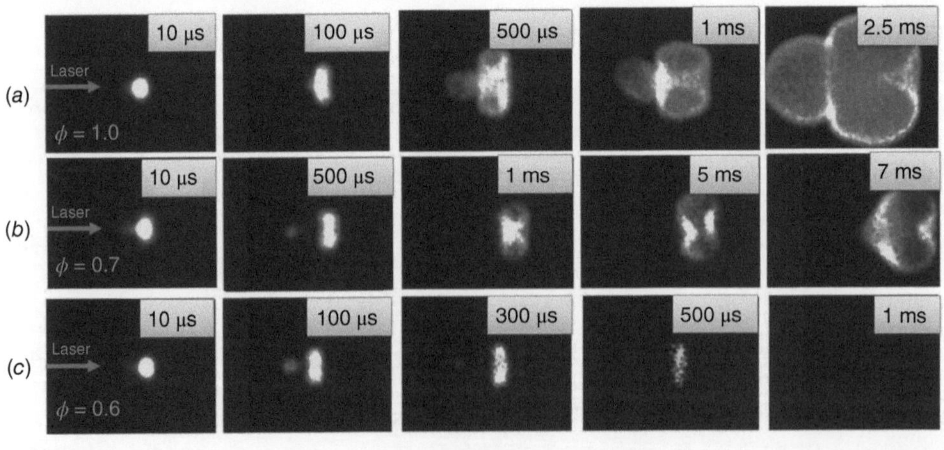

Figure 6.28 Flame kernel development during single-pulse laser ignition of propane–air mixtures at $\phi = 1.0$ (*a*), $\phi = 0.7$ (*b*), and $\phi = 0.6$ (*c*). *Source*: Adapted from Dumitrache et al. 2017/Springer Nature.

kernel that initially grows as a result of the shock wave and by thermal conduction into the surrounding gases. The high pressure in the plasma results in the formation of a spherical shock or blast wave that has a duration from about 10 to 100 ms. As the shock wave moves outward from the plasma, it creates a toroidal vortex ring at the leading and trailing edges of the plasma. The vorticity generated in the trailing edge toroid in the plasma kernel evolves into a third lobe in the direction of the laser beam. The effect of the third lobe depends on the stoichiometry.

The top row of images shows flame kernel development and propagation at $\phi = 1.0$. Under these conditions, one can clearly observe the toroidal flame structure. At leaner conditions, for example, in the middle row of images for which $\phi = 0.7$, the third lobe detaches causing the flame kernel to break into two separate flamelets. The splitting (and in some cases extinction) of the kernel at lean conditions is attributed to insufficient energy deposition in the third lobe given the flame stretching induced by the vorticity effects of the laser plasma; it can be considered as a consequence of the increase in Karlovitz number with the decrease in equivalence ratio. At even leaner conditions, i.e., the bottom row of images for which $\phi = 0.6$, the flame kernel never develops into a self-propagating flame.

The temperature of the plasma created by the laser pulse is greater than that from a conventional spark, since the mechanism of plasma formation is different. On the other hand, it has been found that laser ignition requires much higher minimum ignition energies than spark plugs, since the fuel–air mixture is transparent to the laser beam until the onset of gas breakdown.

6.10 Flame Quenching

Flame quenching can significantly impair the process of flame propagation. There are several mechanisms for flame quenching, including narrow openings, flame cooling, insufficient oxygen, dilution of fuel or air, and reactions with other chemicals such as halogens. The study of flame quenching was initiated by Sir Humphry Davy (1778–1829), an English scientist, when he became interested preventing explosions in coal mines. He invented a safety lamp for use for illumination in flammable atmospheres, known as the Davy lamp. The Davy lamp incorporates a cylindrical wire mesh screen surrounding the lamp's flame. The screen acts as a flame arrestor; air can pass through the mesh freely enough to support combustion, but the holes are too small to allow a flame to propagate through them. In his studies, Davy also found that with methane–air mixtures no flame could propagate along a metal tube smaller than 3.6 mm, and that the addition of an inert gas such as CO_2 or N_2 reduced the flame propagation.

Let us introduce the quenching diameter in a tube by considering the following experiment. After a flame has been established on a Bunsen burner, if the flow of the fuel–air mixture is suddenly cut off, the flame flashes backward and propagates back down the metal burner tube. The tube is replaced with a narrower one. The experiment is repeated until the tube just barely permits the flame to propagate back. However, the smaller the tube, the greater is the surface area to volume ratio within the tube, increasing the relative heat transfer to the tube wall.

The diameter of this last tube is called the quenching diameter d_q for the given fuel–oxidant mixture under the specified conditions of temperature and pressure. Below this diameter, the flame is quenched by heat transfer to the tube wall and the flame cannot propagate back down the tube.

Another geometry of importance is flame quenching between parallel plates. A simplified 1-D quenching analysis can be developed for this geometry, as follows. In this case, the quenching distance d is defined as the distance between the plates such that the rate of thermal energy generation in the flame, $\dot{Q}_c$, is equal to the rate of heat transfer from the flame to the plates, $\dot{Q}_{\text{loss}}$, resulting in quenching of the flame. For parallel plates, experiments indicate the quenching distance $d \sim 0.65d_q$ (Potter 1960).

$$\dot{Q}_c = \dot{Q}_{\text{loss}} \tag{6.154}$$

The rate of heat generation in the flame, for a small volume V of size $A\,d$, can be expressed as

$$\dot{Q}_c = \dot{\omega}_F V q_c = \dot{\omega}_F A d\, q_c \tag{6.155}$$

Since the flame is assumed to be laminar, the rate of heat transfer from the flame can be found using the Fourier heat conduction equation, assuming negligible radiation effects,

$$\dot{Q}_{\text{loss}} = -kA\frac{dT}{dx} \tag{6.156}$$

where

- d is the distance between the two parallel plates (m).
- A is the contact area between the flame and a plate (m^2).
- q_c is the heat of combustion per unit mass of fuel (kJ/kg).
- k is the flame thermal conductivity (W/m-K).
- $\dot{\omega}_F$ is the fuel mass volumetric reaction rate (kg/m^3-s).

Let T_q be the lowest temperature at which the flame can propagate, and let T_u be the unburned gas temperature at the plate boundary. Approximating the temperature distribution in the gas phase by two linear straight-line segments with the temperature T_q at the center, and equating the heat conduction rate with the heat generation rate, yields

$$2Ak\frac{T_q - T_u}{d/2} = \dot{\omega}_F A d\, q_c \tag{6.157}$$

or

$$d^2 = \frac{4k(T_q - T_u)}{\dot{\omega}_F q_c} \tag{6.158}$$

From Section 2.12, the heat of combustion q_c can be expressed in terms of the adiabatic flame temperature T_f,

$$q_c = (s_A + 1)c_p(T_f - T_u) \tag{6.159}$$

Substituting Eq. (6.159) into Eq. (6.158), and assuming $T_q \sim T_f$, we have

$$d^2 = \frac{4k}{\dot{\omega}_F(s_A + 1)c_p}\left(\frac{T_f - T_u}{T_f - T_u}\right) \sim \frac{4k}{\dot{\omega}c_p} \tag{6.160}$$

It is important to note that Eq. (6.160) is derived under the assumption of negligible heat loss from the flame to the unburned gas. Because of this assumption, Eq. (6.160) cannot satisfy the limiting condition at $\dot{\omega} \to 0$. If one sets the reaction rate equal to zero in Eq. (6.160), the quenching distance approaches infinity; however, from experiments, we know that the quenching distance is finite.

From Eq. (6.6), the laminar flame speed S_L is

$$S_L \sim \left(\frac{\alpha\,\dot{\omega}}{\rho_u}\right)^{1/2}$$

where $\alpha = k/\rho_u c_p$. Upon substitution into Eq. (6.160), the quenching distance d can be expressed as

$$d \sim \frac{2\alpha}{S_L} \tag{6.161}$$

From Eq. (6.5), the flame thickness δ_r is

$$\delta_r = \frac{\rho_u S_L}{\dot{\omega}} \sim \frac{\alpha}{S_L} \tag{6.162}$$

so

$$d \sim 2\delta_r \tag{6.163}$$

indicating that the quenching distance is approximately twice the flame thickness. The typical quenching distance for hydrocarbon fuels between parallel plates is around 1.8 mm.

In general, the quenching distance is smaller for more reactive fuels. For a given fuel/oxidant combination, the quenching distance also depends on the inert diluent used in the mixture in the following order:

$$d_{He} > d_{Ar} > d_{N_2} > d_{CO_2} \tag{6.164}$$

This implies that the flame is, in effect, more insulated from the walls by carbon dioxide than by helium, due to the difference in the thermal diffusivities of the diluents,

$$\alpha_{He} > \alpha_{Ar} \approx \alpha_{N_2} > \alpha_{CO_2} \tag{6.165}$$

As the initial temperature of the unburned mixture is increased, the flame is able to pass through smaller openings, and, hence, the quenching distance is decreased. The quenching distance also depends on pressure; it decreases as pressure increases. We can see this dependency using the following equations:

$$\alpha = \frac{k}{\rho c_p} \sim \frac{T^{1.75}}{p}$$

$$S_L \sim p^{(m-2)/2}$$

The pressure dependence of the quenching distance is therefore

$$d \sim \frac{\alpha}{S_L} \sim p^{-m/2} \tag{6.166}$$

Example 6.3 ***Laminar Flame Quenching***

If the quenching distance for a methane–air parallel plate burner operating stoichiometrically at $p_1 = 6$ atm is $d_1 = 2.0$ mm, (a) What is the laminar flame speed S_{L1}? (b) If the pressure is doubled to $p_2 = 12$ atm, what is the flame speed S_{L2} and the quenching distance d_2?

Solution

(a) From Eq. (6.10), the laminar flame speed at 6 atm is

$$S_{L1} = \frac{43}{p_1^{1/2}} = \frac{43}{6^{1/2}} = 17.6\,\text{cm/s}$$

(b) The laminar flame speed at 12 atm is

$$S_{L2} = \frac{43}{p_2^{1/2}} = \frac{43}{12^{1/2}} = 12.4\,\text{cm/s}$$

Now $d \sim \alpha/S_L$ and $\alpha \sim 1/p$, so forming ratios,

$$d_2 = d_1 \frac{\alpha_2}{\alpha_1}\frac{S_{L1}}{S_{L2}} = d_1 \frac{p_1}{p_2}\frac{S_{L1}}{S_{L2}} = 2\left(\frac{6}{12}\right)\left(\frac{17.6}{12.4}\right) = 1.4\,\text{mm}$$

6.11 Flammability Limits of Premixed Laminar Flames

In very lean or very rich fuel/oxidizer mixtures, it is impossible to propagate a flame because there is too little fuel or oxidant to maintain a steady deflagration wave. Thus, there exist upper and lower flammability limits. The upper fuel concentration limit is called the rich or upper flammability limit. Likewise, the lower fuel concentration limit is called the lean or lower flammability limit. As these flammability limits are approached, the flame velocity decreases drastically and the flame is inevitably extinguished due to the instabilities caused by the combined influences of heat loss, flame stretching, and Lewis number effects.

For repeatable measurements of lean and rich flammability limits, external effects need to be accounted for so that the measurements are properties of only the fuel–oxidant mixture and standard test conditions. In the early 1950s, a standard vertical glass tube of 1.22-m (4-ft) length with an inner diameter of 5 cm was selected as a standard tube for observing whether a mixture is capable of propagating a flame indefinitely away from an ignition source. The ignition source is a spark plug located at the bottom of the tube which generates a strong spark of a few millimeters in length (see Figure 6.29*a*).

Typical variations of laminar-flame speed as functions of fuel–oxidant F/O ratio concentration and initial temperature of the gaseous mixture are shown schematically in Figure 6.29*b*. The upper and lower flammability limits of mixtures of various fuels with air are listed in Table 6.5. Note that the flammability limits of hydrogen are much wider than those of hydrocarbon fuels. According to Lewis and von Elbe (1987), the H_2-rich flammability limit that is observed experimentally corresponds to a mole fraction $X_{H_2} \approx 0.94$.

6.11.1 Effect of Pressure and Temperature on Flammability Limits

Flammability limits can be affected by the mixture pressure and temperature. As shown in Figures 6.30 and 6.31, the rich $\phi > 1$ limits become much wider with increasing pressure, while the lean limits are not appreciably dependent on pressure. A sketch of the pressure dependence is shown in Figure 6.30 as a function of the equivalence ratio. This behavior is especially valid for

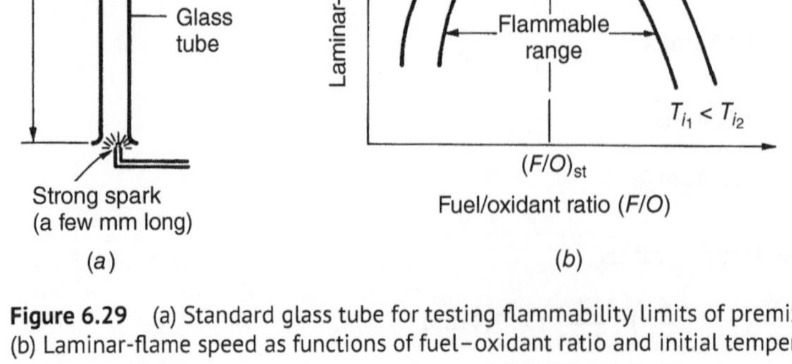

Figure 6.29 (a) Standard glass tube for testing flammability limits of premixed gases and (b) Laminar-flame speed as functions of fuel–oxidant ratio and initial temperature.

Table 6.5 Flammability Limits of Various Fuel–Air Mixtures.

Compound	Lower (Lean) Limits, Vol. (%) in Air	Upper (Rich) Limits, Vol. (%) in Air	Stoichiometric, Vol. (%) in Air
Methane, CH_4	5	15	9.47
Ethane, C_2H_4	2.82	13	5.64
Propane, C_3H_8	2.05	11.38	4.02
Butane, C_4H_{10}	1.86	8.41	3.13
Pentane, C_5H_{12}	1.40	7.80	2.56
Hexane, C_6H_{14}	1.18	7.40	2.16
Heptane, C_7H_{16}	1	6	1.87
Isooctane, C_8H_{18}	0.95	6.0	1.65
Acetylene, C_2H_2	2.5	80	7.72
Hydrogen, H_2	4	74.2	29.2
Carbon Monoxide, CO	12.5	74.2	29.5
Ethylene Oxide, C_2H_4O	3.0	80	7.72
Methanol (g), CH_3OH	5.88	49.94	12.24

Source: Kanury 1975/Gordon and Breach Science Publishers.

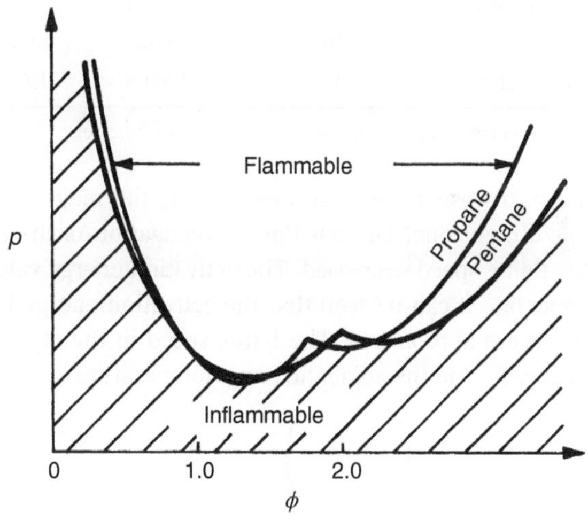

Figure 6.30 Dependence of the flammability limits of a hydrocarbon–air mixture on pressure and equivalence ratio.

hydrocarbon/air mixtures. The change of slope at the upper (rich) limit in Figure 6.31 is caused by the change of a second-order ($m = 2$) reaction at low pressures to a first-order ($m = 1$) reaction as the pressure is increased. For simple hydrocarbons such as ethane, propane, butane, and pentane, it appears that the rich limits extend linearly with increasing pressure. The lean limits, on the other hand, are at first extended slightly and approach asymptotic values thereafter as pressure is increased.

He and Clavin (1993) made an extensive numerical investigation of the flammability limits and structure of adiabatic planar flames in premixed H_2/O_2 gases. Their study used detailed chemical kinetic considerations to aid in the understanding of how the flame speed is affected by changes in the initial unburned mixture temperature T_u, system pressure p, and the equivalence ratio ϕ.

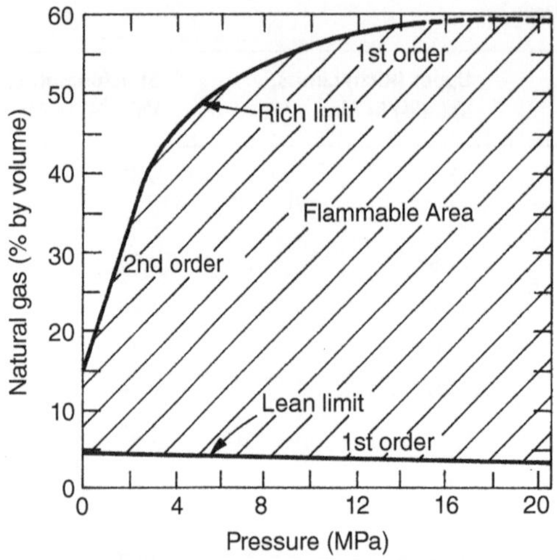

Figure 6.31 Dependence of the width of flammable region on pressure and chemical reaction order *m*. (Adapted from Lewis and von Elbe 1987).

Table 6.6 Global Activation Energy E_a of H_2/O_2 Mixtures at $T_u = 298$ K.

Mole Fraction X_{H_2}	$p = 0.2$ atm E_a (kJ/mol)	1 atm E_a	2 atm E_a	T_f (K)
0.15 (Lean)	61.74	101.72	142.78	1432
0.94	53.16	63.62	88.20	1293
0.952 (rich)	54.04	126.08	196.94	1100

Source: He and Clavin 1993/with permission of Elsevier.

Their analysis focused on determining the major chemical reactions responsible for supporting one-dimensional laminar-flame propagationAs the pressure was increased from 0.2 to 2.0 atm, the flame speed decreased. The activation energy values calculated by He and Clavin are listed in Table 6.6. It can be seen that the activation energy increases as both lean and rich flammability limits are approached. The flame speed of the H_2/O_2 mixture is related to the global activation energy E_a and the adiabatic flame temperature T_f of burned gases by an Arrhenius-type relation,

$$S_L \sim \exp\left(-\frac{E_a}{2R_u T_f}\right) \tag{6.167}$$

6.12 Turbulent Flame Propagation

As the Reynolds number of the mean flow field increases above a critical value, the flow field will become turbulent. Turbulent flow can be envisioned as a mean fluid flow upon which are superimposed eddies of different sizes randomly dispersed in the flow. Due to the presence of the turbulent eddies, the propagation speed of a flame through a turbulent flow field is much greater than the laminar flame speed and increases as the Reynolds number increases.

In turbulent premixed combustion, eddies of unburned mixture are consumed during their flow through the reaction zone. Within each eddy, the burning rate is enhanced by the flow of heat and active species (e.g., radical species) from the enveloping flame front. The turbulent flame surface is

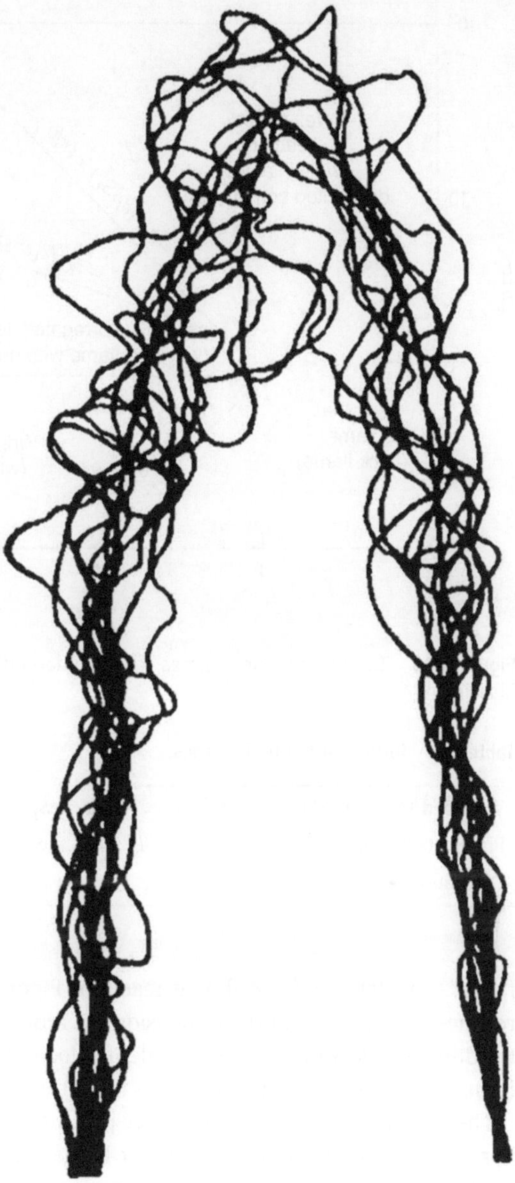

Figure 6.32 Superimposed contours of instantaneous flame boundaries in a wrinkled turbulent flame. *Source*: Adapted from Fox and Weinberg 1962/The Royal Society.

very geometrically complex, and the presence of eddies makes it difficult to locate the instantaneous flame surface. Unlike laminar premixed flames, turbulent premixed flames are usually quite thick due to the fluctuating flame boundary.

The effect of turbulence is to wrinkle and distort the flame front. In the flow field, the turbulent vortices spread ignition sites via a ragged edge wave. As shown in Figure 6.32, a wrinkled turbulent flame has a surface which is constantly changing in time. The position of the flame front moves irregularly, making the time-averaged flame profile appear relatively thick, forming a "turbulent flame brush."

Turbulent flame propagation is generally classified into three overall combustion regimes, depending on the turbulence intensity and related parameters. The regimes are wrinkled laminar

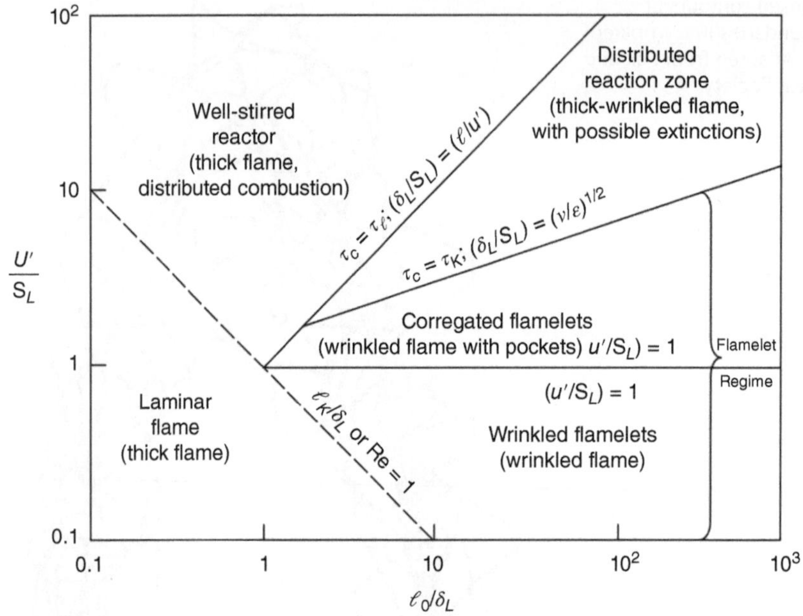

Figure 6.33 Turbulent flame regimes. *Source*: Adapted from Abdel-Gayed et al. 1989/with permission of Elsevier.

Table 6.7 Turbulent Flame Regimes.

Wrinkled Laminar Flame	$\delta_L < \eta$	$u_t \sim S_L$
Flamelets in Eddies	$\eta < \delta_L < l$	$u_t \gg S_L$
Distributed Reactions	$\delta_L > l$	

flame, flamelets, and distributed reaction. Figure 6.33 is a regime map of the turbulent flame regimes, as functions of the dimensionless ratio of the integral scale to flame thickness l/δ_L and the dimensionless ratio of the turbulent velocity to laminar flame speed u_t/S_L. (In Figure 6.33, l_K is the Kolmogorov length scale η, u' is the turbulent velocity u_t, and l_o is the integral length scale l.)

The characteristics of the turbulence parameters in each of the regimes are outlined in Table 6.7. For example, in the wrinkled laminar flame regime, the flame thickness δ_L is thinner than the Kolmogorov microscale η, and the turbulent velocity u_t is the same order as the laminar flame speed S_L. In this regime, chemical reactions occur in thin wrinkled sheets.

The degree to which the fluctuating components of velocity, temperature, density, pressure, and species concentration in a turbulent flame affect the chemical reactions, heat release, and the flame structure depends on various ratios of the characteristic times associated with these parameters. The characteristic chemical reaction time τ_r is the ratio of the laminar flame thickness divided by the laminar flame speed,

$$\tau_r = \frac{\delta_L}{S_L} \tag{6.168}$$

and the turbulent eddy turnover or mixing time τ_i is the integral length scale divided by the turbulent velocity $\tau_i = l/u_t$.

As introduced in Section 5.13, the ratio of the eddy turnover time to the chemical reaction time is defined as the Damköhler number Da.

$$Da = \frac{\tau_i}{\tau_r} = \frac{l}{\delta_L}\frac{S_L}{u_t} \tag{6.169}$$

The Damköhler number is named after G. Damköhler (1908–1944), a German scientist who performed pioneering research on reacting flows. A low Damköhler number ($Da \ll 1$) corresponds to a relatively slow chemical reaction. In this case, reactants and products are mixed by the turbulent eddies before reacting. The formulation of the process becomes that of a system described as a *perfectly stirred reactor*. As the Damköhler number increases, the turbulent eddy turnover time becomes longer than the characteristic chemical reaction time, i.e., the chemical reactions become more rapid than the fluid mixing.

Combustion in most engineering devices occurs with fast chemical reaction rates where $Da \gg 1$. At these large Damköhler numbers, the turbulent eddy time is much longer than the characteristic chemical reaction time, corresponding to a thin reaction zone distorted and convected by the flow field. The turbulent eddies wrinkle and strain the flame surface. The internal structure of the flame is described as a laminar flame element called a "flamelet." The structure of flamelets within premixed turbulent combustion and the effect on the turbulent burning velocity is discussed further in the review by Driscoll (2008).

In the flamelets-in-eddies regime, the flame thickness is greater than the small eddy thickness η, but less than the integral thickness l, and the turbulent intensity is much greater than the laminar flame speed. The increased wrinkling can result in the creation of pockets of unburned gas mixture. Accordingly, in this regime, at higher Damköhler numbers, the burning rate continues to be controlled by the turbulent mixing rate, i.e., the eddy turnover time, not the chemical reaction time.

For a laminar premixed flame, we have defined a flame speed S_L that is independent of geometrical features. It would be equally desirable to define a turbulent flame speed S_t that would be independent of geometry and depends only on the fuel–air ratio and fluid properties such as k, ν, and $\mathcal{D}$. However, in turbulent flow, this is not possible, because the transport properties of turbulent flames are functions of the flow field rather than the fluid. In a turbulent flow field, the turbulent diffusivity can be several orders of magnitude higher than the molecular diffusivity, i.e., $\mathcal{D}_t \gg \mathcal{D}$.

As discussed previously, the laminar flame speed S_L represents the velocity at which a plane laminar flame propagates into the unburned gas under nonturbulent and adiabatic conditions. Similarly, we define the turbulent flame speed S_t in a given turbulent flow as the speed at which an unburned mixture enters the flame zone normal to the flame surface. From the continuity equation,

$$S_t = \frac{\dot{m}}{\rho_u \bar{A}} \tag{6.170}$$

where $\bar{A}$ is a time-averaged flame area, and $\dot{m}$ is the reactant flowrate. This definition of turbulent flame speed is called a consumption speed since it is based on the reactant consumption.

In the wrinkled laminar flame regime, following Damköhler (1947), we can derive a simple turbulence model by assuming that the flamelets propagate with a plane laminar flame speed S_L, so the only effect of turbulence is the wrinkling of the flame. Expanding the continuity equation (6.170), we have

$$\dot{m} = \rho_u \bar{A} S_t = \rho_u A_{\text{flamelets}} S_L \tag{6.171}$$

and therefore

$$\frac{S_t}{S_L} = \frac{A_{\text{flamelets}}}{\bar{A}} \tag{6.172}$$

If we define the term A_{wrinkles} as the additional flame area due to the turbulent wrinkling u_t, then

$$A_{\text{flamelets}} = \bar{A} + A_{\text{wrinkles}} \tag{6.173}$$

and also

$$\frac{A_{\text{wrinkles}}}{\bar{A}} = \frac{u_t}{S_L} \tag{6.174}$$

upon substitution into Eq. (6.172),

$$\begin{aligned} \frac{S_t}{S_L} &= \frac{\bar{A} + A_{\text{wrinkles}}}{\bar{A}} \\ &= 1 + \frac{A_{\text{wrinkles}}}{\bar{A}} \\ &= 1 + \frac{u_t}{S_L} \end{aligned} \tag{6.175}$$

In this simple wrinkled flame model, the turbulent flame speed S_t is simply the sum of the laminar flame speed S_L and the turbulence velocity u_t.

6.13 Example: Turbulent Flame Propagation in a Spark Ignition Engine

A spark ignition engine has a relatively homogeneous turbulent mixture of fuel and air and when ignited by a spark, sustains a reaction process that propagates a turbulent flame in the form of a thin wrinkled sheet through the mixture. Internal combustion engines (and gas turbine combustors) operate in the wrinkled laminar flame and the flamelets-in-eddies regimes, depending on the engine speed. During the combustion process, the energy release rate starts relatively slowly, increases to a maximum value, and then decreases near the end of the combustion process. Accordingly, the cylinder pressure increases to a maximum value after top dead center as the flame propagates across the cylinder and then decreases during the expansion stroke.

The flame speed needs to be rapid enough to propagate across the cylinder during the time available near the end of the compression stroke and the beginning of the expansion stroke. The maximum laminar flame speed is about 0.8 m/s, so that the propagation time for a laminar flame in an engine with a cylinder radius of $b/2 = 0.05$ m is about 60 ms. Since the time for one revolution of a piston operating at 3000 rpm is 20 ms, a factor of 3 smaller, a turbulent fuel–air mixture in the cylinder is required to produce a turbulent flame and increase the flame propagation speed to an acceptable level.

Witze and Vilchis (1981) used a laser shadowgraph technique for combustion visualization in a spark ignition engine. Shadowgraph photography is a method of flow visualization that shows contrasts due to differences in the density of the flow. It does not record light emitted by the flame; rather, it records light transmitted through and refracted by the gases. Figure 6.34 shows laser shadowgraph sequences for lean ($\phi = 0.55$) and slightly rich ($\phi = 1.1$) combustion. A ragged edge wave is seen propagating into the unburned mixture. Ignition delay (in degrees of crank angle) is on the order of 10° for the rich case and 20° for the lean case. At 20° and 25° after ignition in the rich case, the width of the flame front is clearly discernible. The width is more difficult to discern in the lean case because it is two or three times thicker. Thus, a completely burned region does not appear until approximately 40° after ignition. At this time, the whitest region is burned gas, the greyish region in front of the flame is unburned gas and the highly convoluted dark, and white

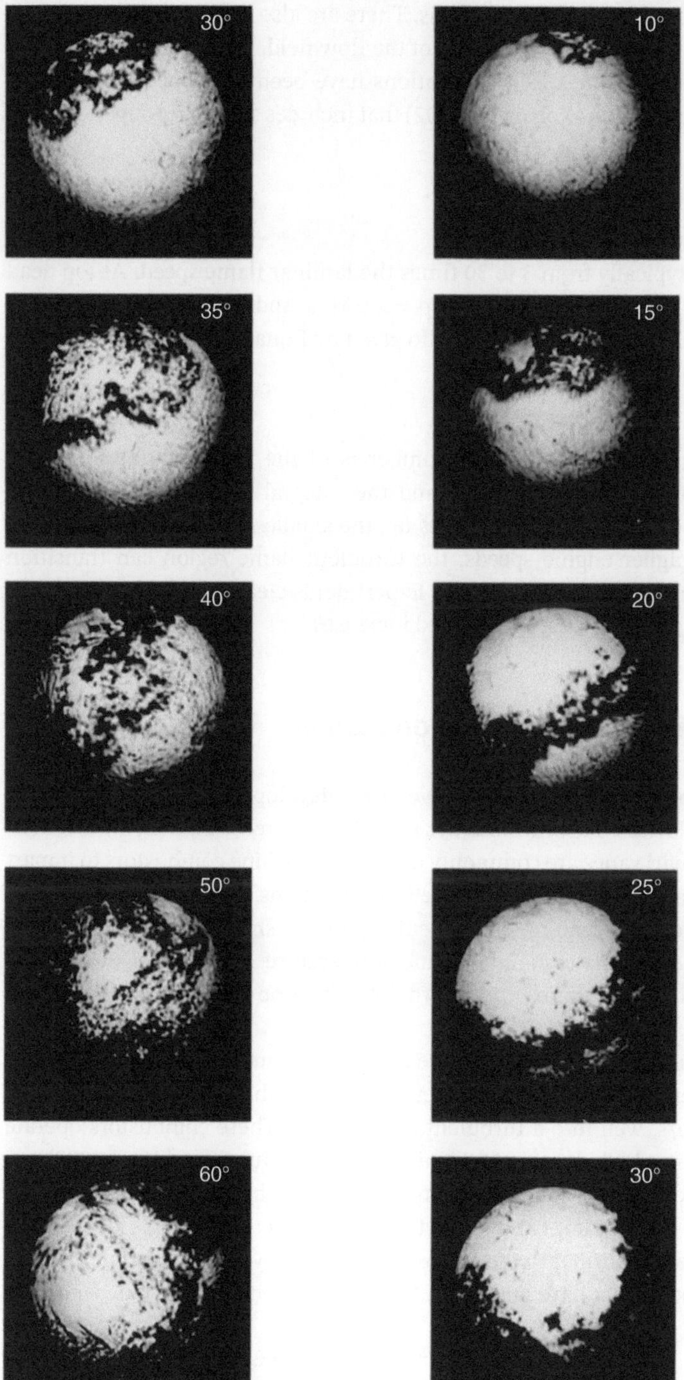

(*a*) (*b*)

Figure 6.34 Laser shadowgraph of combustion in a spark ignition engine. (a) Lean mixture $\phi = 0.55$. (b) Rich mixture $\phi = 1.1$. *Source*: Adapted from Witze and Vilchis 1981/with permission of SAE International.

region is a mixture of burned, burning, and unburned gas. There are also cycle-to-cycle variations in the flame propagation caused by the random features of the flow field.

A number of empirical turbulent flame speed correlations have been developed from engine experiments. A representative correlation (Kobayashi 2002) that includes a pressure dependency is

$$\frac{S_t}{S_L} = 2.9\left(\frac{u_t}{S_L}\frac{p}{p_{ref}}\right)^{0.38} \tag{6.176}$$

where $p_{ref} = 0.1$ MPa.

The turbulent flame speed is typically from 3 to 30 times the laminar flame speed. At top dead center, if the turbulent velocity $u_t = 2.0$ m/s, the pressure $p = 2.0$ MPa, and the laminar flame speed $S_L = 0.5$ m/s, then the turbulent/laminar flame speed ratio given by Equation (6.176) is

$$\frac{S_t}{S_L} = 2.9\left(\frac{2.0}{0.5}\frac{2.0}{0.1}\right)^{0.38} = 15.3$$

For internal combustion engines, the Damköhler number is of the order of 100. Since the turbulence in an engine increases with piston speed and the integral scale is independent of engine speed, we should expect that as engine speed goes up, the smallest eddies of the turbulent flow will decrease in size. At higher engine speeds, the turbulent flame region can transition from a wrinkled sheet to the flamelets-in-eddies regime. Experiments clearly show that the flame wrinkling due to turbulence increases as the engine speed increases.

6.14 Stabilization of Turbulent Combustion Flames

Various flame holder and stabilization techniques have been developed to produce a stable turbulent flame in combustion devices. These techniques include swirl, recirculation, bluff bodies, and pilot flames. For example, swirl vanes are commonly used in gas turbine combustors to impart swirling motion to the primary airflow in order to reduce NO_x emissions. Swirl reduces the flame length by producing higher rates of entrainment, particularly in the shear layer region, which improves flame stability and reduces the maximum flame temperature. A review of the physics of lean premixed combustion for gas turbines and internal combustion engines is presented in Swaminathan and Bray (2011).

In a gas turbine or industrial furnace combustor, we need to keep the premixed flame stable over a variety of operational conditions. In ramjet and turbojet combustors, the velocity of the fuel–air mixture can be as great as 50 m/s, well into a turbulent flow regime. These combustors operate close to a lean blowoff limit using a high degree of fuel–air premixing (Huang and Yang 2009).

As shown in Figure 6.35, recirculation of combustion products can be obtained by several means: (1) by inserting bluff bodies in the stream; (2) by directing part of the flow or one of the flow constituents, usually air, opposed or normal to the mainstream; or (3) by using a step in the wall enclosure. The types of bluff bodies that have been used for recirculation include rods, v-gutters,

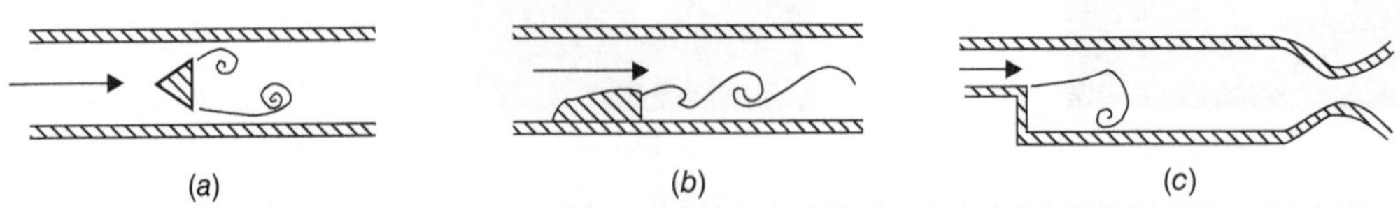

Figure 6.35 Stabilization of turbulent combustion flames. (*a*) Bluff body in free-stream, (*b*) Bluff body attached to wall, and (*c*) Step in wall.

toroids, disks, and spheres. Ramjet-type burners use rods as bluff bodies, which recirculate part of the reacting flow to form hot combustion products, forming a flame in the wake region of the rod. The combustion products ignite the free-stream gases by transferring thermal energy from the hot recirculation wake to the free-stream gases. For Reynolds numbers greater than 10^4, a fully developed turbulent wake exists during combustion. The length of the turbulent wake has been found to be directly proportional to the diameter of the bluff body.

A blowoff velocity V_{bo} correlation with laminar flame speed S_L in Peclet number form can be written as (Spalding 1955):

$$\frac{V_{\text{bo}} d}{\alpha} = C\left(\frac{S_L d}{\alpha}\right)^2 \tag{6.177}$$

where d is the characteristic diameter of the bluff body, α is the thermal diffusivity, and the constant $C \sim 3.5$, depending on the bluff body geometry. The correlation Eq. (6.177) indicates that the blowoff velocity is proportional to the bluff body diameter d, the combustion pressure p, and the square of the laminar flame speed S_L.

6.15 Turbulent Flame Models

6.15.1 Bray–Moss–Libby Model

A widely used flamelet model for turbulent premixed flames was developed by Bray et al. (1985). The model is based on a combustion progress variable c, where $c = 0$ denotes reactants in an unburned (u) state, $c = 1$ represents products in a burned (b) state, and any intermediate values between 0 and 1 denote a partially burned gas mixture with temperatures and composition between those of the reactants and products. For adiabatic flames with Lewis number equal to one, the combustion progress variable c is

$$c = \frac{T - T_u}{T_b - T_u} \tag{6.178}$$

Since from the ideal-gas law,

$$\rho = \frac{p}{RT} \tag{6.179}$$

We can express the density $\rho(c)$ as

$$\rho(c) = \rho_u (c\tau + 1)^{-1} \tag{6.180}$$

where τ is a heat release parameter defined as

$$\tau = \frac{T_b}{T_u} - 1 = \frac{\rho_u}{\rho_b} - 1 \sim 6 \tag{6.181}$$

The Favre averaged density and temperature are

$$\frac{\overline{\rho}}{\rho_u} = (\tilde{c}\tau + 1)^{-1}; \quad \tilde{T} = T_u(\tilde{c}\tau + 1) \tag{6.182}$$

The Bray–Moss–Libby (BML) model introduces a bimodal approximation of the joint probability density function (PDF) that characterizes the probability of finding a value of the combustion progress variable c at location $\mathbf{x}$ and at time t. The PDF $P(c; \mathbf{x}, t)$ is composed of three terms, representing pockets of unburned (u) reactants and fully burned (b) products, with probabilities $\alpha(\mathbf{x}, t)$ and $\beta(\mathbf{x}, t)$, respectively, and a partially burned mixture with probability $\gamma(\mathbf{x}, t)$.

$$P(c; \mathbf{x}, t) = \alpha(\mathbf{x}, t)\delta(c) + \beta(\mathbf{x}, t)\delta(1 - c) + \gamma(\mathbf{x}, t)f(c; \mathbf{x}) \tag{6.183}$$

In Eq. (6.183), δ is the Dirac-delta function, and $\alpha(\mathbf{x}, t)$ and $\beta(\mathbf{x}, t)$ are the probabilities of finding the unburned and burned mixture, respectively, at location $\mathbf{x}$ and time t. The high probability of finding either reactants or products is represented by two Dirac delta functions at $c = 0$ and at $c = 1$ (see Figure 5.25). The function $f(c; \mathbf{x})$ represents the interior distribution of the progress variable c between $0 < c < 1$ and can be considered as a continuous function of product mass fraction. The intermittency between unburned and fully burned gases at a given location $\mathbf{x}$ in a reaction zone is indicated by the graph shown in Figure 6.36.

This formulation of the PDF implies that all mean quantities are the sum of three contributions: the means of reactants and products; weighted by their probabilities, α and β, respectively; and a burning zone contribution weighted by γ. The mean values of quantities such as velocity or progress variables are simply weighted averages of their means in reactants and products and are independent of γ. Reaction rates and reactant concentration gradients are zero everywhere, except in the thin-flame reaction zones, so their means are proportional to γ.

Upon integration of Eq. (6.183) over the combustion progress variable c from 0 to 1, the normalized BML PDF $P(c; \mathbf{x}, t)$ is equal to 1,

$$\alpha(\mathbf{x}, t) + \beta(\mathbf{x}, t) + \gamma(\mathbf{x}, t) = 1 \tag{6.184}$$

The main assumption of the BML model is that the chemistry is "fast"; thus, the intermediate values of c associated with $\gamma(\mathbf{x}, t)$ are relatively infrequent. Based on this assumption, the probability $\gamma(\mathbf{x}, t)$ of finding intermediate states of a gas mixture is much, less than unity at any location $\mathbf{x}$ at time t. In the fast-chemistry combustion regime, in which, a turbulent flame brush consists of unburned and fully burned gases separated by thin burning zones, it is found that $\gamma \sim \mathcal{O}(1/\mathrm{Da})$, where Da is the Damköhler number, the ratio of the turbulent mixing time to the chemical reaction time.

For high Damköhler number flow, $\gamma \ll 1$, so the bimodal PDF distribution of the progress variable c can be expressed as

$$P(c; \mathbf{x}, t) = \alpha(\mathbf{x}, t)\delta(c) + \beta(\mathbf{x}, t)\delta(1 - c) \tag{6.185}$$

where

$$\alpha(\mathbf{x}, t) + \beta(\mathbf{x}, t) = 1 \tag{6.186}$$

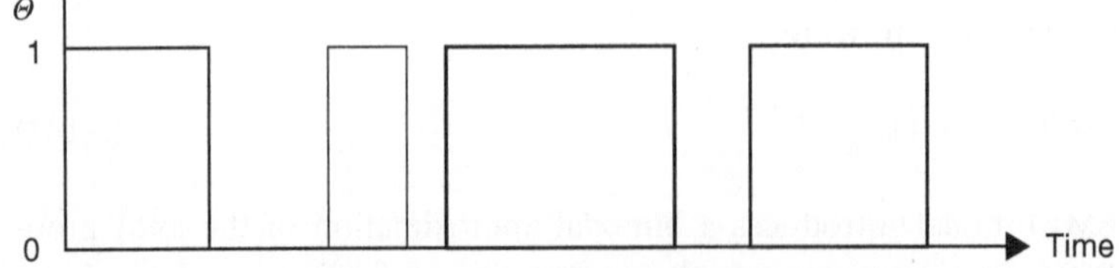

Figure 6.36 Intermittency between unburned and fully burned gases at location **x** in a reaction zone.

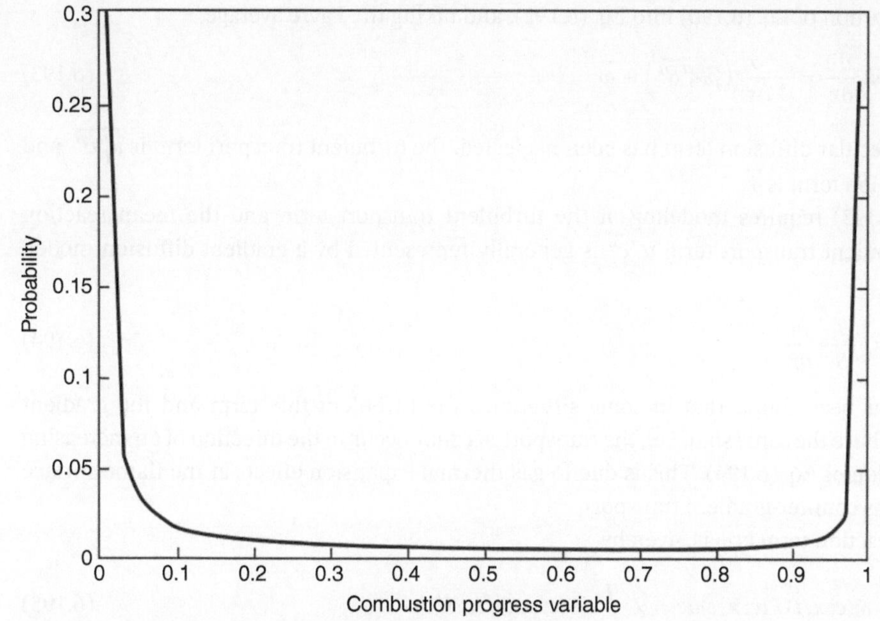

Figure 6.37 Direct numerical simulation (DNS) calculations of the probability density function (PDF) of the combustion progress variable *c*. (Adapted from Chakraborty and Cant 2009).

The bimodal PDF distribution of the progress variable c has been verified by the direct numerical simulation (DNS) results of Chakraborty and Cant (2009), as shown in Figure 6.37.

Recall from Section 5.16, the mean value $\overline{q}$ of any variable q can be determined from the variable's PDF by ensemble averaging, therefore,

$$\overline{q} = \int_0^1 qP(c;\mathbf{x},t)dc = \alpha(\mathbf{x},t)q_u + \beta(\mathbf{x},t)q_b \tag{6.187}$$

where

$$q_u = \int_0^1 q\,\delta(c)dc; \quad \text{and} \quad q_b = \int_0^1 q\,\delta(1-c)dc \tag{6.188}$$

The time-averaged mean progress variable $\overline{c}$ is therefore

$$\overline{c} = \int_0^1 cP(c;\mathbf{x},t)dc = \alpha(\mathbf{x},t)c_u + \beta(\mathbf{x},t)c_b = \beta(\mathbf{x},t) \tag{6.189}$$

since $c_u = 0$ and $c_b = 1$.

The Favre averaged equations for the mean and variance of the reactive scalars can be developed from the mean $\tilde{c}$ and fluctuation c'' components of the progress variable c,

$$c = \tilde{c} + c'' \tag{6.190}$$

Referring to Section 5.14, the Favre mean progress variable $\tilde{c}$ is

$$\tilde{c} = \frac{\overline{\rho c}}{\overline{\rho}} = \int_0^1 \left(\frac{\rho c}{\overline{\rho}}\right) P(c;\mathbf{x},t)dc = \frac{\rho_b}{\overline{\rho}}\beta(\mathbf{x},t) \tag{6.191}$$

The transport equation for c is

$$\frac{\partial \rho c}{\partial t} + \frac{\partial \rho u_i c}{\partial x_i} = \frac{\partial}{\partial x_i}\left(\rho D\frac{\partial c}{\partial x_i}\right) + \dot{\omega}_c \tag{6.192}$$

Upon substitution of Eq. (6.190) into Eq. (6.192), and taking the Favre average,

$$\bar{\rho}\frac{\partial \tilde{c}}{\partial t} + \bar{\rho}\tilde{u}_i\frac{\partial \tilde{c}}{\partial x_i} = -\frac{\partial}{\partial x_i}\left(\bar{\rho}\widetilde{u_i''c''}\right) + \bar{\dot{\omega}} \tag{6.193}$$

where the molecular diffusion term has been neglected, the turbulent transport term is $\widetilde{u_i''c''}$, and the mean reaction term is $\bar{\dot{\omega}}$.

Equation (6.193) requires modeling of the turbulent transport term and the mean reaction term. The turbulent transport term $\widetilde{u_i''c''}$ is generally represented by a gradient diffusion model, Eq. (6.194),

$$\bar{\rho}\widetilde{u_i''c''} = -\frac{\mu_t}{Sc_t}\frac{\partial \tilde{c}}{\partial x_i} \tag{6.194}$$

However, it has been found that in some situations, the turbulent flux term and the gradient diffusion term have the same sign, i.e., the transport of c may occur in the direction of an increasing $\tilde{c}$, a contradiction of Eq. (6.194). This is due to gas thermal expansion effects at the flame surface and is known as counter-gradient transport.

The mean reaction term is $\bar{\dot{\omega}}$ is given by

$$\bar{\dot{\omega}} = \int_0^1 \dot{\omega}(c;\mathbf{x},t)\,P(c;\mathbf{x},t)dc = \gamma\int_0^1 \dot{\omega}(c;\mathbf{x},t)f(c;\mathbf{x})dc \tag{6.195}$$

which requires the specification of the interior distribution function $f(c;\mathbf{x})$.

6.15.2 G-Equation Model

The G-equation model proposed by Williams (1985) is based on flamelet modeling assumptions and uses a level-set method to describe the evolution of the flame front as an interface between the unburned and the burned gases. The level-set function G is a nonreactive scalar which defines the flame surface, i.e., the level-set is the set of all points that lie on the flame surface. Since the level-set function is a nonreactive scalar, it does not have any source terms. It also avoids issues associated with counter-gradient diffusion.

The governing equation for the level-set function describes the propagation of the flame surface. An implicit representation of the flame surface is expressed as

$$G(\mathbf{x},t) = G_o \tag{6.196}$$

where G_o is an arbitrary fixed real number, typically set to zero. The unburned and the burned regions are separated by the flame surface, where $G < G_o$ in the unburned region and $G > G_o$ in the burned region, as shown in Figure 6.38. The applicability of a level-set function requires the flame to be a surface, so the flame needs to be in the thin wrinkled or corrugated flamelet regime.

In general, G is a field variable in 3-D space, and the level-set $G = G_o$ defines a 2-D surface within that field. The 2-D surface is assumed to be smooth and continuous. A flame surface that is convex toward the unburned mixture has a positive curvature and thus is positively stretched. The unit vector $\mathbf{n}$ normal to the surface shown in Figure 6.38 is defined in terms of G by

$$\mathbf{n} = -\frac{\nabla G}{|\nabla G|} \tag{6.197}$$

The negative sign indicates the convention that the unit normal is directed into the unburned mixture from the flame front.

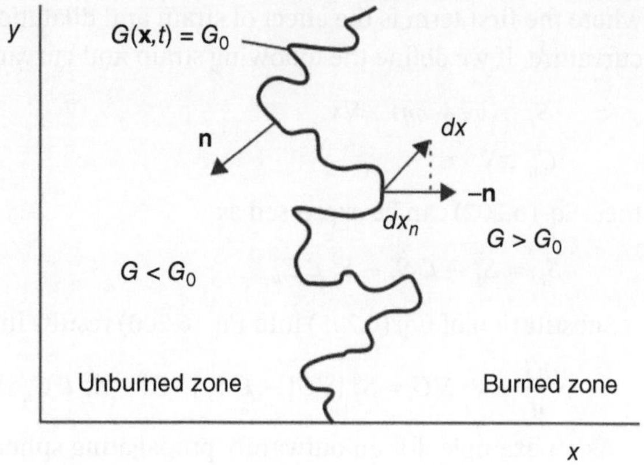

Figure 6.38 Schematic of flame front where $G(\mathbf{x}, t) = G_o$.

On the surface, the total derivative of G is zero,

$$\frac{\partial G}{\partial t} + \frac{d\mathbf{x}_f}{dt} \cdot \nabla G = 0 \tag{6.198}$$

where $\mathbf{x}_f$ is the flame front location. The flame front propagation speed is given by

$$\frac{d\mathbf{x}_f}{dt} = \mathbf{w} = \mathbf{v} + S_L \mathbf{n} \tag{6.199}$$

In Eq. (6.186), $\mathbf{w}$ is the flame surface velocity, $\mathbf{v}$ is the local flow velocity at the flame front, and S_L is the laminar flame speed. This equation is similar to Eq. (6.90). Substituting Eq. (6.199) into Eq. (6.185), and using Eq. (6.197), the G equation becomes

$$\frac{\partial G}{\partial t} + \mathbf{v} \cdot \nabla G = S_L\,|\nabla G| \tag{6.200}$$

The G-equation contains local unsteadiness and advection terms on the left-hand side of the equation and a propagation term with the laminar flame speed on the right-hand side. Note that there is no diffusion term in Eq. (6.200). The location of G_o can be defined to be anywhere in the flame, for instance, at a given temperature iso-surface either immediately ahead of the flame in the unburned gases or immediately behind the flame in the burned gases. Then, in Eq. (6.200), the velocity $\mathbf{v}$ and the laminar flame speed S_L are defined with respect to that location. The burning velocities defined with respect to the unburned and burned gases are denoted as $S_{L,u}$ and $S_{L,b}$, respectively.

The probability density of finding the flame surface $G(\mathbf{x}, t) = G_o$ at $\mathbf{x}$ and t is given by

$$P(G_o, \mathbf{x}, t) = \int_{-\infty}^{+\infty} \delta(G - G_o) P(G; \mathbf{x}, t)\, dG \tag{6.201}$$

Recall from Section 6.12 that the effect of flame stretch on the laminar flame speed is given by Eq. (6.89), restated below as Eq. (6.188),

$$S_L = S_L^o - \mathcal{L}\kappa \tag{6.202}$$

where S_L^o is the unstretched laminar flame speed, and $\mathcal{L}$ is the Markstein length.

The flame stretch factor κ from Eq. (6.111) is

$$\kappa \equiv \frac{1}{\delta A}\frac{d(\delta A)}{dt} = (\vec{v}\vec{v} + \vec{\eta}\vec{\eta}) : \nabla \mathbf{v} + S_L \nabla \cdot \mathbf{n} \tag{6.203}$$

where the first term is the effect of strain and dilatation and the second term is the effect of flame curvature. If we define the following strain and curvature parameters, S_r and C_u, respectively,

$$\begin{aligned} S_r &\equiv (\vec{\nu}\vec{\nu} + \vec{\eta}\vec{\eta}) : \nabla \mathbf{v} \\ C_u &\equiv \nabla \cdot \mathbf{n} \end{aligned} \tag{6.204}$$

then Eq. (6.202) can be expressed as

$$S_L = S_L^o - \mathcal{L}\, S_r - S_L^o\, \mathcal{L}\, C_u \tag{6.205}$$

Substitution of Eq. (6.205) into Eq. (6.200) results in the modified G equation,

$$\frac{\partial G}{\partial t} + \mathbf{v} \cdot \nabla G = S_L^o\, |\nabla G| - \mathcal{L}\, S_r\, |\nabla G| - S_L^o\, \mathcal{L}\, C_u\, |\nabla G| \tag{6.206}$$

As an example, for an outwardly propagating spherical flame with an instantaneous radius r_f, the flame curvature is $C_u = 2/r_f$, the bulk motion $\mathbf{v}$ of the burned products inside the flame is zero, and there is no strain rate effect, i.e., $S_r = 0$. The G differential equation, Eq. (6.206), for a spherical flame is therefore

$$\frac{\partial G}{\partial t} = S_b^o \left(\left| \frac{\partial G}{\partial r} \right| + \frac{2\mathcal{L}_b}{r_f} \frac{\partial G}{\partial r} \right) \tag{6.207}$$

where the laminar burning velocity S_b^o and the Markstein length $\mathcal{L}_b$ are based on the burned gas properties. Choosing a linear dependence of G on r,

$$G = r_f(t) - r \tag{6.208}$$

and upon substitution of Eq. (6.208) into Eq. (6.207),

$$\frac{dr_f}{dt} = S_b^o \left(1 + \frac{2\mathcal{L}_b}{r_f} \right) \tag{6.209}$$

Note that if $r_f < 2\mathcal{L}_b$, then there are no physically meaningful Solutions to Eq. (6.209), indicating there needs to be a minimum radius of the flame kernel for flame propagation to occur.

Further Reading

For more detailed information about premixed combustion, the reader is referred to texts by Kanury (1975), Williams (1985), Swaminathan and Bray (2011), Kuo and Acharya (2012), and Glassman et al. (2014).

References

Abdel-Gayed, R., D. Bradley, and F. Lung (1989), Combustion regimes and the straining of turbulent premixed flames, *Combust. Flame*, Vol. 76, pp. 213–218.

Andrews, G., and D. Bradley (1972a), Determination of burning velocities: a critical review, *Combust. Flame*, Vol. 18, pp. 133–153.

Andrews, G., and D. Bradley (1972b), The burning velocity of methane-air mixtures, *Combust. Flame*, Vol. 19, pp. 275–288.

Berlad, A., and A. Potter (1957), Relation of boundary velocity gradient for flash-back to burning velocity and quenching distance, *Combust. Flame*, Vol. 1, pp. 127–128.

Bosschaart, K., and L. de Goey (2003), Detailed analysis of the heat flux method for measuring burning velocities, *Combust. Flame*, Vol. 132, pp. 170–180.

Botha, J., and D. Spalding (1954), The laminar flame speed of propane/air mixtures with heat extraction from the flame, *Proc. R. Soc. London, Ser. A*, Vol. 225, pp. 71–96.

Bray, K., P. Libby, and J. Moss (1985), Unified modeling approach for premixed turbulent combustion–Part I: General formulation, *Combust. Flame*, Vol. 61, pp. 87–102.

Buckmaster, J. (1993), The structure and stability of laminar flames, *Annu. Rev. Fluid Mech.*, Vol. 25, pp. 21–53.

Candel, S., and T. Poinsot (1990), Flame stretch and the balance equation for the flame area, *Combust. Sci. Technol.*, Vol. 70, pp. 1–15.

Chakraborty, N., and R. Cant (2009), Effects of Lewis number on turbulent scalar transport and its modelling in turbulent premixed flames, *Combust. Flame*, Vol. 156, pp. 1427–1444.

Chung, S., and C. K. Law (1988), An integral analysis of the structure and propagation of stretched premixed flames, *Combust. Flame*, Vol. 72, pp. 325–336.

Clingman, W., R. Brokaw, and R. Pease (1953), Burning velocities of methane with nitrogen-oxygen, argon-oxygen, and helium-oxygen mixtures, *Proc. Combust. Inst.*, Vol. 4, pp. 310–313.

Damköhler, G. (1940), The effect of turbulence on the flame velocity in gas mixtures, *Z. Elektrochem. Angew. Phys. Chem.*, Vol. 46, No. 11, translated as NACA Tech. Memo. TM-1112 (1947).

Dixon-Lewis, G. (1972), Flame structure and flame reaction kinetics. VII. Reactions of traces of heavy water, deuterium and carbon dioxide added to rich hydrogen+ nitrogen+ oxygen flames, *Proc. R. Soc. London, Ser. A*, Vol. 330, pp. 219–245.

Dixon-Lewis, G. (1983), Spherically symmetric flame propagation in hydrogen-air mixtures, *Combust. Sci. Technol.*, Vol. 43, pp. 1–29.

Dowdy, D., D. Smith, S. Taylor, and A. Williams (1990), The use of expanding spherical flames to determine burning velocities and stretch effects in hydrogen/air mixtures, *23rd Symp. (Intl.) on Combust.*, pp. 325–332.

Driscoll, J. (2008), Turbulent premixed combustion: flamelet structure and its effect on turbulent burning velocities, *Prog. Energy Combust. Sci*, Vol. 34, pp. 91–134.

Dumitrache, C., R. Van Osdol, C. Limbach, and A. Yalin (2017), Control of early flame kernel growth by multi-wavelength laser pulses for enhanced ignition, *Nature*, Vol. 7, p. 10239.

Edse, R., and L. Lawrence Jr. (1969), Detonation induction phenomena and flame propagation rates in low temperature hydrogen-oxygen mixtures, *Combust. Flame*, Vol. 13, pp. 479–486.

Egolfopoulos, F., and C. Law (1990), An experimental and computational study of the burning rates of ultra-lean to moderately rich $H_2/O_2/N_2$ laminar flames with pressure variations, *23rd Symp. (Intl.) on Combust.*, pp. 333–340.

Fox, M., and F. Weinberg (1962), An experimental study of burner-stabilized turbulent flames in premixed reactants, *Proc. R. Soc. London, Ser. A*, Vol. 268, pp. 222–239.

Fristrom, R. (1965), Definition of burning velocity and a geometric interpretation of the effects of flame curvature, *Phys. Fluid.*, Vol. 8, pp. 273–280.

Glarborg, P., J. Miller, and R. Kee (1986), Kinetic modeling and sensitivity analysis of nitrogen oxide formation in well-stirred reactors, *Combust. Flame*, Vol. 65, pp. 177–202.

Glassman, I., R. Yetter, and N. Glumac (2014), *Combustion*, 5th ed., Academic Press, Cambridge, MA.

Gupta, V., ed. (2022), *Molecular and Laser Spectroscopy*, Elsevier, New York.

He, L., and P. Clavin (1993), Premixed hydrogen-oxygen flames. Part I: Flame structure near the flammability limits, *Combust. Flame*, Vol. 93, pp. 391–407.

Heimerl, J., and T. Coffee (1980), The detailed modeling of premixed, laminar steady-state flames. I. Ozone, *Combust. Flame*, Vol. 39, pp. 301–315.

Huang, Y., and V. Yang (2009), Dynamics and stability of lean-premixed swirl-stabilized combustion, *Prog. Energy Combust. Sci.*, Vol. 35, pp. 293–364.

Kanury, A. (1975), *Introduction to Combustion Phenomena*, Gordon and Breach, New York.

Karlovitz, B., D. Denniston Jr, D. Knapschaefer, and F. Wells (1953), Studies on turbulent flames: A. Flame propagation across velocity gradients B. turbulence measurement in flames, *Symp. (Intl.) Combust.*, Vol. 4, pp. 613–620.

Kee, R., J. Kee, J. Miller, and T. Jefferson (1980), Chemkin: A general-purpose, problem-independent, transportable, Fortran chemical kinetics code package, *Sandia Report SAND80-8003.*

Kee, R., J. Great, M. Smooke, and J. Miller (1985), A Fortran program for modeling steady laminar one-dimensional premixed flames, *Sandia Report SAND85-8240.*

Kuo, K. and R. Acharya (2012), *Fundamentals of Turbulent and Multiphase Combustion*, John Wiley and Sons, New York, NY.

Kobayashi, H. (2002), Experimental study of high-pressure turbulent premixed flames, *Exp. Therm. Fluid Sci.*, Vol. 26, pp. 375–387.

Kwon, S., and G. Faeth (1992), Stochastic simulation of free turbulent premixed flames in isotropic turbulence, *24th Symp. (Intl.) on Combust.*, pp. 451–460.

Kwon, S., L.-K. Tseng, and G. Faeth (1992), Laminar burning velocities and transition to unstable flames in $H_2/O_2/N_2$ and $C_3H_8/O_2/N_2$ mixtures, *Combust. Flame*, Vol. 90, pp. 230–246.

Law, C. K., D. Zhu, and G. Yu (1986), Propagation and extinction of stretched premixed flames, *21st Symp. (Intl.) on Combust.*, pp. 1419–1426.

Lewis, G., and B. von Elbe (1987), *Combustion, Flames, and Explosions of Gases*, 3rd ed., Elsevier, New York.

Lu, T., Y. Ju, and C. K. Law (2001), Complex CSP for chemistry reduction and analysis, *Combust. Flame*, Vol. 126, pp. 1445–1455.

Maas, U., and J. Warnatz (1988), Ignition processes in hydrogen-oxygen mixtures, *Combust. Flame*, Vol. 74, pp. 53–69.

Mallard, E., and H. LeChatelier (1883), Thermal model for flame propagation, *Ann. Mines*, Vol. 4, No. 18, pp. 379–568.

Markstein, G. (1964), *Nonsteady Flame Propagation*, Macmillan, New York.

Metghalchi, M., and J. Keck (1982), Burning velocities of mixtures of air with methanol, isooctane, and indolene at high pressure and temperature, *Combust. Flame*, Vol. 48, pp. 191–210.

Nilsson, E., L. de Goey, and A. Konnov (2013), Laminar burning velocities of acetone in air at room and elevated temperatures, *Fuel*, Vol. 105, pp. 496–502.

Palm-Leis, A., and R. Strehlow (1969), On the propagation of turbulent flames, *Combust. Flame*, Vol. 13, pp. 111–129.

Potter, A. (1960), Flame quenching, *Prog. Combust. Fuel Technol.*, Vol. 1, pp. 145–182.

Powling, J. (1949), A new burner method for the determination of low burning velocities and limits of inflammability, *Fuel*, Vol. 28, pp. 25–28.

Soid, S., and Z. Zainal (2011), Spray and combustion characterization for internal combustion engines using optical measuring techniques-a review, *Energy*, Vol. 36, pp. 724–741.

Spalding, B. (1955), *Some Fundamentals of Combustion*, Chapter 5. Butterworth, London, England.

Strehlow, R., and L. Savage (1978), The concept of flame stretch, *Combust. Flame*, Vol. 31, pp. 209–211.

Streng, A., and A. Grosse (1957), The ozone to oxygen flame, *6th Symp. (Intl.) on Combust.*, pp. 264–273.

Swaminathan, N., and K. Bray (2011), *Turbulent Premixed Flames*, Cambridge University Press, Cambridge, England.

Taylor, S. (1991), *Burning velocity and the influence of flame stretch*, PhD thesis, University of Leeds, Leeds, UK.

Tsang, W., and R. Hampson (1986), Chemical kinetic data base for combustion chemistry. Part I. Methane and related compounds, *J. Phys. Chem. Ref. Data*, Vol. 15, pp. 1087–1279.

Tseng, L.-K., A. Ismail, and G. Faeth (1993), Laminar burning velocities and Markstein numbers of hydrocarbon/air flames, *Combust. Flame*, Vol. 95, pp. 410–426.

Warnatz, J. (1977), *Berechnung der flammengeschwindigkeit und der struktur von laminaren flachen flammen*, PhD diss., Technische Universitat Darmstadt.

Warnatz, J. (1981), The structure of laminar alkane, alkene and acetylene flames, *18th Symp. (Intl.) on Combust*, pp. 369–384.

Warnatz, J. (1992), ReSolution of gas phase and surface combustion chemistry into elementary reactions, *24th Symp. (Intl.) on Combust.*, pp. 553–579.

Warnatz, J., and N. Peters (1984), Dynamics of flames and reactive systems, *AIAA Prog. Astron. Aeronaut.*, Vol. 95, p. 61.

Weinberg, F. (1963), *Optics of Flames*, Butterworths, London, England.

Williams, F. (1985), *Combustion Theory*, Addison Wesley, New York.

Witze, P., and F. Vilchis (1981), Stroboscopic laser shadowgraph study of the effect of swirl on homogeneous combustion in a spark ignition engine, SAE Technical Paper 810226.

Wohl, K., N. Kapp, and C. Gazley (1949), The stability of open flames, *3rd Symp. (Intl.) on Combust.*, pp. 3–21.

Wu, Y., V. Modica, B. Rossow, and F. Grisch (2016), Effects of pressure and preheating temperature on the laminar flame speed of methane/air and acetone/air mixtures, *Fuel*, Vol. 185, pp. 577–588.

Yetter, R., F. Dryer, and H. Rabitz (1991), A comprehensive reaction mechanism for carbon monoxide/hydrogen/oxygen kinetics, *Combust. Sci. Technol.*, Vol. 79, pp. 97–128.

Zeldovich, Y. (1948), Theory of flame propagation, *Zh. Fiz. Khim.*, Vol. 22, translated as NACA Tech. Memo. TM-1282 (1951).

Homework Problems

6.1 A long cylindrical tube of length L and diameter D is initially filled with a premixed combustible gaseous mixture of methane and oxygen. Before ignition, the mixture is stagnant with a pressure of 1 atm and a temperature of 298 K. Ignition at the left end of the tube is achieved by a spark. The flame front then moves toward the right. If the heat loss to the tube wall is *not* negligible, the flame front will not be a planar surface. The shape of the flame front is approximately parabolic, as shown in Figure 6.39.

(a) Sketch the following: the axial temperature and concentration profiles at various times along the centerline of the tube, and the radial temperature and concentration profiles near the flame front.

(b) If the tube diameter becomes very small, what will happen to the flame propagation process?

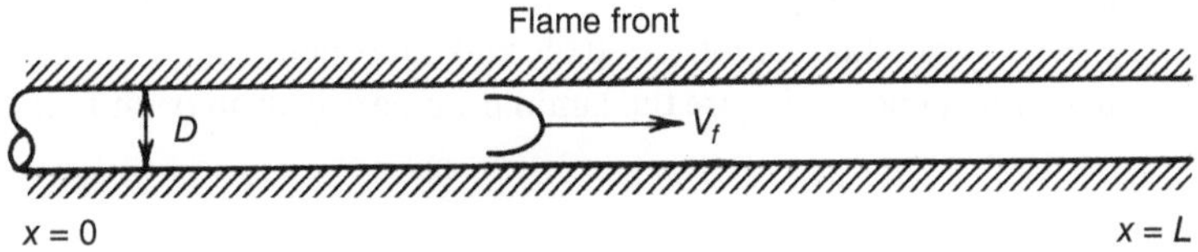

Figure 6.39 Parabolic flame front (Problem 6.1).

6.2 Using the thermal flame theory detailed in Section 6.2, and a single-step reaction model (Table 4.2), estimate the laminar flame speed S_L and thickness δ_r of a stoichiometric mixture of propane in air at 1 atm and $T_o = 300$ K.

6.3 Assume that the temperature profile across a premixed laminar flame of overall thickness δ can be approximated by a linear relationship between T and x. The unburned gas temperature is T_u at $x = 0$, and the fully reacted gas temperature is T_b at $x = \delta$. By dividing the flame into preheat and reaction zones, (a.) Derive expressions for δ and S_L in terms of heat of reaction per molecule of reactant $\tilde{q}$ and temperature difference $(T_b - T_u)$ using the energy equation (6.22) reproduced below, and (b) Describe their functional relationships with pressure.

$$\frac{k}{c_p}\frac{d^2T}{dx^2} - \rho u\frac{dT}{dx} + \frac{\dot{\omega}_r\tilde{q}}{c_p} = 0$$

6.4 A constant-volume spherical combustion chamber of radius R is filled initially with premixed combustible gaseous mixture. At time zero, a spark is produced at the center of the combustion chamber and a spherical flame propagates radially outward. The instantaneous pressure is measured from a pressure transducer mounted on the wall of the spherical combustion chamber. The compression of the unburned gas by the burned gas can be considered to be adiabatic. If the instantaneous radius of the flame is r, derive the laminar-flame speed S_L given by Eq. (6.64) reproduced below.

$$S_L = \frac{dr}{dt} - \frac{R^3 - r^3}{3p\gamma_u r^2}\frac{dp}{dt}$$

6.5 A gas mixture is contained in a soap bubble and ignited by a spark so that a spherical flame spreads radially. It is assumed that the soap bubble can expand. The growth of a flame front along a radius is followed by some photographic means. Relate the velocity of the flame front as determined from the photographs to the laminar-flame speed. If this method were used to measure flame speeds, what would you consider its advantages and disadvantages?

6.6 In each of the cases below, properly order the flame speeds:
(a) Stoichiometric H_2 with
  i. 21 % O_2 – 79 % He
  ii. 21 % O_2 – 79 % N_2
  iii. 21 % O_2 – 70 % Ar
(b) Wet CO–O_2 and dry CO–O_2, both at the same mixture ratio.
(c) A monopropellant decomposition flame and a hydrocarbon–oxygen flame.
(d) A hydrocarbon–air flame at high pressure and the same flame at low pressure.

6.7 Using the Metghalchi and Keck (1982) correlation in Section 6.5, (a) compare the laminar flame speed S_L of iso-octane, propane, and methanol for pressure $p = 20$ atm, unburned mixture temperature $T_u = 600$ K, residual fraction $f_r = 0.05$, and equivalence ratio $\phi = 0.9$, 1.0, 1.1, and 1.2. (b) At what equivalence ratios are the laminar flame speeds maximum for these fuels?

6.8 A stoichiometric fuel–air mixture flowing in a Bunsen burner (see Section 6.4) forms a well-defined conical flame. If the mixture is made leaner, holding the flow velocity in the tube constant, does the cone angle become larger or smaller than the angle for the stoichiometric mixture?

6.9 The laminar flame speed of a gas mixture at $T = 273$ K is to be measured by the Bunsen burner tube method. If the mixture is 9% methane in air, what size should the tube diameter be? (Hint: the laminar flame speed of the mixture is 34 cm/s at $T = 298$ K.)

6.10 What is the critical combustion radius r_c and MIE for ignition of a stoichiometric propane–air mixture at 1 atm and 298 K?

6.11 A ramjet has a flame stabilized in its combustion chamber by a 1.25-cm diameter rod. The mass flow of the fuel–air mixture entering the combustion chamber is 20.0 kg/s, which is the maximum flowrate that can be stabilized at a combustor pressure of 2.8 atm for a cylindrical configuration. The ramjet combustor is redesigned to fly with the same cross-sectional area, the same fuel–air mixture, and a doubling of the mass flow rate to 40.0 kg/s. The inlet diffuser is such that the temperature entering the combustor is the same as in the original case, but the inlet pressure has dropped to 1.8 atm. Assume the reaction is second order. (a) What is the minimum size rod that will stabilize the flame under these new conditions? (b) If the reaction is first order, what is the minimum rod size?

6.12 A straight cylindrical ramjet-type combustion duct is supplied with a stream of premixed fuel and air at 300 K, 1 atm, and inlet velocity of 10 m/s. A spherical bluff body flameholder is used to anchor the flame. Calculate the smallest diameter d of the spherical flameholder that will be able to stabilize the flame in the duct, using Eq. (6.177) correlation reproduced below, where $C = 0.80$.

$$\frac{V_{\text{bo}}d}{\alpha} = C\left[\frac{S_L d}{\alpha}\right]^2$$

The fuel is methane and the flame temperature can be taken as 2213 K for a lean mixture at an 80% stoichiometric condition. The laminar flame speed S_L is 20 cm/s. The thermal properties of gas mixtures before and after combustion can be treated as those of air.

6.13 A laminar flame propagates through a combustible mixture at 1 atm pressure, has a speed of 50 cm/s, and has a mass burning rate of 0.1 g/s-cm^2. The overall chemical reaction rate is second order in that it depends only on the fuel and oxygen concentrations. Now, consider a turbulent flame propagating through the same combustible mixture at a pressure of 10 atm. The integral length scale is $l = 5.0$ mm, and the turbulent diffusivity is 50 times the laminar diffusivity of $\nu = 100 \times 10^{-6}$ m^2/s. Estimate (a) the turbulent flame propagation speed S_t, and (b) the turbulent mass burning rate.

7

Detonation Waves

7.1 Introduction

The main topic of this chapter is detonation of a combustible gas mixture. Detonation results from chemical reactions in a gas mixture which produce pressure waves that propagate through the gas mixture at supersonic speeds. A detonation may be considered to be a shock wave generated by a chemical reaction and the corresponding heat release. Depending on the thermodynamic conditions, a gas mixture may support either a supersonic detonation wave or a subsonic deflagration wave, also called a flame. In this chapter, we will examine the general characteristics of both detonation and deflagration waves and derive the Rankine–Hugoniot equations which relate the fluid properties on either side of a combustion wave. We also outline methods for calculating the speed and structure of a detonation wave.

Detonations were first studied in relation to detonations in coal mines. In practical conditions, confined detonations are much more likely to occur than unconfined detonations. Confinement allows gas pressure buildup and the development of viscous boundary layers on a wall surface. A turbulent boundary layer can be formed on the wall surface, which promotes the transition from deflagration to detonation. Unconfined detonations can be produced due to leakage of explosive gases into the atmosphere. For example, the transportation of large quantities of liquified natural gas (LNG) uses large confinement vessels. The loss of integrity of a confinement vessel will result in a large unconfined cloud of natural gas, and it is possible to detonate this mixture if a sufficiently energetic ignition source is employed.

7.2 Qualitative Differences Between Detonation and Deflagration Waves

The combustion waves we will first consider will be modeled as one-dimensional (1-D) planar waves. A schematic diagram of a stationary 1-D combustion wave in a reference frame moving at the speed of the wave is shown in Figure 7.1 to help visualize the physical situation. In this reference frame following the wave motion, the unburned gases ahead of the wave move at velocity u_1 toward the wave front, and the wave is stationary. In the laboratory reference frame, the combustion wave in Figure 7.1 is moving to the left at a constant velocity u_1. The subscript (1) designates conditions of the unburned gases ahead of the wave, and subscript (2) indicates conditions of the burned gases behind the wave. Note that velocities u_1 and u_2 are defined with respect to the coordinate system relative to the moving reference frame following the wave.

Principles of Combustion, Third Edition. Allan T. Kirkpatrick and Kenneth K. Kuo.

Companion website: www.wiley.com/go/kirkpatrick/principlesofcombustion3e

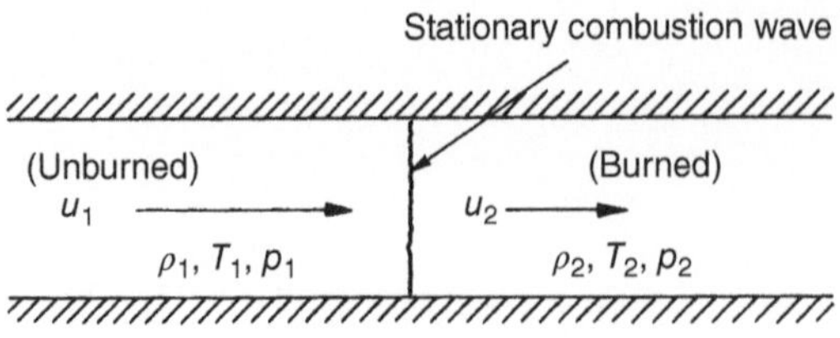

Figure 7.1 Schematic diagram of a one-dimensional deflagration or detonation wave.

Table 7.1 Qualitative Differences Between Detonation and Deflagration in Gases.

	Detonation	Deflagration
$Ma_1 = u_1/c_1$	5–10	0.0001–0.03
u_2/u_1	0.4–0.7 (deceleration)	4–6 (acceleration)
p_2/p_1	13–55 (compression)	≈ 0.98 (slight expansion)
T_2/T_1	8–21 (heat addition)	4–16 (heat addition)
ρ_2/ρ_1	1.7–2.6	0.06–0.25

Source: Adapted from Friedman 1953.

In this figure, we are following the motion of a 1-D combustion wave in a constant area duct closed at one end and open at the other end. The wave was initiated by igniting a combustible mixture in the duct. If ignited at the closed end of the duct, the flame is accelerated as it propagates through the mixture due to the expansion of the burned gas in the region between the flame and the closed end. The rapidly expanding gases produce a shock wave that interacts with the duct walls, sending a forward propagating shock that rapidly ignites the fuel ahead and a backward moving shock that dies out. The forward moving shock–combustion complex is called a detonation wave, and the rearward moving shock is called a retonation. If the wave is ignited at an open end of the duct, it remains at a steady subsonic velocity and does not accelerate to a supersonic detonation wave.

Typical physical values of the upstream (1) and downstream (2) conditions give us a good sense of the physical situation associated with combustion waves. A comparison of the values of typical deflagration and detonation waves is shown in Table 7.1. The upstream Mach *Ma* number of a detonation wave is >1 (supersonic) and that of a deflagration wave is <1 (subsonic). Under ambient conditions, detonation waves have speeds of the order of 1000 m/s or greater, and deflagration waves have speeds under 1 m/s, so detonation speeds are orders of magnitude greater than those of deflagration wave speeds.

The upstream and downstream fluid properties for a detonation wave are very similar to that of a normal shock. The flow velocity decreases across a detonation wave and increases across a deflagration wave. Detonation waves have a high compression ratio, with pressure and density increasing across a detonation wave, and decreasing across a deflagration wave. There is a large increase in temperature across the wave for both detonation and deflagration waves.

In deflagrations, chemical reactions are initiated due to heat and mass diffusion, and in detonations, chemical reactions are activated due to the compression caused by the leading shock. The reaction zone in a detonation wave supplies the energy to sustain the shock. Therefore, in detonations, there is a strong coupling between the shock wave and the zone of chemical energy release, which allows the detonation to maintain itself. As we will see in the next section, the detonation wave speed depends primarily on the heat release per unit mass of reactants, which in turn depends on the heating value of the fuel and the air–fuel ratio.

A detonation wave can also be produced in a long shock cylinder with a combustible gaseous mixture loaded in the low-pressure (driven) side of the shock cylinder. At time zero, the diaphragm separating the driver and driven sides of the shock cylinder is ruptured and a shock wave of known strength propagates into the combustible mixture. The passage of the incident shock raises the temperature and pressure of the reactive gas. When the shock hits the wall at the closed end of the cylinder, it reflects and raises the temperature and pressure of the gas behind the shock even higher. The reaction zone behind the shock sends forth a continuous compression wave that keeps the shock front from decaying. Just prior to the onset of detonation, a highly turbulent flame propagates behind a train of intense shock waves. The origin of the detonation occurs in a localized region in the unburned gases between the shock and the turbulent flame. The detonation wave then propagates upstream into the combustible mixture at a supersonic speed.

7.3 The Hugoniot Curve

In this section, we derive some basic relationships between the unburned and burnt properties of premixed gaseous mixtures contained in a constant-area duct. The analysis uses the 1-D mass, energy, and momentum conservation equations in which the burned region is assumed to be in thermodynamic equilibrium. The combustion wave is assumed to be infinitely thin and propagate at a steady-state speed with no heat loss to the surrounding cylinder wall. The reaction rate is assumed to be infinitely fast. Although these idealized conditions are difficult to achieve, the results derived from the basic 1-D conservation equations are important. In actual detonations, the wave structure is multidimensional due to hydrodynamic instabilities.

The solutions of steady-state detonation and deflagration waves lie on a Hugoniot curve, which is divided into several branches and regimes, corresponding to the different types of combustion waves. The Hugoniot curve is named for the French scientist Pierre Hugoniot (1851–1887) who made important contributions to compressible fluid mechanics.

The conservation equations for steady one-dimensional flow, with no body forces, no external heat addition or heat loss, and negligible Dufour and species interdiffusion effects, are as follows:

Continuity

$$\frac{d(\rho u)}{dx} = 0 \tag{7.1}$$

Momentum

$$\rho u \frac{du}{dx} = -\frac{dp}{dx} + \frac{d}{dx}\left[\left(\frac{4}{3}\mu + \mu'\right)\frac{du}{dx}\right] \tag{7.2}$$

Energy

$$\rho u \left[\frac{d}{dx}\left(h + \frac{u^2}{2}\right)\right] = -\frac{d}{dx}q_{\text{cond}} + \frac{d}{dx}\left[u\left(\frac{4}{3}\mu + \mu'\right)\frac{du}{dx}\right] \tag{7.3}$$

where

$$q_{\text{cond}} = -k\frac{dT}{dx} \tag{7.4}$$

$$h = c_p T + h_f^o \tag{7.5}$$

The viscosity $\left(\frac{4}{3}\mu + \mu'\right)$ in the momentum equation is (see Chapter 5)

$$\tau_{ij} = \mu\left(\frac{\partial u_i}{\partial x_j} + \frac{\partial u_j}{\partial x_i}\right) + \left(\mu' - \frac{2}{3}\mu\right)\frac{\partial u_k}{\partial x_k}\delta_{ij} \tag{7.6}$$

$$\tau_{11} = \left(2\mu + \mu' - \frac{2}{3}\mu\right)\frac{du}{dx} = \left(\frac{4}{3}\mu + \mu'\right)\frac{du}{dx} \tag{7.7}$$

The bulk viscosity μ' is usually very small and can be neglected.

By integrating Eqs. (7.1)–(7.3), we have

$$\rho u = \text{constant} \equiv \dot{m}'' \tag{7.8}$$

$$\rho u\frac{du}{dx} + u\frac{d}{dx}(\rho u) = -\frac{dp}{dx} + \frac{d}{dx}\left(\frac{4}{3}\mu\frac{du}{dx}\right) \tag{7.9}$$

Using the continuity equation (7.1), the momentum equation (7.2) becomes

$$\frac{d}{dx}\left[\rho u^2 + p - \frac{4}{3}\mu\frac{du}{dx}\right] = 0 \tag{7.10}$$

After integration with respect to x, we have

$$\rho u^2 + p - \frac{4}{3}\mu\frac{du}{dx} = \text{constant} \tag{7.11}$$

The energy equation (7.3) after integrating once with respect to x becomes

$$\rho u\left(c_p T + h_f^o + \frac{1}{2}u^2\right) - k\frac{dT}{dx} - u\left(\frac{4}{3}\mu\frac{du}{dx}\right) = \text{constant} \tag{7.12}$$

If we assume that both du/dx and dT/dx are equal to zero in the fully burned and unburned regions, then the following conservation equations provide the relationships between the flow properties in these two regions:

$$\rho_1 u_1 = \rho_2 u_2 = \dot{m}'' \tag{7.13}$$

$$p_1 + \rho_1 u_1^2 = p_2 + \rho_2 u_2^2 \tag{7.14}$$

$$c_{p1} T_1 + \frac{1}{2}u_1^2 + q = c_{p2} T_2 + \frac{1}{2}u_2^2 \tag{7.15}$$

or

$$h_1 + \frac{1}{2}u_1^2 = h_2 + \frac{1}{2}u_2^2 \tag{7.16}$$

The ideal gas equation of state for both regions is

$$p_1 = \rho_1 R_1 T_1 \tag{7.17}$$

$$p_2 = \rho_2 R_2 T_2 \tag{7.18}$$

where the mixture heat release q is

$$q \equiv h_{f1}^o - h_{f2}^o \tag{7.19}$$

and

$$h_f^o = \sum_{i=1}^{N} Y_i h_{f,i}^o \tag{7.20}$$

Equation (7.18) is simply the equation of state of the burned mixture, assuming the ideal-gas law is applicable. There are four equations [Eqs. (7.13)–(7.18)] relating the five unknowns: u_1, u_2, ρ_2, T_2, and p_2. Through algebraic substitutions and manipulations, these four equations can be reduced to a single equation involving only two unknowns, p_2 and ρ_2. This final result is called the Rankine–Hugoniot relation, named after William Rankine (1820–1872), a Scottish engineering professor who made seminal contributions to thermodynamics.

Before presenting the Rankine–Hugoniot relation, it is useful to examine some intermediate relationships. The combining of Eqs. (7.13) and (7.14) gives

$$p_2 - p_1 = \rho_1 u_1^2 - \rho_2 u_2^2 = \frac{(\rho_1 u_1)^2}{\rho_1} - \frac{(\rho_2 u_2)^2}{\rho_2} = \left(\frac{1}{\rho_1} - \frac{1}{\rho_2}\right) \dot{m}''^2 \tag{7.21}$$

$$\text{So} \quad \rho_1^2 u_1^2 = \frac{p_2 - p_1}{1/\rho_1 - 1/\rho_2} = \dot{m}''^2 \tag{7.22}$$

Equation (7.22) is generally referred to as the *Rayleigh-line relation*. This can also be written in terms of the Mach number *Ma*,

$$Ma_1 \equiv u_1/c_1 \tag{7.23}$$

$$c_1 \equiv \sqrt{\gamma R_1 T_1} = \sqrt{\gamma \left(\frac{p_1}{\rho_1}\right)} \tag{7.24}$$

$$\frac{\gamma}{\gamma} \frac{\rho_1^2 u_1^2}{\rho_1 p_1} = \frac{p_2/p_1 - 1}{1 - \rho_1/\rho_2} \tag{7.25}$$

Combining the above equations, the Rayleigh-line relation becomes

$$\gamma Ma_1^2 = \frac{p_2/p_1 - 1}{1 - \rho_1/\rho_2} \tag{7.26}$$

Using the definitions for c_p and γ, we have

$$c_p - c_v = R \tag{7.27}$$

or

$$c_p = \frac{\gamma}{\gamma - 1} R \tag{7.28}$$

Substituting the equation of state and Eq. (7.28) into Eq. (7.15), the conservation of energy equation can be rewritten in the following form:

$$\frac{\gamma}{\gamma - 1}\left(\frac{p_2}{\rho_2} - \frac{p_1}{\rho_1}\right) - \frac{1}{2}(u_1^2 - u_2^2) = q \tag{7.29}$$

Rearranging Eq. (7.14) to obtain expressions for u_1^2 and u_2^2 and substituting these expressions into Eq. (7.29), we have

$$\frac{\gamma}{\gamma - 1}\left(\frac{p_2}{\rho_2} - \frac{p_1}{\rho_1}\right) - \frac{1}{2}\Bigg(\underbrace{\frac{p_2 - p_1}{\rho_1} + \frac{\rho_2}{\rho_1} u_2^2}_{u_1^2} + \underbrace{\frac{p_2 - p_1}{\rho_2} - \frac{\rho_1}{\rho_2} u_1^2}_{-u_2^2}\Bigg) = q \tag{7.30}$$

Rearranging Eq. (7.13) and combining with Eq. (7.30), we have

$$\frac{\gamma}{\gamma - 1}\left(\frac{p_2}{\rho_2} - \frac{p_1}{\rho_1}\right) - \frac{1}{2}(p_2 - p_1)\left(\frac{1}{\rho_1} + \frac{1}{\rho_2}\right) = q \tag{7.31}$$

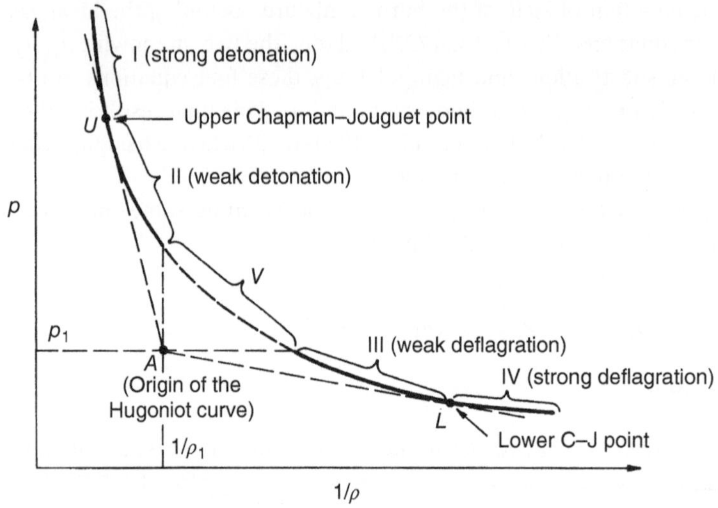

Figure 7.2 Hugoniot curve on p-versus-$1/\rho$ plane for given values of ρ_1, p_1, and q.

Equation (7.31) is known as the *Rankine–Hugoniot relation*. The plot of p_2 versus $1/\rho_2$ for a fixed heat release per unit mass, q, is called the *Hugoniot curve* and is shown in Figure 7.2.

The Hugoniot relation can also be expressed in terms of thermal plus chemical enthalpy h, by combining Eq. (7.5) with Eq. (7.28) and the ideal-gas law,

$$h = \left(\frac{\gamma}{\gamma - 1}\right)\frac{p}{\rho} + h_f^o \tag{7.32}$$

Combining Eqs. (7.19) and (7.32) gives

$$h_2 - h_1 = \frac{\gamma}{\gamma - 1}\frac{p_2}{\rho_2} + h_{f2}^o - \frac{\gamma}{\gamma - 1}\frac{p_1}{\rho_1} - h_{f1}^o = \frac{\gamma}{\gamma - 1}\left(\frac{p_2}{\rho_2} - \frac{p_1}{\rho_1}\right) - q \tag{7.33}$$

Substituting into Eq. (7.31) gives the equivalent Hugoniot relation:

$$h_2 - h_1 = \frac{1}{2}(p_2 - p_1)\left(\frac{1}{\rho_1} + \frac{1}{\rho_2}\right) \tag{7.34}$$

7.3.1 Properties of the Hugoniot Curve

The Hugoniot curve is essentially a plot of all the possible values of $(1/\rho_2, p_2)$ for a given value of $(1/\rho_1, p_1)$ and q. The point $(1/\rho_1, p_1)$, which is usually called the *origin* of the Hugoniot plot, is designated by the symbol A. For different values of q, one obtains a whole family of curves. When $q = 0$, the Hugoniot curve represents an adiabatic shock. Point A is on the $q = 0$ curve, and as q increases, the curve moves outward. Regions of possible solutions are constructed by drawing tangents to the curve through the point A, and vertical and horizontal lines from A. The curve is thus divided into five regions, specified as Roman numerals in Figure 7.2.

The two tangent points to the curve, points labeled U and L, are called Chapman—Jouguet (C–J) points (Chapman 1899; Jouguet 1905), generally referred to as C–J points, and denoted by U for the upper C–J point and L for the lower C–J point. It must be noted that although the curve represents all the *possible* solutions of the Hugoniot equation for the burned mixture, not all the solutions are valid, for physical reasons.

As we shall see, the C–J solutions require no knowledge of the structure of the detonation wave, unlike the deflagration wave, as equilibrium thermodynamic calculations for the C–J state suffice. The energy set free in the chemical reaction is utilized for the propagation of the shock wave in accordance with the Rankine–Hugoniot equations.

We shall establish which regions of the curve represent valid solutions. In region V, $p_2 > p_1$ and $1/\rho_2 > 1/\rho_1$. The Rayleigh-line expression, Eq. (7.22), then implies that u_1 is imaginary. Thus, region V is shown to be a physically impossible region.

To study the characteristics of the C–J points, let us consider q to be fixed and differentiate the Hugoniot relation, Eq. (7.31), with respect to $1/\rho_2$:

$$\frac{1}{\rho_2}\left(\frac{\gamma}{\gamma-1}\right)\frac{dp_2}{d(1/\rho_2)}+\left(\frac{\gamma}{\gamma-1}\right)p_2-\frac{1}{2}(p_2-p_1)$$
$$-\frac{1}{2}\frac{dp_2}{d(1/\rho_2)}\left(\frac{1}{\rho_1}+\frac{1}{\rho_2}\right)=0 \tag{7.35}$$

After rearranging, we have

$$\frac{dp_2}{d(1/\rho_2)}=\frac{(p_2-p_1)-\left(\frac{2\gamma}{\gamma-1}\right)p_2}{\left(\frac{2\gamma}{\gamma-1}\right)\frac{1}{\rho_2}-\left(\frac{1}{\rho_1}+\frac{1}{\rho_2}\right)} \tag{7.36}$$

The slopes at the tangent points U and L can also be represented as

$$\left[\frac{dp_2}{d(1/\rho_2)}\right]_{\text{C-J}}=\frac{p_2-p_1}{1/\rho_2-1/\rho_1} \tag{7.37}$$

Equating the right-hand side of Eq. (7.36) to that of Eq. (7.37) and rearranging, we have

$$\frac{p_2-p_1}{(1/\rho_2-1/\rho_1)}=-\gamma\rho_2p_2 \qquad \text{at C–J points} \tag{7.38}$$

Combining Eq. (7.38) with Eq. (7.22), we have

$$u_2^2=\frac{\gamma p_2}{\rho_2}=c_2^2, \qquad \text{or} \qquad |u_2|=c_2 \tag{7.39}$$

Therefore, at the C–J points, the velocity u_2 of the burned gases is equal to the speed of sound c_2 in the burned gases, i.e., the Mach number Ma of the burned gas is

$$Ma_2=1 \tag{7.40}$$

In the detonation branch of the Hugoniot curve (regions I and II), $1/\rho_2 < 1/\rho_1$; therefore,

$$u_2-u_1=\dot{m}''\left(\frac{1}{\rho_2}-\frac{1}{\rho_1}\right)<0 \tag{7.41}$$

or

$$u_1>u_2 \tag{7.42}$$

where u_1 and u_2 are the velocities of the unburned and burned gases relative to the detonation wave (see Figure 7.1). If we now consider a laboratory coordinate system in which the detonation wave is moving at velocity V_w relative to the stationary duct, the direction of the absolute velocities of the gases before v_1 and behind v_2 the detonation wave is shown in Figure 7.3. The algebraic relationships between the absolute and relative velocities are

$$v_1=V_w-u_1 \tag{7.43}$$

$$v_2=V_w-u_2 \tag{7.44}$$

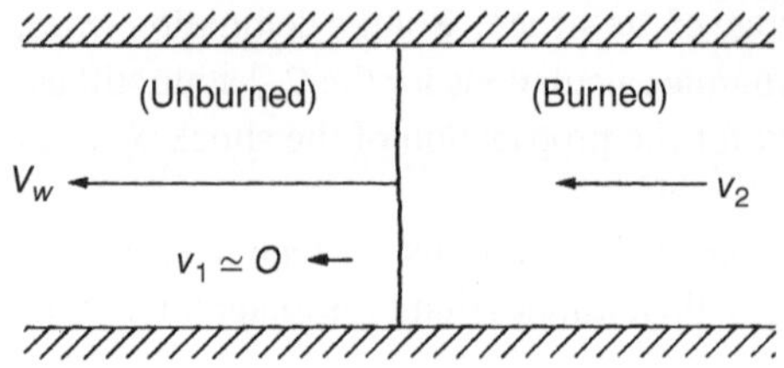

Figure 7.3 Detonation wave in the laboratory coordinate system.

While u_1 and u_2 are positive in the direction shown in Figure 7.1, v_1 and v_2 are positive in the direction shown in Figure 7.3.

Since the absolute velocity of the unburned gases is assumed to be zero, $V_w = u_1$, and from the inequality (7.42), we have

$$v_2 = V_w - u_2 = u_1 - u_2 > 0 \tag{7.45}$$

Physically, this means that the burned gases behind a detonation wave try to follow the wave motion. We shall next examine the magnitude of v_2 with respect to V_w to see whether the burned gases can catch up with the wave motion or not. First of all, the detonation-wave speed V_w can be expressed as the sum of u_2 and v_2 by rearranging Eq. (7.44), i.e.,

$$V_w = u_2 + v_2 \tag{7.46}$$

For the upper C–J point, $u_2 = c_2$ and we have

$$V_w = c_2 + v_2 > c_2 \tag{7.47}$$

This implies that the C–J detonation wave is traveling at a supersonic speed. Also, because $V_w > v_2$, the burned gases, although traveling in the same direction as the detonation wave, can never catch up with it.

Region I is called the strong-detonation region; within this region, the pressure p_2 of the burned gases is greater than the pressure p_U of the C–J detonation wave, i.e., $p_2 > p_U$. In passing through a strong detonation wave, the gas velocity relative to the wave front is slowed down substantially from supersonic speed to subsonic, i.e., $u_2 < c_2$. At the same time, the pressure and density increase significantly. In general, a strong detonation wave is seldom observed, since it requires a special experimental setup for generating overdriven shock waves within very strong confinement.

Region II is called the weak-detonation region; within it, the pressure p_2 of the burned gases is less than that of the C–J detonation wave, i.e., $p_2 < p_U$. In passing through a weak detonation wave, the gas velocity relative to the wave front is slowed down, but the burned mixture still has a velocity greater than that of sound. It is interesting to note that the isochoric $(1/\rho_2 \approx 1/\rho_1)$ weak detonation corresponding to infinite wave velocity [see Eq. (7.22)] is physically unattainable. In general, a weak detonation wave is seldom observed; it requires a gas mixture chosen specially to have extremely fast chemical kinetics.

Under most experimental conditions, detonations are manifested as Chapman–Jouguet waves (at the upper C–J point U). A part of the reason for ruling out regions I and II also stems from the structure of the detonation wave, which will be discussed in a later section. The upper C–J point U in Figure 7.4, therefore, represents the usual solution on the detonation branch of the Hugoniot curve. The wave speed at U corresponds to the minimum detonation wave speed, since a straight line through the origin $(1/\rho_1, p_1)$ will not intersect the detonation branch anywhere if the magnitude of its slope is less than that of the tangent line shown in Figure 7.4.

In contrast to the upper C–J point, the lower C–J deflagration (corresponding to point L on the Hugoniot curve) has the maximum wave speed of all deflagrations [see Eq. (7.22)]. This is because

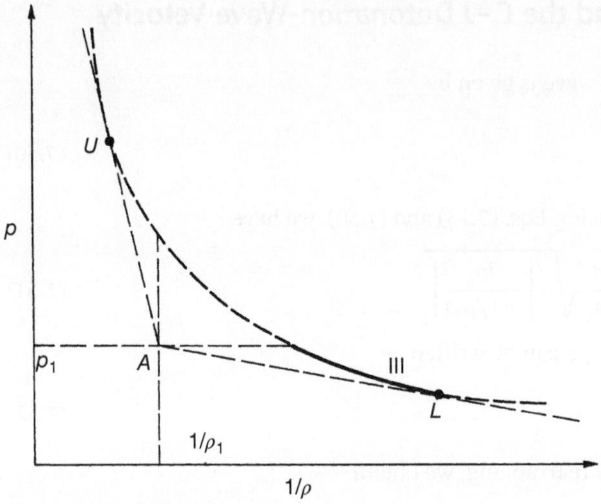

Figure 7.4 Solution regions on the Hugoniot curve.

a straight line through the origin A fails to intersect the deflagration branch if the magnitude of its slope exceeds that of the tangent line, and it intersects the deflagration branch at two points (once in region III and once in region IV) if its slope is less than that of the tangent line (see Figure 7.2).

Region III is called the weak-deflagration region; within it $p_1 \geq p_2 > p_L$ and $1/\rho_2 < 1/\rho_L$. In passing through a weak deflagration wave, the gas velocity relative to the wave front is accelerated from a subsonic velocity to a higher subsonic velocity. The detailed wave structure and methods for determining the laminar wave speed have been discussed in detail in Chapter 6. The isobaric weak deflagration ($p_s = p_1$) corresponds to zero propagation velocity. The solution in region III is often observed; in most experimental conditions, the pressure in the burned-gas zone is slightly lower than that of the unburned gases.

Region IV is called the strong-deflagration region. In passing through a strong deflagration wave, the gas velocity relative to the wave front must be accelerated substantially from subsonic to supersonic. Considerations of wave structure forbid a change from subsonic to supersonic flow in a constant-area duct (Williams 1985). Consequently, strong deflagration is never observed experimentally; the possible solution in the deflagration branch is shown in Figure 7.4 as the solid line on the Hugoniot curve. The C–J deflagration, corresponding to point L, is also not observed experimentally; hence, it is excluded from the solid curve.

In the realizable deflagration-solution region, the burned gases move away from the combustion wave front. This can be shown by using the continuity equation (7.13) and the conditions $1/\rho_2 > 1/\rho_1$ and $\dot{m}'' \geq 0$; we have

$$u_2 - u_1 = \dot{m}''\left(\frac{1}{\rho_2} - \frac{1}{\rho_1}\right) > 0 \tag{7.48}$$

Using Eq. (7.44) and substituting V_w for u_1, the above inequality becomes

$$v_2 < 0 \tag{7.49}$$

This shows that the burned gases, unlike those shown in Figure 7.3 for a detonation wave, flow away from the deflagration wave. This is one of the characteristic differences between deflagration and detonation waves.

7.4 Burned Gas Properties and the C–J Detonation-Wave Velocity

The speed of sound behind a detonation wave is given by

$$c_2 = \sqrt{\left[\frac{\partial p_2}{\partial \rho_2}\right]_s} \tag{7.50}$$

Since $u_2 = c_2$ at the upper C–J point, using Eqs. (7.13) and (7.50), we have

$$u_1 = \frac{1}{\rho_1} c_2 \rho_2 = \frac{1}{\rho_1}\sqrt{\rho_2^2\left[\frac{\partial p_2}{\partial \rho_2}\right]_s} = \frac{1}{\rho_1}\sqrt{-\left[\frac{\partial p_2}{\partial(1/\rho_2)}\right]_s} \tag{7.51}$$

The isentropic relation for the burned gas can be written as

$$p_2\left(\frac{1}{\rho_2}\right)^{\gamma_2} = \text{constant} \tag{7.52}$$

Differentiating the above equation and rearranging, we obtain

$$-\left[\frac{\partial p_2}{\partial(1/\rho_2)}\right]_s = \frac{\gamma_2 p_2}{1/\rho_2} \tag{7.53}$$

Substituting this expression into Eq. (7.51), we have

$$u_1 = \frac{1/\rho_1}{1/\rho_2}\sqrt{\gamma_2 p_2 \frac{1}{\rho_2}} = \frac{\rho_2}{\rho_1}\sqrt{\gamma_2 R_2 T_2} \tag{7.54}$$

A different form of Eq. (7.54) can be written as

$$\rho_1^2 u_1^2 = \gamma_2 p_2 \rho_2 \tag{7.55}$$

Substituting into the Rayleigh-line relation (7.22), we get

$$\frac{1}{\rho_1} - \frac{1}{\rho_2} = \frac{p_2 - p_1}{\gamma_2 p_2 \rho_2} \tag{7.56}$$

If we assume that $p_2 \gg p_1$ (see Table 7.1), then Eq. (7.56) reduces to

$$\frac{1}{\rho_1} - \frac{1}{\rho_2} = \frac{1}{\gamma_2 \rho_2} \tag{7.57}$$

$$\frac{\rho_2}{\rho_1} \approx \frac{\gamma_2 + 1}{\gamma_2} \tag{7.58}$$

The density ratio ρ_2/ρ_1 has maximum values that are close to 1.8 for combustion in air and 1.85 for combustion in oxygen at stoichiometric conditions. The density ratio then decreases on either side of stoichiometric conditions.

Upon substitution of Eq. (7.58) into Eq. (7.54), the detonation velocity u_1 can be expressed as

$$u_1 = \frac{\gamma_2 + 1}{\gamma_2}\sqrt{\gamma_2 R_2 T_2} \tag{7.59}$$

To determine the detonation velocity u_1, the properties of the burned gas at state 2 are needed. Solving the energy equation (7.15) for the burned gas temperature T_2,

$$T_2 = \frac{c_{p1}}{c_{p2}} T_1 + \frac{1}{2c_{p2}}(u_1^2 - u_2^2) + \frac{q}{c_{p2}} \tag{7.60}$$

Using the continuity equation (7.13), Eq. (7.58), and the ideal-gas relation

$$\frac{\gamma_2 R_2}{c_{p2}} = \gamma_2 - 1 \tag{7.61}$$

the temperature T_2 can be expressed as

$$T_2 = \left[\frac{c_{p1}}{c_{p2}} T_1 + \frac{q}{c_{p2}}\right] \frac{2\gamma_2^2}{\gamma_2 + 1} \tag{7.62}$$

Note that an initial estimate of the temperature T_2 is needed for Eq. (7.62), since the specific heat at state 2 is a function of temperature. Also, using the equation of state, Eq. (7.18), p_2 is related to ρ_2/ρ_1 and T_2 by

$$p_2 = \left(\frac{\rho_2}{\rho_1}\right)\left(\frac{R_2 T_2}{R_1 T_1}\right) p_1 \tag{7.63}$$

Gordon and McBride (1994) developed an iterative Newton–Raphson procedure for calculation of the C–J detonation wave velocity using alternative forms of the above equations and included dissociation reactions.

The dependence of the C–J detonation velocity u_1 on the equivalence ratio ϕ for various fuels in air and oxygen at $p_1 = 1$ atm and $T_1 = 298$ K is presented in Figure 7.5. The hydrocarbon detonation velocities in air are of the order of 1800 m/s and in oxygen are of the order of 2500 m/s. Note that hydrogen detonation velocities in air or oxygen are much greater compared to hydrocarbon detonation velocities, due to the difference in their molecular masses. The hydrocarbon detonation velocities are largest near stoichiometric conditions, in comparison, the hydrogen velocities continue to increase as the equivalence ratio is increased beyond stoichiometric.

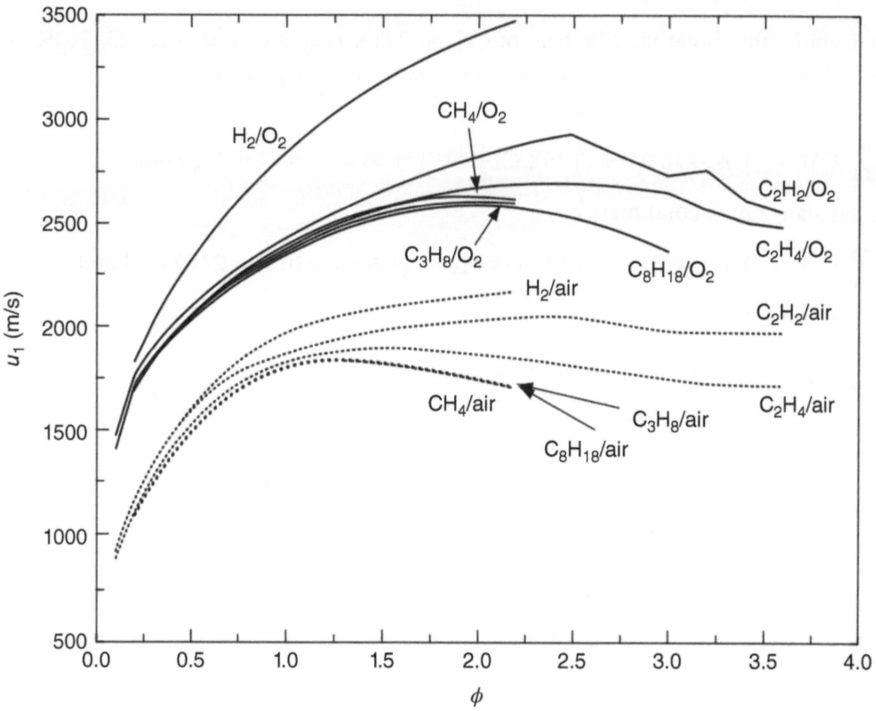

Figure 7.5 Dependence of C–J detonation velocities on equivalence ratio ϕ for $p_1 = 1$ atm and $T_1 = 298$ K. (Adapted from Glassman et al. 2014).

Table 7.2 Reactant and Product Property Table for Example 7.1

	M_i	n_i	$X_i = n_i/n$	$Y_i = X_iM_i/M_{mix}$	h_f^o (kJ/kg)
Reactants					
CH_4	16.04	1	0.095	0.055	−4665
O_2	32.00	2.0	0.19	0.220	0
N_2	28.01	7.52	0.715	0.725	0
Products					
CO_2	44.01	1	0.095	0.151	−8942
H_2O	18.016	2	0.19	0.124	−13, 424
N_2	28.013	7.52	0.715	0.725	0

Example 7.1 ***Detonation Velocity***

A detonation wave is propagating through a stoichiometric mixture of methane CH_4 and air at a temperature of 298 K and pressure of 1 atm. What is the C–J detonation velocity and the burned gas temperature, pressure, and density?

Solution

The stoichiometric methane reaction in air is

$$1CH_4 + 2(O_2 + 3.76N_2) \longrightarrow 1CO_2 + 2H_2O + 7.52N_2$$

From the Appendix, the molar specific heat of CH_4 at 298 K is $\bar{c}_p = c_p * M = (2.225)(16.04) =$ 35.69 kJ/kmol-K. The required reactant and product property data are given in Table 7.2.

The unburned gas state 1 molecular mass is

$$M_1 = \sum X_i M_i = (0.095)(16.04) + (0.19)(32) + (0.715)(28.013) = 27.63\,\text{kg/kmol}$$

The burned gas state 2 molecular mass is

$$M_2 = \sum X_i M_i = (0.095)(44.011) + (0.19)(18.016) + (0.715)(28.013) = 27.63\,\text{kg/kmol}$$

The state 1 specific heat is

$$\bar{c}_{p1} = \sum X_i \bar{c}_{p,i} = (0.095)(35.69) + (0.19)(29.315) + (0.715)(29.071) = 29.75$$

$$c_{p1} = \bar{c}_{p1}/M_1 = 29.75/27.63 = 1.076\,\text{kJ/kgK}$$

and the state 1 density is

$$\rho_1 = \frac{p_1}{R_1 T_1} = \frac{101.3}{(8.314/27.63)298} = 1.13\,\text{kg/m}^3$$

Assume $T_2/T_1 \simeq 9$, so an initial estimate of $T_2 = 9 \cdot 298 \simeq 2700$ K. Since $c_p \sim \Delta h/\Delta T$, using interpolation of the $h(T)$ values in the Appendix at $T = 2600$ K and 2800 K indicates that the state 2 specific heat is

$$\bar{c}_{p2} = \sum X_i \bar{c}_{p,i} = (0.095)(61.839) + (0.19)(54.742) + (0.715)(36.82) = 42.60$$

$$c_{p2} = \bar{c}_{p2}/M_2 = 42.60/27.63 = 1.542\,\text{kJ/kgK}$$

The gas constant R_2 is

$$R_2 = R_u/M_2 = 8.314/27.63 = 0.3009\,\text{kJ/kgK} = 300.9\,\text{J/kgK}$$

and the specific heat ratio γ_2 is

$$\gamma_2 = \frac{c_{p2}}{c_{p2} - R_2} = \frac{1.542}{1.542 - 0.301} = 1.24$$

The heat addition q is

$$\begin{aligned} q &= \sum_1 Y_i h^o_{f,i} - \sum_2 Y_i h^o_{f,i} \\ &= (0.055)(-4665) - (0.151)(-8942) - (0.124)(-13,424) \\ &= 2758\,\text{kJ/kg} \end{aligned}$$

The temperature T_2 is

$$\begin{aligned} T_2 &= \left[\frac{c_{p1}}{c_{p2}} T_1 + \frac{q}{c_{p2}}\right] \frac{2\gamma_2^2}{\gamma_2 + 1} \\ &= \left[\frac{1.076}{1.542} 298 + \frac{2758}{1.542}\right] \frac{(2)1.24^2}{1.24 + 1} \\ &= 2741\,\text{K} \end{aligned}$$

If the calculated value of T_2 is "close enough" to the estimated value, no further iteration is needed.

The detonation velocity u_1 is then

$$\begin{aligned} u_1 &= \frac{\gamma_2 + 1}{\gamma_2} \sqrt{\gamma_2 R_2 T_2} \\ &= \frac{1.24 + 1}{1.24} \sqrt{(1.24)(300.9)(2741)} \\ &= 1827\,\text{m/s} \end{aligned}$$

The density and pressure at state 2 are

$$\rho_2 = \frac{\gamma_2 + 1}{\gamma_2} \rho_1 = \frac{1.24 + 1}{1.24} (1.13) = 2.04\,\text{kg/m}^3$$

$$\begin{aligned} p_2 &= \frac{\rho_2}{\rho_1} \frac{M_1}{M_2} \frac{T_2}{T_1} p_1 \\ &= \left(\frac{2.04}{1.13}\right) \left(\frac{27.63}{27.63}\right) \left(\frac{2741}{298}\right) (101.3) \\ &= 1682\,\text{kPa} \end{aligned}$$

Comment: In air, the temperature ratio T_2/T_1 has a maximum of about 9–10 for hydrocarbon and hydrogen combustion at stoichiometric conditions and decreases on either side of $\phi = 1$.

7.4.1 Comparison of C–J Detonation-Wave Velocities with Experimental Data

The first critical test of the C-J theory for its prediction of detonation velocity was carried out by Lewis and Friauf (1930) on mixtures of hydrogen and oxygen diluted with various gases. The dissociation equilibria considered were

$$H_2O \rightleftharpoons H_2 + \frac{1}{2}O_2 \tag{7.64}$$

$$H_2O \rightleftharpoons \frac{1}{2}H_2 + OH \tag{7.65}$$

$$H_2 \rightleftharpoons 2H \tag{7.66}$$

Table 7.3 gives a summary of the Lewis and Friauf results at $p_1 = 1$ atm, $T_1 = 298$ K. It is quite evident that except for condition with a large excess of hydrogen, the agreement between the calculated and experimental velocities is very good.

It is seen in Table 7.3 that dilution of the stoichiometric mixture with hydrogen results in an increase in the detonation velocity, despite the reduction in temperature T_2. This reflects the influence of the much-reduced density of the mixture, a result that suggests the performance of experiments with helium and argon as diluents. The addition of helium to a stoichiometric mixture should increase the velocity because of the decreased density of the initial mixture.

On the other hand, the addition of argon should decrease the detonation velocity. These two inert gases are identical in their thermal effect, but affect the detonation velocity in opposite directions, since their molecular masses are very different ($M_{He} = 4, M_{Ar} = 40$). How well this prediction is fulfilled can be seen from Table 7.4, obtained by Lewis and Friauf (1930). The experimental fact that dilution with the inert gas helium increases the detonation velocity provides a particularly clear confirmation of the predictability of the detonation wave velocity by the C–J theory.

The influence of species dissociation on the detonation velocity is an important factor. If dissociation is neglected, i.e., no H or OH species present, the calculated velocities are considerably larger,

Table 7.3 Comparison of Experimental Detonation Velocity Data with C–J Theory

Explosive Mixture	p_2 (atm)	T_2 (K)	u_1 (calc.) (m/s)	u_1 (expt.) (m/s)	Dissociation (mol%)
$(2H_2 + O_2)$	18.0	3583	2806	2819	32
$(2H_2 + O_2) + O_2$	17.4	3390	2302	2314	30
$(2H_2 + O_2) + N_2$	17.4	3367	2378	2407	18
$(2H_2 + O_2) + 5N_2$	14.4	2685	1850	1822	2
$(2H_2 + O_2) + 6H_2$	14.2	2650	3749	3532	1

Source: Adapted from Lewis and Friauf (1930).

Table 7.4 Comparison of Experimental Detonation Velocity Data with C–J Theory.

Explosive Mixture	p_2 (atm)	T_2 (K)	u_1 (calc.) (m/s)	u_1 (expt.) (m/s)
$(2H_2 + O_2) + 5\,Ar$	16.3	3097	1762	1700
$(2H_2 + O_2) + 3\,Ar$	17.1	3265	1907	1800
$(2H_2 + O_2) + 1.5\,Ar$	17.6	3412	2117	1950
$(2H_2 + O_2)$	18.0	3583	2806	2819
$(2H_2 + O_2) + 1.5\,He$	17.6	3412	3200	3010
$(2H_2 + O_2) + 3\,He$	17.1	3265	3432	3130
$(2H_2 + O_2) + 5\,He$	16.3	3097	3617	3160

Source: Lewis and Friauf (1930) and Lewis and von Elbe (1987).

by several hundred meters per second, than the experimental data. The equilibrium concentrations of radicals (the amount of dissociation is shown in the last column of Table 7.3) suggest that the chain-carrying concentration in the reaction zone must be very large indeed. This, together with the high temperature in the reaction zone, provides an explanation for the extreme rapidity of the reaction which results in the formation of the detonation wave.

The dependence of detonation wave speed on mixture composition is illustrated in Figures 7.6 and 7.7. It is interesting to note that hydrocarbon mixtures such as propane–oxygen $C_3H_8 - O_2$ (see Figure 7.6) and acetylene–oxygen $C_2H_2 - O_2$ mixtures (see Figure 7.7) show maximum detonation velocities that are different from those for the stoichiometric composition; instead, they are fairly close to rich compositions corresponding to combustion to CO and H_2O. This is because the reaction time scale in the detonation process is very short. The time to form stoichiometric CO and H_2O mixtures is much less than that for stoichiometric CO_2 and H_2O since the oxidation process from CO to CO_2 is relatively slow.

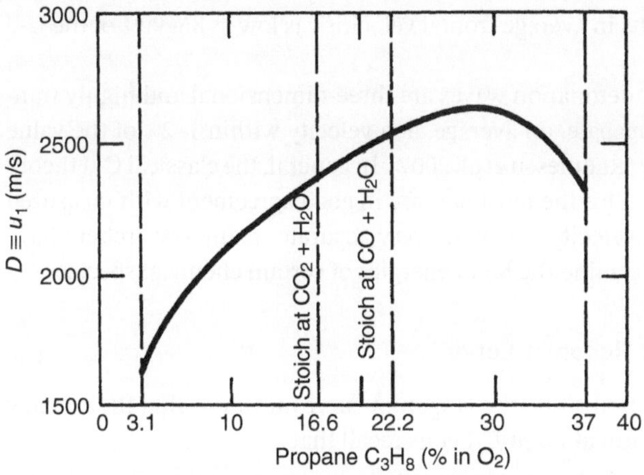

Figure 7.6 Detonation wave velocity as a function of the percentage of propane in O_2.

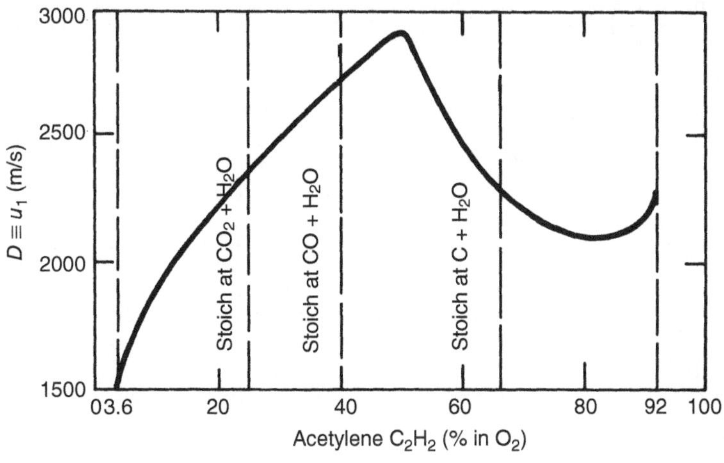

Figure 7.7 Detonation wave velocity as a function of the percentage of acetylene in O_2.

Berets et al. (1950) repeated the calculations made by Lewis and Friauf using more accurate spectroscopic data and obtained good agreement with the theoretical values of detonation velocities (about 1% difference). They further showed that the measured and calculated detonation velocity agree closely except for lean mixtures in the neighborhood of the limits of detonability. Under those conditions, the measured data are significantly below the theoretical values. These discrepancies were attributed to the lateral heat losses, which are not considered in the one-dimensional theory. More comparisons between the experimentally measured detonation velocities and the theoretical results based upon the classical C–J theory can be found in the books by Lewis and von Elbe (1987) and Strehlow (1984).

Poor agreement has been found for initial pressures, mixture compositions, and cylinder dimensions that are near the detonation limit values for the occurrence of a constant-velocity propagating wave (Strehlow 1971). This disagreement has been attributed to a multidimensional transverse structure, which is much more pronounced under these marginal conditions. Recently, Tofaili et al. (2021) studied 1-D unstable detonations in the Chapman–Jouguet regime using high-order numerical simulations and asymptotic techniques in the limit of low heat release. Such unstable detonation is termed *marginal* and results in average frontal velocities as low as 85–90% of the C–J velocity.

To summarize, in spite of the fact that detonation waves are three-dimensional and highly transient, experiments suggest that they propagate, on average, at a velocity within 1–2% of the value given by the one-dimensional C–J theory (Radulescu et al. 2007). In general, the classical C–J theory predicts properties of the flow field which for the most part are in good agreement with measured values. Since the calculated detonation velocity is usually very accurate, many researchers have used the detonation wave velocity to determine the bond energies of certain chemical species.

7.4.2 Entropy Distribution along the Hugoniot Curve

Now we shall consider the entropy variation along the Hugoniot curve and show that the entropy has a minimum at point U and a maximum at point L. Let us recall that

$$h \equiv e + \frac{p}{\rho} \tag{7.67}$$

where e represents internal energy including chemical energy:

$$h_2 - h_1 = (e_2 - e_1) + \left(\frac{p_2}{\rho_2} - \frac{p_1}{\rho_1}\right) \tag{7.68}$$

From Eq. (7.34), we know that

$$e_2 - e_1 = \frac{1}{2}(p_2 - p_1)\left(\frac{1}{\rho_1} + \frac{1}{\rho_2}\right) - \frac{p_2}{\rho_2} + \frac{p_1}{\rho_1} \tag{7.69}$$

After rearranging, we have

$$e_2 - e_1 = \frac{1}{2}(p_2 + p_1)\left(\frac{1}{\rho_1} - \frac{1}{\rho_2}\right) \tag{7.70}$$

From the first and second laws of thermodynamics along the Hugoniot curve, we have

$$T_2 ds_2 = de_2 + p_2 d\left(\frac{1}{\rho_2}\right) \tag{7.71}$$

Differentiating Eq. (7.70), we have

$$de_2 = \frac{1}{2}(dp_2)\left(\frac{1}{\rho_1} - \frac{1}{\rho_2}\right) - \frac{1}{2}(p_2 + p_1)d\left(\frac{1}{\rho_2}\right) \tag{7.72}$$

Substituting Eq. (7.72) into Eq. (7.71) and rearranging, we have

$$T_2 ds_2 = \frac{1}{2}\left(\frac{1}{\rho_1} - \frac{1}{\rho_2}\right) dp_2 + \frac{1}{2}(p_2 - p_1) d\left(\frac{1}{\rho_2}\right) \tag{7.73}$$

or along the Hugoniot curve,

$$T_2 \frac{ds_2}{d(1/\rho_2)} = \frac{1}{2}\left(\frac{1}{\rho_1} - \frac{1}{\rho_2}\right)\left(\frac{dp_2}{d(1/\rho_2)} - \frac{p_2 - p_1}{1/\rho_2 - 1/\rho_1}\right) \tag{7.74}$$

Using Eq. (7.37) for the slopes at the C–J points, Eq. (7.74) gives

$$\left[\frac{ds_2}{d(1/\rho_2)}\right]_{\text{C–J}} = 0 \tag{7.75}$$

Thus, we can conclude that the entropy has a maximum or minimum at the C–J points. In order to determine which it is, we have to obtain the second derivative of s_2 with respect to $1/\rho_2$. Differentiating Eq. (7.74), we have

$$\left[\frac{d^2 s_2}{d(1/\rho_2)^2}\right]_{\text{CJ}} = \frac{1}{2T_2}\left(\frac{1}{\rho_1} - \frac{1}{\rho_2}\right)\left[\frac{d^2 p_2}{d(1/\rho_2)^2}\right]_{\text{CJ}} + \frac{1}{2}\left\{\frac{dp_2}{d(1/\rho_2)} - \frac{p_2 - p_1}{1/\rho_2 - 1/\rho_1}\right\}_{\text{C–J}} \frac{1}{T_2} \tag{7.76}$$

Using Eq. (7.37), the last term in the above equation vanishes, so we have

$$\left[\frac{d^2 s_2}{d(1/\rho_2)^2}\right]_{\text{CJ}} = \frac{1}{2T_2}\left(\frac{1}{\rho_1} - \frac{1}{\rho_2}\right)\left[\frac{d^2 p_2}{d(1/\rho_2)^2}\right]_{\text{C–J}} \tag{7.77}$$

From the shape of the Hugoniot curve (Figure 7.2), we observe that

$$\frac{d^2 p_2}{d(1/\rho_2)^2} > 0 \quad \text{for every point} \tag{7.78}$$

Therefore, for the upper C–J point U, $1/\rho_1 > 1/\rho_2$, and Eq. (7.77) gives

$$\left[\frac{d^2 s_2}{d(1/\rho_2)^2}\right]_U > 0 \tag{7.79}$$

which indicates that the entropy is a minimum at point U.

For the lower C–J point L, $1/\rho_1 < 1/\rho_2$, and Eq. (7.77) gives

$$\left[\frac{d^2 s_2}{d(1/\rho_2)^2}\right]_L < 0 \tag{7.80}$$

indicating that the entropy has a maximum at point L. The entropy distribution thus appears as shown in Figure 7.8.

7.4.3 Comparison of the Burned-Gas Velocity with the Local Speed of Sound

Since the entropy s is a function of p and $1/\rho$,

$$ds = \left[\frac{\partial s}{\partial(1/\rho)}\right]_p d\left(\frac{1}{\rho}\right) + \left[\frac{\partial s}{\partial p}\right]_{1/\rho} dp \tag{7.81}$$

The speed of sound of the burned gases can be obtained from the consideration of a reversible adiabatic (isentropic) process. By setting $ds = 0$, the partial derivatives in Eq. (7.81) can be expressed as

$$\left[\frac{\partial p}{\partial(1/\rho)}\right]_s = -\frac{[\partial s/\partial(1/\rho)]_p}{[\partial s/\partial p]_{1/\rho}} \tag{7.82}$$

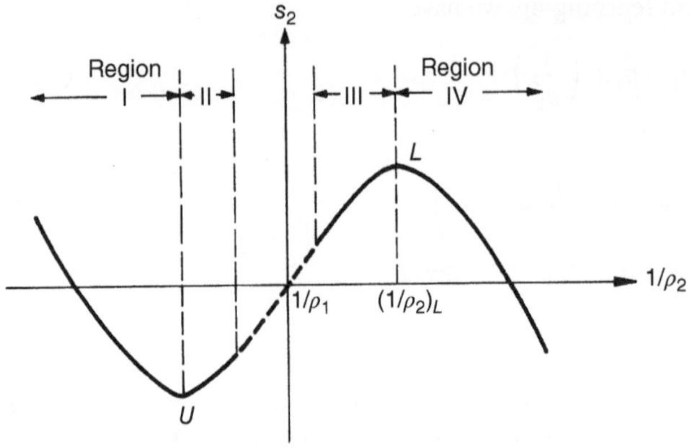

Figure 7.8 Entropy distribution on the Hugoniot curve.

We use the subscript HC to represent that the derivative is being taken along the Hugoniot curve. From Eq. (7.81), we know that

$$\left[\frac{\partial s}{\partial(1/\rho)}\right]_{\mathrm{HC}} = \left[\frac{\partial s}{\partial(1/\rho)}\right]_p + \left[\frac{\partial s}{\partial p}\right]_{1/\rho}\left[\frac{\partial p}{\partial(1/\rho)}\right]_{\mathrm{HC}} \tag{7.83}$$

After rearranging, we have

$$\left[\frac{\partial p}{\partial(1/\rho)}\right]_{\mathrm{HC}} = \frac{1}{[\partial s/\partial p]_{1/\rho}}\left\{\left[\frac{\partial s}{\partial(1/\rho)}\right]_{\mathrm{HC}} - \left[\frac{\partial s}{\partial(1/\rho)}\right]_p\right\} \tag{7.84}$$

Subtracting Eq. (7.82) from Eq. (7.84) yields

$$\left[\frac{\partial p}{\partial(1/\rho)}\right]_{\mathrm{HC}} - \left[\frac{\partial p}{\partial(1/\rho)}\right]_s = \frac{[\partial s/\partial(1/\rho)]_{\mathrm{HC}}}{[\partial s/\partial p]_{1/\rho}} \tag{7.85}$$

At the C–J points along the Hugoniot curve, from Eq. (7.75), we have

$$\left[\frac{\partial s}{\partial(1/\rho)}\right]_{\mathrm{HC}} = 0 \tag{7.86}$$

Thus, Eq. (7.85) implies that

$$\left[\frac{\partial p}{\partial(1/\rho)}\right]_{\mathrm{HC}} = \left[\frac{\partial p}{\partial(1/\rho)}\right]_s \quad \text{at C–J points} \tag{7.87}$$

This means that the tangent lines to the Hugoniot curve on a p-versus-$1/\rho$ plot at C–J points are also the tangents to the constant-entropy line.

After rearranging Eq. (7.74) and combining with Eq. (7.85), we have

$$\frac{2T_2}{1/\rho_1 - 1/\rho_2}\left[\frac{\partial s}{\partial(1/\rho)}\right]_{\mathrm{HC}} + \frac{p_2 - p_1}{1/\rho_2 - 1/\rho_1} - \left[\frac{\partial p}{\partial(1/\rho)}\right]_s = \frac{[\partial s/\partial(1/\rho)]_{\mathrm{HC}}}{[\partial s/\partial p]_{1/\rho}} \tag{7.88}$$

Using Eq. (7.22) and rearranging the above equation, we get

$$-\rho_2^2 u_2^2 + \rho_2^2\left[\frac{\partial p_2}{\partial \rho_2}\right]_s = \left[\frac{\partial s}{\partial(1/\rho)}\right]_{\mathrm{HC}}\left[\frac{-2T_2}{1/\rho_1 - 1/\rho_2} + \frac{1}{[\partial s/\partial p]_{1/\rho}}\right] \tag{7.89}$$

Based on the definition of the speed of sound, we know that

$$\rho_2^2\left[\frac{\partial p_2}{\partial \rho_2}\right]_s = \rho_2^2 c_2^2 \tag{7.90}$$

Substituting Eq. (7.90) into Eq. (7.89), we have

$$\rho_2^2\left(c_2^2 - u_2^2\right) = \left[\frac{\partial s}{\partial(1/\rho)}\right]_{\text{HC}}\left[\frac{-2T_2}{1/\rho_1 - 1/\rho_2} + \frac{1}{[\partial s/\partial p]_{1/\rho}}\right] \tag{7.91}$$

We note that if the right-hand side of Eq. (7.91) is positive in region I, then

$$u_2^2 < c_2^2 \tag{7.92}$$

which implies that the burned-gas velocity is subsonic.

From Figure 7.8, we know that in region I

$$\left[\frac{\partial s}{\partial(1/\rho)}\right]_{\text{HC}} < 0 \tag{7.93}$$

and the right-hand side of Eq. (7.91) will be positive if

$$\frac{2T_2}{1/\rho_1 - 1/\rho_2} > \frac{1}{[\partial s/\partial p]_{1/\rho}} \tag{7.94}$$

In order to prove the above inequality, the partial derivative term on the right-hand side of inequality (7.94) must be first replaced by an algebraic term involving the properties of the gases at the burned end, since $[\partial s/\partial p]_{1/\rho}$ represents the change of entropy of the burned gases with respect to p_2 when $1/\rho_2$ is held constant. Differentiating Eq. (7.67), we have

$$dh = de + p\,d\left(\frac{1}{\rho}\right) + \frac{1}{\rho}dp \tag{7.95}$$

The first law of thermodynamics gives

$$dh = T\,ds + \frac{1}{\rho}dp \tag{7.96}$$

For mixtures with constant c_p, we have

$$c_p dT = Tds + \frac{1}{\rho}dp \tag{7.97}$$

From the perfect gas law,

$$T = \frac{p}{\rho R} \tag{7.98}$$

$$dT = \frac{1}{\rho R}dp + \frac{p}{R}d\left(\frac{1}{\rho}\right) \tag{7.99}$$

Substituting Eqs. (7.99) and (7.28) into Eq. (7.97) yields

$$\frac{\gamma}{\gamma - 1}R\left[\frac{1}{\rho R}dp + \frac{p}{R}d\left(\frac{1}{\rho}\right)\right] = Tds + \left(\frac{1}{\rho}\right)dp \tag{7.100}$$

Substituting Eq. (7.98) into Eq. (7.100) to eliminate T, and solving for ds, we have

$$ds = \frac{\gamma R}{\gamma - 1}\rho d\left(\frac{1}{\rho}\right) + \frac{1}{(\gamma - 1)\rho T}dp \tag{7.101}$$

$$\left[\frac{\partial s}{\partial p}\right]_{1/\rho} = \frac{1}{(\gamma - 1)\rho T} \tag{7.102}$$

Using Eq. (7.102), the second factor on the right-hand side of Eq. (7.91) becomes

$$\frac{-2T_2}{1/\rho_1 - 1/\rho_2} + \frac{1}{[\partial s/\partial p]_{1/\rho}} = \frac{-2T_2}{1/\rho_1 - 1/\rho_2} + (\gamma - 1)\rho_2 T_2 \tag{7.103}$$

It can also be written in the following form after some algebraic manipulations:

$$\frac{-2T_2}{1/\rho_1 - 1/\rho_2} + \frac{1}{[\partial s/\partial p]_{1/\rho}} = \frac{p_2}{R}\left[\gamma - \frac{1 + (\rho_1/\rho_2)}{1 - (\rho_1/\rho_2)}\right] \tag{7.104}$$

In order to prove that $u_2 < c_2$ in region I, we need to show that

$$\gamma - \frac{1 + (\rho_1/\rho_2)}{1 - (\rho_1/\rho_2)} < 0 \tag{7.105}$$

The Rankine–Hugoniot relation given by Eq. (7.31) can be rearranged to give

$$\frac{\gamma+1}{2(\gamma-1)}\frac{p_2}{\rho_2} - \frac{1}{2}\frac{p_2}{\rho_1} + \frac{p_1}{2\rho_2} - \frac{\gamma+1}{2(\gamma-1)}\frac{p_1}{\rho_1} - q = 0 \tag{7.106}$$

This is the equation of a hyperbola in the $p_2 - 1/\rho_2$ plane. To represent p_2 in terms of $1/\rho_2$, we have

$$p_2 = \frac{q + \frac{\gamma+1}{2(\gamma-1)}\frac{p_1}{\rho_1} + \left[-\frac{p_1}{2}\right]\frac{1}{\rho_2}}{-\frac{1}{2}\frac{1}{\rho_1} + \left[\frac{\gamma+1}{2(\gamma-1)}\right]\frac{1}{\rho_2}} = \frac{a + \frac{b}{\rho_2}}{c + \frac{d}{\rho_2}} \tag{7.107}$$

where

$$a = q + \frac{\gamma+1}{2(\gamma-1)}\frac{p_1}{\rho_1} \tag{7.108}$$

$$b = -\frac{p_1}{2} \tag{7.109}$$

$$c = -\frac{1}{2\rho_1} \tag{7.110}$$

$$d = \frac{\gamma+1}{2(\gamma-1)} \tag{7.111}$$

Solving Eq. (7.107) for $1/\rho_2$, we have

$$\frac{1}{\rho_2} = \frac{a - cp_2}{-b + dp_2} \tag{7.112}$$

The vertical asymptote of the Hugoniot curve is determined by setting the denominator of Eq. (7.107) equal to zero, i.e.,

$$c + \frac{d}{\rho_2} = 0 \tag{7.113}$$

or

$$\left(\frac{1}{\rho_2}\right)_{\min} = -\frac{c}{d} = \frac{(\gamma-1)}{(\gamma+1)}\frac{1}{\rho_1} \tag{7.114}$$

Similarly, the horizontal asymptote is determined by $p_2 = b/d$, or

$$(p_2)_{\min} = -\left(\frac{\gamma-1}{\gamma+1}\right)p_1 \tag{7.115}$$

From the vertical asymptote, we know that any solution on the Hugoniot curve must satisfy the following inequality:

$$\frac{1}{\rho_2} > \left(\frac{\gamma-1}{\gamma+1}\right)\frac{1}{\rho_1} \tag{7.116}$$

or

$$(\gamma+1)\frac{\rho_1}{\rho_2} > \gamma - 1 \tag{7.117}$$

or

$$\left(1+\frac{\rho_1}{\rho_2}\right) > \gamma\left(1-\frac{\rho_1}{\rho_2}\right) \tag{7.118}$$

For region I,

$$1-\frac{\rho_1}{\rho_2} > 0 \tag{7.119}$$

Dividing both sides of Eq. (7.118) by this positive quantity, we have

$$\frac{1+(\rho_1/\rho_2)}{1-(\rho_1/\rho_2)} > \gamma \quad \text{for region I} \tag{7.120}$$

The above inequality together with Eqs. (7.104) and (7.91) shows that

$$u_2 < c_2 \quad \text{for region I} \tag{7.121}$$

This indicates that the velocity of the burned gas at any point above the upper C–J point U is subsonic. In the fixed laboratory coordinates, V_w is the speed of the detonation wave relative to the cylinder, and from Eq. (7.44), $V_w = \upsilon_2 + u_2$. Using this relationship, Eq. (7.121) becomes

$$V_w - \upsilon_2 < c_2 \tag{7.122}$$

or

$$c_2 + \upsilon_2 > V_w \quad \text{for region I} \tag{7.123}$$

This equation implies that the resultant propagation velocity of any disturbance in the wave motion direction (as the sum of gas particle velocity υ_2 and any disturbance propagation velocity c_2) can catch up with the wave propagation velocity, V_w. Physically, it means that any rarefaction disturbances behind the detonation wave can propagate at a higher speed than the wave. Therefore, they catch up with the detonation wave and reduce its strength. Thus, any solution in region I is not stable and falls back to the upper C–J point, U.

In the previous derivation, we assumed that the shape of the Hugoniot curve satisfies the inequality given by Eq. (7.78). This inequality can be proved by using the asymptotic limits. By differentiating (7.107) with respect to $1/\rho_2$, we have

$$\frac{dp_2}{d(1/\rho_2)} = \frac{bc-ad}{(c+d/\rho_2)^2} \tag{7.124}$$

and

$$\frac{d^2p_2}{d(1/\rho_2)^2} = \frac{-2d(bc-ad)}{(c+d/\rho_2)^3} \tag{7.125}$$

From the vertical asymptote (7.114), it can be shown that $c/d = -(1/\rho_2)_{\min}$; therefore,

$$\left(\frac{c}{d}+\frac{1}{\rho_2}\right) > 0 \quad \text{or} \quad \left(c+\frac{d}{\rho_2}\right)^3 > 0 \tag{7.126}$$

Also,

$$bc - ad = \frac{p_1}{4\rho_1} - \frac{\gamma+1}{2(\gamma-1)}q - \frac{(\gamma+1)^2}{4(\gamma-1)^2}\frac{p_1}{\rho_1} = -\left[\frac{\gamma}{(\gamma-1)^2}\frac{p_1}{\rho_1} + \frac{\gamma+1}{2(\gamma-1)}q\right] \tag{7.127}$$

Since the quantity in the square brackets is always positive for exothermic reactions, $bc - ad$ is less than zero. Based on Eqs. (7.126), (7.127), and (7.111), we have

$$\frac{d^2p_2}{d(1/\rho_2)^2} > 0 \tag{7.128}$$

The solution in region II is ruled out in view of the Zel'dovich–von Neumann–Döring (ZND) detonation wave structure which will be discussed later in this chapter. In summary, commonly observed detonation waves usually correspond to the upper C–J point, and commonly observed deflagration waves lie in region III of the Hugoniot curve as shown in Figure 7.4.

7.5 Zel'dovich–von Neumann–Döring (ZND) One-Dimensional Wave Model

Extending the classical C–J theory, Zel'dovich (1950), von Neumann (1942), and Döring (1943) independently developed a wave model (known as the ZND wave) in which the detonation wave was assumed to consist of a thin 1-D shock wave moving at the detonation velocity, with chemical reactions occurring in a thicker region behind the shock. They assumed that the shock wave initially heated the reactants to a temperature at which the rate of reaction would be high enough for the ensuing detonation to propagate as fast as the shock.

In the ZND model, shown schematically in Figure 7.9, after an induction period, the chemical reaction starts; and as the reaction progresses, the temperature rises and the density and pressure fall until they reach the C–J values and the reaction attains equilibrium. The pressure, density, and temperature profiles are very flat in an induction zone behind the shock front and then change rapidly in a reaction zone as the reaction goes to completion.

The assumption of very limited reactions in the shock wave region itself was based on the fact that the thickness of a shock wave is of the order of a few mean free paths of the gas molecules.

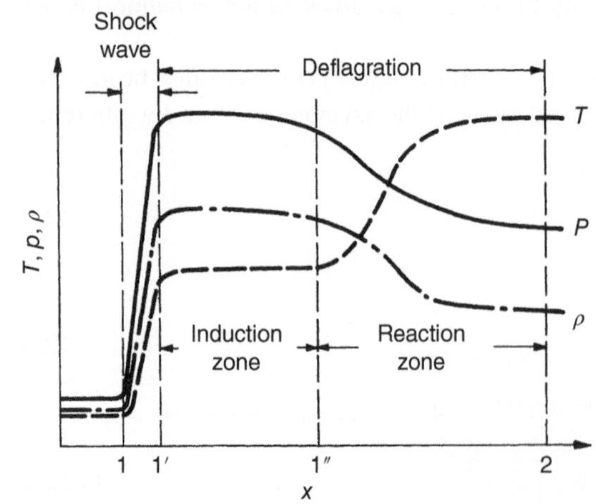

Figure 7.9 Variation of physical properties through a ZND detonation wave.

In view of the very small number of collisions between molecules within the shock wave, the major amount of heat release in a detonation wave was assumed to occur in the thick region behind the shock wave. A rarefaction wave whose steepness depends on the distance traveled by the wave then sets in. Thus, behind the C–J shock, energy is generated by thermal reaction.

Figure 7.9 shows the variation of physical properties through the ZND detonation wave model. The magnitude of the pressure, temperature, and density change in the induction zone (stations $1' - 1''$) immediately behind the shock depends on the fraction of gaseous mixture reacted. If the reaction rate follows the Arrhenius law, then the rate of reaction increases slowly in the region immediately behind the shock, where the temperature is not extremely high. As a result, the pressure, temperature, and density profiles are relatively flat in the induction zone. This region is therefore called the induction zone. In the reaction zone (stations $1'' - 2$), the gas properties change sharply as the reaction rate increases. When the reaction is completed, the thermodynamic properties reach their equilibrium values. The physical distance between the shock front and the fully reacted station is on the order of 1 cm.

The pressure p and temperature T behind a shock are given by the ideal-gas normal shock equations for compressible flow,

$$p = \frac{p_1}{\alpha}\left(\frac{Ma^2}{\beta} - 1\right) \tag{7.129}$$

$$T = \frac{T_1}{\alpha^2 \beta Ma^2}\left(\frac{Ma^2}{\beta} - 1\right)\left(\beta Ma^2 + \frac{1}{\gamma}\right) \tag{7.130}$$

where

$$\alpha \equiv \frac{\gamma + 1}{\gamma - 1}, \quad \beta \equiv \frac{\gamma - 1}{2\gamma} \tag{7.131}$$

and Ma is the Mach number, and T_1 and p_1 are the initial temperature and pressure ahead of the shock. Equations (7.129) and (7.130) apply to the ideal-gas case in which c_p = constant.

The locus of the reacting mixture is shown on the Hugoniot plot in Figure 7.10. As shown in the figure, there are many paths which a reacting mixture can take through the detonation wave from a unreacted state to a fully reacted state. All paths (a, b, c, etc.) can satisfy the conservation equations.

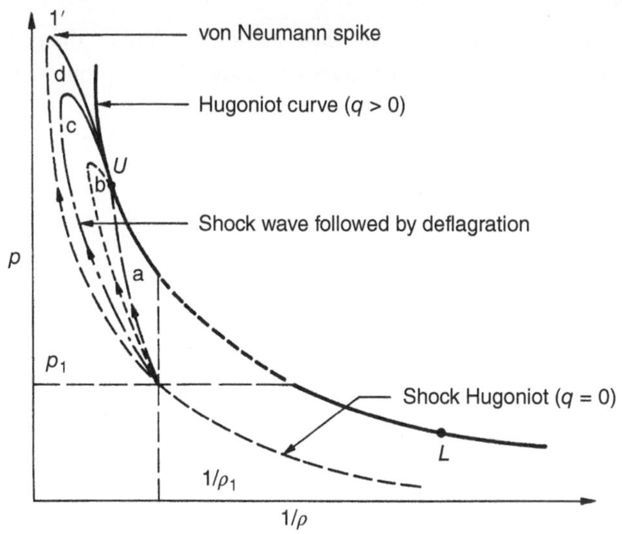

Figure 7.10 ZND detonation structure on $(p, 1/\rho)$ diagram.

A gaseous mixture may enter the detonation wave in the state corresponding to the initial point $(1/\rho_1, p_1)$ and move directly to the upper C–J point by path a. However, this path would demand that reaction occurs everywhere along the path. Since there is no significant compression in the beginning part of this path, there cannot be sufficient temperature to initiate any reaction. Thus, path a is very unlikely to have enough energy release to sustain the wave. Path b represents a possible locus for mixtures with fast chemical kinetics. For slow chemical kinetics, the path may be c. In the limit of zero chemical-energy release in the shock wave, the initial portion of path d will coincide with the shock Hugoniot curve. The peak pressure behind the shock wave in the ZND model is called the *von Neumann spike*.

According to the ZND model, sharp pressure and density gradients exist at the front of the detonation wave. For detonations propagating into subatmospheric initial pressures, the sharp rises in pressure and density have been confirmed experimentally by several investigators, see, for example, Kistiakowsky and Kydd (1956) and Fay (1959). In general, the experimental results agree with a hydrodynamic theory of the detonation wave. There is a sharp increase of the gas density in the wave front, followed by a gradual decrease to essentially constant values. The observed densities at the end of the reaction zone (the Chapman–Jouguet state) compare well with the results of approximate equilibrium calculations, and the observed shape of the density profile qualitatively confirms kinetic predictions.

However, the question of whether the so-called von Neumann spike has been attained or not could not be answered by these experiments, because their effective spatial resolution was insufficient. An extrapolation of the density profile, measured by the X-ray absorption technique to the front of the reaction zone, demonstrated that a peak density of about 70% of the expected von Neumann spike was probably attained.

It is worth pointing out that the state at the von Neumann spike is usually evaluated assuming complete equilibrium for different degrees of freedom. However, it is known in particular that the vibrational relaxation times of several species appearing in the vicinity of the von Neumann spike are on the order of a microsecond. In these cases, lower peak densities and higher pressures than those of the equilibrium von Neumann state can be expected to occur.

Example 7.2 ***ZND Detonation Wave***

For the methane detonation wave of Example 7.1, use the ZND wave model to calculate the gas mixture properties at state 1' immediately after the shock wave passes. Compare with the properties at state 2, the end of the reaction zone.

Solution

Calculating additional properties of the unburned gas at state 1,

$$R_1 = \frac{R_u}{M_1} = \frac{8.314}{27.63} = 0.3009\,\text{kJ/kgK}$$

$$\gamma_1 = \frac{c_{p1}}{c_{p1} - R_1} = \frac{1.076}{1.076 - 0.3009} = 1.388$$

$$c_1 = (\gamma_1 R_1 T_1)^{1/2} = ((1.388)(300.9)(298))^{1/2} = 352.8\,\text{m/s}$$

$$Ma_1 = \frac{u_1}{c_1} = \frac{1827}{352.8} = 5.18$$

Applying the normal shock relations (Eqs. (7.129)–(7.131)) between states 1 and 1′,

$$\alpha = \frac{\gamma_1 + 1}{\gamma_1 - 1} = \frac{1.388 + 1}{1.388 - 1} = 6.155$$

$$\beta = \frac{\gamma_1 - 1}{2\gamma_1} = \frac{1.388 - 1}{2(1.388)} = 0.140$$

The state 1′ pressure p_1' is

$$p_1' = \frac{1}{\alpha}\left(\frac{Ma_1^2}{\beta} - 1\right)p_1 = \frac{1}{6.155}\left(\frac{5.18^2}{0.14} - 1\right)(101.325)$$
$$= 3136\,\text{kPa}$$

and the state 1′ temperature T_1' is

$$T_1' = \frac{1}{\alpha^2 \beta Ma_1^2}\left(\frac{Ma_1^2}{\beta} - 1\right)\left(\beta Ma_1^2 + \frac{1}{\gamma_1}\right)T_1$$
$$= \frac{1}{(6.155)^2(0.14)(5.18)^2}\left(\frac{5.18^2}{0.14} - 1\right)\left((0.14)(5.18^2) + \frac{1}{1.388}\right)(298)$$
$$= 1787\,\text{K}$$

Assuming that the gas constant is unchanged across the shock, $R_1' = R_1$, and the state 1′ density ρ_1' is

$$\rho_1' = \frac{p_1'}{R_1' T_1'} = \frac{3136}{(0.3009)(1787)} = 5.83\,\text{kg/m}^3$$

The state 1′ Mach number $Ma_{1'}$ is

$$Ma_{1'} = \frac{Ma_1^2 + \frac{1}{\beta\gamma_1}}{\frac{Ma_1^2}{\beta} - 1} = \frac{5.18^2 + \frac{1}{(0.14)(1.388)}}{\frac{5.18^2}{0.14} - 1}$$
$$= 0.17$$

Comment: Comparison with the state 2 properties indicates that in the induction/reaction zones from state 1' to state 2, the temperature increased from 1787 to 2741 K, the pressure decreased from 3136 to 1682 kPa, the density decreased from 5.83 to 2.04 kg/m^3, and the Mach number increased from 0.17 to 1.

7.6 Multidimensional Detonation Wave Structure

The motion and structure of an actual detonation front are more complex than the steady one-dimensional wave motion given by the ZND model. Over the range of detonable concentrations of a given fuel–oxidizer mixture, the physical structure of the shock wave is in the form of a multicellular wave front. The cellular nature of the detonation front is caused by the rapid heat release that warps the front, thus causing curved shocks, which interact by means of triple-shock interactions. The shock front is composed of a three-dimensional grid of shock intersections that move transverse to the wave front as the front moves forward.

The transverse waves are relatively weak in comparison to the axial component, which allows one-dimensional modeling of the detonation. The one-dimensional wave structure is generally unstable and evolves into a variety of cellular detonation structures that range from diamond patterns to a complex of shockwaves, reaction regions and shear layers behind leading shock fronts. Inside the multicellular structure, the shock velocity fluctuates within a cell of size λ about the equilibrium Chapman–Jouguet value.

The first evidence of a multidimensional wave structure was obtained by Campbell and Woodhead (1927). By observing spinning detonation in lean mixtures in circular cylinders, they showed that detonation waves could travel in a manner that is locally three-dimensional and unsteady. For more than 25 years after their discovery, this spinning detonation was thought to be a phenomenon peculiar to lean mixtures.

The detailed wave structure of detonation waves was not studied until (Denisov and Troshin 1959) adapted an experimental technique involving soot-coated plates located near a spark discharge which recorded many thin lines on the smoked foil. These lines were interpreted as being caused by the intersection of "sound waves." Denisov and Troshin first applied it to the observation of transverse waves in detonations. In effect, the smoked-foil method exploits the fact that a detonation, unlike any other singular wave process, can actually "write on the walls"; that is, it can leave a record of its passage in the form of an imprint on a film-coated wall, the most convenient material for this purpose being a thin layer of soot.

Since 1959, the smoked-foil method has been extensively applied to the study of detonation structures. Knystautas et al. (1984) measured the cell size in a variety of stoichiometric fuel–oxygen mixtures at initial pressures ranging from 10 to 200 torr. The minimum cell size usually occurred at an equivalence ratio of $\phi \approx 1$. A few examples of smoked-foil records obtained with self-sustaining propagating detonation are shown in Figures 7.11, 7.12, and 7.13.

The reason that a propagating detonation wave is able to write on the walls is the presence of triple points at the intersecting lines of triple shock waves. The triple point is the intersection point of transmitted, incident, and reflected shocks. A slipstream is also generated in this situation. The triple point is described using Figure 7.14.

In Figure 7.14, a steady supersonic two-dimensional flow, passing through a convergent ramp section from right to left, produces the shock-wave pattern shown in the accompanying schematic diagram. The triple-shock structure involves incident, transmitted, and reflected shock waves. The incident shock is oblique and directs the flow toward the point of intersection. The transmitted shock, also called a Mach-stem shock, with which the incident shock intersects, is nearly normal to the flow and hence is stronger.

The overall detonation-wave structure is thus characterized by a nonplanar leading shock wave, which at every instant consists of many curved shock sections that are convex toward the incoming flow. The lines of intersection of these curved shock segments propagate in various directions at high velocities. The reflected shock (R in Figure 7.14) from these intersections extends back into the reactive flow regime and is required for the flow to be balanced at the intersection of two convex leading shock waves.

In agreement with the above wave-structure description, flash schlieren pictures taken by Fay (1959) demonstrated also that the shape of the reaction zone is irregular, and an interferogram of a self-sustained detonation (see Figure 7.15) obtained by White (1961) shows an irregular pattern of the fringes in and behind the reaction zone.

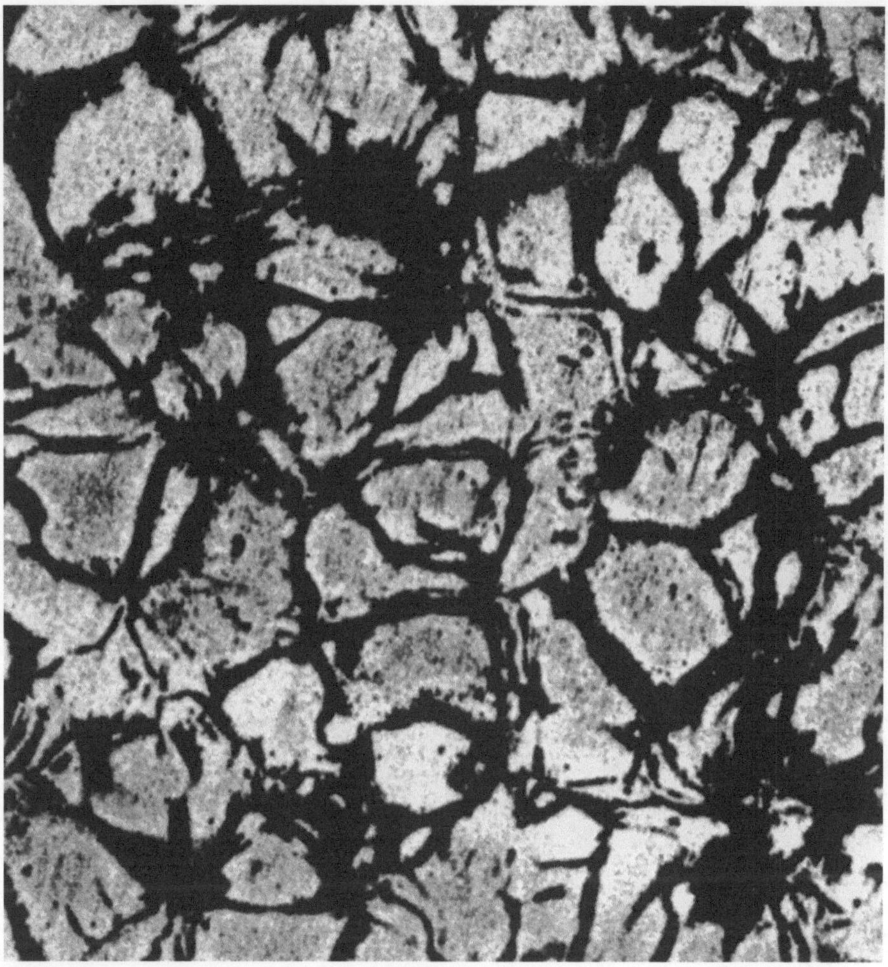

Figure 7.11 Reflection of a detonation from a soot-covered end wall. (Shchelkin and Troshin 1963/NASA/Public domain.)

7.6.1 Numerical Simulation of Detonations

The transition from classical one-dimensional analyses to multicellular wave models with reaction kinetics can be found in comprehensive reviews, for example, see Lee (2008) and Oran and Williams (2012). Numerous theoretical works on 1-D and 2-D analysis of detonation wave instability predict that virtually all waves with realistic reaction kinetics are unstable and develop a spinning or multi-head structure.

Taki and Fujiwara (1978) performed the first numerical simulations of the evolution to the triple point structure in detonations for low-pressure, stoichiometric H_2–O_2 mixtures heavily diluted with argon gas. Using a plane steady Chapman–Jouguet detonation as an initial condition, their calculated results demonstrated that computer simulations could reproduce the types of transverse shock waves and reaction-front structures characterizing multidimensional detonations. An analysis by

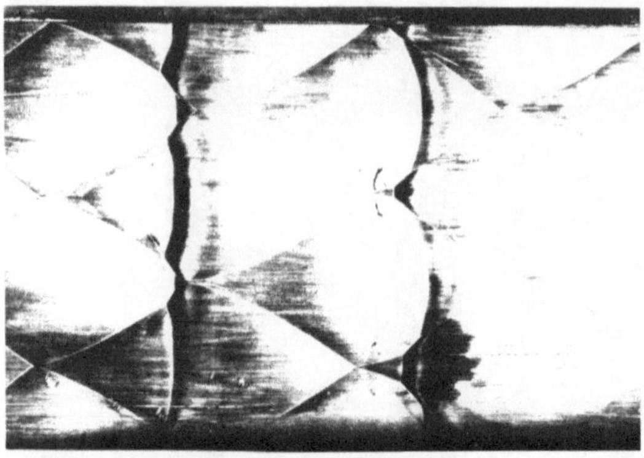

Figure 7.12 Smoked-foil record of H_2-O_2 equilibrium configuration detonation (rectangular mode). The detonation propagation is from left to right. (Strehlow 1984/McGraw Hill Education, Inc.)

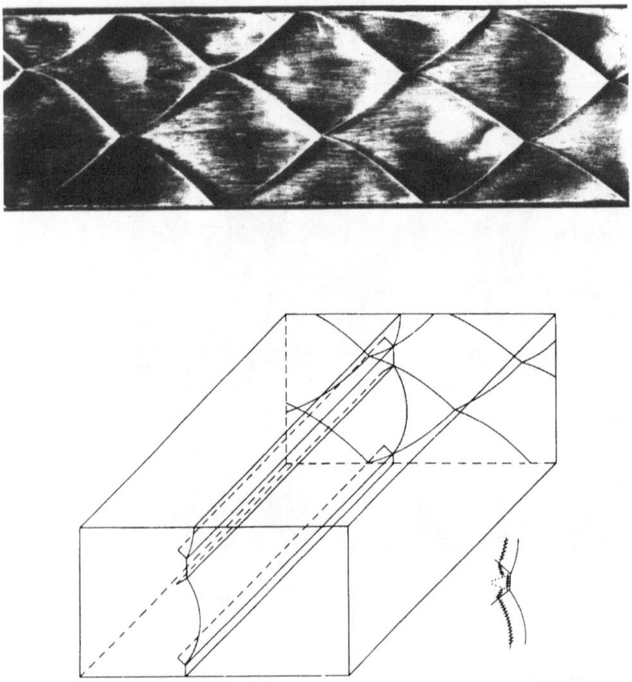

Figure 7.13 Smoked-foil record and schematic diagram of symmetric planar interaction. (Strehlow 1984/McGraw Hill Education, Inc.)

Oran et al. (1987) predicted that there could be some unreacted pockets of material behind the detonation front. Experiments also showed such pockets. The existence of unreacted pockets provides a multidimensional fluid-dynamic mechanism by which a detonation wave can become weaker or even collapse in a situation where, in a 1-D model, it might be expected to propagate. The cellular structure of the computed detonation front is shown in Figure 7.16, based on the calculated pressure, temperature, density, and reaction variable contours (Oran et al. 1987).

The propagation and stability of gaseous detonations at elevated pressures were investigated by Taileb et al. (2021). Their numerical results indicated that at elevated pressures, real gas effects altered the multicellular structure of gaseous detonations. The source of the perceived instabilities was identified as a Mach bifurcation due to jetting and the instability growth was related to a shear layer triple point interaction, giving birth to new triple points.

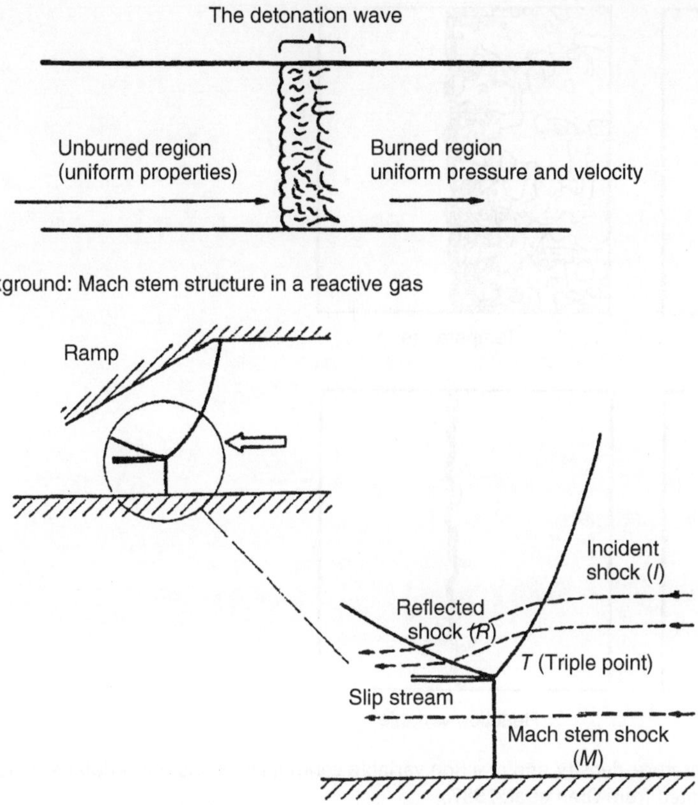

Figure 7.14 Schematic diagram showing the shock wave pattern and triple point in a 2-D supersonic flow passing through a convergent ramp section.

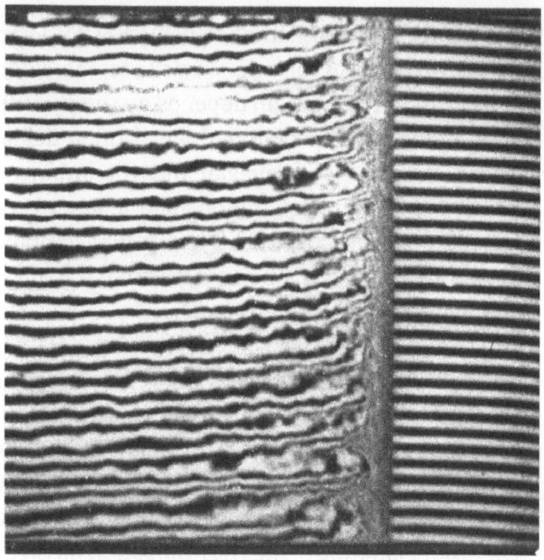

Figure 7.15 Interferogram of a self-sustained detonation in a $2H_2 + O_2 + 2CO$ mixture initially at 0.3-atm pressure and room temperature. (White 1961/with permission of AIP Publishing LLC.)

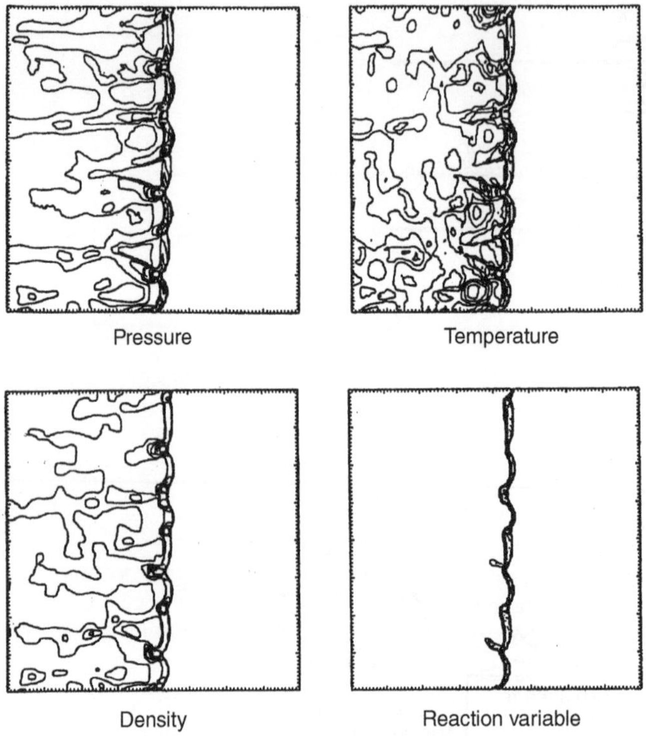

Figure 7.16 Pressure, temperature, density, and reaction variable contours showing the cellular structure of the detonation front. (Adapted from Oran et al. 1987.)

7.7 Deflagration-to-Detonation Transition (DDT) in Gaseous Mixtures

A deflagration-to-detonation transition (DDT) can occur in diverse environments ranging from industrial systems to stars. As we will see, there are several different modes of transition processes, and the onset of detonation depends on the particular pattern of shock fronts created by the accelerating flame. Mixture autoignition is often detected ahead of an accelerating flame giving rise to shock waves which propagate into the unburned and burned gases. The former shock wave is attributed to mixture detonation and the latter is called a retonation shock. It has been found that detonation in tubes and knock in internal combustion engines are essentially the same phenomena. In a tube, the transition length from a deflagration to a detonation wave is of the order of a meter for highly reactive fuels such as acetylene, hydrogen, and ethylene, and longer for most other hydrocarbon–air mixtures.

The transition from deflagration to detonation in confined systems is presented in a review by Oppenheim et al. (1963), with application to enhanced rocket thrust. In addition, direct numerical simulations (DNSs) show that for high enough turbulent intensities that unconfined, subsonic, premixed, turbulent flames are also inherently unstable to DDT (Poludnenko et al. 2011). The occurrence of the DDT in vapor cloud explosions was studied by Oran et al. (2020).

Our understanding of the DDT transition has been furthered due to the use of laser schlieren photography, which has permitted the attainment of extremely short (less than 0.01 μs) light pulses at a high (up to 10^6/s) repetition rate. This technique generates a stroboscopic set of still photographs

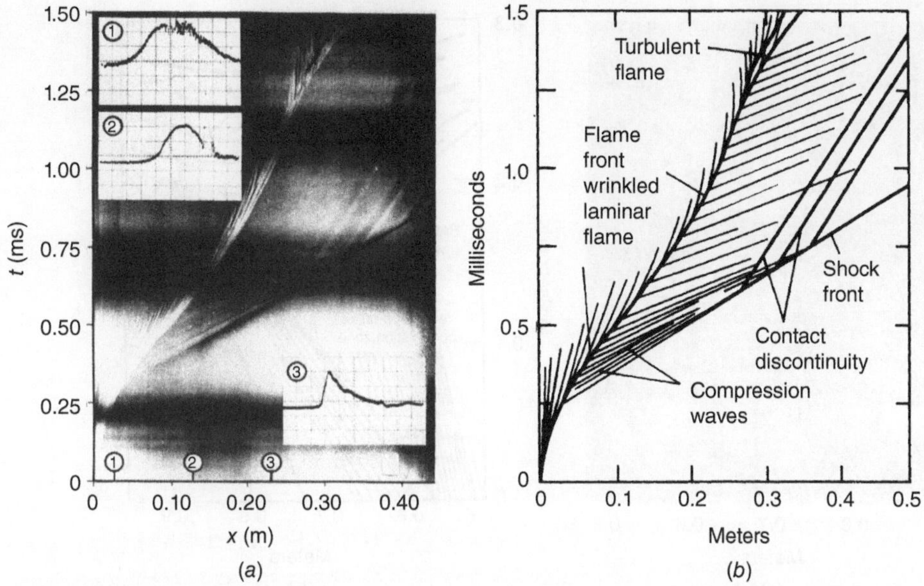

Figure 7.17 (a) Streak schlieren photograph of the development of detonation in stoichiometric hydrogen–oxygen mixture initially at normal temperature and pressure. (b) Generation of pressure waves ahead of the accelerating flame. (Oppenheim et al. 1963/with permission of American Institute of Aeronautics and Astronautics.)

that reveal many of the complex details of the DDT events. The sequence of events leading to detonation in a cylinder containing explosive gases can be seen in Figures 7.17–7.20, which are streak and flash schlieren photographs with explanatory side panels.

The stoichiometric hydrogen–oxygen test mixture of Oppenheim et al. (1963) was initially at normal temperature and pressure. Figure 7.17 shows the generation of compression waves ahead of the accelerating flame. Pressure records at positions 1, 2, and 3 are shown as insets. Figure 7.18 is a streak schlieren photograph of the onset of detonation. The x coordinate denotes the distance from the end of the cylinder. Symbols A and B mark the instant at which the accompanying flash schlieren photographs were obtained.

The complex development of a detonation wave in a cylinder can be summarized as follows:

1. Generation of compression waves ahead of an accelerating laminar flame (see Figure 7.17). The laminar flame front is wrinkled at this stage
2. Formation of a shock front due to coalescence of compression waves (see Figure 7.17).
3. Movement of gases induced by shock, causing the flame to break into a turbulent brush (see Figure 7.17).
4. Onset of "an explosion in an explosion" at a point within the turbulent reaction zone, producing two strong shock waves in opposite directions and transverse oscillations in between. These oscillations are called transverse waves (see Figure 7.18). The forward shock is referred to as a *superdetonation* shock and moves into unburned gases. In the opposite direction, a *retonation* shock moves into the burned gases.
5. Developments of a spherical shock wave at the onset of the "explosion in an explosion," with a center located in the vicinity of the boundary layer (see Figure 7.19).
6. Interaction of transverse waves with the shock front, the retonation wave, and the reaction zone (see Figure 7.20).

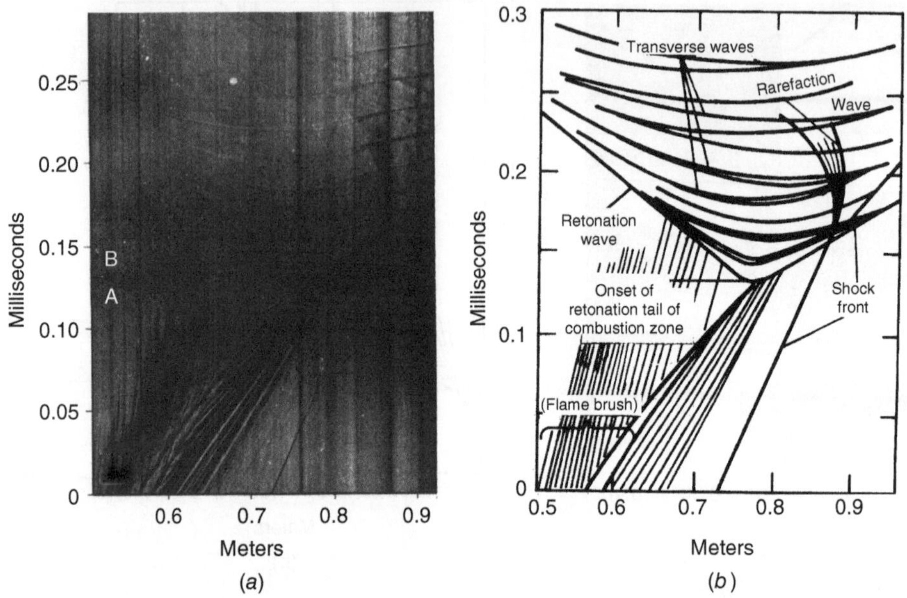

Figure 7.18 Streak schlieren photograph of the onset of *retonation* in a stoichiometric hydrogen–oxygen mixture. (Oppenheim et al. 1963/with permission of American Institute of Aeronautics and Astronautics.)

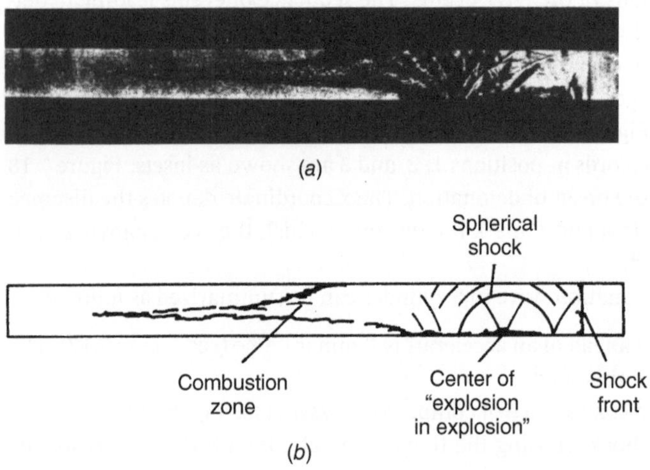

Figure 7.19 Flash schlieren photograph of the onset of retonation in a stoichiometric hydrogen–oxygen mixture at an instant marked by *A* in Figure 7.18. (Oppenheim et al. 1963/with permission of American Institute of Aeronautics and Astronautics.)

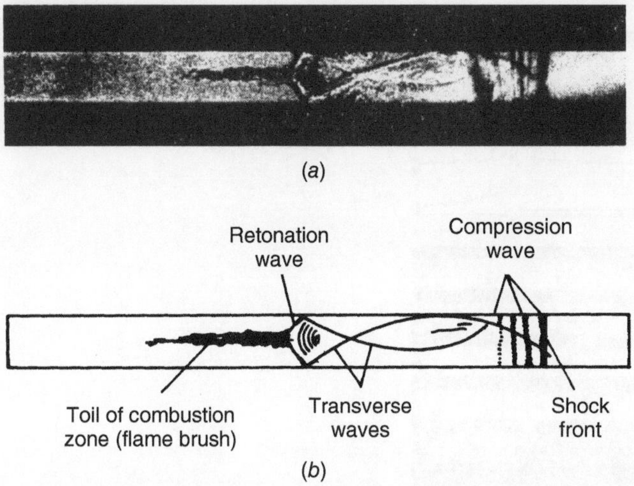

Figure 7.20 Flash schlieren photograph of the onset of retonation in a stoichiometric hydrogen–oxygen mixture at an instant marked by *B* in Figure 7.18. (Oppenheim et al. 1963/with permission of American Institute of Aeronautics and Astronautics.)

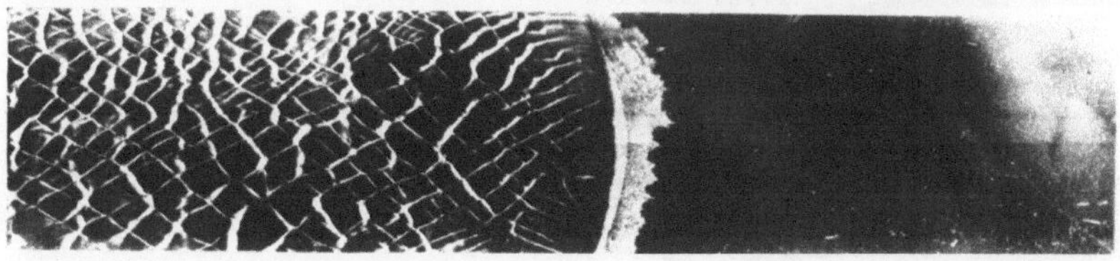

Figure 7.21 Wall imprints of the transition process. (Urtiew and Oppenheim 1966/with permission of The Royal Society.)

7. Establishment of a final "steady wave" as a result of a long sequence of wave interaction processes that lead finally to the shock-deflagration ensemble: the self-sustained C–J detonation wave.

A typical example of the DDT in a $2H_2 + O_2$ mixture ignited by a glow coil in a 1-m-long cylinder is shown in Figure 7.21, obtained by Urtiew and Oppenheim (1966). The forward-moving wave was studied using wall imprints of the detonation process, an example of which is shown in Figure 7.21. The characteristic fish-scale pattern, which corresponds to inception of the forward shock, is a distinguishing feature of a self-sustained detonation front.

Four different types of detonation are generally observed, based on the location of the onset of an "explosion in an explosion," as follows:

1. Between flame and shock front (see Figures 7.22 and 7.23*a*).
2. At flame front (see Figure 7.23*b*).
3. At shock front (see Figure 7.23*c*).
4. At contact discontinuity (see Figure 7.23*d*).

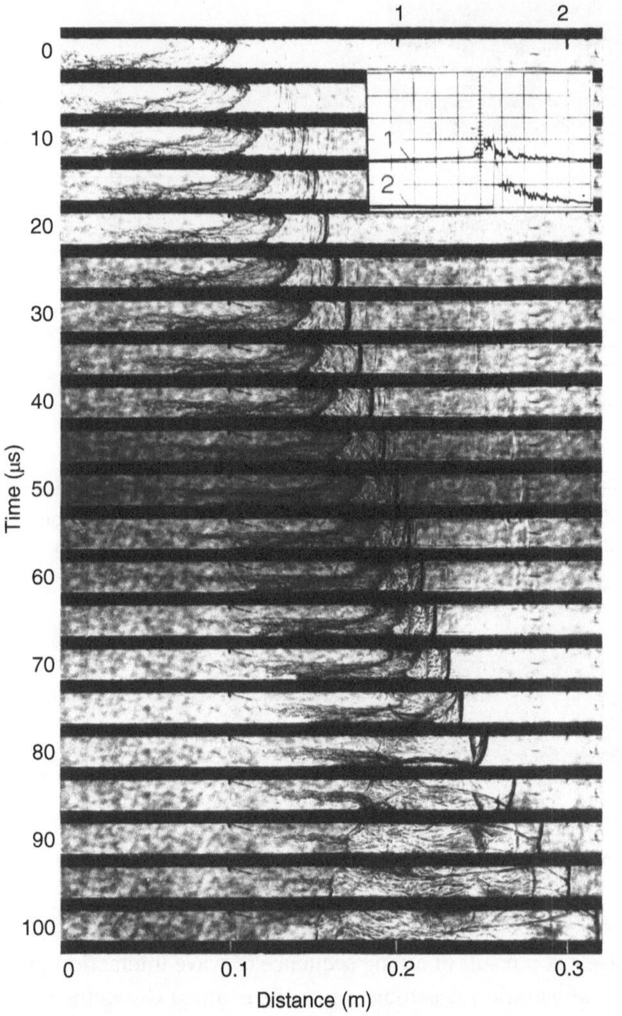

Figure 7.22 Schlieren record of the transition to detonation with onset between flame and shock. $2H_2 + O_2$ initially at a pressure of 0.74 atm. (Adapted from Urtiew and Oppenheim 1966.)

7.8 Detonability Limits

The detonability limits of a reactive mixture are "the critical conditions for the propagation of a self-sustained detonation." Detonability limits are subject to variations in experimental setup, apparatus, and ignition source strength. Despite this lack of standardization, the detonability limits are generally defined as the critical fuel concentrations beyond which no detonation can propagate independently.

The critical conditions are the initial and boundary conditions of the explosive mixture. The initial conditions refer to the fuel's initial characteristics, concentration, thermodynamic state, fluid mechanical state (mean flow and turbulence characteristics), ignition source properties, and all other relevant parameters that characterize the explosive mixture prior to ignition. The boundary conditions refer to the size and geometry of the mixture volume, the degree of confinement, the

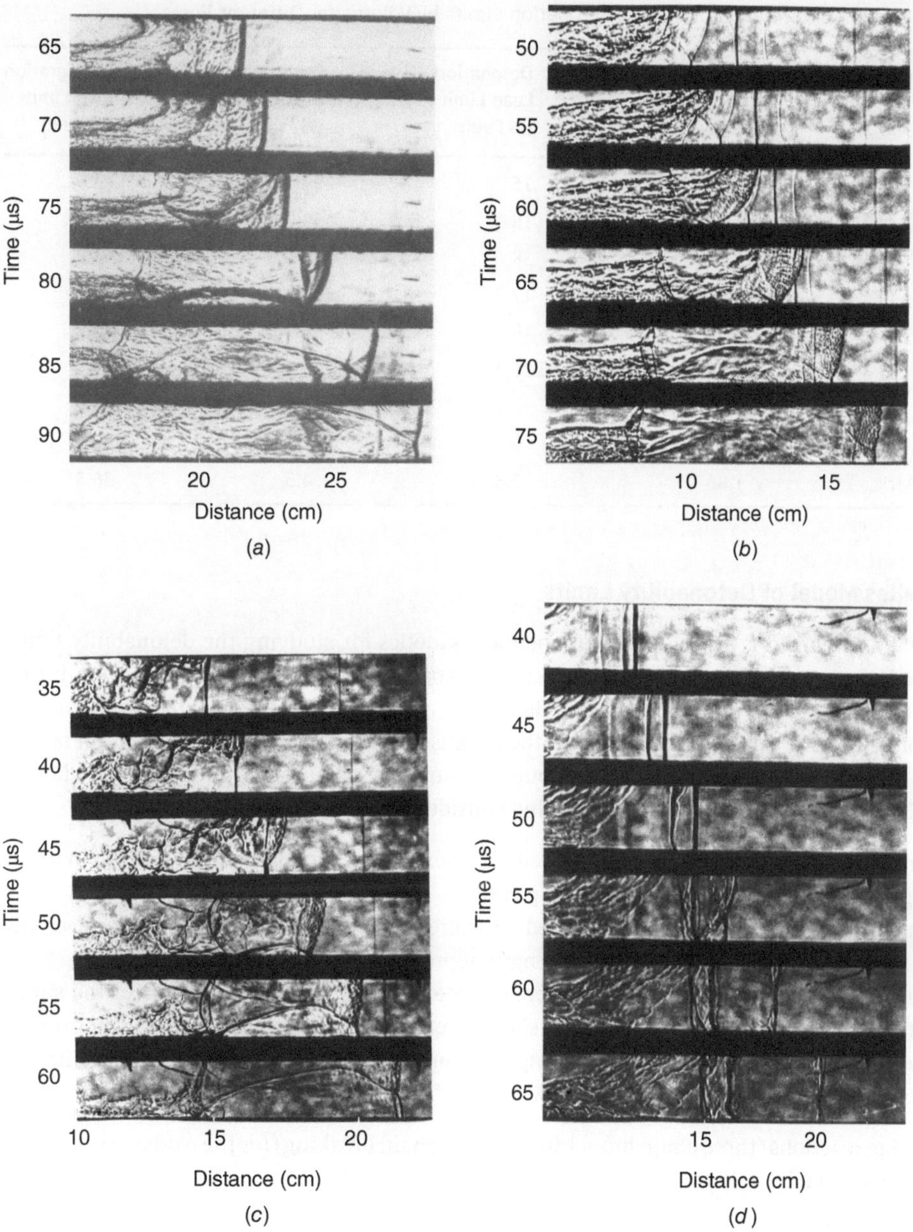

Figure 7.23 Various modes of transition to detonation observed in $2H_2 + O_2$ mixtures. (*a*) Mode 1 with onset occurring between flame and shock, (*b*) mode 2 with onset occurring at flame front, (*c*) mode 3 with onset occurring at shock front, and (*d*) mode 4 with onset occurring at contact discontinuity. (Urtiew and Oppenheim 1966.)

surface characteristics of the walls, and all other relevant parameters that specify the boundary of the explosive mixture.

As shown in Table 7.5, there exist limits of detonability, analogous to limits of flammability, beyond which a stable detonation is not observed. In comparison with the deflagration limits, one can see that the deflagration limits are wider than the detonation limits.

Table 7.5 Experimental Deflagration and Detonation Limits by Volume for Different Fuels

Mixture	Deflagration Lean Limit (% Fuel)	Detonation Lean Limit (% Fuel)	Detonation Rich Limit (% Fuel)	Deflagration Rich Limit (% Fuel)
H_2–O_2	4.6	15	90	93.9
H_2–Air	4	18.3	59	74
CO–O_2 (moist)	15.5	38	90	93.9
(CO + H_2)–O_2	12.5	17.2	91	92
(CO + H_2)–Air	6.05	19	59	71.8
NH_3–O_2	13.5	25.4	75	79
C_3H_8–O_2	2.4	3.2	37	55
C_2H_2–O_2	2.8	3.5	92	93
$C_4H_{10}O$–Air	1.85	2.8	4.5	36.5

7.8.1 Belles Model of Detonability Limits

Belles (1959) developed a model based on chemical kinetics for studying the detonability limits of hydrogen–oxygen mixtures. He combined kinetic data with the observed wave behavior during the transition from flame to detonation and developed a useful technique for predicting the limits of detonability. His analysis was based on the premise that the transition to detonation occurs when the preflame shock becomes strong enough to cause the mixture to explode. The prediction of detonability limits was based on the following considerations and procedures:

1. The chemical kinetics condition for branched-chain explosion of hydrogen was defined in terms of temperature, pressure, and mixture composition.
2. Standard shock-wave equations were used to express the temperature and pressure of the shock-heated gas in terms of the shock strength, represented by the Mach number.
3. Steps 1 and 2 were combined so as to express the explosion condition in terms of the shock strength and mixture composition. Critical shock strengths for explosion were then determined.
4. The detonability limits were predicted using an energy criterion. Some mixtures, on burning, produce enough energy to generate a shock with a critical strength, whereas others cannot.

The following reactions (three chain branching and one chain breaking) for the oxidation of hydrogen were considered by Belles:

$$\mathrm{OH} + \mathrm{H_2} \xrightarrow{k_1} \mathrm{H_2O} + \mathrm{H} \tag{7.132}$$

$$\mathrm{H} + \mathrm{O_2} \xrightarrow{k_2} \mathrm{OH} + \mathrm{O} \tag{7.133}$$

$$\mathrm{O} + \mathrm{H_2} \xrightarrow{k_3} \mathrm{OH} + \mathrm{H} \tag{7.134}$$

$$\mathrm{H} + \mathrm{O_2} + \mathrm{M} \xrightarrow{k_4} \mathrm{HO_2} + \mathrm{M} \tag{7.135}$$

where k_1, k_2, k_3, and k_4 are the rate constants for the above reactions represented by Eqs. (7.132)–(7.135) respectively, and M represents the third body in the formation of HO_2.

Since the diffusion time of a free radical to the cold wall is much longer than the characteristic time of detonation, the chain-termination reactions on the wall were neglected.

The steady-state assumption was used for the free radicals. Thus,

$$\frac{d[H_2O]}{dt} = k_1 \frac{[H_2]^2[M]}{k_4[M] - 2k_2} \tag{7.136}$$

where $[M]$ is the effective concentration of third bodies for the formation of HO_2 in the fourth reaction. The branched-chain explosion limit condition is when the denominator of Eq. (7.136) is zero, or

$$\frac{2k_2}{k_4[M]} = 1 \tag{7.137}$$

The Belles model assumes that an explosion occurs when the rate of the chain-branching reaction (7.133) is equal to half the rate of the chain-breaking reaction (7.135). Using the rate constants of Lewis and von Elbe (1987) and expressing the third body concentration $[M]$ in terms of temperature and pressure by means of the ideal-gas law, Eq. (7.137) becomes

$$\frac{(3.11)T\exp(-8550/T)}{pX_M} = 1 \tag{7.138}$$

where X_M is the effective mole fraction of the third bodies in the formation of HO_2 in Eq. (7.135). Recall that the mole fraction X_M is related to the concentration $[M]$ by the following ideal-gas equation:

$$[M] = \frac{X_M p}{R_u T} \tag{7.139}$$

The mole fraction X_M is given by the following empirical equation, Lewis and von Elbe (1987):

$$X_M = X_{H_2} + 0.35X_{O_2} + 0.43X_{N_2} + 0.2X_{Ar} + 1.47X_{CO_2} \tag{7.140}$$

The numerical factors in Eq. (7.140) express ratios of the rates of reaction for Eq. (7.135) with various third bodies, relative to the rate with hydrogen H_2 as a reference third body.

Rewriting Eq. (7.138) in a logarithmic form, we have

$$\frac{3.71}{T} - \log_{10}\left(\frac{T}{p}\right) = \log_{10}\left(\frac{3.11}{X_M}\right) \tag{7.141}$$

As soon as a given H_2–O_2 mixture, having a characteristic value of X_M, is raised to a temperature and pressure that satisfy Eq. (7.141), the mixture is predicted to explode.

The normal shock equations for the temperature and pressure have been presented previously in the chapter (see Eqs. (7.129) and (7.130)). The critical Mach number Ma_c may be found by combining Eq. (7.141) above with Eq. (7.130). One obtains

$$\begin{aligned} &\frac{3.710\,\alpha^2\beta Ma_c^2}{T_1(Ma_c^2/\beta - 1)(\beta Ma_c^2 + 1/\gamma)} - \log_{10}\left[\frac{T_1(\beta Ma_c^2 + 1/\gamma)}{p_1\alpha\beta Ma_c^2}\right] \\ &= f(T_1, p_1, \gamma, Ma_c) = \log_{10}\left(\frac{3.11}{X_M}\right) \end{aligned} \tag{7.142}$$

Equation (7.142) may be solved graphically. Figure 7.24 shows typical plots of the left-hand side of Eq. (7.142) against the critical Mach number Ma_c, for different hydrogen–oxygen mixtures.

Conservation of energy provides the link between the calculated critical Mach numbers Ma_c for explosion and the prediction of limits of detonability. The shock provides the mechanism whereby the combustion process is continuously sustained. Some of the energy released by combustion is

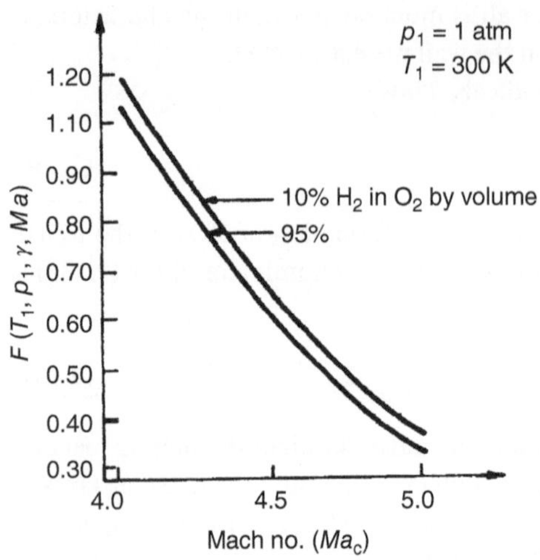

Figure 7.24 Critical Mach number Ma_c as a function of T_1, P_1, and γ.

distributed to unburned gas ahead of the flame. Ultimately, this energy is recovered by the burned gas, when the flame consumes the shock-heated mixture. But, if the enthalpy increases in the shock-heated state corresponding to the critical shock Mach number is greater than the heat of combustion, an impossible situation arises. No transition to detonation in this model could ever occur, even if the duct were infinitely long. The criterion for detonability is therefore

$$\Delta h_s \leq q_c \tag{7.143}$$

where Δh_s is the enthalpy rise across a shock of critical Mach number, and q_c the heat of combustion. Both Δh_s and q_c are expressed as the energy per unit mass of mixture. The energy increase of the shocked state is obtained from

$$\Delta h_s = c_p(T_t - T_0) \tag{7.144}$$

where the total temperature T_t is given by

$$T_t = T_0\left(1 + \frac{\gamma - 1}{2}Ma_c^2\right) \tag{7.145}$$

and c_p, the average specific heat of the mixture. The average heat capacity of the mixture can be evaluated from

$$c_p = \sum_{i=1}^{N} Y_i c_{p,i} \tag{7.146}$$

where $c_{p,i}$ is the average heat capacity at constant pressure of species i.

Figure 7.25 is a plot of Δh_s and q_c versus the percentage of H_2 in the mixture. The two curves intersect at two concentrations. The criterion of detonability, Eq. (7.132), is satisfied only between these two concentrations. Hence, they are the predicted limits of detonability. The calculated results of Belles' model showed good agreement with measured detonability limits by volume (see Table 7.6). This implies that the limits of detonability of H_2 are well accounted for by the kinetic requirements for explosion.

However, one interesting difference between the theoretical model and the experimental treatment is the method of ignition. The theoretical predictions were based on flame transformation to a detonation. The experimentally observed detonations were initiated by an explosive charge or a detonation in an adjoining stoichiometric H_2–O_2 mixture. Despite this discrepancy, the calculated

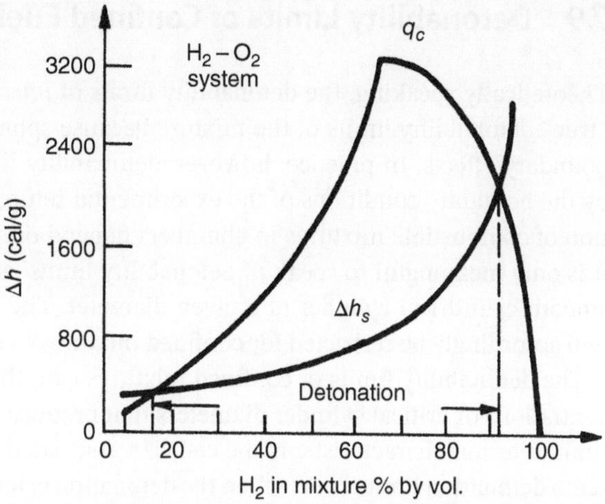

Figure 7.25 Δh_s and q_c versus percentage of H_2 in mixture and the limits of detonability.

Table 7.6 Comparison of Theory and Experiment.

	Lean Limit (% H_2)		Rich Limit (% H_2)	
	Exp.	**Theory**	**Exp.**	**Theory**
Detonation in Air	18.3	15.8	59	59.7
Flammability Limit	14	—	70	—
Detonation in O_2	15	16.3	90	92.3

Source: Adapted from Belles 1959.

and observed results agree and no ignition source dependence seems to be influential. Another feature of the Belles model worth noting is the assumption of no dissociation. This assumption allowed for a simplification of calculations due to the utilization of constant specific heats and the absence of dissociation kinetic equations from the reaction scheme. This assumption proved to be valid due to the occurrence of the detonability limits at low temperatures and pressures. It should be noted that the departure between theory and experiment becomes significant when the concentration of the fuel is decreased below the lean limit as shown in Figure 7.26.

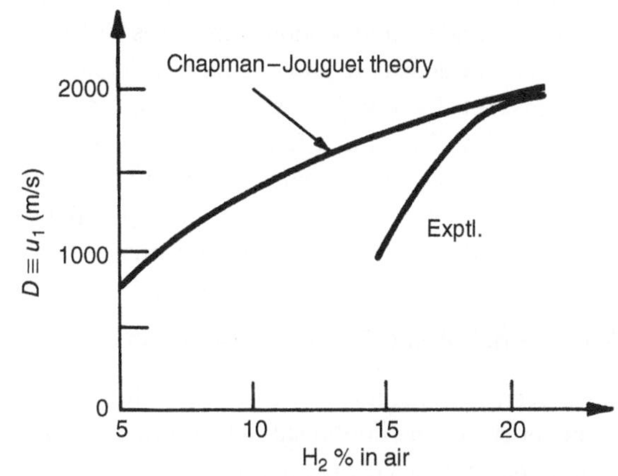

Figure 7.26 Comparison of calculated detonation-wave velocity with experiment as a function of H_2 concentration.

7.9 Detonability Limits of Confined Fuel Mixtures

Theoretically speaking, the detonability limits of *unconfined* mixtures should be regarded as the "true" detonability limits of the mixture because spherical detonations are supposedly free from boundary effects. In practice, however, detonability limits of any mixture are always influenced by the boundary conditions of the experimental setup. It is well known that the limits of detonation of combustible mixtures in chambers depend on the dimensions of the chamber. Therefore, it is only meaningful to speak of detonability limits associated with a given apparatus, such as a smooth cylindrical cylinder of a given diameter. The following discussion of detonability limits will accordingly be restricted for confined mixtures in smooth-walled circular cylinders.

The detonability limits of confined mixtures can, therefore, be defined as the critical fuel concentrations or critical cylinder diameters that produce single-head spinning detonations. Another limit criterion characteristic is the cell size associated with the detonation structure. As we shall see, a detonation limit occurs when the detonation cell width is approximately equal to the cylinder circumference. If the cylinder diameter is less than this critical diameter, no detonation wave can propagate. It is also known that the critical diameter increases with percent dilution.

One limit criterion is the transition from a multiheaded detonation to a single-headed spin detonation as the detonation limits are approached. The detonation limits are also coupled to the cylinder diameter. At each fuel concentration, there is a specific cylinder diameter at which a multiheaded detonation becomes a single-headed spin detonation. The detonation structure of the single-headed spin detonation is unstable to finite perturbations as it loses and then recovers its structure immediately after a perturbation obstacle. Mixtures slightly leaner or richer than that for the onset of the single-headed spin demonstrate the destruction-reformation phenomenon, known as galloping, when detonated.

As the cylinder diameter increases, the detonation limits become wider. Wendlant (1924) showed that for cylinder diameter of 1.4 cm, the detonation limits of H_2–O_2 system range from 18.3 to 59% of H_2 by mole. Lee (1984) used a larger cylinder with a diameter of 43 cm and found that detonation limits of H_2–O_2 system ranged from 13.6 to 70.0% of H_2 by mole. Therefore, detonability limits of confined mixtures are not unique, but strongly dependent on the geometry, i.e., the boundary conditions of the gas mixture in which the detonation propagates.

In order to demonstrate the detonability limit dependence on the initial conditions, such as mixture concentration and the ignition process, the results of Wolanski et al. (1981) on the detonation of methane–air mixtures are considered. In their study, the amount of ignition energy density or critical energy density (MJ/m^2), defined as the total chemical energy in the initiator divided by the main cylinder cross-sectional area, was varied. As the detonation limits were approached, the critical energy density required to initiate detonation of the mixture increased exponentially.

Wolanski et al. showed a U-shaped curve dependency of critical energy density on percentage of CH_4 in the initial mixtures. In their experiments using a 6.35-cm cylinder, they found that the limits of detonation were associated with 8 and 14.5% methane in the CH_4–air mixture. More recently, Matsui (2002) experimentally measured relationships between the initial pressure, cylinder diameter, and oxygen concentration for various fuel (acetylene and ozone)–oxygen mixtures.

7.9.1 Single-Head Spinning Detonation

It has been observed that, in a given smooth circular cylinder, as the mixture composition becomes leaner or richer, the multihead, self-sustained detonation becomes a single-head spinning detonation propagating axially at approximately the C–J velocity. At the first appearance of single-head spin, the characteristic transverse wave is approximately tuned to the lowest acoustic mode of the cylinder which is principally determined by the transverse dimensions of the cylinder. This single-head spinning detonation is characterized by an increase in the transverse wave strength, a three-dimensional wave structure, and mixture gases that rotate at a uniform angular velocity about the cylinder axis.

A detonation cell width is defined as the width of one of the cells formed by the slipstream associated with the interaction of the transverse and longitudinal waves of the detonation. The cellular structure can be mapped during a detonation with the use of soot-covered sheets. A typical cellular structure left on a smoke foil with a corresponding cell width, λ, is shown in Figure 7.29. In these cases, single-head spinning and "galloping" detonations which depend on the coupling between the combustion zone instabilities and the acoustic modes of the cylinder are observed over a fairly wide range of compositions.

As early as 1948, Kogarko and Zel'dovich (1948) proposed that at the onset of single-head spin in a smooth circular cylinder, the detonation cell width, λ, is equal to the cylinder circumference, with a minimum cylinder diameter $d = d^*$,

$$\lambda = \pi d^* \tag{7.147}$$

Lee (2008) provided a derivation of the Kogarko and Zel'dovich (1948) criterion. Because the cylinder circumference, πd, represents the largest characteristic dimension for the cylinder, the longest characteristic acoustic time is $\pi d/c$, where c is the sound speed of the detonation products. Assuming a resonant coupling between the acoustic vibration and the periodic chemical processes in the detonation cell, the characteristic acoustic time, $\pi d^*/c$, must be equal to the characteristic chemical timescale, λ/c. Therefore, the detonation cell width is predicted to be proportional to the cylinder circumference, $\lambda \cong \pi d^*$.

Dove and Wagner (1961) proposed that the single-head spinning detonation is the lowest possible stable wave mode in a given cylinder as the detonation limits of confined mixtures are approached. Figure 7.27 shows a typical path of the single-head detonation wave propagation.

Not only are the single-head detonations associated with the detonability limits for a given fuel–oxidizer mixture coupled to fuel concentration, they are also coupled to cylinder diameter. In an experimental study performed by Moen et al. (1981), three cylinders of diameters 28, 48, and 145 mm were used to detonate ethylene–air mixtures at 1-atm pressure near their lean limits. They showed that for each fuel concentration, there is a specific cylinder diameter at which the multihead detonation becomes a single-head detonation. For a given fuel concentration, that specific diameter was defined as the "critical cylinder diameter, d^*."

The general trend obtained by Moen et al. (1981) follows that shown in Figure 7.28. In Figure 7.28, the cylinder diameter d is plotted against fuel concentration around the lean limit. In the region

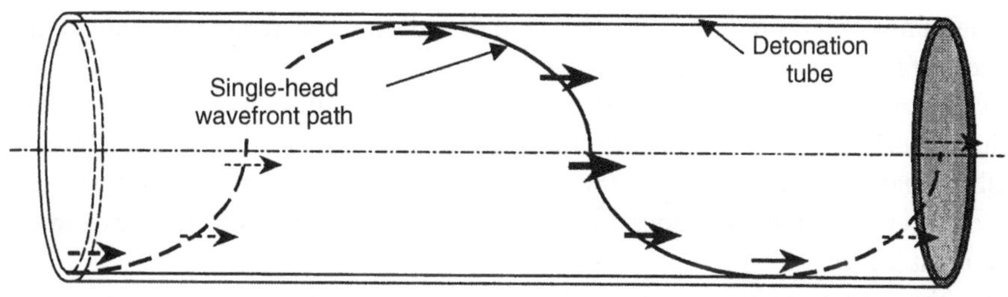

Figure 7.27 Path of a single-head detonation wave front in a straight cylinder.

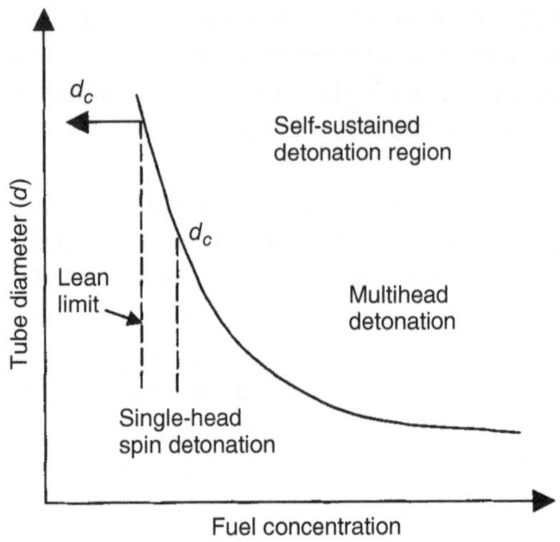

Figure 7.28 Effects of fuel concentration and cylinder diameter on the onset of single-head spin detonation.

above the curve, known as the "self-sustained" detonation region, multihead detonations readily propagate. In the region below the curve, known as the "detonation-like" region, single-head detonations can propagate, but only with powerful igniters.

Moen et al. (1981) also showed that the detonation structure of the single-head detonation is unstable to finite perturbations. In their experiments, a short length of surface obstacle called Shchelkin spiral was used to perturb the detonation. It was observed that for mixture compositions near the onset of single-head spinning, the detonation recovered its structure immediately after the obstacle. For mixtures slightly leaner or richer than that for the onset of the single-head spin detonation, the detonation failed and then recovered its spinning structure far downstream of the Shchelkin spiral.

A schematic illustration of wave motion in a detonation cell is shown in Figure 7.29. In this figure, the detonation wave is propagating upward and the transverse waves are indicated; the reaction-zone thicknesses are given by the widths of the shaded areas.

Moen et al. (1981) formulated a limit criterion based on acoustics (see Fay 1952), which predicts the ratio of spin pitch P to the limiting cylinder diameter,

$$\frac{P}{d^*} = \frac{\pi u}{K_1 c} \cong 1.7 \tag{7.148}$$

Here, u is the C–J detonation velocity and the eigenvalue of the first tangential (1T) wave mode $K_1 = 1.841$. In order to arrive at a limit criterion, $\lambda \cong 1.7d^*$, Moen et al. assumed that the pitch P equals the cell length λ and $u/c \cong 1$ for most fuel–air mixtures. Therefore, it can be inferred from the aforementioned limit criteria that, in smooth round cylinders, the detonability limits occur when the cell size is of the order of the cylinder circumference.

Adopting the criterion that the outset of single-head spinning waves in a given cylinder should correspond to the detonability limits for that cylinder, the limits for circular cylinders can be estimated when cell size data are available. Since the onset of single-head spin corresponds to a cell size of the order of the cylinder circumference, the composition limits for a given circular cylinder can be estimated if λ is known.

Figure 7.30 gives the limiting cylinder diameter d^* as a function of equivalence ratio for various fuels studied by Knystautas et al. (1984). For a given fuel composition of known equivalence ratio,

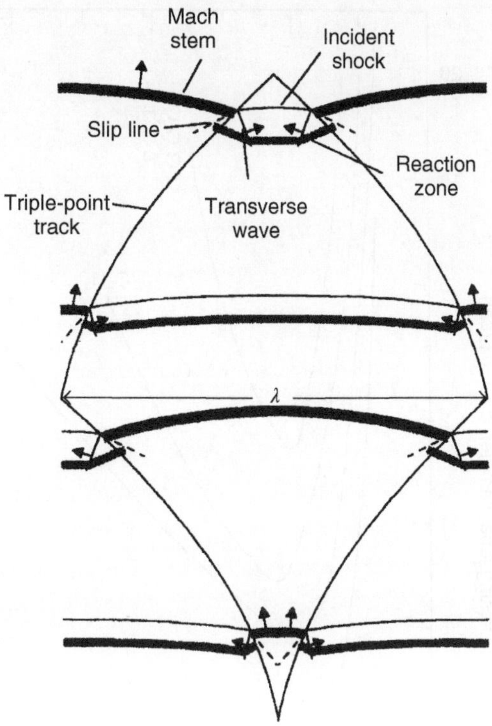

Figure 7.29 Schematic illustration of wave motion in a detonation cell. (Gavrikov et al. 2000/with permission of Elsevier.)

stable detonations cannot be propagated in cylinders with a diameter $d < d^*$. For the case of stoichiometric H_2—air mixtures, $d^* \approx 0.5$ cm, compared to $d^* \approx 0.2$ cm for C_2H_2–air mixtures and $d^* \approx 1.7$ cm for alkane–air mixtures.

7.10 Engineering Applications of Detonation

Attempts have been made to utilize the high combustion pressures and temperatures of a detonation to produce propulsion or power in an engine, but practical applications of detonations are still in the development stage. In a detonation engine, a detonation wave rapidly compresses a fuel/oxidizer mixture and adds heat to the engine at constant volume, which is thermodynamically a more efficient process than the constant pressure heat addition of deflagration-based system such as a Brayton cycle gas turbine. Broadly, three applications of interest have been studied in utilizing detonation engines: gas-turbines, ramjets, and rocket engines.

In a detonation engine, the detonation is initiated in a fuel/oxidizer mixture by an igniter. After the engine is started, the detonations are self-sustaining. One detonation ignites fuel/oxidizer mixture in the combustion chamber, which releases the energy necessary to sustain the detonation. The detonation wave rapidly traverses the chamber resulting in the nearly constant-volume heat addition process. The resulting combustion greatly increases the pressure of the mixture, which then expands through a nozzle producing thrust. The combustion products and any remaining unburnt charge are then pushed out of the engine chamber by the next incoming fuel/oxidizer mixture. Fuel/oxidizer injection and mixing are perhaps the most important operational parameters that affect detonation engine behavior.

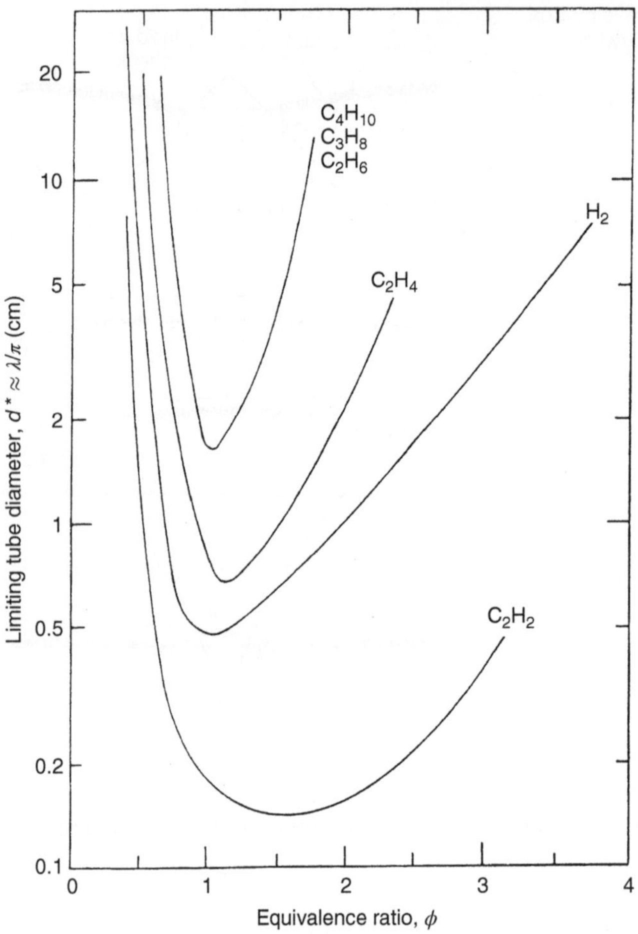

Figure 7.30 Limiting cylinder diameter d^* as a function of equivalence ratio. (Adapted from Knystautas et al. 1984.)

Two geometric configurations of detonation engines that have been developed are the pulse detonation engine (PDE) and the rotating detonation engine (RDE). In a PDE, a spinning detonation wave moves helically in an enclosed chamber with an axial velocity component, similar to the detonation wave shown in Figure 7.27. The operation of multi-tube PDE configurations at high detonation-initiation frequency (about 100 Hz and over) can produce a near-constant thrust. The difficulties in PDEs are achieving detonation without requiring a tube long enough to make it impractical. The concept of PDE is attractive for both subsonic and supersonic flight with PDE as a main propulsion unit or as an afterburner in turbojet or turbofan propulsion system. A comprehensive review of the history and development of PDEs is given in Roy et al. (2004).

The basic concept of an RDE is a detonation wave that travels around a circular annulus. The rotating detonation is axially fixed, and fuel and oxidizer are injected into the annulus through small holes or slits. A rotating detonation wave is initiated by a pulse of hot, high-pressure gas which is introduced tangentially into the annulus. The high pressure behind the wave front gradually decays, allowing fresh reactants to flow into the combustion chamber and exhaust products to flow out of the chamber. The current state of RDEs is given in the review by Anand and Gutmark (2019).

Detonation engines have a number of significant engineering challenges. Some of these include the ability to achieve consistent and reliable detonations in a practical arrangement, understanding of the unsteadiness of the wave motion, and integration with ancillary propulsion systems, and noise.

Internal combustion and rocket engines are typically designed to avoid detonation regimes. Otto cycle internal combustion engines are limited to a maximum compression ratio of about 10:1 due to possible detonation of the fuel/air mixture ahead of the deflagration flame during the compression stroke. This IC engine detonation is also termed *knock*. The resulting high pressures and temperatures cause erosion of the piston head and valves. Rocket engines (both liquid and solid propellant) have observed high-frequency tangential waves (1T) that result in unstable combustion. These rocket combustion instabilities are mitigated through the use of baffles and Helmholtz resonators.

7.11 Possible Measures for Reducing Potential of Detonation

There are a number of gases and vapors that decompose exothermically and can support a detonation wave even in the absence of an oxidant. Examples of such gases are acetylene, ozone, ethylene, and high-concentration hydrogen peroxide. There are common features in the decomposition of derivatives of acetylene; hence, they can also be detonable. In general, all compounds that decompose exothermically should be given due consideration. It is worth noting that there are very few recorded examples of detonations of self-decomposing materials in an unconfined medium.

The minimal conditions required for the development of a detonation wave are the occurrence of a flammable mixture and an ignition source in the case of fuel–oxidant mixtures, or sufficiently high pressures with an ignition source for exothermically self-decomposing fuels. It is not possible to altogether eliminate such conditions in chemical plants. Hence, several measures have been developed to inhibit flames from spreading.

Halogen-containing compounds, products of their pyrolysis, sulfur dioxide, etc., inhibit flame spreading by combination with radicals such as H, OH, and O. These chemical species halt the chain-branching reactions and stop flame spreading. The mechanism involved in the flame inhibition by sodium or potassium salts of organic materials is not fully known. It could be due to a combination of endothermic decomposition and reactions with radicals such as H and OH.

Spraying of water curtains can be effective in many cases and offers the obvious advantage of cost-effectiveness; thus, it is widely used. In the inhibition of accelerating flames, the time factor is of importance. For flames with high velocities, the time available for the suppressants to act is indeed very small. There are several different strategies for the prevention of detonation:

Inhibition of flames of normal burning velocity This is performed mostly by injection of a flame suppressant into the pipes. This requires some kind of flame detection mechanism, which triggers the injection of the suppressants. As the length of the pipes is usually very large in industrial plants, an array of such units is required. For slow-burning flames with burning velocity not exceeding 10 m/s, there is ample time for flame detection and suppression injection. The most widely used inhibitor is mono-ammonium phosphate. Halogenated hydrocarbons are generally used in commercially supplied suppressant equipment. The disadvantage of this family of suppressants is that the products of decomposition of such materials are highly corrosive. Thus, a decision has to be made between the effectiveness of the suppressant and the damage caused due to an accidental injection. Suppression of flames in spherical or cylindrical vessels is more difficult as the time available is much less. Precautions should be taken against central ignition, which is the most harmful. Flames can also be arrested by insertion of sintered metal discs, honeycombs, etc. The choice depends on the allowable pressure drop. These are usually cost effective, as they do not require much maintenance. But, the positioning could prove crucial, as in the later stages of flame spread, such structures could actually enhance the flame due to generation of turbulence.

Venting in the early stages of an explosion An alternative form of prevention of a detonation wave is some form of pressure relief. Bursting diaphragms and hinged or spring-loaded explosion doors can be used. These are much cheaper. An unwanted consequence of such pressure release could be that turbulence could be created that could actually accelerate the flame. This design option could introduce pressure oscillations in the vessel, which could be equally harmful. There are many suggestions available for the placing of pressure reliefs in pipes. In general, it can be said that pressure relief is favored in systems with low working pressures.

Quenching of flame-shock complex Most methods used to quench low-velocity flames are also effective at a stage when the flame has accelerated to near-detonation velocity. An example is the use of ceramic packings. This method has been adopted by some plants in the German chemical industry. The pipelines were designed to sustain the maximum pressure developed, and the packings prevented the transition to detonation.

In addition, there are other methods for reducing the chance for DDT to occur. For example, in mine galleries, trays containing limestone dust or water were supported on ledges on walls that could be easily dislodged by turbulence, thus introducing the inert material just before the flame zone. Injection of a suppressant is perhaps the most effective in all the stages of flame spreading. The choice and design of the system depend on the type and dimensions of the plant and also the nature of the detonable media.

References

Anand, V., and E. Gutmark (2019), Rotating detonation combustors and their similarities to rocket instabilities, *Prog. Energy Combust. Sci.*, Vol. 73, pp. 182–234.

Belles, F. (1959), Detonability and chemical kinetics: prediction of limits of detonability of hydrogen, *Proc. Combust. Inst.*, Vol. 7, pp. 745–751.

Berets, D., E. Greene, and G. Kistiakowsky (1950), Gaseous detonations I. Stationary waves in hydrogen—oxygen mixtures, *J. Am. Chem. Soc.*, Vol. 72, pp.1080–1086.

Campbell, C., and D. Woodhead (1927), CCVII. Striated photographic records of explosion-waves, *J. Chem. Soc.*, pp. 1572–1578.

Chapman, D. (1899), On the rate of explosion in gases, *Philos. Mag.*, Vol. 47, pp. 90–104.

Denisov, Y., and Y. Troshin (1959), Pulsating and spinning detonation of gaseous mixtures in cylinders, *Dokl. Akad. Nauk*, Vol. 125, pp. 110–113.

Döring, W. (1943), On detonation processes in gases, *Ann. Phys.*, Vol. 43, pp. 421–436.

Dove, J., and H. Wagner (1961), A photographic investigation of the mechanism of spinning detonation. *Symp. (Intl.) Combust.*, Vol. 8, No. 1, pp. 589–600.

Fay, J. (1952), A mechanical theory of spinning detonation, *J. Chem. Phys.*, Vol. 20, No. 6, pp. 942–950.

Fay, J. (1959), Two dimensional gaseous detonations: velocity deficit, *Phys. Fluids*, Vol. 2, No. 3, pp. 283–289.

Friedman, R. (1953), Kinetics of the combustion wave, *J. Am. Rocket Soc.*, Vol. 23, pp. 349–354.

Gavrikov, A., A. Efimenko, and S. Dorofeev (2000), A model for detonation cell size prediction from chemical kinetics, *Combust. Flame*, Vol. 120, pp. 19–33.

Glassman, I., R. Yetter, and N. Glumac (2014), *Combustion*, 5th ed., Academic Press. Cambridge, MA.

Gordon, S., and B. McBride (1994), Computer program for calculation of complex chemical equilibrium composition, and applications, NASA RP-1311.

Jouguet, E. (1905), On the propagation of chemical reactions in gases, *J. Math. Pures Appl.*, Vol. 6, pp. 347–425.

Kistiakowsky, G., and J. Kydd (1956), Gaseous detonations IX. A study of the reaction zone by gas density measurements, *J. Chem. Phys.*, Vol. 25, pp. 824–835.

Knystautas, R., C. Guirao, J. Lee, and A. Sulmistras (1984), Measurement of cell size in hydrocarbon-air mixtures and predictions of critical cylinder diameter, critical initiation energy and detonation limits, *Prog. Astron. Aeronaut.*, Vol. 94, pp. 23–37.

Kogarko, S., and Y. Zel'dovich (1948), On detonation of gaseous mixtures, *Dokl. Akad. Nauk SSSR*, Vol. 63, pp. 553–556.

Lee, J. (1984), Dynamic parameters of gaseous detonations, *Annu. Rev. Fluid Mech.*, Vol. 16, pp. 311–336.

Lee, J. (2008), *The Detonation Phenomenon*, Cambridge University Press, Cambridge.

Lewis, B., and J. Friauf (1930), Explosions in detonating gas mixtures. I. Calculation of rates of explosions in mixtures of hydrogen and oxygen and the influence of rare gases, *J. Am. Chem. Soc.*, Vol. 52, pp. 3905–3920.

Lewis, B., and G. von Elbe (1987), *Combustion, Flames and Explosions of Gases*, 3rd ed., Academic Press, New York.

Matsui, H. (2002), Detonation propagation limits in homogeneous and heterogeneous systems. *J. Phys. IV Fr.*, Vol. 12, pp. 11–17.

Moen, I., M. Donato, R. Knystautas, and J. Lee (1981), The influence of confinement on the propagation of detonations near the detonability limits, *Symp. (Intl.) Combust.*, Vol. 18, No. 1, pp. 1615–1622.

von Neumann, J. (1942), Theory of detonation waves, Report No. 549, Office of Scientific Research and Development (OSRD), Washington, DC.

Oppenheim, A., N. Manson, and H. G. Wagner (1963), Recent progress in detonation research, *AIAA J.*, Vol. 1, No. 10, pp. 2243–2252.

Oran, E., and F. Williams (2012), The physics, chemistry and dynamics of explosions, *Phil. Trans. R., Soc. A*, Vol. 370, pp. 534–543.

Oran, E., K. Kailasanath, and R. Guirguis (1987) The structure of detonation waves, *J. Phys. Colloq.*, Vol. 48 (C4), pp. 105–117.

Oran, E., G. Chamberlain, and A. Pekalski (2020), Mechanisms and occurrence of detonations in vapor cloud explosions, *Prog. Energy Combust. Sci.*, Vol. 77, p. 100804.

Poludnenko, A., T. Gardner, and E. Oran (2011), Spontaneous transition of turbulent flames to detonations in unconfined media, *Phys. Rev. Lett.*, Vol. 107, pp. 054501–054514.

Radulescu, M., G. Sharpe, C. Law, and J. Lee (2007), The hydrodynamic structure of unstable cellular detonations, *J. Fluid Mech.*, Vol. 580, pp. 31–81.

Roy, G., S. M. Frolov, A. Borisov, and D. Netzer (2004), Pulse detonation propulsion: challenges, current status, and future perspective, *Prog. Energy Combust. Sci.*, Vol. 30, pp. 545–672.

Shchelkin, K., and Y. Troshin (1963), Gas dynamics of combustion, *Gazodinamika Goreniya, Moscow Izd. Akad. Nauk SSSR*, translated as NASA-TT-F-231 (1964).

Strehlow, R. (1971), Detonation structure and gross properties, *Combust. Sci. Technol.*, Vol. 4, pp. 65–71.

Strehlow, R. (1984), *Combustion Fundamentals*, McGraw-Hill, New York.

Taileb, S., J. Melguizo-Gavilanes, and A. Chimmayya (2021), The influence of the equation of state on the cellular structure of gaseous detonations, *Phys. Fluids*, Vol. 33, p. 036105.
Taki, S., and T. Fujiwara (1978), Numerical analysis of two-dimensional nonsteady detonations, *AIAA J.*, Vol. 16, No. 1, pp. 73–77.
Tofaili, H., G. Lodato, L. Verisch, and P. Clavin (2021), One-dimensional dynamics of gaseous detonations revisited, *Combust. Flame*, Vol. 232, p. 111535-1-13.
Urtiew, P., and A. Oppenheim (1966), Experimental observations of the transition to detonation in an explosive gas, *Proc. Roy. Soc. London, Ser. A.*, Vol. 295, pp. 13–28.
Wendlant, R. (1924), Experimentelle untersuchungen zur detonationsgrenze gasförmiger gemische, *Z. Phys. Chem.*, Vol. 110, pp. 637–655.
White, D. (1961), Turbulent structure of gaseous detonations, *Phys. Fluids*, Vol. 4, pp. 465–480.
Williams, F. (1985), *Combustion Theory*, 2nd ed., Taylor and Francis, Boca Raton, FL.
Wolanski, P., C. Kauffman, M. Sichel, and J. Nicholls (1981), Detonations of methane-air mixtures, *Symp. (Intl.) Combust.*, Vol. 18, No. 1, pp. 1651–1660.
Zeldovich, Y. (1940), The theory of the propagation of detonation in gaseous systems, *Expt. Theor. Phys. S.S.S.R.*, Vol. 10, p. 542; translated as NACA Tech. Memo. TM-1261 (1950).

Homework Problems

7.1 A detonation wave is propagating through a stoichiometric mixture of acetylene C_2H_2 and air at a temperature of 298 K and pressure of 1 atm. What is the detonation velocity and the burned gas temperature, pressure, and density?

7.2 For the acetylene C_2H_2 detonation wave of Problem 7.1, use the ZND wave model to determine the gas mixture properties at state $1'$ immediately after the shock wave passes. Compare with the properties at the end of the reaction zone.

7.3 Suppose the perfect-gas law is replaced by the Noble–Abel dense-gas law

$$p\left(\frac{1}{\rho}-b\right)=RT$$

where b represents the volume occupied by gaseous molecules themselves. Develop a Rankine–Hugoniot relation using the above equation of state for the burned gases behind the detonation wave front.

7.4 A mixture of hydrogen, oxygen, and nitrogen, having partial pressures in the ratio 2:1:5 in the order listed, is observed to detonate with a speed of 1890 m/s when the initial temperature is 298 K and the initial pressure is 1 atm. Using the ZND model, calculate the peak pressure in the detonation wave and calculate the pressure and temperature of the burned gas just after the passage of the wave. Compare the observed speed with the C–J condition. Neglect dissociation, assume that the pressure after the wave passes is much greater than the initial pressure and that the specific heats are independent of pressure.

7.5 Estimate the steady-state detonation-wave velocity for the premixed gaseous mixture $2H_2 + O_2 + 3N_2$ (assuming no dissociation of the product gases). The initial temperature and pressure of the reactants are $T_1 = 298.15$ K and $p_1 = 1$ atm, respectively.

7.6 Beginning with the continuity and momentum equations, Eqs. (7.13) and (7.14), derive equations expressing the pressure ratio p_2/p_1, the temperature ratio T_2/T_1, and the velocity ratio u_2/u_1 in terms of the Mach numbers Ma_1, Ma_2, and the specific-heat ratio γ for the combustion wave shown in Figure 7.1. Assume (1) ideal-gas behavior, (2) steady state, (3) no body force, (4) no external heat addition or heat loss, (5) negligible Dufour effect, and (6) negligible interdiffusion.

7.7 Verify the following C–J condition: *The C–J condition implies that when the Hugoniot curve is tangent to the Rayleigh line passing through the origin, it is also tangent to the isentrope. Hence, the local Mach number at the upper C–J point and the lower C–J point is unity irrespective of the thermodynamic properties of the medium.*

7.8 Assume the conditions at the unburned side of a stationary one-dimensional combustion wave are

$$p_1 = 1\,\text{atm},\ \rho_1 = 1.185\,\text{kg/m}^3,\ \text{and}\ \gamma = 1.40$$

(a) Express the Rankine–Hugoniot relation, Eq. (7.31), in terms of the density ratio, ρ_2/ρ_1, and the pressure ratio, p_2/p_1.
(b) Calculate the heat release per unit mass, assuming $p_2/p_1 = 30$, $\rho_2/\rho_1 = 2.1$ for detonation waves, and $p_2/p_1 = 0.98$, $\rho_2/\rho_1 = 0.15$ for deflagration waves, and compare the two results.

7.9 Interferograms, see Figure 7.15, of both Chapman–Jouguet detonations and slightly overdriven detonations show that the detonation front structure is not planar and that the structure of the following flow has variations in refractive index that we can interpret as being due to turbulence. Derive the form of the Rayleigh and Rankine–Hugoniot relations when turbulence is considered by resolving the burned gas properties of velocity U_2, density ρ_2, pressure p_2, and total enthalpy h_{t2} into mean and fluctuating quantities, and then time averaging. Make the following assumptions:
(a) The detonation wave occurs as a statistically steady process in a constant-area cylinder.
(b) The flow ahead of the detonation wave is nonturbulent.
(c) The mean gas velocity U is parallel to the axis of the cylinder (x-axis), and the transverse derivatives of any time-averaged quantity vanish.
(d) Neglect velocity fluctuation terms of third order or higher.
(e) Assume isotropic turbulence $\overline{u_i' u_j'} = \frac{1}{3}u^2\,\delta_{ij}$.

8
Non-premixed Flames

8.1 Introduction

In this chapter, we examine gaseous non-premixed flames. A non-premixed flame is also called a diffusion flame and is defined as any flame in which the fuel and oxidizer are initially separated, then mix through either molecular or turbulent diffusion. When the fuel and the oxidizer enter a combustion system separately, they must mix on a molecular level before a reaction can take place. Unlike premixed flames, which have a very narrow reaction zone, diffusion flames have a wider region over which the composition changes and chemical reactions can take place.

The rate and extent of reaction are strongly influenced by the amount of mixing that has occurred prior to combustion. If the diffusion rates of fuel and oxidizer to the flame surface are in stoichiometric proportion, then the concentrations of fuel and oxidizer approach zero at the flame surface. Little or no oxidizer is present on the fuel side of the flame, and similarly, there is little or no fuel on the oxidizer side of the flame.

With jet diffusion flames, the fuel flow is in the form of a gas jet issuing from a nozzle. As the fuel jet flows axially downstream, it also diffuses radially outward, mixing with the surrounding gases. The annular flame where combustion takes place is generally quite thin. On either side of the diffusion flame, the fuel–oxidizer mixture is either too rich or too lean to burn. For vertical flames, the buoyancy body force on the hot gas accelerates the flow, narrowing the flame in the axial direction. In order to properly design furnaces and propulsive devices, one needs to predict the shape and structure of the jet flame resulting from the injection of fuel gases into an oxidizing atmosphere.

Diffusion combustion is a common technique used in many combustion devices such as industrial furnaces, gas stoves, natural gas engines, gas-turbine engines, and ramjet engines. A gas jet flame from a cook stove, a flame from natural gas engine fuel injector, a pan of oil burning in air, a fuel droplet burning in oxygen, a solid particle burning in oxidizer-rich environment, and pure solid fuel burning in a ramjet engine all produce diffusion flames. In the next chapter, we discuss diffusion flames originating from liquid fuels.

Many of these applications require some partial premixing with air, a technique called aeration, to reduce or eliminate soot production. The aeration is about 50% of the stoichiometric air flow. For hydrocarbon flames, the oxidation of the soot will result in the flame having an orange or yellow color. Relative to premixed combustion, diffusion flames are safer since they do not propagate as a wave through the combustion system and thus are easier to control.

An important dimensionless number associated with diffusion flames is the Damköhler number, which is a dimensionless number that relates the chemical reaction rates to the rates of species diffusion or mixing. The Damköhler number (Da) is a very useful ratio for determining whether

Principles of Combustion, Third Edition. Allan T. Kirkpatrick and Kenneth K. Kuo.

Companion website: www.wiley.com/go/kirkpatrick/principlesofcombustion3e

the diffusion rates or the reaction rates are more important. It is named after the German chemist G. Damköhler (1908–1944). The general definition of Da is:

$$Da \equiv \frac{\text{Rate of chemical reaction}}{\text{Rate of mixing by species diffusion}}$$

For $Da \gg 1$, the chemical reaction rate is much greater than the diffusion rate, and the reactions are characterized as diffusion limited. Diffusion flames have high Damköhler numbers, since in a diffusion flame, the chemical reaction rates are typically much greater than the species mixing rates. Thus, the overall features of diffusion flames are often approximated by assuming a thin reaction surface in which the chemical reactions occur instantaneously.

As the jet Reynolds number ($Re = ud_i/\nu$) is increased, diffusion flames transition from a laminar flow to a turbulent flow. The jet features of interest are the flame speed, length, and the various flow parameters which govern the jet behavior. Non-premixed flames in a turbulent environment are different from non-premixed laminar flames due to turbulence–chemistry interactions and enhanced mixing by turbulent eddies. Turbulent flows have a large spectrum of length and time scales. The largest scales are based on the physical size of the combustion chamber, and, at the smallest scales, the turbulence energy is dissipated by viscosity.

We start this chapter with a scale analysis of laminar and turbulent jet flames, review the flow and combustion behavior of laminar jets, and then examine turbulent jet mixing and combustion.

8.2 Scale Analysis of Laminar and Turbulent Jet Flames

Prediction of the length of a laminar or turbulent jet flame is an important consideration in the design of combustors that rely on diffusion or non-premixed combustion. Some applications will require short flame lengths and others will require longer flame lengths. A classic plot (Hottel and Hawthorne 1948) of the flame length of a combusting fuel jet as a function of nozzle outlet velocity u_i is shown in Figure 8.1. The laminar regions of the jet are shown as smooth lines, and the turbulent eddy regions are shown as curved lines.

Figure 8.1 has several very interesting features. For laminar flow with low flow rates, the flame length is proportional to the jet velocity or volume flow rate. The transition to turbulence begins near the top of the flame, then as the velocity increases, the turbulence moves upstream to the nozzle exit. In the transition region where turbulent eddies begin to affect the jet mixing, the flame length actually decreases. The transition Reynolds number depends on the fuel type, as shown in Table 8.1. This indicates that both fluid mechanics and chemical kinetics play a role in the combustion of the fuel jet.

In the turbulent regime, the flame length in Figure 8.1 becomes independent of the jet nozzle velocity. For a fuel fraction of unity, the non-dimensional turbulent flame length $h/d_i \sim 125$ (Hottel and Hawthorne 1948). As the flowrate is increased further, a point is reached where the flame lifts off from the nozzle producing a non-burning area near the exit. The lift-off length increases approximately linearly with jet velocity. With further increases in flowrate, the flame blows off completely.

These flame length curve profiles can be explained using a simple scaling analysis in which we equate two time scales. As indicated in Figure 8.2, the length of the flame x_f is defined as the location on the centerline axis where combustion is complete. The nozzle or injector radius is r_i, the outlet velocity of the jet is u_i, and the outlet volumetric flow rate $Q_f = \pi r_i^2 u_i$. The potential core shown in the figure is the constant velocity jet region that is unaffected by the mixing with the ambient gases.

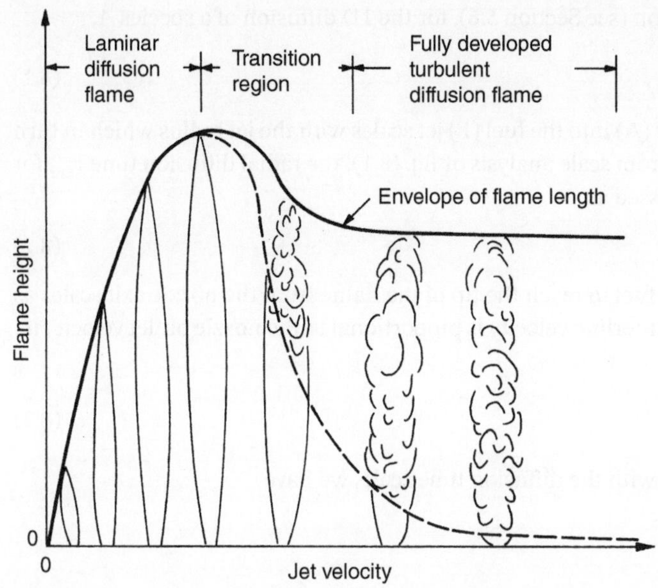

Figure 8.1 Variation of flame length and flow regime as a function of jet velocity. (Hottel and Hawthorne 1948/with permission of Elsevier.)

Table 8.1 Transition Reynolds Number for Jet Diffusion Flames.

Fuel into Air	Formula	Transition *Re*
Hydrogen	H_2	2000
City Gas	H_2 and CO	3500
Carbon Monoxide	CO	4800
Propane	C_3H_8	9000–10,000
Acetylene	C_2H_2	9000–10,000

Source: Hottel and Hawthorne (1948).

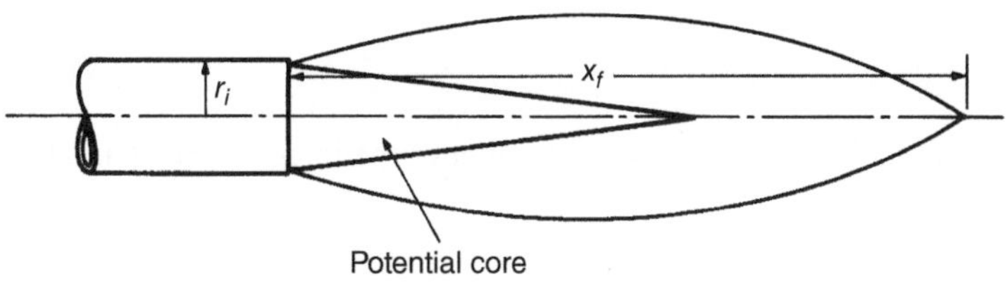

Figure 8.2 Laminar flame length x_f.

In the scaling analysis, there are two related characteristic times. The time τ_{diff} is the time scale of the diffusion of the surrounding air in the radial direction from the edge of the jet to the jet centerline. The other time τ_{conv} is the time scale of the convection of the fuel from the exit nozzle to the tip of the flame. We assume that the combustion process does not affect the mixing rate between the fuel jet and the surrounding air.

From Fick's second law of diffusion (see Section 5.6), for the 1D diffusion of a species A,

$$\frac{\partial[A]}{\partial t} = \mathcal{D}\frac{\partial^2[A]}{\partial r^2} \tag{8.1}$$

The penetration depth δ of the air (A) into the fuel (F) jet scales with the jet radius which in turn scales with the nozzle radius r_i, so from scale analysis of Eq. (8.1), the radial diffusion time τ_{diff} for the fuel and air to mix can be expressed as

$$\tau_{\text{diff}} \sim \delta^2\mathcal{D} \sim r_i^2\mathcal{D} \tag{8.2}$$

The convection time τ_{conv} for the fuel to reach the tip of the flame from the nozzle exit scales as distance/velocity. The average jet centerline velocity is proportional to the nozzle outlet velocity u_i, so we can approximate τ_{conv} as

$$\tau_{\text{conv}} = \frac{x_f}{u_i} \tag{8.3}$$

Equating the convection time scale with the diffusion time scale, we have

$$\begin{aligned}
&\tau_{\text{conv}} \sim \tau_{\text{diff}} \\
&\frac{x_f}{u_i} \sim \frac{r_i^2}{\mathcal{D}} \\
&x_f \sim \frac{r_i^2 u_i}{\mathcal{D}} \sim \frac{\pi r_i^2 u_i}{\pi\mathcal{D}} \sim \frac{Q_f}{\mathcal{D}}
\end{aligned} \tag{8.4}$$

This result indicates that the length of a laminar diffusion flame x_f is proportional to the outlet jet velocity and volumetric flow rate Q_f and inversely proportional to the mass diffusivity $\mathcal{D}$.

Let us now consider the extension of the above relationship to turbulent fuel jets. First, let us recall that the Schmidt number is defined as $Sc = \nu/\mathcal{D}$. For constant Schmidt number, we have $\mathcal{D} \sim \nu$. Hence, Eq. (8.4) can be expressed as

$$x_f \sim \frac{r_i^2 u_i}{\nu} \tag{8.5}$$

For the turbulent case, we can use the same reasoning to arrive at a relationship similar to the above, except that instead of the molecular viscosity ν, we must use the turbulent eddy viscosity ν_t in the denominator, since $\nu_t \gg \nu$.

Thus,

$$x_f \sim \frac{r_i^2 u_i}{\nu_t} \tag{8.6}$$

From Section 5.13, the turbulent eddy viscosity ν_t is the product of the integral scale l and the turbulent velocity u_t,

$$\nu_t = u_t l \tag{8.7}$$

Assuming the turbulent velocity scales linearly with the nozzle velocity u_i and that the integral scale is linearly proportional to the nozzle radius r_i,

$$\nu_t \sim u_i r_i \tag{8.8}$$

Combining Eqs. (8.6) and (8.8), we have

$$x_f \sim \frac{r_i^2 u_i}{r_i u_i} \sim r_i \tag{8.9}$$

This relationship indicates the air entrainment increases at the same rate as the volumetric fuel flowrate, resulting in the turbulent flame length being proportional to the nozzle radius, and independent of volumetric flowrate.

8.3 Laminar Jet Flames

Many commercial and industrial combustion devices rely on laminar jet flames, for example, natural gas ranges and ovens require short jet flames. We model the gas jet as a steady, axisymmetric, laminar flow issuing into still air, with an axial coordinate x and velocity u, and radial coordinate r and velocity v. We also assume that the pressure is uniform, the buoyancy effects are negligible, a low Mach number, and rapid chemical reaction rates.

Following Spalding (1979), the governing laminar differential equations are given as follows:

Overall continuity equation:

$$\frac{\partial}{\partial x}(\rho ur) + \frac{\partial}{\partial r}(\rho vr) = 0 \tag{8.10}$$

Axial (x) momentum conservation equation:

$$\frac{\partial}{\partial x}(\rho uru) + \frac{\partial}{\partial r}(\rho vru) = \frac{\partial}{\partial r}\left(\mu r\frac{\partial u}{\partial r}\right) \tag{8.11}$$

Fuel (F) species continuity equation:

$$\frac{\partial}{\partial x}(\rho urY_F) + \frac{\partial}{\partial r}(\rho vrY_F) = \frac{\partial}{\partial r}\left(\rho D_F r\frac{\partial Y_F}{\partial r}\right) + r\dot{\omega}_F \tag{8.12}$$

Oxidant (O) species continuity equation:

$$\frac{\partial}{\partial x}(\rho urY_O) + \frac{\partial}{\partial r}(\rho vrY_O) = \frac{\partial}{\partial r}\left(\rho D_O r\frac{\partial Y_O}{\partial r}\right) + r\dot{\omega}_O \tag{8.13}$$

Energy-conservation equation:

$$\frac{\partial}{\partial x}(\rho urh) + \frac{\partial}{\partial r}(\rho vrh) = \frac{\partial}{\partial r}\left(kr\frac{\partial T}{\partial r}\right) + (q_c)\frac{\partial}{\partial r}\left(\rho D_F r\frac{\partial Y_F}{\partial r}\right) \tag{8.14}$$

where q_c is the heat of combustion per mass of fuel, k is the gas mixture thermal conductivity, and

$$h \equiv c_p T + Y_F q_c \tag{8.15}$$

To facilitate our discussion and theoretical solution, let us define a term introduced initially by Spalding (1979) for many combustion problems: in a *simple chemically reacting system* (SCRS), the pure fuel and pure oxidant are imagined to always unite in fixed stoichiometric proportions, the specific heats of all components are equal, and the transport properties are equal at any point in the mixture but need not be uniform.

For such systems, a sketch of mixture state as a function of mixture fraction f or Z and reactedness $\mathbf{r}$ is shown in Figure 8.3. The symbols $Y_{F,F}$ and $Y_{F,M}$ in Figure 8.3 represent the mass fractions of fuel in the fuel stream and in the burned mixture stream, respectively.

If the fuel and the oxidizer are assumed to form an SCRS, one can relate the rate of consumption of fuel F to that of oxidizer O by

$$\dot{\omega}_F = \frac{\dot{\omega}_O}{s_O} \tag{8.16}$$

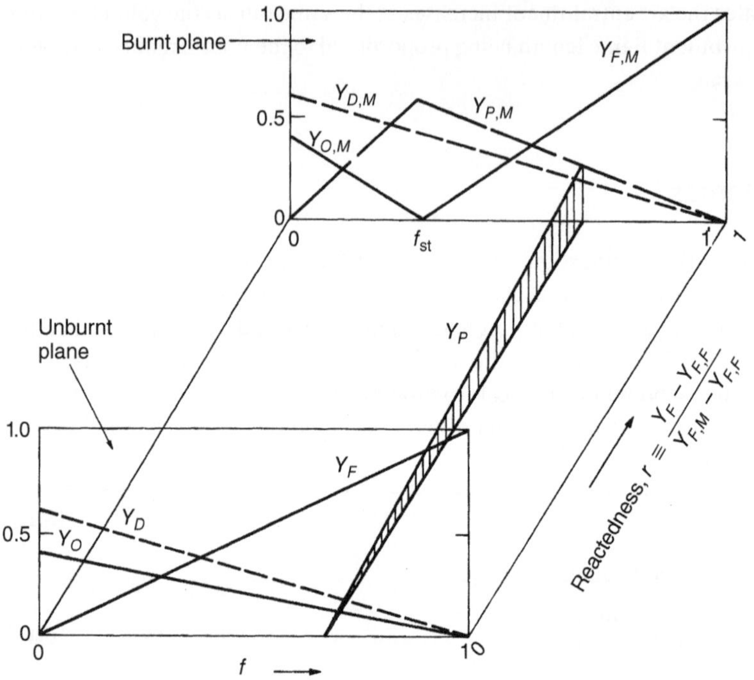

Figure 8.3 Sketch of the mixture state as a function of mixture fraction f and reactedness **r**.

where s_O is the mass stoichiometric ratio of oxidizer to fuel,

$$s_O \equiv \left(\frac{m_O}{m_F}\right)_{st} \tag{8.17}$$

Since the transport properties are equal for an SCRS, the Schmidt number $Sc = 1$, $\nu = \mathcal{D}$, and

$$\rho\mathcal{D} = \rho\nu = \mu \tag{8.18}$$

Equations (8.12) and (8.13) can be combined to form

$$\frac{\partial}{\partial x}(\rho ur\zeta) + \frac{\partial}{\partial r}(\rho vr\zeta) = \frac{\partial}{\partial r}\left(\mu r\frac{\partial \zeta}{\partial r}\right) \tag{8.19}$$

where the conserved property ζ is

$$\zeta \equiv Y_F - Y_O/s_O \tag{8.20}$$

As introduced in Section 2.11, ζ is a conserved property related to the mixture fraction f, so f also satisfies Eq. (8.19), and we can write

$$\frac{\partial}{\partial x}(\rho urf) + \frac{\partial}{\partial r}(\rho vrf) = \frac{\partial}{\partial r}\left(\mu r\frac{\partial f}{\partial r}\right) \tag{8.21}$$

One can also show that Eq. (8.14) can be reduced to

$$\frac{\partial}{\partial x}(\rho urh) + \frac{\partial}{\partial r}(\rho vrh) = \frac{\partial}{\partial r}\left(\mu r\frac{\partial h}{\partial r}\right) \tag{8.22}$$

At the fuel injector nozzle $x = 0$, the fuel jet velocity has a top-hat profile, with a mixture fraction $f = 1$ in the jet, and $f = 0$ outside the jet, i.e.,

$$\begin{aligned} \text{at } x = 0: \quad & u = u_i \\ & r \le r_i, f = 1 \\ & r > r_i, f = 0 \end{aligned} \tag{8.23}$$

The boundary conditions at large r and x are

$$\text{as} \quad r \to \infty: \quad u = f = 0, h = h_\infty \tag{8.24}$$

There are three jet invariants based on the conservation of momentum (I_u), conservation of mass of fuel species (I_f), and conservation of energy (I_h). These jet invariants are as follows:

$$I_u \equiv \frac{1}{\nu}\int_0^\infty u^2 r dr = \frac{1}{\nu}\left(\frac{1}{2}u_i^2 r_i^2\right) \tag{8.25}$$

$$I_f \equiv \frac{1}{\nu}\int_0^\infty ufr dr = \frac{1}{\nu}\left(\frac{1}{2}u_i r_i^2\right) \tag{8.26}$$

$$I_h \equiv \frac{1}{\nu}\int_0^\infty u(h - h_\infty) r dr = \frac{u_i(h_i - h_\infty)r_i^2}{2\nu} \tag{8.27}$$

Upon further analysis (see Schlichting and Gersten 2017), the solutions for the jet velocities and mixture fraction can be written as

$$u(x, r) = \frac{3}{4}\frac{I_u}{x}\left(1 + \frac{\xi^2}{4}\right)^{-2} \tag{8.28}$$

$$v(x, r) = \left(\frac{3}{8}I_u \nu\right)^{1/2} \frac{\xi}{x}\left(1 - \frac{\xi^2}{4}\right)\left(1 + \frac{\xi^2}{4}\right)^{-2} \tag{8.29}$$

$$f(x, r) = \frac{3}{4}\frac{I_f}{x}\left(1 + \frac{\xi^2}{4}\right)^{-2} \tag{8.30}$$

where the self-similar variable ξ is defined as

$$\xi \equiv \left(\frac{3}{8}\frac{I_u}{\nu}\right)^{1/2} \frac{r}{x} \tag{8.31}$$

The similarity of the equations for u, f, and h permits a single form of solution to serve for all three equations,

$$\frac{ux}{I_u} = \frac{fx}{I_f} = \frac{(h - h_\infty)x}{I_h} = \frac{3}{4}\frac{1}{\left(1 + \frac{1}{4}\xi^2\right)^2} \tag{8.32}$$

In terms of the Reynolds number ($Re_{d_i} = u_i d_i/\nu$) of the jet at the fuel injector nozzle $x = 0$, Eq. (8.32) can be written as

$$\begin{aligned} \frac{ux}{u_i r_i}\frac{1}{Re_{d_i}} = \frac{fx}{r_i}\frac{1}{Re_{d_i}} &= \frac{h - h_\infty}{h_i - h_\infty}\frac{x}{r_i}\frac{1}{Re_{d_i}} \\ &= \frac{3/16}{[1 + (3/256)(Re_{d_i}^2)(r^2/x^2)]^2} \end{aligned} \tag{8.33}$$

By examining Eq. (8.32), one can see that there is no primary effect of combustion on the distribution of velocity and on conserved properties such as f and h. The secondary effects are caused mainly by the influence of temperature on density and transport properties.

It can be seen from Eq. (8.32) that the centerline ($\xi = r = 0$) values of the jet velocity, mixture fraction, and enthalpy difference $h_{\mathbb{C}} - h_\infty$ fall off as $1/x$:

$$\begin{aligned} u_{\mathbb{C}} &= \frac{3}{4}\frac{I_u}{x} \\ f_{\mathbb{C}} &= \frac{3}{4}\frac{I_f}{x} \\ h_{\mathbb{C}} - h_\infty &= \frac{3}{4}\frac{I_h}{x} \end{aligned} \tag{8.34}$$

Since the virtual origin of the jet is not included in this analysis, the far-field solutions are valid for $x > 3/4 I_u/u_i$, as $u_{\mathbb{C}} \le u_i$.

The radial values of u and f, when normalized by their respective centerline values, are

$$\frac{u}{u_{\mathbb{C}}} = \frac{f}{f_{\mathbb{C}}} = \frac{1}{\left(1 + \frac{1}{4}\xi^2\right)^2} \tag{8.35}$$

The schematic in Figure 8.4 represents the velocity and concentration profiles across the fuel jet. It should be noted that when

$$\frac{u}{u_{\mathbb{C}}} = 0.5 \qquad \text{then} \quad \xi = 1.287 \tag{8.36}$$

It is useful to characterize the width of the jet by R, defined as the radius at which the velocity u is 1/2 of its centerline value. From Eq. (8.35),

$$\begin{aligned} \frac{R}{x} &= 1.287\left(\frac{8\nu}{3I_u}\right)^{1/2} \\ &= 5.94/Re_{d_i} \end{aligned} \tag{8.37}$$

The half-width spreading angle $\theta/2$ is

$$\theta/2 = \tan^{-1}(5.94/Re_{d_i}) \tag{8.38}$$

A laminar jet has a spreading angle that is not constant but is inversely proportional to the nozzle outlet Reynolds number.

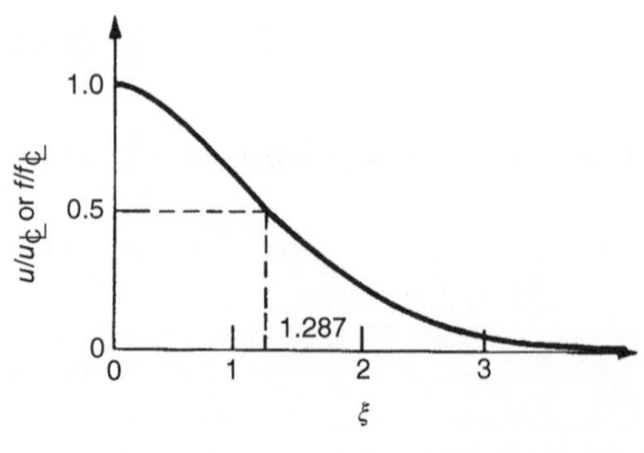

Figure 8.4 Self-similar profiles of velocity and concentration as a function of ξ.

Some other features of laminar flame jets are also of interest. The total mass flow rate in the jet, $\dot{m}$, can be obtained from the following integration:

$$\dot{m} = \int_0^\infty 2\pi r\rho u dr \tag{8.39}$$

Substituting Eq. (8.28) into Eq. (8.39) and carrying out the integration, we have

$$\dot{m} = 2\pi\rho x^2\left(\frac{8}{3}\frac{\nu}{I_u}\right)\left(\frac{3}{4}\frac{I_u}{x}\right)\int_0^\infty \frac{\xi d\xi}{(1+\frac{1}{4}\xi^2)^2} = 8\pi\mu x \tag{8.40}$$

The jet mass flow rate thus increases linearly with the downstream distance x due to the entrainment of the surrounding fluid. The entrainment rate is therefore a constant and can be obtained by differentiating Eq. (8.40),

$$\frac{d\dot{m}}{dx} = 8\pi\mu \tag{8.41}$$

Associated with the entrainment process, an entrainment velocity v_{ent} can be defined by

$$v_{\text{ent}} \equiv -\frac{1}{2\pi r_{jb}\rho}\frac{d\dot{m}}{dx} = -\frac{4\mu}{r_{jb}\rho} = -4\frac{\nu}{r_{jb}} \tag{8.42}$$

where r_{jb} represents the radius at the jet boundary (jb). In dimensionless form, Eq. (8.29) can be written as

$$\frac{v}{v_{\text{ent}}} = -\frac{\sqrt{3}}{16}\frac{u_i r_i}{\nu}\frac{\xi}{x}\left(1-\frac{\xi^2}{4}\right)\left(1+\frac{\xi^2}{4}\right)^{-2} r_{\text{jb}} \tag{8.43}$$

According to Schlichting, the radius at the jet boundary can be expressed as

$$r_{\text{jb}} \cong \frac{16}{\sqrt{3}}\left(\frac{x}{Re_{d_i}}\right) \tag{8.44}$$

Using the above expression for the jet boundary, the radial velocity can be expressed as

$$\frac{v}{v_{\text{ent}}} = -\frac{\xi^2}{4}\left(1-\frac{\xi^2}{4}\right)\left(1+\frac{\xi^2}{4}\right)^{-2} \tag{8.45}$$

If the reaction rates are infinitely fast, fuel and oxidizer cannot be simultaneously present in finite concentration, then the reaction zone is an infinitely thin surface (no overlapping region for Y_F and Y_A or Y_O). For a chemical reaction in air, the stoichiometric reaction surface is located at

$$Y_F - Y_A/s_A = 0 \tag{8.46}$$

and from Section 2.11, the mixture fraction f_{st} at the flame surface is

$$f_{\text{st}} = \frac{1}{1+s_A} \tag{8.47}$$

For a pure oxidizer environment, the stoichiometric reaction surface is located at

$$Y_F - Y_O/s_O = 0 \tag{8.48}$$

where s_O is the mass stoichiometric oxidizer–fuel ratio $(O/F)_{\text{st}}$, and from Eq. (2.193), the mixture fraction f_{st} is

$$f_{\text{st}} = \frac{[Y_O/s_O]_A}{1+[Y_O/s_O]_A} \tag{8.49}$$

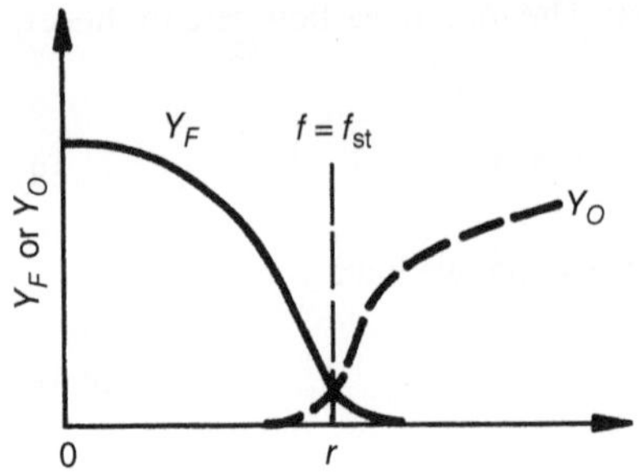

Figure 8.5 Radial distribution of mass fractions.

where $[Y_O]_A$ represents the mass fraction of the oxidizer in the ambient gas. In practice, the overlapping region for many hydrocarbon fuels and air systems is very thin, illustrated schematically in Figure 8.5.

The flame radius r_f at any x location can be found by inserting f_{st} into Eq. (8.33) and rearranging; the result is

$$\frac{r_f}{x} = \frac{16}{\sqrt{3}Re_{d_i}} \sqrt{\left(\frac{3}{16}\frac{Re_{d_i} r_i}{x f_{st}}\right)^{1/2} - 1} \tag{8.50}$$

Curves obeying Eq. (8.50) for fixed values of u/u_i or f have the form shown in Figure 8.6. The flame contour is sketched in Figure 8.7.

The flame length x_f can be obtained by setting r_f to zero in Eq. (8.50),

$$\frac{x_f}{r_i} = \frac{3}{16}\frac{Re_{d_i}}{f_{st}} \tag{8.51}$$

which indicates that the nondimensional flame length x_f/r_i is proportional to the injector Reynolds number.

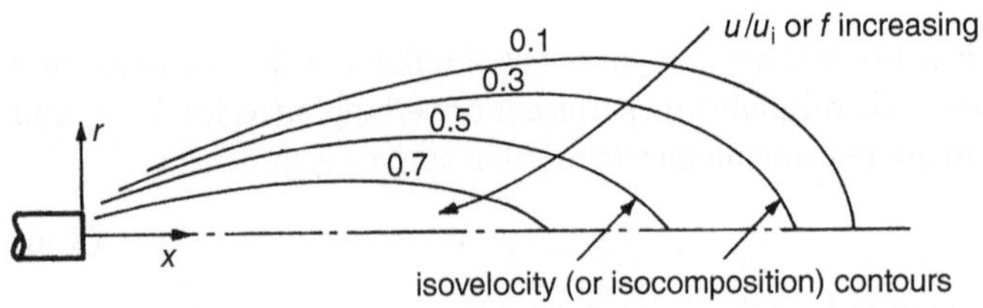

Figure 8.6 Isovelocity or isocomposition contours of a laminar jet.

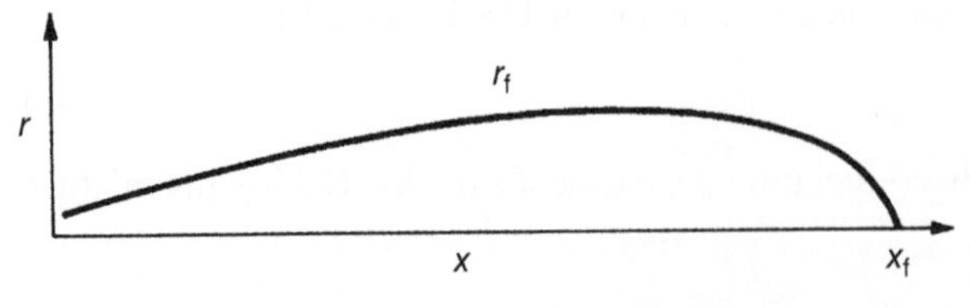

Figure 8.7 Contour of a laminar diffusion flame.

Equation (8.51) implies that

$$\begin{aligned} x_f &= \frac{3}{8}\frac{u_i r_i^2}{f_{st}\nu} \\ &= \frac{3}{8\pi}\frac{1}{f_{st}}\frac{\pi\rho u_i r_i^2}{\mu} \\ &= \frac{3}{8\pi}\frac{1}{f_{st}}\frac{\pi u_i r_i^2}{D} \end{aligned} \tag{8.52}$$

Equation (8.52) indicates that for laminar flows, the flame length x_f is proportional to the injector outlet volumetric flow rate $\pi u_i r_i^2$ and inversely proportional to the mass diffusivity D, consistent with the scale analysis of Section 8.2.

Expressing f_{st} in terms of $[Y_O]_A$, Eq. (8.51) becomes

$$\frac{x_f}{r_i} = \frac{3}{16} Re_{d_i}\left[1 + \frac{1}{[Y_O]_A/s_O}\right] \tag{8.53}$$

This equation says that as $[Y_O]_A$ or $[Y_O]_\infty$ decreases, the flame length increases, which implies at the centerline the fuel has to go farther to reach stoichiometric conditions and form a flame. Short flame lengths are often preferred in rocket motors; to obtain them, Eq. (8.53) suggests the use of a small outlet *Re*, i.e., small injector radius and velocity. This is why many liquid rocket-engine combustors have numerous small diameter fuel-injector nozzles. On the other hand, in industrial furnaces, long flames are often desired for uniformity of heat transfer over the furnace length, and so the fuel is often supplied through a large diameter single nozzle. Since the fuel and oxidizer diffuse together in stoichiometric proportions, the flame temperature closely approaches the adiabatic stoichiometric flame temperature.

Because the effects of buoyancy and the temperature dependence of k, μ, and D have been neglected, the predictions of flame length and shape are necessarily approximate. The effect of buoyancy is to shorten the flame length because it increases the velocity gradients and mixing rates. The increases of properties k, μ, and D with temperature will increase the rate of entrainment, but the increased specific volume widens the jet, reducing the gradients and lengthening the flame.

Example 8.1 ***Laminar Flame Spreading Angle and Length***

A hydrogen H_2 jet diffusion flame is injected into a boiler from a 5-mm diameter nozzle. The injection velocity is 1.0 m/s. The air in the boiler is quiescent and at 300 K and 1 atm. What is the jet spreading angle and the flame length? Assume the fuel fraction is unity at the nozzle exit.

Solution

From the property tables in the Appendix, the dynamic viscosity μ of hydrogen at 300 K and 1 atm is 8.9×10^{-6} Pa-s. From Table 2.8, the stoichiometric mass air–fuel ratio for hydrogen is $s_A = 34.20$.

From the ideal-gas equation, the injection density is

$$\rho = \frac{pM}{R_u T} = \frac{(101.3)(2.016)}{(8.314)(300)} = 0.0819\,\text{kg/m}^3$$

The nozzle Reynolds number is in the laminar flow regime,

$$Re_{d_i} = \frac{\rho u_i d_i}{\mu} = \frac{(0.0819)(1.0)(0.005)}{8.9 \times 10^{-6}} = 46.0$$

From Eq. (8.38), the half-width spreading angle $\theta/2$ is

$$\frac{\theta}{2} = \tan^{-1}\frac{5.94}{Re_{d_i}}$$

$$= \tan^{-1}\frac{5.94}{46.0} = 7.35°$$

The spreading angle is $\theta = 15°$.

From Eq. (8.47), the stoichiometric mixture fraction f_{st} at the flame tip is

$$f_{st} = \frac{1}{1+s_A} = \frac{1}{1+34.20} = 2.84 \times 10^{-2}$$

From Eq. (8.51), the flame length x_f is

$$x_f = \frac{3}{16}\frac{Re_{d_i}}{f_{st}} r_i$$

$$= \frac{3}{16}\frac{46.0}{(2.84 \times 10^{-2})}(2.5 \times 10^{-3})$$

$$= 0.76\,\text{m}$$

8.3.1 Flame Length from Circular Orifices

Roper (1977) and Roper et al. (1977) developed an empirical correlation for the length x_f of laminar jet flames from a variety of geometries. For vertical circular orifices, their correlation is

$$x_f = K\frac{Q_f\,(T_\infty/T_F)}{\ln(1+1/\bar{s}_O)} \tag{8.54}$$

where Q_f is the volumetric fuel flow rate (m^3/s), T_∞ and T_F are the oxidizer and fuel temperatures, $\bar{s}_O$ is the molar stoichiometric oxidizer/fuel ratio, and $K = 1330\ s/m^2$ is a constant.

Diluting the fuel with an inert gas will reduce the flame length since the stoichiometric ratio will be increased. For hydrocarbon fuels, with the chemical formula C_xH_y, it can be shown that $\bar{s}_O$ can be expressed as

$$\bar{s}_O = (x + y/4)\left(\frac{1 - X_{dil}}{X_{O_2}}\right) \tag{8.55}$$

where X_{dil} is the diluent mole fraction in the fuel stream.

8.4 Coaxial Jet Flames

A classic example of a confined laminar jet diffusion flame, which was first described quantitatively by Burke and Schumann (1928), is a system in which fuel and airflow with constant and identical flow velocity in coaxial cylindrical tubes. Burke and Schumann used this configuration to develop the first detailed analysis of non-premixed flames, and it is now considered as a model problem to illustrate the properties of diffusional combustion. The geometry is also relatively straightforward for complementary experimental measurements.

As shown in Figure 8.8, the fuel flows through an inner cylinder, and the air flows through an outer surrounding cylinder, with an axial coordinate z and radial coordinate r. A schematic of the composition profile of a hydrogen–air diffusion flame propagating from a coaxial cylinder geometry

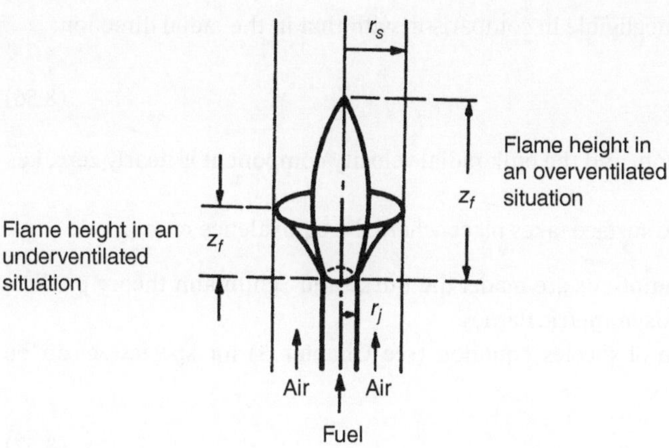

Figure 8.8 The shapes of a coaxial diffusion flame under over- and under-ventilated situations.

Figure 8.9 Radial profiles of species concentration through a coaxial diffusion flame.

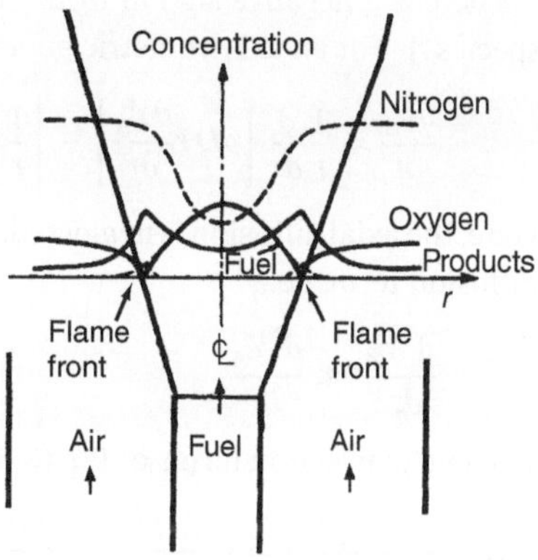

is shown in Figure 8.9. The gas flame at the inner cylinder exit is confined by the outer cylinder. The fuel concentration drops uniformly from the centerline to the edge of the flame, the oxygen concentration drops uniformly from the surroundings to the edge of the flame surface, and the combustion product concentration is maximum at the flame surface.

In general, as shown in Figure 8.11, the observed shapes of coaxial diffusion flames may be divided into two classes. If the ratio of the cylinder outer radius r_s to the inner radius r_j is such that more air is available than what is required for complete combustion, then an *overventilated* flame is formed and the flame boundary converges inward to the cylinder centerline. The flame length z_f is the axial distance from the nozzle where the equivalence ratio $\phi = 1$. On the other hand, if the air supply is insufficient for complete burning, then an *underventilated* flame is produced in which the flame surface expands to the outer cylinder wall (Leonard et al. 1994).

The Burke and Schumann analysis assumed a balance between stream-wise convection and transverse diffusion. Additional assumptions are:

- At the exit nozzle position, the velocities of the air and fuel are constant, equal, and uniform across their respective cylinders. The molar fuel ratio is controlled by varying the radii of the cylinders, and is equal to $r_j^2/\left(r_s^2 - r_j^2\right)$.
- The velocity of the fuel and air in the region of the flame is the same as the velocity at the nozzle.
- The term ρD is constant.

- Diffusion in the axial direction is negligible in comparison with that in the radial direction:

$$\frac{\partial^2 Y_j}{\partial r^2} \gg \frac{\partial^2 Y_j}{\partial z^2} \tag{8.56}$$

- Mixing is caused mainly by diffusion, and the bulk radial velocity component is nearly zero, i.e., $v_r \sim 0$.
- The chemical reaction at the flame surface takes place where the equivalence ratio $\phi = 1$.

Even though many simplifying assumptions are made, the Burke and Schumann theory predicts flame lengths fairly accurately for axisymmetric flames.

The time-dependent conservation-of-species equation (see Chapter 5) for species A can be written as

$$\frac{\partial \rho_A}{\partial t} + \nabla \cdot \rho_A \mathbf{v} = \nabla \cdot \rho \mathcal{D} \nabla Y_A - \dot{\omega}_A \tag{8.57}$$

where Y_A is the mass fraction and ρ_A is the density.

There is a negative sign in front of ω_A, since it now represents the mass rate of consumption of species A. For an axisymmetric geometry, the above species continuity equation (8.57) becomes

$$\frac{\partial \rho_A}{\partial t} = \frac{1}{r}\frac{\partial}{\partial r}\left[\rho \mathcal{D} r \frac{\partial Y_A}{\partial r}\right] - \left[\frac{1}{r}\frac{\partial (r\rho_A v_r)}{\partial r} + \frac{\partial v_z \rho_A}{\partial z}\right] - \dot{\omega}_A \tag{8.58}$$

Here, the axial diffusion is neglected, since it is considered to be small in comparison with the radial diffusion term, i.e.,

$$\frac{\partial^2 Y_A}{\partial z^2} \ll \frac{\partial^2 Y_A}{\partial r^2}$$

The two convection terms of Eq. (8.58) can be rearranged to give

$$\frac{1}{r}\frac{\partial (r\rho_A v_r)}{\partial r} = \frac{1}{r}\frac{\partial [r\rho v_r(\rho_A/\rho)]}{\partial r} = \rho v_r \frac{\partial (Y_A)}{\partial r} + \frac{Y_A}{r}\frac{\partial (r\rho v_r)}{\partial r}$$

$$\frac{\partial v_z \rho_A}{\partial z} = \frac{\partial v_z \rho Y_A}{\partial z} = Y_A \frac{\partial \rho v_z}{\partial z} + \rho v_z \frac{\partial Y_A}{\partial z} \tag{8.59}$$

Since the product $\rho \mathcal{D}$ is a rather weak function of temperature (see Chapter 5), as an approximation, Burke and Schumann considered $\rho \mathcal{D}$ to be constant. Assuming the air density ρ_A to be independent of time, i.e., $\partial \rho_A/\partial t = 0$, and using the overall axisymmetric continuity equation,

$$\frac{1}{r}\frac{\partial (r\rho v_r)}{\partial r} + \frac{\partial (\rho v_z)}{\partial z} = 0$$

Equation (8.58) becomes

$$\frac{\mathcal{D}\rho}{r}\frac{\partial}{\partial r}\left(r\frac{\partial Y_A}{\partial r}\right) - (\rho v_z)\frac{\partial Y_A}{\partial z} = \dot{\omega}_A \tag{8.60}$$

If a single-step forward reaction Air (A) + Fuel (F) → Products (P) is used to represent the overall chemical reaction in the diffusion flame, then the species continuity equation for the fuel F can be written in a similar fashion as Eq. (8.60),

$$\frac{\mathcal{D}\rho}{r}\frac{\partial}{\partial r}\left(r\frac{\partial Y_F}{\partial r}\right) - (\rho v_z)\frac{\partial Y_F}{\partial z} = \dot{\omega}_F = \frac{\dot{\omega}_A}{s_A} \tag{8.61}$$

where $s_A = (m_A/m_F)_{st}$ is the stoichiometric mass air–fuel ratio.

By following the Shvab–Zel'dovich procedure described in Chapter 5, the energy equation can be expressed in the same form as the species conservation equation. The general form of the energy equation (8.62), from Chapter 5, is

$$\rho\frac{Dh}{Dt}-\frac{Dp}{Dt}=-\nabla\cdot\mathbf{q}+\Phi+\dot{Q}+\rho\sum_{k=1}^{N}y_k\mathbf{f}_k\cdot\mathbf{V}_k \tag{8.62}$$

The viscous dissipation term and the body force term are assumed to be small, so

$$\rho\frac{Dh}{Dt}-\frac{Dp}{Dt}=-\nabla\cdot\mathbf{q}+\dot{Q} \tag{8.63}$$

Assuming steady-state conditions where $\partial h/\partial t=0$, and $Dp/Dt=0$, small υ_r, and negligible axial-direction heat conduction, then the energy equation after simplification becomes

$$\rho\upsilon_z\frac{\partial h}{\partial z}=+\frac{k}{r}\frac{\partial}{\partial r}\left(r\frac{\partial T}{\partial r}\right)+\dot{\omega}_F q_c \tag{8.64}$$

where q_c is the heat of reaction per unit mass of fuel (F). Recall that in the Shvab–Zel'dovich formulation, we used

$$Le=1 \qquad \text{i.e.,}\quad \mathcal{D}\rho=\frac{k}{c_p} \tag{8.65}$$

The energy equation, after setting the Lewis number equal to 1, becomes

$$\frac{\mathcal{D}\rho}{r}\frac{\partial}{\partial r}\left[\frac{r\partial(c_pT)}{\partial r}\right]-\rho\upsilon_z\frac{\partial(c_pT)}{\partial z}=-\frac{\dot{\omega}_A}{s_A}q_c \tag{8.66}$$

Multiplying Eq. (8.60) by q_c/s_A and combining with Eq. (8.66), we get

$$\frac{1}{r}\frac{\partial}{\partial r}\left\{\mathcal{D}\rho\left[r\frac{\partial}{\partial r}\left(c_pT+\frac{Y_Aq_c}{s_A}\right)\right]\right\}-\rho\upsilon_z\frac{\partial}{\partial z}\left(c_pT+\frac{Y_Aq_c}{s_A}\right)=0 \tag{8.67}$$

This is a homogeneous differential equation written in terms of the sum of sensible and chemical enthalpy.

For more complex chemical reactions in diffusion flames, one can still follow the Shvab–Zel'dovich procedure to obtain equations of the same form as Eq. (8.67). Some detailed procedures are given below.

In general coordinate systems, the energy and mass equations are

$$\nabla\cdot\left[(\rho\mathbf{v})\left(c_pT\right)-\frac{k}{c_p}\nabla\left(c_pT\right)\right]=\sum_j h^o_{f,j}\dot{\omega}_j \tag{8.68}$$

and

$$\nabla\cdot[(\rho\mathbf{v})Y_j-\rho\mathcal{D}\nabla Y_j]=-\dot{\omega}_j \tag{8.69}$$

where $\dot{\omega}_j$ is the rate of consumption of species j, and $h^o_{f,j}$ is the heat of formation per unit mass of species j.

Let us consider the global reaction of a fuel F with an oxidizer O:

$$\text{Fuel}\,F+\text{Oxidizer}\,O\rightarrow\text{Products}\,P \tag{8.70}$$

including the stoichiometric coefficients for the reactants, we have

$$\nu_FF+\nu_OO\rightarrow\text{Products}\,P \tag{8.71}$$

For example, $CH_4+2\,O_2\longrightarrow CO_2+2\,H_2O$.

If we divide Eq. (8.69) by $M_j(v_j' - v_j)$, we get the general species equation,

$$\nabla \cdot \left[(\rho \mathbf{v}) \frac{Y_j}{M_j(v_j' - v_j)} - (\rho \mathcal{D}) \nabla \frac{Y_j}{M_j(v_j' - v_j)} \right] = -\frac{\dot{\omega}_j}{M_j(v_j' - v_j)} = -\dot{B} \tag{8.72}$$

where v_j is the coefficient for species j appearing as a reactant, and v_j' is the coefficient for species j appearing as a product. Note that in the global reaction, if we let j represent the oxidizer, then

$$v_j' - v_j = -v_O \tag{8.73}$$

and if we let j represent fuel, we have

$$v_j' - v_j = -v_F \tag{8.74}$$

The mass diffusion equation for all the species present in a general system can be put into the same form by defining

$$\alpha_j \equiv \frac{Y_j}{M_j(v_j' - v_j)} \tag{8.75}$$

Then, Eq. (8.72) reduces to

$$\nabla \cdot [(\rho \mathbf{v})\alpha_j - \rho \mathcal{D} \nabla \alpha_j] = -\dot{B} \tag{8.76}$$

Following the Shvab–Zel'dovich formulation, we can arrange the energy equation in the following form:

$$\nabla \cdot \left[\rho \mathbf{v} \frac{c_p T}{q_c M_j(v_j' - v_j)} - \frac{k}{c_p} \nabla \frac{c_p T}{q_c M_j(v_j' - v_j)} \right]$$
$$= \frac{-\dot{\omega}_j}{M_j(v_j' - v_j)} = -\dot{B}$$

Let

$$\alpha_T \equiv \frac{c_p T}{q_c M_j(v_j' - v_j)} \tag{8.77}$$

then

$$\nabla \cdot [(\rho \mathbf{v})\alpha_T - \rho \mathcal{D} \nabla \alpha_T] = -\dot{B} \tag{8.78}$$

Comparing Eq. (8.78) with Eq. (8.76), we notice that both α_j and α_T satisfy the same differential equation.

Equations (8.76) and (8.78) may both be expressed as

$$L(\alpha) = -\dot{B} \tag{8.79}$$

where the linear operator L is defined by

$$L(\alpha) \equiv \nabla \cdot [(\rho \mathbf{v})\alpha - \rho \mathcal{D} \nabla \alpha] \tag{8.80}$$

The inhomogeneous nonlinear rate term may be eliminated from all except one of the relations corresponding to Eq. (8.79). Selecting α_1 to be the dependent variable for the inhomogeneous equation, we have

$$L(\alpha_1) = -\dot{B} \tag{8.81}$$

Other variables are then determined by the linear homogeneous equation through the use of coupling functions β:

$$L(\beta) = 0 \tag{8.82}$$

with $\beta = \alpha_T - \alpha_1 \equiv \beta_T$ or $\beta = \alpha_j - \alpha_1 \equiv \beta_j (j \neq 1)$.

In general, $\rho\mathbf{v}$ and $\rho\mathcal{D}$ do depend on β_j or β_T; thus, the operator L depends implicitly on β, and Eq. (8.82) is actually nonlinear. This Shvab–Zel'dovich linear equation formulation procedure is quite important in solving diffusion-flame problems, since in these problems, one often finds that by solving the linear equations for relations between flow variables, burning rates may eventually be determined without actually solving the nonlinear differential equation.

For the determination of the mass-fraction distribution, the linear differential equation that we need to consider is Eq. (8.82):

$$L(\beta) = 0$$

with

$$\beta = \alpha_F - \alpha_O \tag{8.83}$$

$$\alpha_F = \frac{-Y_F}{\mathrm{M}_F\, \nu_F}, \qquad \alpha_O = \frac{-Y_O}{\mathrm{M}_O\, \nu_O} \tag{8.84}$$

In cylindrical coordinates, this differential equation is

$$\left(\frac{u_z}{\mathcal{D}}\right)\frac{\partial \beta}{\partial z} - \frac{1}{r}\frac{\partial}{\partial r}\left(r\frac{\partial \beta}{\partial r}\right) = 0 \tag{8.85}$$

The boundary conditions for Eq. (8.85) are

$$\beta = -\frac{[Y_F]_0}{\mathrm{M}_F\, \nu_F} \qquad \text{at } z = 0, \quad 0 \le r \le r_j \tag{8.86}$$

$$\beta = +\frac{[Y_O]_0}{\mathrm{M}_O\, \nu_O} \qquad \text{at } z = 0, \quad r_j \le r \le r_s \tag{8.87}$$

and

$$\frac{\partial \beta}{\partial r} = 0 \qquad \text{at } r = 0, \quad z > 0, \text{ and at } r = r_s, z > 0 \tag{8.88}$$

Burke and Schumann utilized the following dimensionless coordinates for r and z:

$$\xi \equiv \frac{r}{r_s} \qquad \text{and} \qquad \eta \equiv \frac{z\mathcal{D}}{u_z r_s^2} \tag{8.89}$$

They also defined several reduced parameters in their solution process for species j,

$$C \equiv \frac{r_j}{r_s} \qquad \text{and} \qquad \nu \equiv \frac{[Y_O]_0\, \mathrm{M}_F\, \nu_F}{[Y_F]_0\, \mathrm{M}_O\, \nu_O} \tag{8.90}$$

The parameter β in the governing equation was replaced by the reduced dependent variable γ, where

$$\gamma \equiv \beta \frac{\mathrm{M}_F\, \nu_F}{[Y_F]_0} \tag{8.91}$$

In terms of these parameters, Eqs. (8.85) through (8.88) become

$$\frac{\partial \gamma}{\partial \eta} = \frac{1}{\xi}\frac{\partial}{\partial \xi}\left(\xi\frac{\partial \gamma}{\partial \xi}\right) \tag{8.92}$$

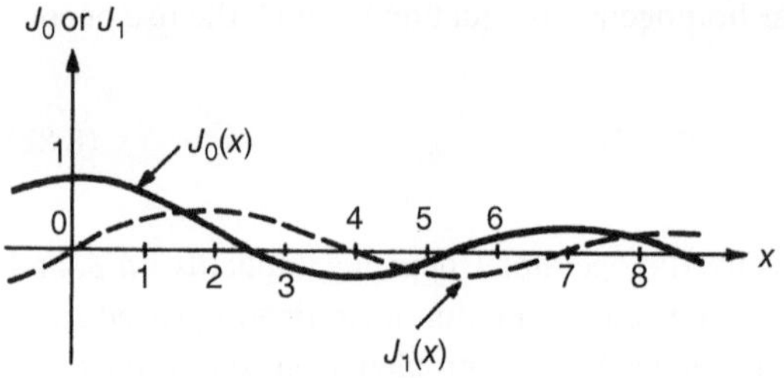

Figure 8.10 Bessel functions of the first kind of order 0 and 1.

$$\gamma = 1 \quad \text{at} \quad \eta = 0, \quad 0 \leq \xi < C \tag{8.93}$$

$$\gamma = -\nu \quad \text{at} \quad \eta = 0, \quad C < \xi < 1 \tag{8.94}$$

$$\frac{\partial \gamma}{\partial \xi} = 0 \quad \text{at}\, \xi = 0 \quad \text{and at}\, \xi = 1, \quad \text{for}\, \eta > 0 \tag{8.95}$$

By the method of separation of variables, it can be shown that a part of the decomposed solution satisfies Bessel's differential equation of order n:

$$\frac{d^2y}{dx^2} + \frac{1}{x}\frac{dy}{dx} + \left(1 - \frac{n^2}{x^2}\right)y = 0 \tag{8.96}$$

The functions $J_0(x)$ and $J_1(x)$ shown in Figure 8.10 are solutions of the above Bessel differential equation; they are the Bessel functions of the first kind (of order 0 and 1, respectively). As indicated in Figure 8.10, the first zero, i.e., root, of $J_0(x)$ is $x = 2.40$, and the second zero is $x = 5.52$. The first zero of $J_1(x)$ is $x = 3.83$, and the second zero is $x = 7.01$. Additional values of $J_0(x)$ and $J_1(x)$ are tabulated in the Appendix.

The mathematical forms of the Bessel functions are

$$J_0(x) = 1 - \frac{x^2}{2^2} + \frac{x^4}{2^4(2!)^2} - \cdots + (-1)^n \frac{x^{2n}}{2^{2n}(n!)^2}$$
$$= \sum_{n=0}^{\infty} (-1)^n \frac{(x/2)^{2n}}{n!\,n!}$$

and the mth-order Bessel function J_m is defined as

$$J_m(x) = \sum_{n=0}^{\infty} \frac{(-1)^n x^{m+2n}}{2^{m+2n}\, n!\, \Gamma(m+n+1)}, \qquad m > 0$$

The final solution for the reduced dependent variable γ in the power-series form is

$$\gamma = (1+\nu)C^2 - \nu + 2(1+\nu)C \sum_{n=1}^{\infty} \frac{1}{x_n} \frac{J_1(Cx_n)}{[J_0(x_n)]^2} J_0(x_n \xi) e^{-x_n^2 \eta} \tag{8.97}$$

where x_n represent successive roots of the equation $J_1(x) = 0$ (with ordering convention $x_n > x_{n-1}$, $x_o = 0$).

To determine the flame shape, Burke and Schumann assumed that the entire reaction occurs on a thin flame surface, then

$$\beta = 0, \qquad \text{or} \quad \gamma = 0$$

defines the flame surface. Hence, setting $\gamma = 0$ in Eq. (8.97) provides a relation between ξ and η that provides the locus of the flame surface [$\eta = f(\xi)$]. The shape of the flame surface obtained in this manner is shown in Figure 8.9 for two different values of the reduced parameter ν.

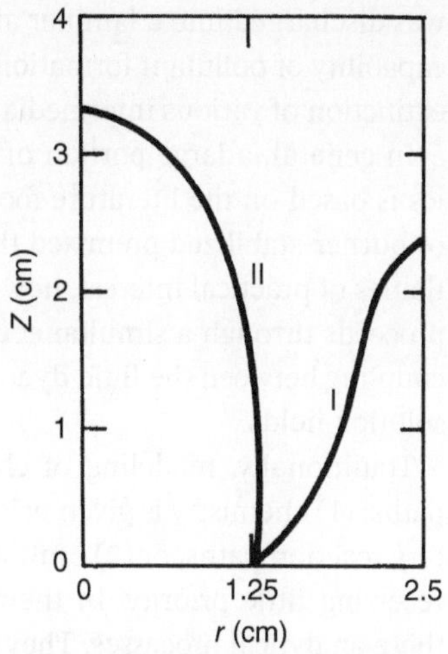

Figure 8.11 Calculated diffusion flame contour by Burke and Schumann: curve I, under-ventilated bowl-shaped flame; curve II, over-ventilated conical flame.

The flame height is obtained by solving Eq. (8.97) for η after setting the dimensionless radius $\xi = 0$ for over-ventilated flames or $\xi = 1$ for under-ventilated flames (and, of course, $\gamma = 0$ in either case). Since the flame heights are generally large enough to cause the factor $e^{-x_n^2\eta}$ to decrease rapidly as n increases at these values of η, it usually suffices to retain only the first few terms of the series in Eq. (8.97) for this calculation. Neglecting all the terms except $n = 1$, we obtain the approximation

$$\eta = \frac{1}{x_1^2} \ln \left\{ \frac{2(1+\nu)CJ_1(Cx_1)}{[\nu - (1+\nu)C^2]x_1 J_0(x_1)} \right\} \tag{8.98}$$

for the dimensionless flame height of an under-ventilated flame.

The flame shapes and flame heights obtained from Eq. (8.98) by Burke and Schumann (Figure 8.11) are in surprisingly good agreement with experiments, considering the many assumptions made in their analysis.

In several subsequent publications, Penner et al. (1983) and Chung and Law (1984) reexamined the Burke–Schumann problem and included the axial diffusion terms. In the Penner et al.'s study, they considered diffusion flames with arbitrary transport coefficients and chemical reactions. The closed-form solution obtained is slightly more complex than that of Burke and Schumann. In Chung and Law's work, the solution was obtained by a perturbation method under the assumption that the Lewis number is very close to unity. The zeroth-order solution was obtained by separation of variables, and the first-order problem was solved using Green's functions. Their results showed that the flame is made longer and narrower by including stream-wise diffusion transport. They also showed that the Burke–Schumann solution is generally valid for large Péclet ($Pe > 10$) numbers. The heat transfer Péclet number is defined as the product of Reynolds number and Prandtl number ($Pe = Re\,Pr$) and is the ratio of the convective heat flux to the conductive heat flux. The Péclet number is named after J. Péclet (1793–1857), a French scientist.

8.5 Combustion Chemistry of Laminar Jet Flames

The detailed combustion processes in axisymmetric laminar diffusion flames were examined numerically and experimentally in a pioneering paper by Smooke et al. (1989). In their theoretical model, the chemical kinetics of the combustion process and the transport phenomena were included. Both unconfined and confined jets were studied. In both cases, the cylindrical fuel jet was discharged into a laminar air stream. Part of their objective was to achieve a better predictive capability of pollutant formation processes and to identify the important reactions controlling the extinction of various intermediate species.

In general, a large portion of the detailed chemistry used in computational combustion studies is based on the literature focusing on one-dimensional configurations, i.e., freely propagating or burner-stabilized premixed flames and counterflow premixed or diffusion flames. In diffusion flames of practical interest, however, the oxidation of the fuel to form intermediates and products proceeds through a simultaneous mixing and diffusion process. In other words, there is a strong coupling between the fluid dynamic and the chemistry kinetics processes in the multidimensional solution fields.

Traditionally, modeling of chemically reacting flows has proceeded along two independent paths: (1) chemistry is given priority over fluid mechanical effects to access the important elementary reaction paths, or (2) multidimensional fluid dynamic effects are emphasized with chemistry receiving little priority. In their study, Smooke et al. were one of the first to combine both of these analytical processes. They considered both the confined and unconfined configurations and utilized the following set of assumptions:

- Two-dimensional axisymmetric flow.
- Combustion takes place at a finite rate.
- Fuel and oxidizer may coexist on either side of the reaction zone.
- Steady flow.
- Newtonian fluid characteristics.
- Gases obey the ideal-gas law.
- Bulk viscosity is negligibly small.
- Dufour and Soret effects are negligible.
- Negligible viscous dissipation.
- Adiabatic system.
- Pressure of the system is constant.
- Infinitely fast kinetics.
- The fuel and oxidizer are separated by a thin exothermic reaction zone.
- Inside the reaction zone, the fuel and oxidizer are in stoichiometric proportion.
- The fuel and oxidizer obey a single-step reaction process.

In their model formulation, they considered a set of 2D axisymmetric governing equations, including the continuity equation, radial-momentum equation, axial-momentum equation, species conservation equations, an energy equation, and an equation of state, written in terms of perfect-gas law. A set of boundary conditions was specified at the injection station ($z = 0$), along the centerline ($r = 0$), at the edge of the outer zone ($r = R_o$), and at the far-field station ($z \to \infty$). To simplify the solution procedure, they considered the stream function ψ, which is defined in terms of the radial and axial components of the velocity vector. Using the stream function as one of the dependent variables, the continuity equation is automatically satisfied.

$$\rho r v_r = -\frac{\partial \psi}{\partial z} \tag{8.99}$$

$$\rho r v_z = \frac{\partial \psi}{\partial r} \tag{8.100}$$

Furthermore, they considered the vorticity (ω) equation in the solution to replace the two momentum equations. In this problem, there is only one component of vorticity, which is defined as

$$\omega \equiv \frac{\partial v_r}{\partial z} - \frac{\partial v_z}{\partial r} \tag{8.101}$$

A partial differential equation for stream function can be obtained by direct substitution of Eqs. (8.99) and (8.100) into Eq. (8.101). Finally, they used $3 + K$ partial differential equations in their solution of ψ, ω, T, and y_i for $i = 1, 2, \ldots, K$ in the chemically reacting flow field. The diffusion velocities in the radial and axial directions are assumed to obey Fick's law. The diffusion coefficients of the ith species against the mixture were calculated using the following relationship given by Curtiss and Hirschfelder (1949).

$$\mathcal{D}_i = \frac{1 - Y_i}{\sum_{j \neq i}^{K} X_j / \mathcal{D}_{ji}} \tag{8.102}$$

The specific rate constants were evaluated using a modified Arrhenius equation with a temperature-dependent pre-exponential term:

$$k_j = A_j T^{\beta_j} \exp\left(-\frac{E_j}{R_u T}\right) \tag{8.103}$$

The pre-exponential factor A_j, the temperature exponent β_j, and the activation energy E_j for a 42-reaction mechanism involving 15 chemical species were obtained from a compilation of published experimental work gathered by Smooke et al. (1989). The binary diffusion coefficients, the viscosity, the thermal conductivity of the mixture, and the chemical production rates are evaluated using the code CHEMKIN developed by Kee et al. (1980).

A discrete solution of the derived system is obtained by applying Newton's method in two dimensions and adaptive gridding to a mesh formed by the intersection of radial and axial lines. For convergence, Newton's method generally requires an initial estimate of flame sheet location that is sufficiently close to the actual solution. An initial estimate of flame sheet location was obtained by assuming a global, infinitely fast reaction model with assumptions that allow the system to be solved using the Shvab–Zel'dovich formulation. Both steady-state and time-dependent solution methods were employed. The time-dependent approach was used to help obtain a converged numerical solution on an initial coarse grid using the initial estimate. Grid points were then inserted adaptively, and the steady-state solution procedure was used to recompute the finer grid. The coarse grid solution was then interpolated linearly to serve as an initial solution estimate for the iteration procedure on the finer grid. The process was continued on finer and finer grids until several termination criteria were satisfied. Detailed transport coefficients and the 42-reaction, 15-species mechanism were used in the calculations. The unconfined fuel exit velocity was 5 cm/s for an exit Reynolds number of 11.2, and the confined fuel exit velocity was 4.5 cm/s for an exit Reynolds number of 30.8.

The measured and calculated temperature fields for the unconfined methane–air laminar diffusion flame are shown in Figure 8.12. The temperature profiles are in good agreement, with a maximum temperature of about 2000 K.

Figure 8.13 is a comparison of the measured and calculated radial temperature profiles at a height of 1.2 cm above the burner. Excellent agreement between the measured and calculated profiles is observed from the axis of symmetry to the low-temperature co-flowing region.

Figure 8.14*a* and *b* compares the measured and calculated radial mole fraction profiles for the major species in the flame (CH_4, O_2, N_2, H_2O, CO_2, CO, and H_2) at a height of 1.2 cm above the burner. The agreement between these computed and measured profiles is very good.

8.5.1 Effect of Preferential Diffusion of Species and Heat in Laminar Diffusion Flames

Takagi and Xu (1993) studied the effect of preferential diffusion of chemical species and thermal energy on the structure of hydrogen–air laminar diffusion flames. They performed numerical computations of axisymmetric laminar jet diffusion flames and included detailed chemical kinetics and

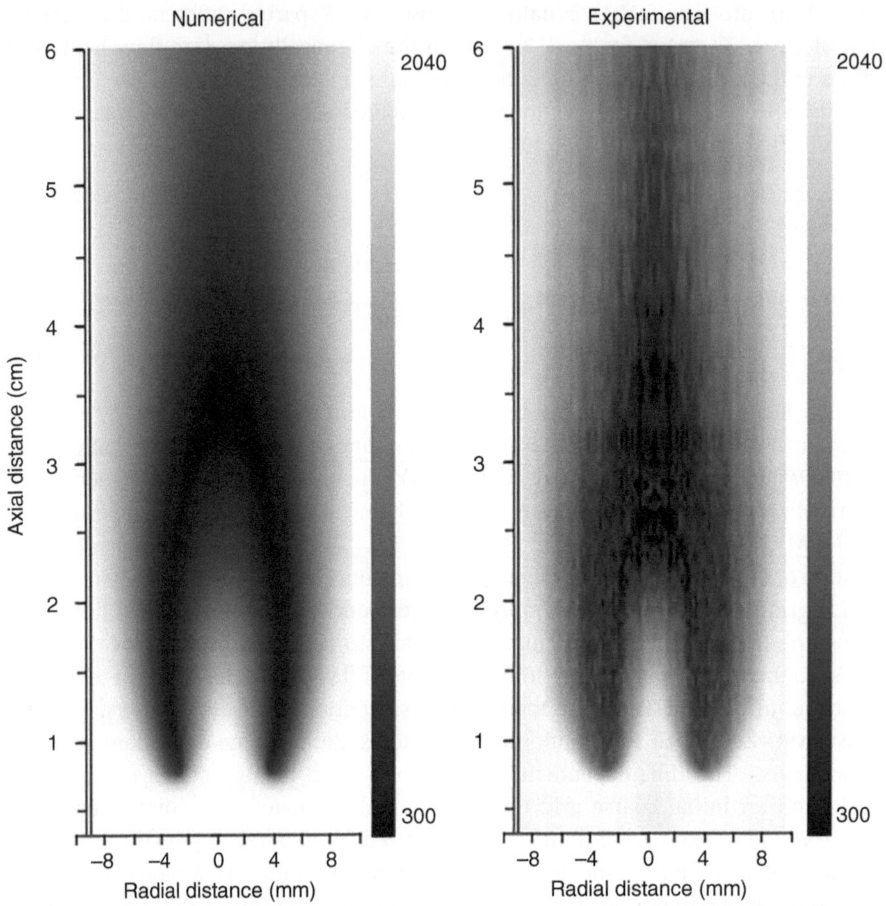

Figure 8.12 Comparison of measured and calculated temperature fields for an unconfined methane–air laminar diffusion flame. (Adapted from Smooke et al. 1989/with permission of Taylor & Francis Group.)

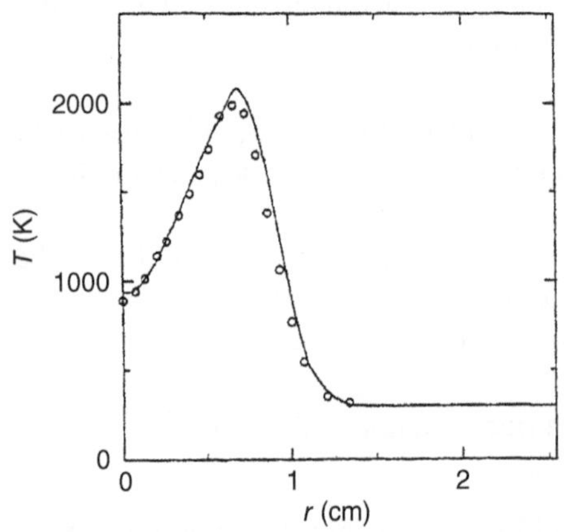

Figure 8.13 Comparison of measured and calculated (solid line) radial temperature profiles for a confined co-flowing methane–air laminar diffusion flame at a height 1.2 cm above burner inlet. (Smooke et al. 1989/with permission of Taylor & Francis Group.)

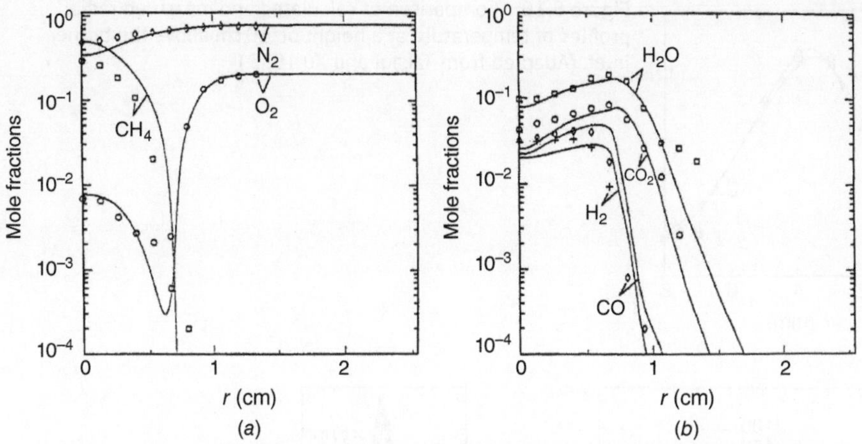

Figure 8.14 Comparison between measured and computed reactant and product profiles for a confined co-flowing methane–air laminar diffusion flame at a height of 1.2 cm above burner inlet. (*a*) CH_4, O_2, and N_2 reactant profiles and corresponding computational values (solid line). (*b*) H_2O, CO_2, CO, and H_2 product profiles and corresponding computational values (solid line). (Smooke et al. 1989/with permission of Taylor & Francis Group.)

multicomponent diffusion. They assumed a fuel jet composition of 30% H_2 and 70% N_2 by mole, with an average velocity of 5.2 m/s at the nozzle. A surrounding air jet was assumed to have an average velocity of 1.7 m/s.

For the hydrogen–air combustion modeling, they considered 10 species: H_2, O_2, H_2O, N_2, O, H, OH, HO_2, N, and NO, and 16 elementary chemical reactions. The mass diffusion coefficients were determined from Eq. (8.102). The comparison of calculated results with experimental data (Takagi et al. 1990) is shown in Figures 8.15 and 8.16. The calculated results are in close agreement with the measured data.

Takagi and Xu showed that preferential diffusion of thermal energy and combustion species resulted in a higher flame temperature than the adiabatic flame temperature for the original fuel. Effects of combustion pressure were also investigated. As shown in Figure 8.17, the temperature profiles at 0.1 and 4.4 MPa are very different, both radially and in peak values,

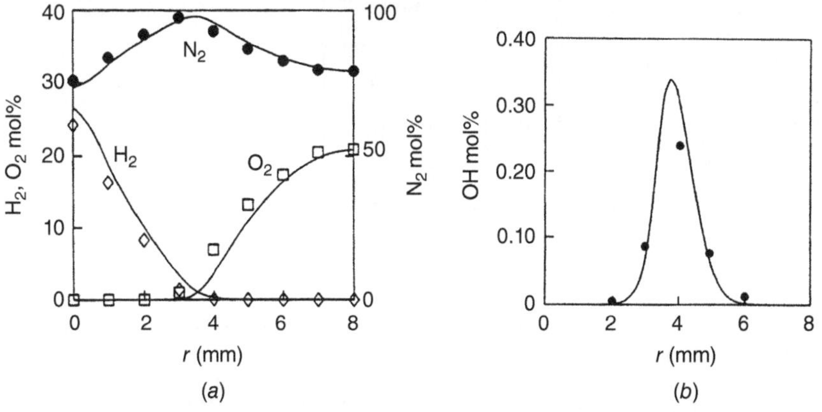

Figure 8.15 (*a*) Comparison of H_2, O_2, and N_2 at a height of 3.0 cm above the burner inlet. (*b*) Comparison of calculated and measured radial profiles of OH. (Adapted from Takagi and Xu 1993.)

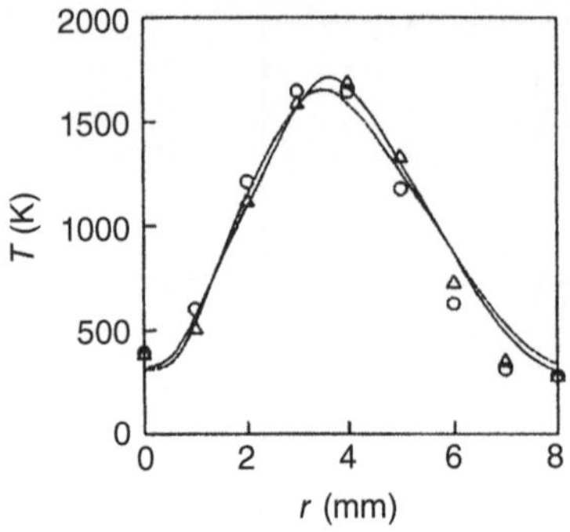

Figure 8.16 Comparison of calculated and measured radial profiles of temperature at a height of 3.0 cm above the burner inlet. (Adapted from Takagi and Xu 1993.)

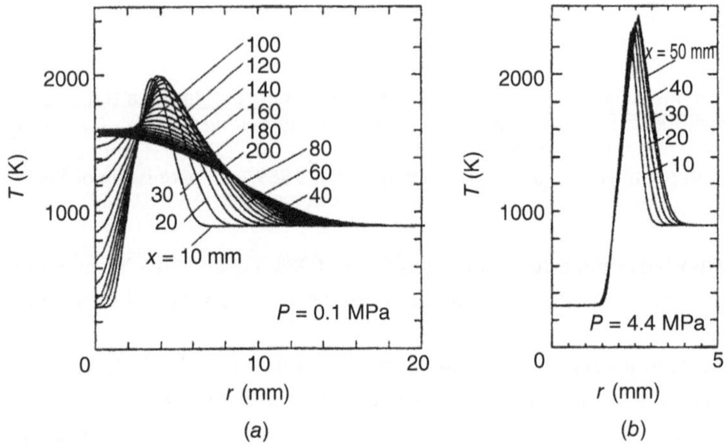

Figure 8.17 Temperature profiles at two different pressures, (*a*) $p = 0.1$ MPa and (*b*) $p = 4.4$ MPa. (Adapted from Takagi and Xu (1993) and Takagi and Xu (1993).)

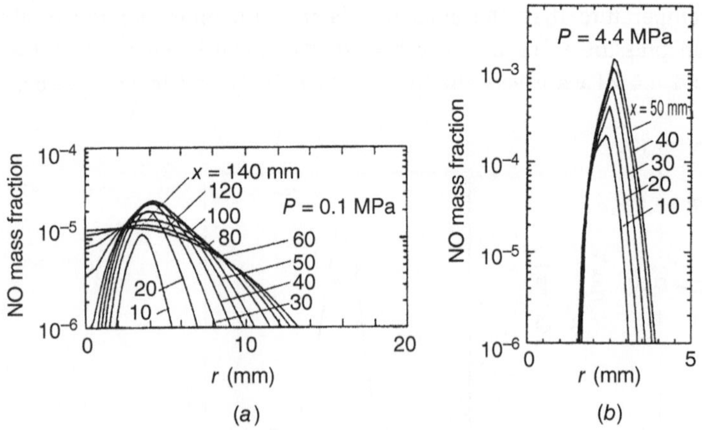

Figure 8.18 NO concentration profiles at two different pressures. (*a*) $p = 0.1$ MPa and (*b*) $p = 4.4$ MPa. (Adapted from Takagi and Xu (1993).)

with greater temperatures at higher pressures. Furthermore, the NO concentration profiles at these two different pressures are very different, as shown in Figure 8.18. An increase in pressure results in higher flame temperatures, which is the main factor responsible for the increase of NO formation.

8.6 Turbulent Gaseous Fuel Jets

As indicated in Table 8.1, when the outlet Reynolds number of a fuel or flame jet is greater than 2000–10,000, the jet flow undergoes a transition from a laminar to a turbulent flow field. As discussed in Section 5.12, a turbulent flow is composed of numerous vortices or eddies of different sizes randomly dispersed in the flow. For a turbulent jet, we will shortly see that neither the velocity decay nor the spreading angle depends on the jet Reynolds number. As a consequence, the mixing of a turbulent jet is independent of the nozzle exit conditions. This is in contrast to the laminar jet, where the velocity decay is directly proportional to the jet Reynolds number and the spreading rate is inversely proportional to the jet Reynolds number.

We now apply a simple analysis to a self-similar turbulent gaseous fuel jet to derive relationships between the jet velocity decay, entrainment, and the outlet nozzle conditions. A schematic of the jet is given in Figure 8.19, a reproduction of Figure 8.3. The injector nozzle diameter is d_i. The outlet velocity u_i is assumed to be constant during the injection duration, and is an effective top hat velocity that represents the injected momentum at the nozzle. The injected gas density is ρ_i, and the cylinder ambient gas density is ρ_g.

At a downstream location x, the axial velocity u_x and fuel concentration Y_F profiles have their maximum values at the centerline $r = 0$ and in the radial direction, the values gradually decay to zero, defining the boundary of the jet. The gas penetration distance is given by S.

The radial velocity profile is assumed to be in the form of Eq. (8.104), a Gaussian distribution where the jet half-width R is defined to be the location where the jet velocity $u(R)$ is equal to 1/2 of the axial centerline velocity u_x:

$$\frac{u(r)}{u_x} = \exp\left[-\ln(2)\,(r/R)^2\right] \tag{8.104}$$

It has been observed that after a short development length of about 10 nozzle diameters, the half-width spreading angle $\theta/2$ is constant, i.e.,

$$R = x\,\tan(\theta/2) \tag{8.105}$$

The momentum rate $\dot{M}n$ of the injected gas jet is constant if the pressure gradients in the cylinder are negligible. The assumption is made that the entrained mass is much greater than the injected

Figure 8.19 Simple model of a turbulent gas jet.

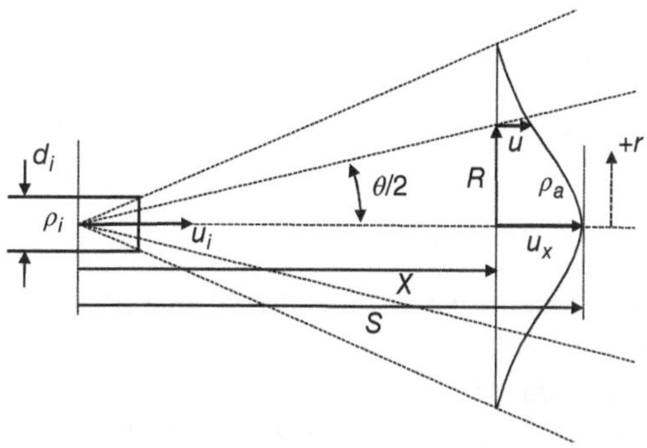

mass, so that the jet density in the self-similar region is assumed to be equal to the cylinder gas density ρ_g.

$$\begin{aligned}\dot{M}n &= \rho_i \frac{\pi}{4} d_i^2 u_i^2 = \int_0^\infty \rho_g 2\pi r u(r)^2 dr \\ &= 2\pi \rho_g u_x^2 \int_0^\infty r \exp\left(-2\ln(2)(r/R)^2\right) dr \\ &= 2\pi \rho_g u_x^2 \frac{R^2}{2\ln(2)} \frac{1}{2}\end{aligned} \tag{8.106}$$

since $\int_0^\infty z \exp(-z^2)\, dz = 1/2$. As indicated by Eq. (8.107) below, the centerline velocity u_x is predicted to have $1/x$ decay profile in the self-similar region:

$$\begin{aligned}\frac{u_x}{u_i} &= \left(\frac{\rho_i}{\rho_g}\right)^{1/2} \left(\frac{\ln(2)}{2}\right)^{1/2} \frac{d_i}{R} \\ &= \left(\frac{\rho_i}{\rho_g}\right)^{1/2} \left(\frac{\ln(2)}{2}\right)^{1/2} \frac{1}{\tan(\theta/2)} \frac{d_i}{x} \\ &= K \left(\frac{\rho_i}{\rho_g}\right)^{1/2} \frac{d_i}{x}\end{aligned} \tag{8.107}$$

where the velocity decay coefficient K is defined as

$$K = \left(\frac{\ln(2)}{2}\right)^{1/2} \frac{1}{\tan(\theta/2)} \tag{8.108}$$

The velocity decay coefficient K for gas jets has been measured to be $K = 5.9$ (Malmstrom et al. 1997), so from Eq. (8.108), the half-width spreading angle is predicted to be $\theta/2 = 5.9°$.

At a distance x downstream of the fuel injector, the gas jet mass flow rate $\dot{m}(x)$ is

$$\begin{aligned}\dot{m}(x) &= \int_0^\infty \rho_g 2\pi r u(r) dr \\ &= 2\pi \rho_g u_x \int_0^\infty r \exp\left(-\ln(2)(r/R)^2\right) dr \\ &= 2\pi \rho_g u_x R^2 \frac{1}{2\ln(2)}\end{aligned} \tag{8.109}$$

Using Eqs. (8.105) and (8.108),

$$u_x R^2 = \left(\frac{\rho_i}{\rho_g}\right)^{1/2} \frac{u_i d_i x}{K} \left(\frac{\ln(2)}{2}\right) \tag{8.110}$$

and upon substitution of Eq. (8.110) into Eq. (8.109), the jet mass flow rate is

$$\frac{\dot{m}(x)}{\dot{m}_i} = \frac{2}{K} \left(\frac{\rho_g}{\rho_i}\right)^{1/2} \frac{x}{d_i} \tag{8.111}$$

Note that the mass flow of the gas jet is predicted to increase linearly with the nondimensional axial distance x/d_i due to entrainment of the surrounding cylinder gases. The entrainment coefficient is predicted to have a value of $2/K = 2/5.9 = 0.34$. By comparison, measurements of the entrainment coefficient result in values around 0.32 (Ricou and Spalding 1961).

8.6.1 Fuel Injection Penetration Distance

The fuel penetration distance has been the subject of extensive experimental and computational research due to its importance in subsequent combustion processes. In an internal combustion engine, if the penetration is too short, air at the edges of the cylinder will not be involved in the combustion process, and if it is too long, the fuel jet will impinge on the cylinder walls, reducing the jet velocity and subsequent entrainment.

In this simple gas jet model, the fuel jet penetration rate dS/dt is assumed to be equal to the axial velocity $u(x)$. This is an approximation, as the transient gas jet has a head vortex with a trailing quasi-steady flow. At $x = S$, from Eq. (8.107),

$$\frac{dS}{dt} = u(x) = K\left(\frac{\rho_i}{\rho_g}\right)^{1/2} \frac{d_i}{S} u_i \tag{8.112}$$

separating variables,

$$S\,dS = K\left(\frac{\rho_i}{\rho_g}\right)^{1/2} d_i\, u_i\, dt \tag{8.113}$$

upon integration,

$$S = (2K)^{1/2}\left(\frac{\rho_i}{\rho_g}\right)^{1/4} (d_i\, u_i\, t)^{1/2} \tag{8.114}$$

$$\frac{S}{d_i} = (2K)^{1/2}\left(\frac{\rho_i}{\rho_g}\right)^{1/4} \left(\frac{u_i\, t}{d_i}\right)^{1/2} \tag{8.115}$$

The nondimensional penetration distance $S(t)/d_i$ is thus predicted by the simple gas model to be proportional to $t^{1/2}$, with a penetration constant $(2K)^{1/2} = (2 \times 5.9)^{1/2} = 3.4$. By way of comparison, the penetration constant measured by Ouellette and Hill (2000) for a transient turbulent gas jet was equal to 3.0, so this simple model overpredicts the penetration distance by about 13%.

The injection velocity u_i can be estimated using Bernoulli's equation applied to the injector nozzle with a discharge coefficient C_d and injector-cylinder pressure difference $\Delta P = P_i - P_c$:

$$u_i = C_d \left(\frac{2\Delta P}{\rho_i}\right)^{1/2} \tag{8.116}$$

Using the ideal-gas law, the injected gas density is

$$\rho_i = \frac{P_i}{R\,T_i} \tag{8.117}$$

and the injector-cylinder pressure ratio P_i/P_c can be expressed as

$$P_i/P_c = \left[1 - \frac{u_i^2}{2R\,T_i\,C_d}\right]^{-1} \tag{8.118}$$

Upon substitution of Eq. (8.116) into Eq. (8.115), the gas jet penetration distance S is predicted to be proportional to the injector-cylinder pressure difference to the 1/4 power, and independent of the injected gas density ρ_i.

$$\begin{aligned}\frac{S}{d_i} &= (2\sqrt{2}K\,C_d)^{1/2}\left(\frac{\Delta P}{\rho_g}\right)^{1/4}\left(\frac{t}{d_i}\right)^{1/2}\\ &= 3.16\left(\frac{\Delta P}{\rho_g}\right)^{1/4}\left(\frac{t}{d_i}\right)^{1/2}\end{aligned} \tag{8.119}$$

for $K = 5.9$ and $C_d = 0.60$. Fuel injection and air–fuel mixing in natural gas engines is discussed in more detail in Kim et al. (2004). Due to the relatively high pressures involved to achieve adequate mixing, consideration needs to be given to the effect of compressible flow features such as Mach disks that occur with high-pressure natural gas fuel injection (Li et al. 2004).

Example 8.2 ***Natural Gas Fuel Injection***

A natural gas fuel injector with a diameter $d_i = 1.0$ mm is being designed for an engine operating at 1500 rpm with a 100-mm bore and stroke. The injection is to begin at $\theta = -90°$ before top dead center where the piston has completed half of the compression stroke. For proper mixing, the gas jet should quickly penetrate to the top of the piston. The design injection velocity $u_i = 100$ m/s, with a centerline velocity decay coefficient $K = 5.9$. Assume an injected natural gas density of $\rho_i = 10$ kg/m^3, temperature $T_i = 350$ K, gas constant $R = 518$ J/kg-K, specific heat ratio $\gamma = 1.3$, dynamic viscosity $\mu = 12.7 \times 10^{-6}$ Pa-s, and an average cylinder gas density $\rho_g = 5$ kg/m^3.

(a) What is the initial Reynolds number of the gas jet?
(b) Estimate the time and the crank angle interval for the penetration of the gas jet onto the piston top.
(c) What is the resulting centerline velocity and half-width of the gas jet?

Solution

(a) The jet Reynolds number is in the turbulent regime,

$$Re_{d_i} = \frac{\rho u_i d_i}{\mu} = \frac{(10)(100)(1.0\times10^{-3})}{12.7\times10^{-6}} = 78,700$$

(b) From Eq. (8.114), assuming $x = S = 50$ mm,

$$\begin{aligned}t &= \frac{1}{2K}\frac{S^2}{d_i\,u_i}\left(\frac{\rho_g}{\rho_i}\right)^{1/2}\\ &= \frac{1}{2(5.9)}\frac{(0.05)^2}{(10^{-3})(100)}\left(\frac{5}{10}\right)^{1/2}\\ &= 1.5\times10^{-3}\,\text{s} \simeq 1.5\,\text{ms}\end{aligned}$$

$$\Delta\theta = 1500\,\text{rev/min}\times 360°/\text{rev}\times 1/60\,\text{min/s}\times 0.0015\,\text{s} = 13.5°$$

The piston continues to move upward when the fuel is initially injected into the cylinder and impacts the piston top, so the penetration distance is somewhat less than $S = 50$ mm. An improved estimate would include the upward piston motion in the analysis.

(c) From Eq. (8.107),

$$\begin{aligned}u_x &= u_i\,K\,d_i/x\\ &= (100)(5.9)(0.001/0.05) = 11.8\,\text{m/s}\end{aligned}$$

and from Eqs. (8.105) and (8.108), the jet half-width R is

$$R = x\left(\frac{\ln 2}{2}\right)^{1/2}(1/K)$$
$$= (0.05)(0.589)(1/5.9) = 5.0 \times 10^{-3}\,\text{m}$$
$$\simeq 5.0\,\text{mm}$$

8.7 Measurements of Non-Premixed Turbulent Flames

Various laser-based combustion diagnostic techniques developed by combustion researchers have been used in order to obtain more detailed and species-specific information about non-premixed combustion processes. The techniques include laser-based Mie, Rayleigh, and Ramán light scattering, as well as laser-induced incandescence and fluorescence. The scattering distribution and intensity depend on the particle size.

Mie scattering, named after Gustav Mie (1869–1957), is defined as elastic scattering from particles whose diameter is of the same order of magnitude or smaller as the light wavelength. For example, liquid droplet spray patterns have been determined via measurements of the Mie scattering. Elastic scattering of light from molecules or small particles with diameters much smaller than the wavelength of the laser light is termed Rayleigh scattering, named after Lord Rayleigh (1842–1919). The Rayleigh scattered light has a wavelength equal to the wavelength of the incident light. Vaporized fuel–air mixture patterns and temperature fields are determined through Rayleigh scattering measurements.

Light that is inelastically scattered with an associated wavelength or frequency change is called Raman scattering, named after the Indian scientist C. Raman (1888–1970). For polarizable molecules, the incident photon energy can excite vibrational modes of the target molecules, yielding scattered photons which have a frequency decrease by the amount of the vibrational transition energies. A spectral analysis of the scattered light under these circumstances will reveal spectral lines (Stokes lines) below the Rayleigh scattering peak at the incident frequency. It is also possible to observe spectral lines (anti-Stokes lines) at frequencies above the incident frequency as the vibrational energy is added to the incident photon energy. Because the frequency change depends on the specific target molecule, it is possible to analyze the composition of combustion products by analyzing the spectrum of the Raman scattered light.

Laser light is also used to induce incandescence and fluorescence of given species. Both relative and absolute soot concentrations have been determined using laser-induced incandescence (LII). Planar laser-induced fluorescence (PLIF) has been used to determine polyaromatic hydrocarbon (PAH) concentrations, which are precursors to soot, OH, CO distributions, and NO distributions. The OH radical distribution provides information about the location and intensity of diffusion and premixed flames. The NO radicals indicate the location of NO_x production in the flame.

An experimental setup for measurements of mixture fraction at several radial and axial locations and subsequent deduction of scalar dissipation rates from these measurements are shown in Figure 8.20. This setup combines line imaging of Raman scattering, Rayleigh scattering, and PLIF to obtain measurements of temperature and the major species concentrations, which enable the calculations of mixture fraction and scalar dissipation rates. Four frequency-doubled Nd:YAG lasers are used in the Raman/Rayleigh system of the diagnostics system. Raman scattered light is collected by using a custom-designed pair of 150-mm diameter achromats which are lens that can bring two wavelengths into focus in the same plane.

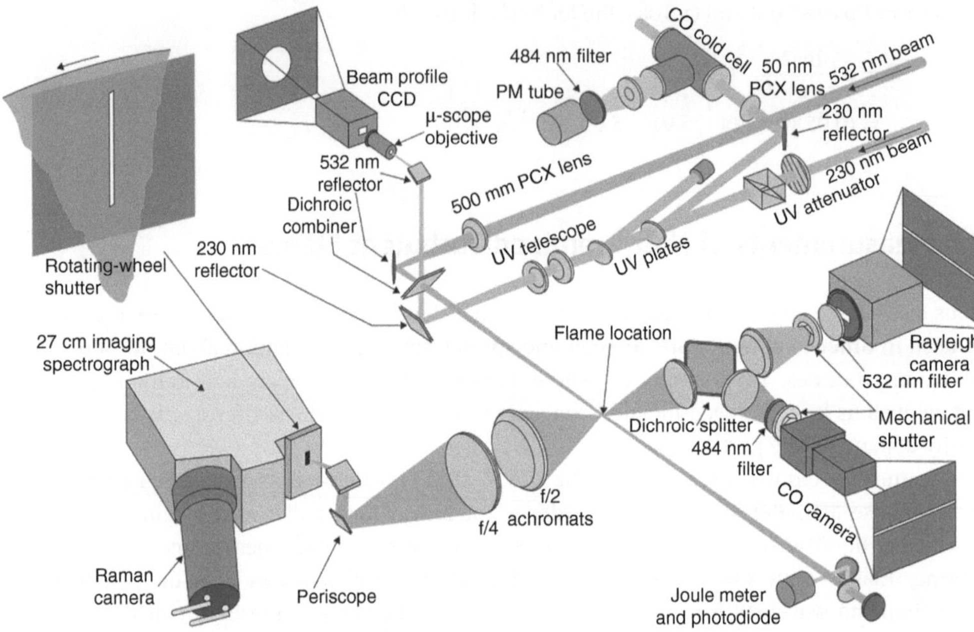

Figure 8.20 Experimental diagnostics setup for simultaneous measurements of Raman scattering, Rayleigh scattering, and carbon monoxide PLIF. (Barlow and Karpetis 2004/Springer Nature.)

Turbulent non-premixed combustion in actual operating combustion systems have complex boundary conditions which make it difficult to implement multiple diagnostic techniques. Because of these issues, measurements in an actual combustion device are not very useful for the study of features of combustion processes such as turbulence, chemical kinetics, radiation, and pollutant formation as well as their interaction with each other. Therefore, research combustion burners are used to perform experimental studies and to obtain detailed benchmark data. A research burner has simple boundary conditions and two, possibly three, of the above features present in the flames.

Simple jet burners are the simplest and most common research combustion devices on which the original studies of turbulent non-premixed flames by Hottel and Hawthorne (1948) and Wohl et al. (1948) were made. A jet flame from a simple jet burner consists of a cylindrical jet of fuel surrounded by a co-flowing stream of air. The fuel jet forms an annular turbulent flame as it issues from a cylindrical tube, mixes with the surrounding air stream, and ignites.

In the classic study by Bergmann et al. (1998), a turbulent jet diffusion flame from an 8-mm diameter tube fueled with a mixture of 22% CH_4, 33% H_2, and 45% N_2 was investigated with various laser techniques: (1) single-pulse point-wise spontaneous Raman/Rayleigh scattering for the determination of the temperature and major species concentrations; (2) 2D Rayleigh scattering for the determination of instantaneous temperature distributions; and (3) 2D PLIF measurements of OH, CH, and NO for the visualization of the structures within the flame. Nitrogen was added to the fuel to provide dilution and increase the signal quality of the Raman scattering technique. Hydrogen was added to stabilize the flame, and a seed amount of NO (70 ppm) was added for PLIF visualization of the fuel constituents. The baseline jet Reynolds number was 15,200.

The fuel jet (with seeded NO) is shown in the left frame of Figure 8.21 for the downstream region from $x/D = 0$ to 20. The images show the mixing of the fuel jet with the surrounding air

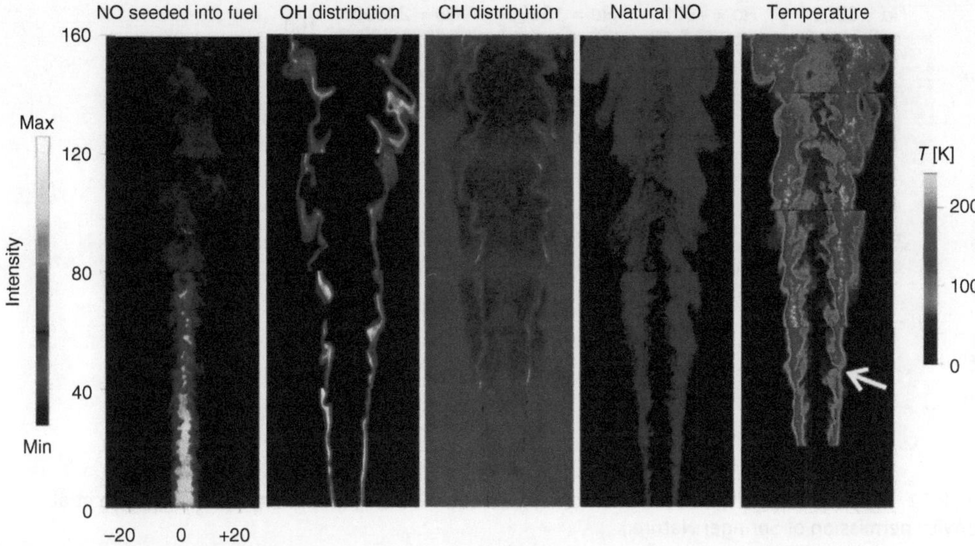

Figure 8.21 PLIF intensity and temperature distributions. Scales are in mm. (Bergmann et al. 1998/with permission of Springer Nature.)

and also the turbulent structures in the inner core of the jet. The bright regions represent zones of pure fuel whereas the contours of the dark regions reflect the entrainment of partially burnt gas into the fuel jet. From $x \approx 40$ mm ($x/D = 5$), downstream combustion products are frequently present on the flame axis and above $x \approx 80$ mm, pockets of pure fuel no longer exist.

The OH fluorescence frame in Figure 8.21 indicates the position and shape of the annular reaction zones and the regions of high temperatures. The OH distribution shows that the reaction zones are thin and rather laminar like up to a downstream position of $x \approx 40$ mm. Above $x \approx 40$ mm ($x/D = 5$), the reaction zones are more and more wrinkled and the effect of laminarization is rapidly reduced. It can be seen from the OH distributions in Figure 8.21 that the reaction zones are not always continuous. The CH distribution shown in Figure 8.21 is very similar to the OH distribution, although the width of the CH distribution is smaller. Throughout the flame, the CH distributions were restricted to thin layers, showing that the fuel-rich side of the reaction zones is thin. An evaluation of the thickness of the fluorescence zones at $x/D = 5$ from 60 single-pulse exposures revealed a mean value of 0.8 mm for CH and 1.0 mm for OH. The small width of the CH distributions supports the concept of combustion taking place in flamelet-like reaction zones throughout the entire flame.

The influence of the Reynolds number on the structure of the flame was investigated by varying the exit velocity and keeping the fuel composition constant. The Rayleigh intensity distributions from flames with four different exit velocities (10.5, 21, 42, and 63 m/s) and Reynolds numbers (3800, 7600, 15,200, and 22,800) were imaged in the stream-wise direction from $x/D = 0.5$ to 20.

As shown in Figure 8.22, the width of the flame zone near the nozzle decreases as the Reynolds number is increased from 3800 to 15,200. Since the time for radial molecular diffusion to the edge of the jet is shorter at higher flow velocities, the radial diffusion path is also shorter, leading to a smaller radial flame zone. For $Re = 22{,}800$, this trend can still be recognized very close to the nozzle ($x < 20$ mm), but further downstream, transport by turbulent diffusion prevails over molecular diffusion and the radial flame zone increases in width. The images show the transition from smooth to wrinkled reaction zones as the Reynolds number is increased. The shape of the flame surface also changes with Reynolds number, due to the development of turbulent vortices. At $Re = 3800$,

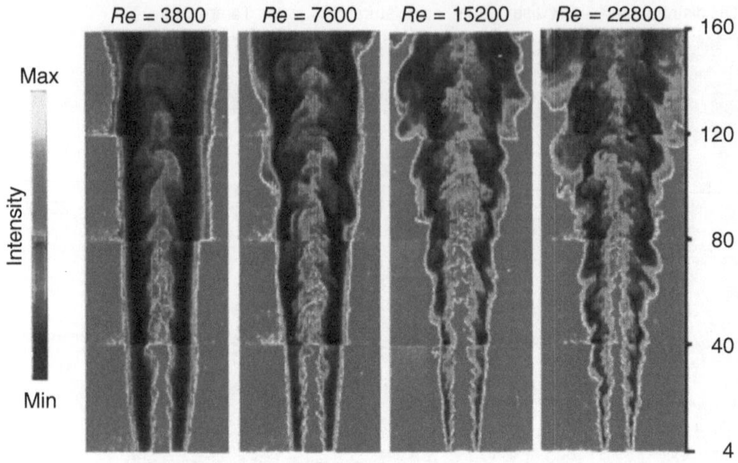

Figure 8.22 Rayleigh intensity distributions for four different exit Reynolds numbers. (Bergmann et al. 1998/with permission of Springer Nature.)

the boundary of the flame is smooth up to $x = 160$ mm, for $Re = 7600$, it becomes slightly curved above $x = 80$ mm, for $Re = 15{,}200$, it is curved and wrinkled above $x = 40$ mm, and for $Re = 22{,}800$, it is wrinkled above $x = 20$ mm.

8.8 Models of Non-Premixed Turbulent Flames

8.8.1 Background

There are a number of models used to calculate non-premixed turbulent flames depending on the particular combustion problem to be solved. The models need to take into account features of non-premixed combustion such as (1) entrainment by the large-scale eddies which create pockets of fuel-rich and fuel-lean mixture, (2) displacement and stretching of iso-concentration surfaces, and (3) interaction between chemistry and turbulence in the flame.

Two turbulent combustion models which account for finite rate chemistry include flamelet models and conditional moment closure (CMC). In these models, chemical reaction effects are included in closed form, and turbulent mixing is represented statistically. Direct numerical stimulation (DNS) has also emerged as a standard tool to study turbulent non-premixed flames.

In a turbulent combustion model, the reactive scalars ψ_i include the mass fractions Y_i of all chemical species and the temperature T,

$$\vec{\psi} = \psi(Y_1, Y_2, Y_3, \dots, Y_N, T) \tag{8.120}$$

where N is the number of reactive species. The flow field equation for a reactive scalar $\psi_i(\mathbf{x}, t)$ is

$$\rho\frac{\partial \psi_i}{\partial t} + \rho\mathbf{v}\cdot\nabla\psi_i = \nabla\cdot(\rho\mathcal{D}_i\nabla\psi_i) + \dot{\omega}_i \tag{8.121}$$

Favre averaged equations for the mean and variance of the reactive scalars are derived by writing $\psi(\mathbf{x}, t)$ as a Favre mean and fluctuation,

$$\psi_i(\mathbf{x}, t) = \tilde{\psi}_i(\mathbf{x}, t) + \psi_i''(\mathbf{x}, t) \tag{8.122}$$

After Favre averaging, the flow field equation is

$$\bar{\rho}\frac{\partial \tilde{\psi}_i}{\partial t} + \bar{\rho}\tilde{\mathbf{v}} \cdot \nabla \tilde{\psi}_i = \nabla \cdot (\overline{\rho \mathcal{D}_i \nabla \psi_i}) - \nabla \cdot (\bar{\rho}\widetilde{\mathbf{v}''\psi_i''}) + \sim \dot{\omega}_i \tag{8.123}$$

As discussed in Chapter 5, density-weighted averaging eliminates the need to explicitly represent the density–momentum and density–scalar correlations, which when kept in the governing equations generate additional closure terms. The three terms on the right-hand side of Eq. (8.123) need to be modeled. In high Reynolds number flows, the first term is normally neglected. Modeling closure is needed for the turbulent transport term $\widetilde{\mathbf{v}''\psi_i''}$ and the mean chemical source term $\tilde{\dot{\omega}}_i$.

8.8.2 Flamelet Models

Flamelet models represent the turbulent flame as a collection of flamelets, which are thin reaction–diffusion layers embedded in a nonreacting turbulent flow field, and are continuously displaced and stretched by the flow field. The modeling of a turbulent diffusion flame as an ensemble of stretched laminar flamelets was first proposed by Williams (1985). The flame is assumed to be wrinkled and thin relative to the Kolmogorov scale.

The flamelet model incorporates finite rate chemistry into a mixture fraction Z framework (Peters 2000). The flame surface is defined as an iso-surface of the mixture fraction Z. Since Z is a nonreacting scalar, its flow field equation, Eq. (8.124), does not contain a chemical source term,

$$\rho\frac{\partial Z}{\partial t} + \rho\mathbf{v} \cdot \nabla Z = \nabla \cdot (\rho \mathcal{D} \nabla Z) \tag{8.124}$$

The mixture fraction is a normalized conserved scalar, with values ranging from $Z = 0$ in the oxidizer and $Z = 1$ in the fuel.

Turbulence equations such as Reynolds Averaged Navier-Stokes (RANS) or Large Eddy Simulation (LES) are used in the flamelet method to solve for the mean and variance of the mixture fraction Z, which allows the development of a presumed probability density function (PDF) of the mixture fraction. Mean values of the reactive scalars can then be calculated by using the presumed PDF of the mixture fraction, whose shape is determined by its statistical moments. For example, in Chapter 5, the first moment of a variable u using a Favre-weighted PDF is

$$\tilde{u} = \int u\tilde{P}(u)du \tag{8.125}$$

and more generally,

$$\tilde{q}(\mathbf{x}) = \int_0^1 q(f)\tilde{P}(f;\mathbf{x})df \tag{8.126}$$

A coordinate transformation of the governing equations, such as the reactive scalar field equation, Eq. (8.123), using the mixture fraction Z as an independent coordinate, yields flamelet equations which express the reactive scalars ψ_i as a function of the mixture fraction Z. More precisely, the derivation of the flamelet equations involves (1) a coordinate transformation of the coordinate normal to the flame surface x_1 to the coordinate Z_1, and the tangential coordinates x_2 and x_3 to the coordinates Z_2 and Z_3, and then (2) an asymptotic boundary layer expansion in which the gradients of the reactive scalars ψ_i in the tangential directions are assumed small relative to the gradients in the normal direction.

The resulting flamelet equations describe the reactive–diffusive field around the flame surface as a function of $Z(\mathbf{x}, t)$. The flamelet equation for a reactive scalar ψ_i is

$$\rho\frac{\partial \psi_i}{\partial t} = \frac{\rho}{Le_i}\frac{\chi}{2}\frac{\partial^2 \psi_i}{\partial Z^2} + \dot{\omega}_i \tag{8.127}$$

where χ is the instantaneous scalar dissipation rate. For example, the flamelet equation for the temperature T is

$$\frac{\partial T}{\partial t} = \frac{\chi}{2}\frac{\partial^2 T}{\partial Z^2} + \frac{\dot{\omega}_T}{\rho c_p} \tag{8.128}$$

The instantaneous scalar dissipation rate χ is defined as

$$\chi = 2D|\nabla Z|^2 = 2D\sum_{i=1}^{3}\left(\frac{\partial Z}{\partial x_i}\right)^2 \tag{8.129}$$

The scalar dissipation rate has the dimension of t^{-1}. It is a measure of how quickly scalar fluctuations decay and can be regarded as the inverse of a diffusion time scale. At the stoichiometric flame surface,

$$\chi = \chi_{\text{st}} \tag{8.130}$$

The mixture fraction Z and scalar dissipation χ are fluctuating quantities and their statistical distribution in space and time needs to be included. If the joint PDF $\tilde{P}(Z, \chi_{\text{st}}; \mathbf{x}, t)$ is known, and the flamelet equations are solved to obtain ψ_i as a function of Z and χ_{st}, then the Favre mean $\tilde{\psi}_i(\mathbf{x}, t)$ is

$$\tilde{\psi}_i(\mathbf{x}, t) = \int_0^1 \int_0^\infty \psi_i(Z, \chi_{\text{st}})\tilde{P}(Z, \chi_{\text{st}}; \mathbf{x}, t)\, d\chi_{\text{st}}\, dZ \tag{8.131}$$

8.8.3 Conditional Moment Closure Models

CMC models were independently developed by Klimenko (1990) and Bilger (1993), who proposed that the reactive scalars ψ_i should be conditioned on a given value of the mixture fraction $Z = Z^*$, since most of the fluctuations of the reactive scalars are associated with fluctuations of the mixture fraction. The CMC model results in conditional means of reactive scalars that respond to spatial and temporal variations in the scalar dissipation χ. It has been used successfully in a wide range of non-premixed reacting flow problems, including extinction and ignition.

Instead of considering flame surface statistics and the laminar reactive–diffusive flow fields adjacent to the flame surface as in the flamelet method, the CMC method is based on conditional moments at a fixed $\mathbf{x}$ and time t in the flow field. This approach is useful for comparison with measured mass fractions Y_i.

As introduced in Chapter 5, the notation for a conditional PDF is $P(u|v)$, which describes the probability density of u for a fixed value of v, and the notation for the conditional average or mean of u, for a fixed value of v, is given by angular brackets $\langle u|v\rangle$, for example,

$$\langle u|v\rangle = \int_{-\infty}^{+\infty} uP(u|v)du \tag{8.132}$$

In the CMC method, a reactive scalar ψ_i is decomposed into a conditional mean Q_i and conditional fluctuation y_i,

$$\psi_i(\mathbf{x}, t) = Q_i(Z^*; \mathbf{x}, t) + y_i(Z^*; \mathbf{x}, t) \tag{8.133}$$

The first conditional moment Q_i, i.e., the mean of ψ_i, is

$$Q_i(Z^*; \mathbf{x}, t) = \langle \psi_i|Z^*\rangle = \int_0^1 \psi_i P(\psi_i|Z^*; \mathbf{x}, t)\, d\psi_i \tag{8.134}$$

Upon substitution of Eq. (8.133) into the governing equations, Eq. (8.121), for the reactive scalars and determination of their conditional average, the governing balance equations for the mean values Q_i of the conditional reactive scalars are expressed as (Peters 2000),

$$\langle \rho|Z^*\rangle \frac{\partial Q_i}{\partial t} + \langle \rho|Z^*\rangle \tilde{\mathbf{v}}_{Z^*} \cdot \nabla Q_i = \langle \rho|Z^*\rangle \tilde{\chi}_{Z^*} \frac{\partial^2 Q_i}{\partial Z^{*2}} + \langle \dot{\omega}_i|Z^*\rangle \tag{8.135}$$

The Favre conditional velocity $\tilde{\mathbf{v}}_{Z^*}$, the Favre conditional scalar dissipation rate $\tilde{\chi}_{Z^*}$, and the chemical source term $\dot{\omega}_i$ are unclosed and need to be modeled. Since there are the second-order derivatives with respect to mixture fraction, a sufficiently large number of conditioning Z^* values are required. The PDF $P(\psi_i|Z^*; \mathbf{x}, t)$ is presumed from the mean and variance of the mixing fraction, $\tilde{Z}^*$ and $\widetilde{Z''}^2$.

8.9 Buoyant Flames

Buoyancy forces on a flame can be an important factor in the flame behavior. For vertical flames, the buoyancy body force on the hot gas accelerates the flow, narrowing the flame in the axial direction. A common example of a buoyant laminar diffusion flame is a candle flame. The paraffin of the candle first melts due to radiative heating from the flame to the candle, driven by capillary forces into the wick where it evaporates to become paraffin vapor, which serves as the gaseous fuel in the buoyant combustion process.

A schematic of a single fuel jet flame is shown in Figure 8.23*a*. For upward vertical jets with the density of the injected fluid lower than that of the surroundings, buoyancy effects can cause the momentum flux in the jet to increase with height. When the buoyancy or crossflow effect is significant, the flame geometry and the boundary of the hot gases will not be axisymmetric (see Figure 8.23*b*). The solutions of these physical situations become more complex and one usually has to rely on numerical methods.

The transition from laminar to turbulent flow is due to either shear laminar type instabilities or buoyancy-induced instabilities, depending on the magnitude of the buoyancy forces relative to the inertia forces near the nozzle exit. Buoyancy-induced transition instability in jets is associated with the formation of toroidal vortices, and shear transition instability is associated with a sinusoidal motion of the jet column which transitions into turbulent vortices.

The nondimensional Froude number *Fr* is used to represent the ratio of fluid momentum to buoyancy or potential energy in a fluid flow. The Froude number is named after W. Froude (1810–1879) an English engineer who is known for his contributions to ship hydrodynamics. The basic Froude number in fluid mechanics is defined as

$$Fr = \frac{u_i}{(gd_i)^{1/2}} \tag{8.136}$$

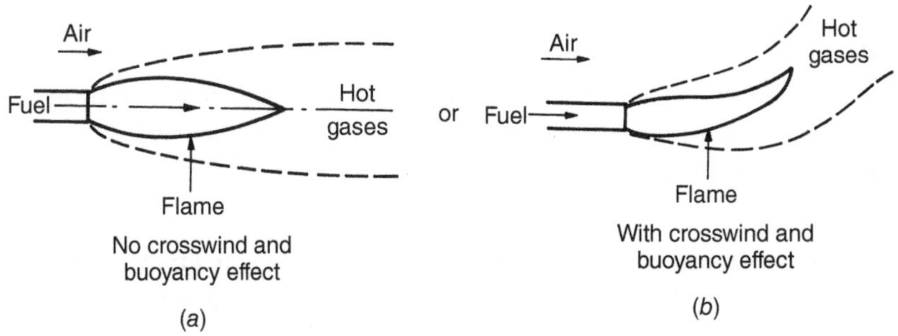

Figure 8.23 Diffusion flame jets (*a*) without and (*b*) with crosswind and buoyancy effect.

where g is the gravitational acceleration (9.81 m/s^2), u_i is a fluid characteristic velocity, and d_i is a characteristic length. For small values of Fr, the flow field is dominated by buoyancy, and for large values of Fr, the flow field is momentum dominated.

When modeling turbulent flames, the formulation of the Froude number is more complex, as additional combustion-related parameters such as flame temperature and density need to be included. A turbulent flame Froude number Fr_f has been developed in order to account for the combustion of the fuel jet and to distinguish between the momentum-controlled and the buoyancy-controlled flow regimes (Delichatsios 1993). This flame Froude number is

$$Fr_f = \frac{u_i f_{\mathrm{st}}^{3/2}}{\left[\rho_i/\rho_\infty\right)^{1/4}\left(g\,d_i(T_f - T_\infty)/T_\infty\right]^{1/2}} \tag{8.137}$$

The flame Froude number parameters in Eq. (8.137) are:

u_i = jet exit velocity (m/s)
f_{st} = stoichiometric mixture fraction = $(1 + s_A)^{-1}$
ρ_i = jet exit density (kg/m^3)
ρ_∞ = ambient air density (kg/m^3)
T_f = adiabatic flame temperature (K)
T_∞ = ambient temperature (K)
g = gravitational acceleration
d_i = nozzle diameter (m)

A correlation for a dimensionless turbulent flame length L^* as a function of the turbulent Froude number is (Delichatsios 1993),

$$L^* = \frac{13.5\,Fr_f^{2/5}}{[1 + 0.07\,Fr_f^2]^{1/5}} \tag{8.138}$$

where the flame length x_f is defined as

$$x_f = L^* \frac{d_i(\rho_i/\rho_\infty)^{1/2}}{f_{st}} \tag{8.139}$$

Note that for buoyancy-controlled flames, where $Fr_f < 5$, Eq. (8.138) approaches $L^* \sim 3.5\,Fr_f^{2/5}$. With momentum-controlled flames, where $Fr_f > 5$, Eq. (8.138) approaches $L^* \sim 23.0$.

Example 8.3 ***Turbulent Flame Length***

What is the outlet Reynolds number and length L_f of a turbulent methane jet flame burning in air at T_∞ = 300 K and p = 1 atm? The exit nozzle diameter d_i is 5 mm. The methane exit density is ρ_i = 1.00 kg/m^3, dynamic viscosity $\mu = 11.0 \times 10^{-6}$ Pa-s, and flowrate $\dot{m}_i = 1.0 \times 10^{-3}$ kg/s.

Solution

From Table 2.8, the stoichiometric mass air–fuel ratio of methane is s_A = 17.20. From Table 2.10, the adiabatic flame temperature of methane is T_f = 2227 K. The ambient air density ρ_∞ is

$$\rho_\infty = \frac{p}{RT_\infty} = \frac{101.2}{(0.287)(300)} = 1.17\,\mathrm{kg/m^3}$$

The mixture fraction f_{st} is

$$f_{st} = \frac{1}{1+s_A} = \frac{1}{1+17.20} = 0.055$$

The jet exit velocity u_i is

$$u_i = \frac{\dot{m}_i}{\rho_i \pi r_i^2} = \frac{1.0\times10^{-3}}{(1.00)\pi(2.5\times10^{-3})^2} = 64\,\mathrm{m/s}$$

The jet Reynolds number is in the turbulent regime,

$$Re_{d_i} = \frac{\rho_i u_i d_i}{\mu} = \frac{(1.0)(64)(5.0\times10^{-3})}{11.0\times10^{-6}} = 28,865$$

The flame Froude number is

$$Fr_f = \frac{u_i f_{st}^{3/2}}{(\rho_i/\rho_\infty)^{1/4}\left[g\,d_i(T_f - T_\infty)/T_\infty\right]^{1/2}}$$

$$= \frac{(64)\,(0.055)^{3/2}}{(1.00/1.17)^{1/4}\left[(9.81)\,(5.0\times10^{-3})(2227-300)/300\right]^{1/2}} = 1.51$$

Since $Fr_f < 5$, the flame is primarily buoyancy controlled. The nondimensional flame length L^* is

$$L^* = \frac{13.5\,Fr_f^{2/5}}{[1+0.07\,Fr_f^2]^{1/5}}$$

$$= \frac{13.5\,(1.51)^{2/5}}{[1+0.07\,(1.51)^2]^{1/5}} = 15.4$$

From Eq. (8.139), the flame length x_f is

$$x_f = \frac{d_i}{f_{st}}(\rho_i/\rho_\infty)^{1/2}L^*$$

$$= \frac{5.0\times10^{-3}}{0.055}(1.00/1.17)^{1/2}(15.4) = 1.29\,\mathrm{m}$$

The nondimensional length L_f/d_i is

$$\frac{x_f}{d_i} = \frac{1.29}{5.0\times10^{-3}} = 258$$

References

Barlow, R., and A. Karpetis (2004), Measurements of scalar variance, scalar dissipation, and length scales in turbulent piloted methane/air jet flames, *Flow, Turbul. Combust.*, Vol. 72, pp. 427–448.

Bergmann, V., W. Meier, D. Wolff, and W. Stricker (1998), Application of spontaneous Raman and Rayleigh scattering and 2D LIF for the characterization of a turbulent $CH_4/H_2/N_2$ jet diffusion flame, *J. Appl. Phys. B*, Vol. 66, No. 4, pp. 489–502.

Bilger, R. (1993), Conditional moment closure for turbulent reacting flow, *Phys. Fluids A: Fluid Dyn.*, Vol. 5, No. 2, pp. 436–444.

Burke, S., and T. Schumann (1928), Diffusion flames, *Ind. Eng. Chem.*, Vol. 20, No. 10, pp. 998–1004.

Chung, S., and C. K. Law (1984), Burke-Schumann flame with streamwise and preferential diffusion, *Combust. Sci. Technol.*, Vol. 37, No. 1-2, pp. 21–46.

Curtiss, C., and J. Hirschfelder (1949), Transport properties of multicomponent gas mixtures, *J. Chem. Phys.*, Vol. 17, No. 6, pp. 550–555.

Delichatsios, M. (1993), Transition from momentum to buoyancy-controlled turbulent jet diffusion flames and flame height relationships, *Combust. Flame*, Vol. 92, pp. 349–364.

Hottel, H., and W. Hawthorne (1948), Diffusion in laminar flame jets, *3rd Symp. on Comb. and Flame, and Explosion Phenomena*, pp. 254–266.

Kee, R., J. Miller, and T. Jefferson (1980), CHEMKIN: A general-purpose, problem-independent, transportable, FORTRAN chemical kinetics code package, SAND–80-8003, Sandia Nat. Lab.

Kim, G., A. Kirkpatrick, and C. Mitchell (2004), Computational modeling of natural gas injection in a large bore engine, *J. Eng. Gas Turbines Power*, Vol. 126, No. 3, pp. 656–664.

Klimenko, A. (1990), Multicomponent diffusion of various admixtures in turbulent flow, *Fluid Dyn.*, Vol. 25, No. 3, pp. 327–334.

Leonard, S., G. Mulholland, R. Puri, and R. Santoro (1994), Generation of CO and smoke during underventilated combustion, *Combust. Flame*, Vol. 98, pp. 20–34.

Li, Y., A. Kirkpatrick, C. Mitchell, and B. Willson (2004), Characteristic and computational fluid dynamics modeling of high pressure gas jet injection, *J. Eng. Gas Turbines Power*, Vol. 126, No. 1, pp. 1–6.

Malmstrom, T., A. Kirkpatrick, B. Christensen, and K. Knappmiller (1997), Centreline velocity decay measurements in low-velocity axisymmetric jets, *J. of Fluid Mech.*, Vol. 246, pp. 363–377.

Ouellette, P., and P. Hill (2000), Turbulent transient gas injections, *J. Fluids Eng.*, Vol. 122, pp. 743–753.

Penner, S., M. Bahadori, and E. Kennedy (1983), Laminar diffusion flames with cylindrical symmetry, arbitrary values of diffusion coefficients and inlet velocities, and chemical reactions in the approach streams, in *Dynamics of Flames and Reactive Systems*, Progress in Aeronautics and Astronautics, AIAA, pp. 261–292.

Peters, N. (2000), *Turbulent Combustion*, Cambridge University Press, Cambridge, UK.

Ricou, F., and D. Spalding (1961), Measurements of entrainment by axisymmetrical turbulent jets, *J. Fluid Mech.*, Vol. 11, pp. 21–32.

Roper, F. (1977), The prediction of laminar jet diffusion flame sizes: Part I. Theoretical model, *Combust. Flame*, Vol. 29, pp. 219–226.

Roper, F., C. Smith, and A. Cunningham (1977), The prediction of laminar jet diffusion flame sizes: Part II. Experimental verification, *Combust. Flame*, Vol. 29, pp. 227–234.

Schlichting, H., and K. Gersten (2017), *Boundary-Layer Theory*, 9th ed., Springer, New York.

Smooke, M., R. Mitchell, and D. Keyes (1989), Numerical solution of two-dimensional axisymmetric laminar diffusion flames, *Combust. Sci. Technol.*, Vol. 67, pp. 85–122.

Spalding, D. (1979), *Combustion and Mass Transfer*, Pergamon Press, New York.

Takagi, T., and Z. Xu (1993), Effects of preferential diffusion of heat and species in diffusion flames, in *Turbulence and Molecular Processes in Combustion*, Elsevier Science Publishers, pp. 393–402.

Takagi, T., K. Tada, and M. Komiyama (1990), Numerical simulation of the hydrogen diffusion flame, *Bull. JSME*, Vol. 56, pp. 2109–2114.

Williams, F. (1985), *Combustion Theory*, 2nd ed., CRC Press, Boca Raton, FL.

Wohl, K., C. Gazley, and N. Kapp (1948), Diffusion flames, *3rd Symp. on Comb. and Flame, and Explosion Phenomena*, pp. 288–300.

Homework Problems

8.1 Integrate the x-momentum equation (8.11) from $r = 0$ to $r = \infty$ and show that the parameter I_u, defined in Eq. (8.25), of an axisymmetric jet is an invariant. Discuss the physical meaning of I_u.

8.2 Show that the parameter I_f, defined in Eq. (8.26), of an axisymmetric jet is an invariant by integrating the mixture fraction equation (8.21), from $r = 0$ to $r = \infty$. What is the physical meaning of I_f?

8.3 A laminar fuel jet issues from a tube into air and is ignited to form a flame. The length of the flame is 10 cm. Using the same fuel, the diameter of the tube is increased by 50% and the velocity leaving the jet is decreased by 50%. (a) What is the length of the flame after the changes are made? (b) Suppose the measurement is repeated with grids inserted in the fuel tube so that the jet flow is turbulent. What is the length of the turbulent flame with the larger diameter tube? Assume the flame length is also 10 cm for the initial turbulent condition.

8.4 A laminar jet of ethylene C_2H_4 flows from a 10-mm diameter nozzle into standard air at 298 K. What is the spreading angle and the axial location where the jet centerline mass fraction is stoichiometric? (a) The initial velocity is $u_i = 0.1$ m/s. (b) The initial velocity is $u_i = 0.01$ m/s.

8.5 Due to wall friction, the exit velocity from a circular nozzle is parabolic, $u(r) = 2u_m[1 - (r/R)^2]$, where u_m is the mean velocity of the nozzle flow, and R is the nozzle radius. How does the resulting momentum flow compare with a frictionless top hat velocity profile that has the same mean velocity u_m?

8.6 Estimate the length of an ethane C_2H_6–air diffusion flame from a 10-mm diameter nozzle if the outlet velocity is 0.05 m/s. The air and ethane are at 300 K and 1 atm. The dynamic viscosity μ of ethane is 9.51×10^{-6} Pa-s at 300 K.

8.7 Propane fuel can be diluted with nitrogen to reduce soot formation. If the propane fuel flow is 50% N_2 by mass, what is the change in the flame length compared with an undiluted flame? The fuel, nitrogen, and air are at 300 K and 1 atm. Assume the total fuel flowrate in both cases is $\dot{m}_F = 6.0 \times 10^{-6}$ kg/s.

8.8 What is the length of a turbulent methane jet flame burning in air at $T_\infty = 300$ K and p = 1 atm? The exit nozzle diameter is 3 mm. The methane exit density is $\rho_i = 1.00$ kg/m^3, and flowrate $\dot{m}_i = 0.4 \times 10^{-3}$ kg/s.

9

Droplet Combustion

9.1 Introduction

In this chapter, we study diffusion flames originating from liquid fuels. Liquid fuels and/or oxidizers are usually delivered to combustion chambers as a spray of droplets, which then undergo a sequence of processes including vaporization, mixing, ignition, and flame spreading. With liquid droplet combustion, there are several important questions to consider. How long is the lifetime of a droplet? Which parameters govern the mass burning rate of a fuel droplet? What are the effects of relative motion between the droplet and surrounding gases? What is the effect of buoyancy? Is there an internal recirculation within the droplet? What is the effect of ambient pressure on the lifetime of the droplet and other parameters?

For typical conditions in a combustion chamber, the evaporation time is on the order of milliseconds and depends on the droplet temperature, size, velocity, and composition. In this chapter, we start with the evaporation and combustion of a fuel droplet in a quiescent environment, followed by analysis of liquid fuel injection, droplet formation, evaporation, and combustion in convective environments. Finally, we study droplet vaporization and combustion in high pressure, i.e., supercritical conditions.

Liquid fuel vaporization and combustion processes have attracted the interest and attention of many researchers due to their direct application to combustion devices such as boilers, gasoline and diesel engines, liquid-propellant rockets, gas-turbine engines, pulse-detonation engines, and liquid-fuel ramjets. For example, Figure 9.1 shows the gasoline fuel injector in a direct injection internal combustion engine producing a spray of liquid droplets. The droplets vaporize and the fuel–air gas mixture is ignited by a spark plug as the piston approaches top dead center.

9.2 Droplet Evaporation in Quiescent Environments

We shall follow Spalding's classic analysis (Spalding 1979) of the evaporation of a single fuel droplet in a quiescent, i.e., non-convective, environment at a given temperature and pressure. A *quasi-steady assumption* is utilized in Spalding's analysis for determining the distributions of fuel vapor mass fraction and temperature around the evaporating spherical droplet. The main advantage of this assumption is to temporarily avoid the consideration of time variations associated with the droplet regression process. The justification of this assumption is based on the relatively slow regression rate of the fuel droplet. For a spherical droplet with a radial gas velocity u at a

Principles of Combustion, Third Edition. Allan T. Kirkpatrick and Kenneth K. Kuo.

Companion website: www.wiley.com/go/kirkpatrick/principlesofcombustion3e

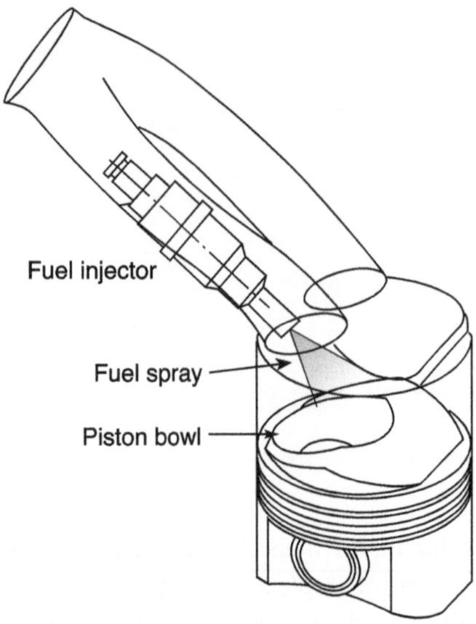

Figure 9.1 Schematic of gasoline direct fuel injection.

radius r, the 1-D steady-state species and energy differential equations in Shvab–Zeldovich form (see Example 5.2) are:

Fuel species continuity equation:

$$r^2 \rho u \frac{dY_F}{dr} = \frac{d}{dr}\left(r^2 \mathcal{D} \rho \frac{dY_F}{dr}\right) \tag{9.1}$$

Energy equation:

$$r^2 \rho u \frac{dc_p T}{dr} = \frac{d}{dr}\left(\frac{k}{c_p} r^2 \frac{dc_p T}{dr}\right) \tag{9.2}$$

Let us further assume that the ambient gases do not dissolve into the fuel droplet; just below the droplet surface, only fuel (F) exists. From Fick's law (see Chapter 5) for the vaporizing droplet, the total fuel mass flux $\dot{m}_F''$ leaving the droplet surface (s) must be equal to the sum of the fuel mass flux due to the bulk convective motion of the gaseous mixture at the surface and the mass diffusion rate of the fuel vapor in the radial direction due to the fuel concentration gradient above the droplet surface. Fick's Law is

$$\dot{m}_F'' = Y_F \dot{m}'' - \rho \mathcal{D} \nabla Y_F \tag{9.3}$$

We assume the outward flux of the gas mixture at the surface is composed only of fuel, so $\dot{m}'' = \dot{m}_F''$, with no contribution to the flux from the other species present. With a radial concentration gradient, species conservation at the droplet surface can be written as

$$\dot{m}_F'' = Y_{F,s} \dot{m}_F'' - \rho_s \mathcal{D}_s \frac{dY_{F,s}}{dr} \tag{9.4}$$

$$\rho_s u_s = Y_{F,s} \rho_s u_s - \rho_s \mathcal{D}_s \frac{dY_{F,s}}{dr} \tag{9.5}$$

At the droplet surface r_s, ρ_s is the density of the fuel vapor, u_s is the fuel radial velocity, and $Y_{F,s}$ is the fuel mass fraction. After rearranging Eq. (9.5), we have

$$u_s = \frac{\mathcal{D}_s (dY_{F,s}/dr)}{Y_{F,s} - 1} \tag{9.6}$$

Spalding defined a dimensionless b parameter as

$$b(r) \equiv \frac{Y_F(r)}{Y_{F,s} - 1} \equiv -\frac{\text{mass fraction of } F \text{ at any radius } r}{\text{mass fraction of other species at surface } r = r_s} \tag{9.7}$$

u_s, the fuel radial velocity, can be expressed as

$$u_s = \mathcal{D}_s \left(\frac{db}{dr} \right)_s \tag{9.8}$$

Since $db/dr = dY_F/dr$, the fuel species conservation equation Eq. (9.1) becomes

$$r^2 \rho u \frac{db}{dr} = \frac{d}{dr} \left(r^2 \mathcal{D} \rho \frac{db}{dr} \right) \tag{9.9}$$

The boundary condition at $r = \infty$ is $Y_F = Y_{F,\infty}$,

$$r = \infty : \qquad b = b_\infty = \frac{Y_{F,\infty}}{Y_{F,s} - 1} \tag{9.10}$$

Overall mass conservation implies

$$\dot{m}(r) = 4\pi r^2 \rho u = \text{constant} \tag{9.11}$$

Therefore,

$$r^2 \rho u = r_s^2 \rho_s u_s = \text{constant} \tag{9.12}$$

Using Eq. (9.12) and integrating Eq. (9.9) with respect to r, we get

$$r^2 \rho u b = r^2 \rho \mathcal{D} \frac{db}{dr} + \text{constant} \tag{9.13}$$

Applying the boundary condition (9.8) at the surface, we have

$$r_s^2 \rho_s u_s (b - b_s + 1) = r^2 \rho \mathcal{D} \frac{db}{dr} \tag{9.14}$$

By separating variables, we have

$$\frac{r_s^2 \rho_s u_s}{r^2 \rho \mathcal{D}} dr = \frac{db}{b - b_s + 1} \tag{9.15}$$

Taking the product $\rho\mathcal{D}$ to be constant ($= \rho_s \mathcal{D}_s$) and integrating again, we have

$$-\frac{r_s^2 u_s}{r \mathcal{D}_s} = \ln(b - b_s + 1) + \text{constant} \tag{9.16}$$

Using the boundary condition (9.10), we have

$$\frac{r_s^2 u_s}{r \mathcal{D}_s} = \ln \left(\frac{b_\infty - b_s + 1}{b - b_s + 1} \right) \tag{9.17}$$

At $r = r_s$,

$$\frac{r_s u_s}{\mathcal{D}_s} = \ln[1 + (b_\infty - b_s)] \tag{9.18}$$

The difference $b_\infty - b_s$ is called the *Spalding transfer number B or B_M*, a dimensionless mass fraction difference, which provides the driving potential for mass transfer. The subscript M stands for mass, i.e.,

$$B \equiv B_M \equiv b_\infty - b_s = \frac{Y_{F,s} - Y_{F,\infty}}{1 - Y_{F,s}} \tag{9.19}$$

Then, Eq. (9.18) becomes

$$r_s u_s = \mathcal{D}_s \ln(1 + B_M) \tag{9.20}$$

Since the mass flowrate from the droplet surface is

$$\dot{m}_F = 4\pi r_s^2 \rho_s u_s \tag{9.21}$$

we can express the mass evaporation rate as

$$\dot{m}_F = 4\pi r_s \rho_s \mathcal{D}_s \ln(1 + B_M) \tag{9.22}$$

When the transfer number B_M is zero, there is no evaporation, and as the transfer number increases, the evaporation rate increases. In order to calculate the mass evaporation rate $\dot{m}_F$ from Eq. (9.22), the value of B_M must be known; before that, the fuel mass fraction Y_F (or fuel vapor partial pressure p_F) must be determined. A reasonable assumption would be that the gas surrounding the droplet is saturated by the fuel vapor at the surface temperature. Since the vapor-pressure data for most fuels are available, the problem now becomes to determine the surface temperature T_s.

In order to determine T_s, we use the energy equation, Eq. (9.2). Using the integrated continuity equation of (9.11) and assuming that the specific heat c_p of the gas surrounding the droplet is independent of temperature, Eq. (9.2) becomes

$$r_s^2 \rho_s u_s c_p \frac{dT}{dr} = \frac{d}{dr}\left(r^2 k \frac{dT}{dr}\right) \tag{9.23}$$

Integrating the above equation with respect to r, we have

$$r_s^2 \rho_s u_s c_p T = r^2 k \frac{dT}{dr} + \text{constant} \tag{9.24}$$

The integration constant is determined from the boundary condition at the droplet surface (see Figure 9.2):

$$k\left(\frac{dT}{dr}\right)_s = \rho_s u_s h_{fg}, \qquad T = T_s \ @\ r = r_s \tag{9.25}$$

After applying the boundary condition at the droplet surface, we have

$$r_s^2 \rho_s u_s c_p \left(T - T_s + \frac{h_{fg}}{c_p}\right) = r^2 k \frac{dT}{dr} \tag{9.26}$$

After separating the variables and integrating the above equation, we have

$$-\frac{r_s^2 \rho_s u_s c_p}{rk} = \ln\left(T - T_s + \frac{h_{fg}}{c_p}\right) + \text{constant}' \tag{9.27}$$

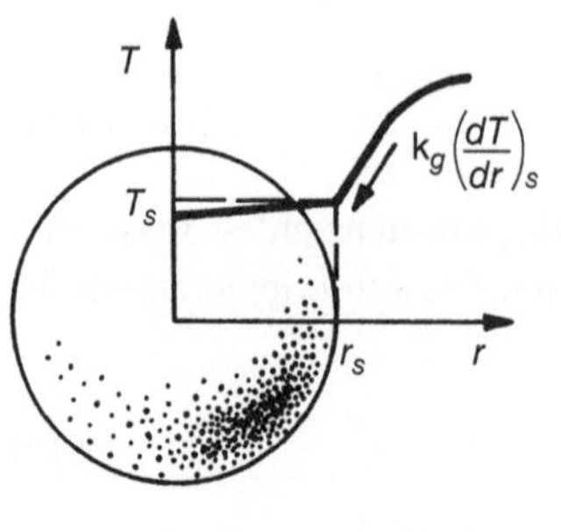

Figure 9.2 Temperature distribution of an evaporating liquid droplet.

Using the boundary condition that $T \to T_\infty$ as $r \to \infty$, we have

$$\frac{r_s^2 \rho_s u_s c_p}{rk} = \ln\left(\frac{T_\infty - T_s + h_{fg}/c_p}{T - T_s + h_{fg}/c_p}\right) \tag{9.28}$$

Setting r equal to r_s at the surface, we have

$$\frac{r_s \rho_s u_s c_p}{k} = \ln\left(1 + \frac{c_p(T_\infty - T_s)}{h_{fg}}\right) \tag{9.29}$$

Since the thermal diffusivity $\alpha \equiv k/\rho c_p$,

$$r_s u_s = \alpha_s \ln\left[1 + \frac{c_p(T_\infty - T_s)}{h_{fg}}\right] \equiv \alpha_s \ln[1 + B_T] \tag{9.30}$$

Comparing Eq. (9.30) with Eq. (9.20), we have

$$r_s u_s = \alpha_s \ln\left[1 + \frac{c_p(T_\infty - T_s)}{h_{fg}}\right] \equiv \mathcal{D}_s \ln\left[1 + \frac{Y_{F,s} - Y_{F,\infty}}{1 - Y_{F,s}}\right] \tag{9.31}$$

or

$$r_s u_s = \alpha_s \ln[1 + B_T] \equiv \mathcal{D}_s \ln[1 + B_M] \tag{9.32}$$

If $\alpha_s = \mathcal{D}_s$ (i.e., $Le = 1$), then the energy and the mass transfer numbers are identical,

$$B_T = B_M \tag{9.33}$$

or

$$B = B_T = \frac{c_p(T_\infty - T_s)}{h_{fg}} = B_M = \frac{Y_{F,s} - Y_{F,\infty}}{1 - Y_{F,s}} \tag{9.34}$$

This equation relates the two unknowns T_s and $Y_{F,s}$; one additional equation is needed to solve for them. The following equation relates the fuel mass fraction Y_F to the partial pressure p_F:

$$Y_F = \frac{\rho_F}{\rho} = \frac{n_F \mathrm{M}_F}{n\mathrm{M}} = \frac{p_F}{p}\frac{\mathrm{M}_F}{\mathrm{M}} \quad \text{or} \quad Y_{F,s} = \frac{p_{F,s}}{p}\frac{\mathrm{M}_F}{\mathrm{M}} \tag{9.35}$$

This equation is helpful, since we can further relate the partial pressure of the fuel $p_{F,s}$ to the surface temperature T_s using the Clausius–Clapeyron vapor-pressure equation introduced in Chapter 2. The Clausius–Clapeyron equation is

$$\frac{d \ln p_{F,s}}{dT_s} = \frac{h_{fg}}{RT_s^2} \tag{9.36}$$

Integrating Eq. (9.36), we have

$$\ln \frac{p_{F,s}}{p_{Fs,\mathrm{ref}}} = \frac{h_{fg}}{R}\left(\frac{1}{T_{s,\mathrm{ref}}} - \frac{1}{T_s}\right) \tag{9.37}$$

Now, we have three equations [Eqs. (9.34), (9.35), and (9.37)] for three unknowns: $Y_{F,s}$, T_s, and $p_{F,s}$. Once the solution is obtained, the transfer number B_M and mass evaporation rate can be calculated from Eqs. (9.19) and (9.22).

The lifetime or evaporation time, t_v, of a liquid droplet can be calculated using Eq. (9.22). The droplet evaporation time is an important parameter in combustion-chamber design, since the lifetime of the largest droplet in a spray determines the minimum time the droplet must be allowed to reside in the combustion chamber. Also, the residence time is related to other design parameters

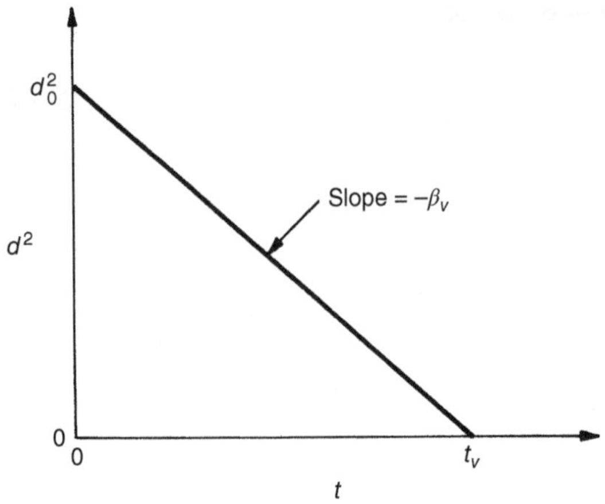

Figure 9.3 The d^2-evaporation law for liquid fuel droplets.

such as the air–stream velocity through the combustor, the velocity of the spray injection, the angle of injection, and the combustor geometry.

Using the mass continuity at the droplet surface ($-\rho_l\, dr_s/dt = \rho_s u_s$) and rearranging Eq. (9.22), we have

$$\frac{dr_s}{dt} = -\frac{\rho_s \mathcal{D}_s}{\rho_l r_s}\ln(1+B) \tag{9.38}$$

After integration, expressing the result in terms of droplet diameter, and since $\alpha_s = \mathcal{D}_s$, we have

$$d^2 = d_o^2 - \left[\frac{8\rho_s\alpha_s}{\rho_l}\ln(1+B)\right]t \tag{9.39}$$

We define an evaporation coefficient β_v as

$$\beta_v = \frac{8\rho_s\alpha_s}{\rho_l}\ln(1+B) = \frac{8k}{\rho_l c_p}\ln(1+B) \tag{9.40}$$

The evaporation coefficient β_v represents the magnitude of the negative slope of the straight line on the d^2—t plot shown in Figure 9.3. The lifetime of the droplet is therefore equal to

$$t_v = \frac{d_o^2}{\beta_v} = \frac{\rho_l\, d_o^2}{8\rho_s\alpha_s\ln(1+B)} \tag{9.41}$$

This implies that t_v is longer for larger droplets and also that lighter fuels evaporate faster. The effect of gas density and thermal diffusivity on droplet evaporation time can also be seen clearly from Eq. (9.41). When the temperature of the ambient atmosphere is substantially the same as that of the drop and is low in relation to the boiling point of the material of the drops, then the evaporation rate is determined by diffusion processes. The predominant fuel parameter is the vapor pressure of the liquid. In this situation, the mass rate of evaporation of a drop into quiescent air is proportional to the first power of the drop diameter and, when the vapor pressure is a small fraction of the total pressure, to the first power of the difference in the vapor pressure at the drop surface. Thus, under these conditions, the drop lifetime is proportional to the square of the drop diameter. It will be shown later that even though Eq. (9.41) was derived for evaporating droplets, it is also applicable to burning droplets. Godsave (1953) and many other experimentalists have found in droplet-burning studies that the time variation of the droplet diameter is similar to that of evaporation.

Some discussion about the physical properties, namely, the vapor thermal conductivity, and specific heat, needed for the above equations is in order. The physical properties need to be evaluated at a reference temperature. At this level of modeling, the properties are typically determined using the average boundary layer temperature. Following Law and Williams (1972), the vapor thermal conductivity k is a weighted average of the fuel vapor and air thermal conductivity, shown in Eq. (9.42),

$$k = 0.4k_F(\overline{T}) + 0.6k_{\text{air}}(\overline{T}) \tag{9.42}$$

and the vapor-specific heat is based on the fuel vapor-specific heat,

$$c_p = c_{pF}(\overline{T}) \tag{9.43}$$

Example 9.1 ***Droplet Evaporation***

What is the evaporation time t_v of a $d_o = 100\,\mu\text{m}$ diameter n-hexane C_6H_{14} droplet in quiescent air at $T_\infty = 800$ K and $p = 1$ atm? Hexane is a component of crude oil and gasoline. Assume $Le = 1$ and the droplet is at a uniform saturation temperature T_{sat}.

Solution

Using webbook.nist.gov properties for n-hexane, at 1 atm pressure, the boiling temperature of n-hexane is $T_{\text{sat}} = T_s = 342$ K and liquid density $\rho_l = 664\,\text{kg/m}^3$. At $T_s = 342$ K, the n-hexane heat of vaporization $h_{\text{fg}} = 3.35 \times 10^5$ J/kg.

The average vapor boundary layer temperature is

$$\overline{T} = (T_{\text{sat}} + T_\infty)/2 = (342 + 800)/2 = 571\,K$$

From the property data tables for n-hexane vapor and air in the Appendix, the n-hexane vapor-specific heat and the thermal conductivity are $c_{pF}\,(571\text{ K}) = 2772$ J/kg-K and $k_F\,(571\text{ K}) = 0.0456$ W/m-K. The air thermal conductivity is $k_{\text{air}}(571\text{ K}) = 0.0443$ W/m-K. From Eqs. (9.42) and (9.43),

$$\begin{aligned} c_p &= c_{pF}\,(571\text{ K}) = 2772\,\text{J/kg} - \text{K} \\ k &= 0.4\,k_F(571\text{ K}) + 0.6\,k_{\text{air}}(571\text{ K}) \\ &= 0.4(0.0456) + 0.6(0.0443) = 0.0448\,\text{W/m} - \text{K} \end{aligned}$$

Since $Le = 1$, the transfer number B is

$$B = B_T = \frac{c_p(T_\infty - T_s)}{h_{\text{fg}}} = \frac{(2772)(800 - 342)}{3.35 \times 10^5} = 3.79$$

The evaporation coefficient β_v is

$$\beta_v = \frac{8k}{\rho_l c_p} \ln(1 + B) = \frac{8(0.0448)}{(664)(2772)} \ln(1 + 3.79) = 3.05 \times 10^{-7}\text{m}^2/\text{s}$$

Therefore, the n-hexane droplet evaporation time is

$$t_v = \frac{d_o^2}{\beta_v} = \frac{(100 \times 10^{-6})^2}{3.05 \times 10^{-7}} = 0.0328\,\text{s} \sim 33\,\text{ms}$$

9.3 Combustion of a Fuel Droplet in a Quiescent Environment

In the combustion of a single fuel droplet in a quiescent atmosphere, fuel is evaporated from the liquid surface and the fuel vapor diffuses to the flame front, while the oxygen moves from the surroundings to the flame front. The rate at which the droplet evaporates and burns is generally

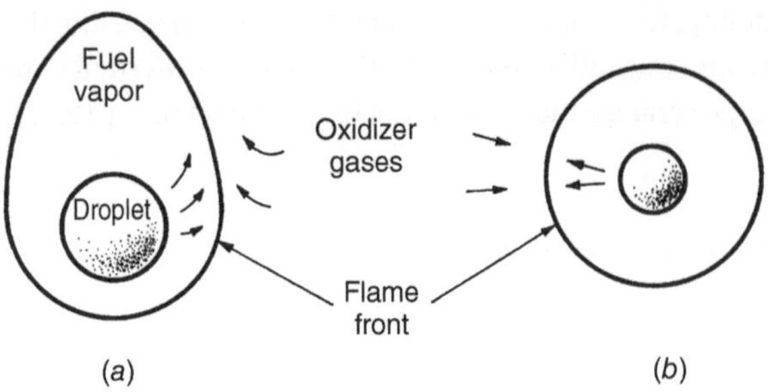

Figure 9.4 Shapes of diffusion flames surrounding a burning spherical fuel droplet: (*a*) nonspherical and (*b*) spherical.

considered to be determined by the rate of heat transfer from the flame front to the fuel surface. The shape of the flame front can be either spherical or nonspherical (see Figure 9.4). Nonspherical flames are generally caused by the convection effect due to the relative motion between the surrounding gases and droplet. When the droplet is very small, the surrounding gases can entrain it easily and the relative velocity between the droplet and nearby gases becomes small; then, the diffusion flame surrounding the droplet becomes nearly spherical.

In a pioneering development of a theoretical model describing the gaseous diffusion flames surrounding the burning droplet, Spalding (1979) proposed a *double-film model* for the temperature and chemical species distributions around the single droplet. One film covers the region between the droplet surface and the infinitely thin flame front, and the other film lies between the flame front and the surrounding oxidizer (see Figure 9.5). The chemical reaction processes in the gaseous flame are assumed to occur so rapidly that the droplet burning rate is determined solely by the rates of heat and mass transfer.

In most model developments, the liquid surface is assumed to be at the normal boiling point temperature of the fuel. Surveys of the temperature fields in burning liquids indicate that in fact, the temperature is only slightly below the boiling temperature (Law 1982). Many fuels have multiple components with a wide range of volatilities. At this level of analysis, we assume the fuel is composed of a single component with lumped properties.

In the following analysis, the temperature distribution within the droplet is assumed to be uniform and is slightly below the normal boiling point. In film I, heat is conducted from the flame front to the liquid, where it vaporizes the fuel. Most analyses assume that the fuel is heated to the

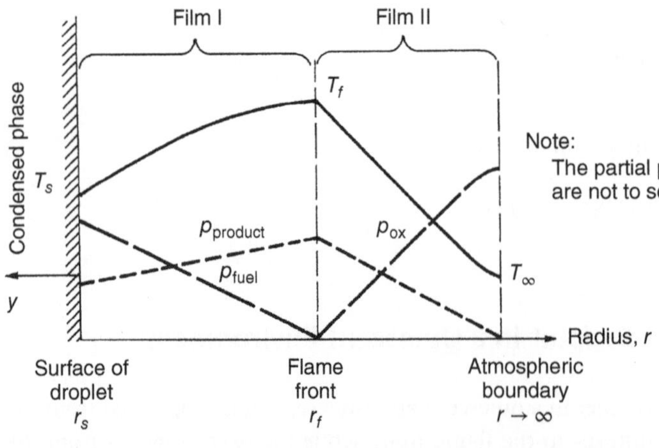

Figure 9.5 Radial variations of temperature and species partial pressures in double-film model.

flame temperature before it chemically reacts and that the vaporized fuel species does not react until it reaches the flame front. It is interesting to note that it was not necessary to determine the flame temperature T_f in the process to obtain the droplet burning rate.

The value of T_f can, however, be determined by assuming an infinitely thin reaction zone at the stoichiometric position. In film II, oxygen diffuses to the flame front, while combustion products and heat are transported to the surrounding atmosphere. The position of this far-field boundary, designated by ∞, is determined by the outward bulk flow condition. Strictly speaking, for a droplet burning in a stagnant environment, this boundary could be located at very far distance from the fuel surface. However, for practical purpose, this boundary can be quite close to the fuel surface, as long as the temperature is asymptotically close to the ambient hot-gas temperature, T_∞.

At low pressures, the liquid droplet vaporization process is one of the controlling factors governing the duration of the droplet combustion. For a burning droplet, the fuel vapor species conservation equation is the same as Eq. (9.1), except that we have to add a source term on the right-hand side, which corresponds to the generation of fuel species by chemical reaction. Since the fuel species is consumed by chemical reactions, this source term is generally negative (see Example 5.2).

$$r^2 \rho u \frac{dY_F}{dr} = \frac{d}{dr}\left(r^2 \rho D \frac{dY_F}{dr}\right) + r^2 \dot{\omega}_F \tag{9.44}$$

Similarly, one can write a species conservation equation for the oxidizer, i.e.,

$$r^2 \rho u \frac{dY_O}{dr} = \frac{d}{dr}\left(r^2 \rho D \frac{dY_O}{dr}\right) + r^2 \dot{\omega}_O \tag{9.45}$$

Again, the source term for this equation is generally negative due to the consumption of oxidizer species by chemical reactions. The energy conservation equation can be written as

$$r^2 \rho u \frac{dc_p T}{dr} = \frac{d}{dr}\left(\frac{k}{c_p} r^2 \frac{dc_p T}{dr}\right) - r^2 \dot{\omega}_F q_c \tag{9.46}$$

where q_c represents the heat of combustion per unit mass of fuel.

In the following droplet-burning analysis, we shall assume that the flame reactions are stoichiometric, i.e., the fuel (F) and oxidant (O) depletion rates are related in stoichiometric proportions,

$$\frac{\dot{\omega}_F}{\dot{\omega}_O} = (F/O)_{\text{st}} = \frac{1}{s_O} \tag{9.47}$$

where s_O is the stoichiometric oxidant/fuel ratio, as defined in Section 2.10. If one defines the following Shvab–Zel'dovich variables

$$b_{F,O} \equiv \frac{Y_F - Y_O(1/s_O)}{(Y_{F,s} - 1) + Y_{O,s}(1/s_O)} \tag{9.48}$$

$$b_{F,T} \equiv \frac{Y_F q_c + c_p T}{h_{\text{fg}} + q_c(Y_{F,s} - 1)} \tag{9.49}$$

$$b_{O,T} \equiv \frac{Y_O(1/s_O) q_c + c_p T}{h_{\text{fg}} + (1/s_O) Y_{O,s} q_c} \tag{9.50}$$

then the energy and species conservation equations can be combined (assuming $Le = 1$) to give the following homogeneous equation without any source terms:

$$r^2 \rho u \frac{db}{dr} = \frac{d}{dr}\left(r^2 \rho D \frac{db}{dr}\right) \tag{9.51}$$

where b can be either $b_{F,O}$ or $b_{F,T}$ or $b_{O,T}$ defined by Eqs. (9.48) through (9.50). Since the product $\rho\mathcal{D}$ is considered to be independent of temperature and $r^2\rho u$ = constant, Eq. (9.51) can be integrated twice with the boundary conditions

$$r = r_s: \qquad u_s = \mathcal{D}\left(\frac{db}{dr}\right)_s \tag{9.52}$$

$$r \to \infty: \qquad b = b_\infty \tag{9.53}$$

to obtain the solution

$$\frac{r^2\rho u}{\rho\mathcal{D}r} = \ln\left(\frac{b_\infty - b_s + 1}{b - b_s + 1}\right) \tag{9.54}$$

At the droplet surface $r = r_s$, we have

$$r_s\rho_s u_s = \rho_s\mathcal{D}_s \ln[1 + (b_\infty - b_s)] = \rho_s\mathcal{D}_s \ln(1 + B) \tag{9.55}$$

So the droplet burning rate $\dot{m}_F$ is

$$\begin{aligned} \dot{m}_F &= 4\pi r_s^2\rho_s u_s \\ &= 4\pi r_s\rho_s\mathcal{D}_s \ln(1 + B) \\ &= 4\pi r_s \frac{k}{c_p} \ln(1 + B) \end{aligned} \tag{9.56}$$

Note that this equation is identical to Eq. (9.22) for the evaporation case. The only difference is that the transfer number B in Eq. (9.56) can be obtained from one of the following:

$$B_{F,O} = \underbrace{\frac{(Y_{F,\infty} - Y_{F,s}) + (Y_{O,s} - Y_{O,\infty})(1/s_O)}{(Y_{F,s} - 1) + (1/s_O)(Y_O)_s}}_{\text{with or without combustion}} = \underbrace{\frac{Y_{O,\infty}(1/s_O) + Y_{F,s}}{1 - Y_{F,s}}}_{\text{with combustion}} \tag{9.57}$$

$$B_{F,T} = \underbrace{\frac{q_c(Y_{F,\infty} - Y_{F,s}) + c_p(T_\infty - T_s)}{h_{\text{fg}} + q_c(Y_{F,s} - 1)}}_{\text{with or without combustion}} = \underbrace{\frac{c_p(T_\infty - T_s) - Y_{F,s}q_c}{h_{\text{fg}} + q_c(Y_{F,s} - 1)}}_{\text{with combustion}} \tag{9.58}$$

$$\begin{aligned} B_{O,T} &= \underbrace{\frac{(1/s_O)(Y_{O,\infty} - Y_{O,s})q_c + c_p(T_\infty - T_s)}{h_{\text{fg}} + (1/s_O)Y_{O,s}q_c}}_{\text{with or without combustion}} \\ &= \underbrace{\frac{c_p(T_\infty - T_s) + Y_{O,\infty}q_c(1/s_O)}{h_{\text{fg}}}}_{\text{with combustion}} \end{aligned} \tag{9.59}$$

where for the combustion case $Y_{O,s} = Y_{F,\infty} = 0$, since it is assumed that neither fuel nor oxidizer can penetrate through the reaction zone.

Because $B_{F,O} = B_{O,T}$, we have

$$\frac{Y_{O,\infty}(1/s_O) + Y_{F,s}}{1 - Y_{F,s}} = \frac{c_p(T_\infty - T_s) + Y_{O,\infty}(1/s_O)q_c}{h_{\text{fg}}} \tag{9.60}$$

The unknowns in this equation are T_s and $Y_{F,s}$. These two unknowns together with the vapor pressure of the fuel can be calculated by the same method outlined in Section 9.2 for a non-combusting

Table 9.1 Transfer Number Values for Various Fuels.

Substance	B
iso-Octane	6.41
Benzene	5.97
n-Heptane	5.82
Toluene	5.69
Aviation Gasoline	≈5.5
Automobile Gasoline	≈5.3
Kerosene	≈3.4
Gas Oil	≈2.5
Light Fuel Oil	≈2.0
Heavy Fuel Oil	1.7
Carbon	0.12

Source: Adapted from Glassman et al. 2014.

liquid droplet. Once these parameters are known, the droplet burning rate $\dot{m}_F$ can be determined from Eq. (9.56).

Values of B taken from Spalding (1979) and Glassman et al. (2014) for various condensed combustible substances are given in Table 9.1. The most convenient form of B is $B_{O,T}$, which in the case of relatively small subcooling, can be expressed as

$$B_{O,T} = \frac{c_p(T_\infty - T_s) + Y_{O,\infty}(1/s_O)q_c}{h_{fg}} \approx \frac{Y_{O,\infty}(1/s_O)q_c}{h_{fg}} \tag{9.61}$$

As soon as $Y_{O,\infty}$ is known, $B_{O,T}$ can be calculated from Eq. (9.61) and the burning rate of the droplet can be determined from Eq. (9.56). For a droplet burning in air, on a molar basis, air (A) can be modeled as 1 O_2 + 3.762 N_2, so the far-field oxidizer mass fraction $Y_{O,\infty}$ is

$$Y_{O,\infty} = \frac{n_{O_2} M_{O_2}}{n_A M_A} = \frac{(1)(32.00)}{(4.762)(28.97)} = 0.232 \tag{9.62}$$

Note that

$$Y_{O,\infty}(1/s_O) = (F/A)_{st} \tag{9.63}$$

where $(F/A)_{st}$ is the stoichiometric mass fuel–air ratio (see Table 2.9).

By following the development of the droplet evaporation time t_v in Section 9.2, we can also define a burning rate coefficient β_b,

$$\beta_b = \frac{8k}{\rho_l c_p} \ln(1 + B) \tag{9.64}$$

and a droplet burning time t_b,

$$t_b = \frac{d_o^2}{\beta_b} \tag{9.65}$$

In general, the surface temperature has less effect on the burning rate than that of the mass fractions $Y_{F,s}$ and $Y_{O,\infty}$. It is interesting to note that for most hydrocarbon fuels, B does not change very much and is in the range of $1.7 < B < 6.5$.

It can be seen clearly from Eqs. (9.56) and (9.64) that the controlling term for the difference in mass burning rates between various fuels is due to the magnitude of the product $\mathcal{D}_s\rho_s$, i.e., the ratio k/c_p. Since $(1+B)$ is inside a natural logarithm, a tenfold variation in B results in only an approximately twofold variation in burning rate. Therefore, the transfer number does not have a strong influence on the magnitude of the mass burning rate.

If it is assumed that the flame exists at the position where $Y_O = Y_F = 0$, then the temperature of the flame T_f and its position r_f can be determined. Experimentally, it has been found that $r_f/r_s \approx 10$. However, note that the burning rate can be determined *without* knowing exactly where the flame front r_f is.

Equation (9.54) may be written in the following form:

$$\frac{r^2\rho u}{\rho \mathcal{D} r} = \frac{r_s^2\rho_s u_s}{d_s\rho_s r} = \ln\left[\frac{Y_{F,\infty} - Y_{F,s} - (1/s_O)(Y_{O,\infty} - Y_{O,s}) + (Y_{F,s} - 1)}{Y_F - Y_{F,s} - (1/s_O)(Y_O - Y_{O,s}) + (Y_{F,s} - 1)}\right] \tag{9.66}$$

At the flame surface $Y_F = Y_O = 0$ and $Y_{F,\infty} = Y_{O,s} = 0$, thus Eq. (9.66) becomes

$$\frac{r_s^2\rho_s u_s}{\rho_s \mathcal{D}_s r_f} = \ln[1 + (1/s_O)Y_{O,\infty}] \tag{9.67}$$

Since $4\pi r_s^2\rho_s u_s = \dot{m}_F$ is known, the flame front r_f can be solved for. The flame temperature at r_f can be obtained by setting $b = b_{O,T}$ in Eq. (9.54) and making use of Eq. (9.67). It is interesting to note that the flame temperature T_f can be *higher* than the adiabatic flame temperature, since the nitrogen diffused into film I shown in Figure 9.5 is preheated, which can in turn heat the fuel vapor before its reaction with the oxidizer.

Example 9.2 ***Droplet Combustion***

What is the initial mass burning rate $\dot{m}_F$ and the burning time t_b of an *n*-hexane C_6H_{14} droplet burning in quiescent air at $T_\infty = 300$ K and 1 atm when its initial diameter is $d_o = 100\ \mu$m? Assume $Le = 1$ and the droplet is at a uniform saturation temperature T_{sat}.

Solution

The stoichiometric reaction at the flame front is

$$C_6H_{14} + 9.5(O_2 + 3.76\ N_2) \longrightarrow 6\ CO_2 + 7\ H_2O + 35.7\ N_2$$

The molecular mass of hexane is $M_F = 86.18$ kg/kmol, and the molecular mass of oxygen is $M_{O_2} = 32.00$ kg/kmol. From webbook.nist.gov, at $p = 1$ atm, the saturation (boiling) temperature of *n*-hexane is $T_{sat} = T_s = 342$ K, liquid density $\rho_l = 664\ \text{kg/m}^3$, the heat of vaporization $h_{fg}(342\ \text{K}) = 3.35 \times 10^5$ J/kg, and the lower heat of combustion $q_c = 4.51 \times 10^7$ J/kg.

Assuming the flame temperature T_f is 2200 K, the reference film temperature is

$$\overline{T} = (T_{sat} + T_f)/2 = (342 + 2200)/2 = 1271\ \text{K}$$

From the property data tables for *n*-hexane vapor and air in the Appendix, the *n*-hexane vapor-specific heat and the thermal conductivity are $c_{pF}(1271\ K) = 4232$ J/kg-K and $k_F(1271\ K) = 0.2308$ W/m-K. The air thermal conductivity is $k_{air}(1271\ K) = 0.08097$ W/m-K. From Eqs. (9.42) and (9.43),

$$c_p = c_{pF}(1271\ K) = 4232\ \text{J/kg} - \text{K}$$

$$\begin{aligned} k &= 0.4\,k_F(1271\ K) + 0.6\,k_{air}(1271\ K) \\ &= 0.4(0.2308) + 0.6(0.0897) = 0.146\ \text{W/m} - \text{K} \end{aligned}$$

The $Y_{O,\infty}(1/s_O)$ term is

$$Y_{O,\infty}(1/s_O) = (0.232)\left[\frac{n_F M_F}{n_{O_2} M_{O_2}}\right]$$

$$= (0.232)\left[\frac{(1)(86.18)}{(9.5)(32.00)}\right]$$

$$= 6.58 \times 10^{-2}$$

The transfer number B is

$$B = B_{O,T} = \frac{c_p(T_\infty - T_s) + Y_{O,\infty}(1/s_O)q_c}{h_{fg}}$$

$$= \frac{4232(300 - 342) + (6.58 \times 10^{-2})(4.51 \times 10^7)}{3.35 \times 10^5}$$

$$= \frac{-1.78 \times 10^5 + 2.968 \times 10^6}{3.35 \times 10^5}$$

$$= 8.32$$

The initial droplet burning rate $\dot{m}_F$ is

$$\dot{m}_F = 4\pi r_s \frac{k}{c_p} \ln(1 + B)$$

$$= 4\pi(50 \times 10^{-6})\frac{(0.146)}{(4232)} \ln(1 + 8.32) = 4.84 \times 10^{-8}\ \text{kg/s}$$

The burning rate coefficient β_b is

$$\beta_b = \frac{8k}{\rho_l c_p} \ln(1 + B)$$

$$= \frac{8(0.146)}{(664)(4232)} \ln(1 + 8.32) = 9.28 \times 10^{-7}\ \text{m}^2/\text{s}$$

and the droplet burning time t_b is

$$t_b = \frac{d_o^2}{\beta_b}$$

$$= \frac{(100 \times 10^{-6})^2}{9.28 \times 10^{-7}} = 0.0108\,\text{s} \sim 11\,\text{ms}$$

Comment: Note that the burning time of 11 ms is less than the 33 ms evaporation time for the same droplet of Example 9.1.

9.4 Droplet Formation in Convective Environments

The droplet formation process has been modeled with two sequential phases, the *liquid jet breakup* phase followed by the *droplet breakup* phase. In the jet breakup phase, the cylindrical liquid fuel jet breaks up into large droplets and ligaments. The breakup of the fuel jet occurs near the injector nozzle and is due to shear forces acting on the circumference of the jet, cavitation, and turbulence. With the decrease in static pressure in the nozzle accompanying the fluid acceleration, it is also possible for cavitation to occur inside the nozzle, producing a bubbly two-phase injection flow from the nozzle.

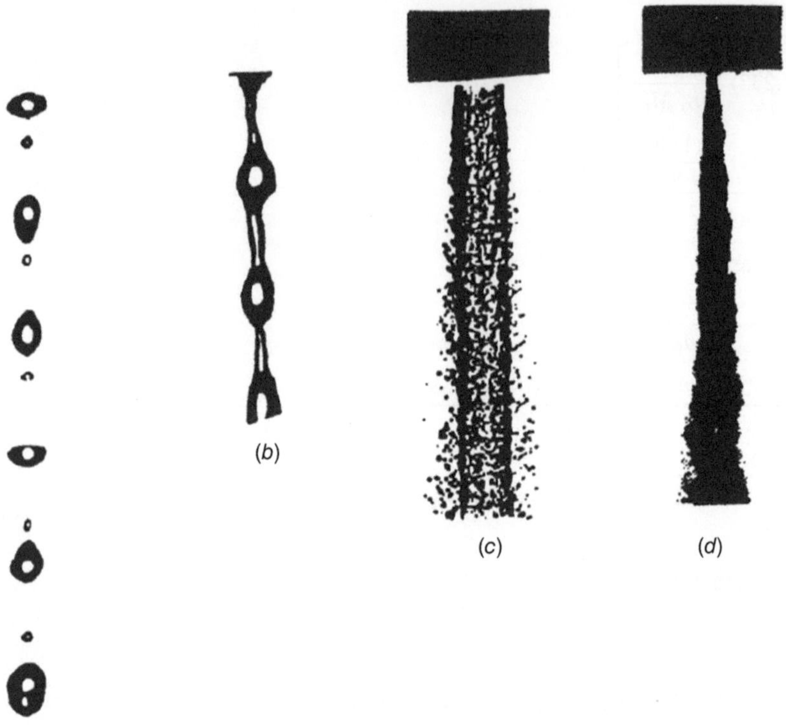

Figure 9.6 Liquid jet breakup regimes: (a) Rayleigh, (b) first wind induced, (c) second wind induced, and (d) atomization. (Adapted from Lin and Reitz 1998.)

9.4.1 Liquid Jet Breakup

As indicated in Figure 9.6, four general liquid jet breakup regimes have been classified: the Rayleigh, wind induced (first and second), and atomization regimes. Rayleigh breakup is caused by surface tension and yields a stream of drops having diameters larger than the jet diameter. First and second wind-induced breakups are due to shear instabilities caused by the relative motion of the gas and liquid. The atomization regime is characterized by jet breakup immediately at the jet exit. Flow in the atomization regime yields drops whose average diameter is much smaller than the jet diameter.

With high injection pressures or short nozzles, the breakup length decreases, and breakup can occur at the nozzle exit, leading directly to the atomization regime. In this case, the virtual origin of the spray is inside the injector. For diesel fuels, the high injection pressures result in an outlet jet relative velocity of the order of 100–200 m/s.

The two major dimensionless parameters governing the fuel jet breakup are the liquid Weber number We_l, which is the ratio of the shear and the surface tension forces; and the liquid Reynolds number Re_l, the ratio of the inertial and the viscous forces,

$$We_l = \frac{\rho_l u_f^2 d_i}{\sigma} \tag{9.68}$$

$$Re_l = \frac{\rho_l u_f d_i}{\mu_l} \tag{9.69}$$

where ρ_l is the liquid fuel density, μ_l is the fuel dynamic viscosity, u_f is the injection velocity of the jet, d_i is the injector nozzle diameter, and σ is the fuel surface tension.

For a simple orifice, the droplet spray is initially conical in shape, with the virtual origin of the jet inside the nozzle. The spreading angle of the spray, θ, depends on the ratio of the cylinder gas and the liquid fuel densities. Experimental data for diesel fuel in the atomization regime has been correlated by Reitz and Bracco (1979) with the following expression:

$$\tan\frac{\theta}{2} = \frac{4\pi}{A}\left(\frac{\rho_g}{\rho_l}\right)^{1/2}\frac{\sqrt{3}}{6} \tag{9.70}$$

where the parameter A is a function of the injector nozzle diameter d_i and length L_i:

$$A = 3.0 + 0.28(L_i/d_i) \tag{9.71}$$

For values of $\rho_g/\rho_l = 25\times10^{-3}$, and $A = 5$, the spread angle $\theta = 13°$. As the cylinder gas density increases, for example, during the compression stroke, the spreading angle of the spray increases.

The spray penetration distance S has been the subject of extensive experimental and computational research due to its importance in the behavior of subsequent combustion processes. If the penetration is too short, air at the edges of the combustion chamber will not be involved in the combustion process, and if the penetration is too long, the jet spray will impinge on the chamber walls, which will reduce the jet velocity and subsequent entrainment. Depending on the droplet Weber number, the impact of a droplet on the cylinder wall can result in the formation of a liquid film on the surface, breakup and vaporization, or rebound back into the cylinder.

Prior to the fuel jet breakup, $t < t_b$, the liquid jet is assumed to penetrate at the initial injection velocity u_i, so the penetration distance S increases linearly.

$$S = C_d\left(\frac{\Delta P}{\rho_l}\right)^{1/2} t \tag{9.72}$$

where ΔP is the injector nozzle pressure drop, and t is the time after start of injection (SOI). The distance from the nozzle at which the spray penetration changes from being proportional to time t to being proportional to $t^{1/2}$ is considered the jet breakup length.

After jet breakup, $t \geq t_b$, a widely used correlation by Jung and Assanis (2001) for the spray penetration distance is

$$\frac{S}{d_i} = 2.95\left(\frac{\Delta P}{\rho_g}\right)^{1/4}\left(\frac{t}{d_i}\right)^{1/2} \tag{9.73}$$

where the breakup time t_b of the liquid jet is

$$t_b = 4.35\frac{\rho_l d_i}{C_d^2(\rho_g\Delta P)^{1/2}} \tag{9.74}$$

As the injection pressure or the nozzle diameter is increased, the penetration distance is increased. Once the initial transient phase of the injection is complete, the length of the liquid core remains fairly constant until the end of injection.

A representative droplet size is given by the Sauter mean diameter, d_{sm}, named after a German scientist, J. Sauter, who in 1928 developed a measure of average particle size for a distribution of particles. The Sauter mean diameter, d_{sm}, as defined in Eq. (9.75), is the diameter of a model drop whose volume to surface ratio is equal to that of the total spray. Typical values of d_{sm} for diesel spray range between 10 and 50 μm. As the gas density and nozzle diameter decreases, d_{sm} decreases, decreasing the volume/surface ratio.

$$d_{sm} = \left(\sum_{j=1}^{n} d_j^3\right) \Big/ \left(\sum_{j=1}^{n} d_j^2\right) \tag{9.75}$$

A two-equation correlation developed by Hiroyasu et al. (1996) for diesel injectors accounts for observed differences in breakup mechanisms at different injection velocities:

$$\begin{aligned} d_{sm}^{k} &= 4.12\, d_i\, Re_l^{0.12}\, We_l^{-0.75} \left(\frac{\mu_l}{\mu_g}\right)^{0.54} \left(\frac{\rho_l}{\rho_g}\right)^{0.18} \\ d_{sm}^{h} &= 0.38\, d_i\, Re_l^{0.25}\, We_l^{-0.32} \left(\frac{\mu_l}{\mu_g}\right)^{0.37} \left(\frac{\rho_l}{\rho_g}\right)^{-0.47} \\ d_{sm} &= \max\left(d_{sm}^{k}, d_{sm}^{h}\right) \end{aligned} \tag{9.76}$$

There is a distribution of droplet sizes about the Sauter mean diameter in the spray. The cumulative χ^2 function is used to model droplet size distribution of hollow cone sprays and is represented by

$$f_V(d) = 1 - \exp\left(\frac{d}{\overline{d}}\right) \left[1 + \left(\frac{d}{\overline{d}}\right) + \frac{1}{2}\left(\frac{d}{\overline{d}}\right)^2 + \frac{1}{6}\left(\frac{d}{\overline{d}}\right)^3\right] \tag{9.77}$$

where $\overline{d} = 1/3\, d_{sm}$, and $f_V(d)$ is the fraction of the total volume contained in droplets of diameter less than d.

9.4.2 Droplet Breakup

The droplet breakup phase involves subsequent disintegration of the larger droplets and ligaments into smaller droplets. The droplet breakup process is due to the shear forces between the droplets and the surrounding gases, which distort the droplet by creating surface waves that are sheared off to form smaller droplets. Collisions between droplets can result in droplet coalescence. This droplet breakup phase is characterized by the dimensionless gas Weber and Reynolds numbers, *We* and *Re*, respectively, defined as follows,

$$We = \frac{\rho_g u_d^2 d}{\sigma} \tag{9.78}$$

$$Re = \frac{\rho_g u_d d}{\mu_g} \tag{9.79}$$

where ρ_g is the surrounding gas density, μ_g is the gas dynamic viscosity, u_d is the relative velocity between the droplet and the surrounding gas, and d is the droplet diameter. Wierzba (1990) noted that there is a critical gas Weber number, $We_{cr} \approx 10$, below which a droplet is stable, nearly spherical, and does not break up.

Different types of breakup modes occur as the Weber number is increased. The three general regimes are termed *bag*, *stripping*, and *catastrophic* and described below. Additional droplet breakup nomenclature is given in Figure 9.7.

1. *Bag type* The droplet flattens perpendicular to the flow and eventually forms a shroud. This shroud either breaks off, forming a group of small droplets, or several parachutes are formed

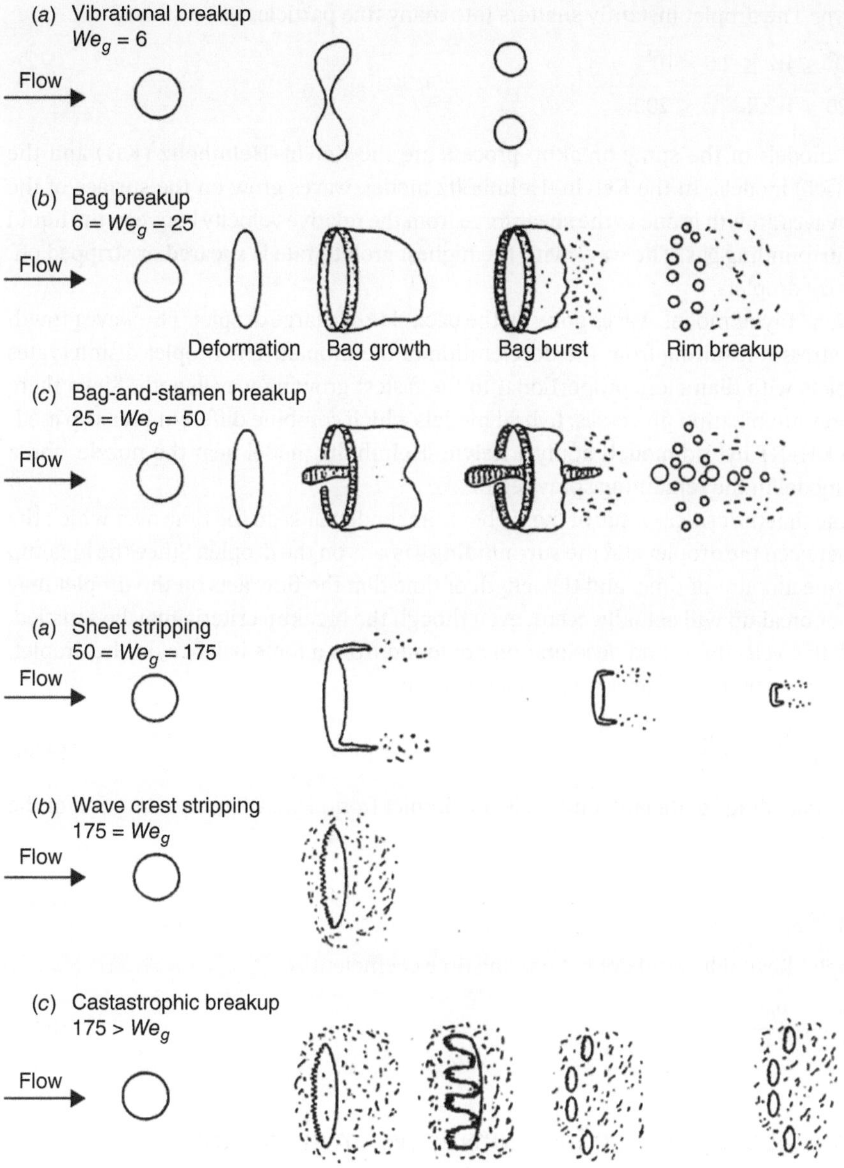

Figure 9.7 Liquid droplet breakup regimes. (Adapted from Pilch and Erdman 1987.)

from one droplet. In the final stage, the breakup process produces a group of fine droplets. This usually happens when

$$8 \leq We \leq 40$$
$$0.2 \leq We\,Re^{-0.5} \leq 1.6$$

2. *Stripping type* The droplet flattens perpendicular to the flow, and the gas flow tears off shrouds from the flattened droplet, followed by the disk-shaped droplet reaching a critical deformation and breaking into smaller droplets.

$$20 \leq We \leq 2.0 \times 10^4$$
$$1.0 \leq We\,Re^{-0.5} \leq 20$$

3. *Catastrophic type* The droplet instantly shatters into many fine particles.

$$2.0 \times 10^3 \leq We \leq 2.0 \times 10^5$$

$$20 \leq We\, Re^{-0.5} \leq 200$$

Two analytical models of the spray breakup process are the Kelvin–Helmholtz (KH) and the Rayleigh–Taylor (RT) models. In the Kelvin–Helmholtz model, waves grow on the surface of the fuel droplet. The waves growth is due to the shear force from the relative velocity between the liquid surface and the surrounding gas. The wave with the highest growth rate is sheared or stripped off, forming smaller size droplets.

With the Rayleigh–Taylor model, waves grow on the back side of a large droplet. The wave growth is due to normal stresses resulting from the deceleration of the droplet. The droplet disintegrates into smaller droplets with diameters proportional to the fastest growing wavelength. Since there are a variety of spray atomization processes, hybrid models which combine different breakup models are used. The KH–RT hybrid models apply a Kelvin–Helmholtz model near the nozzle, and a Rayleigh–Taylor model in the remaining spray region.

It is useful to note that the critical value of the Weber number depends on the time over which the relative velocity between the droplet and the surrounding gas acts on the droplet. Since the breakup process takes a finite amount of time, and the length of time that the flow acts on the droplet may determine whether breakup will actually occur, even though the breakup criteria may be satisfied.

The droplet relative velocity u_d and deceleration are found from a force balance on the droplet, with the drag force F_d acting to decrease the droplet velocity,

$$F_d = -C_d A_d \rho_g \frac{u_d^2}{2} = m\frac{du_d}{dt} \tag{9.80}$$

where C_d is the droplet drag coefficient, and A_d is the droplet frontal area. The deceleration of the droplet is therefore

$$\frac{du_d}{dt} = -\frac{3}{8} C_d \frac{\rho_g}{\rho_l} \frac{u_d^2}{r} \tag{9.81}$$

For spherical droplet Reynolds numbers < 1000, the drag coefficient is

$$C_d = \frac{24}{Re}\left(1 + \frac{Re^{2/3}}{6}\right) \tag{9.82}$$

9.5 Fuel Droplet Evaporation in a Convective Environment

When the droplets are not at rest, as is the case for flow from liquid fuel injectors, the results for quiescent flow will no longer hold true and effects of forced convection must be considered. When the Reynolds number $Re > 20$, there exists a boundary-layer flow region around the front of the droplet and a wake region behind the droplet. Determination of the vaporization rate of fuel droplets in a convective environment involves solutions of the energy equation and the coupled mass and heat transfer rate equations. The energy for evaporation is provided by the convective heat transfer from the surrounding hot gas. The heat transfer from the ambient gas will increase the droplet temperature and result in increased evaporation from the droplet.

In the control volume analysis in this section, we assume the fuel (F) droplet has a spatially uniform temperature $T_F(t)$ equal to the surface temperature $T_s(t)$, with no internal motion, an instantaneous radius r_s, and is surrounded by a vapor film of thickness δ. The Lewis number is assumed to be equal to one, i.e., the droplet has identical thermal and concentration boundary layer thicknesses, δ.

The rate of increase in temperature of the fluid droplet is found from an energy balance on the droplet, Eq. (9.83), where the time rate of energy storage equals the net heat transfer rate to the droplet minus the rate of vaporization.

$$m_F c_{p,F} \frac{dT_F}{dt} = \dot{Q}(r_s) - \dot{m}_F h_{fg} \tag{9.83}$$

The term $\dot{Q}(r_s)$ is the heat transfer to the droplet surface at temperature T_s, m_F is the fuel droplet instantaneous mass, $c_{p,F}$ is the fuel-specific heat, and h_{fg} is the fuel latent heat. A portion of the heat transfer from the ambient is used to heat the vapor film above the droplet temperature T_s, since the temperature of the vapor boundary layer increases from $T = T_s$ at $r = r_s$ to the ambient gas temperature $T = T_\infty$ at $r = r_s + \delta$. It is assumed that the maximum droplet temperature is the saturation temperature T_{sat} which increases with increasing ambient pressure p.

A spherical control volume in the vapor film is used to obtain the temperature profile through the vapor boundary layer. The control volume inner radius is the droplet surface r_s and the outer radius is at a radius r. The heat transfer from the ambient into the control volume is denoted $\dot{Q}_{in}$, and $\dot{Q}(r_s)$ is the heat transfer out of the control volume into the fluid droplet at r_s. There is also vapor flow into the control volume from the fluid droplet at the surface r_s, and vapor flow out of the control volume at radius r.

$$\dot{Q}_{in} - \dot{Q}(r_s) = \dot{m}_F[h(r) - h(r_s)] \tag{9.84}$$

We assume a laminar boundary layer, so the heat transfer from the ambient into the fuel vapor boundary layer is by conduction. Accordingly, $\dot{Q}_{in} = kA\,dT/dr$ and $h = c_p T$. Solving for the temperature gradient dT/dr in the vapor film in Eq. (9.84),

$$\frac{dT}{dr} = \frac{\dot{Q}(r_s) + \dot{m}_F c_p (T(r) - T_s)}{4\pi k r^2} \tag{9.85}$$

Separating variables and integrating Eq. (9.85) across the boundary layer from $r = r_s$ to $r = r_s + \delta$,

$$\ln\left[1 + \frac{\dot{m}_F c_p (T_\infty - T_s)}{\dot{Q}(r_s)}\right] = \frac{\dot{m}_F c_p}{4\pi} \frac{\delta}{r_s (r_s + \delta) k} \simeq \frac{\dot{m}_F c_p}{\mathbf{h} A_s} \tag{9.86}$$

In Eq. (9.86), we define the laminar boundary layer heat transfer coefficient **h**,

$$\mathbf{h} = \frac{k}{\delta} \tag{9.87}$$

We have also assumed $r_s \gg \delta$, and incorporated the droplet surface area $A_s = 4\pi r_s^2$. Defining the parameter z as

$$z = \frac{\dot{m}_F c_p}{\mathbf{h} A_s} \tag{9.88}$$

and solving Eq. (9.86) for the heat transfer to the droplet $\dot{Q}(r_s)$ results in

$$\dot{Q}(r_s) = \frac{z}{e^z - 1} \mathbf{h} A_s (T_\infty - T_s) \tag{9.89}$$

Note that the superheating of the vapor by the hot ambient gas as the vapor moves from the droplet surface out to the ambient reduces the heat transfer to the droplet $\dot{Q}(r_s)$ by the fraction $z/(e^z - 1)$.

The rate of vaporization $\dot{m}_F$ of the fuel droplet is a function of the bulk gas motion and the fuel vapor diffusion. Fick's law of vapor diffusion is given in Eq. (9.90). The first term is the bulk motion of the gas, and the second term is an additional component representing the fuel vapor diffusion proportional to the concentration gradient.

$$\dot{m}_F = Y_F \dot{m} - \rho_g \mathcal{D} A_s \frac{dY_F}{dr} \tag{9.90}$$

The parameter Y_F is the fuel vapor mass fraction, $\mathcal{D}$ is the mass diffusivity of fuel vapor in air, $\rho_g = P/(RT)$ is the gas density surrounding the droplet, and A_s is the droplet surface area equal to $4\pi r_s^2$. If we assume the bulk motion $\dot{m}$ is due to the fuel vapor motion $\dot{m}_F$, i.e., $\dot{m} = \dot{m}_F$, the fuel species conservation equation becomes

$$\begin{aligned}\dot{m}_F &= Y_F \dot{m}_F - \rho_g \mathcal{D} 4\pi r^2 \frac{dY_F}{dr} \\ &= -\frac{4\pi r^2 \rho_g \mathcal{D}}{1 - Y_F} \frac{dY_F}{dr}\end{aligned} \tag{9.91}$$

Separating variables,

$$\frac{dY_F}{1 - Y_F} = -\left[\frac{\dot{m}_F}{4\pi \rho_g \mathcal{D}}\right] \frac{dr}{r^2} \tag{9.92}$$

Equation (9.92) can be integrated across the boundary layer from the surface of the droplet at $r = r_s$ to the ambient free stream gas at $r = r_s + \delta$, resulting in

$$\ln\left[\frac{1 - Y_{F,\infty}}{1 - Y_{F,s}}\right] = \left[\frac{\dot{m}_F}{4\pi \rho_g \mathcal{D}}\right] \frac{\delta}{r_s(r_s + \delta)} \tag{9.93}$$

If the mass transfer coefficient $\mathbf{h}_m$ for the vapor laminar boundary layer is defined as

$$\mathbf{h}_m = \frac{\mathcal{D}}{\delta} \tag{9.94}$$

and assuming $\delta / r_s \ll 1$, then Eq. (9.93) becomes

$$\dot{m}_F = 4\pi r_s^2 \rho_g \mathbf{h}_m \ln\left[\frac{1 - Y_{F,\infty}}{1 - Y_{F,s}}\right] \tag{9.95}$$

Again defining the Spalding mass transfer number B_M (see Eq. ((9.19)) as

$$B_M \equiv \frac{Y_{F,s} - Y_{F,\infty}}{1 - Y_{F,s}} \tag{9.96}$$

and

$$1 + B_M = \frac{1 - Y_{F,\infty}}{1 - Y_{F,s}} \tag{9.97}$$

Upon substitution of Eq. (9.97), the rate of evaporation, Eq. (9.95), becomes

$$\begin{aligned}\dot{m}_F &= A_s \rho_g \mathbf{h}_m \ln[1 + B_M] \\ &= 2\pi r_s \rho_g \mathcal{D}\, Sh \ln[1 + B_M] \\ &= 2\pi r_s (k/c_p)\, Nu \ln[1 + B_T]\end{aligned} \tag{9.98}$$

Since the Lewis number is assumed to be equal to one, the Sherwood and Nusselt numbers are equal, and $B = B_T = B_M$. Note that the droplet evaporation mass flow rate in a convective flow differs from that in the quiescent case, Eq. (9.22), by the inclusion of the Sherwood or Nusselt number.

Similarly, the droplet evaporation coefficient β_v for a convective environment can be expressed using either the Sherwood or Nusselt numbers as

$$\beta_v = \frac{4k\,Sh}{\rho_l c_p}\ln(1+B_M) = \frac{4k\,Nu}{\rho_l c_p}\ln(1+B_T) \tag{9.99}$$

and the droplet lifetime t_v is

$$t_v = \frac{d_o^2}{\beta_v} \tag{9.100}$$

For a combustion environment, the droplet combustion coefficient β_b and the droplet lifetime t_b are expressed as

$$\beta_b = \frac{4k\,Nu}{\rho_l c_p}\ln(1+B_T) \tag{9.101}$$

and the droplet lifetime t_b is

$$t_b = \frac{d_o^2}{\beta_b} \tag{9.102}$$

The heat and mass transfer coefficients $\mathbf{h}$ and $\mathbf{h}_m$ are obtained from the Nusselt number Nu, and the Sherwood number Sh, respectively, where

$$\begin{aligned} Nu &= \frac{2\mathbf{h}\,r_s}{k} \\ Sh &= \frac{2\mathbf{h}_m\,r_s}{D} \end{aligned} \tag{9.103}$$

Correlations for the Nusselt and Sherwood numbers are given by Ranz and Marshall (1952) as

$$\begin{aligned} Nu &= 2 + 0.6Re^{1/2}Pr^{1/3} \\ Sh &= 2 + 0.6Re^{1/2}Sc^{1/3} \end{aligned} \tag{9.104}$$

where $d = 2r_s$, and from Eq. (9.79), $Re = \rho_g u_d d/\mu_g$ is the Reynolds number evaluated at the relative velocity u_d between the drop and the gas, and $Pr = Sc$ are the molecular Prandtl and Schmidt numbers, respectively. Note that in a pure conduction situation where $Sh = Nu = 2$, the convective mass flow equation, Eq. (9.98), reduces to the quiescent environment mass flow equation, Eq. (9.22).

Following Law and Williams (1972), the Reynolds and Prandtl numbers are evaluated using properties at ambient conditions. The Reynolds number decreases during droplet evaporation and combustion due to the decrease in the droplet diameter and velocity. In Example 9.3, the average Reynolds number is based on half the initial diameter d_o, and the velocity is assumed to remain constant. In Example 9.4, the Reynolds number is based on the instantaneous droplet diameter and velocity.

Example 9.3 ***Droplet Evaporation and Combustion in a Convective Environment***

Estimate the evaporation and combustion times for the 100 μm diameter n-hexane C_6H_{14} droplet of Examples 9.1 and 9.2 if it has a constant velocity of 10 m/s in a combustion chamber. Assume $Le = 1$.

Solution

(a) Evaporation with $T_\infty = 800$ K.

From Example 9.1, the reference thermal conductivity $k = 0.0448$ W/m-K and specific heat $c_p = 2772$ J/kg-K. From the tables in Appendix B, the ambient air properties at $T_\infty = 800$ K are: gas density $\rho = 0.4354$ kg/m^3, dynamic viscosity $\mu = 3.698 \times 10^{-5}$ Pa-s, and Prandtl number $Pr = 0.709$.

The Reynolds number is based on half the initial diameter d_o, accounting for the droplet diameter decrease with vaporization,

$$Re = \frac{\rho_g u_d d_o/2}{\mu} = \frac{(0.4354)(10)(50 \times 10^{-6})}{3.698 \times 10^{-5}} = 5.88$$

The Nusselt number is

$$Nu = 2 + 0.6Re^{1/2}Pr^{1/3} = 2 + 0.6\,(5.88)^{1/2}(0.709)^{1/3} = 3.3$$

The transfer number B_T is

$$B_T = \frac{c_p(T_\infty - T_s)}{h_{\text{fg}}} = \frac{(2772)(800 - 342)}{3.35 \times 10^5} = 3.79$$

The evaporation coefficient β_v is

$$\beta_v = \frac{4k\,Nu}{\rho_l c_p}\ln(1 + B_T) = \frac{4(0.0448)(3.3)}{(664)(2772)}\ln(1 + 3.79) = 5.0 \times 10^{-7}\text{m}^2/\text{s}$$

Therefore, the 100 μm *n*-hexane droplet evaporation time is

$$t_v = \frac{d_o^2}{\beta_v} = \frac{(100 - 10^{-6})^2}{5.0 \times 10^{-7}} = 19.8 \times 10^{-3}\,\text{s} \sim 20\,\text{ms}$$

Comment: The convective environment has reduced the evaporation time from 33 to 20 ms relative to a quiescent environment.

(b) Combustion with $T_\infty = 300$ K, $T_f = 2200$ K

From Example 9.2, the reference thermal conductivity $k = 0.146$ W/m-K and specific heat $c_p = 4232$ J/kg-K. From the tables in Appendix B, the ambient air properties at $T_\infty = 300$ K are: gas density $\rho = 1.161$ kg/m^3, dynamic viscosity $\mu = 1.846 \times 10^{-5}$ Pa-s, and Prandtl number $Pr = 0.707$.

The Reynolds number is based on half the initial diameter d_o, accounting for the droplet diameter decrease,

$$Re = \frac{\rho_g u_d d_o/2}{\mu} = \frac{(1.161)(10)(50 - 10^{-6})}{1.846 \times 10^{-5}} = 31$$

The Nusselt number is

$$Nu = 2 + 0.6Re^{1/2}Pr^{1/3} = 2 + 0.6\,(31)^{1/2}(0.707)^{1/3} = 4.97$$

The transfer number B is

$$B = B_{O,T} = \frac{c_p(T_\infty - T_s) + Y_{O,\infty}(1/s_O)q_c}{h_{\text{fg}}}$$

$$= \frac{4232(300 - 342) + (6.58 \times 10^{-2})(4.51 \times 10^7)}{3.35 \times 10^5}$$

$$= \frac{-1.78 \times 10^5 + 2.968 \times 10^6}{3.35 \times 10^5}$$

$$= 8.32$$

The combustion coefficient β_b is

$$\beta_b = \frac{4k\,Nu}{\rho_l c_p}\ln(1 + B) = \frac{4(0.146)(4.97)}{(664)(4232)}\ln(1 + 8.32) = 2.32 \times 10^{-6}\text{m}^2/\text{s}$$

Therefore, the 100 μm *n*-hexane droplet combustion time is

$$t_b = \frac{d_o^2}{\beta_v} = \frac{(100 - 10^{-6})^2}{2.32 \times 10^{-6}} = 4.33 \times 10^{-3}\ \text{s} \sim 4\ \text{ms}$$

Comment: The convective environment has reduced the combustion time from 10 to 4 ms relative to a quiescent environment.

Example 9.4 ***Numerical Modeling of Droplet Evaporation in a Convective Environment***

What is the evaporation time for *n*-octane and *n*-heptane fuel sprays, given an initial droplet diameter of 20 μm, initial droplet velocity of 10 m/s, initial temperature of 295 K, with a combustion cylinder temperature and pressure $T_g = 600$ K, $p = 1000$ kPa? Assume the density ρ_l of *n*-octane is 690 kg/m^3 and that of *n*-heptane is 585 kg/m^3. At 1000 kPa, T_{sat}, the saturation (boiling) temperature of *n*-octane is 507 K and for *n*-heptane is 475 K.

Solution

The droplet evaporation equations are incorporated into the Matlab program `Droplet.m` listed in the text web page. Accounting for the change in droplet velocity, temperature, and diameter, the program calculates the evaporating droplet mass and temperature versus time. The inputs to the program are the fuel type (*n*-octane and *n*-heptane) which are surrogates for gasoline and diesel fuels, and a cylinder pressure *p* and temperature T_g, which are assumed to remain constant.

Given an initial droplet diameter, velocity, and temperature; cylinder pressure and temperature, the Matlab program integrates in time Eq. (9.81) for droplet velocity,

$$\frac{du_d}{dt} = -\frac{3}{8} C_d \frac{\rho_g}{\rho_l} \frac{u_d^2}{r}$$

Equation (9.83) for droplet temperature,

$$m_F c_{p,F} \frac{dT_F}{dt} = \dot{Q}(r_s) - \dot{m}_F h_{fg}$$

and Eq. (9.98) for the evaporated mass from the droplet,

$$\frac{dm_F}{dt} = A_s \rho_g \mathbf{h}_m \ln[1 + B_M]$$

Curvefits for the fuel droplet thermal transport property values k, c_p, and μ of *n*-octane and *n*-heptane as a function of temperature are given in the Appendix. The vaporized mass, diameter, and temperature profiles in time for the *n*-octane and *n*-heptane droplets are plotted in Figure 9.8. The 20 μm *n*-octane droplet evaporates in about 0.8 ms, with the *n*-heptane droplet evaporating slightly later in about 0.9 ms. For the initial conditions given, neither droplet reaches its saturation temperature before fully evaporating, with a maximum temperature of the *n*-octane droplet of about 460 K, and the maximum temperature of the *n*-heptane droplet of about 435 K.

Jeng and Deng (1996) developed a comprehensive computational model for the numerical simulation of nonspherical evaporating droplet dynamics in convective flows. Their model simulated deforming, evaporating droplet dynamic behavior when surface tension, phase change, and mass and heat transfer are influential. Their computational results on evaporating droplets indicated that the dynamics of droplet deformation are basically unaffected by the evaporation process. However, deformation effects on the evaporation rate are significant. The evaporation rates per unit area of deformed droplets are smaller (e.g., ~85%) than those of non-deformed droplets. This is due to the larger recirculation zone at the back of the deformed droplet, where

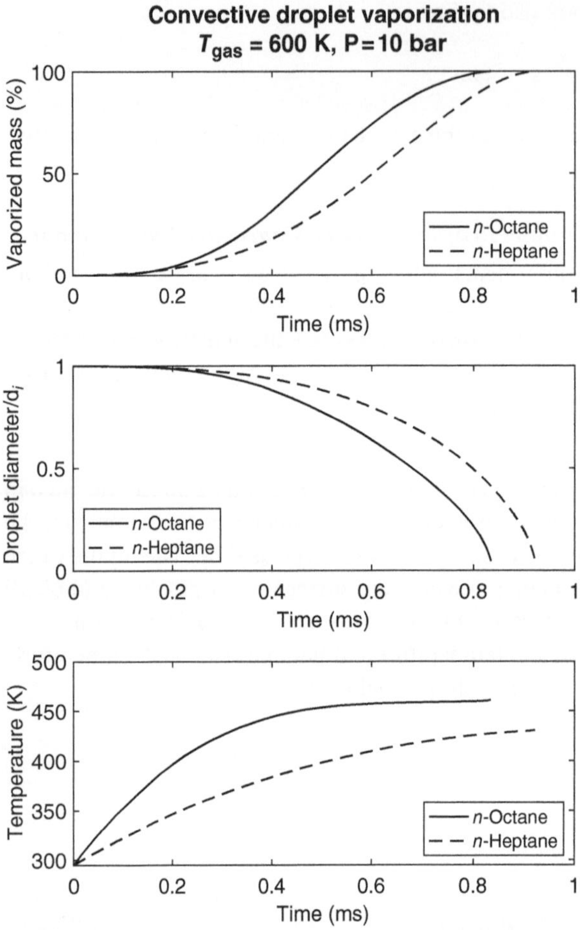

Figure 9.8 Droplet vaporization (Example 9.4).

the evaporation rate is significantly decreased. For droplets with severe deformation, more and more energy must be fed into the liquid droplet to raise its temperature. Nevertheless, the total evaporation rate of the severely deformed droplet can be larger than that of the undeformed spherical droplet, since the surface area of the deformed droplet is much larger than that of undeformed spherical droplet. A set of typical calculated results of Jeng and Deng for stripping-type breakup (Re = 36 and We = 298) are shown in Figure 9.9. In this figure, the gases are flowing from right to left.

9.5.1 Effect of Internal Circulation on Droplet Evaporation

In many practical high-pressure combustors, the Reynolds number based on the relative velocity between gas and droplet is on the order of 100 or higher. This implies that the shear stress at the gas–liquid interface can be large enough to induce internal liquid-phase circulation. The liquid motion can then be important in determining the heat and mass transfer processes within the droplet and thereby modify the vaporization rate. Prakash and Sirignano (1980) developed a

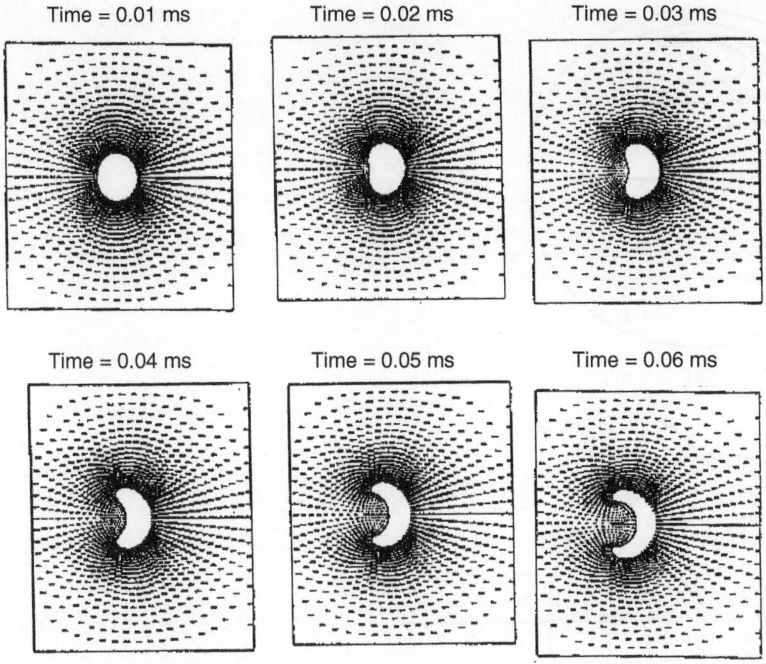

Figure 9.9 Calculated droplet deformation when it is undergoing stripping-type breakup. (Adapted from Jeng and Deng 1996.)

theoretical model for the prediction of droplet vaporization rate, considering the effect of internal circulation within a droplet.

They demonstrated that boundary layers exist in both the gas and liquid phases near the droplet surface over most of its lifetime and that convection is an important transport mode within the droplet. They considered a thin boundary layer near the surface of the droplet in the liquid phase. The liquid motion in the core region was approximated by Hill's spherical vortex as shown in Figure 9.10*a*. The third liquid-phase region considered in their physical model was an inviscid wake near the axis of symmetry shown in Figure 9.10*b*. The gas-phase motion near the surface was analyzed for three different regions: a front stagnation region, a boundary-layer region, and a separation region. For the boundary layer, the Karman–Pohlhausen integral approach (see Schlichting and Gersten 2017) was used. The matching with the front stagnation region was performed by requiring that the gradient of momentum and energy along the droplet surface be continuous. In the separation region, the boundary-layer equations cannot be used. The heat flux in this region was neglected in comparison with the fluxes in the boundary layer.

The gas-phase flow over a vaporizing droplet, rigorously speaking, is unsteady due to the temporal change in the size of the droplet. However, the characteristic time for changes in the gas phase is the residence time in the neighborhood of the droplet, and the time is usually much shorter than the droplet lifetime. Therefore, the quasi-steady gas-phase assumption was employed in Prakash and Sirignano's model. It should be noted that although the conditions in many combustors are turbulent, a locally laminar boundary layer exists over the spherical droplet surface, since the typical size of a droplet is much smaller than a typical large eddy size. This is why laminar gas-phase equations are considered in their model.

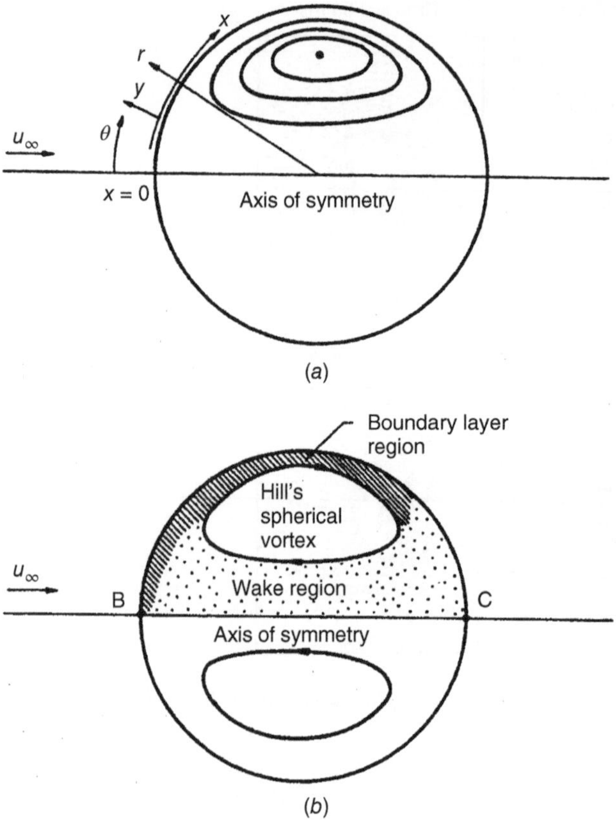

Figure 9.10 (*a*) Hill's spherical vortex and orthogonal boundary-layer coordinate system. (*b*) Modeled droplet regions. (Prakash and Sirignano 1980/with permission of Elsevier.)

In brief, for evaluating the heating of the droplet, the thermal-boundary-layer equations were solved in a way similar to the viscous-boundary-layer equations. The heating of the core has been shown to be essentially normal to the closed streamlines of the vortex, and the heating process is unsteady. In the solution of this coupled gas–liquid problem, the gas-phase boundary layer is first treated using the initial guess or the relaxed previous iterated values for surface velocity and surface temperature. With the shear stress and the heat flux from the gas-phase solution, the liquid-phase viscous and thermal boundary layer and then the thermal core are resolved. The gas-phase and liquid-phase solutions are iterated until convergence is reached. The new droplet radius corresponding to the next time step is calculated, and the process is repeated at the advanced time. Since the droplet becomes heated as time increases, the liquid viscosity and thus the strength of the Hill's vortex are updated in time. The overall vaporization rate at any instant is calculated by integrating the vaporizing flux over the droplet surface.

Three hydrocarbon fuels (n-hexane, n-decane, and n-hexadecane) evaporating in air were considered for $T = 1000$ K and $p = 10$ atm. The results for n-decane are shown in Figure 9.11, which shows the variation of $(r_s/r_{so})^{3/2}$ and the dimensionless average vaporizing mass flux. A comparison of the calculated results of Prakash and Sirignano with those based on the Ranz and Marshall (1952) correlation and the Spalding correlation (1979) is shown in this figure. The Ranz–Marshall correlation is

$$\dot{m} = \dot{m}_{ss}(1 + 0.276\,Re^{1/2}Pr^{1/3}) \tag{9.105}$$

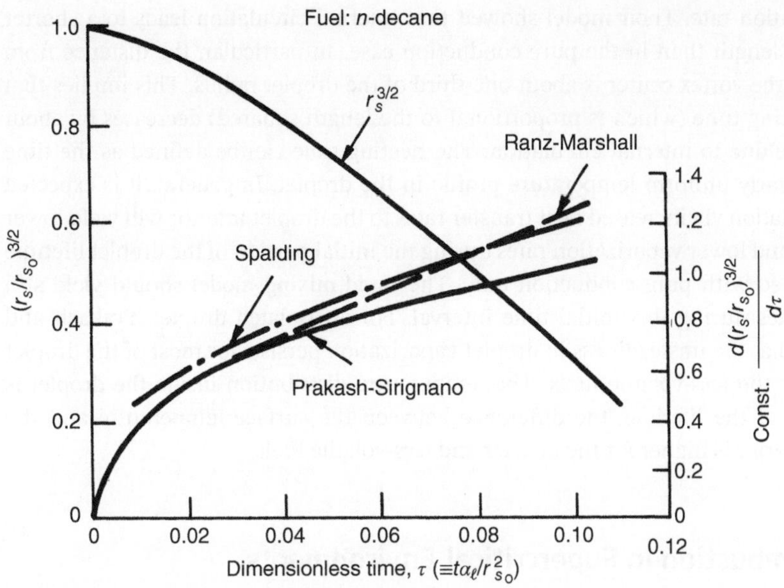

Figure 9.11 Droplet size and vaporization rate versus time. (Prakash and Sirignano 1980/with permission of Elsevier.)

where $\dot{m}_{ss}$ is the mass vaporization rate with spherically symmetric vaporization. The Spalding correlation is given in terms of the Spalding transfer number B and the Reynolds number Re,

$$\frac{\dot{m}}{4\pi r_s \mu} = 0.265 B^{3/5} Re^{1/2} \quad \text{for} \quad 0.6 < B < 5 \tag{9.106}$$

Spalding's correlation was based on experiments with Reynolds numbers between 800 and 4000, and the scatter of the experimental results was about 15–20%.

In using these correlations, the heat of vaporization was modified by Prakash and Sirignano to take into account the heat flux into the liquid phase. The modified heat of vaporization was calculated using their results for the average heat flux into the liquid phase. Figure 9.11 shows that the solution of Prakash and Sirignano is in good agreement with the Ranz–Marshall correlation during the initial part of the lifetime when the vaporization rate is small. The numerical results are lower than Spalding's correlation by about 15–20% toward the end of the lifetime of the droplet. According to Prakash and Sirignano, the difference could be due to the wake region's contribution to vaporization, which was neglected in their model. The contribution may be important when the droplet is heated. A part of the disagreement could also be due to the experimental error in the correlation.

The reason for the small variation in the overall lifetime of a single droplet is that the surface temperature for the less-volatile fuel rises quickly, during which the quasi-steady thermal boundary layer is established. After the initial establishment of the thermal boundary layer, the surface temperature rises slowly, so that the less-volatile fuel is always at a higher temperature at any time, and, hence, the droplet lifetime is not significantly affected by the differences in volatility.

According to Prakash and Sirignano (1980), the internal circulation model lies between two extreme models: (*a*) a pure conduction model without any circulation and (*b*) a rapid-mixing model. The pure conduction model has a zero vortex strength, while the rapid-mixing model has an infinite circulation rate. Their model showed that internal circulation leads to a shorter characteristic heating length than in the pure conduction case. In particular, the distance from the droplet surface to the vortex center is about one-third of the droplet radius. This implies that the characteristic heating time (which is proportional to the length squared) decreases by about an order of magnitude due to internal circulation. The heating time can be defined as the time required to reach a nearly uniform temperature profile in the droplet. In general, it is expected that the internal circulation via increased heat transfer rates to the droplet interior will yield lower surface temperatures and lower vaporization rates during the initial portion of the droplet lifetime than would be achieved with pure conduction only. The rapid-mixing model should yield still lower vaporization rates during this initial time interval. For an isolated droplet, Prakash and Sirignano concluded that the unsteadiness in droplet vaporization persists for most of the droplet lifetime, especially for the less-volatile fuels. The temperature distribution inside the droplet is nonuniform for most of the lifetime; the difference between the surface temperature and the temperature in the interior is higher for the heavier and less-volatile fuels.

9.6 Droplet Combustion in Supercritical Environments

9.6.1 Background

Very often, combustion processes take place at pressure levels well above the thermodynamic critical points of the reactant fluids. Even though the reactant liquids are usually injected initially at subcritical temperatures, they can be heated up quickly and experience a thermodynamic state transition into the supercritical regime during their lifetimes. A thermodynamic phase diagram schematic indicating the critical pressure p_c, critical temperature T_c, and the supercritical combustion regime are shown in Figure 9.12. The critical temperature and pressure values for selected fluids including cryogenic hydrogen and oxygen are given in Table 9.2.

The supercritical vaporization process exhibits many characteristic differences from those of subcritical conditions, so conventional quasi-steady analysis and results for low-pressure applications are inappropriate for supercritical combustion at high pressures. These are caused by the (1) thermodynamic non-ideality at high pressures, (2) significant transport and thermal property variations with pressure and temperature beyond the critical point, (3) ambient gas solubility into the dense "liquid" material, and (4) transient diffusion process of gas-phase species. These distinct factors render conventional approaches developed for low-pressure applications invalid.

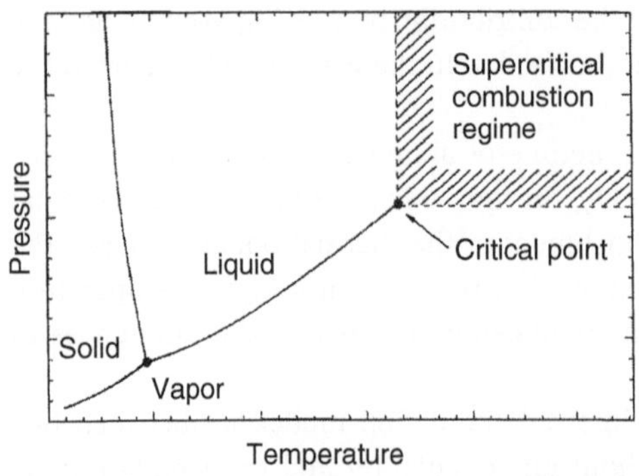

Figure 9.12 Phase diagram of a liquid fuel.

Table 9.2 Critical Properties of Selected Fluid.

Species	Chemical Formula	Critical Pressure p_c (bar)	Critical Temperature T_c (K)
Hydrogen	H_2	13.00	33.18
Nitrogen	N_2	33.98	126.19
Oxygen	O_2	50.43	154.58
Methane	CH_4	45.99	190.56
Carbon Dioxide	CO_2	73.77	304.13
Ammonia	NH_3	112.8	405.50
Ethanol	C_2H_6	63.90	516.25

Source: https://.webbook.nist.gov/

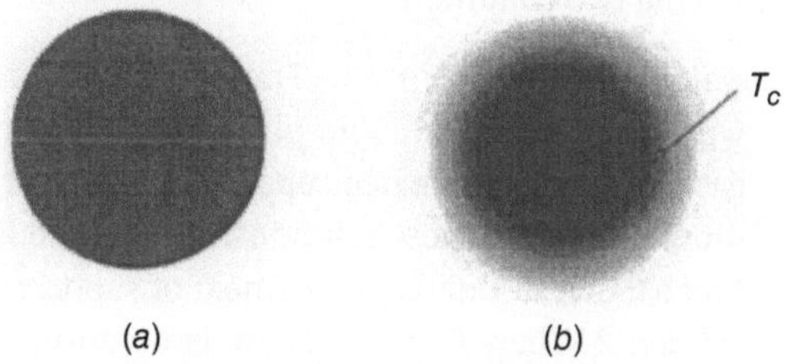

Figure 9.13 Subcritical (*a*) and supercritical (*b*) droplet vaporization. (Adapted from Yang 2000.)

As shown in Figure 9.13, the surface of a droplet under supercritical vaporization is very different from that of subcritical evaporation. The abrupt change of liquid phase to gas phase at the droplet outer radius for the subcritical condition is replaced by a gradual density change for the supercritical condition. Subcritical atomization and vaporization processes transition under supercritical conditions to diffusive mixing processes where surface tension forces diminish with increasing pressure and temperature.

The different thermodynamic behavior of two species that are being mixed also contributes to the differences between subcritical and supercritical vaporization. In the subcritical regime, the amount of hydrogen dissolved in the liquid oxygen (LOX) is quite limited, decreasing progressively with increasing temperature and reducing to zero at the boiling point of oxygen. At supercritical pressures, however, the hydrogen gas solubility becomes substantial and increases with temperature. Because of the distinct differences in the physical properties between the two species, the dissolved hydrogen modifies the liquid properties and, subsequently, the vaporization behavior. The phase equilibrium results also indicate that the critical mixing temperature T_c decreases with pressure. The peak temperature of each curve (called the critical mixing point) decreases with the increase of pressure for $p_r > 1$.

The pressure–temperature diagram in Figure 9.14 shows the phase equilibrium behavior of an O_2/H_2 mixture under different thermodynamic conditions. The boiling line is made up of the boiling points for subcritical pressures. At a fixed pressure, as the temperature increases, an equilibrium vapor–liquid mixture will transition to superheated vapor across this line. The critical mixing line indicates the variation of the critical mixing temperature T_c with pressure. It intersects the boiling line at the critical point of oxygen, the highest temperature at which the vapor and liquid phases of an O_2/H_2 binary system can coexist in equilibrium.

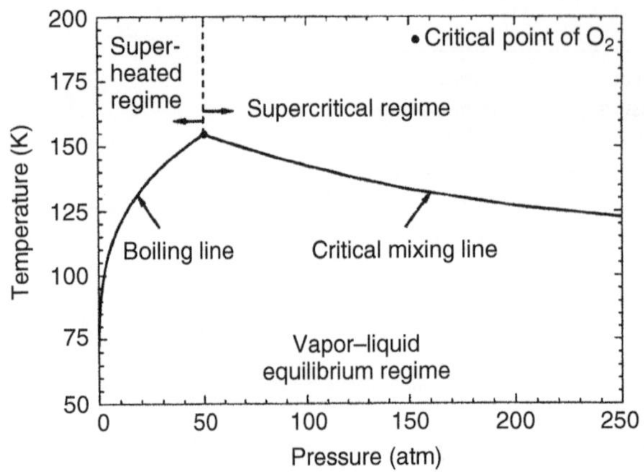

Figure 9.14 Pressure–temperature diagram for an O_2/H_2 system in equilibrium. (Yang 2000/with permission of Elsevier.)

To better understand the significance of the critical mixing temperature T_c, we can consider the following two conditions:

Condition 1 When $T_s < T_c$ – The droplet surface provides a well-defined interfacial boundary, which separates the liquid phase from the ambient gases. Under this condition, the liquid and gas phases can be treated separately and then coupled together through the relationships at the droplet surface. These relationships include liquid–vapor phase equilibrium, mass, and energy flux balance at interface. The heat of vaporization is nonzero as shown in Figure 9.15.

Condition 2 When $T_s \geq T_c$ – There is no abrupt phase change near the so-called droplet surface. In fact, the definition of the interfacial boundary is very vague. The heat of vaporization is zero as shown by the heavy dots in Figure 9.15. Under this condition, temperature and density vary continuously across the interfacial region. The interior of the droplet could still be at a subcritical condition. The regression of the droplet is best characterized by the motion of the "surface," which has a temperature equal to the critical mixing temperature of the system. The surface tension σ also vanishes under supercritical conditions.

As shown in Figure 9.15, the enthalpy of vaporization is a function of both temperature and pressure. At a given temperature, the enthalpy of vaporization decreases as P_r increases. Also, at a

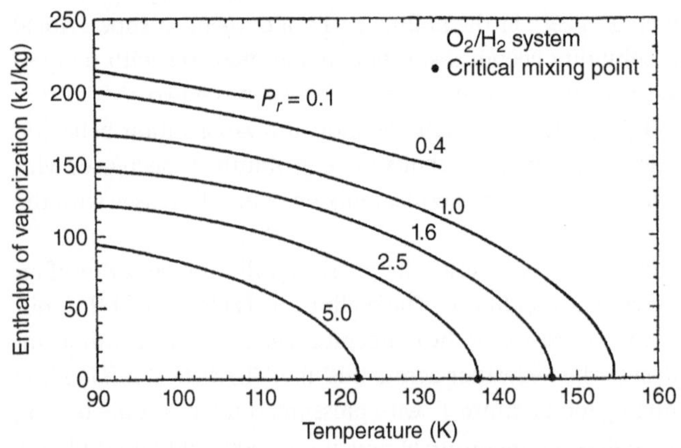

Figure 9.15 Effect of pressure on enthalpy of vaporization of O_2 in an equilibrium mixture of O_2 and H_2. (Yang et al. 1994/with permission of Taylor & Francis Group.)

fixed value of P_r, the enthalpy of vaporization (h_{fg}) decreases to zero when the temperature reaches the critical mixing point.

Spalding (1959) developed a theory to predict the rate of combustion of liquid droplets under high pressure. Rosner (1967) extended Spalding's analysis by taking into account the finite dimensions of the droplet. The basic model consists of a pocket of fuel vapor surrounded by a spherical flame zone. Outside the flame zone is an environment containing an oxidant. Fuel diffuses outward to the flame zone, and oxidant diffuses inward to the flame, where they react instantly to form combustion products.

The influences of convection, density variation, and finite-rate chemical kinetics on supercritical combustion were studied by Brzustowski (1965), Chervinsky (1969), and Polymeropoulos and Peskin (1972). Manrique and Borman (1969) found that the effects of thermodynamic non-idealities, property variations, and high-pressure corrections for phase equilibrium could influence the vaporization mechanisms significantly. Based on these findings, Canada and Faeth (1975) conducted a series of experimental and theoretical studies on droplet combustion in both stagnant and forced convective environments, with special attention focused on the high-pressure phenomena of phase equilibrium.

Reviews of supercritical droplet vaporization and combustion are given in Yang (2000) and Bellan (2000). Dahms and Oefelein (2015) and Nguyen et al. (2022) recently made important contributions to increasing our understanding the transition from a classical spray regime to a supercritical diffusion-controlled mixing regime.

As described in Appendix F, the fluid density ρ can be derived from a modified Benedict–Webb–Rubin (BWR) equation of state proposed by Jacobsen and Stewart (1973) due to its superior performance over conventional cubic equations of state. The BWR equation of state is extremely valuable in correlating both liquid and vapor thermodynamic and volumetric data; however, the temperature constants involved have been available only for a limited number of pure substances. To overcome this constraint, an extended corresponding state (ECS) principle developed by Ely and Hanley (1983) can also be used.

The basic idea of ECS is to assume that the properties of a single-phase fluid can be evaluated via conformal mappings of temperature and density to those of a given reference fluid. As a result, only the BWR constants for the reference fluid are needed. For a multicomponent system, accounting for changes in properties due to mixing is much more complicated. A pseudo pure-substance model is adopted to evaluate the properties of a mixture, treating the mixture as a single-phase pure substance with its own set of properties.

9.6.2 Droplet Vaporization in Supercritical Quiescent Environments

Research on droplet vaporization under supercritical conditions in quiescent environments has been conducted to better understand the effects of pressure and temperature on transient diffusion and interfacial thermodynamics. Both hydrocarbon droplets in air (e.g., see Harstad and Bellan 2000), and LOX droplets in hydrogen (e.g., see Yang 2000) have been the subject of extensive research. In contrast to the low-pressure case, where the enthalpy of vaporization is rate controlling, it has been found that thermal diffusion processes prevail at high pressures.

Figure 9.16 shows representative curves of droplet diameter versus time for a spherical oxygen droplet evaporating in a supercritical quiescent hydrogen environment at various pressures. These data were obtained by means of direct numerical simulations at conditions with $T_\infty = 1000\,\text{K}$; $T_0 = 90\,\text{K}$; and $D_0 = 100\,\mu\text{m}$. Note the deviation from the d^2 evaporation law as the pressure increases.

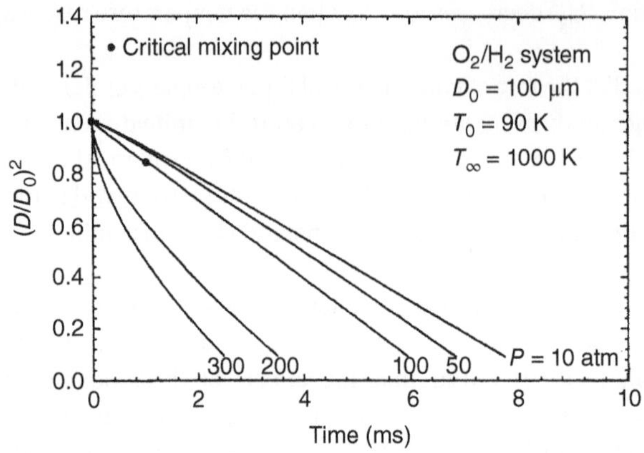

Figure 9.16 Droplet diameter versus time for a spherical oxygen droplet evaporating in a supercritical quiescent hydrogen environment at various pressures. (Adapted from Lafon et al. 1995.)

Numerical calculations of LOX droplet lifetimes in pure hydrogen have been correlated using an approximate analysis that takes into account the effect of transient heat diffusion in terms of a reduced critical temperature T_c^* (Yang 2000).

$$T_c^* \equiv \frac{T_\infty - T_c}{T_\infty - T_0} \tag{9.107}$$

The resultant expression of the droplet lifetime t_v takes the form

$$t_v = [0.0115 + 0.542(1 - T_c^*)]\frac{r_o^2}{\alpha_o}\mathcal{F}\left(\frac{\alpha_\infty}{\alpha_o}\right) \tag{9.108}$$

where r_o is the initial droplet radius, and the term α_∞/α_o represents the ratio of the thermal diffusivity of the gas mixture to that of the droplet.

The correction factor $\mathcal{F}\left(\frac{\alpha_\infty}{\alpha_o}\right)$ is

$$\mathcal{F}\left(\frac{\alpha_\infty}{\alpha_o}\right) = 1 + 3.9\left\{1 - \exp\left[-0.035\left(\frac{\alpha_\infty}{\alpha_o} - 1\right)\right]\right\} \tag{9.109}$$

The high-pressure Spalding transfer number, B_T, can be related to the mixture of critical temperature T_c and the reduced critical temperature T_c^* by the following algebraic equation:

$$B_T = \frac{T_\infty - T_c}{T_c - T_0} = \frac{T_c^*}{1 - T_c^*} \tag{9.110}$$

Thus, the correlation given in Eq. (9.108) clearly shows that pressure affects the droplet lifetime through its influence on the mixture critical temperature T_c and the ambient thermal diffusivity, α_∞. Figure 9.17 shows the dependence of oxygen droplet lifetimes on the reduced critical mixing temperature in a quiescent hydrogen environment at various pressures. At high pressures, T_c^* decreases and therefore the lifetimes of the droplets are shorter.

In general, the droplet lifetime decreases gradually with increasing pressure, and no discernible variation occurs across the critical transition. In the subcritical regime, as pressure increases, the enthalpy of vaporization decrease thereby facilitates the vaporization process and consequently leads to a decrease in droplet lifetime. The situation is different at supercritical conditions, at which the interfacial boundary disappears and the enthalpy of vaporization is zero. In this case, the decrease in droplet lifetime is mainly attributed to the increase in thermal diffusivity of the surrounding gas near the droplet surface.

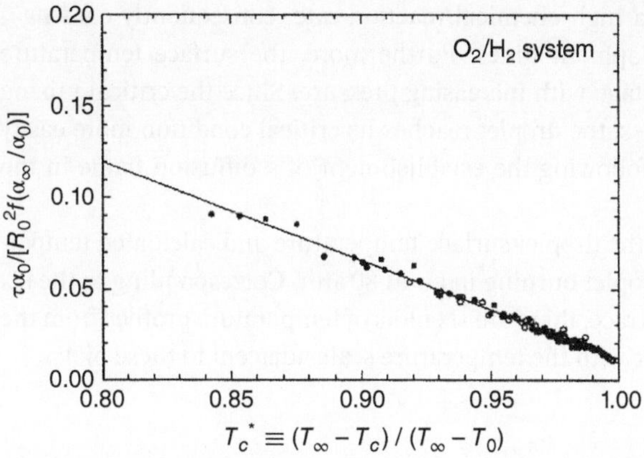

Figure 9.17 Dependency of oxygen droplet lifetime on critical mixing temperature in a quiescent hydrogen environment. (Adapted from Lafon et al. 1995.)

9.6.3 Droplet Combustion in Supercritical Quiescent Environments

As discussed in previous sections, the combustion processes of a burning droplet are very similar to those involving vaporization only. It has been found that at low pressures, the vaporization process is the primary factor controlling the duration of the combustion of a liquid droplet, whereas in a supercritical environment, the gas-phase diffusion process plays a more important role.

The combustion of a hydrocarbon (e.g., *n*-pentane) fuel droplet at various pressures in air was studied by Shuen et al. (1992) assuming an initial droplet temperature of 300 K and ambient air temperature of 1000 K. Figure 9.18 shows the time history of the calculated droplet surface temperature for a 100 μm droplet. Once ignition is achieved in the gas phase, energy feedback from the flame causes a rapid increase in droplet surface temperature.

At low pressures ($p \leq 20$ atm), the surface temperature varies very slowly following the onset of flame development and approximately levels off at the pseudo wet-bulb temperature. As the ambient pressure increases, the high concentrations of oxygen in the gas phase and the fuel

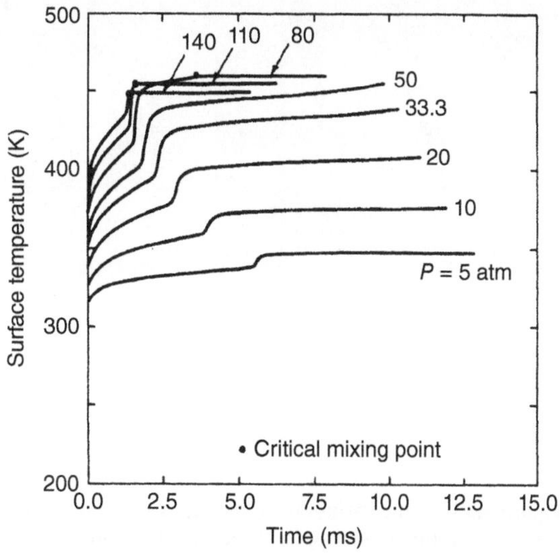

Figure 9.18 Time variations of *n*-pentane droplet surface temperature at various pressures, combustion occurring in a hot-air environment. (Shuen et al. 1992/with permission of Elsevier.)

vapor from the droplet surface result in a high chemical reaction rate, consequently causing a decrease in the ignition and flame-development times. Furthermore, the surface temperature increases during the flame-development stage with increasing pressure. Since the critical mixing temperature decreases as pressure increases, the droplet reaches its critical condition more easily at higher pressures, almost immediately following the establishment of a diffusion flame in the gas phase for $p \geq 110$ atm.

Figure 9.19 shows the time variation of the droplet surface temperature and calculated temperature distribution around an n-pentane droplet burning in air at 80 atm. Corresponding to the six time locations on the surface temperature trace, there are six plots of temperature profiles from the center of the droplet to the surrounding air with the temperature scale adjacent to these plots.

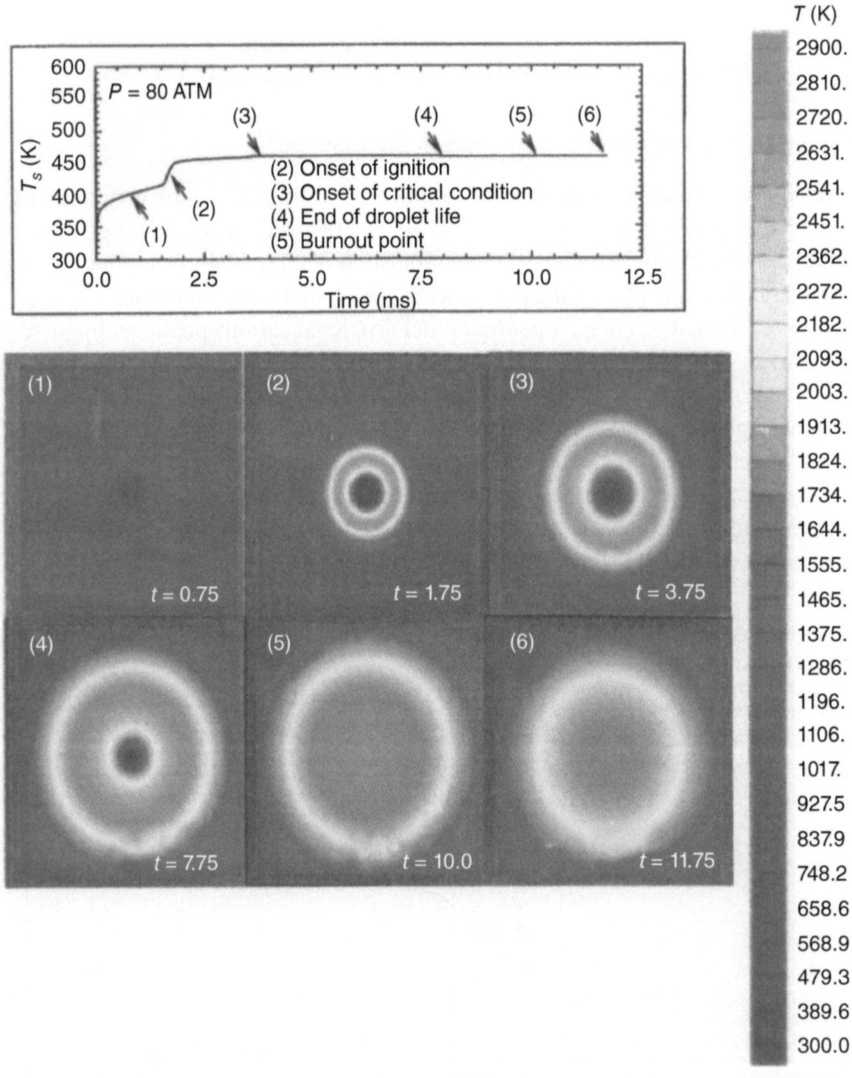

Figure 9.19 Time variations of droplet surface temperature and temperature distributions around an n-pentane droplet burning in air at 80 atm. (Shuen et al. 1992/with permission of Elsevier.)

The vaporization lifetime decreases continuously as pressure increases partly due to the decay of the critical mixing line, whereas the single-phase vapor combustion lifetime increases progressively with pressure due to the reduction of species mass diffusion at higher pressures. Furthermore, as discussed above, the enthalpy of vaporization of the droplet is a function of pressure and temperature. Due to these opposite effects, the pressure dependence of the burnout time and combustion lifetime both exhibit non-monotonic behavior.

The combustion behavior of a large droplet has several characteristics distinct from those of a small droplet. First, as mentioned above, ignition for large droplets occurs in the very early stage of the entire droplet lifetime. The influence of vaporization prior to ignition on the overall burning mechanism appears to be quite limited. Second, the combustion lifetime decreases with increasing pressure at lower pressures, reaching a minimum near the critical pressure of the liquid fuel (~33 atm). As the pressure further increases, the combustion time increases due to the reduction of mass diffusivity at high pressures. The droplet burning lifetime decreases continuously with pressure, whereas the single vapor-phase combustion lifetime increases progressively with pressure.

9.6.4 Droplet Vaporization in Supercritical Convective Environments

As discussed earlier in this chapter, when a droplet is introduced into a convective environment, the forced convection results in increases of heat and mass transfer between the droplet and surrounding gases, which consequently enhances the droplet vaporization process. Extensive research has been conducted to examine droplet vaporization in supercritical forced convection, including the effects of pressure and free-stream velocities on droplet dynamics. Hsiao et al. (1995) and Yang (2000) developed a comprehensive model of liquid-oxygen (LOX) droplet vaporization in a supercritical hydrogen stream, covering a pressure range of 100–400 atm. The model takes into account multidimensional flow motions and enables a thorough examination of droplet behavior during its entire lifetime, including dynamic deformation, viscous stripping, and secondary breakup. Detailed flow structures and thermodynamic properties are obtained to reveal mechanisms underlying droplet vaporization as well as deformation and breakup dynamics.

Figure 9.20 shows six frames of isotherms and isopleths of oxygen concentration in a droplet in a hydrogen stream with a velocity of 2.5 m/s and an ambient pressure of 100 atm. The free-stream Reynolds number $Re_d = 31$ based on the initial droplet diameter. Soon after the introduction of the droplet into the hydrogen stream, the flow adjusts to form a boundary layer near the surface. The gasified oxygen is carried downstream both by convection and mass diffusion.

The evolution of the temperature field exhibits features distinct from that of the concentration field due to the different timescales associated with the thermal and mass diffusion processes (i.e., Lewis number $\neq 1$). The thermal wave penetrates into the droplet interior faster than the surrounding hydrogen species. Since the liquid core possesses a greater inertia and moves slower compared to the gasified oxygen, at $t = 0.79$ ms, the droplet (delineated by the dark region in the temperature contours) reveals an olive shape while the oxygen concentration contours deform into a crescent shape with the edge bent to the streamwise direction. At $t = 1.08$ ms, the subcritical liquid core disappears.

Figure 9.21 shows a numerical simulation of the streamline patterns and oxygen concentration contours of four different modes commonly observed in supercritical droplet vaporization (Hsiao et al. 1995). Depending on the local flow conditions, the droplet can remain in a spherical configuration, deform to an olive shape, or even break up into fragments. Unlike the low-pressure cases in which the large shear stress at the gas–liquid interface induces internal flow circulation in the liquid core (Prakash and Sirignano 1980), no discernible recirculation takes place in the droplet

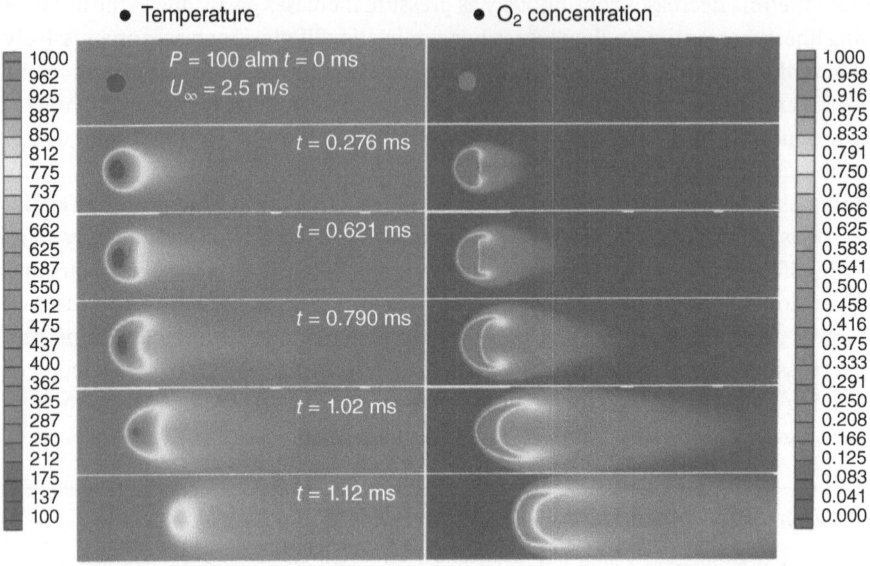

Figure 9.20 Calculated LOX droplet vaporization in a supercritical hydrogen flow, $p = 100$ atm, $u_\infty = 2.5$ m/s. (Yang 2000/with permission of Elsevier.)

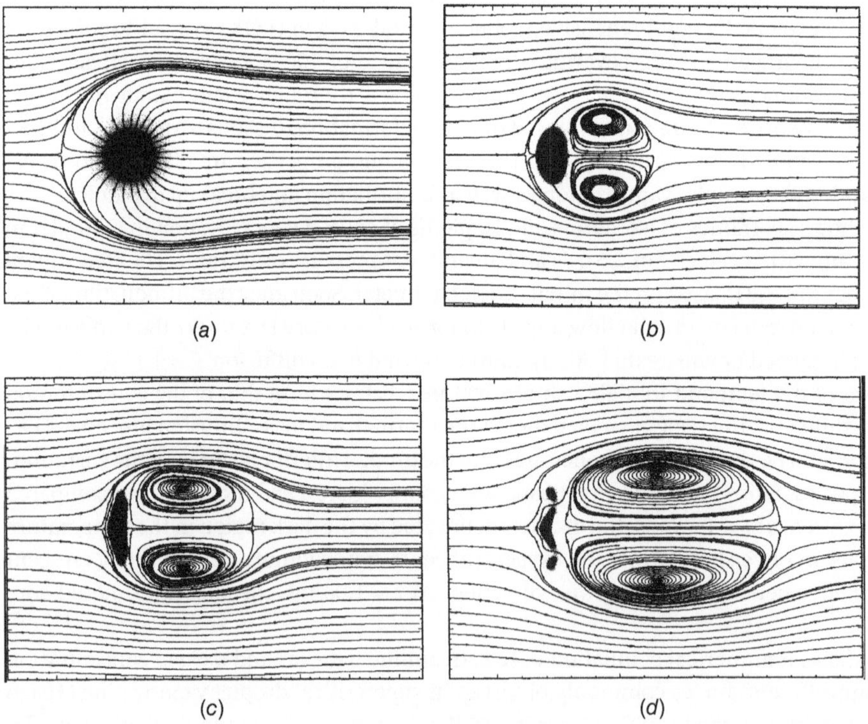

Figure 9.21 LOX Droplet vaporization in supercritical hydrogen flow at $p = 100$ atm. (*a*) Spherical mode; $u_\infty = 0.2$ m/s, $t = 0.61$ ms. (*b*) Deformation mode; $u_\infty = 1.5$ m/s, $t = 0.61$ ms. (*c*) Stripping mode; $u_\infty = 5$ m/s, $t = 0.17$ ms. (*d*) Breakup mode; $u_\infty = 15$ m/s, $t = 0.17$ ms. (Adapted from Hsiao et al. 1995.)

interior (shown as a dark region), regardless of the Reynolds number and deformation mode. This may be attributed to the diminishment of surface tension at supercritical conditions. In addition, the droplet regresses so fast that there may not be enough time to develop an internal vortical flow before the droplet evaporates completely. The rapid deformation of the droplet configuration further precludes the existence of stable shear stress in the liquid core and consequently obstructs the formation of recirculation.

The spherical mode shown in Figure 9.21*a* typically occurs at very low Reynolds numbers. No recirculating eddy is found in the wake behind the droplet. The vorticity generated is too weak to form any confined eddy. When the ambient velocity increases to 1.5 m/s, the droplet deforms into an olive shape with a recirculating ring attached behind it, as shown in Figure 9.21*b*. Owing to the droplet deformation and vaporization, the threshold Reynolds number above which the recirculating eddy forms is considerably lower than that for a hard sphere. Figure 9.21*c* depicts the flow structure with viscous stripping at an ambient velocity of 5 m/s, showing an oblate droplet with a stretched vortex ring. The flattened edge of the droplet enhances the strength of the recirculating eddies and as such increases the viscous shear stress drastically. Consequently, a thin sheet of fluid is stripped off the edge of the droplet and swept toward the outer boundary of the recirculating eddy. At a very high ambient velocity of 15 m/s, droplet breakup takes place, as clearly shown in Figure 9.21*d*. The hydrogen flow penetrates through the liquid phase and divides the droplet into two parts: the core disk and surrounding ring. The vortical structure in the wake region expands substantially as a result of the strong shear stress.

The effect of ambient pressure and Reynolds number on droplet lifetime t_v has been correlated with respect to a reference value at zero Reynolds number (Hsiao et al. 1995). The result takes the form

$$\frac{t_v}{t_{Re_d} = 0} = \frac{1}{1 + 0.17\, Re_d{}^{1.1}\, (p_{r,O_2})^{-0.88}} \tag{9.111}$$

where Re_d and p_{r,O_2} refer to the Reynolds number based on the initial droplet diameter and the reduced pressure ($p_r = p/p_c$) of oxygen, respectively. This correlation bears a resemblance to the popular (Ranz and Marshall 1952) correlation for droplet convection heat transfer. However, the Ranz and Marshall correlation applies only to low-pressure flows and has a weaker Reynolds number dependency.

The drag coefficient has been generally adopted as a dimensionless parameter to measure the drag force acting on a droplet. Yuen and Chen (1976) found that the drag coefficient of an evaporating droplet is smaller than that of a non-vaporizing solid sphere at the same Reynolds numbers. Several researchers, Renksizbulut and Haywood (1988), and Chiang et al. (1983), numerically analyzed evaporating droplet motion and proposed the following correlation,

$$C_d = \frac{C_d^o}{(1+B)^b} \tag{9.112}$$

where C_d^o denotes the drag coefficient for a hard sphere, and the exponent b has a value of 0.2 for Renksizbulut and Haywood's model and 0.32 for Chiang's correlation. For droplet vaporization at low to moderate reduced pressures ($p_r \leq 0.5$), the Spalding transfer number B is widely used to account for the effect of evaporation on the momentum transfer to the droplet,

$$B = \frac{c_p(T_\infty - T_s)}{h_{fg}} \tag{9.113}$$

However, the enthalpy of vaporization h_{fg} becomes zero at the critical point, rendering an infinite value for the transfer number. This deficiency may be remedied by introducing a transfer number B_d

for supercritical droplet vaporization, similar to the non-convective transfer number, B_T, previously introduced in Eq. (9.110) (Hsiao et al. 1995).

$$B_d = \frac{T_\infty - T_c}{T_c - T_d} \tag{9.114}$$

where T_d is the instantaneous average temperature of the droplet, and T_c the critical mixing temperature. Since B_d diverges at $T_c = T_d$ at the end of the droplet lifetime, the calculation of drag force was terminated when $(T_c - T_d)$ becomes less than 1 K, at which the droplet residual mass is usually less than one-thousandth of the initial mass. Following the procedure leading to Eq. (9.112), Hsiao et al. obtained a correlation for an LOX droplet drag coefficient,

$$C_d = \frac{C_d^o}{(1 + aB_d)^b} \tag{9.115}$$

where a and b are selected to be 0.05 and $1.592(p_{r,O_2})^{-0.7}$, respectively. The data clusters along the classical drag curve, Eq. (9.112), in the low Reynolds-number region but deviates considerably at high Reynolds numbers (i.e., $Re_d > 10$). Although a shape factor may be employed to account for this phenomenon arising from the increased form drag due to droplet deformation, the difficulty of calculating this factor and conducting the associated data analysis preclude its use in correlating the drag coefficient herein. Instead, a simple correction factor $Re_d^{0.3}$ is incorporated into Eq. (9.115), i.e.,

$$C_d = \frac{C_d^o Re_d^{0.3}}{(1 + aB_d)^b}, \qquad 0 < Re_d < 300 \tag{9.116}$$

9.7 Application: Compression Ignition Engine Combustion

9.7.1 Background

A compression ignition (Diesel) engine has separate fuel and air streams that combust as they are mixed together at temperatures greater than the autoignition temperature. In this type of internal combustion engine, a low volatility fuel must be converted from a liquid state into finely atomized droplets, which are vaporized, mixed with air, and raised to a temperature to support autoignition. The autoignition temperature of vehicular diesel fuel (2-D) is about 530 K, about 110 K less than that of gasoline. One of the limiting features of compression ignition engines is the finite mixing rate of the fuel and air, especially for higher speed diesel engines, so a variety of direct injection and indirect injection systems have been developed.

With direct injection, the fuel is injected into the cylinder after the intake air has been compressed to about 50 bar and a temperature high enough (>850 K) for autoignition of the fuel–air mixture. The fuel flows through very small orifices in the injector tip, forming a liquid jet which subsequently breaks up into droplets that evaporate and mix with the surrounding air. Direct injection systems operate at high pressures, up to 3000 bar, to ensure that the fuel velocity will be high enough to penetrate deeply into the cylinder, and the atomized droplets will be small enough for rapid evaporation and subsequent ignition during the time available before the piston reaches top dead center. Distribution of the fuel in the cylinder is accomplished by both penetration along the stream-wise flow direction and turbulent entrainment perpendicular to the flow.

The compression engine combustion process has been classified into three phases: ignition delay, premixed combustion, and mixing-controlled combustion. The ignition delay period τ_{id} is defined as the time or crank angle between the SOI and the start of combustion, which includes the droplet formation, atomization, and mixing processes. The ignition delay depends not only on the fluid mechanics of atomization, vaporization, and mixing but also on the chemical characteristics of the fuel.

Aromatic hydrocarbons and alcohols have strong chemical bonds, resulting in a long ignition delay. If these fuels mix completely with air before autoignition occurs, when ignition does occur, they will all burn rapidly in the premixed phase, producing a large rate of change of pressure and a high peak pressure. On the other hand, the chemical bonds of some fuels, such as alkanes (straight chain paraffins), are easily broken. Ignition delay is then short, and with a long injection, most of the fuel to be burned is injected after the start of ignition occurs. Relatively little fuel burns in the premixed phase and most of the fuel burns in the diffusion phase at a rate limited by the rate of mixing with the cylinder air.

Ignition delay data from have been correlated by Arrhenius equations of the general form given by Eq. (9.117):

$$\tau_{\text{id}} = Ap^{-n} \exp\left(\frac{E_a}{R_u T}\right) \tag{9.117}$$

The terms A and n are constants which depend on the fuel type, injection parameters, and cylinder airflow characteristics.

A correlation, Eq. (9.118), by Hardenberg and Hase (1979) for direct injection engines is widely referenced. The correlation includes the cetane number CN, cylinder pressure p_{tdc} and temperature T_{tdc} at top dead center, and mean piston speed $\overline{U}_p$. In Eq. (9.118), the ignition delay θ_{id} is in units of crank angle degrees, piston speed in m/s, pressure in bars, and temperature in degrees Kelvin.

$$\theta_{\text{id}} = (0.36 + 0.22\overline{U}_p) \exp\left[E_a\left(\frac{1}{R_u T_{\text{tdc}}} - \frac{1}{17,190}\right)\left(\frac{21.2}{p_{\text{tdc}} - 12.4}\right)^{0.63}\right] \tag{9.118}$$

The term E_a is the apparent activation energy in J/mol,

$$E_a = \frac{618,840}{\text{CN} + 25} \tag{9.119}$$

The mean piston speed $\overline{U}_p$ is twice the product of the engine speed N (rpm) and stroke s (m),

$$\overline{U}_p = 2\frac{N}{60}s \tag{9.120}$$

Assuming a polytropic process with exponent n and compression ratio r, the temperature and pressure at top dead center are:

$$\begin{aligned} T_{\text{tdc}} &= T_i\, r^{n-1} \\ p_{\text{tdc}} &= p_i\, r^n \end{aligned} \tag{9.121}$$

The premixed combustion phase begins once the regions of fuel vapor–air mixture that forms around the fluid jet as it is first injected into the cylinder are at or above the autoignition temperature. The combustion heat release of this initial transient vapor–air mixture depends on the length of the ignition delay.

The mixing-controlled combustion reaction, which produces a diffusion flame, takes place at the interface between the fuel and the air. Combustion in this phase occurs at a rate limited by the rate at which the fuel can be mixed with the entrained air. The energy release begins at a relatively high value and then decreases as the available oxygen is depleted.

Compression engine energy release profiles (see Figure 9.22) will typically have two maxima, resulting from the premixed and the diffusion combustion phases. The first peak occurs during the

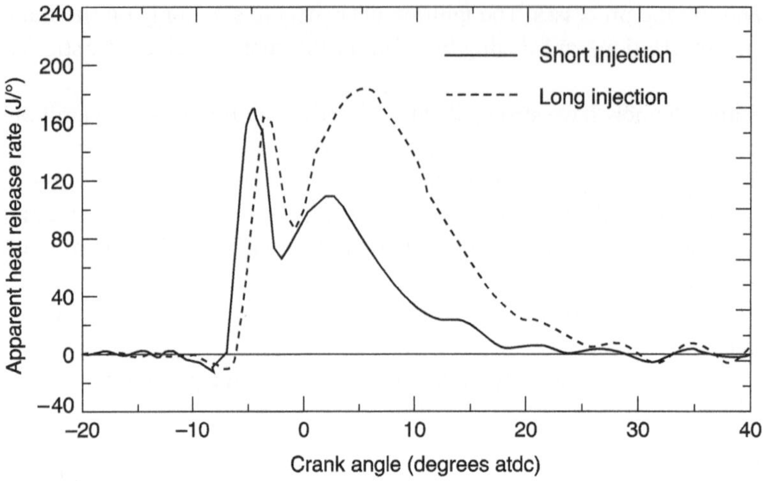

Figure 9.22 Energy release profiles for short- and long-duration fuel injection. (Adapted from Dec 1997.)

premixed combustion phase and results from the rapid combustion of the portion of the injected fuel that has vaporized and mixed with the air during this period. The energy release curve in the premixed combustion phase is relatively independent of the load, since the initial mixing is independent of the injection duration. The second peak occurs during the diffusion combustion. The energy release and burn duration during this phase depends on the injection duration. As the injection duration is increased to meet an increased engine load, the amount of fuel injected increases, thus increasing the magnitude and duration of the diffusion combustion energy release.

A single-cylinder diesel engine modified for optical access by researchers at Sandia National Laboratory (Dec and Espey 1995) was used for visualization of the compression engine combustion process. Figure 9.23 shows a high-speed cinematography sequence of the combustion process in the above test engine. The high-speed photo sequence shows the fuel jet penetration and spread of the luminous combustion zones. The SOI is at 8.8°btdc, and each image is about 2.4°apart. The first luminosity is seen about 5°after the start of injection (ASI). The rapid appearance of widespread combustion indicates that the ignition occurs at multiple points throughout the jet. The luminosity is yellow, indicating the presence of hot soot particles and suggesting fuel-rich combustion. Between 2.9°and 1.8°btdc, the burning fuel jets contact the edge of the combustion bowl, and then spread along the circumference into the space between the jets and downward into the bottom of the bowl. The combustion heat release ends at about 17°atdc. In this case, about half of the fuel is injected after the burning fuel jet reaches the edge of the bowl.

Rapid compression machines (RCMs) have also been used to study compression ignition processes and verify proposed chemical kinetic mechanisms. The RCM has been used for many years to study autoignition phenomena and simulate the compression stroke of a compression ignition combustion engine (Lee and Hochgreb 1998). With an RCM, a reactive mixture is introduced into a chamber, rapidly compressed by a moving piston, and the resulting temperature and pressure profiles measured.

9.7.2 Diesel Combustion Analysis

The fuel–air mixture in compression ignition engines is not homogeneous, since the injected fuel is in two phases and is distributed nonuniformly within the cylinder. Multi-zone models

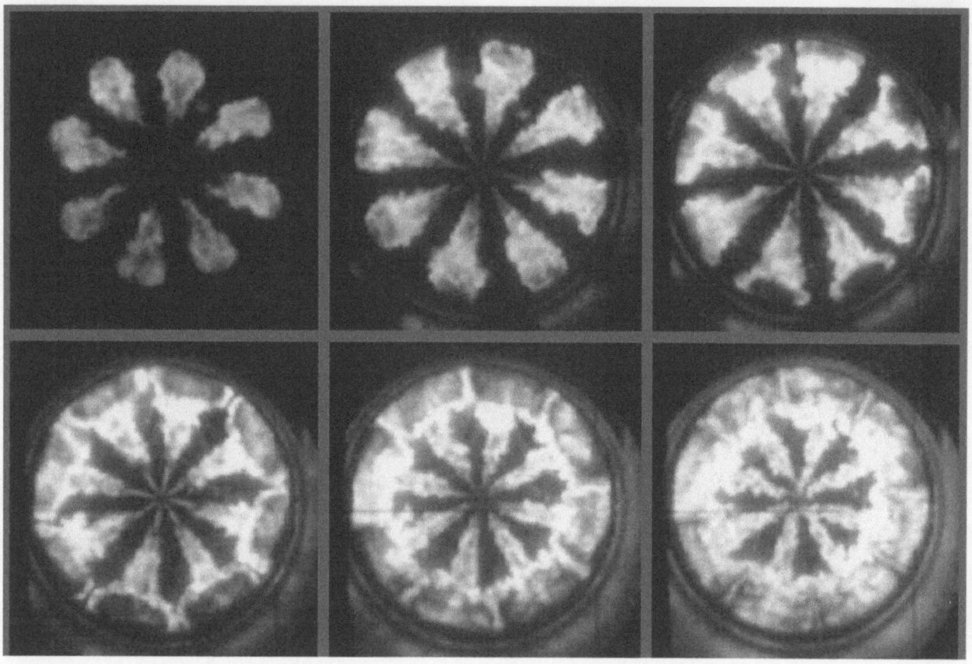

Figure 9.23 High-speed photographic sequence of the luminosity of a diesel flame. (Adapted from Espey and Dec 1993.)

(see Dobos and Kirkpatrick 2017) are used to take the heterogeneous character of diesel spray combustion into account. The models discretize the fuel spray into a series of finite packets that break into droplets, evaporate, mix with combustion air, and ignite. These models are also designated as quasi-dimensional, since the exact cylinder geometry is not modeled.

With multi-zone modeling, momentum conservation is used to determine the entrainment rate, and empirical correlations are used to calculate the penetration and breakup of the diesel spray. Chemical kinetic models or empirical correlations for ignition delay are used to calculate the start of combustion. With the use of chemical kinetic equations, no model calibrations are needed to match heat release and pressure data, and the evolution of pollutant species such as NO_x and soot during the combustion of the fuel spray can also be calculated.

Early analyses of diesel combustion assumed that a burning diesel jet was composed of a dense fuel-rich core surrounded by a uniformly leaner fuel–air mixture, as shown in Figure 9.24. With reference to steady-state spray combustion in furnaces and gas turbines, the diesel fuel autoignition and premixed combustion phases were also assumed to occur in a diffusion flame in the near stoichiometric ($\phi \sim 1$) regions between the rich $\phi > 1$ and the lean $\phi < 1$ limits, at the inner and outer edges of the jet. Soot was assumed to form in a narrow region on the fuel-rich side of the diffusion flame.

Recent laser sheet diagnostic experiments in diesel engines have indicated that the combustion process in diesel engines is different than that in furnaces and gas turbines. Dec (1997) has proposed an alternative conceptual model based on laser sheet experimental results. The Dec model features two stages of fuel oxidation for both the premixed and mixing-controlled combustion phases. The first stage is partial oxidation of the fuel in a rich premixed reaction, and the second stage is combustion of the fuel rich, partially oxidized products of the first stage in a near stoichiometric diffusion flame.

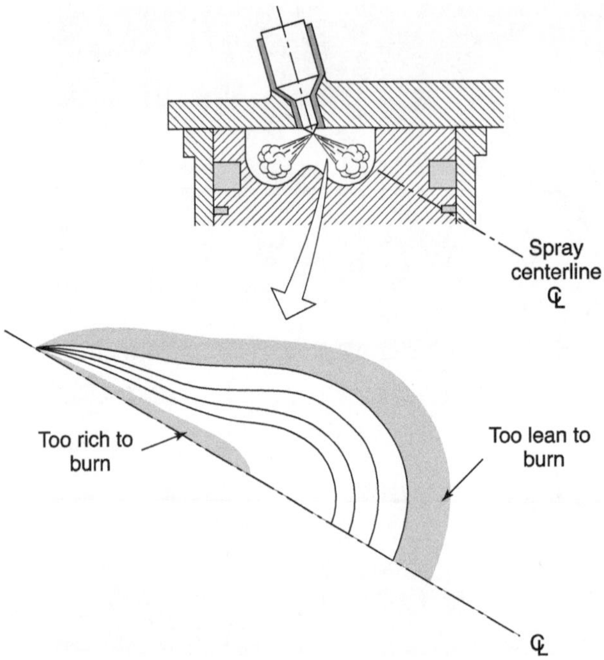

Figure 9.24 Simple model of diesel combustion.

This conceptual model is shown schematically in Figure 9.25, a temporal sequence showing the progressive changes during the injection process. Significant events in the evolution of the jet state are drawn at successive degrees ASI. Six parameters are shown in Figure 9.25: the liquid fuel, the vapor–air mixture, the PAHs, the diffusion flame, the chemiluminescence emission region, and the soot concentration.

At 1.0°in Figure 9.25, near the beginning of the ignition delay phase, as the liquid fuel is injected into the cylinder, it entrains hot cylinder air along the sides of the jet, leading to fuel evaporation. Note that throughout the injection process, the liquid length portion of the jet remains relatively constant. There is limited penetration of the fuel droplets into the combustion chamber. The penetration depth of the liquid jet has been found to be dependent on the volatility of the fuel, injector hole size, fuel and cylinder air temperature, and relatively insensitive to the injection pressure.

At about 4.0°ASI, a vapor head vortex is beginning to form in the leading portion of the jet downstream of the liquid jet. The bulk of the vaporized fuel is in the head of the jet. The fuel vapor–air mixture region in the head vortex is relatively uniform, has a well-defined boundary separating it from the surrounding air, and has an equivalence ratio between 2 and 4 throughout its cross section. At about 5.0°ASI, premixed combustion begins in the head vortex. As a consequence of the high equivalence ratio, the initial premixed combustion is fuel rich with a temperature of about 1600 K, and produces PAHs and soot. The soot concentration is fairly uniform throughout the jet cross section.

At about 6.5°ASI, a turbulent diffusion flame forms at the edge of the jet around the products of the initial premixed stage. This turbulent diffusion flame begins the transition to the mixing-controlled phase, and is near stoichiometric. The diffusion flame causes the formation of

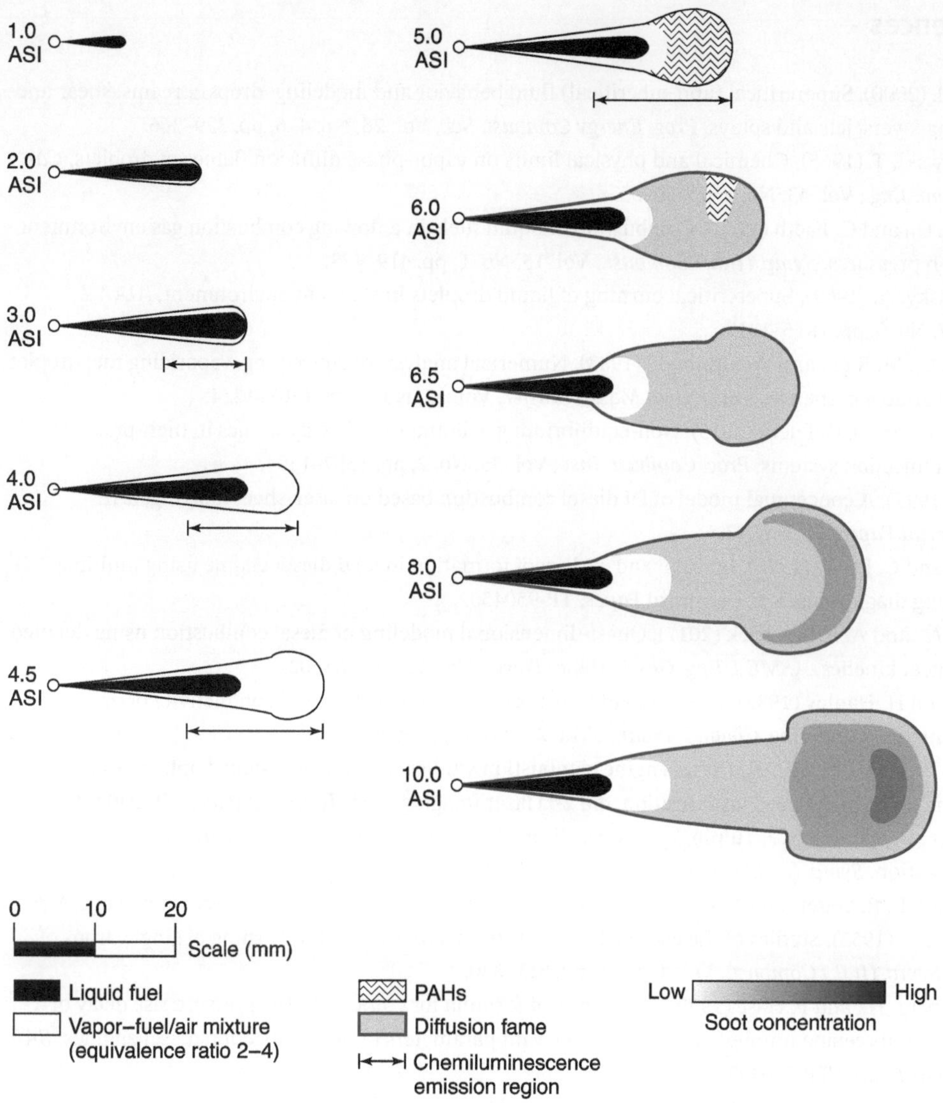

Figure 9.25 Detailed model of diesel combustion. (Dec 1997/SAE International.)

larger soot particles at the jet periphery. The soot concentration continues to increase throughout the head vortex region at the head of the jet. Since the head vortex of the jet is composed of recirculating gases, the soot particles also recirculate and grow in size.

At about 8° ASI, the jet reaches a quasi-steady condition in which the general features of the jet do not change significantly as it expands across the combustion chamber. The combustion is in the mixing controlled phase. The fuel first passes through a very fuel-rich ($\phi > 4$) premixed reaction stage and then burns out in the turbulent diffusion flame at the edge of the jet. Most of the soot is burned with the fuel in the diffusion flame. The fraction of soot that is not oxidized becomes an exhaust emission. NO_x is formed in the high-temperature regions in the diffusion flame, where both oxygen and nitrogen are available, and in the post-combustion hot gas regions.

References

Bellan, J. (2000), Supercritical (and subcritical) fluid behavior and modeling: drops, streams, shear and mixing layers, jets and sprays, *Prog. Energy Combust. Sci.*, Vol. 26, No. 4–6, pp. 329–366.

Brzustowski, T. (1965), Chemical and physical limits on vapor-phase diffusion flames of droplets, *Can. J. Chem. Eng.*, Vol. 43, No. 1, pp. 30–35.

Canada, G., and G. Faeth (1975), Combustion of liquid fuels in a flowing combustion gas environment at high pressures, *Symp. (Intl.) Combust.*, Vol. 15, No. 1, pp. 419–428.

Chervinsky, A. (1969), Supercritical burning of liquid droplets in stagnant environment, *AIAA J.*, Vol. 7, No. 9, pp. 1815–1817.

Chiang, C., M. Raju, and W. Sirignano (1983), Numerical analysis of convecting, vaporizing fuel droplet with variable properties, *Int. J. Heat Mass Transfer*, Vol. 35, No. 5, pp. 1307–1324.

Dahms, R., and J. Oefelein (2015), Non-equilibrium gas-liquid interface dynamics in high-pressure liquid injection systems, *Proc. Combust. Inst.*, Vol. 35, No. 2, pp. 1587–1594.

Dec, J. (1997), A conceptual model of DI diesel combustion based on laser-sheet imaging, *SAE Technical Paper*, TP-970873.

Dec, J., and C. Espey (1995), Ignition and early soot formation in a DI diesel engine using multiple 2-D imaging diagnostics, *SAE Technical Paper*, TP-950456.

Dobos, A., and A. Kirkpatrick (2017), Quasi-dimensional modeling of diesel combustion using detailed chemical kinetics, *ASME J. Eng. Gas Turbines Power*, Vol. 139, p. 081502.

Ely, J., and H. Hanley (1983), Prediction of transport properties. 2. Thermal conductivity of pure fluids and mixtures, *Ind. Eng. Chem. Fundam.*, Vol. 22, No. 1, pp. 90–97.

Espey, C. and J. Dec. (1993), Diesel engine combustion studies in a newly designed optical-access engine using high-speed visualization and 2-D laser imaging, *SAE Technical Paper*, TP-930971.

Faeth, G., D. Dominicis, J. Tulpinsky, and D. Olson (1969), Supercritical bipropellant droplet combustion, *Symp. (Intl.) Combust.*, Vol. 12, No. 1, pp. 9–18.

Glassman, I., R. Yetter, and N. Glumac (2014), *Combustion*, 5th ed., Academic Press, Cambridge, MA.

Godsave, G. (1953), Studies of the combustion of drops in a fuel spray–the burning of single drops of fuel, *Symp. (Intl.) Combust.*, Vol. 4, No. 1, pp. 818–830.

Hardenberg, H., and F. Hase (1979), An empirical formula for computing the pressure rise delay of a fuel from its cetane number and from the relevant parameters of direct-injection diesel engines, *SAE Technical Paper* TP-790493.

Harstad, K., and J. Bellan (2000), An all-pressure fluid drop model applied to a binary mixture: heptane in nitrogen, *Int. J. Multiphase Flow*, Vol. 26, No. 10, pp. 1675–1706.

Hiroyasu, H., M. Arai, and M. Tabata (1989), Empirical equations for the Sauter mean diameter of a diesel spray, *SAE Trans.*, Vol. 98, pp. 868–877.

Hsiao, G., V. Yang, and J. Shuen (1995), Supercritical vaporization and dynamics of liquid oxygen (LOX) droplet in hydrogen stream, AIAA paper 95-0383, *33rd Aerospace Sciences Meeting*, Reno, NV.

Jacobsen, R., and R. Stewart (1973), Thermodynamic properties of nitrogen including liquid and vapor phases from 63 K to 2000 K with pressures to 10,000 bar, *J. Phys. Chem. Ref. Data*, Vol. 2, No. 4, pp. 757–922.

Jeng, S., and Deng, Z. (1996), Numerical simulation of deformed droplet dynamics and evaporation, in *Recent Advances in Spray Combustion, Progress in Astronautics and Aeronautics*, Vol. 171, pp. 305–328.

Jung, D., and D. Assanis (2001), Multi-zone DI diesel spray combustion model for cycle simulation studies of engine performance and emissions, *SAE Transactions*, pp. 1510–1532.

Lafon, P., V. Yang, and M. Habiballah (1995), Pressure-coupled vaporization and combustion responses of liquid oxygen (LOX) droplets in supercritical hydrogen environments, *31st Joint Propulsion Conference*, p. 2432.

Law, C. K. (1982), Recent advances in droplet vaporization and combustion, *Prog. Energy Combust. Sci.*, Vol. 8, pp. 171–201.

Law, C. K., and F. Williams (1972), Kinetics and convection in the combustion of alkane droplets, *Combust. Flame*, Vol. 19, No. 3, pp. 393–405.

Lee, D., and S. Hochgreb (1998), Rapid compression machines: heat transfer and suppression of corner vortex, *Combust. Flame*, Vol. 114, pp. 531–545.

Lin, S. P., and R. Reitz (1998), Drop and spray formation from a liquid jet, *Annu. Rev. Fluid Mech.*, Vol. 30, No. 1, pp. 85–105.

Manrique, J., and G. Borman (1969), Calculations of steady state droplet vaporization at high ambient pressures, *Int. J. Heat Mass Trans.*, Vol. 12, No. 9, pp. 1081–1095.

Nguyen, T., R. Dahms, L. Pickett, and F. Tagliante (2022), The corrected distortion model for Lagrangian spray simulation of transcritical fuel injection, *Int. J. Multiphase Flow*, Vol. 148, p. 103927.

Pilch, M. and C. Erdman (1987), Use of breakup time data and velocity history data to predict the maximum size of stable fragments for acceleration-induced breakup of a liquid drop, *Intl. J. of Multiphase Flow*, Vol. 13, pp. 741–757.

Polymeropoulos, C., and R. Peskin (1972), Combustion of fuel vapor in a hot, stagnant oxidizing environment, *Combust. Sci. Technol.*, Vol. 5, No. 1, pp. 165–174.

Prakash, S., and W. Sirignano (1980), Theory of convective droplet vaporization with unsteady heat transfer in the circulating liquid phase, *Int. J. Heat Mass Trans.*, Vol. 23, No. 3, pp. 253–268.

Ranz, W., and W. Marshall (1952), Evaporation from droplets, *Chem. Eng. Prog.*, Vol. 48, No. 3, pp. 141–146.

Reitz, R., and F. Bracco (1979), On the dependence of spray angle and other spray parameters on nozzle design and operating conditions. *SAE Technical Paper*, TP-790494.

Renksizbulut, M., and R. Haywood (1988), Transient droplet evaporation with variable properties and internal circulation at intermediate Reynolds numbers, *Int. J. Multiphase Flow*, Vol. 14, No. 2, pp. 189–202.

Rosner, D. (1967), On liquid droplet combustion at high pressures, *AIAA J.*, Vol. 5, No. 1, pp. 163–166.

Sato, J. (1993), Studies on droplet evaporation and combustion in high pressure, AIAA Paper 93-0813, *31st Aerospace Sciences Meeting*, pp. 11–14.

Schlichting, H., and K. Gersten (2017), *Boundary-Layer Theory*, 9th ed., Springer, New York.

Shuen, J., V. Yang, and C. Hsiao (1992), Combustion of liquid-fuel droplets in supercritical conditions, *Combust. Flame*, Vol. 89, No. 3-4 pp. 299–319.

Spalding, D. (1959), Theory of particle combustion at high pressures, *Am. Rocket Soc. J.*, Vol. 29, No. 11, pp. 828–835.

Spalding, D. (1979), *Combustion and Mass Transfer*, Pergamon Press, New York.

Wierzba, A. (1990), Deformation and breakup of liquid drops in a gas stream at nearly critical Weber numbers, *Exp. Fluids*, Vol. 9, pp. 59–64.

Yang, V. (2000), Modeling of supercritical vaporization, mixing, and combustion processes in liquid-fueled propulsion systems, *Proc. Combust. Inst.*, Vol. 28, No. 1, pp. 925–942.

Yang, V., and J. Shuen (1994), Vaporization of liquid oxygen (LOX) droplets in supercritical hydrogen environments, *Combust. Sci. Technol.*, Vol. 97, pp. 247–270.

Yuen, M. and Chen, L. (1976), On drag of evaporating liquid droplets, *Comb. Sci. and Tech.*, Vol. 14, pp. 147–154.

Homework Problems

9.1 What is the evaporation time t_v of a $d_o = 100\ \mu m$ diameter *n*-hexane C_6H_{14} droplet in quiescent air at $T_\infty = 1000$ K and 1 atm? Assume $Le = 1$ and the droplet is at a uniform saturation temperature T_{sat}.

9.2 For an octane C_8H_{18} droplet with an initial diameter of 1 mm in quiescent air at 1000 K and 1 atm, (a) What is the time t_v to vaporize 50% and 90% of the droplet? (b) What is the mass flow $\dot{m}_F$ from the droplet when the droplet is 50% and 90% vaporized?

9.3 Suppose fuel droplets of various sizes are formed at one end of a combustor and move with the gas through the combustor at a velocity of 25 m/s. From experiment, the 50 μm diameter droplets completely vaporize in 5 ms. How far will 100 μm diameter droplets travel before they are completely vaporized?

9.4 Starting with the energy equation, Eq. (9.86), show that for a convective environment that the droplet evaporation coefficient β_v can be expressed as

$$\beta_v = \frac{4k\,Nu}{\rho_l c_p} \ln(1 + B_T)$$

and the droplet lifetime t_v is

$$t_v = \frac{d_o^2}{\beta_v}$$

where d_o is the initial droplet diameter, Nu is the Nusselt number, and B_T is the Spalding energy transfer number.

9.5 Find the evaporation time t_v for a 2-mm-diameter water droplet evaporating in dry air at a 1 atm pressure and ambient temperature of 1000 K.

9.6 What is the mass burning rate $\dot{m}_F$ of an *n*-hexane C_6H_{14} droplet burning in quiescent air at $T_\infty = 350$ K and 1 atm when its diameter is $d = 1.25$ mm? Assume $Le = 1$ and the droplet is at a uniform saturation temperature T_{sat}.

9.7 The velocity of *n*-pentane C_5H_{12} droplets in a combustor is 200 m/s. The combustion chamber length is 2 m and the gas temperature in the combustor is 1000 K. What is the maximum allowable size of the droplets if they have to be totally consumed by (a) vaporization and (b) combustion, before they exit the combustor?

9.8 A benzene C_6H_6 fuel spray is burned in quiescent air at 1 atm and $T_\infty = 298$ K. The spray combustion process can be assumed to be characterized by single-droplet burning. The Sauter mean diameter of the spray is 100 μm. (a) Calculate a mean burning time for the spray? (b) What is the steady-state temperature of the fuel droplet as it is being consumed?

9.9 Two liquid sprays are evaluated in a single-cylinder fuel–air test engine at T=1000 K. The first is liquid methanol CH_3OH, which has a transfer number $B = 2.9$. The second is a regular diesel fuel (heptane C_7H_{16}), which has a transfer number $B = 3.9$. The diesel spray has a

Sauter mean diameter of 150 μm which is 1.5 times that of methanol. Which spray requires the longer burning time and how much longer is it than the other?

9.10 A turbocharged diesel engine has a compression ratio of 16:1 and stroke of 150 mm, with an inlet manifold conditions of 2 bar and 380 K. The cetane number of the diesel fuel is 40. What is the ignition delay θ_{id} both in crank angle degrees and in milliseconds, if $N = 2500$ rpm? Assume a polytropic coefficient $n = 1.35$.

10
Emissions

10.1 Introduction

In this chapter, we discuss how pollutants are formed during a combustion process and examine measures that have been taken to reduce emissions. The major emissions from combustion processes include nitrogen oxides (NO_x), carbon monoxide (CO), particulates (PM), hydrocarbons (HC), and carbon dioxide (CO_2). These emissions are a significant source of environmental air pollution and are responsible for a variety of adverse health and environmental effects, as there is a strong correlation between air pollution levels and human health effects. The emissions irritate the respiratory passages, increasing inflammation and lowering the resistance to respiratory illnesses such as influenza, and increase the risk of serious illness or death from asthma attacks and chronic obstructive pulmonary disease. Additional human health effects include cancer, reduced lung function, cardiovascular issues, asthma, bronchitis, emphysema, and eye irritation.

Nitrogen oxides (nitric oxide NO and nitrogen dioxide NO_2) are formed during the combustion process, due to the dissociation of nitrogen N_2 and oxygen O_2 into their atomic states, and subsequent reactions with molecular oxygen and nitrogen. In the atmosphere, nitrogen oxides react with water vapor and solar radiation to form nitric acid, a component of acid rain, and ground-level ozone, O_3, another component of smog. In addition to creating significant respiratory system problems, acid rain and smog damage forests, streams, and agricultural products. As strong oxidizing agents, nitrogen oxides will react with unburned hydrocarbons and volatile organic compounds (VOCs) in the presence of sunlight to form ozone and smog. High levels of nitrogen oxide emissions from natural gas stoves are also an indoor air quality concern. Another motivation for the study of NO_x formation mechanism arises from the fact that energetic materials, such as explosives, always contain nitrogen compounds, and nitrogen oxides are always present in their combustion products.

Carbon monoxide is a product of rich combustion and reacts with oxygen and nitrogen oxides in the exhaust stream and atmosphere to form smog. The term "smog" was coined in 1905 by the English scientist H. Des Voeux from the combination of the words soot and fog. When inhaled, carbon monoxide interferes with oxygen distribution throughout the circulatory system due to its high affinity for hemoglobin, about 200 times that of oxygen. Carbon monoxide poisoning is the most common type of fatal air poisoning worldwide. The US Environmental Protection Agency (EPA) first set air quality standards for CO in 1971, specifying an eight-hour maximum of 9 parts per million (ppm) and a one-hour maximum of 35 ppm.

Particulate matter, also called soot or black carbon, emitted from combustion processes such as internal combustion engines and solid fuel combustion is a health issue since inhalation of particulates can create respiratory problems and increased risk of lung cancer. Formation of particulates starts with condensation of the gas phase hydrocarbons, which is followed by surface

Principles of Combustion, Third Edition. Allan T. Kirkpatrick and Kenneth K. Kuo.

Companion website: www.wiley.com/go/kirkpatrick/principlesofcombustion3e

growth and agglomeration of the nuclei particles. The particles pose a health hazard because they are too small to be trapped by the nose and large enough that some particle deposition in the lungs occurs. Because of their small size, particulates on the order of ~2.5 μm or less ($PM_{2.5}$) can penetrate the deepest part of the lungs such as the bronchioles or alveoli. In addition, large-scale deposition of soot on snow or ice increases the net absorption of solar radiation and thus has global climate change implications.

Combustion of solid biomass fuel is especially prone to incomplete combustion, often resulting in harmful emissions. These emissions, when combined with poor ventilation, often lead to poor indoor air quality and a number of adverse health effects (Eilenberg et al. 2018). Over 1.6 million deaths a year are attributed to poor indoor air quality as a result of domestic biomass combustion products. Poor indoor air quality is thought to be the leading cause of death worldwide for children under the age of five. In addition, recent studies have found that black carbon or soot is the second largest cause of global warming, accounting for approximately 18% of global warming compared with 40% for carbon dioxide. Household use of biofuels has been estimated to account for approximately 18% of these black carbon emissions (Bond et al. 2004).

Hydrocarbon emissions result from release of unburned or partially combusted gas phase hydrocarbon fuels. Polycyclic aromatic hydrocarbons (PAHs) are hydrocarbon rings containing only carbon and hydrogen resulting from partial combustion. Cancer is a primary human health risk of exposure to PAHs. Exposure to PAHs has also been linked with cardiovascular disease and poor fetal development. Various other hydrocarbon compounds can cause increased incidence of respiratory problems and lung cancer. Hydrocarbon emissions also contribute to the chemical reactions that form ground-level ozone.

Carbon dioxide, a primary gaseous combustion product of hydrocarbons, is also a greenhouse gas and is in the process of being regulated as well, due to its increasingly adverse influence on climate and weather patterns. The combustion of fossil fuels such as petroleum and coal is the leading cause of increasing CO_2 concentration in the atmosphere. Currently, about 33 billion tons of CO_2 is released into the atmosphere each year. About 60% of the CO_2 produced by combustion remains in the atmosphere, the remainder is removed from the atmosphere by plant photosynthesis and by diffusion into ocean water. Before the advent of the Industrial Revolution, the average atmospheric concentration of CO_2 was about 280 ppm. Long-term precision monitoring began in 1960, when the average atmospheric concentration of CO_2 was measured at 316 ppm. By 1970, it was 325 ppm, in 1990, the concentration had risen to 354 ppm, and in 2022, the concentration was 418 ppm, increasing at a rate of about 1–2 ppm per year. The current CO_2 concentration above 400 ppm is the highest in recorded history.

10.2 Nitrogen Oxides

The formation mechanisms of oxides of nitrogen have been a topic of intensive research for many decades. The formation of NO is a complex combustion process that involves many elementary chemical reactions. It involves both equilibrium and nonequilibrium processes that take place in the precombustion, combustion, and post-flame regions. The reactions forming NO_x are highly temperature dependent. The primary nitrogen oxide formed in combustion systems is NO, although in some systems, appreciable NO_2 is produced, usually because of $NO \rightarrow NO_2$ conversion in low-temperature mixing regions of non-premixed systems.

The formation of NO is for the most part determined by the peak temperature achieved during combustion. The nitrogen molecule N_2 has a triple bond, with a bond dissociation energy of

945 kJ/mol, and the oxygen molecule O_2 has a double-bond dissociation energy of 498 kJ/mol, so a relatively high combustion temperature is required to break the N_2 and O_2 bonds and begin the NO_x formation reactions. In turn, the peak temperature depends on a number of variables, including equivalence ratio ϕ, quantity of diluents present, fuel composition, and the initial temperature of the fuel–air mixture. It should be noted that the combustion of hydrogen produces higher adiabatic flame temperatures, resulting in greater NO_x generation relative to other fuels.

Detailed reaction pathways for NO_x creation and decay have been formulated, and the corresponding rate parameters for the reactions have been established. In what follows, the equal sign indicates both forward and backward reactions need to be considered. A detailed study of nitrogen chemistry in combustion is contained in Miller and Bowman (1989), and experimental studies and models of NO_x formation in simple non-premixed flames were reviewed by Turns (1995).

The global reaction for NO formation is

$$N_2 + O_2 \rightarrow 2\,NO \tag{10.1}$$

The major NO_x formation routes are categorized below:

- Thermal NO route, also known as the Zel'dovich NO mechanism
- Prompt NO route, also known as the Fenimore NO mechanism
- N_2O route
- NO_2 route
- Fuel-bound nitrogen (FN) route

Zel'dovich (1946) developed the thermal NO mechanism, now known as the Zel'dovich NO mechanism. Later on, Fenimore (1970) proposed the prompt NO mechanism to explain the additional NO being produced over and above the Zel'dovich thermal NO in hydrocarbon flames. The prompt mechanism occurs within a relatively thin combustion flame front. Since the volume of the burned gases which are at a high temperature is much larger than the instantaneous volume of the flame front, the amount of NO formed from the prompt mechanism is relatively small compared to that formed from the thermal or Zel'dovich mechanism. Fuel-bound NO is formed from nitrogen in the fuel. Recently, researchers have identified an additional route, the NNH mechanism through which NO and other oxides of nitrogen are formed. The radical NNH or N_2H is also called diazenylium.

In laminar flames at low pressures, the NO yield is dominated by the Fenimore mechanism and a super-equilibrium O/OH Zel'dovich route. At high pressures, the equilibrium concentrations of O, OH, and H have strong influence on the formation rate, and the Zel'dovich thermal-NO mechanism dominates. For fuel-rich mixtures, Fenimore prompt-NO mechanism dominates. For sufficiently fuel-rich mixtures, however, this route no longer dominates. Lean conditions can suppress the formation of CH radicals, hence leading to less Fenimore prompt NO, and low temperature can suppress the Zel'dovich NO. What remains is NO generated via a N_2O route. The fuel-bound nitrogen route is also important in non-premixed flames.

10.2.1 Zel'dovich Thermal NO Mechanism

The Zel'dovich mechanism is the dominant mechanism for NO_x formation from combustion of fuels which are nitrogen free. This mechanism consists of three principal reactions as follows (Miller and Bowman 1989):

$$O + N_2 \underset{k_{1b}}{\overset{k_1}{\rightleftharpoons}} NO + N \tag{10.2}$$

$$\mathrm{N} + \mathrm{O_2} \underset{k_{2b}}{\overset{k_2}{\rightleftharpoons}} \mathrm{NO} + \mathrm{O} \tag{10.3}$$

$$\mathrm{N} + \mathrm{OH} \underset{k_{3b}}{\overset{k_3}{\rightleftharpoons}} \mathrm{NO} + \mathrm{H} \tag{10.4}$$

The rate constants for the Zel'dovich reaction mechanism are given in Eq. (10.5) (Hanson and Salimian 1984). The rate constants have units of m^3/kmol-s, and the temperature T is in degrees Kelvin:

$$\begin{aligned} k_1 &= 1.82 \times 10^{11} \exp(-38,370/T) \\ k_{1b} &= 3.8 \times 10^{10} \exp(-425/T) \\ k_2 &= 1.80 \times 10^{7} T \exp(-4,680/T) \\ k_{2b} &= 3.8 \times 10^{6} T \exp(-20,820/T) \\ k_3 &= 7.1 \times 10^{10} \exp(-450/T) \\ k_{3b} &= 1.7 \times 10^{11} \exp(-24,560/T) \end{aligned} \tag{10.5}$$

The chain reaction nature of this mechanism can be observed in Figure 10.1. The Zel'dovich mechanism is also known as the "thermal" mechanism, since reaction (10.2) has a very high activation energy (319 kJ/mol), and so it is sufficiently fast only at high temperatures. Thermal NO is generally considered to be formed in the high-temperature post-flame gases. As a rule-of-thumb, the thermal mechanism is usually unimportant at temperatures below 1800 K.

Because of the low reaction rate coefficient k_1, the reaction (10.2) is the rate-determining step in the Zel'dovich reaction mechanism. The concentration of NO cannot be predicted accurately by assuming it to be in equilibrium in reaction (10.2), because the reaction (10.2) is so slow that equilibrium is reached only for times that are much longer than typical combustion chamber residence times in the high-temperature range.

In general, this mechanism is coupled with the fuel combustion chemistry through the O_2, O, and OH species. Since the overall rate of NO formation by the thermal mechanism generally is slow compared with the fuel oxidation reactions, it often is assumed, following the suggestion of Zel'dovich, that the NO formation reactions can be decoupled from the fuel oxidation process.

Following Heywood (1976), one can write the following expression for the rate of change of NO concentration, with the brackets denoting molar concentrations in units of kmol/m^3:

$$\begin{aligned} \frac{d}{dt}[\mathrm{NO}] = &+ k_1[\mathrm{O}][\mathrm{N_2}] - k_{1b}[\mathrm{NO}][\mathrm{N}] + k_2[\mathrm{N}][\mathrm{O_2}] \\ &- k_{2b}[\mathrm{NO}][\mathrm{O}] + k_3[\mathrm{N}][\mathrm{OH}] - k_{3b}[\mathrm{NO}][\mathrm{H}] \end{aligned} \tag{10.6}$$

To apply Eq. (10.6) two approximations are introduced. First, that the C–O–H system is in equilibrium and is not perturbed by N_2 dissociation, and second, the N atoms change concentration by

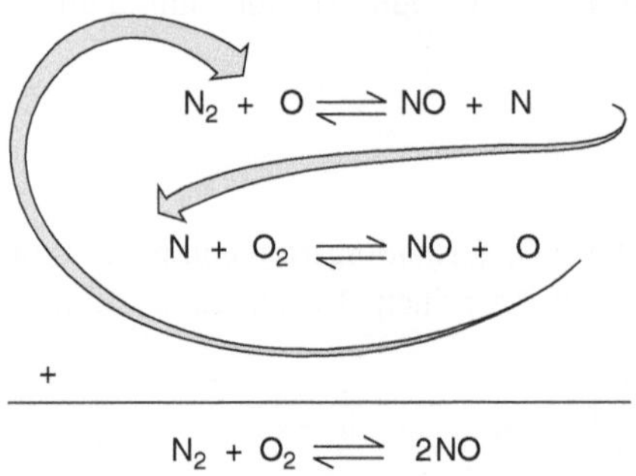

Figure 10.1 Chain reaction nature of the thermal NO formation mechanism.

a quasi-steady process. The first approximation means simply that given the pressure, temperature, equivalence ratio, and residual fraction of a fluid element, one simply computes the equilibrium composition to determine the concentrations of N_2, O_2, O, OH, and H. The second approximation means that one can solve for the N atom concentration by setting the rate of change of N atoms to zero:

$$\begin{aligned}\frac{d}{dt}[\mathrm{N}] &= +k_1[\mathrm{O}][\mathrm{N_2}] - k_{1b}[\mathrm{N}][\mathrm{NO}] - k_2[\mathrm{N}][\mathrm{O_2}] \\ &\quad + k_{2b}[\mathrm{NO}][\mathrm{O}] - k_3[\mathrm{N}][\mathrm{OH}] + k_{3b}[\mathrm{NO}][\mathrm{H}] \\ &= 0\end{aligned} \tag{10.7}$$

With these two approximations, it can be shown that

$$\frac{d[\mathrm{NO}]}{dt} = \frac{2R_1(1-\alpha^2)}{1+\alpha R_1/(R_2+R_3)} \tag{10.8}$$

where α is the ratio of the nitrogen oxide concentration to its equilibrium value:

$$\alpha = \frac{[\mathrm{NO}]}{[\mathrm{NO}]_e} \tag{10.9}$$

and $R_i(i = 1, 2, 3)$ are various rates of reaction, with the equilibrium concentrations labeled with the subscript e:

$$R_1 = k_1[\mathrm{O}]_e[\mathrm{N_2}]_e$$

$$R_2 = k_{2b}[\mathrm{NO}]_e[\mathrm{O}]_e$$

$$R_3 = k_{3b}[\mathrm{NO}]_e[\mathrm{H}]_e$$

The concentration of N_2 can be accurately measured with a probe or can be estimated assuming equilibrium in the burnt gases. If the relevant timescales are sufficiently long, one can assume that the N_2, O_2, O, and OH concentrations are at their equilibrium values and N atoms are in quasi-steady state. However, in the flame zone proper and in certain short timescale post-flame processes, the equilibrium assumption is not truly valid. According to Warnatz et al. (1982), *super-equilibrium concentrations* of O atoms, up to several orders of magnitude greater than equilibrium, can greatly increase NO formation.

A *partial-equilibrium* assumption for O atoms at high temperatures (> 1600 K) gives a better estimate of O atom concentration, as has been verified from experiments. Consider partial equilibrium for the following reactions:

$$\mathrm{H} + \mathrm{O_2} \underset{k_{4b}}{\overset{k_4}{\rightleftharpoons}} \mathrm{OH} + \mathrm{O}$$

$$\mathrm{O} + \mathrm{H_2} \underset{k_{5b}}{\overset{k_5}{\rightleftharpoons}} \mathrm{OH} + \mathrm{H}$$

$$\mathrm{OH} + \mathrm{H_2} \underset{k_{6b}}{\overset{k_6}{\rightleftharpoons}} \mathrm{H_2O} + \mathrm{H}$$

For each of the reactions, the forward and backward reaction rates are assumed to be equal. Therefore,

$$k_4[\mathrm{H}][\mathrm{O_2}] = k_{4b}[\mathrm{OH}][\mathrm{O}]$$

$$k_5[\mathrm{O}][\mathrm{H_2}] = k_{5b}[\mathrm{OH}][\mathrm{O}]$$

$$k_6[\mathrm{OH}][\mathrm{H_2}] = k_{6b}[\mathrm{H_2O}][\mathrm{H}]$$

Solving for the O-atom concentration,

$$[O] = \frac{k_4 k_6 [O_2][H_2]}{k_{4b} k_{6b} [H_2O]} \tag{10.10}$$

Therefore, the O atom concentration can be calculated from the concentrations of H_2O, O_2, and H_2, which can be measured or estimated easily, since they are stable species.

10.2.2 Prompt NO Mechanism (Fenimore Mechanism)

The prompt NO mechanism is of importance in rich conditions with hydrocarbon fuels. Fenimore (1970, 1979) discovered that NO was produced in the flame zone of laminar premixed flames before there would be enough time to form NO by the thermal mechanism, and he termed this rapidly formed NO the "prompt" NO.

Numerous studies have shown that in hydrocarbon flames NO can be formed at low temperatures (<1000 K) by a reaction sequence that results from the rapid reaction of hydrocarbon radicals – CH, methylene CH_2, dicarbon C_2, ethynyl C_2H, and atomic carbon C, with molecular nitrogen N_2, leading to the formation of amines or hydrogen cyanide HCN that then react to form NO. The activation energy of the rate-limiting reaction of the prompt-NO mechanism, $CH + N_2 \rightarrow HCN + N$, is about 92 kJ/mol, compared with 319 kJ/mol for the rate-limiting reaction of the thermal-NO mechanism.

The principal reactions for the prompt NO mechanism are,

$$CH + N_2 \underset{k_{7b}}{\overset{k_7}{\rightleftharpoons}} HCN + N \tag{10.11}$$

$$CH_2 + N_2 \underset{k_{8b}}{\overset{k_8}{\rightleftharpoons}} HCN + NH \tag{10.12}$$

$$CH_2 + N_2 \underset{k_{9b}}{\overset{k_9}{\rightleftharpoons}} H_2CN + N \tag{10.13}$$

$$C + N_2 \underset{k_{10b}}{\overset{k_{10}}{\rightleftharpoons}} CN + N \tag{10.14}$$

Reaction (10.11) is the primary path and is the rate-limiting step in the sequence. The rate constant k_7 correlated from many sets of experimental data can be expressed as

$$k_7 = 4.4 \times 10^9 \exp(-11,060/T) \quad \text{m}^3/(\text{kmol} \cdot \text{s}) \tag{10.15}$$

Overall, prompt NO formation involves three separate kinetic issues, including (1) the CH concentration and how it is established; (2) the rate of molecular nitrogen N_2 fixation, i.e., the value of the rate coefficient k_7; and (3) the rate of inter-conversion among fixed nitrogen fragments.

From experiments and calculations, it has been found that at low temperatures ($T < 2000$ K), the hydrocarbon–nitrogen prompt-NO channel dominates the rate of NO formation. As the temperature increases, the relative importance of the hydrocarbon–nitrogen prompt-NO channel decreases, so that for temperatures above 2500 K, NO formation is controlled mainly by the thermal mechanism.

From sensitivity analysis conducted by Miller and Bowman (1989), it was found that reaction (10.11) is an important rate-limiting step for NO, HCN, and TFN (total fixed nitrogen, TFN = NO + HCN + NH_3) production under all conditions investigated. The reaction has sensitivity coefficients for these species approaching unity in rich mixtures. Even under lean conditions ($\phi < 0.8$), it plays a major role in NO formation. There have been indirect measurements of k_7 from flame, shock tube, and stirred-reactor data and one direct shock-tube measurement. The estimated accuracy of k_7 using Eq. (10.15) is within about a factor of 2 at the present time. The activation energy for reaction (10.11) is between 75 and 92 kJ/mol, which is much lower than the 319 kJ/mol required for the formation of thermal NO. Thus, reaction (10.11) can be very important for NO formation through the production of the HCN species.

Now, let us consider the endothermic reaction (10.12) in the prompt mechanism. The suggested activation energy for this reaction is approximately 356 kJ/mol, a very high value. A rate coefficient expression of $k_8 = 10^9 \exp(-356,000/R_u T)$ m^3/kmol-s is suggested by Miller and Bowman (1989). Due to the high value of the activation energy, this reaction is considered to be an insignificant contributor to prompt NO.

The reaction (10.13) is endothermic by 330.5 kJ/mol. The rate coefficient expression k_9 is essentially the same as k_1 of the Zel'dovich mechanism. This reaction is also an insignificant contributor to prompt NO. The last prompt-NO reaction considered in the mechanism is reaction (10.14). Experiments reported by Miller and Bowman indicate that the reverse of this reaction is very fast; hence, reaction (10.14) is a minor, but non-negligible, contributor to prompt NO under most conditions. This reaction is also highly endothermic. Thus, its importance with respect to reaction (10.11) increases with increasing temperature.

The thermal NO mechanism is the dominant source of NO only in the lean equivalence-ratio range ($\phi = 0.8 - 1.0$). Miller and Bowman showed that, for a series of premixed rich methane/air flames ($1.37 \geq \phi > 1.06$), the transition from the hydrocarbon–nitrogen prompt-NO mechanism to the thermal mechanism occurs as conditions become leaner. For lean mixtures with $\phi < 0.8$, the temperature becomes sufficiently low and the following mechanism becomes an important source of prompt NO formation:

$$O + N_2 + M = N_2O + M \tag{10.16}$$

$$H + N_2O = NO + NH \tag{10.17}$$

$$O + N_2O = NO + NO \tag{10.18}$$

The mechanism of CH formation depends slightly on the fuel, since CH is formed as an intermediate at the flame front only. In all cases, CH is formed from methylene CH_2 by reaction with H and OH, i.e.,

$$H + CH_2 = CH + H_2 \tag{10.19}$$

$$OH + CH_2 = CH + H_2O \tag{10.20}$$

CH_2 can be formed either from a methyl radical CH_3 or from CH. For equivalence ratios $\phi < 1.5$, CH_3 is the dominant source.

$$CH_3 + OH = CH_2 + H_2O \tag{10.21}$$

$$CH_3 + H = CH_2 + H_2 \tag{10.22}$$

The most important CH-consuming steps are the following reactions:

$$CH + H = C + H_2 \tag{10.23}$$

$$CH + H_2O = CH_2O + H \tag{10.24}$$

$$CH + CO_2 = HCO + CO \tag{10.25}$$

Up to an equivalence ratio $\phi \approx 1.2$, the hydrogen cyanide HCN and N produced by reaction (10.11) are converted rapidly to NO, primarily by the following reactions:

$$HCN + O = NCO + H \tag{10.26}$$

$$NCO + H = NH + CO \tag{10.27}$$

$$NH + H = N + H_2 \tag{10.28}$$

$$N + OH = NO + H \tag{10.29}$$

Above an equivalence ratio $\phi = 1.2$, several factors (Miller and Bowman 1989) combine to cause the NO concentration to decrease relative to hydrogen cyanide HCN:

- The conversion of HCN to NO by the above mechanism is no longer rapid.
- The "recycle" of NO to HCN by the mechanism, reactions (10.30)–(10.32), below begins to inhibit NO production.
- The reaction $N + NO \Leftrightarrow N_2 + O$ shifts direction from reverse to forward.

A decrease in the O atom concentration contributes to all three of the above occurrences. Under rich conditions, reactions that convert CH_3 to CH_2 become more competitive with the oxidation reaction of CH_3 according to $CH_3 + O \Leftrightarrow CH_2O + H$, thereby producing the radicals C, CH, and CH_2 and allowing the following recycle reactions to be more effective:

$$C + NO = CN + O \tag{10.30}$$

$$CH + NO = HCN + O \tag{10.31}$$

$$CH_2 + NO = HCNO + H \tag{10.32}$$

From their calculations, Miller and Bowman found that at an equivalence ratio $\phi = 1.4$, the peaks in the CH and TFN concentrations occur simultaneously. Above $\phi = 1.4$, the fixed nitrogen concentration is limited by the availability of the chain carriers (H and OH) required for producing CH from methane. The following chain-branching reactions occur

$$H + O_2 = OH + O \tag{10.33}$$

$$OH + CH_3 = CH_2OH + H \tag{10.34}$$

$$CH_2OH + M = CH_2O + H + M \tag{10.35}$$

In this regime, the following reaction

$$CH_3 + N = H_2CN + H \tag{10.36}$$

appears as a competitor for nitrogen N atoms, competing with

$$N + OH = NO + H \tag{10.37}$$

In fossil fuels such as coal, nitrogen is present as the chemically bound nitrogen. The fuel-bound nitrogen route is important in non-premixed flames. Available data suggest that the gas phase fuel nitrogen reaction sequence is initiated by a rapid conversion of the fuel nitrogen compounds to HCN and NH_3. HCN appears to be the principal product when the fuel nitrogen is bound by an aromatic ring, and NH_2 when the fuel nitrogen is in the form of amines. The mechanism of NO production from fuel-bound nitrogen is the mechanism of oxidation of HCN and NH_3. As the reaction pathways in Figure 10.2 indicate, the dissociation of H_2CN following reaction (10.36) results in the formation of hydrogen cyanide HCN, rather than nitric oxide NO formation in reaction (10.37).

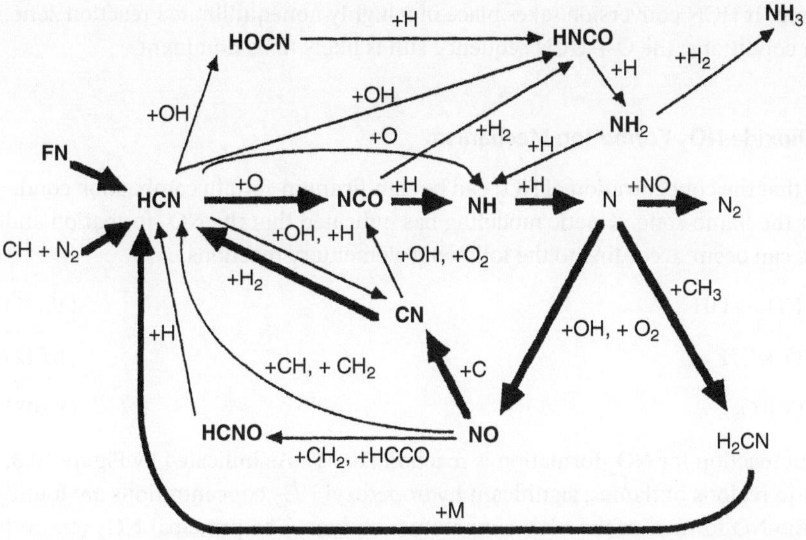

Figure 10.2 Reaction path diagram showing major steps in prompt NO mechanism and conversion of fuel-bound nitrogen to NO. *Source*: Miller and Bowman 1989/with permission of Elsevier.

As shown in Figure 10.2, the consumption of hydrogen cyanide HCN is controlled by the reaction of HCN with oxygen O atoms, even in rich flames.

$$HCN + O = NCO + H$$

$$HCN + O = NH + CO$$

The subsequent reactions of NCO and NH with hydrogen atoms, producing N, are relatively fast and have little influence on the observed species profiles. The distribution of NO and N_2 in the flame is governed by the N atom reactions presented earlier in this section,

$$N + OH = NO + H$$

$$N + NO = N_2 + O$$

In the post-flame gases of very rich atmospheric-pressure flames, HCN is removed primarily by reaction with OH. Two routes have been suggested by Miller and Bowman, one that is the first order in OH,

$$OH + HCN = HOCN + H \tag{10.38}$$

and one that is the second order,

$$OH + HCN = CN + H_2O \tag{10.39}$$

$$CN + OH = NCO + H \tag{10.40}$$

under the partially equilibrated conditions prevalent in the post-flame gases of atmospheric-pressure flames. The first route is dominant below 2300 K and the second route is dominant above 2300 K.

Accumulated experience by chemists in modeling a variety of flames, stirred reactors, and flow reactors indicates that the O + HCN sequence always plays a major role in the conversion to NO and N_2. The OH + HCN reactions normally come into play only under conditions that are both rich and nearly equilibrated. If HCN conversion takes place in a highly nonequilibrated reaction zone, even under very rich conditions, the O + HCN sequence still is likely to be dominant.

10.2.3 Nitrogen Dioxide NO_2 Formation Mechanism

It has been observed that the concentration of NO_2 can be significant in certain combustion conditions, especially near the flame zone. Kinetic modeling has indicated that the NO_2 formation and destruction in flames can occur according to the following elementary reactions,

$$NO + HO_2 = NO_2 + OH \tag{10.41}$$

$$NO_2 + H = NO + OH \tag{10.42}$$

$$NO_2 + O = NO + O_2 \tag{10.43}$$

The most important reaction for NO_2 formation is reaction (10.41). As indicated by Figure 10.3, in the low-temperature regions of flames, significant hydroperoxyl HO_2 concentrations are found that can react with the NO formed in the high-temperature regions. The principal NO_2 removal step is reaction (10.42), while reaction (10.43) is not as important. The NO_2 removal reactions are rapid, and in the presence of high radical concentrations, NO_2 will be converted rapidly back to NO. Since the rate of this reaction depends on the HO_2 concentration, NO_2 formation is also sensitive to reactions forming and removing HO_2. The following is the principal HO_2 formation reaction in the low-temperature region of the flame.

$$H + O_2 + M = HO_2 + M \tag{10.44}$$

The NO_2 removal rate is sensitive to the hydrogen H atom concentration. The H atoms needed can be transported from the high-temperature regions by molecular diffusion. This phenomenon can be observed in Figure 10.3 for an atmospheric pressure 10% methane–air flame.

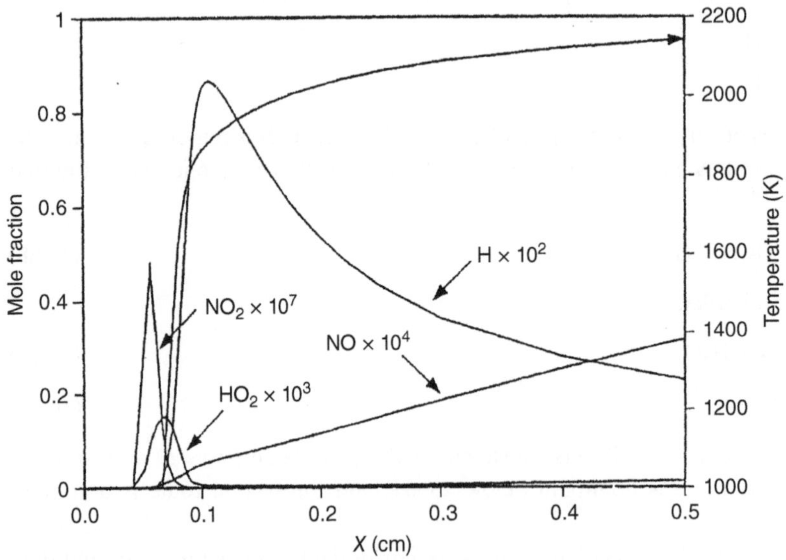

Figure 10.3 NO_2 flame structure and profiles of H, NO, and HO_2. *Source*: Miller and Bowman 1989/with permission of Elsevier.

10.2.4 Nitrous Oxide N_2O Mechanism

Nitrous oxide N_2O is an important nitrogen oxide under fuel-lean conditions. This route is the major source of NO in lean premixed combustion in devices such as gas turbine engines. Studies indicate that N_2O is a very short-lived species in hot combustion gases and that the principal N_2O formation reactions involve NO and various nitrogen-containing radicals such as NH and NCO,

$$NH + NO = N_2O + H \tag{10.45}$$

$$NCO + NO = N_2O + CO \tag{10.46}$$

The principal N_2O formation reaction is reaction (10.45). For rich mixtures, both reactions (10.45) and (10.46) are important in N_2O formation. The N_2O formed in these reactions rapidly reacts with H atoms to form N_2:

$$N_2O + H = N_2 + OH \tag{10.47}$$

In fuel-lean flames, the primary N_2O removal reaction is reaction (10.47). While the removal is done by reaction (10.47), there is another mechanism for the formation of N_2O and subsequent conversion to NO. This mechanism is analogous to the thermal mechanism in that an O atom attacks the nitrogen molecule N_2. In the presence of a third body, molecule M, the outcome of this reaction is N_2O,

$$N_2 + O + M \rightarrow N_2O + M \tag{10.48}$$

The N_2O may subsequently react with O atoms to form NO

$$N_2O + O \rightarrow NO + NO \tag{10.49}$$

The activation energy of this reaction is around 97 kJ/mol. Under fuel-lean conditions, CH formation is suppressed, leading to less prompt NO, and at low-temperatures, the Zel'dovich thermal NO production is also less. What remains is NO generated from N_2O via reaction (10.49), which is promoted at high pressures because of the three-body reaction (10.48). Since the activation energy for a three-body reaction is relatively low, a low temperature does not affect the formation rate of NO.

10.2.5 NNH Mechanism

Bozzelli and Dean (1995) proposed a new high-temperature pathway for NO formation that involves the reaction of NNH with oxygen atoms. The presence of significant quantities of NNH results from the reaction of H with N_2, reaction (10.50). The reaction forms the HNNO* energized adduct via a rapid combination reaction; HNNO* then rapidly dissociates to NH + NO, reaction (10.51).

$$H + N_2 \Longleftrightarrow NNH \tag{10.50}$$

$$NNH + O \Longleftrightarrow NH + NO \tag{10.51}$$

This pathway has been shown to be important in premixed and non-premixed flames of both hydrogen and hydrocarbon fuels with large carbon/hydrogen ratios (Klippenstein et al. 2011).

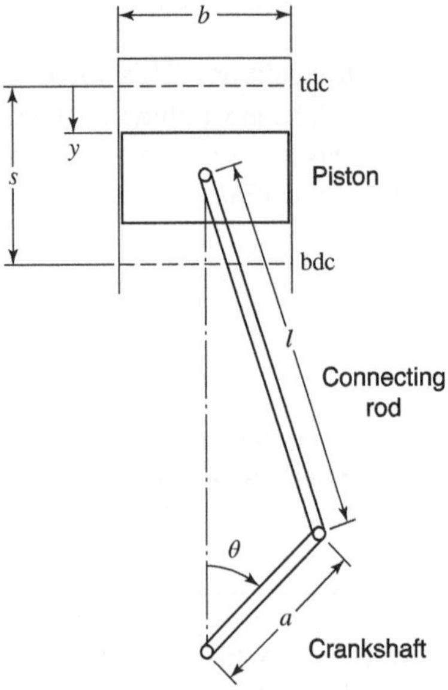

Figure 10.4 Engine slider crank geometry.

10.2.6 Example: NO_x Emissions from Internal Combustion Engines

A simple model of the crankshaft, connecting rod, piston, and head assembly of an internal combustion engine cylinder is the flattop slider crank mechanism shown in Figure 10.4. Of particular interest are the following geometric parameters: bore b, connecting rod length l, crank radius a, stroke s, clearance height c, and crank angle θ. The stroke s is twice the crankshaft radius a.

The term "top dead center" (tdc) of an engine refers to the crankshaft at a position $\theta = 0°$. The cylinder volume at this position is minimum and is called the clearance volume, V_o. The term "bottom dead center" (bdc) refers to the crankshaft at a position $\theta = 180°$. The cylinder volume at bottom dead center is the maximum volume V_1.

The compression ratio r is defined as the ratio of the maximum to minimum volume.

$$r = \frac{V_{\text{bdc}}}{V_{\text{tdc}}} = \frac{V_1}{V_o} \tag{10.52}$$

The displacement volume V_d is the difference between the maximum and the minimum volume; for a single cylinder,

$$V_d = V_1 - V_o = \frac{\pi}{4}\, b^2\, s \tag{10.53}$$

The computation of NO production and subsequent emission from an internal combustion engine is nontrivial, as the instantaneous NO molar production rate $d[\text{NO}]/dt$ (kmol/m^3-s) depends on the varying burned (b) mixture temperature, pressure, and the equilibrium and instantaneous NO molar concentrations. The NO production rate thus follows the increase, then decrease, of the temperature and pressure of the burned mixture from the start to the end of combustion (Lavoie et al. 1970). In general, with spark ignition engine exhaust, the dominant component of NO_x is nitrogen oxide, NO, with concentrations of the order of 1000 ppm, and the concentration of nitrogen dioxide, NO_2, is of the order of 10 ppm, i.e., about 1%. In compression ignition engines, the concentration of NO_2 can be higher, approaching 10–30% of the total NO_x.

The difference between the equilibrium and instantaneous NO emissions from an engine can be illustrated using a simple two-zone model. In this model, the fuel and air are fully mixed prior

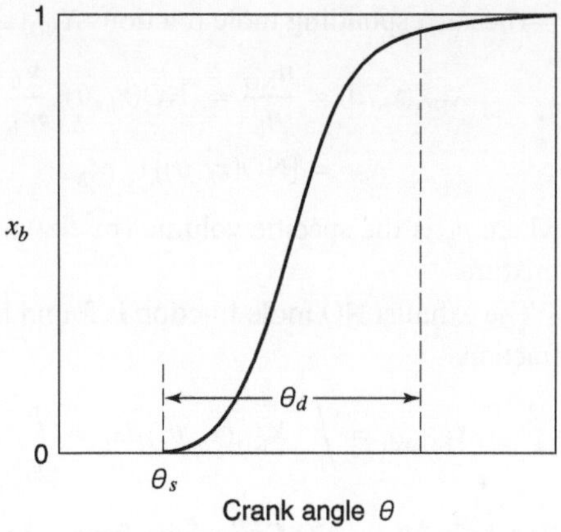

Figure 10.5 Cumulative mass fraction burned versus crank angle.

to combustion, and the combustion chamber is divided into two zones, burned (b) and unburned (u). Differential equations are developed for the change in pressure and temperature in each zone during the compression stroke from −180° to 0°, and the expansion stroke from 0° to +180°. The modeling is based on equations for energy and mass conservation, equation of state, and mass fraction burned. The assumption is made that both zones are at the same pressure p. This is known as a *zero-dimensional* model, since it is a function only of crank angle and not a function of the combustion engine geometry.

A cumulative mass fraction burned curve, i.e., the fraction of fuel energy released, is given in Figure 10.5. The representative figure plots the cumulative mass fraction burned x_b versus crank angle. The characteristic features of the mass fraction burned curve are an initial small slope region beginning with spark ignition and the start of energy release at θ_s, followed by a region of rapid growth, and then a more gradual decay. The three regions correspond to the initial ignition development, a rapid burning region, and a burning completion region. The end of combustion $\theta_e = \theta_s + \theta_d$, where θ_d is the duration of combustion. This S-shaped mass fraction curve can be represented analytically by a cosine function as indicated by Eq. (10.54),

$$x_b(\theta) = \frac{1}{2}\left[1 - \cos\left(\frac{\pi(\theta - \theta_s)}{\theta_d}\right)\right] \tag{10.54}$$

The derivative of Eq. (10.54) with respect to crank angle is the combustion energy release rate per crank angle,

$$\frac{dx_b}{d\theta} = \frac{\pi}{2\theta_d}\left[\sin\left(\frac{\pi(\theta - \theta_s)}{\theta_d}\right)\right] \tag{10.55}$$

In this illustrative model, the assumption is made that there is no turbulent mixing between fluid elements during the combustion process. As a consequence, different NO concentrations for various burn fractions will be established in the burned mixture. Each mass is assumed to burn at its adiabatic flame temperature based on the unburned gas temperature at the time it burned. At a given crank angle θ, the NO concentrations for a range of burn fractions x_b from 0 to $x_b(\theta)$ in the burned mixture can be found by integration of Eq. (10.8).

$$[\mathrm{NO}(x_b, \theta)] = \frac{1}{\omega}\int_{\theta}^{\theta+d\theta} \frac{d[\mathrm{NO}]}{dt}\, d\theta \tag{10.56}$$

The corresponding mole fractions $X_{NO}(x_b, \theta)$ in the burned mixture are

$$\begin{aligned} X_{NO}(x_b, \theta) &= \frac{n_{NO}}{n_b} = [NO(x_b, \theta)]\, \frac{V_b}{m_b}\frac{m_b}{n_b} \\ &= [NO(x_b, \theta)]\, v_b\, M_b \end{aligned} \tag{10.57}$$

where v_b is the specific volume (m^3/kg) and M_b is the molecular mass (kg/kmol) of the burned mixture.

The exhaust NO mole fraction is found by summing the mole fractions $X_{NO}(x_b, \theta_e)$ of each burn fraction,

$$X_{NO,exh} = \int_0^1 X_{NO}(x_b, \theta_e)\, dx_b = \int_{\theta_s}^{\theta_e} X_{NO}(x_b, \theta_e) \frac{dx_b}{d\theta} d\theta \tag{10.58}$$

Example 10.1 ***NO Emissions from a Spark Ignition Engine***

A spark ignition engine is operated lean ($\phi = 0.88$) with the following baseline conditions given in the table below. The start of heat release is 15° before top dead center and the burn duration is 45°. The inlet conditions are $p_1 = 100$ kPa, $T_1 = 350$ K. What is the burn fraction, burned and unburned temperatures, equilibrium, and rate-limited NO values as a function of crank angle?

Solution

The NO thermal mechanism has been incorporated into a two-zone (burned and unburned zones) heat release program `TwoZoneNO.m` listed in the text website. The program computes and plots both the equilibrium and the rate-limited NO concentrations as a function of crank angle and also calculates NO concentration in the exhaust.

Parameter	Value
Bore (m)	0.0825
Stroke (m)	0.1143
Half Stroke/Rod Ratio	0.25
Compression Ratio	7
Engine Speed (rpm)	1200
Equivalence Ratio	0.88
Residual Fraction	0.05
Cylinder Heat Transfer Coeff. (W/m^2K)	500
Mass Blowby Coeff. (s^{-1})	0.8
Cylinder Wall Temp. (K)	400

The burn fraction x_b profile for the engine is plotted in Figure 10.6. The increased cylinder pressure (see Figure 10.7) due to the combustion results in compression of both the burned and unburned gas. The calculated unburned T_u and burned T_b temperature profiles are plotted in Figure 10.8.

The unburned gases have the same uniform temperature and pressure profiles and are compressed approximately isentropically. The burned gases are compressed both before and after combustion and so have different temperatures for different burn fractions. Once burned, the temperature of a gas element tracks the pressure, as it also is more or less isentropically compressed or expanded. Notice that the first mass to burn is compressed significantly. Each subsequent element

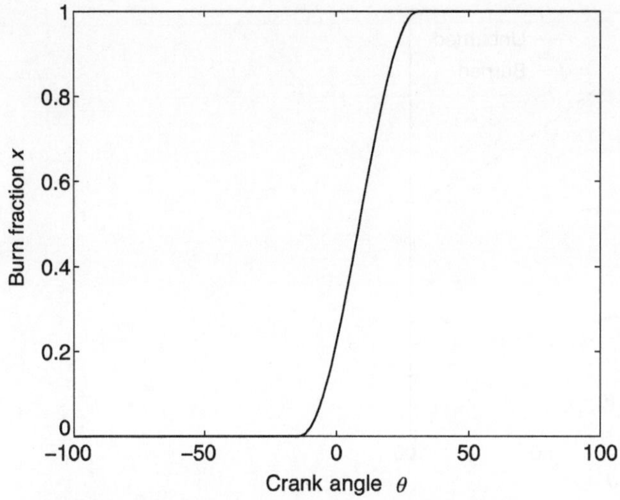

Figure 10.6 Mass fraction burned versus crank angle.

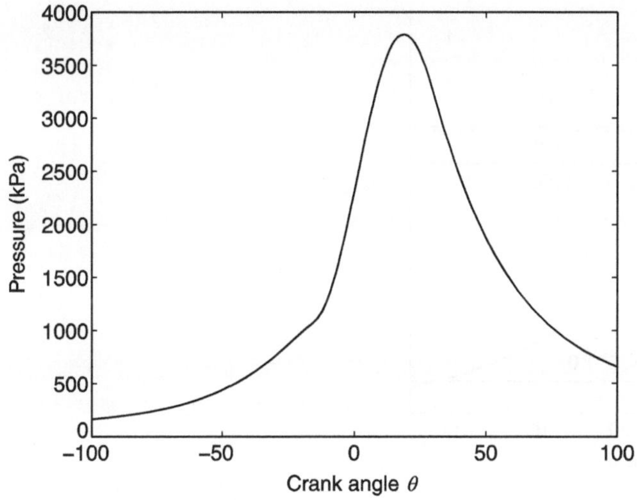

Figure 10.7 Cylinder pressure versus crank angle.

to burn is compressed less, and the last element to burn undergoes no compression. As a result, the first mass to burn is hotter than all the rest, and the last mass to burn is the coolest.

The resulting NO profiles are plotted in Figure 10.9 for lean combustion with $\phi = 0.88$, and Figure 10.10 for rich combustion with $\phi = 1.14$. The equilibrium concentration is computed based on the local temperature, pressure, and equivalence ratio during the expansion stroke.

The plots illustrate some important points about NO combustion in engines. First, there is a significant difference between the equilibrium and the rate-limited concentrations of NO during the combustion process, as the rate-limited NO concentrations lag behind the equilibrium NO concentrations. Second, since the reaction rates increase strongly with temperature, there are large differences between the nitrogen oxide concentrations in the first mass to burn ($x = 0$) and last mass to burn ($x = 1$).

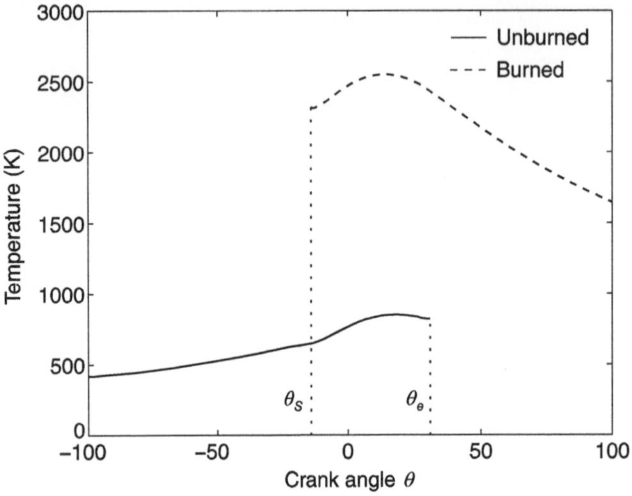

Figure 10.8 Temperature of burned gas T_b and unburned gas T_u.

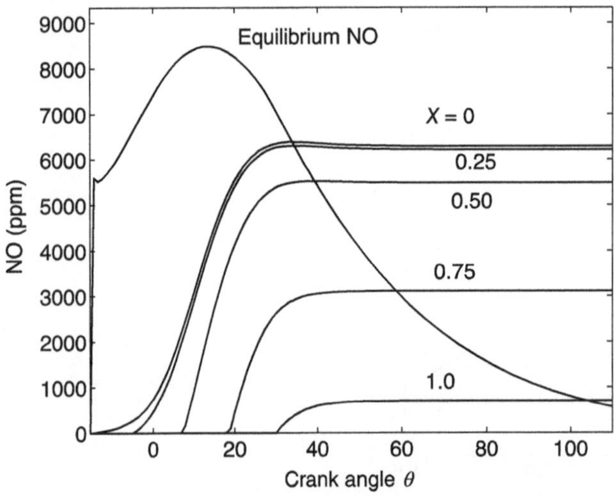

Figure 10.9 Computed equilibrium and rate-limited NO concentrations at $\phi = 0.88$.

Furthermore, it can be seen by comparison of Figure 10.8 with Figures 10.9 and 10.10 that when the temperatures drop to about 2000 K, the decomposition rate becomes very slow and for practical purposes it may be said that the nitrogen oxides freeze at a concentration greater than the equilibrium values. The relatively high values of the "frozen" concentrations of nitrogen oxides in the exhaust are thus a function of the gas temperatures during combustion, not temperatures in the exhaust.

The model also indicates that as the mixture becomes richer, the exhaust NO concentration freezes later in the exhaust stroke. Therefore, the exhaust NO concentration for lean mixtures depends on the peak burned gas temperatures early in the exhaust stroke, and for rich mixtures, depends on the gas temperatures later in the exhaust stroke.

Figure 10.11 shows a result typical of all internal combustion engines that nitrogen oxides are maximized with mixtures slightly lean of stoichiometric. Recall that increased temperatures favor nitrogen oxide formation and that burned gas temperatures are maximized with mixtures that are

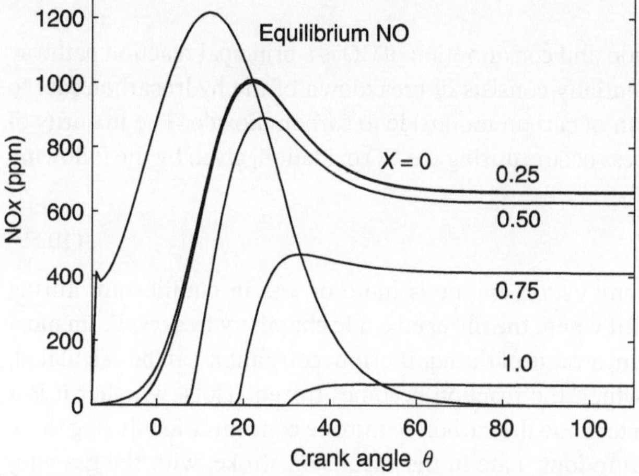

Figure 10.10 Computed equilibrium and rate-limited NO concentrations at $\phi = 1.14$.

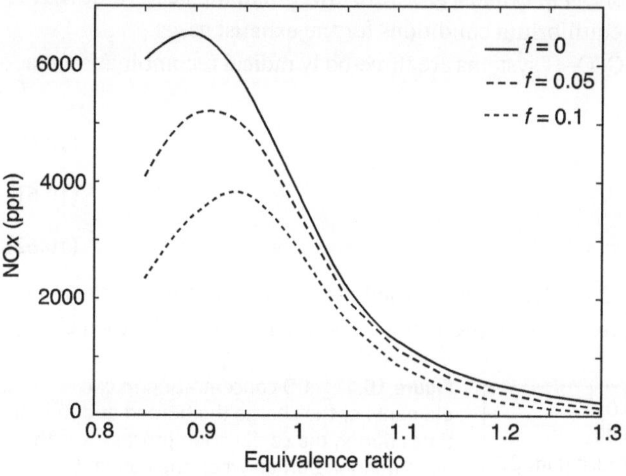

Figure 10.11 Computed exhaust NO concentration versus equivalence ratio ϕ and residual fraction f.

slightly rich. On the other hand, in rich mixtures, there is little excess oxygen to dissociate and attach to nitrogen atoms to form nitrogen oxide. The interplay between these two effects results in maximum nitrogen oxides occurring in slightly lean mixtures, where there is a slight excess of oxygen atoms to react with the nitrogen atoms.

As the residual fraction is increased, the NO levels decrease. Since the effective specific heat of the combusting mixture increases with increasing residual fraction, the peak combustion temperatures are lowered, resulting in a decrease in NO formation.

10.3 Carbon Monoxide

Carbon monoxide is a product of rich combustion since there is insufficient oxygen to convert all the carbon in the fuel to carbon dioxide. Another CO source is the dissociation of CO_2 at high combustion temperatures. CO subsequently reacts with oxygen and nitrogen oxides in the exhaust stream and atmosphere to form smog.

As discussed in Chapter 3, the formation and consumption of CO is a principal reaction pathway in hydrocarbon combustion, which essentially consists of breakdown of the hydrocarbon fuel to carbon monoxide, and then the oxidation of carbon monoxide to carbon dioxide. The majority of the heat release for the combustion process occurs during the CO oxidation, given by the following global reaction:

$$CO + OH \rightleftharpoons CO_2 + H \tag{10.59}$$

The C–O–H system in an internal combustion engine is more or less in equilibrium during combustion and expansion up to the point where the nitrogen oxide chemistry freezes. From measurements of the exhaust gas composition, a value of the equilibrium constant K can be calculated, which determines the temperature at which the reaction becomes frozen. Thus, whether it is a lean- or rich-running engine, one can determine the carbon monoxide concentration during these times using equilibrium chemistry assumptions. Late in the expansion stroke, with the gas temperatures down to about 1800 K, the chemistry in C–O–H systems starts to become controlled by chemical kinetics and is generally frozen by the time blowdown finishes, and the exhaust valve opens. Therefore, the measured values of CO in exhaust gases are lower than the peak values in the combustion chamber, but greater than equilibrium conditions for the exhaust gases.

The rate-controlling reactions in the C–O–H systems are three-body radical recombination reactions with molecule M:

$$H + H + M \rightleftharpoons H_2 + M \tag{10.60}$$

$$H + OH + M \rightleftharpoons H_2O + M \tag{10.61}$$

$$H + O_2 + M \rightleftharpoons HO_2 + M \tag{10.62}$$

Results obtained by using an unmixed model for the burned gas and accounting for these rate limiting reactions are illustrated in Figure 10.12. In these plots, x is the fraction of the total charge

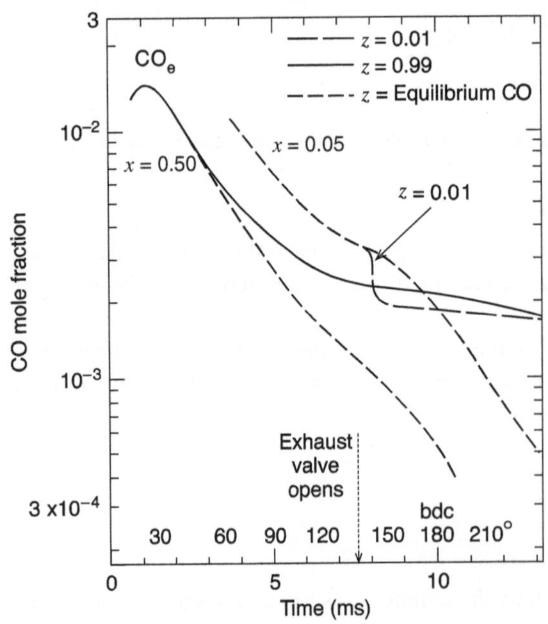

Figure 10.12 CO concentration in two elements of the charge the burned at different times during the combustion process; x is the mass fraction burned when the element burned, and z is the fraction of gas that has left the cylinder during the exhaust process. *Source*: Heywood 1976/with permission of Elsevier.

burned when an element is burned and z is the mass fraction that has left the cylinder at the time an element leaves the cylinder. Because that time is unknown, results are given for several values of z for each element. Gas that leaves early ($z \gg 1$) cools more rapidly than gas that leaves last ($z \simeq 1$).

The results show that gases that burn early carry more CO into the exhaust than gases that burn later. They also show the fortuitous fact that the frozen concentrations are close to the equilibrium concentrations that exist in the cylinder at the time the exhaust valve opens. This suggests an approximation that is often used in practice, to assume that the C–O–H system is in equilibrium until the exhaust valve opens, at which time, it then freezes instantaneously.

In lean-running engines, there appears to be an additional source of CO caused by the flame–fuel interaction with the walls, the oil films, and the deposits. Under these circumstances, the exhaust concentrations are so low they are not a practical problem, and thus details of these interactions remain largely unexplored. Thus, the key to minimizing CO emissions is to minimize the times an internal combustion engine must run rich (such as during start-up). Since diesel engines run lean overall, their emissions of carbon monoxide are low and generally not considered a problem. However, it does appear that direct-injection diesel engines emit relatively more CO than indirect-injection diesel engines.

10.4 Particulate Matter

Another common pollutant resulting from the combustion of hydrocarbons is particulate matter, i.e., soot. Particulates are primarily composed of carbon and hydrogen, with a C/H ratio of about 8:1. The yellow luminosity of a wood flame is due to the incandescence of the burning soot particles.

The size distribution of particulates is of importance since they consist of a mixture of fine, ultra-fine, and nanoparticles. Particles less than 2.5 μm in diameter ($PM_{2.5}$) are referred to as "fine" particles and are believed to pose the largest health risks. Particles with diameters between 2.5 and 10 μm are referred to as "coarse." Particles with diameters less than 0.1 μm (100 nm) are referred to as "ultra-fine," and with diameters less than 0.05 μm (50 nm) are referred to as "nanoparticles."

The chemical processes that occur in the formation of particulates are very complex and include pyrolysis, inception, surface growth, coalescence, graphitization, agglomeration, and oxidation. There is much unknown about the various reaction pathways for the formation and the oxidization of soot and their dependence on the combustion processes such as fuel atomization, combustion, turbulence, as well as overall cylinder pressure, temperature and equivalence ratio. A general mechanism of soot formation is shown in Figure 10.13. The organic fraction of a soot particle results from all the processes that generate hydrocarbons and their partial oxidation products and is also influenced by the turbulent fluid mixing during combustion.

The vast majority of the burning fuel undergoes oxidation to final products, indicated by the bold arrow in Figure 10.13. However, a portion of the fuel in fuel-rich zones in the premixed flames (or diffusion flames) follows the reaction path to soot via pyrolysis reactions. These reactions break down the fuel molecules into unsaturated radicals and intermediates like acetylene (C_2H_2), known as soot precursors. These soot precursors then react to form small polycyclic aromatic hydrocarbons (PAH). PAH is composed of multiple aromatic rings with carbon–carbon bonds internal to the ring structure, as shown in Figure 10.14. Five- and six-ring structures are the most common. Note that carbon atoms are shared between ring structures. The growth of the PAH species continues through reactions in which acetylene is added to aromatic ring structures. These incipient soot particles continue to grow by reactions with acetylene and smaller PAH, a process known as surface growth. They also collide and coalesce, resulting in the formation of soot particles.

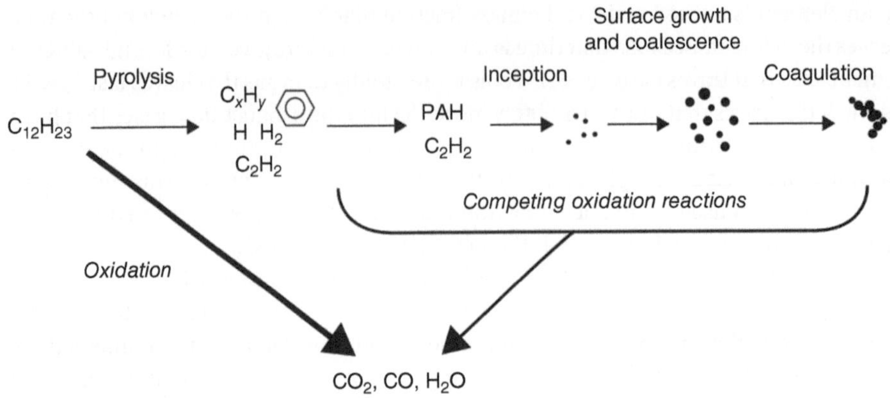

Figure 10.13 Soot formation process and oxidation pathways during combustion of hydrocarbon–air mixtures. (Adapted from Kuo and Acharya 2012.)

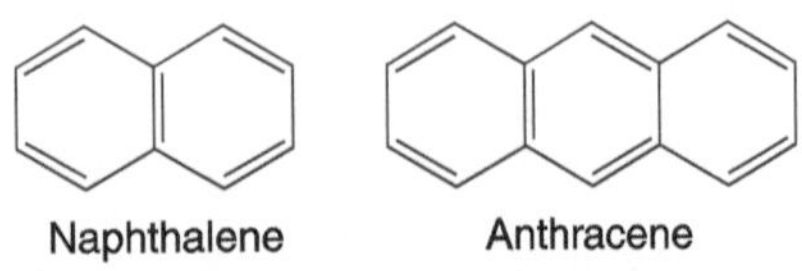

Figure 10.14 Two- and three-ring polycyclic aromatic hydrocarbon (PAH) structures.

Inspection of the soot particles under an electron microscope, as shown in Figure 10.15, reveals them to be an agglomerate of spherical clusters or spherules, each approximately 0.02–0.04 μm (20–40 nm) in diameter. Each spherule contains about 10^6 carbon atoms, in a concentric shell layering similar to the layers of an onion. The carbon atoms in each layer are bonded together in planes in a fractal cluster hexagonal network. Each agglomerate particle can contain as many as 4000 spherules in a variety of configurations. The agglomerate configuration can resemble a bunch of grapes or can be branched and chainlike in character.

10.4.1 Soot Combustion Modeling

Current soot combustion models assume fuel combustion with fuel-rich conditions will form fragmented hydrocarbons such as acetylene (C_2H_2) that then react to form PAH-type molecules. Formation of particulates starts with condensation of the gas phase hydrocarbons, which is followed by surface growth and agglomeration of the nuclei particles.

The soot particle inception is modeled as a condensation and dehydrogenation reaction of intermediate gas phase species,

$$C_2H_2 \longrightarrow C(s) + H_2$$

The incipient soot particle is typically assumed to be a nanoparticle 2.4 nm in diameter with about 700 carbon atoms. The soot particle surface growth involves the gas-phase deposition of hydrocarbon intermediates on the surfaces of the spherules. The surface growth, a heterogeneous surface reaction, results from acetylene and PAH molecules colliding and reacting with the soot carbon surface, increasing the soot particle volume.

$$nC(s) + C_2H_2 \longrightarrow (n+2)C(s) + H_2$$

The second stage of combustion burns the CO, soot, and other first-stage products to completion in a diffusion flame. The dominant oxidants are O_2 and OH, which react with the surface area

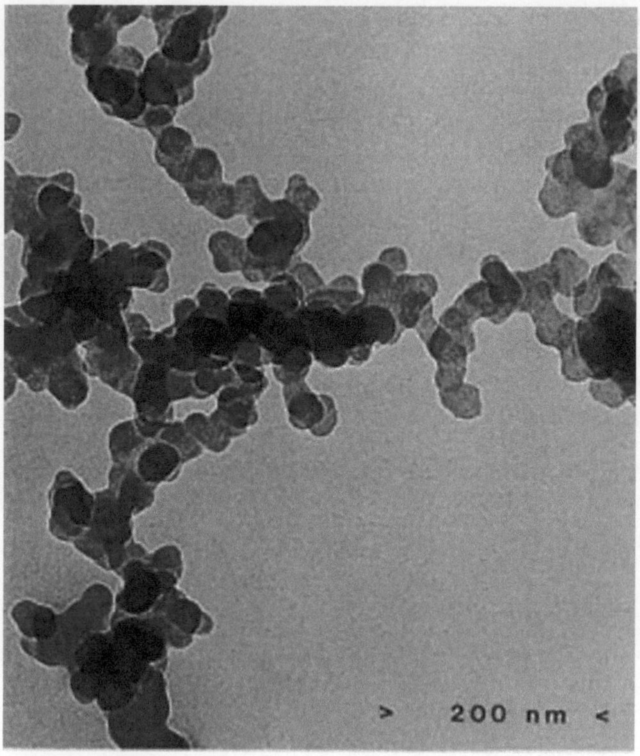

Figure 10.15 Electron microscope photograph of soot particle structure. *Source*: Dobbins et al. 1994/with permission of Springer Nature.

of the soot particles, reducing the soot particle size. However, due to the inhomogeneous mixture distribution, not all of the soot will be oxidized. The rate of soot oxidation is kinetically controlled. If there is not enough oxygen present, incomplete oxidation will result in a sooting flame.

OH and O concentrations in flames are determined by detailed chemical kinetics and cannot be accurately predicted from simple equilibrium at the local temperature and stoichiometry. This is particularly the case when active soot oxidation is occurring and the local temperature is decreasing with flame residence time. As a consequence, most attempts to model soot oxidation in flames have used a relation based on oxidation by O_2 and then applied a correction factor to augment the rate to include the effect of oxidation by radicals.

The net soot production rate during the above combustion stages is the difference between the soot formation (sf) and oxidation (so) rates:

$$\frac{dm_s}{dt} = \frac{dm_{\text{sf}}}{dt} - \frac{dm_{\text{so}}}{dt} \tag{10.63}$$

For the soot formation rate, many diesel engine simulations use a single-step Arrhenius-type expression (Hiroyasu and Kadota 1976) in which the formation rate is proportional to the mass of fuel vapor m_{fv}:

$$\frac{dm_{\text{sf}}}{dt} = A_p m_{\text{fv}} P^{0.5} \exp(-E_a/R_u T) \tag{10.64}$$

where A_p is the pre-exponential factor, and E_a is the activation energy. For diesel combustion, representative values are $A_p = 200$ 1/s, and $E_a = 52.30$ kJ/mol. More recently acetylene (C_2H_2) and even benzene (C_6H_6) are used as the inception species instead of the fuel vapor since acetylene is a dominant precursor in the soot formation process, and the fuel mass reactions include other precursor species that do not participate in soot formation.

A model widely used for the soot oxidation rate is the Nagle and Strickland-Constable (1962) semiempirical model of soot surface kinetics. Nagle and Strickland-Constable (NSC) observed at temperatures greater than 2000 K, a decrease in the oxidation rate until the surface reached a higher temperature. To explain this phenomenon, they proposed that carbon oxidation proceeds through two types of surface sites, types A and B, that both react with oxygen to produce CO and generate a new type A site. In this model, the soot oxidation occurs by two types of oxidation reactions (A and B) on the particle surface. The "A" and "B" reaction sites have different reactivities, k_a, k_z, and k_b, k_t, respectively. The "A" sites are more reactive, and their oxidation rate is proportional to the fraction of sites not covered by surface oxides. The "A" sites can convert to "B" sites with rate constant k_t. The surface oxidation reactions are

$$\mathrm{A} + \mathrm{O_2} \xrightarrow{k_a} \text{surface oxide}$$

$$\text{surface oxide} \xrightarrow{k_a/k_z} 2\mathrm{CO} + \mathrm{A}$$

$$\mathrm{B} + \mathrm{O_2} \xrightarrow{k_b} 2\mathrm{CO} + \mathrm{A}$$

$$\mathrm{A} \xrightarrow{k_t} \mathrm{B}$$

The soot oxidation rate is proportional to the soot mass m_s:

$$\frac{dm_{so}}{dt} = \frac{6.0\,\mathrm{M}}{\rho_s d_s} m_s R_{ox} \tag{10.65}$$

where M is the carbon molecular mass, ρ_s is the soot density, and d_s is the soot diameter. The term R_{ox} is the net surface oxidation rate, given by Eq. (10.66).

$$R_{ox} = \left(\frac{k_a P_{O_2}}{1 + k_z P_{O_2}}\right) x_a + k_b P_{O_2}(1 - x_a) \tag{10.66}$$

and x_a is the fraction of the soot particle surface populated by A reaction sites, given by Eq. (10.67).

$$x_a = \frac{P_{O_2}}{P_{O_2} + (k_t/k_b)} \tag{10.67}$$

For a given temperature and relatively low (<1 atm) oxygen partial pressures, the soot oxidation rate is the first order with oxygen partial pressure, and at higher (>1 atm) partial pressures, the rate asymptotically approaches a zero-order limit with oxygen partial pressure.

The rate constants in Eqs. (10.66) and (10.67) are:

$$\begin{aligned} k_a &= 20 \exp(-15,100/T) \quad (\mathrm{g/cm^2\text{-}s\text{-}atm}) \\ k_b &= 4.66 \times 10^{-3} \exp(-7640/T) \quad (\mathrm{g/cm^2\text{-}s\text{-}atm}) \\ k_t &= 1.51 \times 10^5 \exp(-48,800/T) \quad (\mathrm{g/cm^2\text{-}s}) \\ k_z &= 21.3 \exp(+2060/T) \quad (1/\mathrm{atm}) \end{aligned} \tag{10.68}$$

The above soot formation-oxidation model is also known as the Hiroyasu two-step soot model. It has been used extensively to compute soot distribution and transport in conventional high-temperature diesel combustion simulations (e.g., see Dempsey et al. 2018).

For more complex combustion situations, additional phenomenological models are required. The models include additional considerations such as soot particle inception, polymerization, turbulence, particle coagulation, and soot oxidation by other species such as OH radicals. For example, Frenklach (2019) has developed a soot surface oxidation model that incorporates more rigorous multisite kinetics and surface reactivity. Sun and Rigopoulos (2022) used a combination of gas-phase chemistry, soot kinetics, and aerosol dynamics to determine soot-forming zones and profiles of soot volume fraction in diffusion flames.

Example 10.2 ***Soot Formation and Oxidation Rates***

A diesel fuel combustion process occurs at a constant pressure of 100 atm. Assume the acetylene (C_2H_2) mass is 1.0×10^{-3} g, the soot mass is 1.0×10^{-5} g, soot particle diameter is 2.5×10^{-6} cm, and soot density is 2.0 g/cm^3. Using the Hiroyasu two-step soot model, calculate and plot the soot formation and oxidation rates for an engine operating at 1800 rpm for temperatures ranging from 1200 to 2750 K.

(a) At what temperature will the soot formation and oxidation rates be the same?
(b) At what temperature will the net soot production be a maximum? Assume the mole fraction of oxygen in this process is 0.20.

Solution

The above Hiroyasu two-step equations have been incorporated into the Matlab program `Soot.m` listed in the text website. The inputs to the program are the pressure and temperature and soot particle properties. The program computes the soot formation and oxidation rates, and the net soot production rate. The results of the computations are presented in Figure 10.16.

(a) At lower temperatures, the soot formation rate is greater than the oxidation rate, and as the combustion temperature increases, soot oxidation increases at a greater rate than soot formation. Above 2500 K, the oxidation rate is greater than the formation rate.
(b) The maximum net soot production is about 5.7×10^{-6} g/degree at a temperature of 2085 K.

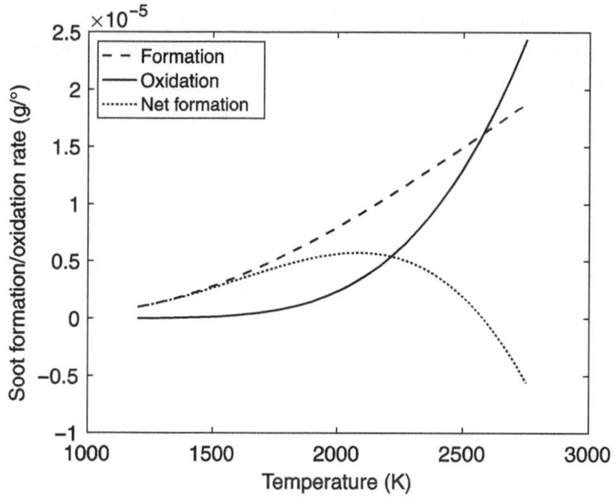

Figure 10.16 Soot formation and oxidation versus temperature. (Example 10.2).

10.5 Soot Radiation

10.5.1 Radiation Background

During the combustion process of a hydrocarbon fuel in an engine, the high-temperature gases and soot particles radiate through the combusting fuel–air mixture to the cylinder walls. The portion of total chemical energy released during combustion that is lost to radiation heat transfer to the cylinder walls varies with crank angle and depends on the specific engine parameters and the particulars of the combustion process.

In a spark ignition engine, the thin flame front propagates quickly across the combustion chamber through a nearly homogeneous fuel–air mixture, and so the gaseous and particulate matter radiation is very small in comparison to the convection heat transfer to the cylinder wall.

In a compression ignition engine, fuel burns in a turbulent diffusion flame formed around the entire fuel spray in the region where the equivalence ratio is close to 1. As a consequence, the radiative heat transfer to the cylinder walls in compression ignition engines is greater than that in spark ignition engines and can be as high as 10–30% of the total wall heat loss, increasing with engine operating load and equivalence ratio. As the equivalence ratio is increased, the flame temperature and concentration of combustion products increase.

The propagation of radiation through participating media, such as the case in a combusting gas–particulate mixture in an engine, involves radiation absorption, emission, and scattering over a wavelength spectrum throughout the combustion chamber. The radiation originating from CO_2 and H_2O gas-phase molecules is primarily in narrow spectral bands. The CO_2 bands are centered at 2.0, 2.7, 4.3, 10.4, and 15 μm, and the H_2O bands are centered at 1.38, 1.88, 2.7, 6.3, and 20–200 μm. In contrast, during the soot oxidation process, the radiation generated from soot particles occurs over a continuous spectrum. The majority of the emitted radiation is reabsorbed by the combusting mixture before reaching the cylinder walls.

The relative contributions of the molecular gas and soot radiation are an active area of research. Recent research on radiation in diesel engines has found that the radiation from the CO_2 and H_2O molecules can contribute a significant portion to the net radiation heat transfer to the cylinder walls in comparison with the soot radiation, especially at low values of the soot volume fraction. Since the combustion of non-carbon fuels such as hydrogen and ammonia does not produce soot and CO_2, their thermal radiation originates only from H_2O gas molecules.

The absorption of radiation produced during the combustion process is modeled with a spectral absorption coefficient K_λ. The decrease in the radiation intensity of a given wavelength λ over a distance ds along the radiation path due to absorption has been found to depend on the following relationship,

$$dI_\lambda = -K_\lambda I_\lambda ds \tag{10.69}$$

where K_λ is the spectral absorption coefficient, and I_λ is the spectral radiation intensity, defined as the radiant energy per unit area, per unit solid angle, per unit wavelength (W/m^2-sr-μm). Integrating over a distance L for a constant K_λ,

$$I_\lambda(L) = I_\lambda(0)\exp(-K_\lambda L) \tag{10.70}$$

which is called Beer's law, named after the German scientist A. Beer (1825–1868). The spectral transmissivity τ_λ is defined as

$$\tau_\lambda = \frac{I_\lambda(L)}{I_\lambda(0)} = \exp(-K_\lambda L) \tag{10.71}$$

and the spectral absorptivity α_λ is defined as

$$\alpha_\lambda = 1 - \tau_\lambda = 1 - \exp(-K_\lambda L) \tag{10.72}$$

The parameter $K_\lambda L$ can be interpreted physically as the optical depth of the combustion gas. If $K_\lambda L < 0.1$, the gas is considered "optically thin," with high transmissivity, and if $K_\lambda L > 10$, the gas is considered "optically thick," with high absorptivity.

A blackbody is a theoretically perfect emitter and used as a reference for actual radiation emission. A blackbody at temperature T emits radiant energy over a continuous spectrum, described by the following Planck intensity distribution,

$$I_{b\lambda}(T) = \frac{C_1}{\lambda^5[\exp(C_2/\lambda T) - 1]} \tag{10.73}$$

where $C_1 = 1.191 \times 10^{-16}$ W-m^2, and $C_2 = 1.439 \times 10^{-2}$ m-K. The blackbody spectral emissive power $E_{b\lambda}$ is the integral over all solid angles,

$$E_{b\lambda} = \pi I_{b\lambda} \tag{10.74}$$

The blackbody emissive power E_b (W/m^2) is found by integrating Eq. (10.74) over all wavelengths yielding

$$E_b(T) = \int_0^\infty E_{b\lambda} d\lambda = \sigma T^4 \tag{10.75}$$

where the Stefan–Boltzmann constant $\sigma = 5.67 \times 10^{-8}$ W/m^2-K^4.

The spectral emissivity ϵ_λ is the ratio of the radiant energy emitted at wavelength λ into a solid angle $d\Omega$ to the radiant energy emitted into that solid angle by a blackbody at the same temperature and wavelength. From Kirchhoff's law, $\epsilon_\lambda = \alpha_\lambda$, so the emissivity is

$$\epsilon_\lambda = 1 - \exp(-K_\lambda L) \tag{10.76}$$

The total radiative emissivity depends on the optical depth $K_\lambda L$, temperature T, and pressure P, as indicated by Eq. (10.77),

$$\begin{aligned} \epsilon(T, L, P) &= \frac{1}{E_b}\int_0^\infty \epsilon_\lambda E_{b\lambda} d\lambda \\ &= \frac{\pi}{\sigma T^4}\int_0^\infty I_{b\lambda}[1 - \exp(-K_\lambda L)] d\lambda \end{aligned} \tag{10.77}$$

An expression similar to Eq. (10.69) can be written for the increase in intensity due to radiation emission. Kirchhoff's law for optically thick gases in which the emission sources can be modeled as blackbodies with intensity $I_{b\lambda}$ implies

$$dI_\lambda = +K_\lambda I_{b\lambda} ds \tag{10.78}$$

Combining the absorption and emission equations, Eqs. (10.69) and (10.78), yields the radiative transfer equation,

$$\frac{dI_\lambda}{ds} = +K_\lambda(-I_\lambda + I_{b\lambda}) \tag{10.79}$$

One can add a radiation scattering term to the radiative transfer equation, but in internal combustion engine applications, radiation scattering is usually considered to be negligible. There are a variety of numerical methods for solution of the radiative transfer equation including flux methods, Monte Carlo techniques, the discrete ordinates method (DOM), and the discrete transfer method (DTM). The DOM discretizes the radiative transfer equation for a set of finite solid angle directions. The resulting discrete ordinates equations are solved along the solid angle directions using a control volume technique.

10.5.2 Soot Emissivity and Emissive Power

Soot particles are formed in the fuel-rich zone of the diffusion flame as an intermediate combustion step in a compression ignition engine and radiate as they are oxidized in the high-temperature region of the flame. The spectral absorption coefficient, K_λ, of a soot cloud has been found to scale with ratio of the soot volume fraction and the wavelength,

$$K_\lambda \simeq \frac{f_v}{\lambda} = \frac{kf_v}{\lambda} \tag{10.80}$$

where f_v is the soot (s) volume fraction,

$$f_v = \frac{V_s}{V} = \frac{m_s}{\rho_s V} \tag{10.81}$$

and k is a constant of proportionality depending on the H/C ratio of the soot particle. Following Yuen and Tien (1977), upon substitution of Eqs. (10.73) and (10.80) into Eq. (10.77), and subsequent integration,

$$\begin{aligned}\epsilon_s(T,L) &= 1 - \frac{1}{\left(1+\frac{k}{C_2}f_v LT\right)^4} \\ &\simeq 1 - \exp\left(-3.6\frac{k}{C_2}f_v LT\right)\end{aligned} \tag{10.82}$$

The emissive power E_s of a soot cloud is

$$E_s = \epsilon_s E_b(T) = \epsilon_s \sigma T^4 \tag{10.83}$$

Example 10.3 ***Thermal Radiation***

Compare the total emissivity and emissive power of (a) the gaseous (g) thermal radiation and (b) the soot (s) thermal radiation during combustion of cetane $C_{16}H_{34}$ in a compression ignition engine for both light and heavy soot masses m_s of 1.0×10^{-5} g and 1.0×10^{-4} g, respectively. Assume the following instantaneous conditions: combustion temperature $T = 2000$ K, combustion chamber volume $V = 17\,\text{cm}^3$ and pressure $P = 15$ bar, characteristic path length $L = 7.0$ cm, and soot particle density $\rho_s = 2\,\text{g/cm}^3$.

Solution

(a) The gaseous radiation from the CO_2 and H_2O gas-phase molecules can be found using the procedure developed by Hottel and Egbert (1942), a methodology included in most heat transfer books. Due to the numerous charts and graphs used, it is not reproduced here. With the Hottel procedure, the total gas emissivity ϵ_g is expressed as

$$\epsilon_g = \epsilon_w + \epsilon_c - \Delta\epsilon \tag{10.84}$$

where ϵ_w is the water emissivity, ϵ_c is the carbon dioxide emissivity, and $\Delta\epsilon$ is the wavelength overlap correction factor. The water and carbon dioxide emissivities are a function of the gas temperature T_g and the product of their partial pressure and characteristic length $P_i L$. For combustion of cetane at a pressure of $P = 15$ bar and temperature $T = 2000$ K, the partial pressures of CO_2, and H_2O are each approximately 2 bar. Using the Hottel procedure, the predicted emissivities are $\epsilon_w = 0.11$, $\epsilon_c = 0.084$, and $\Delta\epsilon = 0.035$, so

$$\epsilon_g = 0.11 + 0.084 - 0.035 = 0.16 \tag{10.85}$$

and the gas radiation emissive power E_g is

$$E_g = \epsilon_g \sigma T^4 = (0.16)(5.67 \times 10^{-8})2000^4 = 1.45 \times 10^5\ \mathrm{W/m^2} \tag{10.86}$$

(b) From Eq. (10.81), the soot volume fractions f_v are

$$f_v = \frac{m_s}{\rho_s V} = \frac{1 \times 10^{-5}}{(2)(17)} = 2.9 \times 10^{-7}\ \text{(light)}$$

$$= \frac{1 \times 10^{-4}}{(2)(17)} = 2.9 \times 10^{-6}\ \text{(heavy)}$$

and from Eq. (10.82), the total soot emissivity ϵ_s is

$$\epsilon_s = 1 - \exp\left(-3.6 \frac{k}{C_2} f_v \mathrm{LT}\right)$$

$$= 1 - \exp\left[-3.6 \frac{5}{1.439} (2.9 \times 10^{-7})(7.0)(2000)\right] = 0.051\ \text{(light)}$$

$$= 1 - \exp\left[-3.6 \frac{5}{1.439} (2.9 \times 10^{-6})(7.0)(2000)\right] = 0.40\ \text{(heavy)}$$

The soot emissive power E_s is

$$E_s = \epsilon_s \sigma T^4 = (0.051)(5.67 \times 10^{-8})2000^4 = 4.64 \times 10^4\ \mathrm{W/m^2}\ \text{(light)}$$

$$= (0.40)(5.67 \times 10^{-8})2000^4 = 3.61 \times 10^5\ \mathrm{W/m^2}\ \text{(heavy)}$$

Note: Depending on the soot volume fraction, the soot thermal radiation can be less or greater than the gas thermal radiation.

10.5.3 Diesel Engine Particulates

Particulates are a major emissions consideration for diesel engines, as their performance is soot or smoke limited. Particulate emissions in spark-ignition engines are comparatively much less and occur with direct fuel injection combustion at high equivalence ratios. A high concentration of particulate matter is manifested as visible smoke or black soot in the exhaust gases. Particulates form in diesel engines because diesel combustion relies on diffusion flames, and there is incomplete combustion of the fuel in the fuel-rich zone of the diffusion flame. Even though the diesel engine combustion process is lean overall, there are regions of fuel-rich combustion, both in the premixed phase and in the mixing-controlled combustion phases of combustion. Most diesel engines require after-treatment devices to meet regulated levels for particulate matter.

In-cylinder techniques used to decrease soot in diesel engines are to increase the turbulence intensity, advance the injection timing, decrease the amount of exhaust gas recirculation (EGR), and increase the premixed combustion phase. Enhancement of the turbulence intensity will promote combustion and suppress soot formation. The premixed combustion phase can be increased by increased injection pressure, finer fuel spray, and increased swirl. High-pressure injection will increase atomization and entrainment, making the fuel–air mixture leaner. The advanced injection timing and decreased EGR will result in higher temperatures and thus greater oxidation during the expansion stroke. An increase in turbulence can also be accomplished through the use of auxiliary gas injection during the late stages of combustion.

It is challenging to diesel engine designers that, generally, when a reduction in soot has been achieved it is at the expense of an increase in NO_x. This is due to the fact that an increase in

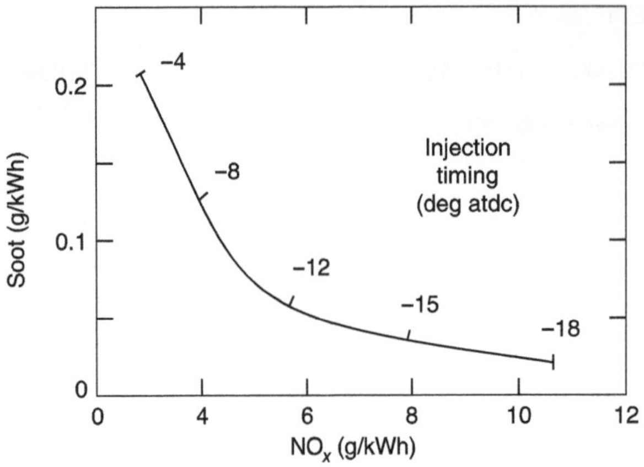

Figure 10.17 Representative plot of soot and NO_x tradeoff versus injection timing.

the temperature of the diffusion flame will increase the oxidization of soot but also increase the NO_x formation. Figure 10.17 is an example plot indicating that as the timing is advanced, the soot decreases, but the NO_x increases, creating a tradeoff between NO_x and smoke.

Equivalence ratio–temperature ($\phi - T$) plots are used to show contour plots of NO_x and soot concentrations during the diesel engine combustion process. A representative $\phi - T$ diagram is shown in Figure 10.18. The diagram indicates that soot is formed in regions with $\phi > 2$ at temperatures of approximately 1700 K, while NO_x is formed in regions with $\phi < 2$ at temperatures above approximately 2000 K. The amount of soot created is highly dependent on the equivalence ratio. As the equivalence ratio is doubled, increasing fuel-rich combustion zones, the soot increases by an order of magnitude.

Also plotted in Figure 10.18 are the evolution of packets of injected diesel fuel at 5, 20, and 35° after start of injection (ASI) as computed by a quasi-dimensional diesel combustion model (Dobos and Kirkpatrick 2017). The fuel and air begin to react at +5° ASI at 1000 K with a rich mixture. Note that there are fuel packets that are in a soot formation region at +20° ASI.

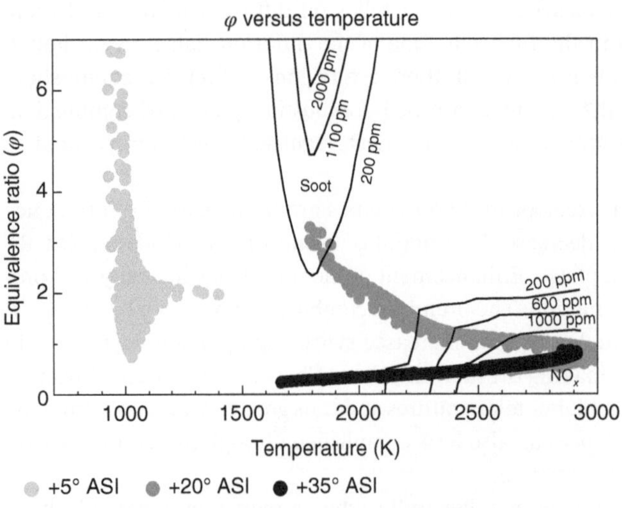

Figure 10.18 Soot and NO_x formation on a $\phi - T$ plot.

At +35° ASI, as mixing and combustion continue, the packets enter the NO_x formation region as a lean or stoichiometric high-temperature mixture.

10.6 Emissions Control in Internal Combustion Engines

Meeting emission requirements has been a major challenge and also an opportunity for combustion engineers. Prior to the mid-1960s, exhaust emissions were uncontrolled. In 1968, the United States adopted the Clean Air Act which regulated vehicular and stationary emissions on a nationwide basis. Since then exhaust emissions from internal combustion engines have been regulated worldwide by governmental agencies. In the United States, exhaust regulations are set by the U.S. EPA. The 1968 Clean Air Act required the EPA to set national ambient air quality standards for "criteria pollutants." Currently, there are six "criteria pollutants":

- carbon monoxide
- nitrogen oxides
- sulfur oxides
- ozone
- lead
- particulate matter

The Clean Air Act also defined a list of hazardous air pollutants (HAPs). Currently, there are 188 HAP's identified, such as benzene, formaldehyde, and numerous other VOCs. During the intervening years since the mid-1960s, emission requirements have become increasingly rigorous, and internal combustion engines today generate significantly less pollution than their 1968 counterparts. Large displacement engines used in construction, agriculture, locomotive, and marine applications, stationary engines, and small displacement engines used in non-road applications, such as lawn mowers, snow blowers, and chainsaws, are also now regulated, since they also have been found to be significant sources of emissions.

There are two basic methods used to control engine emissions: control of the combustion process and the use of after-treatment devices in the exhaust system. Application of technological advances in fuel injectors, oxygen sensors, and onboard computers to engines has increased the control and subsequent optimization of the engine combustion process. Alternative and oxygenated fuels have also been used to reduce emissions.

Two NO_x control measures that have been used in vehicular engines since the 1970s are retard of the ignition spark and EGR. The aim of these measures is to reduce the peak combustion temperature and thus the formation of NO_x. Retarding the spark timing lowers the NO_x since a greater fraction of the combustion occurs in an expanding volume, lowering the peak cylinder pressure and temperature. With the use of EGR, some fraction of the exhaust gas is routed back into the intake manifold. The nonreactive exhaust gas acts as a diluent in the fuel–air mixture, lowering the combustion temperature, and the NO_x formation. In spark ignition engines, the EGR fraction increases with engine load up to the lean limit, which is about 15–20% of the fuel–air flow rate. Greater EGR fractions are used in compression ignition engines, since the combustion is heterogeneous.

10.6.1 Exhaust After-Treatment Devices

A variety of exhaust after-treatment devices have been added to vehicles to meet emission requirements. These include catalytic converters, NO_x traps, and particulate filters. Currently, the most important after-treatment device is the three-way heterogenous catalyst, invented in 1950 by the American engineer Eugene Houdry, and first installed on the exhaust systems in passenger cars in 1975. It derives its name from the fact that it works on all three of the gaseous pollutants of concern: carbon monoxide, hydrocarbons, and nitrogen oxides. Heterogenous catalysts provide a solid surface for gaseous reactions and reduce the activation energy required for the reaction. When the reactants are adsorbed on the catalyst surface, their internal molecular bonds are weakened, new bonds are more easily formed between fragments of different molecular species, forming reaction products which then desorb from the catalyst surface back into the exhaust stream.

The surface area for reaction is important as it determines the number of catalytic sites. All catalytic converters are built in a porous honeycomb or pellet geometry to expose the exhaust gases to a larger surface made of small particles (< 50 nm) of one or more of the noble metals, platinum (Pt), palladium (Pal), rhodium (Rh), and cerium oxide (CeO_2). Platinum is the principal metal used to remove HC and CO, and rhodium is the principal metal used to remove NO.

Figure 10.19 is a schematic of a three-way honeycomb catalyst. In the converter shown, a thin layer of the noble metals covers a washcoat of inert alumina Al_2O_3 on a cordierite honeycomb foundation. The operation of the catalytic converter is severely inhibited by lead and sulfur compounds in the exhaust so that vehicular fuels have been reformulated to eliminate their lead and sulfur content.

The reactions removing CO and HC are oxidation reactions as shown in Eqs. (10.87) and (10.88), thus work best in a lean environment with $\phi < 1$, as the excess oxygen will drive the reactions to completion:

$$2CO + O_2 \longrightarrow 2CO_2 \tag{10.87}$$

$$C_xH_{2x+2} + \frac{3x+1}{2}O_2 \longrightarrow xCO_2 + (x+1)H_2O \tag{10.88}$$

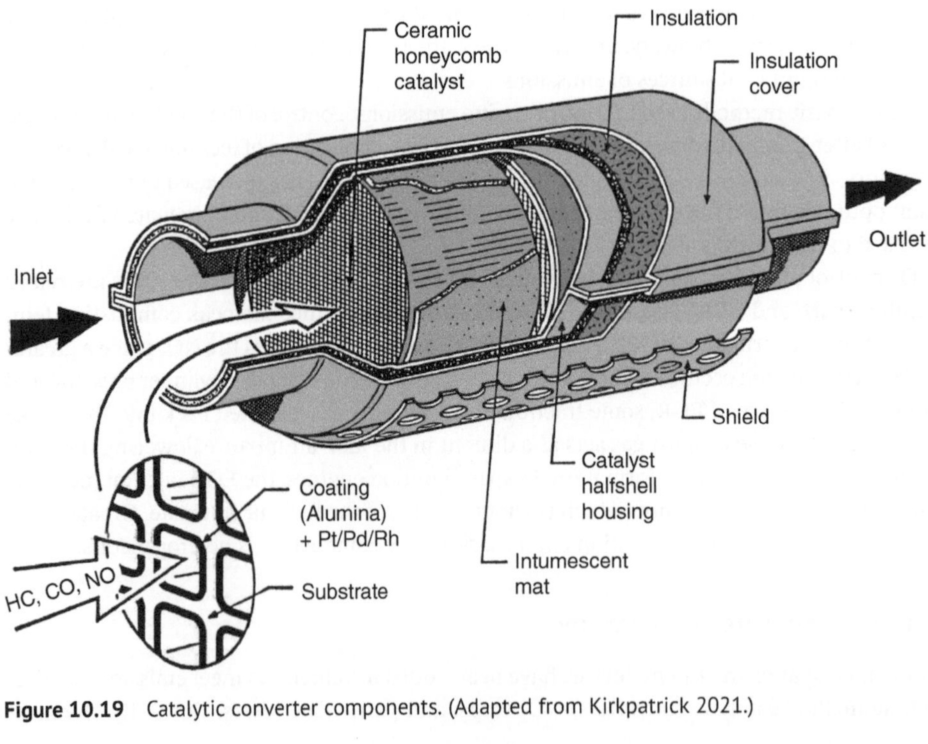

Figure 10.19 Catalytic converter components. (Adapted from Kirkpatrick 2021.)

The reactions removing the NO are reduction reactions involving CO, H_2, and HC. Since these are reduction reactions, they work best in a rich environment with $\phi > 1$. Two representative reactions are given in Eqs. (10.89) and (10.90),

$$NO + CO \longrightarrow \frac{1}{2}N_2 + CO_2 \tag{10.89}$$

$$NO + H_2 \longrightarrow \frac{1}{2}N_2 + H_2O \tag{10.90}$$

As the exhaust gases flow through the catalyst, the CO and hydrocarbons are removed through the above oxidation reactions forming CO_2 and H_2O products. The oxidation rate of hydrocarbons increases with molecular weight, so that the oxidation of low molecular weight fuels such as methane is very slow in the converter. The NO reacts with the CO, the hydrocarbons, and H_2 via reduction reactions on the surface of the catalyst. A three-way catalyst will function correctly only if the exhaust gas composition is within 1% of stoichiometric conditions. If the exhaust is too lean, nitrogen oxides are not destroyed and if it is too rich, carbon monoxide and hydrocarbons are not destroyed (see Figure 10.20). Herein lies one constraint that emission control imposes upon engine operation; to use a three-way catalyst, the engine must operate in a narrow window about stoichiometric fuel–air ratios.

10.6.2 NO_x Control Techniques for Lean Combustion Engines

With lean combustion engines, such as diesel and natural gas engines, reduction of the exhaust nitrogen oxides is poor because of the high oxygen content of the exhaust gases. Two nitrogen oxide control techniques which have been developed for lean combustion conditions are selective catalytic reduction (SCR) and the lean NO_x trap (LNT).

Selective catalytic reduction (SCR) uses liquid ammonia (NH_3) or urea ($CO(NH_2)^2$) sprayed into the exhaust stream upstream of the catalytic converter. As the urea mixes with the hot exhaust

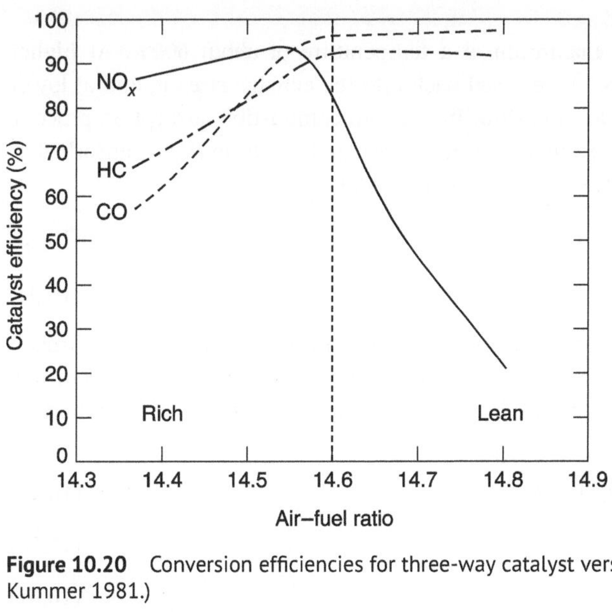

Figure 10.20 Conversion efficiencies for three-way catalyst versus air–fuel ratio. (Adapted from Kummer 1981.)

gases, it decomposes into ammonia and CO_2. Subsequent reactions with NO_x over a Pt/Rh catalyst produce nitrogen and water vapor, as indicated by the global reactions (10.91) and (10.92):

$$4NO + 4NH_3 + O_2 \longrightarrow 4N_2 + 6H_2O \tag{10.91}$$

$$2NO_2 + 4NH_3 + O_2 \longrightarrow 3N_2 + 6H_2O \tag{10.92}$$

The SCR control technique was developed by the Englehard Corporation in 1957. The largest users of SCR are heavy-duty commercial trucks, and off-road equipment that need to meet EPA Tier 4 emission standards. For vehicular applications, a mixture of about 32% urea and 68% water, called DEF (diesel exhaust fluid) is typically used, with an injection rate of about 2% of the diesel fuel flow rate. The urea injection rate needs to be controlled, since the catalyst efficiency, typically about 90%, is a function of temperature. Delivering the proper amount of the reductant fluid to the catalyst is a challenge for engine control systems, especially in transient conditions. If the injection rate is too low, some NO_x will not be reduced, and it the injection rate is too high, some ammonia or urea will pass or "slip" into the exhaust stream.

The disadvantages of SCR include the additional urea or ammonia injection system cost, weight, and maintenance. A separate onboard tank is required for urea storage, and periodic replenishment is required. The storage tank needs to be heated since the urea freezing point is −11° C. The use of ammonia SCR systems is limited to large marine engines and large bore stationary engines, due to the high toxicity of ammonia. Since these applications operate primarily at steady-state conditions, the risk of significant ammonia slip is reduced.

The lean NO_x trap (LNT) operates in a cyclic fashion, in which the trap adsorbs and stores NO_x as a nitrate during lean burn operation and releases it as molecular nitrogen during a short fuel-rich reducing desorption process. The NO_x trap is composed of an alkali earth material and a precious metal catalyst such as platinum washcoated onto a flow-through ceramic monolith. The alkaline earth material, such as barium carbonate, is a storage component that absorbs and forms nitrate species on the surface of the catalyst during the adsorption process. Once these nitrate species saturate the catalyst, the catalyst is regenerated by operating the engine in a rich condition to reduce the NO_2 to N_2. After regeneration, the alkali or alkaline earth materials are again available for NO_2 trapping.

The storage capacity of the LNT is maximum at a temperature of about 600 K. At higher temperatures, the NO_x stored in the LNT is released back into the exhaust stream, and at lower temperatures, the adsorption reactions are too slow. In the lean combustion adsorption process, the NO is oxidized to NO_2 as it flows over the platinum catalyst, and resulting NO_2 is adsorbed as the salt barium nitrate, as indicated by reactions (10.93) and (10.94).

$$NO + \frac{1}{2}O_2 \longrightarrow NO_2 \tag{10.93}$$

$$BaCO_3 + 2NO_2 \longrightarrow Ba(NO_3)^2 + CO \tag{10.94}$$

During the short fuel-rich desorption step, excess fuel is used to produce molecular nitrogen as indicated by reactions (10.95)–(10.97). With these three reactions, the adsorbed barium nitrate reacts to form barium oxide, the NO_2 is reduced to N_2 in the presence of CO, and the barium carbonate is restored.

$$Ba(NO_3)^2 \longrightarrow BaO + 2NO_2 + CO + O \tag{10.95}$$

$$2NO_2 + 2CO \longrightarrow N_2 + 2CO_2 \tag{10.96}$$

$$BaO + CO_2 \longrightarrow BaCO_3 \tag{10.97}$$

The engine control system controls the cycling rate between lean and rich burn conditions to achieve a desired NO_x conversion efficiency. Lean NO_x traps have been used on diesel, gasoline, and natural gas-fueled engines, in which the engine fuel was used as a fuel source for the rich reducing step, so this technique has a fuel penalty.

Example 10.4 ***SCR Injection System***

A 12 l four-stroke diesel truck engine uses an SCR ammonia injection system to remove NO from the engine exhaust. The engine operates at 2000 rpm with a volumetric efficiency of 0.90. If 0.15% of the nitrogen entering the engine has been converted to NO during the combustion process, what is the mass flow (kg/hr) of ammonia needed to reduce the exhaust NO to N_2? Assume air density $\rho_i = 1.17\ \text{kg/m}^3$, air molecular mass $M_a = 28.97\ \text{kg/kmol}$, and ammonia molecular mass $M_{NH_3} = 17.03\ \text{kg/kmol}$.

Solution

The air flowrate into the engine is

$$\begin{aligned}\dot{m}_a &= e_v \cdot \rho_i \cdot V_d \cdot N/2 \\ &= (0.90)(1.17)(12 \times 10^{-3})\left(\frac{2000}{2 \cdot 60}\right) \\ &= 0.21\ \text{kg/s}\end{aligned}$$

so the air molar flowrate is

$$\begin{aligned}\dot{n}_a &= \frac{\dot{m}_a}{M_a} = \frac{0.21}{28.97} \\ &= 7.27 \times 10^{-3}\ \text{kmol/s}\end{aligned}$$

Since air is 79% N_2, the nitrogen molar flowrate is

$$\begin{aligned}\dot{n}_{N_2} &= y_{N_2} \cdot \dot{n}_a = (0.79)(7.27 \times 10^{-3}) \\ &= 5.74 \times 10^{-3}\ \text{kmol/s}\end{aligned}$$

The global NO formation reaction during the combustion process is

$$N_2 + O_2 \rightarrow 2NO$$

so two moles of NO are formed from 1 mol of N_2, and for 0.15% conversion

$$\begin{aligned}\dot{n}_{NO} &= 2\dot{n}_{N_2} \cdot (0.0015) = 2\left(5.74 \times 10^{-3}\right)(0.0015) \\ &= 1.72 \times 10^{-5}\ \text{kmol/s}\end{aligned}$$

In the SCR injection system, as indicated by Eq. (10.91), there are 4 kmols of NO reacting with 4 kmols of NH_3, so they have the same molar flowrate,

$$\dot{n}_{NH_3} = \dot{n}_{NO}$$

The mass flowrate of NH_3 is

$$\begin{aligned}\dot{m}_{NH_3} &= \dot{n}_{NH_3} \cdot M_{NH_3} = \left(1.72 \times 10^{-5}\right)(17.03) \\ &= 2.93 \times 10^{-4}\ \text{kg/s} = 1.05\ \text{kg/h}\end{aligned}$$

10.6.3 Diesel Particulate Filters

Diesel particulate filters (DPFs) are an after-treatment technique used upstream of the catalytic converters to remove particulates or soot from the exhaust stream of a diesel engine. They were introduced for vehicular applications in 2000 to meet increased diesel particulate standards, 0.01 g/bhp-hr for heavy-duty engines, and 0.02 g/mi for light-duty engines. The filter efficiency is in the range of 98–99% or greater for filtration of particulate matter down to 1 μm size. The filters are made from cordierite ceramic, silicon carbide, or ceramic fiber and have the same honeycomb flow geometry as a catalytic converter.

The particulate filters are regenerated by periodically burning off the collected particulates, so the particulate filter must be able to withstand carbon combustion temperatures of about 870 K. The frequency of regeneration is governed by the engine control system, typically by measurement of the pressure drop in the filter. There are a number of regeneration processes used, ranging from fuel injection into the cylinder during the exhaust stroke, resistive heating coils upstream of the filter, or use of a fuel-born catalyst. The regeneration process will continue until the pressure differential across the DPF is decreased to an acceptable level. If the soot level has built up to a point where the DPF cannot be regenerated, the DPF is replaced.

References

Bond, T., D. Streets, K. Yarber, S. Nelson, J. Woo, and Z. Klimont (2004), A technology-based global inventory of black and organic carbon emissions from combustion, *J. Geophys. Res.*, Vol. 109, pp. D14203/1–D14203/43.

Bozzelli, J., and A. Dean (1995), O + NNH: A possible new route for NO_x formation in flames, *Int. J. Chem. Kinet.*, Vol. 27, 11, pp. 1097–1109.

Dempsey, A., P. Seiler, K. Svensson, and Y. Qi (2018), Evaluation of the two-step Hiroyasu soot model over a broad range of diesel combustion systems, *SAE Technical Paper* 2018-01-0242.

Dobbins, R., and H. Subramanisivam (1994), Soot precursor particles in flames, in *Soot Formation in Combustion*, Springer/Berlin, pp. 290–301.

Dobos, A., and A. Kirkpatrick (2017), Quasi-dimensional modeling of diesel combustion using detailed chemical kinetics, *ASME J. Eng. Gas Turbines Power*, Vol. 139, p. 081502.

Eilenberg, S., K. Bilsback, M. Johnson, J. Kodros, E. Lipsky, A. Naluwagga, K. Fedak, M. Benka-Coker, B. Reynolds, J. Peel, M. Clark, M. Shanh, S. Sambandam, C. L'Orange, J. Pierce, R. Subramanian, J. Volckens, and A. Robinson (2018), Field measurements of solid-fuel cookstove emissions from uncontrolled cooking in China, Honduras, Uganda, and India, *Atmos. Environ.*, Vol. 190, pp. 116–125.

Fenimore, C. (1970), Formation of nitric oxide in premixed hydrocarbon flames, *13th Symposium (Intl.) on Combustion*, The Combustion Institute, pp. 373–380.

Fenimore, C. (1979), Studies of fuel-nitrogen species in rich flame gases, *17th Symposium (Intl.) on Combustion*, The Combustion Institute, pp. 661–670.

Frenklach, M. (2019), New form for reduced modeling of soot oxidation: accounting for multi-site kinetics and surface reactivity, *Combust. Flame*, Vol. 201, pp. 148–159.

Hanson, R., and S. Salimian (1984), Survey of rate constants in the N/H/O system, Chapter 6 in *Combustion Chemistry* (W. Gardiner, Jr., ed.), Springer-Verlag, New York, pp. 361–421.

Heywood, J. (1976), Pollutant formation and control in spark ignition engines, *Prog. Energy Combust. Sci.*, Vol. 1, pp. 135–164.

Hiroyasu, H., and T. Kadota (1976), Models for combustion and formation of nitrogen oxide and soot in DI diesel engines, *SAE Technical Paper* 760129.

Hottel, H., and R. Egbert (1942), Radiant heat transmission from water vapor. *Trans. Am. Inst. Chem. Eng.*, Vol. 38, pp. 531–568.

Kirkpatrick, A. (2021), *Internal Combustion Engines*, John Wiley and Sons, New York, pp. 318–329.

Klippenstein, S., L. Harding, P. Glarborg, and J. Miller (2011), The role of NNH in NO formation and control, *Combust. Flame*, Vol. 158, pp. 774–789.

Kummer, J. (1981), Catalysts for automobile emission control, *Prog. Energy Combust. Sci.*, Vol. 6, pp. 177–199.

Kuo, K., and R. Acharya (2012), *Fundamentals of Turbulent and Multiphase Combustion*, John Wiley and Sons, New York.

Lavoie, G., J. Heywood, and J. Keck (1970), Experimental and theoretical study of nitric oxide formation in internal combustion engines, *Combust. Sci. Technol.*, Vol. 1, pp. 313–326.

Miller, J., and C. Bowman (1989), Mechanism and modeling of nitrogen chemistry in combustion, *Prog. Energy Combust. Sci.*, Vol. 15, pp. 287–338.

Nagle, J., and R. Strickland-Constable (1962), Oxidation of carbon between 1000-2000 C., *Proc. Fifth Carbon Conf.*, Pergamon, Oxford, Vol. 1, pp. 154–164.

Sun, B., and S. Rigopoulos (2022), Modelling of soot formation and aggregation in turbulent flows with the LES-PBE-PDF approach and a conservative sectional method, *Combust. Flame*, Vol. 242, p. 112152.

Turns, S. (1995), Understanding NO_x formation in nonpremixed flames: experiments and modeling, *Prog. Energy Combust. Sci.*, Vol. 21, pp. 361–385.

Warnatz, J., H. Bockhorn, A. Moser, and H. Wenz (1982), Experimental investigations and computational simulation of acetylene-oxygen flames from near stoichiometric to sooting conditions, *19th Symposium (Intl.) on Combustion*, The Combustion Institute, pp. 197–209.

Yuen, W., and C. Tien (1977), A Simple calculation scheme for the luminous flame emissivity, *16th Symposium (Intl.) on Combustion*, The Combustion Institute, pp. 1481–1487.

Zel'dovich, Y. (1946), The oxidation of nitrogen in combustion and explosions, *Acta Physiochim. U.R.S.S.*, Vol. 21, pp. 577–628.

Homework Problems

10.1 Consider an equilibrium mixture of exhaust gases composed of O_2, N_2, and NO. Using the equilibrium constant K_p equation for the reaction $\frac{1}{2}O_2 + \frac{1}{2}N_2 \rightleftharpoons NO$ (see Table 2.11), plot the equilibrium mole fraction of NO, $X_{NO,e}$, as a function of temperature from 1000 to 4000 K for (a) $p = 1$ atm and (b) $p = 40$ atm.

10.2 The initial NO formation rate (kmol/m^3-s) kinetics can be estimated (Heywood 1976) given the equilibrium (e) concentrations (kmol/m^3) of oxygen and nitrogen, from the relation

$$\frac{d[\mathrm{NO}]}{dt} = \frac{6 \times 10^{-13}}{T^{1/2}} \exp\left(\frac{-69,090}{T}\right) [\mathrm{O_2}]_e^{1/2} [\mathrm{N_2}]_e$$

(a) Plot the NO formation rate as a function of temperature over a range of temperatures from 1000 to 3000 K at $p = 1000$ kPa and equilibrium mole fractions of $X_{N_2} = 0.79$ and $X_{O_2} = 0.21$.

(b) Estimate the NO concentration (ppm) formed in the first 1 ms at $T = 2500$ K and $p = 2020$ kPa.

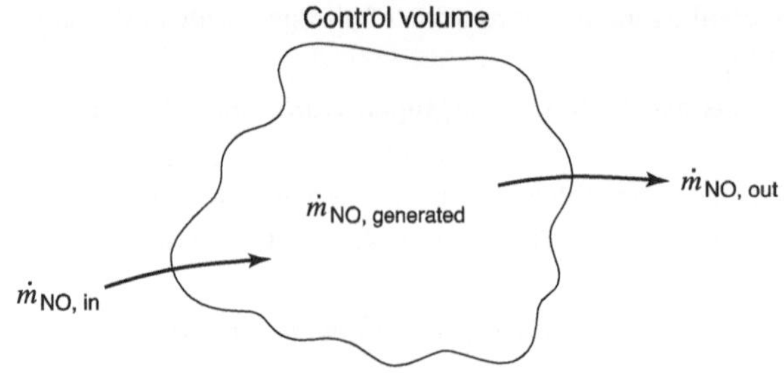

Figure 10.21 Illustration for Homework Problem 10.3.

10.3 The rate of change of the nitrogen oxide mass fraction Y_{NO} in a fluid element is due not only to chemical reactions in the element. The mass fraction can also change because of NO convected in and out of the fluid element. Consider the control volume shown in Figure 10.21. Write an expression for the rate of change of the nitrogen oxide mass fraction for this element assuming the fluid entering is devoid of nitrogen oxides, the fluid leaving has the same properties as fluid in the element, and there is generation of NO within the control volume.

10.4 A spark ignition engine operates with a rich ($\phi = 1.10$) mixture of methane and air at the conditions given in Example 10.1. Plot the burn fraction, pressure, burned and unburned temperatures, equilibrium, and rate-limited NO values as a function of crank angle. Discuss the NO trends, and how they differ from the lean combustion of Example 10.1. (Use the `TwoZoneNO.m` program listed on the text web site.)

10.5 A diesel fuel combustion process occurs at a constant pressure $p = 150$ atm, with the conditions given in Example 10.2. At what temperature will the net soot formation rate be a maximum? (Use the `Soot.m` program listed on the text website.)

10.6 What are the equilibrium constants $\ln K_p$ at 650 K for the two ammonia reactions given by Eqs. (10.91) and (10.92)? At 650 K, the ammonia properties are $\overline{h}_f^o = -45,920$ kJ/kmol, $\Delta\overline{h} = 15,519$ kJ/kmol, and $\overline{s}^o = 224$ kJ/kmol-K. The properties of the other reactants and products are given in the Appendix. (Hint: Start with Eq. (2.225).)

10.7 A large bore two-stroke stationary engine uses an ammonia injection system to remove NO from the exhaust. The engine has a displacement of 300 l and speed of 400 rpm. It is fueled with methane directly injected into the cylinder and operates with a delivery ratio of 0.92. If 0.10% of the nitrogen entering the engine has reacted to form NO during the combustion process, what is the mass flow (kg/hr) of ammonia needed to reduce the NO to N_2?

10.8 An automobile is driving at 50 mph. The gasoline-fueled engine is operating at stoichiometric A/F ratio and uses a three-way catalyst. The engine power is 40 kW and the brake efficiency is 22%. Measured pre-catalyst emissions are 2300 ppm NO and 4400 ppm CO. The catalyst reduction efficiencies for NO and CO are 95 and 92%, respectively. The molecular mass of the exhaust gas is 28.50 kg/kmol. Determine the following:
(a) Total flowrate of exhaust gas in kg/hr.
(b) Pre-catalyst emissions of NO and CO in g/kW-hr.
(c) Post-catalyst emissions in grams/mile.

10.9 For gas turbines, the NO and CO emissions are normalized as ppm of NO and % of CO adjusted to a 15% mole fraction of O_2. The normalization procedure is performed by virtually adding (or subtracting) air to the exhaust mixture to produce a 15% O_2 mole fraction in the exhaust gas.

(a) What is the normalized equivalence ratio ϕ at which the exhaust gases contain 15% O_2?

(b) If the measured exhaust gases contain 2.5% O_2, 0.15% CO, and 80 ppm NO, what are the normalized NO and CO emissions?

10.10 If a vehicle mileage standard is 54.5 miles per gallon, estimate the corresponding carbon dioxide emissions in grams/mile for a vehicle with a gasoline-fueled engine.

11

Solid Fuel Combustion

11.1 Introduction

The combustion processes occurring with a solid fuel are more complex in comparison to gaseous and liquid droplet combustion. The burning of a solid fuel requires chemical decomposition to produce volatiles which flow out from the solid and burn in a flame. When the temperature of a small fuel particle or a layer within a large particle is greater than about 470 K, solid fuels begin to decompose, releasing volatiles. The volatiles flow out of the solid through pores in the solid, preventing ambient oxygen from penetrating into the solid.

This devolatilization process is known as the pyrolysis stage. The volatile gases from the pyrolysis of wood are a mixture of CO, CO_2, H_2O, and short-chain hydrocarbons. If enough ambient air is available and the gaseous ignition temperature is reached, the volatile gases leaving the solid fuel particle will burn as a diffusion flame. The combustion process will be self-sustaining if the heat released by the flame provides adequate thermal energy feedback for subsequent combustion of the unburned solid region adjacent to the flame.

Combustion of a solid biomass particle thus involves the following processes: drying, distillation, pyrolysis, volatiles combustion, and char oxidation. For a small particle, these combustion processes occur in a sequential manner, resulting in the production of a flame. For larger fuel particles, some of the processes will occur simultaneously. A cross-section of a combusting wood slab is shown in Figure 11.1 indicating the char, pyrolysis, and unburned layers. In the following sections, we discuss the drying, pyrolysis, and char oxidation processes in more detail.

11.2 Drying of Solid Fuels

As described in Chapter 1, wood is a complex mixture of polymers of high molecular weight, primarily cellulose, hemicellulose, and lignin. The proportions of these polymers vary by species. Heating of the solid surface will initiate the drying and distillation steps in which water and liquid hydrocarbons such as alcohols, aldehydes, and terpenes in the solid are vaporized. The water content of wood can range from 50% for green wood to 5% for dried wood.

Water can exist within a wood matrix as water vapor, chemically bound (adsorbed) water, or free liquid water. The heating value of a given wood species is strongly dependent on its free and bound water content. Free water is water that is absorbed by capillary uptake and resides in the wood pores. Bound water is water that is adsorbed into the wood or coal cell wall and is held by hydrogen bonds. As such, the energy required to dry a given species of wood depends on the energy needed to heat the wood and the subcooled water to 373 K, the heat of vaporization (~2400 kJ/kg) of the

Principles of Combustion, Third Edition. Allan T. Kirkpatrick and Kenneth K. Kuo.

Companion website: www.wiley.com/go/kirkpatrick/principlesofcombustion3e

Figure 11.1 Burning charred wood with pyrolysis zone. (Courtesy of Dr. O. Masek).

free water, and the heat of sorption (~200–300 kJ/kg) of the bound water, all of which also affect the rate of combustion.

Within the wood, water is first taken up as bound water until the fiber saturation point is reached (all available adsorption sites are occupied), and then it becomes free water within the pores. The driving force for the free water vaporization is the difference between the surface equilibrium vapor pressure and the vapor partial pressure. Bound water is released as a result of chemical reactions in the cell wall, with a chemical potential gradient. For small thermally thin wood particles of the order of 100 μm or less, the temperature profile is spatially uniform, and water is vaporized and released through the pores of the particle before volatiles are produced.

For larger thermally thick particles, such as wood chips, a temperature profile exists within the particle. Due to the temperature gradient, moisture is being produced from inside the particle, while volatiles are being generated nearer the surface of the particle. Due to the greater pressure in the fuel pores during the volatilization at the surface, some of the moisture is forced to the center of the particle until the pressure builds up throughout the particle. Therefore, drying of larger wood particles involves inward vapor migration as well as outward vapor flow to the surface of the particle.

The combined outward and inward flow of vapor reduces the heat and mass transfer to the surface, thus reducing the burning rate, similar to the droplet vaporization processes analyzed in Chapter 9. A schematic of the burning rate versus mass fraction burned is shown in Figure 11.2 for three values (dry, 15%, and 200%) of moisture content for 10 mm pine wood cubes burned in air at 1100 K and $Re = 120$. Figure 11.2*a* is the normalized mass and Figure 11.2*b* is normalized burning

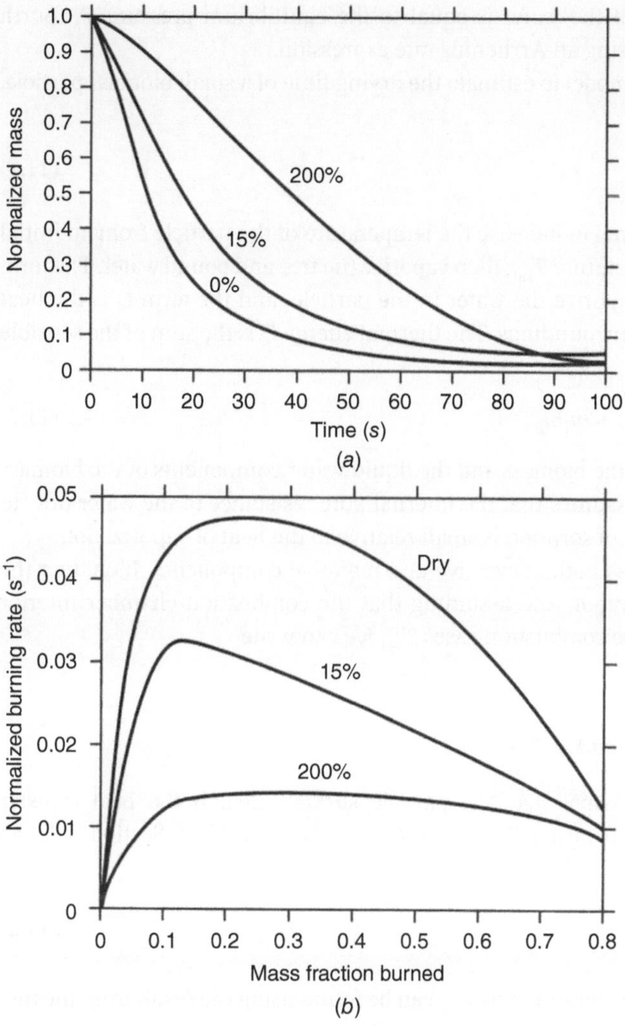

Figure 11.2 Effect of moisture content for three values (dry, 15%, and 200%) on burning rate (*a*) Normalized mass (*b*) Normalized burning rate. *Source*: Adapted from Simmons et al. 1986)/with permission of Taylor & Francis Group.

rate. The dry particle has the greatest burning rate, and as the moisture content is increased the burning rate accordingly decreases. Measurements of H_2O, CO_2, and CO indicate that pyrolysis products are released while the fuel is drying. Drying and pyrolysis are complete at 120 seconds for the 200% moisture case, at which point, some of the char has burned (as indicated by the mass remaining), and the remaining char burns in air generating CO and CO_2.

There are a number of models of wood drying in the literature that attempt to include the complex physics of moisture transport within the anisotropic porous solid particle (Peters and Bruch 2003). The simplest model is a basic thermodynamics model in which the drying front is assumed to be at 100°C, infinitely thin, and an energy sink. A second method uses an algebraic expression for temperature in an atmosphere of superheated steam as a function of moisture content to describe the equilibrium moisture content in wood. Another method models the evaporation of bound water with a diffusion expression, and recondensation is accounted for by assuming that the partial pressure of the water is equal to the equilibrium pressure. A fourth method models the drying of wood using an Arrhenius rate expression.

Applying a basic thermodynamics model to estimate the drying time of a small biomass particle, we can write

$$t_{\text{dry}} = \frac{Q}{\dot{Q}} \tag{11.1}$$

The drying time t_{dry} is the time required to increase the temperature of the particle from an initial temperature T_i to the saturation temperature T_{sat}, then vaporize the free and bound water. The term Q is the thermal energy needed to vaporize the water in the particle, and the term $\dot{Q}$ is the heat transfer rate to the particle from the surroundings. The thermal energy Q is the sum of the sensible and latent energies,

$$Q = \left[(mc)_b + (mc_p)_l\right](T_{\text{sat}} - T_i) + m_l h_{\text{fg}} \tag{11.2}$$

where the subscripts b and l indicate the biomass and the liquid water components of the biomass particle. Note that this basic model assumes that the internal pore resistance to the water flow to the surface is negligible, and the heat of sorption is small relative to the heat of vaporization.

The heat transfer to the particle has both convective and radiative components. Including the effect of superheating of the water vapor, and assuming that the combustion chamber interior surface is the same temperature as the combustion gases T_∞, we can write

$$\begin{aligned}\dot{Q} &= (\dot{Q}_{\text{conv}} + \dot{Q}_{\text{rad}}) \\ &= \left(\frac{z}{e^z - 1}\right)\mathbf{h}A_p(T_\infty - T_s) + \epsilon\sigma A_p(T_\infty^4 - T_s^4)\end{aligned} \tag{11.3}$$

where ϵ is the average particle emissivity, A_p the particle surface area, $\mathbf{h}$ the heat transfer coefficient, and average surface temperature $T_s = (T_i + T_{\text{sat}})/2$. As derived in Section 9.5, the superheating effect z is

$$z = \frac{\dot{m}_w c_{\text{pv}}}{\mathbf{h}A_p} \tag{11.4}$$

An estimate of the average water evaporation rate $\dot{m}_w$ can be found using the result from the fuel droplet analysis in Chapter 9, assuming the Nusselt number $Nu = 2$.

$$\dot{m}_w = 2\pi d \frac{k_g}{c_{\text{pl}}} \ln(1 + B_m) \tag{11.5}$$

where the transfer number B_m is

$$B_m = \frac{c_{\text{pl}}(T_\infty - T_{\text{sat}})}{h_{\text{fg}}} \tag{11.6}$$

and the terms c_{pl} and c_{pv} are the liquid and vapor-specific heats of water, respectively.

Example 11.1 ***Drying of a Biomass Particle***

A combustion chamber at atmospheric pressure and temperature $T_\infty = 1500$ K is supplied with ground biomass (b) particles $d = 100\,\mu$m in diameter at a temperature $T_i = 300$ K and a moisture content of 30% on a dry basis. The biomass particles have a thermal conductivity $k_b = 0.3$ W/m-K, dry density $\rho_b = 200$ kg/m^3, specific heat $c_{pb} = 1.30$ kJ/kg-K, and emissivity $\epsilon = 0.9$. Estimate the drying time t_{dry} of a biomass particle, assuming a quiescent environment.

Solution

At the saturation temperature $T_{sat} = 373$ K, the specific heat of the liquid water $c_{pl} = 4.22$ kJ/kg-K and the specific heat of the water vapor $c_{pv} = 2.22$ kJ/kg-K, with a heat of vaporization $h_{fg} = 2257$ kJ/kg. Since the particle is in a non-convective environment, $Nu = 2$. Using an average boundary layer temperature of $(1500 + 300)/2 = 900$ K, the surrounding air thermal conductivity $k_g = 0.062$ W/m-K, and the heat transfer coefficient **h** is

$$\mathbf{h} = \text{Nu}\frac{k_g}{d} = 2\frac{0.062}{100 \times 10^{-6}} = 1240\ \text{W/m}^2\text{K} = 1.24\ \text{kW/m}^2\text{K}$$

The particle surface area $A_p = \pi d^2 = (3.14)(100 \times 10^{-6})^2 = 3.14 \times 10^{-8}\ \text{m}^2$. The mass m_b of a dry biomass particle is

$$m_b = \rho_b V_b = 200\frac{\pi}{6}(100 \times 10^{-6})^3 = 1.05 \times 10^{-10}\ \text{kg}$$

and the mass of the water in the particle is

$$m_l = (0.30)1.05 \times 10^{-10} = 3.15 \times 10^{-11}\ \text{kg}$$

The transfer number B_m is

$$B_m = \frac{c_{pl}(T_\infty - T_{sat})}{h_{fg}} = \frac{(4.22)(1500 - 373)}{2257} = 2.11$$

The evaporation rate $\dot{m}_w$ is

$$\begin{aligned}\dot{m}_w &= 2\pi d\frac{k_g}{c_{pl}}\ln(1 + B_m)\\ &= 2\pi(100 \times 10^{-6})\frac{(0.062 \times 10^{-3})}{4.22}\ln(1 + 2.11)\\ &= 1.05 \times 10^{-8}\ \text{kg/s}\end{aligned}$$

The z term is

$$z = \frac{\dot{m}_w c_{pv}}{\mathbf{h}A_p} = \frac{(1.05 \times 10^{-8})(2.22)}{(1.240)(3.14 \times 10^{-8})} = 0.60$$

and

$$\frac{z}{e^z - 1} = \frac{0.60}{e^{0.60} - 1} = 0.73$$

The vaporization energy Q is

$$\begin{aligned}Q &= \left[(mc)_b + (mc_p)_l\right](T_{sat} - T_i) + m_l h_{fg}\\ &= \left[(1.05 \times 10^{-10})(1.3) + (3.15 \times 10^{-11})(4.22)\right](373 - 300)\\ &\quad + (3.15 \times 10^{-11})(2257)\\ &= 1.97 \times 10^{-8} + 7.11 \times 10^{-8}\\ &= 9.08 \times 10^{-8}\ \text{kJ} = 9.08 \times 10^{-5}\ \text{J}\end{aligned}$$

Note that the latent component is about 3.6× the subcooled component.

The average surface temperature T_s is

$$T_s = \frac{T_i + T_{sat}}{2} = \frac{300 + 373}{2} = 336\ \text{K}$$

The heat transfer rate $\dot{Q}$ is

$$\dot{Q} = \left(\frac{z}{e^z - 1}\right)\mathbf{h}A_p(T_\infty - T_s) + \epsilon\sigma A_p(T_\infty^4 - T_s^4)$$
$$= (0.73)(1240)(3.14\times10^{-8})(1500 - 336)$$
$$+ (0.9)(5.67\times10^{-8})(3.14\times10^{-8})(1500^4 - 336^4)$$
$$= 33.1\times10^{-3} + 8.11\times10^{-3} = 41.2\times10^{-3}\ \mathrm{W}$$

Therefore, the drying time is estimated to be

$$t_{\mathrm{dry}} = \frac{Q}{\dot{Q}} = \frac{9.08\times10^{-5}}{41.2\times10^{-3}} = 2.2\ \mathrm{ms}$$

Check: using just the mass transfer equations, we can write

$$t_{\mathrm{dry}} = \frac{m_w}{\dot{m}_w} = \frac{3.15\times10^{-11}}{1.05\times10^{-8}} = 3.0\ \mathrm{ms}.$$

which is comparable to the drying time calculated from the heat transfer equations.

11.3 Biomass Pyrolysis

Biomass pyrolysis starts at the surface of a solid particle and moves radially inward with the release of volatiles and formation of char and tar behind the pyrolysis layer, as indicated by the reaction schematic in Figure 11.3. Biomass pyrolysis is typically classified into three separate regimes based on particle thickness and moisture–thermally thin, thermally thick, and wave regimes (Bryden and Hagge 2003). In the thermally thin regime, temperature is nearly constant across the particle; consequently, drying and pyrolysis occur in series. In the thermally thick regime, there are significant temperature gradients, but the reaction rate is slow relative to the heat transfer rate. In the wave regime, wet unreacted wood, the pyrolysis zone, and char exist simultaneously. The drying and pyrolysis regions travel through the particle like a wave.

As the surface of the particle is heated, a pyrolysis layer is initiated at the outer edges of the solid and gradually moves inward while releasing volatiles and forming char. In the pyrolysis step, starting at about 470 K for wood and about 600 K for coal, the solid fuel begins to decompose and releases volatiles. The volatiles form within the solid particle and flow outward through pores in the solid. With decomposition, chemical bonds in the high molecular weight molecules are broken and form lower molecular weight char, tar, and volatile gases. Since wood and coal are heterogenous compounds, there are a variety of chemical reactions that compose the pyrolysis decomposition phase. The chemical restructuring during the pyrolysis reactions also results in shrinkage of the porous particle.

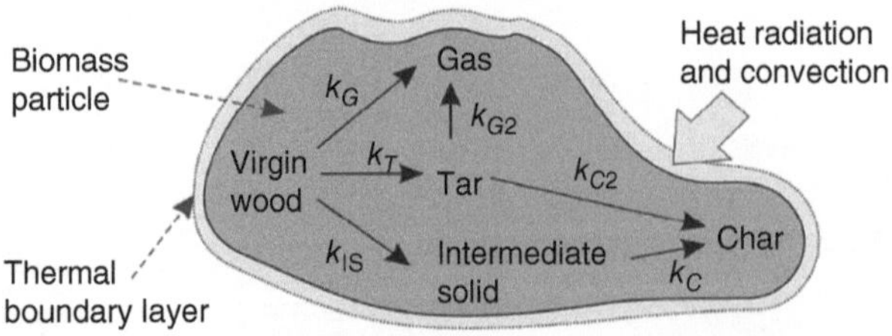

Figure 11.3 Solid fuel pyrolysis reactions. (Adapted from Park et al. 2010).

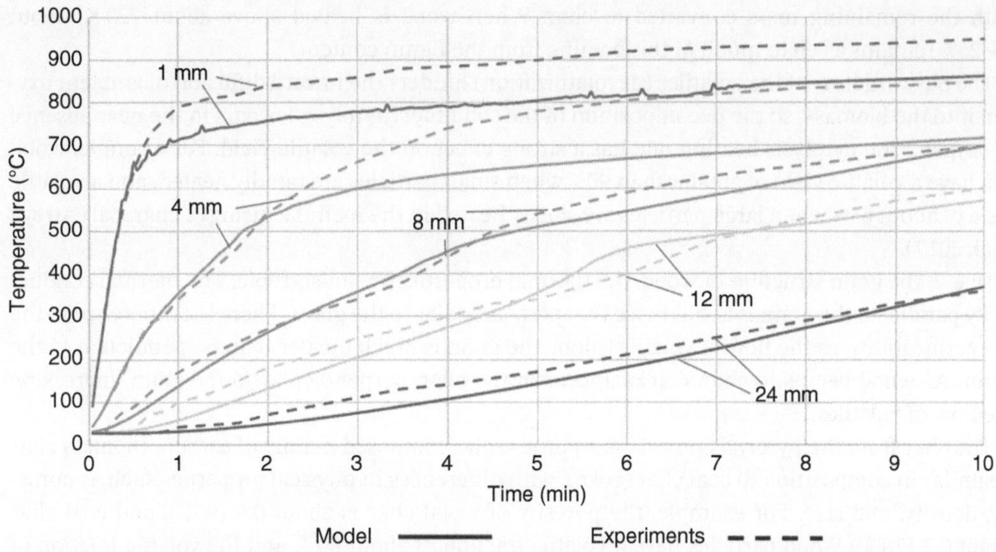

Figure 11.4 Temperature–penetration depth plot for pinewood irradiated with $\dot{Q}'' = 130\,\text{W/m}^2$ radiation flux. *Source*: Adapted from Gomez et al. 2018)/with permission of Elsevier.

A representative temperature–penetration depth plot is shown in Figure 11.4. The plot compares experimental and modeled temperatures in a thick pinewood slab undergoing radiative heating. Thermocouples were placed at 1, 4, 8, 12, and 24 mm from the left side of the pinewood. The figure shows the thermal energy progressing through the solid, and gradually increasing the interior temperatures.

In general, pyrolysis of hemicellulose and lignin is an exothermic process, while cellulose pyrolysis is an endothermic process at low temperatures and becomes an exothermic process at high temperatures. As indicated in Figure 11.5, cellulose pyrolyses in the temperature range of 510–620 K, hemicellulose pyrolyses in the temperature range of 470–530 K, and has the highest reactivity of the wood components. Pyrolysis reactions of lignins occur over a wide temperature range of 550–770 K. If lignin is heated to temperatures above 670–720 K, about 50% volatilizes,

Figure 11.5 Devolatilization rates as a function of temperature for beech wood heated at 5 K/min. The solid and dashed lines are the modeled rates and the symbols are the measured rates. *Source*: Adapted from Di Blasi 2008)/with permission of Elsevier.

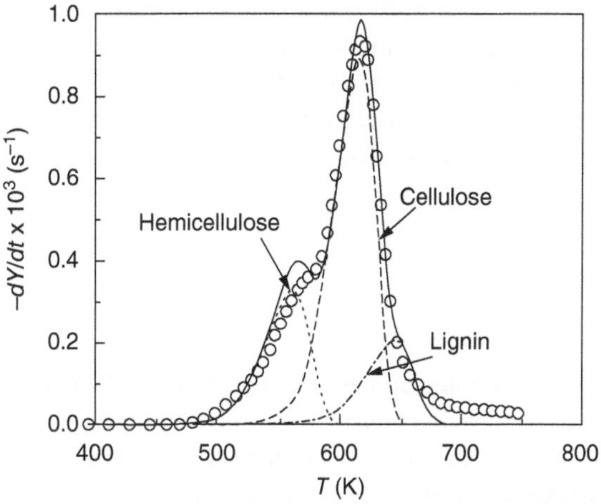

with the remaining mass converted to char. When wood is heated above about 720 K, about 15–25% remains as char, much of this coming from the lignin content.

The outward flow of the volatiles (devolatilization) hinders the inward diffusion of ambient oxygen into the biomass, so the decomposition of the solid fuel (pyrolysis) occurs in the near absence of oxygen. The pyrolysis heating rate has a strong effect on the volatile yield. For example, wood can have a volatile yield of greater than 90% when small particles are rapidly heated, and a volatile yield of about 65% when large particles are slowly heated in the manufacturing of charcoal (Atreya et al. 2017).

Due to the grain structure of wood, the thermal properties are anisotropic. The thermal conductivity parallel to the grain is about twice that perpendicular to the grain. There is a difference in the gas permeability, as the flow of volatiles along the grain is much greater than perpendicular to the grain. As wood begins to char, cracks appear in the char, perpendicular to the grain, increasing the flow of volatiles.

The char from the pyrolysis process is a porous solid composed mainly of carbon. Biomass char is similar in composition to coal char (coke), with differences in physical properties such as porosity, density, and size. For example, the porosity of wood char is about 0.9 (90%), and coal char about 0.7 (70%). Wood particles have a volatile fraction of about 80%, and the volatile fraction of coal is about 35%. The volatile and char fractions depend on the heating rate, with high heating rates (> 60 K/min) producing higher volatile and lower char fractions than indicated by a proximate analysis.

The volatile gases from the pyrolysis of the solid wood are a mixture of CO, CO_2, H_2O, and short-chain hydrocarbons. If enough ambient air is available and the gaseous ignition temperature is reached, the volatile gases leaving the solid fuel particle will burn as a diffusion flame. The ignition temperature for wood volatiles is about 530 K and for coal volatiles about 620 K.

Tar is a volatile heavy hydrocarbon composed primarily of a mixture of condensed-ring aromatic hydrocarbons such as benzene. As the char layer thickens, secondary pyrolysis reactions convert the tar produced by the primary pyrolysis reactions to light hydrocarbons and char. The tar cracks under combustion conditions, producing additional volatile gases, primarily CO, and but can also repolymerize to secondary char.

11.3.1 Physical Modeling of the Pyrolysis Process

A wide variety of physical models of the pyrolysis process have been developed (see Di Blasi 2008; Richter and Rein 2017). There has been recent interest (Ji et al. 2022) in combined neural network and physics-based models, which is a very promising technique. The basic models use a single or multiple Arrhenius rates to describe the rate of evolution of volatiles from solid particles. A single Arrhenius rate model assumes that the devolatilization rate is the first order with respect to the mass of volatile (v) matter remaining in the particle, $m_{v\infty} - m_v$.

$$\frac{dm_v}{dt} = k(m_{v\infty} - m_v) \tag{11.7}$$

with rate constant k,

$$k = A\exp(-E/R_uT) \tag{11.8}$$

The single-rate approach defined by Eq. (11.7) indicates an increasing rate of pyrolysis at higher temperatures but fails to account for the effect of temperature on volatile yields.

In order to account for increased volatile yields with temperature, Kobayashi et al. (1977) proposed the use of a model with two competing pyrolysis reactions, a low-temperature and a high-temperature reaction. Each reaction produces volatile matter and char. The first reaction dominates at relatively low temperatures, and the second reaction is dominant at high temperatures. The volatile yield f is defined as the fraction of the initial particle mass that is released as volatile matter v, and $(1 - f)$ is the fraction of the initial particle mass that is converted to char c. The rate of change of the mass fraction Y_p of the solid particle that has *not* reacted is given by

$$\frac{dY_p}{dt} = -k_1 Y_p - k_2 Y_p \tag{11.9}$$

where

$$\begin{aligned} k_1 &= A_1 \exp(-E_1/R_u T) \\ k_2 &= A_2 \exp(-E_2/R_u T) \end{aligned} \tag{11.10}$$

At high temperatures, the second reaction is faster than the first one, and E_2 is much larger than E_1, producing larger volatile yields. The parameters in the Kobayashi model are determined empirically. The low-temperature volatile yield f_1 is estimated from a proximate analysis (see Section 1.8) of the volatile matter.

Integrating Eq. (11.9) with the initial condition $Y_p = 1$ at $t = 0$,

$$Y_p = \exp(-t/\tau_d) \tag{11.11}$$

where the characteristic time τ_d for devolatilization is

$$\tau_d = (k_1 + k_2)^{-1} \tag{11.12}$$

Since the fraction of the particle mass that is converted to volatile gases by reaction 1 is f_1, and by reaction 2 is f_2, the rate of volatile formation is

$$\begin{aligned} \frac{dY_v}{dt} &= f_1 k_1 Y_p + f_2 k_2 Y_p \\ &= (f_1 k_1 + f_2 k_2) \exp(-t/\tau_d) \end{aligned} \tag{11.13}$$

Integrating Eq. (11.13), and using the initial condition that $Y_v = 0$ at $t = 0$, the total volatile mass fraction Y_v is

$$Y_v = \left(\frac{f_1 k_1 + f_2 k_2}{k_1 + k_2}\right)[1 - \exp(-t/\tau_d)] \tag{11.14}$$

In the asymptotic limit as $t \to \infty$,

$$Y_v = \frac{f_1 k_1 + f_2 k_2}{k_1 + k_2} \tag{11.15}$$

Similarly, since $Y_c + Y_v = 1$ at the end of pyrolysis, the total char mass fraction Y_c is

$$Y_c = 1 - Y_v = \frac{(1 - f_1)k_1 + (1 - f_2)k_2}{k_1 + k_2} \tag{11.16}$$

Example 11.2 ***Solid Fuel Pyrolysis***

The combustion pyrolysis at a temperature of 1500 K of a wood particle is modeled with a two-reaction model with the following parameters. What is the resulting volatile mass fraction Y_v and the char mass fraction Y_c?

$$f_1 = 0.40$$
$$f_2 = 0.80$$

$$A_1 = 2.0 \times 10^5\,\mathrm{s^{-1}}$$
$$A_2 = 1.5 \times 10^{10}\,\mathrm{s^{-1}}$$
$$E_1/R_u = 10,000\,\mathrm{K}$$
$$E_2/R_u = 20,000\,\mathrm{K}$$

Solution

The rate coefficient k_1 is

$$k_1 = A_1 \exp(-E_1/R_uT)$$
$$= (2.0 \times 10^5)\exp(-10,000/1500) = 2.54 \times 10^2\,\mathrm{s^{-1}}$$

and the rate coefficient k_2 is

$$k_2 = A_2 \exp(-E_2/R_uT)$$
$$= (1.5 \times 10^{10})\exp(-20,000/1500) = 2.42 \times 10^4\,\mathrm{s^{-1}}$$

The volatile mass fraction Y_v is

$$Y_v = \frac{f_1k_1 + f_2k_2}{k_1 + k_2}$$
$$= \frac{(0.40)(2.54 \times 10^2) + (0.80)(2.42 \times 10^4)}{(2.54 \times 10^2) + (2.42 \times 10^4)} = 0.79$$

and the char mass fraction Y_c is

$$Y_c = 1 - Y_v = 1 - 0.79 = 0.21$$

Note that about 21% of the original solid material remains as char, and the rest has been converted into volatile gases.

11.4 Char Oxidation

As the rates of local drying and devolatilization decrease in the fuel particle, allowing external oxygen to diffuse and convect into the char layer, char oxidation will begin. Char oxidation is a catalytic process that occurs both at the surface and at the interior of the porous char, involving adsorption and desorption reactions. As carbonaceous particles are composed primarily of carbon, the chemical oxidation of char proceeds in a similar manner to that previously described in Section 10.4 for soot particles.

It is thought that the oxidation of char occurs through adsorption of oxygen, either immediately releasing a CO or CO_2 molecule or forming a stable surface oxygen complex that may later desorb as CO or CO_2.The carbon atoms in the char are oxidized to CO by O_2 molecules that collide with and stick to the surface. The carbon is strongly attached to the oxygen and weakly attached to the surface, creating a vapor pressure, and leaves the surface in a gas phase. When the char oxidation is complete, the remaining material is ash, which is nonorganic and by definition, noncombustible.

In a simple reaction model, let us assume that there are three main global reactions that occur on the carbon surface,

$$\mathrm{C} + \frac{1}{2}\mathrm{O_2} \longrightarrow \mathrm{CO} \tag{11.17}$$

$$\mathrm{C} + \mathrm{CO_2} \longrightarrow 2\mathrm{CO} \tag{11.18}$$

$$\mathrm{C} + \mathrm{H_2O} \longrightarrow \mathrm{CO} + \mathrm{H_2} \tag{11.19}$$

The first reaction, Eq. (11.17), is an oxidation reaction and is much faster than the two reduction reactions, Eqs. (11.18) and (11.19), which become important at low oxygen concentrations at the particle surface.

The surface combustion kinetics have been modeled by numerous researchers using Arrhenius expressions which calculate the particle reaction rate based on the external surface area of the particle and the concentration or partial pressure of oxygen at the particle surface. Reviews of biomass and carbon particle combustion models are given in Caram and Amundson (1977) and Lu et al. (2008).

A single film model was first developed by Nusselt (1924) which assumes that the oxidation of carbon is controlled by diffusion of oxygen through a stationary boundary layer to the surface of the carbon particle where it reacts to form CO and/or CO_2. A double-film model was proposed by Burke and Schuman in 1931 which assumes carbon reacts at the surface with CO_2, and the CO produced diffuses radially outward and subsequently reacts with oxygen in a thin flame sheet in the boundary layer.

In what follows, we present a single film oxidation model with CO formation at the surface of the particle. We neglect the interior diffusion within the carbon particle. A mass balance at the particle surface $r = r_s$ is

$$\dot{m}_C + \dot{m}_{O_2} = \dot{m}_{CO} \tag{11.20}$$

As introduced in Chapter 2, we can express the carbon–oxygen mass balance in terms of the mass stoichiometric ratio s, in this case defined as the mass of oxidizer O_2 consumed per mass of carbon C in the oxidation reaction, Eq. (11.17).

$$\begin{aligned} s \equiv \frac{\dot{m}_{O_2}}{\dot{m}_C} = \frac{m_{O_2}}{m_C} = \frac{n_{O_2} M_{O_2}}{n_C M_C} \\ = \frac{1/2 \cdot 31.999}{1 \cdot 12.011} = 1.332 \end{aligned} \tag{11.21}$$

where

$$\dot{m}_{O_2} = s \cdot \dot{m}_C \tag{11.22}$$

$$\dot{m}_{CO} = (s+1)\,\dot{m}_C \tag{11.23}$$

Fick's law of diffusion (see Chapter 5) states that the mass flux $\dot{\mathbf{m}}''_A$ of a given species A relative to stationary coordinates can be expressed as

$$\dot{\mathbf{m}}''_A = \rho_A \mathbf{v} - \rho \mathcal{D}_{AB} \nabla Y_A \tag{11.24}$$

The first term represents the flux of A resulting from the bulk motion $\mathbf{v}$ of the fluid. The second term with the minus sign represents the relative flux of A resulting from the diffusion of A. Since $\mathbf{v} = \dot{\mathbf{m}}''/\rho$, Fick's law for the mass flux $\dot{\mathbf{m}}''_{O_2}$ can be expressed as

$$\dot{\mathbf{m}}''_{O_2} = Y_{O_2}(\dot{\mathbf{m}}''_{O_2} + \dot{\mathbf{m}}''_{CO}) - \rho \mathcal{D} \frac{dY_{O_2}}{dr} \mathbf{i}_r \tag{11.25}$$

Upon substitution of Eqs. (11.22) and (11.23), and adopting the convention that radially inward flow is negative,

$$-s\dot{m}''_C - Y_{O_2}\dot{m}''_C = -\rho \mathcal{D} \frac{dY_{O_2}}{dr} \tag{11.26}$$

Since

$$\dot{m}_C = 4\pi r_s^2 \dot{m}_C'' \tag{11.27}$$

the carbon mass flow rate $\dot{m}_C$ differential equation is

$$\dot{m}_C = \frac{4\pi r^2 \rho D}{(1 + Y_{O_2}/s)} \frac{d\left(Y_{O_2}/s\right)}{dr} \tag{11.28}$$

The boundary conditions are

$$\begin{aligned} Y_{O_2} &= Y_{O_2,s} \quad \text{at } r = r_s \\ Y_{O_2} &= Y_{O_2,\infty} \quad \text{at } r = \infty \end{aligned} \tag{11.29}$$

Upon integration, the carbon mass flow rate $\dot{m}_C$ is

$$\begin{aligned} \dot{m}_C &= 4\pi r_s \rho D \ln \left[\frac{1 + Y_{O_2,\infty}/s}{1 + Y_{O_2,s}/s}\right] \\ &= 4\pi r_s \rho D \ln \left[1 + \frac{Y_{O_2,\infty} - Y_{O_2,s}}{s + Y_{O_2,s}}\right] \\ &= 4\pi r_s \rho D \ln \left[1 + B_{O,m}\right] \end{aligned} \tag{11.30}$$

where the transfer number $B_{O,m}$ is defined as

$$B_{O,m} \equiv \frac{Y_{O_2,\infty} - Y_{O_2,s}}{s + Y_{O_2,s}} \tag{11.31}$$

Note the similarity with the mass transfer number B_M for liquid droplet evaporation.

We use surface kinetics to obtain an expression for the oxygen mass fraction at the surface $Y_{O_2,s}$. Assuming the first-order kinetics with respect to O_2, the oxygen reaction rate at the particle surface is

$$\dot{m}_{O_2} = 4\pi r_s^2 k_c \rho_{O_2,s} \tag{11.32}$$

with a rate coefficient k_c,

$$k_c = A_c \exp\left[-E_c/R_u T\right] \tag{11.33}$$

Equating the oxygen diffusion rate across the boundary layer to the oxygen reaction rate,

$$\dot{m}_{O_2} = 4\pi r_s^2 \mathbf{h}_m (\rho_{O_2,\infty} - \rho_{O_2,s}) = 4\pi r_s^2 k_c \rho_{O_2,s} \tag{11.34}$$

where $\mathbf{h}_m$ is the mass transfer coefficient. Solving for the oxygen density at the surface $\rho_{O_2,s}$,

$$\rho_{O_2,s} = \frac{\mathbf{h}_m}{k_c + \mathbf{h}_m} \rho_{O_2,\infty} \tag{11.35}$$

The relationship between oxygen mass fraction and gas density is

$$Y_{O_2} = \frac{m_{O_2}}{m} = \frac{n_{O_2} M_{O_2}}{nM} = \frac{p_{O_2} M_{O_2}}{pM} = \rho_{O_2} \frac{R_u T}{pM} \tag{11.36}$$

where $p, T,$ and M are the gas mixture total pressure, temperature, and molecular mass, respectively. Since for an ideal gas

$$\frac{R_u T}{pM} = \text{constant,} \tag{11.37}$$

we can express Eq. (11.35) in terms of the mass fraction Y,

$$Y_{O_2,s} = \frac{\mathbf{h}_m}{k_c + \mathbf{h}_m} Y_{O_2,\infty} \tag{11.38}$$

The oxygen reaction rate, Eq. (11.32), can thus be written in terms of gas density ρ or mass fraction Y,

$$\dot{m}_{O_2} = 4\pi r_s^2 \frac{k_c \mathbf{h}_m}{k_c + \mathbf{h}_m} \rho_{O_2,\infty} = 4\pi r_s^2 \frac{k_c \mathbf{h}_m}{k_c + \mathbf{h}_m} \frac{pM}{R_u T} Y_{O_2,\infty} \tag{11.39}$$

Also note that when $k_c \gg \mathbf{h}_m$,

$$\frac{k_c \mathbf{h}_m}{k_c + \mathbf{h}_m} \to \mathbf{h}_m \tag{11.40}$$

the oxygen consumption is *diffusion controlled*, since the reaction kinetics are rapid, and when $\mathbf{h}_m \gg k_c$,

$$\frac{k_c \mathbf{h}_m}{k_c + \mathbf{h}_m} \to k_c \tag{11.41}$$

the oxygen consumption is *kinetically controlled*, since the boundary layer diffusion is rapid.

Example 11.3 ***Char Oxidation Rate***

What is the oxidation rate at a temperature of 1800 K of a d = 300 μm diameter char particle in a quiescent environment? Assume the far field oxygen mass fraction $Y_{O_2,\infty} = 0.23$, the gas molecular mass M = 29.0 kg/kmol, the O_2–CO kinetic parameters are $A_c = 1.50 \times 10^5$ kJ/kmol and activation energy $E_c = 1.40 \times 10^5$ kJ/kmol, and the binary diffusion coefficient $\mathcal{D}(273\,K) = 1.85 \times 10^{-5}\ \mathrm{m^2/s}$ (CO in air).

Solution

The surface oxidation reaction is

$$\mathrm{C} + \frac{1}{2}\mathrm{O_2} \longrightarrow \mathrm{CO}$$

with the mass stoichiometric ratio $s = 1.33$.

The rate coefficient k_c is

$$k_c = A_c \exp\left[-E_c/R_u T\right] = 1.50 \times 10^5 \exp\left[-1.40 \times 10^5/(8.314)(1800)\right] = 13.0\,\mathrm{m/s}$$

The gas density ρ at the particle surface is

$$\rho = \frac{\mathrm{M}p}{R_u T} = \frac{(29.0)(101.3)}{(8.314)(1800)} = 0.196\,\mathrm{kg/m^3}$$

The binary diffusion coefficient $\mathcal{D}$ at 1800 K is

$$\mathcal{D}(1800\,K) = (1800/273)^{3/2} \cdot 1.85 \times 10^{-5} = 3.13 \times 10^{-4}\,\mathrm{m^2/s}$$

Since the particle is in a quiescent environment, the Sherwood number $Sh = 2$, and the mass transfer coefficient $\mathbf{h}_m$ is

$$\mathbf{h}_m = \frac{\mathrm{Sh}\,\mathcal{D}}{d} = \frac{2(3.13 \times 10^{-4})}{300 \times 10^{-6}} = 2.09\,\mathrm{m/s}$$

The oxygen mass fraction at the particle surface (s) is

$$Y_{O_2,s} = \frac{\mathbf{h}_m}{k_c + \mathbf{h}_m} Y_{O_2,\infty} = \left[\frac{2.09}{13.0 + 2.09}\right](0.23) = 0.032$$

The transfer number $B_{O,m}$ is

$$B_{O,m} = \frac{Y_{O_2,\infty} - Y_{O_2,s}}{s + Y_{O_2,s}} = \frac{0.23 - 0.032}{1.33 + 0.032} = 0.14$$

and the char burning rate is

$$\begin{aligned} \dot{m}_C &= 4\pi r_s \rho D \ln\left[1 + B_{O,m}\right] \\ &= 4\pi(150 \times 10^{-6})(0.196)(3.13 \times 10^{-4}) \ln[1 + 0.14] \\ &= 1.51 \times 10^{-8}\ \text{kg/s} \end{aligned}$$

11.4.1 Char Oxidation Time

The oxidation time for a biomass particle can be found from integration of Eq. (11.30). If we assume a constant density burn with a decreasing particle diameter d, then the char mass m_C at any instant is

$$m_C = \rho_C \frac{\pi d^3}{6} \tag{11.42}$$

and the mass burn rate becomes

$$\begin{aligned} \dot{m}_C = \frac{dm_C}{dt} &= 4\pi r_s \rho D \ln\left[1 + B_{O,m}\right] \\ &= 2\pi d\, \rho D \ln\left[1 + B_{O,m}\right] \\ &= 2\pi \left(\frac{6\, m_C}{\rho_C \pi}\right)^{1/3} \rho D \ln\left[1 + B_{O,m}\right] \end{aligned} \tag{11.43}$$

Separating variables,

$$m_C^{-1/3} dm_C = 2\pi \left(\frac{6}{\rho_C \pi}\right)^{1/3} \rho D \ln\left[1 + B_{O,m}\right] dt \tag{11.44}$$

Upon integration from the initial char mass $m_{C,i}$ to zero, and in terms of the initial particle diameter d_i, we have

$$t = \frac{\rho_C}{8 \rho D \ln\left[1 + B_{O,m}\right]} d_i^2 \tag{11.45}$$

Note the similarity with the d^2 droplet evaporation equation of Chapter 9. If the particle is assumed to remain at a constant diameter with decreasing density, then it can be shown that the burn up time is proportional to d, not d^2.

Example 11.4 ***Char Oxidation Time***

Using a constant density analysis, how long does it take for the char particle in the quiescent environment of Example 11.3 to completely burn up? Assume the char density $\rho_C = 2000\ \text{kg/m}^3$.

Solution

$$\begin{aligned} t &= \frac{\rho_C}{8 \rho D \ln\left[1 + B_{O,m}\right]} d_i^2 \\ &= \frac{2000}{8(0.196)(3.13 \times 10^{-4}) \ln[1 + 0.14]} (300 \times 10^{-6})^2 \\ &= 2.8\ \text{s} \end{aligned}$$

11.5 Solid Fuel Ignition

11.5.1 Introduction

Ignition of solid fuels can occur either by ignition at the surface or by ignition of the volatiles released from the solid. The rate of convective and radiative heat transfer to the particle governs the ignition sequence. When a solid fuel is introduced into a furnace, the surface temperature of the fuel will increase from an initial ambient temperature to its ignition temperature. The ignition will occur first at the surface if the radiative and convection heat transfer is high enough for the surface to heat up to the ignition temperature. If the surface heating rate is low, then the emitted volatiles will ignite first since they have a lower ignition temperature than the surface material.

There are two important nondimensional numbers, the Biot number and the Fourier number, which characterize the surface temperature increase. The Biot number Bi characterizes the degree to which a dry fuel particle or slab is *thermally thin*. It is defined as the ratio of a solid's thermal resistance to the surface resistance. The Biot number is named after the French scientist J. Biot (1774–1862) who made many contributions to the science of optics and magnetism:

$$Bi = \frac{L_c/k}{1/\mathbf{h}} = \frac{\mathbf{h}L_c}{k} \tag{11.46}$$

where L_c is a characteristic length, k is the solid's thermal conductivity, and $\mathbf{h}$ is the convective heat transfer coefficient. The length L_c is defined as the ratio of the solid's volume V to its surface area A, so $L_c = 1/2$ of the thickness of a thin slab, $L_c = r/2$ for a cylinder of radius r, and $L_c = r/3$ for a sphere of radius r. It is important to emphasize that whether or not a particle or slab is thermally thin depends on both the properties of the solid and the ambient conditions.

The Fourier number Fo is a dimensionless time and defined as the ratio of time t to the diffusion time L_c^2/α

$$Fo = \frac{\alpha\, t}{L_c^2} \tag{11.47}$$

The Fourier number provides a stability limit to the size of the time step used in numerical analysis of transient conduction problems. The Fourier number is named after the French scientist J. Fourier (1768–1830) who made seminal contributions to the thermal sciences and mathematics. Representative thermophysical properties of fir, oak, and pine used to determine the Biot and Fourier numbers are given in Table 11.1.

The simplest conduction model is one where the fuel particle is thermally thin, with no internal temperature gradients, and the fuel heats up uniformly. In this case, the internal resistance to heat transfer is much less than the surface resistance. Thin pieces of wood can be ignited easily, since the rate of heat transfer from the surface to the interior is minimal, and the surface temperature increase is rapid.

Table 11.1 Thermophysical Properties of Wood.

Material	Density ρ (kg/m^3)	Thermal Conductivity k (W/m-K)	Specific Heat c (J/kg-K)
Fir	415	0.11	2720
Oak	800	0.17	2385
Pine	640	0.14	2850

Source: Adapted from Incropera et al. (2007).

If the solid is not thermally thin, then there are internal temperature gradients that need to be taken into account that lower the rate of the surface temperature rise, resulting in ignition delay. A thermally thick slab of wood will not burn unless the radiation and convection heat transfer to the surface is increased to account for the conduction heat transfer from the surface of the wood to the interior.

11.5.2 Lumped Capacitance Analysis

We use the criterion $Bi < 0.1$ for classifying a dry solid as thermally thin, a classification also known as *lumped capacitance*. For a solid being convectively heated in a hot environment, the time-dependent temperature profile $T(t)$ is found from the energy equation,

$$\dot{Q} = \frac{dE}{dt}$$
$$\mathbf{h}A(T_\infty - T) = \rho Vc\frac{dT}{dt} \tag{11.48}$$

where T_∞ is the surrounding air temperature, c is the specific heat, and ρ is the density of the solid. Upon integration,

$$\frac{T_\infty - T}{T_\infty - T_o} = \exp\left(-\frac{t}{\tau}\right) \tag{11.49}$$

where T_o is the initial temperature, and the time constant τ is defined as

$$\tau \equiv \frac{\rho Vc}{\mathbf{h}A} \tag{11.50}$$

The time t_{ign} for a spatially isothermal solid to reach an ignition temperature T_{ign} is therefore given by

$$t_{ign} = \frac{\rho c L_c}{\mathbf{h}} \ln\left(\frac{T_\infty - T_o}{T_\infty - T_{ign}}\right) \tag{11.51}$$

In this case, the time to ignition is directly proportional to the characteristic length L_c of the solid and rises logarithmically with the temperature differences.

Example 11.5 ***Thermally Thin Cylinder***

A dry cylindrical pinewood stick at 20°C is placed on a fire and heated by the surrounding heated air at $T_\infty = 600$°C.
(a) What is the maximum diameter d of the stick for modeling as thermally thin?
(b) How long will it take this diameter stick to reach an ignition temperature T_{ign} of 340°C? Assume a convective heat transfer coefficient $\mathbf{h}$ of 25 W/m²K.

Solution
From Table 11.1 for pine, the thermal conductivity $k = 0.14$ W/m-K, density $\rho = 640$ kg/m^3, and specific heat $c = 2850$ J/kg-K.

(a) Using a Biot number criterion of $Bi = 0.1$,

$$Bi = \frac{\mathbf{h}L_c}{k} = \frac{\mathbf{h}\,r/2}{k} = 0.1$$

Therefore,

$$r = \frac{(2)(0.1)k}{\mathbf{h}} = \frac{(2)(0.1)(0.3)}{25} = 2.4 \times 10^{-3}\,\text{m} = 2.4\,\text{mm}$$

The maximum diameter d of the wooden stick is 4.8 mm.

(b) From Eq. (11.51), the time for ignition t_{ign} for a thermally thin cylinder is

$$t_{ign} = \frac{\rho c L_c}{\mathbf{h}} \ln\left(\frac{T_\infty - T_o}{T_\infty - T_{ign}}\right)$$
$$= \frac{(640)(2850)(2.4 \times 10^{-3}/2)}{25} \ln\left(\frac{600 - 20}{600 - 340}\right)$$
$$= 70\,\text{s}$$

11.5.3 Heat Conduction Equation

If the solid fuel is thermally thick, i.e., $Bi > 0.1$, then there are internal temperature gradients that need to be taken into account. The internal temperature profile can be determined from a 1-D energy balance on a small element of size $dx\,dy\,dz$, with heat flow in the x-direction. The net heat flow in the x-direction is

$$Q_x - Q_{x+dx} = mc\frac{dT}{dt} = \rho c(dx\,dy\,dz)\frac{dT}{dt} \tag{11.52}$$

Expanding the heat flow in a Taylor series,

$$Q_{x+dx} = Q_x + \frac{dQ_x}{dx}dx + \cdots \tag{11.53}$$

Therefore,

$$-\frac{dQ_x}{dx}dx = \rho c(dx\,dy\,dz)\frac{dT}{dt} \tag{11.54}$$

The Fourier heat conduction equation in the x-direction is

$$Q_x = -kA\frac{dT}{dx} = -k(dy\,dz)\frac{dT}{dx} \tag{11.55}$$

Upon substitution of Eq. (11.55) into (11.54), and canceling like terms,

$$\frac{d}{dx}\left(k\frac{dT}{dx}\right) = \rho c\frac{dT}{dt} \tag{11.56}$$

If the thermal conductivity k is constant,

$$\frac{d^2T}{dx^2} = \frac{\rho c}{k}\frac{dT}{dt} = \frac{1}{\alpha}\frac{dT}{dt} \tag{11.57}$$

Since the temperature T is a function of both space and time, we need to use partial derivative notation,

$$\frac{\partial^2 T}{\partial x^2} = \frac{1}{\alpha}\frac{\partial T}{\partial t} \tag{11.58}$$

Equation (11.58) is the 1-D heat conduction equation, with no internal heat generation. To solve for $T(x, t)$ it is necessary to specify an *initial* condition and two *boundary* conditions.

11.5.4 Semi-infinite Solid

The semi-infinite solid provides a useful idealization for combustion problems. Convection and flame impingement onto a surface can be modeled with the heat conduction equation applied to a semi-infinite solid, which is a slab heated on one side only. As we will see, concepts such as a critical ignition temperature T_{ign}, and a critical radiant heat flux $\dot{Q}''$ are used to model solid ignition processes.

We can perform a scale analysis of the heat conduction equation for a semi-infinite solid. If at $t = 0$, the boundary condition at $x = 0$ is suddenly raised from T_o to T_s, a thermal wave will be conducted through the solid in the positive x-direction from the boundary at $x = 0$. The length scale is the penetration depth δ at a time scale t, and the temperature scale $\Delta T = T_s - T_o$. Upon substitution into Eq. (11.58),

$$\frac{\Delta T}{\delta^2} \sim \frac{1}{\alpha}\frac{\Delta T}{t} \tag{11.59}$$

or

$$\delta \sim (\alpha\, t)^{1/2} \tag{11.60}$$

This scaling indicates that the penetration distance is proportional to the square root of the thermal diffusivity and the time.

We solve the semi-infinite slab problem by first defining a nondimensional variable η which collapses the partial differential heat conduction equation to an ordinary differential equation. The space variable x is nondimensionalized by the penetration distance δ, and from Eq. (11.60),

$$\eta = \frac{x}{\delta} = \frac{x}{2\sqrt{\alpha\, t}} \tag{11.61}$$

where the factor of 2 is introduced to simplify the subsequent algebra.

Upon substitution into Eq. (11.58), and using the chain rule, the conduction equation can be expressed as

$$\frac{d^2T}{d\eta^2} = -2\eta\frac{dT}{d\eta} \tag{11.62}$$

Integrating twice,

$$T = c_1\int_0^\eta e^{-v^2}\,dv + c_2 \tag{11.63}$$

For a constant surface temperature T_s boundary condition,

$$\begin{aligned} &\text{at } \eta = 0,\ T = T_s \Rightarrow c_2 = T_s \\ &\text{at } \eta = \infty,\ T = T_o \Rightarrow T_o = c_1\int_0^\infty e^{-v^2}\,dv + T_s \end{aligned} \tag{11.64}$$

The integral in the second boundary conditions (b.c.) is equal to $\pi^{1/2}/2$, so

$$c_1 = \frac{2(T_o - T_s)}{\pi^{1/2}} \tag{11.65}$$

and the temperature profile $T(x, t)$ for a constant surface temperature is

$$\frac{T(x,t) - T_s}{T_o - T_s} = \frac{2}{\pi^{1/2}}\int_0^\eta e^{-v^2}\,dv \equiv \text{erf}(\eta) = \text{erf}\left(\frac{x}{2\sqrt{\alpha\, t}}\right) \tag{11.66}$$

The function erf(η) is the *Gaussian error function*, named after J. Gauss (1777–1855), a celebrated German mathematician with over one hundred scientific terms named after him. The erf(η) function is tabulated in Appendix G. It has a value of 0 at $\eta = 0$, is $\sim \eta$ for $\eta < 0.7$, and then asymptotically approaches a value of 1 as η increases. The complementary error function, erfc(η) is defined as

$$\text{erfc}(\eta) = 1 - \text{erf}(\eta) \tag{11.67}$$

The temperature profile in the semi-infinite solid depends on the surface boundary conditions. Three combustion-related boundary conditions are (1) constant surface heat flux, (2) surface convection heating, and (3) combined surface heat flux and convection.

(1) For a boundary condition of constant surface heat flux $\dot{Q}''$, where the solid particle is being heated by radiation from a nearby flame, we assume all of the incident heat flux is conducted into the interior of the solid. A non-unity absorptivity can be included to account for radiation losses from the surface. The solution of the heat conduction equation (11.58) for the temperature profile $T(x, t)$ is

$$T(x,t) - T_o = \frac{2\dot{Q}''(\alpha t/\pi)^{1/2}}{k} \exp\left(\frac{-x^2}{4\alpha t}\right) - \frac{\dot{Q}''x}{k}\text{erfc}\left(\frac{x}{2\sqrt{\alpha t}}\right) \tag{11.68}$$

(2) For a surface convection boundary condition, where the convection heat transfer is from the hot ambient air at T_∞ into the interior of the solid initially at T_o,

$$-k\frac{\partial T}{\partial x}|_{x=0} = \mathbf{h}[T_\infty - T_o] \tag{11.69}$$

the resulting temperature profile $T(x, t)$ is

$$\begin{aligned}\frac{T(x,t) - T_o}{T_\infty - T_o} &= \text{erfc}\left(\frac{x}{2\sqrt{\alpha t}}\right) \\ &- \left[\exp\left(\frac{\mathbf{h}x}{k} + \frac{\mathbf{h}^2\alpha t}{k^2}\right)\right]\left[\text{erfc}\left(\frac{x}{2\sqrt{\alpha t}} + \frac{\mathbf{h}\sqrt{\alpha t}}{k}\right)\right]\end{aligned} \tag{11.70}$$

and the variation of the surface temperature $T_s = T(0, t)$ with time is

$$\frac{T_s - T_o}{T_\infty - T_o} = 1 - \exp\left(\frac{\mathbf{h}^2\alpha t}{k^2}\right) \cdot \text{erfc}\left(\frac{\mathbf{h}\sqrt{\alpha t}}{k}\right) \tag{11.71}$$

If we define the term β as

$$\beta = \frac{\mathbf{h}\sqrt{\alpha t}}{k} = \frac{\mathbf{h}\sqrt{t}}{\sqrt{k\rho c}} = \text{Bi} \cdot \text{Fo}^{1/2} \tag{11.72}$$

Then, Eq. (11.71) can be written as

$$\frac{T_s - T_o}{T_\infty - T_o} = 1 - \exp(\beta^2) \cdot \text{erfc}(\beta) \tag{11.73}$$

(3) Finally, let us consider the case with both radiative heating and surface convection, when the initial particle temperature and the ambient temperature are assumed to be equal, and radiation from the particle to the surroundings is neglected. In this case, the incident radiative energy is converted to both convection from the particle to the ambient and also to conduction into the particle interior.

If we set $T_\infty = T_o$, and $\dot{Q}''$ is the net absorbed radiation flux, the appropriate boundary conditions are

$$\begin{aligned}&\text{at } x = 0,\ \dot{Q}'' - \mathbf{h}(T_s - T_o) = -k\frac{dT}{dx} \\ &\text{at } t = 0,\ T(x, 0) = T_o\end{aligned} \tag{11.74}$$

The resulting temperature distribution is

$$\frac{T - T_o}{\dot{Q}''/\mathbf{h}} = \text{erfc}\left(\frac{x}{2\sqrt{\alpha t}}\right) - \exp\left(\frac{\mathbf{h}x}{k} + \frac{\mathbf{h}^2\alpha t}{k^2}\right)\cdot \text{erfc}\left(\frac{x}{2\sqrt{\alpha t}} + \frac{\mathbf{h}\sqrt{\alpha t}}{k}\right) \tag{11.75}$$

at the surface ($x = 0$), Eq. (11.75) reduces to

$$\frac{T_s - T_o}{\dot{Q}''/\mathbf{h}} = 1 - \exp(\beta^2)\cdot \text{erfc}(\beta) \tag{11.76}$$

Note the similarity of the right-hand sides of Eqs. (11.76) and (11.72). If we non-dimensionalize the radiant energy flux $\dot{Q}''t$, where

$$\gamma \equiv \frac{\dot{Q}''t}{(T_s - T_o)\rho c\sqrt{\alpha t}} \tag{11.77}$$

then Eq. (11.76) can be expressed as

$$\gamma = \frac{\beta}{1 - \exp(\beta^2)\cdot \text{erfc}(\beta)} \tag{11.78}$$

Example 11.6 ***Thermally Thick Slab***

A thick oak wall initially at 20°C is suddenly exposed to hot combustion gases at 600°C. At an elapsed time of 120 seconds, what is the surface temperature of the oak wall? Assume the convection heat transfer coefficient $\mathbf{h} = 20\ \text{W/m}^2\text{-K}$. If the critical ignition temperature is 360°C, will the oak surface ignite at 120 seconds?

Solution

From Table 11.1 for oak, the thermal conductivity is $k = 0.17$ W/m-K, density ρ is 800 kg/m³, and specific heat c is 2385 J/kg-K. The thermal diffusivity α of oak is

$$\alpha = \frac{k}{\rho c} = \frac{0.17}{(800)(2385)} = 8.91\times 10^{-8}\ \text{m/s}$$

From Eq. (11.72), the β term is

$$\beta = \frac{\mathbf{h}\sqrt{\alpha t}}{k} = \frac{(20)\sqrt{(8.91\times 10^{-8})(120)}}{0.17} = 0.385$$

From Eq. (11.73), the nondimensional surface temperature is

$$\frac{T_s - T_o}{T_\infty - T_o} = 1 - \exp(\beta^2)\cdot \text{erfc}(\beta)$$
$$= 1 - \exp(0.385^2)\cdot \text{erfc}(0.385)$$
$$= 0.320$$

The surface temperature T_s is therefore

$$T_s = 0.320(T_\infty - T_o) + T_o$$
$$= 0.320(600 - 20) + 20$$
$$= 206\ \text{C}$$

Since the surface temperature is below the critical ignition temperature of 360°C, the oak surface will not ignite at 120 seconds of exposure to the hot combustion gases.

11.5.5 Cylindrical Solid

The transient 1-D heat conduction equation in cylindrical coordinates is

$$\frac{1}{r}\frac{\partial}{\partial r}\left(r\frac{\partial T}{\partial r}\right) = \frac{1}{\alpha}\frac{\partial T}{\partial t} \tag{11.79}$$

For a uniform initial temperature T_o and a convection boundary condition, the solution (Carslaw and Jaeger 1986) in dimensionless form to Eq. (11.79) for a long cylinder of radius r_o is

$$\theta^*(r^*, Fo) = \sum_{n=1}^{\infty} C_n \exp(-x_n^2 Fo) J_0(x_n r^*) \tag{11.80}$$

where the dimensionless temperature θ^* and radius $r*$ are defined as

$$\theta^* = \frac{T_\infty - T(r,t)}{T_\infty - T_o}$$
$$r^* = r/r_o \tag{11.81}$$

The coefficient C_n is

$$C_n = \frac{2}{x_n}\frac{J_1(x_n)}{J_0^2(x_n) + J_1^2(x_n)} \tag{11.82}$$

The discrete values of x_n are positive roots of the transcendental equation

$$x_n\frac{J_1(x_n)}{J_0(x_n)} = Bi \tag{11.83}$$

For this conduction case, note that the Biot number is defined as $Bi \equiv hr_o/k$.

The terms J_0 and J_1 are Bessel functions of the first kind of order 0 and 1, respectively. Their values are tabulated in Appendix G.

For a Fourier number $Fo \geq 0.2$, the above series solution, Eq. (11.80) can be approximated by the single leading term with $n = 1$,

$$\theta^*(r^*, Fo) = C_1 \exp(-x_1^2 Fo) J_0(x_1 r^*) + \cdots \tag{11.84}$$

Values of the coefficients C_1 and x_1 are tabulated in Table 11.2 for a range of Biot numbers.

Example 11.7 ***Thermally Thick Cylinder***

A long dry cylindrical piece of pinewood 30 mm in diameter initially at 20°C is placed on a fire grate above a hot plume of combustion products at 600°C. How long will it take for the surface of the pine to reach $T_s = 400$°C? The natural convection heat transfer coefficient of the plume is $\mathbf{h} = 5\,\text{W/m}^2\text{-K}$.

Solution

From Table 11.1 for pine, the thermal conductivity $k = 0.14$ W/m-K, density $\rho = 640\,\text{kg/m}^3$, and specific heat $c = 2850$ J/kg-K. The thermal diffusivity α of pine is

$$\alpha = \frac{k}{\rho c} = \frac{0.14}{(640)(2850)} = 7.87 \times 10^{-8}\,\text{m/s}$$

The Biot number Bi test for a spatially isothermal cylinder is

$$Bi = \frac{\mathbf{h}(r_o/2)}{k} = \frac{5(0.015/2)}{0.14} = 0.27$$

Table 11.2 C_1 and x_1 Coefficients for Eq. (11.84).

Bi	x_1	C_1	*Bi*	x_1	C_1
0.01	0.1412	1.0025	0.5	0.9408	1.1143
0.02	0.1995	1.0050	0.6	1.0185	1.1346
0.03	0.2439	1.0075	0.7	1.0873	1.1539
0.04	0.2814	1.0099	0.8	1.1490	1.1725
0.05	0.3142	1.0124	0.9	1.2048	1.1902
0.06	0.3438	1.0148	1.0	1.2558	1.2071
0.07	0.3708	1.0173	2.0	1.5995	1.3384
0.08	0.3960	1.0197	3.0	1.7887	1.4191
0.09	0.4195	1.0222	4.0	1.9081	1.4698
0.10	0.4417	1.0246	5.0	1.9898	1.5029
0.15	0.5376	1.0365	6.0	2.0490	1.5253
0.20	0.6170	1.0483	7.0	2.0937	1.5411
0.25	0.6856	1.0598	8.0	2.1286	1.5526
0.30	0.7465	1.0712	9.0	2.1566	1.5611
0.40	0.8516	1.0932	10.0	2.1795	1.5677

Note: $Bi = \mathbf{h} r_o/k$

Since $Bi > 0.1$, the internal temperature gradients of the cylinder need to be taken into account, and Eq. (11.84) is to be used to calculate the surface temperature T_s.

$$\theta^*(r^*, Fo) = C_1 \exp(-x_1^2 Fo) J_0(x_1 r^*) + \cdots$$

At the cylinder surface, $r^* = 1$, and the dimensionless temperature θ^* is

$$\theta^* = \frac{T_s - T_\infty}{T_o - T_\infty} = \frac{400 - 600}{20 - 600} = 0.34$$

From Table 11.2, if $Bi = \mathbf{h} r_o/k = 0.536$, then $x_1 = 0.969$ and $C_1 = 1.12$. From the Bessel function table in Appendix G, $J_0(x_1) = J_0(0.969) = 0.778$.

Rearranging Eq. (11.84) for the Fourier number *Fo*,

$$\begin{aligned} Fo &= -\frac{1}{x_1^2} \ln\left[\frac{\theta^*}{C_1 J_0(x_1)}\right] \\ &= -\frac{1}{(0.969^2)} \ln\left[\frac{0.34}{(1.12)(0.778)}\right] \\ &= 1.005 \end{aligned}$$

The Fourier number $Fo > 0.2$, so the single-term approximation to Eq. (11.80) is appropriate. Since $Fo \equiv \alpha t/r_o^2$,

$$t = Fo\frac{r_o^2}{\alpha} = (1.005)\frac{(0.015)^2}{7.87 \times 10^{-8}} = 2873 \text{ sec}$$

Additional solutions of the heat conduction equation for a wide variety of geometries and boundary conditions can be found in Carslaw and Jaeger (1986).

11.5.6 Pilot and Spontaneous Ignition

Ignition of a combustible solid from a nearby flame is defined as pilot ignition. Simms (1963) used Eq. (11.78) to correlate data on the piloted ignition of wood in which vertical samples of several species were exposed to radiant heat fluxes $\dot{Q}''$ ranging from 6.3 to 63 kW/m^2. A pilot flame was located 12.5 mm from the wood surface in the plume of vapors emitted from the wood surface and the time to ignition t_{ign} measured. The wood samples tested were cedar, mahogany, oak, freijo, and iroko.

Assuming an initial temperature T_o of 20°C, Simms found that a value of the critical surface temperature $T_{ign} = T_s = 360$°C gave the best correlation for all of the data from the variety of woods. Simms also found that the results depended on the location of the pilot flame. As the pilot flame was placed varying distances away from test samples of oak and pine, the critical surface temperature increased from 320 to 430°C to maintain a correlation with Eq. (11.78).

If a combustible solid is exposed to a high heat flux, the resulting emitted volatiles may ignite spontaneously. Spontaneous ignition requires a higher heat flux than piloted ignition since a higher surface temperature is required as indicated in Table 11.3. The critical radiant heat flux $\dot{Q}''$ was reported by Simms (1963) to be about 12 kW/m^2 for pilot ignition and 28 kW/m^2 for spontaneous ignition. The volatiles released from the surface must be above their autoignition temperature as they mix with the surrounding air to produce a flame. With convective heating, heated air mixes with the volatiles released from the wood surface producing a flammable mixture, so the critical surface temperature with convective heating is less than that for radiative heating.

Moisture contained in the wood affects the ignition process in two ways, by increasing the effective specific heat and by diluting the pyrolysis products which reduces the heat of combustion. As indicated in Table 11.4, the ignition time increases with increasing moisture content.

Table 11.3 Critical Surface Temperature T_{ign} Required for Spontaneous (SI) and Pilot (PI) Ignition.

Mode of Heat Transfer	Surface Temperature (SI)	Surface Temperature (PI)
Radiation	600°C	300–410°C
Convection	490°C	450°C

Source: Adapted from Kanury (1972).

Table 11.4 Effect of Moisture on Time to Ignition. (Horizontal Samples, Douglas Fir).

Moisture Content (%)	Time to Ignition (Seconds)
0	55
11	100
17	145
27	215

Source: Adapted from Atreya and Abu-Zaid (1991).

11.6 Example: Biomass Cookstove

It is estimated that about half the world's population, more than 3 billion people, cook over an open biomass fire. This energy source is especially common in rural areas of developing countries, where an estimated 90% of households rely on biomass fuels for cooking. Despite biomass being a simple, convenient, and potentially carbon-neutral energy source, there are two important environmental issues associated with biomass combustion – inefficiency and air pollution, as mentioned in Chapter 10.

Biomass-based cookstoves are increasingly being used to mitigate these two issues. Cookstove performance is typically characterized by the following two metrics: reduction of harmful emissions, primarily carbon monoxide (CO) and particulate matter (PM), and increased thermal efficiency. The majority of these biomass cooking stoves are wood burning and operate in a natural convection mode – driven by buoyant fluid forces.

The *three stone fire* is the most basic, but still extremely common, method for cooking with biomass fuel, and is often used as a benchmark for comparison. The three-stone fire is simply an open fire which is named from the stones used to hold the cook piece over the fire. However, a three-stone fire is inefficient and polluting. The emissions of soot and other pollutants from a three-stone fire are high due to poor mixing of fuel and air and incomplete combustion, as the combusting gases are quickly quenched by the cool pot and the ambient air. In a three-stone fire, the heat loss to the surroundings by radiation and convection is significant.

Cookstove heat transfer efficiency can be improved by enclosing and insulating the combustion chamber, limiting the size of the combustion chamber, and directing the flow of hot gases next to the pot surface by using a pot skirt or other device. In addition, limiting the flow of air improves stove performance by increasing the temperature of the combustion chamber.

The biomass cookstoves have been able to reduce CO and PM emissions by 50–75%, and increase fuel efficiency by 30–50%, compared to the three-stone fire configuration. These reductions are accomplished largely by enclosing the fire in a combustion chamber. The primary benefits of chamber enclosure are airflow control and reduced radiative and convective heat transfer losses.

The well-known *rocket elbow stove*, shown in Figure 11.6, is a good example of current cookstove technology. A typical rocket stove chimney height h is 200 mm, and chimney diameter is 100 mm. The small chamber limits the amount of fuel being combusted at any given time, ensuring the stove is not over-fueled relative to the cooking task. In addition, the combustion chamber is well

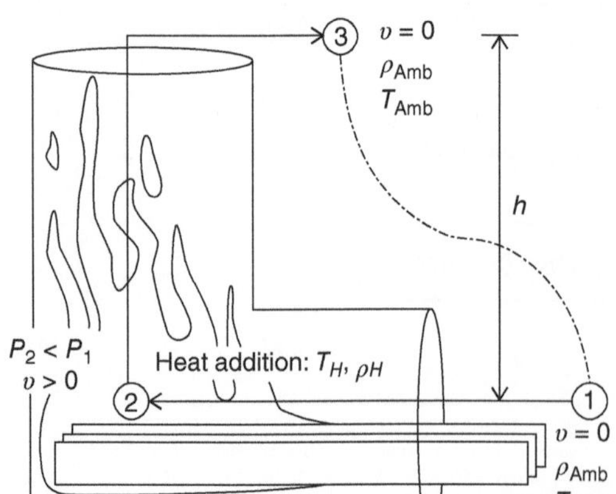

Figure 11.6 Schematic diagram of rocket stove parameters. (Adapted from Agenbroad et al. 2011).

insulated and limits the heat loss from the cooking fire, and it directs the hot gases under the pot, thereby increasing the heat transfer efficiency. The fuel wood sits on a grill located in the horizontal leg of the stove, and air is drawn by natural convection.

11.6.1 Modeling Cookstove Combustion

A simplified model of the stove operation will be developed (following Agenbroad et al. 2011) for predicting air flow rate, temperature, and excess air ratio, based on stove geometry (including chimney height and chimney area) and the operating firepower. The stove firepower $\dot{Q}$, i.e., the amount of firewood supplied to the stove, is controlled by the user and an input to the simple model.

The physics of the stove have been simplified to two basic and fundamental driving processes: heat addition $\dot{Q}$ from wood combustion at point (2) and kinetic energy per unit mass V^2 addition between points (1) and (2) due to the chimney effect, i.e., the buoyancy of the lower-density air in the chimney. The two processes are interconnected and together govern the overall stove operation.

The wood fire transfers thermal energy to the airflow passing through the stove, increasing its temperature from the ambient at T_{amb} to T_H and decreasing its density from ρ_{amb} to ρ_H. The heat addition is assumed to be taking place entirely at point 2. A constant specific heat, evaluated at an average temperature, will be assumed. Heating takes place over a range from approximately 300–1300 K, over which the air-specific heat c_p varies by a maximum of 15%.

For a known mass flow rate, $\dot{m}$, and stove firepower, $\dot{Q}$, the stove temperature T_H can be calculated using the first law of thermodynamics for a control volume as shown in Eq. (11.85),

$$T_H = \frac{\dot{Q}}{\dot{m}c_p} + T_{\text{amb}} \sim \frac{\dot{Q}}{\dot{m}_A c_p} + T_{\text{amb}} \tag{11.85}$$

Since cookstoves operate in a very lean condition, it is assumed that the gas flow is dominated by the airflow, so $\dot{m} \sim \dot{m}_A$.

The higher temperature and lower density air in the chimney between points (2) and (3) create a lower hydrostatic pressure at point (2) compared to the ambient pressure at point (1). Assuming the density ρ_H in the chimney is constant and the ambient pressure in a horizontal plane at the top of the chimney is uniform, the pressure difference Δp_{1-2} between points (1) and (2) can be calculated using the hydrostatic equation, given below in Eq. (11.86), where h is the chimney height, and $g = 9.81\ \text{m/s}^2$ is the gravitational acceleration.

$$\Delta p_{1-2} = (\rho_{\text{amb}} - \rho_H)gh \tag{11.86}$$

The upward chimney flow results from the horizontal pressure difference calculated in Eq. (11.86). The gain in kinetic energy per unit mass of the chimney flow at point (2) relative to the ambient air at point (1) is equated to this pressure difference using Bernoulli's equation, as shown in Eq. (11.87).

$$\frac{\rho_H V^2}{2} = \Delta p_{1-2} = (\rho_{\text{amb}} - \rho_H)gh \tag{11.87}$$

Assuming ideal-gas behavior, the gas density is related to the temperature by the ideal-gas law, Eq. 11.88

$$\rho_{\text{amb}} = \frac{p}{RT_{\text{amb}}} \tag{11.88}$$

So

$$\frac{\rho_{\text{amb}} - \rho_H}{\rho_H} = \frac{T_H - T_{\text{amb}}}{T_{\text{amb}}} \tag{11.89}$$

and the chimney air flow velocity V can be expressed as

$$V = \sqrt{2\left(\frac{T_H - T_{\text{amb}}}{T_{\text{amb}}}\right) gh} \tag{11.90}$$

A loss coefficient C_d can be included to account for friction, and inlet/outlet flow losses, so Eq. (11.90) becomes

$$V = C_d\sqrt{2\left(\frac{T_H - T_{\text{amb}}}{T_{\text{amb}}}\right) gh} \tag{11.91}$$

The cookstove air mass flowrate $\dot{m}_A$ is therefore

$$\begin{aligned} \dot{m}_A &= \rho_H A V \\ &= C_d\left(\frac{pA}{RT_H}\right)\sqrt{2\left(\frac{T_H - T_{\text{amb}}}{T_{\text{amb}}}\right) gh} \end{aligned} \tag{11.92}$$

Eqs. (11.85) and (11.92) constitute a simple model for predicting stove temperature T_H and mass flow $\dot{m}_A$ for a given stove firepower $\dot{Q}$.

The air–fuel ratio A/F and excess air ratio EA are calculated as follows. The fuel mass flowrate $\dot{m}_F$ is

$$\dot{m}_F = \frac{\dot{Q}}{q_{\text{lhc}}} \tag{11.93}$$

where q_{lhc} is the lower heat of combustion or heating value of the wood. The higher heat of combustion (*hhc*) is not used because the latent heat of the water vapor is significant and will not be recovered from the flow. The wood heating value includes the combustion of gaseous volatiles as well as the surface oxidation of leftover carbon char. The two processes require different amounts of oxygen and have different heat release values. Some error results as the proportion of carbon originating from each process is unknown. However, this effect is minimized as wood volatiles contributions (on a mass basis) are much larger than those from char oxidation.

The air–fuel ratio A/F is

$$A/F = \frac{\dot{m}_A}{\dot{m}_F} \tag{11.94}$$

and the equivalence ratio ϕ is

$$\phi = \frac{(A/F)_{st}}{A/F} = s_A\frac{\dot{m}_F}{\dot{m}_A} \tag{11.95}$$

The excess air EA is

$$EA = \frac{1-\phi}{\phi} \tag{11.96}$$

The above equations were solved over a typical range of operating firepowers and the results are shown in Figure 11.7. A value of $q_{\text{lhc}} = 18.28$ MJ/kg for Douglas fir was chosen as a representative value for wood. The loss coefficient C_d was assumed to have a value of 1. Figure 11.7 indicates that increasing firepower leads to increasing stove airflow temperature. However, there is a peak flow rate, after which mass flowrate decreases with increasing firepower, due to the decreasing airflow density. As a consequence, the excess air ratio is large ($EA \sim 1000\%$) while operating at low firepower, and decreases rapidly ($EA \sim 100\%$) as firepower is increased. Note that relative to boiler applications, a 100% excess air is still high. Experimental validation of the simple model is given in Agenbroad et al. (2011).

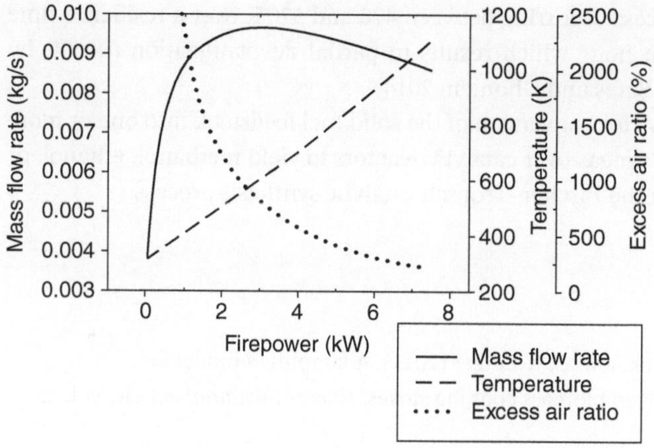

Figure 11.7 Mass flow rate, temperature, and excess air as a function of firepower. (Adapted from Agenbroad et al. 2011).

11.7 Gasification of Solid Fuels

Solid fuels like wood and coal are heterogenous compounds that do not burn completely. Their combustion residue is in the form of inorganic ash particles. As a consequence, they cannot be used in combustion devices like gas turbines or engines that require a cleaner fuel. Gasification is a combustion technique that has been developed to convert a solid fuel to a clean gaseous or liquid fuel that can be used in combustion devices. Coal gasification has many applications, as it has been used for the generation of synthetic natural gas ("Syngas"), hydrogen, ammonia, and specialty chemicals.

A gasifier differs from a combustor as it uses a controlled air or oxygen supply and steam. With gasification, only the drying, distillation, and pyrolysis processes occur. The pyrolysis reactions are controlled by the level of oxygen or air made available for the reactions. A volatile flame is not established. The products are low-molecular-weight hydrocarbons, CO, H_2, CO_2, H_2O, N_2, and tar. The tar, char, and ash particles and any vaporized inorganic compounds are filtered from the pyrolysis products before use in an engine, a process called hot gas cleanup.

Either air or oxygen can be used as the primary oxidant gas, although steam is generally added to assist in char gasification. The additional reactions with steam result in a gaseous fuel primarily composed of carbon monoxide CO and hydrogen H_2, for example,

$$H_2O + C \longrightarrow CO + H_2 \tag{11.97}$$

The generation of CO and CO_2 from the oxygen consumed in the gasifier releases the energy required to devolatilize the solid fuel particles and gasify the resultant char through the endothermic reactions with steam and CO_2.

Biomass gasifiers are typically operated at atmospheric pressure because of the difficulty of feeding biomass particles into a pressurized gasifier and because the high biomass volatile content and high reactivity of biomass chars reduce some of the need for elevated pressures to obtain good carbon conversion. Because CO_2 and H_2O react much slower with char than O_2, coal gasifiers usually operate at elevated pressure, to increase the collision rate of the gasifying species with the char.

Torrefaction is a mild pyrolysis process occurring between 470 and 570 K over a residence time between several minutes to about an hour which results in partial devolatilization (0–60% by weight) of the original dry feedstock (Bates and Ghoniem 2014).

Gasification of solid fuels also allows for conversion of the solid fuel feedstock into one or more liquid fuels by processing the gasifier syngas over catalytic reactors to yield methanol, ethanol, or even gasoline and diesel fuel, through the Fischer–Tropsch catalytic synthesis process.

References

Agenbroad, J., M. DeFoort, A. Kirkpatrick, and C. Kreutzer (2011), A simplified model for understanding natural convection driven biomass cooking stoves, *Energy Sustainable Dev.*, Vol. 15, pp. 160–168.

Atreya, A., and M. Abu-Zaid (1991), Effect of environmental variables on piloted ignition, *Fire Saf. Sci.*, Vol. 3, pp. 177–186.

Atreya, A., P. Olszewski, Y. Chen, and H. Baum (2017), The effect of size, shape and pyrolysis conditions on the thermal decomposition of wood particles and firebrands, *Int. J. Heat Mass Transfer*, Vol. 107, pp. 319–328.

Bates, R., and A. Ghoniem (2014), Modeling kinetics-transport interactions during biomass torrefaction: the effects of temperature, particle size, and moisture content, *Fuel*, Vol. 137, pp. 216–229.

Bryden, K., and M. Hagge (2003), Modeling the combined impact of moisture and char shrinkage on the pyrolysis of a biomass particle, *Fuel*, Vol. 82, pp. 1633–1644.

Caram, H., and N. Amundson (1977), Diffusion and reaction in a stagnant boundary layer about a carbon particle, *Ind. Eng. Chem. Fundam.*, Vol. 16, pp. 171–181.

Carslaw, H., and J. Jaeger (1986), *Conduction of Heat in Solids*, 2nd ed., Oxford University Press, Oxford, UK.

Di Blasi, C. (2008), Modeling chemical and physical processes of wood and biomass pyrolysis, *Prog. Energy Combust. Sci.*, Vol. 34, pp. 47–90.

Gomez, M., J. Porteiro, S. Chapela, and J. Miguez (2018), An Eulerian model for the simulation of the thermal conversion of a single large biomass particle, *Fuel*, Vol. 220, pp. 671–681.

Incropera, F., D. Dewitt, T. Bergman, and A. Levine (2007), *Fundamentals of Heat and Mass Transfer*, 6th ed., John Wiley and Sons, New York.

Ji, W., F. Richter, M. Gollner, and S. Deng (2022), Autonomous kinetic modeling of biomass pyrolysis using chemical reaction neural networks, *Combust. Flame*, Vol. 240, p. 111992.

Kanury, A. (1972), Ignition of cellulosic materials: a review, *Fire Res. Abstr. Rev.*, Vol. 14, pp. 24–52.

Kobayashi, H., J. Howard, and A. Sarofim (1977), Coal devolatilization at high temperatures, *Symp. (Intl.) Combust.*, Vol. 16, pp. 411–425.

Lu, H., W. Robert, G. Peirce, B. Ripa, and L. Baxter (2008), Comprehensive study of biomass particle combustion, *Energy Fuels*, Vol. 22, pp. 2826–2839.

Nusselt, W. (1924), The process of combustion in powdered coal firing, *Ver. Z. Deut. Ing.*, Vol. 68, pp. 124–128.

Park, W., A. Atreya, and H. Baum (2010), Experimental and theoretical investigation of heat and mass transfer processes during wood pyrolysis, *Combust. Flame*, Vol. 157, no. 3, pp. 481–494.

Peters, B., and C. Bruch (2003), Drying and pyrolysis of wood particles: experiments and simulation, *J. Anal. Appl. Pyrolysis*, Vol. 70, pp. 233–250.

Richter, F., and G. Rein (2017), Pyrolysis kinetics and multi-objective inverse modeling of cellulose at the microscale, *Fire Saf. J.*, Vol. 91, pp. 191–199.
Simmons, W., and K. Ragland (1986), Burning rate of millimeter sized wood particles in a furnace, *Combust. Sci. Technol.*, Vol. 46, pp. 1–15.
Simms, D. (1963), On the pilot ignition of wood by radiation, *Combust. Flame*, Vol. 7, pp. 253–261.

Homework Problems

11.1 A combustion chamber at atmospheric pressure is supplied with $d = 120\,\mu\text{m}$ diameter wood particles at $T_i = 300$ K. They have a moisture content of 25% on a dry basis. Estimate the drying time t_{dry} of a single wood particle. Use the property data given in Example 11.1.

11.2 The combustion pyrolysis at a temperature of 1500 K of a wood particle is modeled with a two-reaction model with the following parameters. What is the characteristic time τ_d for devolatilization?

$$f_1 = 0.35$$

$$f_2 = 0.75$$

$$A_1 = 2.0 \times 10^5\ \text{s}^{-1}$$

$$A_2 = 1.5 \times 10^{10}\ \text{s}^{-1}$$

$$E_1/R_u = 12{,}000\ \text{K}$$

$$E_2/R_u = 18{,}000\ \text{K}$$

11.3 Using the Nusselt single film model, what is the oxidation rate at a combustion temperature $T = 1500$ K of a $d = 250\,\mu\text{m}$ diameter char particle in a quiescent environment? Assume the far field oxygen mass fraction $Y_{O_2,\infty} = 0.23$, the gas molecular mass M = 29.0 kg/kmol, the O_2–CO kinetic parameters are $A_c = 1.50 \times 10^5$ kJ/kmol and activation energy $E_c = 1.40 \times 10^5$ kJ/kmol, and the binary diffusion coefficient $\mathcal{D}(273\,K) = 1.85 \times 10^{-5}\ \text{m}^2/\text{s}$ (CO in air).

11.4 How long does it take for the char particle in the quiescent environment of Problem 11.3 to completely burn up? Use a constant density model and assume the char density $\rho_c = 2000\ \text{kg/m}^3$.

11.5 Using the Nusselt single film model, what is the surface temperature of a 1-mm diameter carbon particle burning in air, with the air and surrounding surfaces at a temperature of 1500 K? Assume the particle emissivity $\epsilon = 0.9$, and the parameters of Problem 11.3.

11.6 (a) What is the oxidation rate of a $d = 50\,\mu\text{m}$ carbon particle at a combustion temperature of 1500 K in air at 1 atm? Assume the far field oxygen mass fraction $Y_{O_2,\infty} = 0.23$, the gas molecular mass M= 30.0 kg/kmol, and the kinetic constant $k = 1.9$ m/s.
(b) If the burning rate is constant, and the particle density is 2200 kg/m^3, how long will it take for the particle to completely burn up?

11.7 A 150 μm oak particle initially at 300 K and containing 5% moisture is placed in a furnace at 1500 K. What is a) the drying time, and b) the time needed to reach the ignition temperature of 360° C? The convection coefficient **h** is 25 W/m^2. The particle density $\rho = 800\,\text{kg/m}^3$, and specific heat $c = 2385\,\text{J/kg-K}$.

11.8 A vertical slab of 40 mm thick pine initially at 20° C is exposed to a plume of 600°C hot air above a fire. The ignition temperature of the wood is 320°C. How long will it take the wood surface to reach this temperature and ignite? The convection coefficient h is 20 W/m^2.

11.9 With reference to problem 11.8, compare the conduction heat transfer into the wood interior with the radiative losses from the wood surface at the time the wood ignites. Assume the emissivity $\epsilon = 0.7$, and the surroundings are at 20°C.

11.10 A vertical strip of cotton fabric 0.6 mm thick and 1 m long initially at 20°C is exposed to a radiant heat flux of 20 W/m^2. The ignition temperature of the fabric is 300°C. The natural convection heat transfer coefficient is 15 W/m^2-K and the thermal conductivity k = 0.04 W/m-K. The fabric density is $\rho = 300\,\text{kg/m}^3$, specific heat $c = 1400\,\text{J/kg-K}$, and emissivity $\epsilon = 0.9$. How long will it take for the fabric to reach its ignition temperature?

11.11 A rocket elbow cookstove is producing 4 kW firepower. What is the air–fuel ratio of the cookstove? The stove has a diameter of 100 mm and a height of 200 mm.

A

Conversion Factors and Physical Constants

Table A.1 Unit Conversion Factors.

Area	1 m^2	= 1550.0 in^2
		= 10.764 ft^2
Energy	1 J	= 9.4787×10^{-4} Btu
		= 0.239 calories
Energy/mass	1 kJ/kg	= 0.4303 Btu/lb_m
Force	1 N	= 0.22481 lb_f
Heat Transfer Rate	1 W	= 3.4123 Btu/h
Heat Flux	1 W/m^2	= 0.3171 Btu/h-ft^2
Heat Transfer Coefficient	1 W/m^2-K	= 0.17612 Btu/h-ft^2 °F
Thermal Diffusivity	1 m^2/s	= 3.875×10^4 ft^2/h
Length	1 m	= 39.370 in.
		= 3.2808 ft.
	1 km	= 0.62137 mile
Mass	1 kg	= 2.2046 lb_m
Mass Density	1 kg/m^3	= 0.062428 lb_m/ft^3
Mass Flow Rate	1 kg/s	= 7936.6 lb_m/h
Mass Transfer Coefficient	1 m/s	= 1.1811×10^4 ft/h
Power	1 kW	= 1.341 hp
Pressure and Stress	1 Pa (1 N/m^2)	= 0.020886 lb_f/ft^2
	1.01325×10^5 N/m^2	= 1 standard atmosphere
		= 760 mmHg
	1×10^5 N/m^2	= 1 bar
		= 750.06 mmHg
Rotational Speed	1 rev/min	= 0.10472 rad/s
Specific Heat	1 J/kg-K	= 2.3886×10^{-4} Btu/lb_m- °F
Temperature	K	= (5/9) °R
		= (5/9) (°R + 459.67)
		°C + 273.15

Principles of Combustion, Third Edition. Allan T. Kirkpatrick and Kenneth K. Kuo.

Companion website: www.wiley.com/go/kirkpatrick/principlesofcombustion3e

Table A.1 (Continued)

Temperature Difference	1 K	= 1 °C
		= (9/5) °R = (9/5) °F
Thermal Conductivity	1 W/m-K	= 0.57782 Btu/h ft- °F
Thermal Resistance	1 K/W	= 0.52750 °F/h-Btu
Torque	1 Nm	= 0.73756 lb_f ft
Viscosity (dynamic)	1 kg/m-s^2 (Pa-s)	= 2419.1 lb_m/ft-h
Viscosity (kinematic)	1 m^2/s	= 3.875×10^4 ft^2/h
Volume	1 m^3 (10^3 L)	= 6.1023×10^4 $in.^3$
		= 35.314 ft^3
		= 264.17 gal
Volume Flow Rate	1 m^3/s	= 2.1189×10^3 ft^3/min
		= 1.5850×10^4 gal/min

Source: Adapted from Incropera et al. (2007).

Table A.2 Physical Constants.

Universal Gas Constant	R_u	= 8.3145 kJ/kmol-K
		= 8.3145 Pa m^3/mol-K
		= 8.3145×10^{-2} m^3 bar/kmol-K
		= 8.2057×10^{-2} m^3 atm/kmol-K
		= 1545.35 ft lb_f/lbmole-°R
		= 1.986 Btu/lbmole-°R
Avogadro's Number	N_A	= 6.0221×10^{23} molecules/mol
Planck's Constant	h	= 6.6261×10^{-34} J-s
Boltzmann's Constant	k	= 1.3806×10^{-23} J/K
Speed of Light in Vacuum	c_o	= 2.9979×10^8 m/s
Stefan–Boltzmann Constant	σ	= 5.6704×10^{-8} W/m^2-K^4
		= 0.1714×10^{-8} Btu/h ft^2-°R^4
Gravitational Acceleration	g	= 9.8067 m/s^2
Standard Atmospheric Pressure	P	= 101,325 N/m^2
		= 101.325 kPa

Reference

Incropera, F., D. DeWitt, T. Bergman, and A. Lavine (2007), *Fundamentals of Heat and Mass Transfer*, John Wiley & Sons, New York.

B

Thermal and Transport Properties of Air

B.1 Introduction

In the simulation of many combustion processes, it is important to evaluate accurately the thermal and transport properties of both pure substances and mixtures of multiple chemical species. The thermal properties include density and specific heat, while the transport properties include viscosity, thermal conductivity, and binary mass diffusivity. The accuracy of the numerical simulation depends strongly on how well these thermal and transport properties are evaluated. Mixing rules have also been developed to evaluate these properties when multiple chemical species are present.

For more detailed information about property evaluation of gases and liquids, the readers are recommended to consult an excellent reference book by Elliott et al. (2023). Thermophysical properties of a number of selected gases and liquids can also be obtained using the NIST website: http://webbook.nist.gov/chemistry/fluid/.

Principles of Combustion, Third Edition. Allan T. Kirkpatrick and Kenneth K. Kuo.

Companion website: www.wiley.com/go/kirkpatrick/principlesofcombustion3e

Table B.1 Properties of Air at Atmospheric Pressure.

T (K)	ρ (kg/m^3)	c_p (kJ/kg-K)	$\mu \times 10^7$ (Pa-s)	$\nu \times 10^6$ (m^2/s)	$k \times 10^3$ (W/m-K)	$\alpha \times 10^6$ (m^2/s)	Pr
100	3.5562	1.032	71.1	2.00	9.34	2.54	0.786
150	2.3364	1.012	103.4	4.426	13.8	5.84	0.758
200	1.7458	1.007	132.5	7.590	18.1	10.3	0.737
250	1.3947	1.006	159.6	11.44	22.3	15.9	0.720
300	1.1614	1.007	184.6	15.89	26.3	22.5	0.707
350	0.9950	1.009	208.2	20.92	30.0	29.9	0.700
400	0.8711	1.014	230.1	26.41	33.8	38.3	0.690
450	0.7740	1.021	250.7	32.39	37.3	47.2	0.686
500	0.6964	1.030	270.1	38.79	40.7	56.7	0.684
550	0.6329	1.040	288.4	45.57	43.9	66.7	0.683
600	0.5804	1.051	305.8	52.69	46.9	76.9	0.685
650	0.5356	1.063	322.5	60.21	49.7	87.3	0.690
700	0.4975	1.075	338.8	68.10	52.4	98.0	0.695
750	0.4643	1.087	354.6	76.37	54.9	109	0.702
800	0.4354	1.099	369.8	84.93	57.3	120	0.709
850	0.4097	1.110	384.3	93.80	59.6	131	0.716
900	0.3868	1.121	398.1	102.9	62.0	143	0.720
950	0.3666	1.131	411.3	112.2	64.3	155	0.723
1000	0.3482	1.141	424.4	121.9	66.7	168	0.726
1100	0.3166	1.159	449.0	141.8	71.5	195	0.728
1200	0.2902	1.175	473.0	162.9	76.3	224	0.728
1300	0.2679	1.189	496.0	185.1	82	238	0.719
1400	0.2488	1.207	530	213	91	303	0.703
1500	0.2322	1.230	557	240	100	350	0.685
1600	0.2177	1.248	584	268	106	390	0.688
1700	0.2049	1.267	611	298	113	435	0.685
1800	0.1935	1.286	637	329	120	482	0.683
1900	0.1833	1.307	663	362	128	534	0.677
2000	0.1741	1.337	689	396	137	589	0.672
2100	0.1658	1.372	715	431	147	646	0.667
2200	0.1582	1.417	740	468	160	714	0.655
2300	0.1513	1.478	766	506	175	783	0.647
2400	0.1448	1.558	792	547	196	869	0.630
2500	0.1389	1.665	818	589	222	960	0.613
3000	0.1135	2.726	955	841	486	1570	0.536

Source: Adapted from Incropera et al. (2007).

Table B.2 Physical Properties of Air at Atmospheric Conditions ($T = 298$ K, $P = 1$ atm $= 1.0133$ bar).

Molecular Weight	$M = 28.966$ kg/kmol
Gas Constant	$R = 0.28704$ kJ/kg-K
Speed of Sound	$c = 345.9$ m/s
Binary Diffusion with Octane	$D = 5.68 \times 10^{-6}$ m^2/s
Schmidt Number	$Sc = \nu/D = 2.77$
Lewis Number	$Le = \alpha/D = 3.91$

References

Elliott, J., V. Diky, T. Knotts, and W. Wilding (2023), *The Properties of Gases and Liquids*, 6th ed., McGraw-Hill, New York.

Incropera, F., D. DeWitt, T. Bergman, and A. Lavine (2007), *Fundamentals of Heat and Mass Transfer*, John Wiley & Sons, New York.

C

Thermodynamic Property Tables for Various Ideal Gases

Table C.1 Properties of Various Ideal Gases at 298 K (SI Units).

Gas	Chemical Formula	M (kg/kmol)	R (kJ/kg-K)	c_{po} (kJ/kg-K)	c_{vo} (kJ/kg-K)	γ
Air		28.97	0.287	1.004	0.717	1.400
Acetylene	C_2H_2	26.038	0.3193	1.699	1.380	1.231
Ethane	C_2H_6	30.07	0.27650	1.7662	1.4897	1.186
Ethanol	C_2H_5OH	46.069	0.18048	1.427	1.246	1.145
Ethylene	C_2H_4	28.054	0.29637	1.5482	1.2518	1.237
Helium	He	4.003	2.07703	5.1926	3.1156	1.667
Hydrogen	H_2	2.016	4.12418	14.2091	10.0849	1.409
Methane	CH_4	16.04	0.51835	2.2247	1.7354	1.299
Methanol	CH_3OH	32.042	0.25948	1.4050	1.1455	1.227
Nitrogen	N_2	28.013	0.29680	1.0416	0.7448	1.400
Nitrous Oxide	N_2O	44.013	0.18891	0.8793	0.6904	1.274
n-Octane	C_8H_{18}	114.23	0.07279	1.7113	1.6385	1.044
Oxygen	O_2	31.999	0.25983	0.9216	0.6618	1.393
Propane	C_3H_8	44.097	0.18855	1.6794	1.4909	1.126
Steam	H_2O	18.015	0.46152	1.8723	1.4108	1.327

Source: Adapted from Sonntag et al. (2003).

Principles of Combustion, Third Edition. Allan T. Kirkpatrick and Kenneth K. Kuo.

Companion website: www.wiley.com/go/kirkpatrick/principlesofcombustion3e

Table C.2 Ideal-Gas Properties of N_2 and N (SI Units), Entropies at 0.1-MPa (1-bar) Pressure.

	Nitrogen, Diatomic (N_2) $\bar{h}_f^o = 0$ kJ/kmol $M = 28.013$		Nitrogen, Monatomic (N) $\bar{h}_f^o = 472{,}680$ kJ/kmol $M = 14.007$	
T (K)	$\Delta\bar{h}$ (kJ/kmol)	$\bar{s}^o$ (kJ/kmol-K)	$\Delta\bar{h}$ (kJ/kmol)	$\bar{s}^o$ (kJ/kmol-K)
0	−8670	0	−6197	0
100	−5768	159.812	−4119	130.593
200	−2857	179.985	−2040	145.001
298	0	191.609	0	153.300
300	54	191.789	38	153.429
400	2971	200.181	2117	159.409
500	5911	206.740	4196	164.047
600	8894	212.177	6274	167.837
700	11,937	216.865	8353	171.041
800	15,046	221.016	10,431	173.816
900	18,223	224.757	12,510	176.265
1000	21,463	228.171	14,589	178.455
1100	24,760	231.314	16,667	180.436
1200	28,109	234.227	18,746	182.244
1300	31,503	236.943	20,825	183.908
1400	34,936	239.487	22,903	185.448
1500	38,405	241.881	24,982	186.883
1600	41,904	244.139	27,060	188.224
1700	45,430	246.276	29,139	189.484
1800	48,979	248.304	31,218	190.672
1900	52,549	250.234	33,296	191.796
2000	56,137	252.075	35,375	192.863
2200	63,362	255.518	39,534	194.845
2400	70,640	258.684	43,695	196.655
2600	77,963	261.615	47,860	198.322
2800	85,323	264.342	52,033	199.868
3000	92,715	266.892	56,218	201.311
3200	100,134	269.286	60,420	202.667
3400	107,577	271.542	64,646	203.948
3600	115,042	273.675	68,902	205.164
3800	122,526	275.698	73,194	206.325
4000	130,027	277.622	77,532	207.437
4400	145,078	281.209	86,367	209.542
4800	160,188	284.495	95,457	211.519
5200	175,352	287.530	104,843	213.397
5600	190,572	290.349	114,550	215.195
6000	205,848	292.984	124,590	216.926

Source: Adapted from Sonntag et al. (2003).

Table C.3 Ideal-Gas Properties of O_2 and O (SI Units), Entropies at 0.1-MPa (1-bar) Pressure.

	Oxygen, Diatomic (O_2) $\bar{h}_f^o = 0$ kJ/kmol $M = 31.999$		Oxygen, Monatomic (O) $\bar{h}_f^o = 249{,}170$ kJ/kmol $M = 16.00$	
T (K)	$\Delta\bar{h}$ (kJ/kmol)	$\bar{s}^o$ (kJ/kmol-K)	$\Delta\bar{h}$ (kJ/kmol)	$\bar{s}^o$ (kJ/kmol-K)
0	−8683	0	−6725	0
100	−5777	173.308	−4518	135.947
200	−2868	193.483	−2186	152.153
298	0	205.148	0	161.059
300	54	205.329	41	161.194
400	3027	213.873	2207	167.431
500	6086	220.693	4343	172.198
600	9245	226.450	6462	176.060
700	12,499	231.465	8570	179.310
800	15,836	235.920	10,671	182.116
900	19,241	239.931	12,767	184.585
1000	22,703	243.579	14,860	186.790
1100	26,212	246.923	16,950	188.783
1200	29,761	250.011	19,039	190.600
1300	33,345	252.878	21,126	192.270
1400	36,958	255.556	23,212	193.816
1500	40,600	258.068	25,296	195.254
1600	44,267	260.434	27,381	196.599
1700	47,959	262.673	29,464	197.862
1800	51,674	264.797	31,547	199.053
1900	55,414	266.819	33,630	200.179
2000	59,176	268.748	35,713	201.247
2200	66,770	272.366	39,878	203.232
2400	74,453	275.708	44,045	205.045
2600	82,225	278.818	48,216	206.714
2800	90,080	281.729	52,391	208.262
3000	98,013	284.466	56,574	209.705
3200	106,022	287.050	60,767	211.058
3400	114,101	289.499	64,971	212.332
3600	122,245	291.826	69,190	213.538
3800	130,447	294.043	73,424	214.682
4000	138,705	296.161	77,675	215.773
4400	155,374	300.133	86,234	217.812
4800	172,240	303.801	94,873	219.691
5200	189,312	307.217	103,592	221.435
5600	206,618	310.423	112,391	223.066
6000	224,210	313.457	121,264	224.597

Source: Adapted from Sonntag et al. (2003).

Table C.4 Ideal-Gas Properties of CO_2 and CO (SI Units), Entropies at 0.1-MPa (1-bar) Pressure.

	Carbon Dioxide (CO_2) $\bar{h}_f^o = -393{,}522$ kJ/kmol $M = 44.01$		Carbon Monoxide (CO) $\bar{h}_f^o = -110{,}527$ kJ/kmol $M = 28.01$	
T (K)	$\Delta\bar{h}$ (kJ/kmol)	$\bar{s}^o$ (kJ/kmol-K)	$\Delta\bar{h}$ (kJ/kmol)	$\bar{s}^o$ (kJ/kmol-K)
0	−9364	0	−8671	0
100	−6457	179.010	−5772	165.852
200	−3413	199.976	−2860	186.024
298	0	213.794	0	197.651
300	69	214.024	54	197.831
400	4003	225.314	2977	206.240
500	8305	234.902	5932	212.833
600	12,906	243.284	8942	218.321
700	17,754	250.752	12,021	223.067
800	22,806	257.496	15,174	227.277
900	28,030	263.646	18,397	231.074
1000	33,397	269.299	21,686	234.538
1100	38,885	274.528	25,031	237.726
1200	44,473	279.390	28,427	240.679
1300	50,148	283.931	31,867	243.431
1400	55,895	288.190	35,343	246.006
1500	61,705	292.199	38,852	248.426
1600	67,569	295.984	42,388	250.707
1700	73,480	299.567	45,948	252.866
1800	79,432	302.969	49,529	254.913
1900	85,420	306.207	53,128	256.860
2000	91,439	309.294	56,743	258.716
2200	103,562	315.070	64,012	262.182
2400	115,779	320.384	71,326	265.361
2600	128,074	325.307	78,679	268.302
2800	140,435	329.887	86,070	271.044
3000	152,853	334.170	93,504	273.607
3200	165,321	338.194	100,962	276.012
3400	177,836	341.988	108,440	278.279
3600	190,394	345.576	115,938	280.422
3800	202,990	348.981	123,454	282.454
4000	215,624	352.221	130,989	284.387
4400	240,992	358.266	146,108	287.989
4800	266,488	363.812	161,285	291.290
5200	292,112	368.939	176,510	294.337
5600	317,870	373.711	191,782	297.167
6000	343,782	378.180	207,105	299.809

Source: Adapted from Sonntag et al. (2003).

Table C.5 Ideal-Gas Properties of H_2O and OH (SI Units), Entropies at 0.1-MPa (1-bar) Pressure.

	Water (H_2O) $\bar{h}_f^o = -241{,}826$ kJ/kmol $M = 18.015$		Hydroxyl (OH) $\bar{h}_f^o = 38{,}987$ kJ/kmol $M = 17.007$	
T (K)	$\Delta\bar{h}$ (kJ/kmol)	$\bar{s}^o$ (kJ/kmol-K)	$\Delta\bar{h}$ (kJ/kmol)	$\bar{s}^o$ (kJ/kmol-K)
0	−9904	0	−9172	0
100	−6617	152.386	−6140	149.591
200	−3282	175.488	−2975	171.592
298	0	188.835	0	183.709
300	62	189.043	55	183.894
400	3450	198.787	3034	192.466
500	6922	206.532	5991	199.066
600	10,499	213.051	8943	204.448
700	14,190	218.739	11,902	209.008
800	18,002	223.826	14,881	212.984
900	21,937	228.460	17,889	216.526
1000	26,000	232.739	20,935	219.735
1100	30,190	236.732	24,024	222.680
1200	34,506	240.485	27,159	225.408
1300	38,941	244.035	30,340	227.955
1400	43,491	247.406	33,567	230.347
1500	48,149	250.620	36,838	232.604
1600	52,907	253.690	40,151	234.741
1700	57,757	256.631	43,502	236.772
1800	62,693	259.452	46,890	238.707
1900	67,706	262.162	50,311	240.556
2000	72,788	264.769	53,763	242.328
2200	83,153	269.706	60,751	245.659
2400	93,741	274.312	67,840	248.743
2600	104,520	278.625	75,018	251.614
2800	115,463	282.680	82,268	254.301
3000	126,548	286.504	89,585	256.825
3200	137,756	290.120	96,960	259.205
3400	149,073	293.550	104,388	261.456
3600	160,484	296.812	111,864	263.592
3800	171,981	299.919	119,382	265.625
4000	183,552	302.887	126,940	267.563
4400	206,892	308.448	142,165	271.191
4800	230,456	313.573	157,522	274.531
5200	254,216	318.328	173,002	277.629
5600	278,161	322.764	188,598	280.518
6000	302,295	326.926	204,309	283.227

Source: Adapted from Sonntag et al. (2003).

Table C.6 Ideal-Gas Properties of H_2 and H (SI Units), Entropies at 0.1-MPa (1-bar) Pressure.

	Hydrogen (H_2) $\bar{h}_f^o$ = 0 kJ/kmol M = 2.016		Hydrogen, Monatomic (H) $\bar{h}_f^o$ = 217,999 kJ/kmol M = 1.008	
T (K)	$\Delta\bar{h}$ & $\bar{s}^o$ (kJ/kmol)	$\Delta\bar{h}$ (kJ/kmol-K)	$\bar{s}^o$ (kJ/kmol)	 (kJ/kmol-K)
0	−8467	0	−6197	0
100	−5467	100.727	−4119	92.009
200	−2774	119.410	−2040	106.417
298	0	130.678	0	114.716
300	53	130.856	38	114.845
400	2961	139.219	2117	120.825
500	5883	145.738	4196	125.463
600	8799	151.078	6274	129.253
700	11,730	155.609	8353	132.457
800	14,681	159.554	10,431	135.233
900	17,657	163.060	12,510	137.681
1000	20,663	166.225	14,589	139.871
1100	23,704	169.121	16,667	141.852
1200	26,785	171.798	18,746	143.661
1300	29,907	174.294	20,825	145.324
1400	33,073	176.637	22,903	146.865
1500	36,281	178.849	24,982	148.299
1600	39,533	180.946	24,060	149.640
1700	42,826	182.941	29,139	150.900
1800	46,160	184.846	31,218	152.089
1900	49,532	186.670	33,296	153.212
2000	52,942	188.419	35,375	154.279
2200	59,865	191.719	39,532	156.260
2400	66,915	194.789	43,689	158.069
2600	74,082	197.659	47,847	159.732
2800	81,355	200.355	52,004	161.273
3000	88,725	202.898	56,161	162.707
3200	96,187	205.306	60,318	164.048
3400	103,736	207.593	64,475	165.308
3600	111,367	209.773	68,633	166.497
3800	119,077	211.856	72,790	167.620
4000	126,864	213.851	76,947	168.687
4400	142,658	217.612	85,261	170.668
4800	158,730	221.109	93,576	172.476
5200	175,057	224.379	101,890	174.140
5600	191,607	227.447	110,205	175.681
6000	208,332	230.322	118,519	177.114

Source: Adapted from Sonntag et al. (2003).

Table C.7 Ideal-Gas Properties of NO and NO_2 (SI Units), Entropies at 0.1-MPa (1-bar) Pressure.

	Nitric Oxide (NO) $\bar{h}_f^o = 90{,}291$ kJ/kmol $M = 30.006$		**Nitrogen Dioxide (NO_2)** $\bar{h}_f^o = 33{,}100$ kJ/kmol $M = 46.005$	
T (K)	$\Delta\bar{h}$ (kJ/kmol)	$\bar{s}^o$ (kJ/kmol-K)	$\Delta\bar{h}$ (kJ/kmol)	$\bar{s}^o$ (kJ/kmol-K)
0	−9192	0	−10, 186	0
100	−6073	177.031	−6861	202.563
200	−2951	198.747	−3495	225.852
298	0	210.759	0	240.034
300	55	210.943	68	240.263
400	3040	219.529	3927	251.342
500	6059	226.263	8099	260.638
600	9144	231.886	12,555	268.755
700	12,308	236.762	17,250	275.988
800	15,548	241.088	22,138	282.513
900	18,858	244.985	27,180	288.450
1000	22,229	248.536	32,344	293.889
1100	25,653	251.799	37,606	298.904
1200	29,120	254.816	42,946	303.551
1300	32,626	257.621	48,351	307.876
1400	36,164	260.243	53,808	311.920
1500	39,729	262.703	59,309	315.715
1600	43,319	265.019	64,846	319.289
1700	46,929	267.208	70,414	322.664
1800	50,557	269.282	76,008	325.861
1900	54,201	271.252	81,624	328.898
2000	57,859	273.128	87,259	331.788
2200	65,212	276.632	98,578 3	37.182
2400	72,606	279.849	109,948	342.128
2600	80,034	282.822	121,358	346.695
2800	87,491	285.585	132,800	350.934
3000	94,973	288.165	144,267	354.890
3200	102,477	290.587	155,756	358.597
3400	110,000	292.867	167,262	362.085
3600	117,541	295.022	178,783	365.378
3800	125,099	297.065	190,316	368.495
4000	132,671	299.007	201,860	371.456
4400	147,857	302.626	224,973	376.963
4800	163,094	305.940	248,114	381.997
5200	178,377	308.998	271,276	386.632
5600	193,703	311.838	294,455	390.926
6000	209,070	314.488	317,648	394.926

Source: Adapted from Sonntag et al. (2003).

Reference

Sonntag, R., C. Borgnakke, and G. Van Wylen (2003), *Fundamentals of Thermodynamics*, John Wiley & Sons, New York.

D

Curve-Fit Equations for Thermodynamic Properties of Gaseous Fuels and Combustion Products

Specific heats of fuels and ideal gases are curve-fitted to polynomials of the form below. For any given species, the specific heat is approximated by

$$\frac{c_p}{R} = \frac{\bar{c}_p}{R_u} = a_1 + a_2 T + a_3 T^2 + a_4 T^3 + a_5 T^4$$

For an ideal gas, $dh = c_p dT$ and $ds = (c_p/T)dT$. It follows that the enthalpy and entropy at atmospheric pressure are

$$\frac{h}{RT} = \frac{\bar{h}}{R_u T} = a_1 + \frac{a_2}{2}T + \frac{a_3}{3}T^2 + \frac{a_4}{4}T^3 + \frac{a_5}{5}T^4 + \frac{a_6}{T}$$

$$\frac{s^o}{R} = \frac{\overline{s^o}}{R_u} = a_1 \ln T + a_2 T + \frac{a_3}{2}T^2 + \frac{a_4}{3}T^3 + \frac{a_5}{4}T^4 + a_7$$

The temperature range 300–1000 K range is useful for the unburned mixture property computation, and the temperature range 1000–3000 K range is useful for the burned mixture property computation, as discussed in Chapter 2.

Principles of Combustion, Third Edition. Allan T. Kirkpatrick and Kenneth K. Kuo.

Companion website: www.wiley.com/go/kirkpatrick/principlesofcombustion3e

Table D.1 Curve-Fit Coefficients for Ideal-Gas Properties of Selected Fuels in Vapor Phase ($300 \leq T \leq 1000$ K).

Fuel	Formula	a_1	a_2	a_3	a_4	a_5	a_6	a_7
Methane	CH_4	5.14987613	−1.36709788E−02	4.91800599E−05	−4.84743026E−08	1.66693956E−11	−1.02466476E+04	−4.64130376
Gasoline	C_7H_{17}	4.0652	6.0977E−02	−1.8801E−05	0	0	−3.5880E+04	1.545E+01
Diesel	$C_{14.4}H_{24.9}$	7.9710	1.1954E−01	−3.6858E−05	0	0	−1.9385E+04	−1.7879
Methanol	CH_3OH	5.71539582	−1.52309129E−2	6.52441155E−5	−7.10806889E−8	2.61352698E−11	−2.56427656E4	−1.50409823
Ethanol	CH_5OH	4.85868178	−3.74006740E−03	6.95550267E−05	−8.86541147E−8	3.51684430E−11	−2.99961309E+04	4.80192394
Nitromethane	CH_3NO_2	1.412633	2.087101E−02	−8.142134E−06	0	0	−1.026351E+04	1.917126E+01

Source: Gordon and McBride (1971) and McBride et al. (1993).

Table D.2 Curve-Fit Coefficients for Thermodynamic Properties of Combustion Products ($300 \leq T \leq 1000$ K).

Species	a_1	a_2	a_3	a_4	a_5	a_6	a_7
CO_2	0.24007797E+01	0.87350957E−02	−0.66070878E−05	0.20021861E−08	0.63274039E−15	−0.48377527E+05	0.96951457E+01
H_2O	0.40701275E+01	−0.11084499E−02	0.41521180E−05	−0.29637404E−08	0.80702103E−12	−0.30279722E+05	−0.32270046E+00
N_2	0.36748261E+01	−0.12081500E−02	0.23240102E−05	−0.63217559E−09	−0.22577253E−12	−0.10611588E+04	0.23580424E+01
O_2	0.36255985E+01	−0.18782184E−02	0.70554544E−05	−0.67635137E−08	0.21555993E−11	−0.10475226E+04	0.43052778E+01
CO	0.37100928E+01	−0.16190964E−02	0.36923594E−05	−0.20319674E−08	0.23953344E−12	−0.14356310E+05	0.2955535E+01
H_2	0.30574451E+01	0.26765200E−02	−0.58099162E−05	0.55210391E−08	−0.18122739E−11	−0.98890474E+03	−0.22997056E+01
H	0.25000000E+01	0	0	0	0	0.25471627E+05	−0.46011762E+00
O	0.29464287E+01	−0.16381665E−02	0.24210316E−05	−0.16028432E−08	0.38906964E−12	0.29147644E+05	0.29639949E+01
OH	0.38375943E+01	−0.10778858E−02	0.96830378E−06	0.18713972E−09	−0.22571094E−12	0.36412823E+04	0.49370009E+00
NO	0.40459521E+01	−0.34181783E−02	0.79819190E−05	−0.61139316E−08	0.15919076E−11	0.97453934E+04	0.29974988E+01

Source: Adapted from Gordon and McBride (1971).

Table D.3 Curve-Fit Coefficients for Thermodynamic Properties of Combustion Products ($1000 \leq T \leq 3000$ K).

Species	a_1	a_2	a_3	a_4	a_5	a_6	a_7
CO_2	0.446080E+01	0.309817E−02	−0.123925E−05	0.227413E−09	−0.155259E−13	−0.489614E+05	−0.986359
H_2O	0.271676E+01	0.294513E−02	−0.802243E−06	0.102266E−09	−0.484721E−14	−0.299058E+05	0.663056E+01
N_2	0.289631E+01	0.151548E−02	−0.572352E−06	0.998073E−10	−0.652235E−14	−0.905861E+03	0.616151E+01
O_2	0.362195E+01	0.736182E−03	−0.196522E−06	0.362015E−10	−0.289456E−14	−0.120198E+04	0.361509E+01
CO	0.298406E+01	0.148913E−02	−0.578996E−06	0.103645E−09	−0.693535E−14	−0.142452E+05	0.634791E+01
H_2	0.310019E+01	0.511194E−03	0.526442E−07	−0.349099E−10	0.369453E−14	−0.877380E+03	−0.196294E+01
H	0.25E+01	0	0	0	0	0.254716E+05	−0.460117
O	0.254205E+01	−0.275506E−04	−0.310280E−08	0.455106E−11	−0.436805E−15	0.292308E+05	0.492030E+01
OH	0.291064E+01	0.959316E−03	−0.194417E−06	0.137566E−10	0.142245E−15	0.393538E+04	0.544234E+01
NO	0.3189E+01	0.133822E−02	−0.528993E−06	0.959193E−10	−0.648479E−14	0.982832E+04	0.674581E+01

Source: Adapted from Gordon and McBride (1971).

References

Gordon, S., and B. McBride (1971), Computer program for calculation of complex chemical equilibrium compositions, rocket performance, incident and reflected shocks, and Chapman–Jouquet detonations, NASA SP-273.

McBride, B., S. Gordon, and M. Reno (1993), Coefficients for calculating thermodynamic and transport properties of individual species, NASA TM-4513.

E

Transport Properties of Gases: Mass Diffusion, Viscosity, and Thermal Conductivity

E.1 Introduction

Mass diffusion, viscosity, and thermal conductivity in a gas are very similar processes, since mass diffusion is the transfer of molecular mass with a diffusion coefficient, $\mathcal{D}_{AB}$, viscous diffusion is the transfer of molecular momentum with a viscosity μ and thermal diffusion is the transfer of molecular energy with a thermal conductivity k.

E.2 Binary Diffusion Coefficients $\mathcal{D}_{AB}$

According to Fick's law of diffusion, described in Chapter Five, the binary gas diffusion coefficient, $\mathcal{D}_{AB}$, for a binary mixture of molecules A and B is the proportionality constant between the mass flux and the driving potential. Some basic diffusion relationships derived from the kinetic theory of gases between the temperature, pressure, and the molecular dimensions have been given in Section 5.5. A table of diffusion coefficients for common gases at 1 atm and 273 K is given in Table E.1.

By solving the Boltzmann equation, Chapman and Enskog (Chapman and Cowling 1970) arrived at the following relationship:

$$\mathcal{D}_{AB} = \frac{3}{16}\frac{(4\pi kT/\mathrm{M}_{AB})^{1/2}}{n^*\pi\sigma^2\Omega_D}f_D \tag{E.1}$$

The term M_{AB} is the combined molecular mass of species A and B, according to

$$\mathrm{M}_{AB} = 2.0\left(\frac{1}{\mathrm{M}_A}+\frac{1}{\mathrm{M}_B}\right)^{-1} \tag{E.2}$$

In Eq. (E.1), k_B is Boltzmann's constant, and σ is the interaction characteristic length between A and B, expressed in Å and defined as

$$\sigma = \frac{\sigma_A+\sigma_B}{2} \tag{E.3}$$

where σ_A and σ_B are the characteristic Lennard–Jones lengths of species A and B.

The dimensionless parameter, Ω_D, represents the collision integral for mass diffusion and is a function of temperature. The value of Ω_D depends on the intermolecular force between the colliding molecules.

The f_D parameter in Eq. (E.1) represents a correction term that is on the order of unity. According to Elliott et al. (2023), f_D lies between 1.0 and 1.02 if M_A and M_B are of the same order. If one

Principles of Combustion, Third Edition. Allan T. Kirkpatrick and Kenneth K. Kuo.

Companion website: www.wiley.com/go/kirkpatrick/principlesofcombustion3e

Table E.1 Binary Diffusion Coefficients at 1 atm.

Substance A	Substance B	T (K)	$D_{AB} \times 10^5$ (m^2/s)
Benzene	Air	273	0.77
Carbon Dioxide	Air	273	1.38
Cyclohexane	Air	318	0.86
n-Decane	Nitrogen	363	0.84
n-Dodecane	Nitrogen	399	0.81
Ethanol	Air	273	1.02
n-Hexane	Nitrogen	288	0.757
Hydrogen	Air	273	0.611
Methanol	Air	273	1.32
n-Octane	Air	273	0.505
Water	Air	273	2.2

Source: Adapted from Perry et al. (1984).

component has a much lower molecular weight than the other, then f_D can vary between 1.0 and 1.1. Taking f_D as 1.0 and replacing the number density n^* with the ideal-gas law relationship, Eq. (E.1) becomes

$$\mathcal{D}_{AB} = \frac{0.0266\, T^{3/2}}{p\, \mathrm{M}_{AB}^{1/2}\, \sigma^2\, \Omega_D} \tag{E.4}$$

The $\mathcal{D}_{AB}$ calculated from this equation is in m^2/s, pressure p is in Pa, temperature T in K, and σ_{AB} in Å. Note that $D_{AB} \sim T^{3/2}/p$.

Based on Neufeld et al. (1972), the dimensionless collision integral Ω_D can be calculated from

$$\Omega_D = \frac{A}{T^{*B}} + \frac{C}{\exp(DT^*)} + \frac{E}{\exp(FT^*)} + \frac{G}{\exp(HT^*)} \tag{E.5}$$

where

$$T^* = \frac{kT}{\varepsilon} \tag{E.6}$$

The parameter ε depends on the characteristic Lennard–Jones energies ε_A and ε_B, of species A and B respectively, according to the following equation:

$$\varepsilon = (\varepsilon_A \varepsilon_B)^{1/2} \tag{E.7}$$

The constants in Eq. (E.5) are defined as follows:

$A = 1.06036 \quad B = 0.15610 \quad C = 0.19300 \quad D = 0.47635$

$E = 1.03587 \quad F = 1.52996 \quad G = 1.76474 \quad H = 3.89411$

The values of ε/k_B in the tabulated Lennard–Jones potential for various gases are presented in Table E.2.

Table E.2 Lennard–Jones Parameters.

Species	σ (Å)	ε/k_B (K)	Species	σ (Å)	ε/k_B (K)
Air	3.711	78.6	C_5H_{12}	5.784	341.1
Al	2.655	2750	C_6H_6	5.349	412.3
Ar	3.542	93.3	C_6H_{12}	6.182	297.1
B	2.265	3331	C_6H_{14}	5.949	399.3
BO	2.944	596	H	2.708	37.0
B_2O_3	4.158	2092	H_2	2.827	59.7
C	3.385	0.56	H_2O	2.641	809.1
CH	3.370	30.6	H_2O_2	4.196	389.3
CH_3OH	3.626	481.8	He	2.551	10.22
CH_4	3.758	148.6	N	3.298	71.4
CN	3.856	75	NH_3	2.900	558.3
CO	3.690	91.7	NO	3.492	116.7
CO_2	3.941	195.2	N_2	3.798	71.4
C_2H_2	4.033	231.8	N_2O	3.828	232.4
C_2H_4	4.163	224.7	O	3.050	106.7
C_2H_6	4.443	215.7	OH	3.147	79.8
C_3H_8	5.118	237.1	O_2	3.467	106.7

Source: Adapted from Svehla (1995).

E.3 Dynamic Viscosity μ

The viscosity of a fluid is a measure of the resistance between the shear stress and the rate of strain. The resistance is due to cohesion of the molecules and the molecular transfer from one layer to another, setting up a shear stress. In gases, cohesion is weak, but molecular transfer between molecules is significant. Since molecular transfer in gases increases with temperature, gas viscosity increases with temperature. In general, liquid viscosity decreases as temperature increases due to weakened cohesion of the molecules. The viscosity of a dilute gas is independent of pressure due to two competing effects. As the pressure increases, the density of the gas increases, but the mean free path decreases, causing the viscosity to remain unchanged.

The dynamic viscosity μ is defined as the ratio of shear stress to the rate of strain with dimensions $Pa - s$, $N \cdot s/m^2$, $kg/(m \cdot s)$, or centipoise ($\equiv 0.001\ N \cdot s/m^2$). The kinematic viscosity ν is defined as the ratio of the dynamic viscosity to the density, i.e., $\nu \equiv \mu/\rho$ with dimensions m^2/s.

The viscosity of a gas can be modeled in a simple manner using kinetic theory. The kinetic model makes use of the following assumptions: (*a*) all molecules are noninteracting, rigid spheres of diameter σ, (*b*) each molecule has a mass of m, (*c*) molecules move randomly with mean velocity v, and (*d*) the number density of the gas is n molecules per unit volume.

The rate of momentum transfer per unit cross-sectional flow area is proportional to the velocity gradient in the transverse direction. The proportionality constant μ can be expressed as

$$\mu = \frac{vL}{3} mn \tag{E.8}$$

where L is the mean free path; L is proportional to $(n\sigma^2)^{-1}$. The random velocity v is proportional to $(R_u T/\mathrm{M})^{1/2}$, where R_u is the universal gas constant, and M is the molecular mass of the gas. After substituting these basic kinetic theory relationships, the dynamic viscosity can be expressed as

$$\mu = 26.69 \frac{(\mathrm{M}T)^{1/2}}{\sigma^2} \tag{E.9}$$

In the above equation, μ is given in μP (micropoise), M in g/mol, T in K, and σ, the hard sphere diameter, in Å (ångström ≡ 10^{-10} m). Note the similarity with the equation for mass diffusivity $\mathcal{D}_{AB}$, Eq. (E.1), also derived using kinetic theory.

The rigid, noninteracting sphere model is the simplest gas viscosity model. It shows the general relationship between viscosity, temperature, and molecule size. Typically, correction factors are required to determine the actual viscosity. The Chapman–Enskog approach (Chapman and Cowling 1970) uses the collision diameter σ and the collision integral Ω_v, which is a function of the dimensionless temperature T^*, where k_B is Boltzmann's constant and ε is the characteristic energy.

$$\Omega_v = \Omega_v(T^*) \quad \text{and} \quad T^* \equiv \frac{T}{\varepsilon/k_B} \tag{E.10}$$

The interactions between colliding molecules are described by the intermolecular potential function $\psi(r)$ for gas molecules separated by a distance of r. For example, the Lennard–Jones 12-6 potential is

$$\psi(r) = 4\varepsilon \left[\left(\frac{\sigma}{r}\right)^{12} - \left(\frac{\sigma}{r}\right)^{6} \right] \tag{E.11}$$

There are many other forms of $\psi(r)$ one can utilize to determine the collision integral Ω_v. In general, Ω_v is not very sensitive to the form of $\psi(r)$. Many scientists have used Eq. (E.11) to obtain Ω_v. Neufeld et al. (1972) suggested the following empirical equation for Ω_v:

$$\Omega_v = \lfloor A(T^*)^{-B} \rfloor + C[\exp(-DT^*)] + E[\exp(-FT^*)] \tag{E.12}$$

where $A = 1.16145, B = 0.14874, C = 0.52487, D = 0.77320, E = 2.16178$, and $F = 2.43787$. Chung et al. (1984) suggested the use of

$$\frac{\varepsilon}{k_B} = \frac{T_c}{1.2593} \quad \text{and} \quad \sigma = 0.809 V_c^{1/3} \tag{E.13}$$

where T_c is the critical temperature, in K, and V_c is the critical volume, in cm^3/mol.

Reichenberg (1975) found Ω_v to be a power law of T^*, i.e.,

$$\Omega_v = a(T^*)^n \tag{E.14}$$

With the consideration of the interactions between the molecules, Eq. (E.9) is modified as

$$\mu = 26.69 \frac{(\mathrm{M}\,T)^{1/2}}{\sigma^2 \Omega_v} \tag{E.15}$$

Using either Eqs. (E.13) or (E.14), the gas viscosity can be calculated from Eq. (E.15). Note that the collision integral Ω_v is equal to 1 if the molecules do not attract each other. To account for molecular shapes and polarities, Chung et al. (1984) suggested the following equation for viscosity calculations:

$$\mu = 40.785\frac{F_c(M\,T)^{1/2}}{V_c^{2/3}\Omega_v} \tag{E.16}$$

where the factor F_c is related to the acentric factor ω, a correction factor for highly polar substances κ, and the dimensionless dipole moment η_r, according to

$$F_c = 1 - 0.2756\,\omega + 0.059035\,\eta_r^4 + \kappa \tag{E.17}$$

where the ω, κ, and η_r values for many hydrocarbon fuels and combustion products are given in Elliott et al. (2023).

E.4 Thermal Conductivity *k*

The German physicist A. Eucken (1884–1950) developed a simple model to relate the thermal conductivity k to the viscosity μ and specific heat c_v (Eucken 1913). The Prandtl number is defined as

$$Pr = \frac{c_p\mu}{k} \tag{E.18}$$

If $Pr \sim 1$, then $k \sim c_p\mu \sim c_v\mu$. Eucken separated the thermal conductivity k into a component k_{tr} for the translational energy and a component k_{in} for the remaining internal energies. The specific heat term c_v was accordingly divided into two parts, one for the translational energy and the other for the internal molecular energy, and the term f is a constant.

$$\begin{aligned} k &= k_{\text{tr}} + k_{\text{in}} \\ &= (f_{\text{tr}}c_{v,\text{tr}} + f_{\text{in}}c_{v,\text{in}})\mu \end{aligned} \tag{E.19}$$

Using kinetic theory, for purely translational motion $f_{\text{tr}} = 5/2$, and for internal motion $f_{\text{in}} = 1$. The specific heat terms are $c_{v,\text{tr}} = \frac{3}{2}R$, and so $c_{v,\text{in}} = c_v - \frac{3}{2}R$. The ideal gas constant is $R = c_p - c_v = (\gamma - 1)c_v$. Upon substitution, Eucken's formula for thermal conductivity can be expressed as

$$\begin{aligned} k &= \left(\frac{5}{2}\cdot\left[\frac{3}{2}(\gamma-1)\right] + 1\cdot\left[1 - \frac{3}{2}(\gamma-1)\right]\right)\mu\,c_v \\ &= \frac{1}{4}(9\gamma - 5)\mu\,c_v \end{aligned} \tag{E.20}$$

E.5 Curve-Fits and Coefficients for Viscosity, Thermal Conductivity, and Specific Heat

The temperature dependence of the thermal conductivity k, dynamic viscosity μ, and specific heat c_p of selected fuels are curve-fitted to polynomials in Table E.3. For any given species, k, μ, and c_p are approximated by the following sets of a_i coefficients.

$$k(T) = a_1 + a_2T + a_3T^2 + a_4T^3 + a_5T^4 + a_6T^5 + a_7T^6$$

$$\mu(T) = a_1 + a_2T + a_3T^2 + a_4T^3 + a_5T^4 + a_6T^5 + a_7T^6$$

$$c_p(T) = a_1 + a_2T + a_3T^2 + a_4T^3 + a_5T^4 + a_6T^5 + a_7T^6$$

Table E.3 Curve-Fit Coefficients for Fuel Vapor Thermal Conductivity k (W/m-K), Dynamic Viscosity μ (Pa-s), and Specific Heat (J/kg-K).

Fuel	Formula	Property	a_1	a_2	a_3	a_4	a_5	a_6	a_7
Hydrogen	H_2	k (100–500) K	2.009705E−02	3.234622E−04	2.1637249E−06	−6.49151204E−09	5.52407932E−12	0	0
		k(500–1500) K	0.1083105	2.21163789E−04	2.26380948E−07	−1.74258636E−10T	4.6468625E−14	0	0
		μ (10–500) K	−0.135666E−06	6.84115878E−08	−3.928747E−10	1.8996E−12	−5.23104E−15	7.4490972E−18	−4.250937E−21
		μ (500–2000) K	2.72941E−06	2.3224377E−08	−7.6287854E−12	2.92585E−15	−5.2889938E−19		0
		c_p (100–425) K	5006.6253	101.569422	−0.602891517	2.7375894E−03	−8.4758275E−06	1.43800374E−08	9.8072403E−12
		c_p (490–1365) K	14920.082	−1.996917584	2.540615E−03	−4.7588954E−07	0	0	0
Methane	CH_4	k (100–1000) K	−1.3401499E−02	3.6630706E−04	−1.8224861E−06	5.9398800E−09	−9.1405505E−12	6.7896889E−15	−1.9504874E−18
		μ (70–1000) K	2.9682670E−07	3.7112010E−09	1.2182980E−11	−7.02426E−14	7.543269E−17	−2.7237166E−20	0
n-Hexane	C_6H_{14}	k (150–1000) K	1.287757E−03	−2.0049944E−05	2.378588E−07	−1.609445E−10	7.710272E−14	0	0
		μ (270–900) K	1.545412E−06	1.150809E−08	2.722165E−11	−3.26900E−14	1.245459E−17	0	0
n-Heptane	C_7H_{16}	k (250–1000) K	−4.606147E−2	5.9565222E−4	−2.988931E−6	8.4461287E−9	−1.22927E−11	9.0127E−15	−2.6296E−18
		μ (270–580) K	1.540097E−06	1.095157E−08	1.800664E−11	−1.36379E−14	0	0	0
		c_p (300–755) K	9.4626E+1	5.860997	−1.98231E−3	−6.886993E−8	−1.93795E−10	0	0
		c_p (755–1365) K	−7.40308E+2	1.089354E+1	−1.265124E−2	9.843763E−6	−4.3228296E−9	7.863665E−13	
n-Octane	C_8H_{18}	k (250–500) K	−4.013919E−03	3.387960E−05	8.192918E−08	0	0	0	0
		μ (300–650) K	8.324354E−07	1.400450E−09	−8.793765E−12	−6.840300E−15	0	0	0
		c_p (275–755) K	2.144198E+2	5.3569	−1.17497E−3	−6.991155E−7	0	0	0
		c_p (755–1365) K	2.4359686E+3	−4.4681947	−1.6684329E−2	−1.7885605E−5	8.6428202E−9	−1.614265E−12	
Methanol	CH_3OH	k (300–550) K	−2.0298675E−02	1.2191097E−04	−2.237484E−08	0	0	0	0
		μ (250–650) K	1.197900E−06	2.450280E−08	1.861627E−11	−1.306748E−14	0	0	0
Ethanol	CH_5OH	k (250–550) K	−2.466630E−02	1.558925E−04	−8.229548E−08	0	0	0	0
		μ (270–600) K	−6.335950E−08	3.207134E−08	−6.250795E−14	0	0	0	0

Source: Adapted from Andrews and Biblarz (1981).

References

Andrews, J., and O. Biblarz (1981), *Temperature Dependence of Gas Properties in Polynomial Form*, Naval Postgraduate School, Monterey, CA.

Chapman, S., and T. Cowling (1970), *The Mathematical Theory of Non-uniform Gases*, Cambridge University Press, New York.

Chung, T., L. Lee, and K. Starling (1984), Applications of kinetic gas theories and multi-parameter correlation for prediction of dilute gas viscosity and thermal conductivity, *Ind. Eng. Chem. Fundam.*, Vol. 23, pp. 8–13.

Elliott, J., V. Diky, T. Knotts, and W. Wilding (2023), *The Properties of Gases and Liquids*, 6th ed., McGraw-Hill, New York.

Eucken, A. (1913), Über das Wärmeleitvermögen, die spezifische Wärme und die innere Reibung der Gase, *Phys. Z.*, Vol. 14, no. 8, pp. 324–332.

Neufeld, P., A. Janzen, and R. Aziz (1972), Empirical equations to calculate 16 of the transport collision integrals $\Omega^{(l,s)*}$ for the Lennard-Jones (12–6) potential, *J. Chem. Phys.*, Vol. 57, pp. 1100–1102.

Perry, R. H., D. Green, and J. O. Maloney (1984), *Perry's Chemical Engineers' Handbook*, McGraw-Hill, New York.

Reichenberg, D. (1975), New methods for the estimation of the viscosity coefficients of pure gases at moderate pressures (with particular reference to organic vapors), *AIChE J.*, Vol. 21, pp. 181–183.

Svehla, R. (1995), Transport coefficients for the NASA Lewis chemical equilibrium program, NASA-TM-4647.

F

Equations of State

F.1 Introduction

The density of gaseous mixtures can be calculated from several accurate equations of state, which relate density to temperature and pressure. Most equations of state are accurate only when the density is substantially lower than the critical density of the substance, although several have been developed to maintain accuracy up to 2.5 times the critical density. Most equations fall into one of the three categories: generalized, empirical, or theoretical. p–v–T equations of state may be of two types: *pressure-explicit*, in which case p is given as a function of v, T, and composition, and *volume-explicit*, in which v is given as a function of T, p, and composition.

F.1.1 Ideal-Gas Equation of State

The simplest equation of state is that for an ideal gas. The ideal-gas law, which applies to both pure components and mixtures, has been established from empirical observation and is accurate for gases at low density or up to about 2 atm for most compounds. For nondissociating molecules, this relationship holds for low to moderate pressures.

$$pV = mRT = m\frac{R_u}{M}T = nR_uT \qquad \text{or} \qquad pv = RT \qquad \text{or} \qquad p = \rho RT \tag{F.1}$$

F.1.2 Van der Waals Equation of State

The van der Waals equation of state (introduced in 1873) is one of the best-known generalized equations of state. It is essentially a modified version of the ideal-gas law, expressed by Eq. (F.1), except that it accounts for the intermolecular forces that exist between molecules (represented by the term a/v^2) and also corrects for the volume, b, occupied by the molecules themselves. The van der Waals equation of state is

$$\left(p + \frac{a}{v^2}\right)(v - b) = RT \tag{F.2}$$

where a and b are evaluated from the general behavior of gases. These constants are related to the critical temperature T_c and pressure p_c of pure substances by

$$a = \frac{27}{64}\frac{R^2T_c^2}{p_c} \qquad \text{and} \qquad b = \frac{RT_c}{8p_c} \tag{F.3}$$

Principles of Combustion, Third Edition. Allan T. Kirkpatrick and Kenneth K. Kuo.

Companion website: www.wiley.com/go/kirkpatrick/principlesofcombustion3e

F.1.3 Redlich–Kwong Equation of State

The Redlich–Kwong equation of state (introduced in 1949) is representative of the commonly used *empirical* cubic equations of state. It is considerably more accurate than the van der Waals equation and has been shown to be very successful for not only pure substances but also mixture calculations and phase equilibrium correlations. The original Redlich–Kwong equation is given as

$$p = \frac{R_u T}{\overline{v} - b} - \frac{a}{\overline{v}(\overline{v} + b)T^{1/2}} \tag{F.4}$$

where $\overline{v}$ is the molar specific volume (m^3/kmol), and where

$$a = \frac{0.42748 R_u^2 T_c^{2.5}}{p_c} \quad \text{and} \quad b = \frac{0.08664 R_u T_c}{p_c} \tag{F.5}$$

F.1.4 Soave (SRK) and Peng–Robinson Equations of State

The Soave (SRK) equation and the Peng–Robinson equations of state are also "cubic" equations of state developed to improve the Redlich–Kwong form. Both approaches have used the same method to set the parameters a and b. That is, both the first and second partial derivatives of pressure with respect to specific volume are set to zero, as is done for the Redlich–Kwong equation of state. Both of these equations rely heavily on computations based on the compressibility factor, $Z = pv/RT$. The cubic form of the equations and their coefficients are provided in the summary table (Table F.1).

The parameter ω in the Soave and Peng–Robinson equations in Table F.1 is the acentric factor, a measure of the forces between the noncentral portions of molecules, i.e., their nonsphericity, and is defined as

$$\omega = -\log_{10}(p_r^{\text{sat}}) - 1, \text{ at } T_r = 0.7 \tag{F.6}$$

where $p_r^{\text{sat}} = p^{\text{sat}}/p_c$ is the reduced saturation vapor pressure and $T_r = T/T_c$ is the reduced temperature (Pitzer et al. 1955). The f_1 and f_2 terms are

$$f_1(\omega) = 0.48 + 1.574\,\omega - 0.176\,\omega^2 \tag{F.7}$$

and

$$f_2(\omega) = 0.37464 + 1.5423\,\omega - 0.26992\,\omega^2 \tag{F.8}$$

Representative acentric values are given in ascending order in Table F.2.

The four equations of state (van der Waals, Redlich–Kwong, Soave, and Peng–Robinson) in Table F.1 are classified as "cubic equations of state," since the equations contain volume terms

Table F.1 Summary of Four Common "Cubic" Equations of State and Their Constants.

Equation	*u*	w	*b*	*a*
van der Waals	0	0	$\frac{R_u T_c}{8P_c}$	$\frac{27}{64}\frac{R_u^2 T_c^2}{P_c}$
Redlich–Kwong	1	0	$\frac{0.08664 R_u T_c}{P_c}$	$\frac{0.42748 R_u^2 T_c^{2.5}}{P_c T^{0.5}}$
Soave	1	0	$\frac{0.08664 R_u T_c}{P_c}$	$\frac{0.42748 R_u^2 T_c^2}{P_c}[1 + f_1(\omega)(1 - T_r^{0.5})]^2$
Peng–Robinson	2	−1	$\frac{0.07780 R_u T_c}{P_c}$	$\frac{0.42748 R_u^2 T_c^2}{P_c}[1 + f_2(\omega)(1 - T_r^{0.5})]^2$

Table F.2 Representative Acentric Factors.

Molecule	Formula	ω
Hydrogen	H_2	−0.216
Methane	CH_4	0.011
Oxygen	O_2	0.025
Nitrogen	N_2	0.039
Carbon monoxide	CO	0.066
Propane	C_3H_8	0.153
Nitrous oxide	N_2O	0.165
Carbon dioxide	CO_2	0.239
Ammonia	NH_3	0.250
Water	H_2O	0.344
Octane	C_8H_{18}	0.398
Methanol	CH_3OH	0.556
Nitric oxide	NO	0.588

Source: Elliott et al. (2023)/with permission of McGraw-Hill Education.

raised to either the first, second, or third power. The two-parameter (containing a and b) equations are represented by the following general equation:

$$p = \frac{R_u T}{\overline{v} - b} - \frac{a}{\overline{v}^2 + ub\overline{v} + wb^2} \tag{F.9}$$

F.1.5 Virial Equation of State

The virial equation of state, expressed in a polynomial series in reciprocal molar specific volume, is a theoretical equation of state derived from kinetic theory and statistical mechanics.

$$p = \frac{R_u T}{\overline{v}} + \frac{R_u TB}{\overline{v}^2} + \frac{R_u TC}{\overline{v}^3} + \cdots \tag{F.10}$$

where B, C, … are called the second, third, … virial coefficients and are functions of temperature and type of pure substance. The number of terms required for accurate predictions is highly dependent on pressure: as pressure increases, more terms must be used. Considerable amounts of information exist on the second virial coefficients; however, very little is known about the third coefficients, and essentially nothing about the fourth and higher coefficients.

F.1.6 Beattie–Bridgeman Equation of State

The Beattie–Bridgeman equation of state (Beattie and Bridgeman 1928) is an empirical, pressure-explicit equation requiring five constants and is reasonably accurate for densities of less than about 0.8 times the critical density.

$$p = \frac{R_u T(1-\varepsilon)}{v^2}(v + B) - \frac{A}{v^2} \tag{F.11}$$

where $A = A_0(1 - a/v)$, $B = B_0(1 - b/v)$, $\varepsilon = c/vT^3$, and a, b, and c are constants for different gases. The values of these constants for several substances are given in Table F.3.

Table F.3 Constants for Beattie–Bridgeman Equation of State (Pressure in kPa; Specific Volume in m^3/kmol; Temperature in Kelvin; $R_u = 0.08206$ atm-l/g-mol · K).

Gas	A_0	a	B_0	b	$10^{-4}\ c$
Helium	0.0216	0.05984	0.01400	0.0	0.0040
Argon	1.2907	0.02328	0.03931	0.0	5.99
Hydrogen	0.1975	−0.00506	0.02096	−0.04359	0.0504
Nitrogen	1.3445	0.02617	0.05046	−0.00691	4.20
Oxygen	1.4911	0.02562	0.04624	0.004208	4.80
Air	1.3012	0.01931	0.04611	−0.001101	4.34
CO_2	5.0065	0.07132	0.10476	0.072350	66.00

F.1.7 Benedict–Webb–Rubin Equation of State

The Benedict, Webb, and Rubin (BWR) equation of state is an extension of the Beattie–Bridgeman equation of state through the addition of high-density terms (Benedict et al. 1951). It requires eight empirical constants. Its popularity stems from its ability to calculate both liquid and vapor properties, and it is accurate for those compounds and mixtures for which constants have been determined. Expressed in terms of molar density ρ, the BWR equation is

$$\begin{aligned} p = {} & R_u T\rho + \left(B_0 R_u T - A_0 - \frac{C_0}{T^2} \right) \rho^2 + (bR_u T - a)\rho^3 + a\alpha\rho^6 \\ & + \left(\frac{c\rho^3}{T^2} \right) (1 + \gamma\rho^2) \exp(-\gamma\rho^2) \end{aligned} \tag{F.12}$$

where A_0, B_0, C_0, b, a, α, c, and γ are the eight constants. Values for these constants can be found in Perry and Green (2019). When solved for density, this equation can have up to six roots. The lowest value is the vapor density, and the highest is the liquid density.

F.1.8 Ideal-Gas Mixing Rules

The mixing rule most commonly used for the ideal-gas equation is Dalton's law of partial pressures, which is applicable only to ideal gases. In Dalton's model, the properties of each component are evaluated as though each component existed separately at the volume and temperature of the mixture.

Considering a mixture consisting of two components, A and B, where both the mixture and each individual component behave as ideal gases, then, for the mixture,

$$pV = nR_u T \tag{F.13}$$

where

$$n = n_A + n_B \tag{F.14}$$

For the components,

$$p_A V = n_A R_u T, \qquad p_B V = n_B R_u T \tag{F.15}$$

On substituting Eq. (F.15) into Eq. (F.14), we have

$$\frac{pV}{R_uT} = \frac{p_AV}{R_uT} + \frac{p_BV}{R_uT} \tag{F.16}$$

which yields

$$p = p_A + p_B \tag{F.17}$$

Thus, for a mixture of ideal gases, the pressure is the sum of the partial pressures of the individual components. It must be stressed that this approximation only holds for ideal gases and assumes that no interactions between different components take place.

F.1.9 Mixing Rules for Redlich–Kwong Types of Equation of State

The mixing rules recommended by Redlich–Kwong for all two-constant cubic equations of state (i.e., van der Waals, Redlich–Kwong, Soave, and Peng–Robinson) are broadly used. The constants a_m and b_m for the mixture (m) with n components can be calculated from

$$a_m = \sum_i^n \sum_j^n X_i X_j (a_i a_j)^{1/2} (1 - \overline{k}_{ij}) \tag{F.18}$$

$$b_m = \sum_i^n X_i b_i \tag{F.19}$$

where a_i, b_i, and the binary interaction coefficients $\overline{k}_{ij}$ for the Soave and Peng–Robinson equations are given in Perry and Green (2019). For hydrocarbon pairs, $\overline{k}_{ij}$ is usually taken as zero.

F.1.10 Benedict-Webb-Rubin Mixing Rules

For mixtures of n components, the eight constant parameters in Eq. (F.12) are functions of composition in the mixture. The mixture parameter φ_{mix} can be calculated from

$$\varphi_{\text{mix}} = \sum_{i=1}^n X_i \varphi_i^{1/r} \tag{F.20}$$

where r, the inverse exponential integer, is based on the type of constant represented by φ.

F.1.11 Virial Equation Mixing Parameters

Because the second virial coefficient is the most used, it will be the only one treated here, as very little is known about the third and fourth virial coefficients. The second virial coefficient for mixtures of n components, B_{mix}, is obtained from the following expression:

$$B_{\text{mix}} = \sum_{i=1}^n \sum_{j=1}^n X_i X_j B_{ij} \tag{F.21}$$

where B_{ii} represents the pure-component second virial coefficient of component i, and B_{ij} is the binary-interaction second virial coefficient for components i and j (note that $B_{ij} = B_{ji}$).

For example, in a ternary system with mixture components 1, 2, and 3,

$$B_{\text{mix}} = X_1^2 B_1 + X_2^2 B_2 + X_3^2 B_3 + 2X_1X_2B_{12} + 2X_1X_3B_{13} + 2X_2X_3B_{23} \tag{F.22}$$

F.2 Thermodynamic Derivatives

As shown by Bridgman (1914), any thermodynamic derivative can be expressed in terms of three independent derivatives. These derivatives are

$$\begin{aligned} &\left(\frac{\partial h}{\partial T}\right) = c_p \\ &\left(\frac{\partial v}{\partial P}\right)_T \\ &\left(\frac{\partial v}{\partial T}\right)_P \end{aligned} \tag{F.23}$$

For example, γ, the ratio of specific heats, is given by

$$\gamma = \frac{c_p}{c_v} = \frac{c_p}{c_p + T\left(\frac{\partial v}{\partial T}\right)_P^2 / \left(\frac{\partial v}{\partial P}\right)_T} \tag{F.24}$$

These derivatives were chosen by Bridgman as they can be determined directly by experiment. The term $\left(\frac{\partial v}{\partial P}\right)_T$ is the isothermal compressibility, the term $\left(\frac{\partial v}{\partial T}\right)_P$ is the coefficient of volumetric expansion, and the term $\left(\frac{\partial h}{\partial T}\right)$ is the specific heat at constant pressure. These three variables are used in this text to compute fuel–air mixture properties and perform cycle analysis.

For a gas mixture, the derivatives can be expressed as functions of pressure P, specific volume v, temperature T, and molecular mass M. Using the ideal gas law, the specific volume v of a mixture is

$$v = \frac{R_u T}{MP} \tag{F.25}$$

Upon differentiation of Eq. (F.25) with respect to pressure P,

$$\begin{aligned} \left(\frac{\partial v}{\partial P}\right)_T &= \frac{R_u T}{M}\frac{\partial}{\partial P}\left(\frac{1}{P}\right) + \frac{R_u T}{P}\frac{\partial}{\partial P}\left(\frac{1}{M}\right) \\ &= -\frac{v}{P} - \frac{v}{M}\frac{\partial M}{\partial P} \\ &= -\frac{v}{P} - \frac{v}{M}\sum M_i \frac{\partial X_i}{\partial P} \end{aligned} \tag{F.26}$$

where

$$\frac{\partial M}{\partial P} = \sum M_i \frac{\partial X_i}{\partial P} \tag{F.27}$$

and upon differentiation of Eq. (F.25) with respect to temperature T,

$$\begin{aligned} \left(\frac{\partial v}{\partial T}\right)_P &= \frac{R_u}{PM} + \frac{R_u T}{P}\frac{\partial}{\partial T}\left(\frac{1}{M}\right) \\ &= \frac{v}{T} - \frac{v}{M}\frac{\partial M}{\partial T} \\ &= -\frac{v}{T} - \frac{v}{M}\sum M_i \frac{\partial X_i}{\partial T} \end{aligned} \tag{F.28}$$

where

$$\frac{\partial M}{\partial T} = \sum M_i \frac{\partial X_i}{\partial T} \tag{F.29}$$

Note the important role that the changes in mole fraction with respect to pressure and temperature, $\partial X_i/\partial P$ and $\partial X_i/\partial T$, have in the determination of the mixture properties and thermodynamic derivatives.

Adopting the notation used by Bridgeman (1914),

$$\left(\frac{\partial y}{\partial x}\right)_z = \frac{(\partial y)_z}{(\partial x)_z}$$

the following list can be used to construct any other required thermodynamic first derivative in terms of $c_p = \partial h/\partial T, \partial v/\partial P$, and $\partial v/\partial T$:

$$(\partial T)_P = -(\partial P)_T = 1$$

$$(\partial v)_P = -(\partial P)_v = \left(\frac{\partial v}{\partial T}\right)_P$$

$$(\partial s)_P = -(\partial P)_s = \frac{c_p}{T}$$

$$(\partial u)_P = -(\partial P)_u = c_p - P\left(\frac{\partial v}{\partial T}\right)_P$$

$$(\partial h)_P = -(\partial P)_h = c_p$$

$$(\partial v)_T = -(\partial T)_v = -\left(\frac{\partial v}{\partial P}\right)_T$$

$$(\partial s)_T = -(\partial T)_s = \left(\frac{\partial v}{\partial T}\right)_P$$

$$(\partial u)_T = -(\partial T)_u = T\left(\frac{\partial v}{\partial T}\right)_P + P\left(\frac{\partial v}{\partial P}\right)_T$$

$$(\partial h)_T = -(\partial T)_h = -v + T\left(\frac{\partial v}{\partial T}\right)_P$$

$$(\partial s)_v = -(\partial v)_s = \frac{c_p}{T}\left(\frac{\partial v}{\partial P}\right)_T + \left(\frac{\partial v}{\partial T}\right)_P^2$$

$$(\partial u)_v = -(\partial v)_u = c_p\left(\frac{\partial v}{\partial P}\right)_T + T\left(\frac{\partial v}{\partial T}\right)_P^2$$

$$(\partial h)_v = -(\partial v)_h = c_p\left(\frac{\partial v}{\partial P}\right)_T + T\left(\frac{\partial v}{\partial T}\right)_P^2 - v\left(\frac{\partial v}{\partial T}\right)_P$$

$$(\partial u)_s = -(\partial s)_u = \frac{P}{T}\left[c_p\left(\frac{\partial v}{\partial P}\right)_T + T\left(\frac{\partial v}{\partial T}\right)_P^2\right]$$

$$(\partial h)_s = -(\partial s)_h = -\frac{vc_p}{T}$$

$$(\partial h)_u = -(\partial u)_h = -v\left[c_p - P\left(\frac{\partial v}{\partial T}\right)_P\right] - P\left[c_p\left(\frac{\partial v}{\partial P}\right)_T + T\left(\frac{\partial v}{\partial T}\right)_P^2\right]$$

References

Beattie, J. A., and O. Bridgeman (1928), A new equation of state for fluids, *Proc. Am. Acad. Arts Sci.*, Vol. 63, pp. 229–308.

Benedict, M., G. Webb, and L. Rubin (1951), An empirical equation for thermodynamic properties of light hydrocarbons and their mixtures: constants for twelve hydrocarbons, *Chem. Eng. Prog.*, Vol. 47, no. 8, pp. 419–422.

Bridgman, P. (1914), *A Complete collection of thermodynamic formulas*, Phys. Rev., Vol. 3, pp. 273–281.

Elliott, J., V. Diky, T. Knotts, and W. Wilding (2023), *The Properties of Gases and Liquids*, 6th ed., McGraw-Hill, New York.

Perry, R. H., and D. Green (2019), *Perry's Chemical Engineers' Handbook*, 9th ed., McGraw-Hill, New York.

Pitzer, K., D. Lippmann, R. Curl, C. Huggins, and D. Petersen (1955), The volumetric and thermodynamic properties of fluids. II. Compressibility factor, vapor pressure and entropy of vaporization, *J. Am. Chem. Soc.*, Vol. 77, pp. 3433–3440.

G

Mathematical Relations

G.1 Numerical Solution of Equilibrium Combustion Equations

The 4 × 4 nonlinear system of equations developed in Chapter 2 with unknowns X_3, X_4, X_5, X_6, i.e., the mole fractions of N_2, O_2, CO, H_2, respectively, is solved using Newton–Raphson iteration and Gaussian elimination.

$$f_j(X_3, X_4, X_5, X_6) = 0 \quad j = 1, 2, 3, 4 \tag{G.1}$$

A first-order Taylor series expansion of the solution mole fraction vector $[X_3^*, X_4^*, X_5^*, X_6^*]$ is $X_i^* = X_i^{(1)} + \Delta X_i$, where X_i^* is a set of mole fractions reasonably close to the solution vector. The steps ΔX_i are computed from the matrix equation

$$\mathbf{A}\Delta X_i = \mathbf{F} \tag{G.2}$$

where **A** is the Jacobian matrix of the partial derivative of each equation with respect to each mole fraction, $A_{ij} = \partial f_j / \partial X_i$. At each iteration denoted by k, the new X_i solution vector is thus given by

$$X_i^{k+1} = X_i^k - \mathbf{A}^{-1}\mathbf{F} \tag{G.3}$$

Iteration continues until the maximum step change ΔX_i is below a specified tolerance. In certain cases, Newton–Raphson iteration may run into difficulties due to the numerical precision of the computer, and the solution of the matrix equation for ΔX_i may yield singularities. This behavior is reduced by introducing an under-relaxation parameter, which was determined by trial and error to be 0.05.

The elements of the Jacobian matrix **A** are determined by first defining the following partial derivatives:

$$D_{ij} = \frac{\partial X_k}{\partial X_l} \frac{i = 1, 2, 7, 8, 9, 10}{j = 3, 4, 5, 6}$$

$$D_{jj} = \frac{\partial X_j}{\partial X_j} = 1$$

Performing the mole fraction differentiations, where the c_i terms are $K_{p,i}$ terms modified by pressure.

$$D_{76} = \frac{1}{2} \frac{c_1}{X_6^{1/2}} \qquad D_{103} = \frac{1}{2} \frac{c_4 X_4^{1/2}}{X_3^{1/2}}$$

$$D_{84} = \frac{1}{2} \frac{c_2}{X_4^{1/2}} \qquad D_{26} = c_5 X_4^{1/2}$$

Principles of Combustion, Third Edition. Allan T. Kirkpatrick and Kenneth K. Kuo.

Companion website: www.wiley.com/go/kirkpatrick/principlesofcombustion3e

$$D_{94} = \frac{1}{2}\frac{c_3 X_6^{1/2}}{X_4^{1/2}} \qquad D_{24} = \frac{1}{2}\frac{c_5 X_6}{X_4^{1/2}}$$

$$D_{96} = \frac{1}{2}\frac{c_3 X_4^{1/2}}{X_6^{1/2}} \qquad D_{14} = \frac{1}{2}\frac{c_6 X_5}{X_4^{1/2}}$$

$$D_{104} = \frac{1}{2}\frac{c_4 X_3^{1/2}}{X_4^{1/2}} \qquad D_{15} = c_6 X_4^{1/2}$$

The matrix elements A_{ij} are

$$A_{11} = 1 + D_{103} \qquad A_{12} = 1 + D_{14} + D_{24} + D_{84} + D_{104} + D_{94}$$

$$A_{13} = 1 + D_{15} \qquad A_{14} = 1 + D_{26} + D_{76} + D_{96}$$

$$A_{21} = 0 \qquad A_{22} = 2D_{24} + D_{94} - d_1 D_{14}$$

$$A_{23} = -d_1(1 + D_{15}) \qquad A_{24} = 2D_{26} + 2 + D_{76} + D_{96}$$

$$A_{31} = D_{103} \qquad A_{32} = 2 + (2 - d_2)D_{14} + D_{24} + D_{84} + D_{104} + D_{94}$$

$$A_{33} = 1 - d_2 + (2 - d_2)D_{15} \qquad A_{34} = D_{26} + D_{96}$$

$$A_{41} = 2 + D_{103} \qquad A_{42} = D_{104} + d_3 D_{14}$$

$$A_{43} = -d_3(1 + D_{15}) \qquad A_{44} = 0$$

Evaluating the three thermodynamic derivatives $c_p = (\partial h/\partial T)_p$, $(\partial v/\partial T)p$, and $(\partial v/\partial p)_T$ requires the change in the mole fractions X_i due to changes in temperature and pressure. The four independent mole fraction (X_3, X_4, X_5, X_6) derivatives with respect to temperature are

$$\frac{\partial X_3}{\partial T}, \frac{\partial X_4}{\partial T}, \frac{\partial X_5}{\partial T}, \text{ and } \frac{\partial X_6}{\partial T} \tag{G.4}$$

and the remaining derivatives are expressed in terms of this independent set.

$$\frac{\partial X_7}{\partial T} = X_6^{1/2}\frac{\partial c_1}{\partial T} + D_{76}\frac{\partial X_6}{\partial T}$$

$$\frac{\partial X_8}{\partial T} = X_4^{1/2}\frac{\partial c_2}{\partial T} + D_{84}\frac{\partial X_4}{\partial T}$$

$$\frac{\partial X_9}{\partial T} = X_4^{1/2}X_6^{1/2}\frac{\partial c_3}{\partial T} + D_{94}\frac{\partial X_4}{\partial T} + D_{96}\frac{\partial X_6}{\partial T}$$

$$\frac{\partial X_{10}}{\partial T} = X_4^{1/2}X_3^{1/2}\frac{\partial c_4}{\partial T}$$

$$\frac{\partial X_2}{\partial T} = X_4^{1/2}X_6\frac{\partial c_5}{\partial T} + D_{24}\frac{\partial X_4}{\partial T} + D_{26}\frac{\partial X_6}{\partial T}$$

$$\frac{\partial X_1}{\partial T} = X_4^{1/2}X_5\frac{\partial c_6}{\partial T} + D_{14}\frac{\partial X_4}{\partial T} + D_{15}\frac{\partial X_5}{\partial T}$$

These four independent derivatives are found by solution of the matrix equation that results from differentiating Eq. (G.1) with respect to T,

$$\frac{\partial f_j}{\partial T} + \frac{\partial f_j}{\partial X_3}\frac{\partial X_3}{\partial T} + \frac{\partial f_j}{\partial X_4}\frac{\partial X_4}{\partial T} + \frac{\partial f_j}{\partial X_5}\frac{\partial X_5}{\partial T} + \frac{\partial f_j}{\partial X_6}\frac{\partial X_6}{\partial T} = 0 \tag{G.5}$$

$j = 1, 2, 3, 4$

In the matrix form

$$[\mathbf{A}]\left[\frac{\partial \mathbf{X}}{\partial T}\right] + \left[\frac{\partial \mathbf{f}}{\partial T}\right] = 0 \tag{G.6}$$

where the matrix **A** is identical to that used earlier to solve for the mole fractions. To evaluate $\partial f_j/\partial T$, define

$$\begin{aligned} x_1 &= \frac{X_1}{c_6} \\ x_2 &= \frac{X_2}{c_5} \\ x_k &= \frac{X_k}{c_{k-6}} \qquad \text{where } k = 7, 8, 9, 10 \end{aligned} \tag{G.7}$$

Note that the terms x_k are functions of X_3, X_4, X_5, and X_6 only, so upon substitution of x_k into Eq. (G.1) followed by differentiation with respect to T yields the $[\partial f/\partial T]$ terms:

$$\begin{aligned} \frac{\partial f_1}{\partial T} &= \frac{\partial c_6}{\partial T}x_1 + \frac{\partial c_5}{\partial T}x_2 + \frac{\partial c_1}{\partial T}x_7 + \frac{\partial c_2}{\partial T}x_8 + \frac{\partial c_3}{\partial T}x_9 + \frac{\partial c_4}{\partial T}x_{10} \\ \frac{\partial f_2}{\partial T} &= 2\frac{\partial c_5}{\partial T}x_2 + \frac{\partial c_1}{\partial T}x_7 + \frac{\partial c_3}{\partial T}x_9 - d_1\frac{\partial c_6}{\partial T}x_1 \\ \frac{\partial f_3}{\partial T} &= 2\frac{\partial c_6}{\partial T}x_1 + \frac{\partial c_5}{\partial T}x_2 + \frac{\partial c_2}{\partial T}x_8 + \frac{\partial c_3}{\partial T}x_9 + \frac{\partial c_4}{\partial T}x_{10} - d_2\frac{\partial c_6}{\partial T}x_1 \\ \frac{\partial f_4}{\partial T} &= \frac{\partial c_4}{\partial T}x_{10} - d_3\frac{\partial c_6}{\partial T}x_1 \end{aligned} \tag{G.8}$$

From the definitions of c_i, the $\partial c_i/\partial T$ terms are

$$\begin{aligned} \frac{\partial c_1}{\partial T} &= \frac{1}{p^{1/2}}\frac{dK_{p,1}}{dT} & \frac{\partial c_4}{\partial T} &= \frac{dK_{p,4}}{dT} \\ \frac{\partial c_2}{\partial T} &= \frac{1}{p^{1/2}}\frac{dK_{p,2}}{dT} & \frac{\partial c_5}{\partial T} &= p^{1/2}\frac{dK_{p,5}}{dT} \\ \frac{\partial c_3}{\partial T} &= \frac{dK_{p,3}}{dT} & \frac{\partial c_6}{\partial T} &= p^{1/2}\frac{dK_{p,6}}{dT} \end{aligned} \tag{G.9}$$

As discussed in Chapter 2, the curve-fit equations for the equilibrium coefficients $K_{p,i}$ are of the form

$$\log_{10}K_{p,i}(T) = A_i \ln(T/1000) + \frac{B_i}{T} + C_i + D_iT + E_iT^2 \tag{G.10}$$

where T is in Kelvin. Differentiation of Eq. (G.10) with respect to temperature yields

$$\frac{dK_{p,i}}{dT} = 2.302585\, K_{p,i}\left[\frac{A}{T} - \frac{B}{T^2} + D + 2ET\right] \tag{G.11}$$

The evaluation of the mole fraction partial derivatives with respect to pressure parallels the development of the temperature derivatives, with $\partial/\partial p$ replacing $\partial/\partial T$. Note that since neither c_3 nor c_4 depends on pressure, the terms containing $\partial c_3/\partial p$ and $\partial c_4/\partial p$ are zero. The remaining derivatives are

$$\frac{\partial c_1}{\partial p} = -\frac{1}{2}\frac{c_1}{p} \qquad \frac{\partial c_5}{\partial p} = \frac{1}{2}\frac{c_5}{p}$$
$$\frac{\partial c_2}{\partial p} = -\frac{1}{2}\frac{c_2}{p} \qquad \frac{\partial c_6}{\partial p} = -\frac{1}{2}\frac{c_6}{p} \tag{G.12}$$

G.2 Gaussian Error Function

The Gaussian error function is defined as

$$\operatorname{erf} w \equiv \frac{2}{\sqrt{\pi}} \int_0^w e^{-v^2}\, dv \tag{G.13}$$

Table G.1 lists values for erf w from $w = 0$ to 3. The complementary error function is

$$\operatorname{erfc} w \equiv 1 - \operatorname{erf} w \tag{G.14}$$

Table G.1 Gaussian Error Function.

η	erf η	η	erf η	η	erf η
0.00	0.00000	0.36	0.38933	1.04	0.85865
0.02	0.02256	0.38	0.40901	1.08	0.87333
0.04	0.04511	0.40	0.42839	1.12	0.88679
0.06	0.06762	0.44	0.46622	1.16	0.89910
0.08	0.09008	0.48	0.50275	1.20	0.91031
0.10	0.11246	0.52	0.53790	1.30	0.93401
0.12	0.13476	0.56	0.57162	1.40	0.95228
0.14	0.15695	0.60	0.60386	1.50	0.96611
0.16	0.17901	0.64	0.63459	1.60	0.97635
0.18	0.20094	0.68	0.66378	1.70	0.98379
0.20	0.22270	0.72	0.69143	1.80	0.98909
0.22	0.24430	0.76	0.71754	1.90	0.99279
0.24	0.26570	0.80	0.74210	2.00	0.99532
0.26	0.28690	0.84	0.76514	2.20	0.99814
0.28	0.30788	0.88	0.78669	2.40	0.99931
0.30	0.32863	0.92	0.80677	2.60	0.99976
0.32	0.34913	0.96	0.82542	2.80	0.99992
0.34	0.36936	1.00	0.84270	3.00	0.99998

G.3 Bessel Functions of the First Kind

The Bessel functions of the first kind, order 0 and order 1, are defined as $J_0(x)$ and $J_1(x)$. Table G.2 lists values for $J_0(x)$ and $J_1(x)$ from $x = 0$ to 2.4.

Table G.2 Bessel Functions of the First Kind, Order 0 and 1.

x	$J_0(x)$	$J_1(x)$
0.0	1.0000	0.0000
0.1	0.9975	0.0499
0.2	0.9900	0.0995
0.3	0.9776	0.1483
0.4	0.9604	0.1960
0.5	0.9385	0.2423
0.6	0.9120	0.2867
0.7	0.8812	0.3290
0.8	0.8463	0.3688
0.9	0.8075	0.4059
1.0	0.7652	0.4400
1.1	0.7196	0.4709
1.2	0.6711	0.4983
1.3	0.6201	0.5220
1.4	0.5669	0.5419
1.5	0.5118	0.5579
1.6	0.4554	0.5699
1.7	0.3980	0.5778
1.8	0.3400	0.5815
1.9	0.2818	0.5812
2.0	0.2239	0.5767
2.1	0.1666	0.5683
2.2	0.1104	0.5560
2.3	0.0555	0.5399
2.4	0.0025	0.5202

H

Computer Programs

The Matlab programs contained in this section are listed below. Digital copies of the programs are available for downloading at the John Wiley website for the textbook or from the author (allan@engr.colostate.edu).

AdiabaticFlameTemp.m	Computes constant pressure adiabatic flame temperature of fuel–air mixture.
farg.m	Computes mole fractions and thermodynamic properties of a fuel–air–residual gas mixture.
fuel.m	Fuel thermodynamic properties file.
RunEcp.m	Input/output file for program ecp.m.
ecp.m	Computes mole fractions and thermodynamic properties for equilibrium combustion of fuel–air mixture.
Droplet.m	Computes droplet evaporation for two fuels, n-octane and n-heptane.
Kinetic.m	Computes species fractions using a reduced kinetic model of hydrocarbon autoignition.
Soot.m	Computes soot formation and oxidation rates.
TwoZoneNO.m	Computes NO versus crank angle, and state parameters for a two-zone fuel–air cycle.

Principles of Combustion, Third Edition. Allan T. Kirkpatrick and Kenneth K. Kuo.

Companion website: www.wiley.com/go/kirkpatrick/principlesofcombustion3e

Index

Principles of Combustion, Third Edition. Allan T. Kirkpatrick and Kenneth K. Kuo.

Companion website: www.wiley.com/go/kirkpatrick/principlesofcombustion3e